pci design handbook

Precast and Prestressed Concrete

FOURTH EDITION

MNL 120-92

PRECAST/PRESTRESSED CONCRETE INSTITUTE
175 WEST JACKSON BOULEVARD
CHICAGO, ILLINOIS 60604

312 786-0300   FAX 312 786-0353

Copyright© 1992
By Precast/Prestressed Concrete Institute
First edition, first printing, 1971
First edition, second printing, 1972
Second edition, first printing, 1978
Second edition, second printing, 1980
Third edition, first printing, 1985
Fourth edition, first printing, 1992

All rights reserved. This book or any part thereof may not be reproduced in any form without the written permission of the Precast/Prestressed Concrete Institute.

Library of Congress Catalog Card Number 92-61318

ISBN 0-937040-51-7

Printed in U.S.A.

PCI Design Handbook

Precast and Prestressed Concrete

FOURTH EDITION

PCI Industry Handbook Committee

Helmuth Wilden, P.E. Chairperson

Roger J. Becker, P.E.	Harry Gleich, P.E.	Ben G. Olson, P.E.
E. Fred Brecher, P.E.	Edward Gloppen, P.E.	Kim E. Seeber, P.E.
Ian R. Chin, S.E.	Phillip J. Iverson, P.E.	Gordon Samuelson, P.E
Marco Cordero, P.E.	Daniel P. Jenny	A. Fattah Shaikh, P.E.
Sepp Firnkas, P.E.	Rafael A. Magana, P.E.	Irwin J. Speyer, P.E.
Sidney Freedman	Leslie D. Martin, S.E.	
Todd Gerhart, P.E.	Robert F. Mast, P.E.	

Co-Editors

Irwin J. Speyer, P.E.   A. Fattah Shaikh, P.E.

FOREWORD

The Precast/Prestressed Concrete Institute, a non-profit corporation, was founded in 1954 for the purpose of advancing the design, manufacture, and use of structural precast and prestressed concrete and architectural precast concrete in the United States and Canada.

To meet this purpose, PCI continually disseminates information on the latest concepts, techniques and design data to the architectural and engineering professions through regional and national programs and technical publications.

The First Edition of the *PCI Design Handbook* was published in 1971 with its primary focus on structural products and buildings. To fill a void in the design of architectural precast concrete, the *PCI Manual for Structural Design of Architectural Precast Concrete* was published in 1977. In 1978, the Second Edition of the *PCI Design Handbook* was published. In keeping with the tradition of continually updating information, an Industry Handbook Committee was formed in 1979 to develop the Third Edition, which was published in 1985. That edition provided, in a single source, information on the design of both architectural precast concrete and structural precast and prestressed concrete. The Fourth Edition continues to present both architectural and structural products and systems.

Since 1985, the committee has continued to monitor technical progress within the industry, with particular assistance from the many committees of PCI responsible for a variety of specific topics. This Fourth Edition is the culmination of those efforts, and represents current industry practice.

The members of the committee listed on the title page have made significant contributions of their time and expertise. In addition, PCI committees have provided reviews of specific areas. Many individuals within the industry have also provided advice and comment. The Institute offers all those involved in this process a special note of recognition and appreciation.

The final review phase consisted of a Blue Ribbon Review Committee, made up of Plant Engineers and Consulting Engineers from each of the PCI Zones and from Canada. These individuals were Michael M. Armor, P.E., Joel L. Bartlett, P.E., J. Edward Britt, P.E., Anant Y. Dabholkar, P.E., Paul Doak, P.E., John D. Garlich, P.E., Richard Golec, P.E., Keith Kaufman, P.E., Dean C. McKee, P.E., Courtney B. Phillips, P.E., Joseph W. Retzner, P.E., John R. Salmons, P.E., Theodore Spool, P.E., and John R. Spronken, P.E., To them, PCI extends its appreciation for their very important input.

Changes have been made throughout the document. Some of particular importance include:

- Updated to the ACI Building Code 318-89, other current PCI publications, and publications of other technical associations.

- New load tables have been added.

- Guidelines on epoxy coated strand, structural integrity, human response to vibrations, durability, repair and maintenance of structures, and responsibility have been added.

- Chapter One is greatly expanded, to illustrate representative projects using precast products and systems.

- Seismic section is expanded, and now includes additional examples.

- Shrinkage coefficients are modified.

- Stress-strain curves are modified.

- Torsion design procedure has been updated and modified.

- Equations for stud groups are updated.

Substantial effort has been made to ensure that all data and information in this Handbook are accurate. However, PCI cannot accept responsibility for any errors or oversights in the use of material or in the preparation of engineering plans. The designer must recognize that no handbook or code can substitute for experienced engineering judgment. This publication is intended for use by professional personnel competent to evaluate the significance and limitations of its contents and able to accept responsibility for the application of the material it contains.

The Institute considers each new Edition of the *PCI Design Handbook* to be a living document. The user is encouraged to offer comments to PCI on the content and suggestions for improvement to be incorporated in the next edition. Questions concerning the source and derivation of any material in the Handbook should be directed to the Institute.

GUIDE TO CHAPTERS

Chapter 1—Precast and Prestressed Concrete: Applications and Materials
1–1 to 1–30

Chapter 2—Product Information and Capability
2–1 to 2–57

Chapter 3—Analysis and Design of Precast, Prestressed Concrete Structures
3–1 to 3–87

Chapter 4—Design of Precast and Prestressed Concrete Components
4–1 to 4–74

Chapter 5—Product Handling and Erection Bracing
5–1 to 5–35

Chapter 6—Design of Connections
6–1 to 6–82

Chapter 7—Selected Topics for Architectural Precast Concrete
7–1 to 7–25

Chapter 8—Tolerances for Precast and Prestressed Concrete
8–1 to 8–19

Chapter 9—Thermal, Acoustical, Fire and Other Considerations
9–1 to 9–73

Chapter 10—Specifications and Standard Practices
10–1 to 10–45

Chapter 11—General Design Information
11–1 to 11–32

Index

CHAPTER 1
PRECAST AND PRESTRESSED CONCRETE: APPLICATIONS AND MATERIALS

	Page No.
1.1 Introduction	1-2
1.2 Applications	1-3
1.2.1 Building Structures	1-4
1.2.1.1 Residential buildings	1-6
1.2.1.2 Office buildings	1-8
1.2.1.3 Warehouses and industrial buildings	1-10
1.2.1.4 Other building structures	1-11
1.2.2 Parking Structures	1-12
1.2.3 Justice Facilities	1-14
1.2.4 Precast Concrete Cladding	1-15
1.2.5 Stadiums/Arenas	1-17
1.2.6 Bridges	1-19
1.2.7 Other Structures	1-20
1.3 Materials	1-22
1.3.1 Concrete	1-22
1.3.1.1 Compressive strength	1-22
1.3.1.2 Tensile strength	1-23
1.3.1.3 Shear strength	1-23
1.3.1.4 Modulus of elasticity	1-23
1.3.1.5 Poisson's ratio	1-23
1.3.1.6 Volume changes	1-23
1.3.2 Grout, Mortar and Drypack	1-24
1.3.2.1 Sand-cement mixtures	1-25
1.3.2.2 Non-shrink grouts	1-25
1.3.2.3 Epoxy grouts	1-25
1.3.3 Reinforcement	1-25
1.3.3.1 Prestressing tendons	1-25
1.3.3.2 Deformed reinforcing bars	1-26
1.3.3.3 Welded wire fabric	1-26
1.3.4 Durability	1-26
1.3.4.1 Freeze-thaw and chemical resistance	1-26
1.3.4.2 Protection of reinforcement	1-27
1.3.4.3 Sulphate attack	1-29
1.4 References	1-29

PRECAST AND PRESTRESSED CONCRETE: APPLICATIONS AND MATERIALS

1.1 Introduction

The growth of precast and prestressed concrete is a story of the vision and daring of a few notable men. These men took a new idea and maximized its potential by modifying and improving existing methods, conceiving new methods and inventing new devices—all with a focus on mass production techniques. An excellent portrayal of the beginnings and the growth of precast and prestressed concrete in North America and the early pioneers is given in a series of papers[1] developed to commemorate the 25-Year Silver Jubilee of the founding of the Prestressed Concrete Institute (now the Precast/Prestressed Concrete Institute—PCI).

The single most important event leading to the launching of the precast/prestressed concrete industry in North America was the construction in 1950 of the famed Walnut Lane Memorial Bridge (Fig. 1.1.1) in Philadelphia, Pennsylvania. From a technical perspective it is surprising, and from a historical perspective it is fascinating, that the Walnut Lane Memorial Bridge was constructed in prestressed concrete. There was very little published information on the subject and there was a total lack of experience with linear prestressing in this country. Furthermore, the length of bridge spans involved would have been a daring venture in the late 1940's anywhere in the world. The bridge became a reality (Ref. 1) through a fortunate sequence of events, and the vision, courage and persistence of a few extraordinary individuals.

Following completion of the Walnut Lane Memorial Bridge, American engineers and the construction industry enthusiastically embraced prestressed concrete. Many of the early post-Walnut Lane Memorial Bridge applications remained in bridge construction, such as the lower Tampa Bay crossing now known as the Sunshine Skyway. Simultaneously, American engineers and constructors were conceiving new devices, improving techniques and developing new materials.

The 1950's were the years that brought into focus the 7-wire strand, precasting, long-line beds (Fig. 1.1.2), admixtures, high strength concrete, vacuum concrete, steam curing and many other innovations. With these developments coupled with the technical and logistic support provided by the Prestressed Concrete Institute (PCI), which was chartered in 1954, the industry grew and the applications of precast and prestressed concrete began to appear in an impressive variety of structures.

Development of standard products was one of the major activities through the 1950's and the 1960's.

Fig. 1.1.1 Walnut Lane Memorial Bridge: Recipient of the 1978 ASCE's *Outstanding Civil Engineering Achievement Award.*

Fig. 1.1.2 Long-line prestressed double tee casting bed.

Double tee and hollow-core slabs (Fig. 1.1.3) are the most widely used building products. Double tees are efficient for spans up to 70 ft although longer spans are possible with deeper sections. Hollow-core slabs are available in a variety of widths ranging from 16 in. to 8 ft and are used for spans up to about 40 ft. Fig.

Fig. 1.1.3 Erection of hollow-core deck members.

1.1.4 shows cross sections of some other commonly used products. The I-beam, box beam and bulb tee are used in bridge construction. The inverted tee, ledger beam and rectangular beam are used for structural framing to support deck members. Square or rectangular columns, with or without corbels, are an integral part of the column-beam-deck framing that makes rapid, all-weather erection possible. Piles are manufactured in a variety of shapes including round, square, hexagonal and octagonal, as well as rectangular sheet piles. Channel slabs are used to support heavy floor or roof loads in short and medium span ranges.

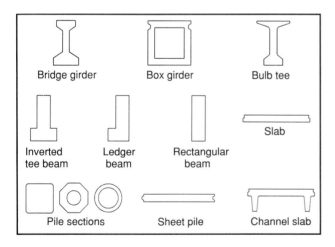

Fig. 1.1.4 Other common precast/prestressed concrete products.

Precast and prestressed concrete products are designed in accordance with the latest engineering standards and produced in plants where PCI's Plant Certification is an integral part of plant production. PCI Plant Certification (Fig. 1.1.5) assures specifiers of a manufacturing plant's audited capability to produce quality products. A minimum of two unannounced inspections each year by specially trained engineers evaluates compliance. Performance standards are found in the PCI quality control manuals, MNL-116[2] and MNL-117.[3] These manuals are considered industry standards for quality assurance and control. A plant that does not achieve a minimum required score loses certification. Plant certification is a prerequisite for PCI membership. Plants in Canada subscribe to a similar Canadian plant certification program based on Canadian Standards Association (C.S.A.) A-251 Standard.

Fig. 1.1.5 Assurance of quality through PCI Plant Certification.

The high quality standard products noted above form the basis for configuring a wide variety of framing systems for buildings, bridges and other structures. The following pages provide an overview of some of the applications. For other applications and also for products and practices suitable in different geographical zones, contact with the local producers is highly recommended. A list of PCI Producer Members is available from the Precast/Prestressed Concrete Institute (PCI).

1.2 Applications

The developments in products, materials and techniques noted in the previous section have made precast/prestressed concrete competitive in a variety of residential, commercial, transportation and many other types of structures. A few examples of applications to different types of structures are given in this section.

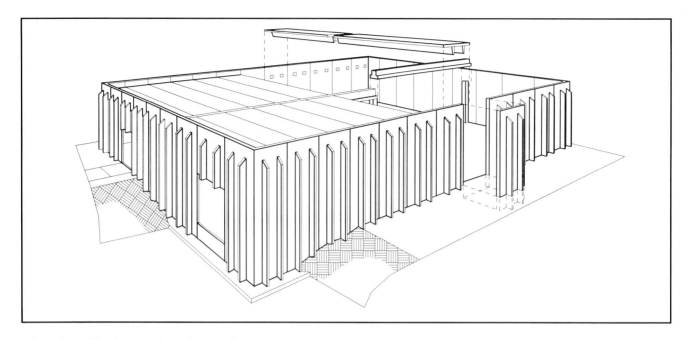

Fig. 1.2.1 Single-story bearing wall construction:
Provides economy by eliminating the need for a structural frame at the perimeter. The wall panels themselves can be selected from a variety of standard sections or flat panels, and specially formed architectural precast shapes. Any of the standard precast deck units can be used for roofs. (Note: This figure and some other revised figures in this Edition were drawn using CADKEY© software version 4.0 by CADKEY, Inc.)

1.2.1 Building Structures

Owners and developers quickly recognize the many inherent qualities of precast/prestressed concrete which make it suitable for many types of building structures. Precast/prestressed concrete structures, assembled from high-quality plant produced products, provide superior flexibility for achieving the required degrees of fire resistance, sound control, energy efficiency and durability. The availability of a variety of materials and finishes makes it possible to render virtually any desired aesthetic character. Furthermore, the construction speed possible with precast/prestressed concrete minimizes on-site labor costs, reduces the cost of interim financing, and thus provides important overall economy to the owner or developer.

Both bearing wall construction (Figs. 1.2.1 and 1.2.2) and beam-column framing (Figs. 1.2.3 and 1.2.4) have been successfully used for various height buildings. Resistance to lateral loads can be provided by interior shear walls (Fig. 1.2.5), exterior shear walls (Fig. 1.2.6), or rigid frame action (Fig. 1.2.7), or some combination of these.

1.2.1.1 Residential buildings

Precast/prestressed concrete enjoys broad acceptance in low-rise and mid-rise apartment buildings, hotels, motels, and nursing homes. The superior fire resistance and sound control features are specifically recognized by owners and developers.

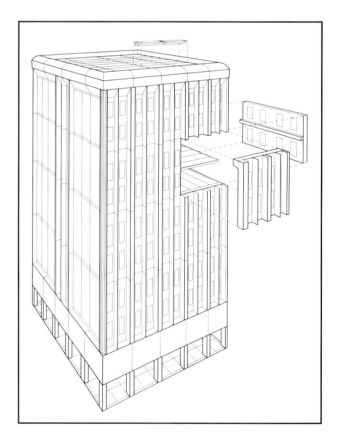

Fig. 1.2.2 Multi-story bearing wall construction:
Precast bearing wall units can be cast in one or multi-story high units. The units may be started at the second floor level with the first floor framing consisting of beams and columns to obtain a more open space on the first level.

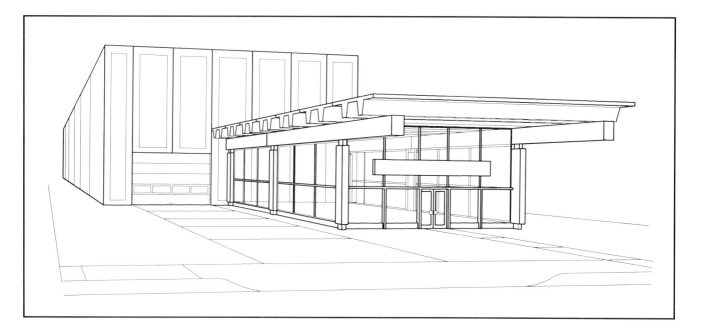

Fig. 1.2.3 Single-story beam-column construction:
Any of the standard precast beam and column sections shown in Chapter 2 can be used for single-story structures. Selection of the type of beam to be used depends on considerations such as span length, level of superimposed loads, and also on depth of ceiling construction and desired architectural expression.

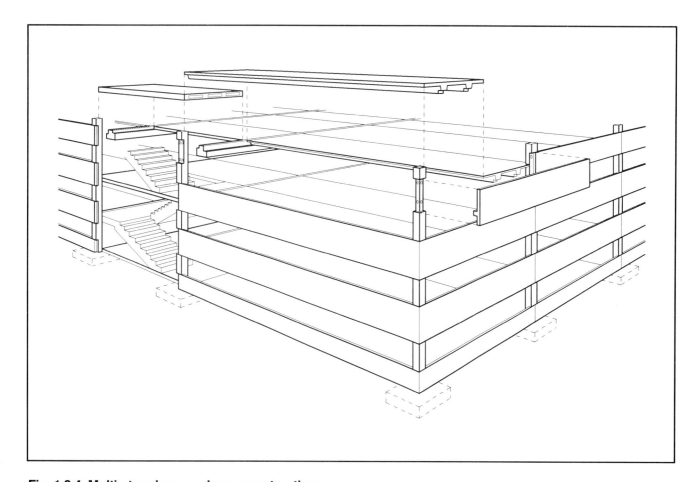

Fig. 1.2.4 Multi-story beam-column construction:
Beam-column framing is suitable for both low-rise and high-rise buildings. Architectural and engineering considerations dictate whether the beams are continuous with single-story columns, or whether multi-story columns are used with single span beams.

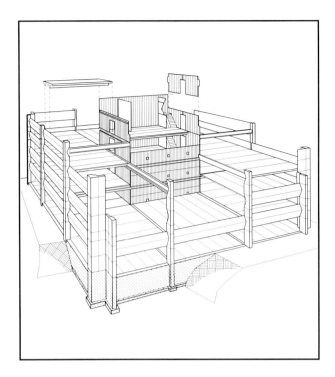

Fig. 1.2.5 Interior shear wall system:
Lateral loads are transmitted by floor diaphragms to a structural core of precast shear walls. The shear walls can be tied together vertically and at corners to form a structural tube that cantilevers from the foundation.

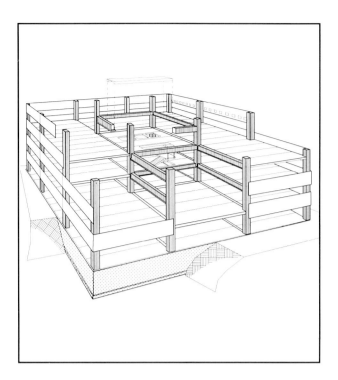

Fig. 1.2.7 Rigid frame system:
All lateral loads are transferred to a moment-resisting frame that ties beams and columns together with rigid connections. The need for shear walls is eliminated.

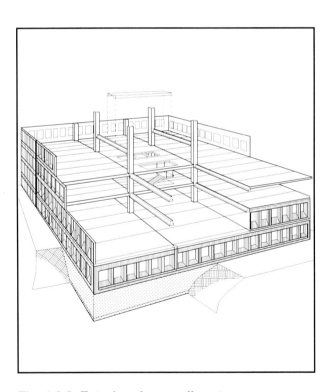

Fig. 1.2.6 Exterior shear wall system:
In general, the exterior shear wall system permits greater design flexibility than the interior shear wall system because it eliminates the need for a structural core. By combining gravity load bearing function with lateral load resistance, the exterior shear wall system is, in general, more economical.

Fig. 1.2.8 Precast concrete walls and floors ensure fire containment and lower insurance rates.

Two-hour fire containment within each living unit provides safety from adjacent units. With this type of high quality precast concrete housing, fire insurance rates are reduced and often higher incomes can be generated because of the safe, soundproof high quality environment and lifestyle offered.

The hollow-core slab is a standard product in this type of construction. Fig. 1.2.9 shows a typical apartment building with hollow-core floors, load bearing precast concrete walls and a durable maintenance free exterior spandrel panel. Details of this type of construction are shown in Fig. 1.2.10.

Fig. 1.2.9 High quality concrete in precast concrete walls and prestressed hollow-core floors provide sound resistance, firesafety and reduced maintenance cost in multi-family housing.

Fig. 1.2.10 Details of a load bearing wall—hollow-core floor building. Spandrels can be load bearing if required.

Fig. 1.2.11 Precast concrete provides a quiet, safe, comfortable and high quality place to live.

1.2.1.2 Office buildings

Significant time savings usually result from the choice of a totally precast concrete structure. The superstructure is prefabricated while the on-site foundations are being built. Potential delays are reduced with the complete building system being supplied under one contract. Erection of large precast concrete components can proceed even during adverse weather conditions to quickly enclose the structure. The prestressed floors provide an immediate working platform to allow the interior tradesmen an early start on the mechanical, electrical and interior finishing work. The quality finishes and fast schedules result in early occupancy, tenant satisfaction and reduced financing costs. These factors make a precast/prestressed concrete building very suitable for office buildings.

The uses of precast/prestressed concrete in office building construction are many, from total building systems to single products like precast concrete stairs. Precast/prestressed beams, columns and floors are used in frame systems; shear walls can be used alone or in conjunction with beams and columns to resist lateral loads. Precast concrete stairs, along with being economical, provide immediate safe use of stairwells.

Architectural precast concrete is used with all types of framing systems. It provides an economical, fireproof, soundproof, durable, maintenance-free curtain wall that allows the architect much freedom of expression and results in beautiful facades. Architectural precast concrete is discussed more fully in Sect. 1.2.4 and in Chapter 7.

Fig. 1.2.12 This total precast concrete office building is an example of the beauty and economy that can be achieved using precast concrete. Real savings can be achieved by allowing the architectural facade to also perform structurally.

Fig. 1.2.13 Architectural precast concrete panels can perform structurally.

Fig. 1.2.14 Using precast concrete beams and columns with hollow-core floor slabs provides a quick economical structure. The addition of an architectural precast concrete facade provides the owner with the security of a single source of responsibility for the structure.

Fig. 1.2.15 By varying minor details like rustication, architectural precast concrete exteriors can be made expressive and economical.

Fig. 1.2.16 Office buildings quite often utilize precast concrete spandrels.

Fig. 1.2.17 The light industrial office structure is erected and put into service quickly using insulated precast concrete sandwich panels.

1.2.1.3 Warehouses and industrial buildings

The ability of precast/prestressed concrete to span long distances with small depths and carry heavy loads is particularly suitable for warehouses and industrial buildings. Standard precast/prestressed concrete walls, insulated or non-insulated, are very economical for warehouse and light manufacturing applications. Total precast systems with precast/prestressed roof diaphragms and shear walls can provide owners with a complete "structural package". A wide variety of wall finishes is available.

In heavy industrial projects, precast floor units capable of carrying the typical heavy floor loads can be combined with other precast components to construct versatile, corrosion-resistant structural systems. The precast framing can be designed to accommodate a variety of mechanical systems and to support bridge cranes for industrial uses. High quality precast concrete provides protection against fire, dampness and a variety of chemical substances. The smooth surfaces achievable in precast concrete make it ideal for food processing, wet operations, computer components manufacturing, as well as many other types of manufacturing and storage operations. Clear spans of 40 ft and 90 ft are possible using hollow-core slabs and double tees, respectively. Even longer spans to about 150 ft can be obtained with bridge-type girders.

Fig. 1.2.18 Precast and prestressed concrete flat panels are used for warehouse walls. Hollow-core slabs are often used in this application.

Fig. 1.2.19 Erection of precast concrete wall panels is fast and can be performed year round even in cold climates. Roof diaphragms and supporting beams and columns are often also precast concrete.

Fig. 1.2.20 Precast concrete insulated sandwich panels provide excellent insulation for cold storage facilities. Smooth interior finishes make these panels an excellent choice for food processing plants.

Fig. 1.2.21 Total precast concrete structures provide many answers for heavy industrial applications. High quality plant produced concrete provides excellent corrosion resistance and durability. Heavy loads are no problem for prestressed concrete beams and slabs.

1.2.1.4 Other building structures

The many benefits of precast/prestressed concrete, make it suitable for many other types of buildings in addition to residential, office, and industrial buildings. Applications abound in educational institutions, commercial buildings such as shopping malls, and public buildings including hospitals, libraries, and airport terminals. Precast/prestressed concrete and glass fiber reinforced concrete have also been effectively used in numerous retrofit projects.

Fig. 1.2.22 Waste water treatment buildings benefit from the durability, economy and aesthetic capability of precast concrete.

Fig. 1.2.23 This airport terminal illustrates the long span capability of precast/prestressed concrete.

Fig. 1.2.24 Long-span perforated precast, prestressed concrete girders provide flexible, column-free interior space in this government administration and public services building.

Fig. 1.2.25 This boathouse incorporates a training center and restaurant.

1.2.2 Parking Structures

Increasingly, architects, engineers, developers and owners are turning to precast/prestressed concrete as the answer to their commercial, municipal and institutional parking needs. Though classified and constructed as buildings, parking structures are unique; in some ways, they may be compared to bridges with multiple decks. They are subjected to moving loads from automobile traffic and the roof level of a parking structure is exposed to weather in much the same way as a bridge deck. Furthermore, they are usually not enclosed and thus the entire structure is subjected to ambient weather conditions. Also, exposure to deicing salts in northern climates or to salt-laden atmospheres in coastal locations requires consideration to ensure long-time performance.

The controlled conditions of a precast concrete plant assures the parking structure owner of the quality concrete and workmanship that provides long term durability. The low water-cement ratio concrete that precast concrete manufacturers use has been proven to increase resistance to corrosion due to chlorides. Studies[4] have also shown that accelerated curing makes precast concrete more resistant to chlorides than field cured concrete.

These inherent durability characteristics along with low-cost, rapid erection in all weather conditions, unlimited architectural expression and long clear spans make precast concrete the natural choice for parking structures.

Through surveys of existing structures and other experiences, and through research and development, significant improvements have been achieved in the engineering state-of-the-art of parking structures. This accumulated experience and knowledge has been assembled in a comprehensive publication[5] by PCI that includes recommendations on planning, design, construction and maintenance.

Figs. 1.2.26 and 1.2.27 show a typical precast concrete parking structure with architectural load bearing spandrels and stair tower walls. Long span double tees are shown in Fig. 1.2.28 bearing on an innovative "light" wall that adds openness and a feeling of security. Fig. 1.2.29 shows a precast, prestressed double tee being erected into a load bearing spandrel which is pocketed to reduce load induced torsion. The double tee is being set down on an interior inverted tee beam spanning from column to column.

Fig. 1.2.26 Long clear spans with architectural load bearing spandrels make precast concrete an economical and aesthetically preferable parking structure solution.

Fig. 1.2.27 Stair towers acting as shearwalls resist lateral loads and provide architectural features.

Fig. 1.2.28 Opened load bearing walls provide the security of visibility and openness as well as carrying vertical loads.

Fig. 1.2.30 Interior double ledger beam is deep enough to also act as the spandrel/car stop. These beams bear on column corbels.

Vertical expansion of precast concrete parking structures can be economically accomplished with special cranes that carry new members over erected decks as shown in Fig. 1.2.32. This erection method also allows precast concrete to be considered for expansion of non-precast concrete structures.

Fig. 1.2.29 Erection of precast concrete components is fast and independent of climate. Pocketed load bearing spandrels accept the double tee, which then bears on the interior double ledger beam. Erection is readily accomplished even at jobsites with limited space.

Fig. 1.2.31 Erecting precast concrete directly from the truck eliminates on-site storage and solves limited space problems.

Fig. 1.2.32 Special erection equipment is available for adding floors to existing parking structures.

1.2.3 Justice Facilities

Justice facilities encompass many types of building occupancies. These include jails, prisons, police stations, courthouses, juvenile halls, and special mental health or drug abuse centers.

Precast/prestressed concrete has proven to be the favored material for justice facilities because it has many inherent benefits which are important to these building types. In addition to the benefits noted previously, namely fire resistance, durability, speed of erection and flexibility for aesthetics, precast/prestressed concrete is ideal for building-in the desired level of physical security coupled with accommodation of security hardware and communication systems. The thickness and reinforcement of precast wall and slab systems, designed for gravity and lateral loads and volume changes, are generally sufficient even for maximum security requirements. Also, the modular nature of precast/prestressed concrete products facilitates pre-installation of necessary security and communication hardware in the plant, greatly simplifying field installation work as well as saving valuable time along the project's critical path.

The use of precast concrete box modules in a one-cell or two-cell format has greatly reduced field labor, erection time, punch list problems, multi-trade confusion and, ultimately—risk. These units can be quickly stacked and can be substantially complete requiring little finishing work (see Fig. 1.2.35).

Given the serious shortage of justice facilities, savings in total project time is often a critical consideration in selection of a structural material and system for these projects. Case histories show that total precast concrete projects have resulted in saving one to two years of construction time over the estimated schedule for competing systems. These experiences and the other considerations noted above have led to rapid growth in the use of precast/prestressed concrete in justice facilities projects.

Fig. 1.2.33 The repetitive nature of cell layout and the typical quick occupancy requirements make precast concrete an ideal solution to prison overcrowding.

Fig. 1.2.34 Precast double-cell module being lifted into place.

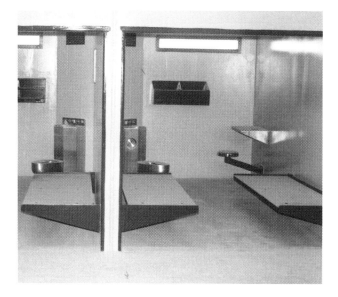

Fig. 1.2.35 Box module cells are often supplied with electric conduit and boxes, furniture, window and door embedments, and mechanical and plumbing chases cast in.

Fig. 1.2.37 The Transamerica Corporation building in San Francisco is 48 stories tall. Floor-height, double-window units, weighing 3½ tons each make up half of the total precast concrete pieces used.

Fig. 1.2.36 Interior finishes are durable and maintenance free. Soundproof and firesafe, precast concrete adds to security and limits vandalism.

1.2.4 Precast Concrete Cladding

Architectural precast concrete cladding provides many degrees of freedom for architectural expression with the economy of mass production of precast elements. The cladding may serve only as an enclosure for the structure, or may be designed to support gravity loads as well. Attention is currently also being given to the use of cladding to contribute to resistance to lateral loads of the structural frame.[6]

Architectural precast concrete can be cast in almost any color, form or texture to meet aesthetic and practical requirements.[7] Special sculptured effects can provide such visual expression as strength and massiveness, or grace and openness. Design flexibility is possible in both color and texture by varying aggregate and matrix color, size of aggregates, finishing processes and depth of exposure. Additional flexibility of aesthetic expression is achieved by casting various other materials as veneers on the face of precast concrete panels. Natural stone, such as polished and thermal-finished granite, limestone and marble, and clay products such as brick, tile and terra cotta have been frequently used as veneer materials.[7]

In addition to the freedom of aesthetic expression achievable with load bearing or non-load bearing architectural precast concrete, there are a number of other important functional and construction advantages. Insulated wall panels, consisting of two concrete wythes with a continuous layer of rigid insulation sandwiched between, contribute substantially to the overall thermal efficiency of a building. In cast-in-place concrete construction, precast concrete cladding panels are sometimes used as permanent concrete formwork, thus becoming an integral part of the structure. Off-site pre-assembly of all components comprising a total wall system, including window sash and glazing, can also be very cost effective. For more comprehensive information on product and design, see Ref. 7.

Fig. 1.2.39 The moldability of architectural precast concrete allows for an infinite number of sizes and shapes.

Fig. 1.2.38 The NBC tower in Chicago has 2500 pieces of precast concrete with limestone veneer as its exterior.

Glass fiber reinforced concrete (GFRC) is a recent innovation in materials technology which has been adopted for use in producing strong, thin, lightweight architectural cladding panels.[8] GFRC is a portland cement-based composite reinforced with randomly dispersed alkali-resistant glass fibers. The fibers serve as reinforcement to enhance flexural, tensile, and impact strength of concrete. A major benefit of GFRC is its light weight which provides for substantial economy resulting from reduced costs of product handling, transportation, and erection, and which also results in lower seismic loads.

Fig. 1.2.40 Repetition of size and shape, which allows multiple form use, is a key to economy.

Fig. 1.2.41 Spandrel panels spanning from column to column can be load bearing or solely architectural.

Fig. 1.2.42 This major sports stadium was built using total precast/prestressed concrete after evaluating other systems.

1.2.5 Stadiums/Arenas

Large stadiums and arenas (Figs. 1.2.42 to 1.2.46) are impressive structures. Often these projects are built on tight schedules to accommodate some important sporting event. Precast/prestressed concrete has been the overwhelming choice for many of these projects. The technique of post-tensioning precast segments together has allowed this versatile material to form complex cantilever arm and ring beam construction which supports the roofs of these structures. Post-tensioning is also commonly employed to minimize the depth of precast concrete cantilevered raker beams which carry the seating and provide unhindered viewing of the playing surface. Mass produced seating units have been manufactured in a variety of configurations and spans to provide for quick installation and long lasting service. The ability to eliminate costly field formwork makes precast/prestressed concrete the best choice for many components of stadium construction, especially seating which can be standardized to take advantage of repeated form utilization. Components that would require tall scaffolding towers to field-form, such as raker beams and ring beams, can be simplified by precasting in a plant and delivering and lifting them into place. Pedestrian ramps, mezzanine floors, concession, toilet, and dressing room areas can all be framed and constructed using precast/prestressed concrete elements.

Fig. 1.2.43 Precast concrete risers and raker beams. Columns and mezzanine floors are also precast concrete components.

PCI Design Handbook/Fourth Edition   1–17

Fig. 1.2.44 Fast track construction allows compressed schedules. Site preparation and component manufacture can take place simultaneously. Multiple erection crews speed erection.

Fig. 1.2.45 Special framing is handled easily with precast concrete, eliminating expensive field formwork.

Fig. 1.2.46 Repetition, simplicity and multiple form use allows this seating to be framed economically with total precast concrete.

Fig. 1.2.47 Smaller grandstands and huge stadiums all benefit from the use of precast/prestressed concrete.

Fig. 1.2.48 Quality plant-produced precast/prestressed concrete results in durability. Low maintenance, economy and the ability to span long distances make precast/prestressed concrete the preferred system for bridges of all spans. Here, precast/ prestressed concrete piling easily resists the marine environment.

1.2.6 Bridges

Bridge construction gave the prestressing industry its start in North America. Precast/prestressed concrete is now the dominant structural material for short-to-medium span bridges. With its inherent durability, low maintenance and assured quality, precast/prestressed concrete is a natural product for bridge construction. The ability to quickly erect precast concrete components in all types of weather with little disruption of traffic adds to the economy of the job. For short spans (spans to 100 ft), use of box sections and double tee sections has proved economical. However, the most common product for short-to-medium spans is the I-girder. Spans to 150 or 160 ft are not uncommon with I-girders and bulb tees. Spliced girders allow spans as much as 300 ft.[9] Even longer spans (300 to 400 ft) can be achieved using precast box girder segments which are then post-tensioned in the field. Using cable stays, the spanning capability of precast/prestressed concrete has been increased to over 1000 ft.

A recent innovation in bridge construction has been the use of precast concrete in horizontally curved bridges. A study commissioned by PCI[10] documents the technical feasibility and the economic viability of this application.

Another growing application of precast/prestressed concrete in bridge construction includes the use of precast deck panels[11]. Used as stay-in-place forms, the panels reduce field placement of reinforcing steel and concrete resulting in considerable savings. The panels become composite with the field-placed concrete for live loads.

Figs. 1.2.48 to 1.2.52 show some of the applications mentioned above.

Fig. 1.2.49 The length of this bridge with extensive repetition, form usage and large quantity of product made it worthwhile to create a special section to span from pile cap to pile cap. Precast concrete pile caps eliminated expensive overwater formwork. Precast/ prestressed concrete piling added speed to the project as well as durability in a harsh marine environment.

Fig. 1.2.50 Standard AASHTO shapes offer immediate availability, economy, durability, speed and low maintenance.

Fig. 1.2.51 Standardization has optimized bridge designs.

Fig. 1.2.52 These shallow depth 35 x 48 in. box girders offer a clean simple design and provide excellent performance.

1.2.7 Other Structures

The inherent qualities of precast/prestressed concrete noted in previous sections and the high degree of design flexibility also make it ideal for a wide variety of special applications. Properties, such as corrosion resistance, fire resistance, durability and fast installation, have been used to good advantage in the construction of poles, piles, storage tanks, retaining walls and sound barriers. Where repetition and standardization exist, precasting components can economically provide the quality of plant manufactured products while eliminating expensive and risky field procedures. These applications are too numerous to categorize here separately; only a few examples are given in Figs. 1.2.53 to 1.2.59.

Fig. 1.2.53 Precasting this arched culvert allows quick erection with little site disruption and no field formwork.

Fig. 1.2.54 Precast concrete sound wall protects residential neighborhood from highway noise.

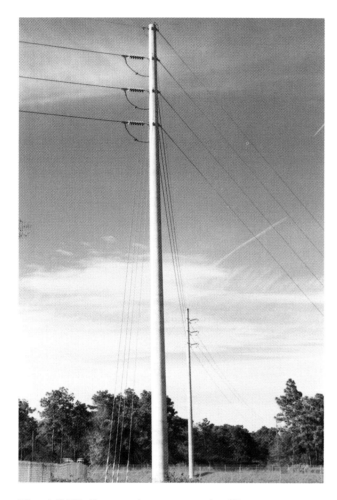

Fig. 1.2.55 Precast/prestressed utility poles provide low maintenance, durable, economical poles capable of carrying heavy line loads.

Fig. 1.2.57 Large circular storage tank under construction. Vertically prestressed precast concrete wall segments are braced temporarily. Cast-in-place concrete joints and circumferential post-tensioning complete the structure.

Fig. 1.2.56 This pier and ammunition loading dock for the US Navy is a total precast concrete structure. Economic analysis shows precast/prestressed concrete to be the best solution.

Fig. 1.2.58 Precast/prestressed concrete has proven to be a viable alternative to timber for railroad ties.

© Walt Disney Productions

Fig. 1.2.59 Famous amusement park building in background has precast/prestressed roof and walls. Monorail guideway is also precast/prestressed concrete.

1.3 Materials

This section provides a brief review of properties of the major materials used in precast and prestressed concrete. Also included is a discussion of the durability of concrete. Refs. 12-23 provide more complete information on these subjects.

1.3.1 Concrete

Concrete properties, such as stress-strain relationship, tensile strength, shear strength and bond strength, are frequently expressed in terms of the compressive strength of concrete. Generally, these expressions have been empirically established based on experimental data of concretes with compressive strengths less than 6000 psi. These expressions are given in this section and are applicable to most precast and prestressed concrete since is it usually specified in the 5000 psi to 6000 psi compressive strength range.

While use of these equations also yields reasonable design values for concrete with compressive strengths up to about 8000 psi, different equations which provide better correlations with experimental data are available and should be considered.[12,13]

Concrete with compressive strengths higher than 8000 psi has been used in columns of high-rise buildings, in precast/prestressed concrete piling and bridge girders. Often, higher strength is a result of using high performance concrete to achieve added durability. Interest in a more widespread use is growing. Extrapolated use of the equations given in this section is not advised for concretes with compressive strengths higher than 8000 psi. Recommendations given in Refs. 12 and 13 should be followed in these cases.

1.3.1.1 Compressive strength

The compressive strength of concrete, made with aggregate of adequate strength, is governed by either the strength of the cement paste or the bond between the paste and the aggregate particles. At early ages the bond strength is lower than the paste strength; at later ages the reverse may be the case. For a given cement and acceptable aggregates the strength that may be developed by a workable, properly placed mixture of cement, aggregates, and water (under the same mixing, curing, and testing conditions) is influenced by (a) the ratio of water to cement, (b) the ratio of cement to aggregate, (c) grading, surface texture, shape and strength of aggregate particles, and (d) maximum size of the aggregate. Mix factors, partially or totally independent of water-cement ratio, which affect the strength are (a) type and brand of cement, (b) amount and type of admixture or pozzolan, and (c) mineral composition of the aggregate.

Compressive strength is measured by testing 6 x 12 in. cylinders in accordance with standard ASTM procedures. The precast concrete industry also uses 4 x 8 in cylinders and 4 in. cube specimens. Adjustment factors need to be applied to these nonstandard specimens to correlate with the standard 6 x 12 in. cylinders.

Because of the need for early strength gain, Type III cement is often used by precasters so that molds may be reused daily. Structural precast concrete and much architectural precast concrete is made with gray cement that meets ASTM C150. Type III and Type I white and buff portland cements are frequently used in architectural products. These are usually assumed to have the same characteristics (other than color) as gray cement. Pigments are also available to achieve colored concrete, and have little effect on strength at the recommended dosages. Cement types and experience with color should be coordinated with the local producers.

High strength concrete mixes (6000 psi or more) are available in some areas. Local suppliers should be contacted to furnish mix and design information.

Initial curing of precast concrete takes place in the form, usually by covering to prevent loss of moisture and sometimes, especially in structural prestressed products, by the application of radiant heat or live steam. Additional curing has been shown to rarely be necessary to attain the specified strength. Control techniques for the most effective and economical accelerated curing are reported in PCI publication TR No. 1.[14]

When concrete is subjected to freezing and thawing and other aggressive environments, air entrainment is often specified (see Sect. 1.3.4.1). In some concrete mixes, reduction of strength may occur with air entrainment.

1.3.1.2 Tensile strength

A critical measure of performance of architectural precast concrete is its resistance to cracking, which is dependant on the tensile strength of concrete. Reinforcement does not prevent cracking, but controls crack widths after cracking occurs. Tensile stresses in prestressed concrete, which could result in cracking, are permitted by ACI 318-89.[15]

Flexural tensile strength is measured by the modulus of rupture. It can be determined by test, but the modulus of rupture for structural design is generally assumed to be a function of compressive strength as given by:

$$f_r = K\lambda\sqrt{f'_c} \qquad \text{(Eq. 1.3.1)}$$

where:

f_r = modulus of rupture, psi
f'_c = compressive strength, psi
K = a constant, usually between 8 and 10 but prescribed to be equal to 7.5 by ACI 318-89
λ = 1.0 for normal weight concrete
 0.85 for sand-lightweight concrete
 0.75 for all-lightweight concrete

1.3.1.3 Shear strength

The shear (or diagonal tension) strength of concrete is also a function of its compressive strength. The equations for shear strength specified in ACI 318-89 are given in Chapter 4. The shear strength of lightweight concrete is a function of the splitting tensile strength, which is determined by test. However, in lieu of test, ACI 318-89 permits the use of the coefficient, λ, as described above.

1.3.1.4 Modulus of elasticity

Modulus of elasticity (E) is the ratio of normal stress to corresponding strain for tensile or compressive stresses. It is the material property which determines the deformability under load. Thus, it is used to calculate deflections, axial shortening and elongation, buckling and relative distribution of applied forces in composite and non-homogeneous structural members.

The modulus of elasticity of concrete and other masonry materials is not as well defined as, for example, steel. It is therefore defined by some approximation, such as the "secant modulus". Thus, calculations which involve its use have inherent imprecision, but this is seldom critical enough to affect practical performance. While it may be desirable in some rare instances to determine modulus of elasticity by test, especially with some lightweight concretes, the equation given in ACI 318-89 is usually adequate for design:

$$f_r = K\lambda\sqrt{f'_c} \qquad \text{(Eq. 1.3.2)}$$

where:

E_c = modulus of elasticity of concrete, psi
w = unit weight of concrete, pcf
f'_c = compressive strength, psi

1.3.1.5 Poisson's ratio

Poissons's ratio is the ratio of transverse strain to axial strain resulting from uniformly distributed axial load. It generally ranges between 0.11 and 0.27, and, for design, is usually assumed to be 0.20 for both normal weight and lightweight concrete.

1.3.1.6 Volume changes

Volume changes of precast concrete are caused by variations in temperature, shrinkage due to air-drying, and by creep caused by sustained stress. If precast concrete is free to deform, volume changes are of little consequence. If these members are restrained by foundations, connections, steel reinforcement, or connecting members, significant stresses may develop over time.

The volume changes due to temperature variations can be positive (expansion) or negative (contraction), while volume changes from shrinkage and creep are only negative.

Precast concrete members are generally kept in yard storage for a period of time. Thus, much of the shrinkage will have taken place by the time of erection. However, connection details and joints must be designed to accommodate the changes which will occur after the precast member is erected and connected to the structure. In most cases, the shortening that takes place prior to making the final connections will reduce the shrinkage and creep strains to manageable proportions.

Typical creep, shrinkage, and temperature strains and design examples are given in Chapter 3.

Temperature effects. The coefficient of thermal expansion of concrete varies with the aggregate used as shown in Table 1.3.1. Range for normal weight concrete is 5 to 7 x 10^{-6} in./in./deg F when made with siliceous aggregates and 3.5 to 5 x 10^{-6} when made with calcareous aggregates. The range for structural lightweight concretes is 3.6 to 6 x 10^{-6} in./in./deg F depending on the type of aggregate and amount of natural sand. For design, coefficients of 6 x 10^{-6} in./in./deg F for normal weight concrete and 5 x 10^{-6} for sand-lightweight concrete are frequently used. If greater accuracy is needed, tests should be made on the specific concrete.

Table 1.3.1 Coefficients of linear thermal expansion of rock (aggregate) and concrete.[16]

Type of rock (aggregate)	Average coefficient of thermal expansion x 10^{-6} in./in./deg F	
	Aggregate	Concrete*
Quartzite, Cherts	6.1 - 7.0	6.6 - 7.1
Sandstones	5.6 - 6.7	5.6 - 6.5
Quartz Sands & Gravels	5.5 - 7.1	6.0 - 8.7
Granites & Gneisses	3.2 - 5.3	3.8 - 5.3
Syenites, Diorites, Andesite, Gabbros, Diabas, Basalt	3.0 - 4.5	4.4 - 5.3
Limestones	2.0 - 3.6	3.4 - 5.1
Marbles	2.2 - 3.9	2.3
Dolomites	3.9 - 5.5	—
Expanded Shale, Clay & Slate	—	3.6 - 4.3
Expanded Slag	—	3.9 - 6.2
Blast-Furnace Slag	—	5.1 - 5.9
Pumice	—	5.2 - 6.0
Perlite	—	4.2 - 6.5
Vermiculite	—	4.6 - 7.9
Barite	—	10.0
Limonite, Magnetite	—	4.6 - 6.0
None (Neat Cement)	—	10.3
Cellular Concrete	—	5.0 - 7.0
1:1 (Cement: Sand) **	—	7.5
1:3 **	—	6.2
1:6 **	—	5.6

*Coefficients for concretes made with aggregates from different sources vary from these values, especially those for gravels, granites, and limestones. Fine aggregates generally are the same material as coarse aggregates.

**Tests made on 2-yr. old samples.

Since the thermal coefficient for steel is also about 6 x 10^{-6} in./in./deg F, the thermal effects on precast/prestresed concrete may be evaluated by treating it as plain concrete.

Shrinkage and creep. Precast concrete members are subject to air-drying as soon as they are removed from the molds or forms. During this exposure to the atmosphere, the concrete slowly loses some of its original water causing a shrinkage volume change to occur.

When concrete is subjected to a sustained load, the deformation may be divided into two parts: (1) an elastic deformation which occurs immediately, and (2) a time-dependent deformation which begins immediately and continues over time. This time-dependent deformation is called creep.

Creep and shrinkage strains vary with relative humidity, volume-surface ratio (or ratio of area to perimeter—see Fig. 1.3.1), level of sustained load including prestress, concrete strength at time of load application, amount and location of steel reinforcement, and other characteristics of the material and design. Typical creep and shrinkage strains are given in Table 3.3.1, along with multipliers in Tables 3.3.2, 3.3.3 and 3.3.4 to account for the effects of the more significant variables mentioned above.

Fig. 1.3.1 Volume surface ratios for precast structural concrete members.

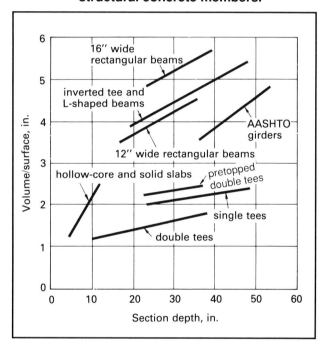

1.3.2 Grout, Mortar and Drypack

When water, sand and a cementitious material are mixed together without coarse aggregate, the result is called grout, mortar or drypack, depending on consistency. These materials have numerous applications with precast concrete: sometimes only for fire or corrosion protection, or for cosmetic treatment, other times to transfer loads through horizontal and vertical joints.

Different cementitious materials are used:

1. Portland cement.
2. Shrinkage-compensating portland cement.
3. Expansive portland cement made with special additives.
4. Gypsum or gypsum/portland cements.
5. Epoxy-cement resins.

In masonry mortar about one-half of the portland cement is replaced with lime. This improves its bonding characteristics but reduces its strength. Such mortar should not be used indiscriminately as a substitute for grout.

Quality control of grout is as important as that of concrete. Site mixed grout should be made and tested at regular intervals according to ASTM C-1019 which parallels ASTM C-39 for concrete. For more general information on grout, see Ref. 17.

1.3.2.1 Sand-cement mixtures

Most grout is a simple mixture of portland cement, sand, and water. Proportions are usually one part portland cement to 2 to 3 parts sand. The amount of water depends on the method of placement of the grout.

Flowable grouts are used to fill voids that are either formed in the field or cast into the precast member. They are used at joints that are heavily congested but not confined, thus requiring some formwork. These grouts usually have a high water-cement ratio (typically about 0.5), resulting in low strength and high shrinkage. There is also a tendency for the solids to settle, leaving a layer of water on the top. Special ingredients or treatments can improve these characteristics.

For very small spaces in confined areas, grout may be pumped or pressure injected. The confinement must be of sufficient strength to resist the pressure. Less water may be used than for flowable grouts, hence less shrinkage and higher strength.

A stiffer grout, or mortar, is used when the joint is not totally confined, for example in vertical joints between wall panels. This material will usually develop strengths of 3000 to 6000 psi, and have much less shrinkage than flowable grouts.

Drypack is the common name used for very stiff sand-cement mixes. They are used if a relatively high-strength is desired, for example, under column base plates. Compaction is attained by hand tamping.

When freeze-thaw durability is a factor, the grout should be air-entrained. Air content of plastic grout or mortar of 9 or 10% is required for adequate protection.

Typical portland cement mortars have very slow early strength gain when placed in cold weather. Heating the material is usually not effective, because the heat is rapidly dissipated to surrounding concrete. Thus, unless a heated enclosure can be provided, special proprietary mixes, usually containing gypsum, may be indicated. Mixes containing a high percentage of gypsum are known to deteriorate under prolonged exposure to water.

1.3.2.2 Non-shrink grouts

Shrinkage can be reduced, or more appropriately, compensated for by the use of commercially available non-shrink, pre-mixed grouts. These mixes expand during the initial hardening to offset the subsequent shrinkage of the grout. Since the non-shrink grouts are primarily proprietary, their chemical composition is usually not available to study their potential effects on the interfacing materials, such as reinforcement and inserts in the connection. Thus, it is advisable that manufacturers' recommendations should be carefully followed. For a general reference on the characteristics and methods of testing of non-shrink grouts, see Ref. 18.

1.3.2.3 Epoxy grouts

Epoxy grouts are mixtures of epoxy resins and a filler material, usually sand. These are used when high strength is desired, or when improved bonding to concrete is necessary. Ref. 19 is a comprehensive report on the subject by Committee 503 of the American Concrete Institute.

The physical properties of epoxy compounds vary widely. Also, the epoxy grouts behave very differently than the sand-cement grouts. For example, the thermal expansion of an epoxy grout can be as much as 7 times the thermal expansion of sand-cement grout. It is therefore important that use of these grouts be based on experience and/or appropriate tests. Recommended tests and methods are given in Ref. 18.

1.3.3 Reinforcement

Reinforcement used in structural and architectural precast concrete includes prestressing tendons, deformed steel bars, and welded wire fabric.

1.3.3.1 Prestressing tendons

Tendons for prestressing concrete may be wires, strands, or bars. In precast/prestressed structural concrete, nearly all tendons are 7-wire strands conforming to ASTM A416. The strands are pretensioned, that is, they are tensioned prior to placement of the concrete. After the concrete has reached a predetermined strength, the strands are cut and the prestress force is transferred to the concrete through bond.

Two types of strand are covered in ASTM A416—"low-relaxation" strand and "stress-relieved" strand. Over the past several years, use of low-relaxation strand has progressively increased to a point that currently the stress-relieved strand is seldom used. Thus, low-relaxation strand is used in the load tables in Chapter 2 and in the various examples throughout this Handbook.

Architectural precast concrete is also sometimes prestressed. Depending on the facilities available at the plant, the prestressing tendons may be either pretensioned or post-tensioned. In post-tensioning, the tendons are either placed in a conduit or are coated so they will not bond to the concrete. The tendons are then tensioned after the concrete has reached the predetermined strength. When the tendons are placed in a conduit, they are usually grouted after tensioning (bonded post-tensioning). When they are greased and wrapped, or coated, they usually are not grouted (unbonded post-tensioning). For more information on post-tensioning in general, see Ref. 20.

Precast products are typically prestressed with 7-wire strand not prestressing bars. Prestressing bars meeting ASTM A722 have been used in connections between members.

The properties of prestressing strand, wire and bars are given in Chapter 11.

1.3.3.2 Deformed reinforcing bars

Reinforcing bars are hot-rolled from steels with varying carbon content. They are usually required to meet ASTM A615, A616, or A617. These specifications are of the performance type, and do not closely control the chemistry or manufacture of the bars. Bars are usually specified to have a minimum yield of 40,000 psi (Grade 40) or 60,000 psi (Grade 60). Grade 40 bars will usually have a lower carbon content than Grade 60, but not necessarily.

ASTM A706 specifies a bar with controlled chemistry that is weldable. For bars that are not to be welded, see Sect. 6.5.1.

In order for the reinforcing bar to develop its full strength in the concrete, a minimum length of embedment is required, or the bars may be hooked. Information on bar sizes, bend and hook dimensions and development length are given in Chapter 11 and Ref. 21.

1.3.3.3 Welded wire fabric

Welded wire fabric is a prefabricated reinforcement consisting of parallel, cold-drawn wires welded together in square or rectangular grids. Each wire intersection is electrically resistance-welded by a continuous automatic welder. Pressure and heat fuse the intersecting wires into a homogeneous section and fix all wires in their proper position.

Plain wires, deformed wires or a combination of both may be used in welded wire fabric. Plain wire sizes are specified by the letter W followed by a number indicating the cross-sectional area of the wire in hundredths of a square inch. For example, W16 denotes a plain wire with a cross-sectional area of 0.16 sq in. Similarly, deformed wire sizes are specified by the letter D followed by a number which indicates area in hundredths of a square inch.

Plain wire fabric bonds to concrete by the mechanical anchorage at each welded wire intersection. Deformed wire fabric utilizes wire deformations plus welded intersections for bond and anchorage.

Welded wire fabric for architectural precast concrete is normally supplied in flat sheets. Use of fabric from rolls, particularly in thin precast sections, is not recommended because the rolled fabric can not be flattened to the required tolerance. In addition to using fabric in flat sheets, many plants have equipment for bending sheets into various shapes, such as U-shaped stirrups, four-sided cages, etc.

Available wire sizes, common stock sizes and other information on welded wire fabric are given in Chapter 11 and Ref. 22.

1.3.4 Durability

Durability is of concern when the structure is exposed to an aggressive environment. The designer must be concerned about the deleterious effects of (a) freeze-thaw, (b) chemical attack, and (c) corrosion of embedded metals. The ideal approach is to make the concrete impermeable, which means to make the concrete uniformly as dense as possible, and designed for adequate crack control. In this respect, precast/prestressed concrete is inherently advantageous since it is produced in a controlled environment which lends itself to high quality concrete, and prestressing leads to effective crack control.

Penetrating surface sealers can improve the durability of concrete by reducing moisture and chloride penetration. Sealers have however, little ability to bridge cracks and should not be expected to provide protection from moisture absorption or chloride penetration at cracks. Some sealers have proven to be more effective than others. Their evaluation should be based on criteria established in National Cooperative Highway Research Program (NCHRP 244)—"Concrete Sealers for Protection of Bridge Structures." Silane based sealers are hydropic and have been demonstrated to reduce chloride penetration into concrete as much as 95%.

1.3.4.1 Freeze-thaw and chemical resistance

The typical dense mixes used for precast products have high resistance to freezing and thawing. Entrained air may be used to further improve freeze-thaw resistance in particularly severe environments. For some of the concrete mixtures, such as low-slump mixtures in extruded products or gap-graded mixtures in architectural precast concrete, it is difficult to measure air content. Thus, it is recommended that a "normal dosage" of the air-entraining agent be used instead of specifying a particular range of air content percentage. For precast concrete elements constructed above grade and oriented in a vertical position, air contents as low as 2 to 3% will usually provide the required durability. The precast/prestressed concrete industry does not use air-entraining portland cements. Instead, admixtures are added to the concrete during the mixing cycle to entrain air. Tolerance on air content is ± 1.5%. ACI 318-89 permits a 1% lower air content for concrete strengths higher than 5000 psi.

In addition to entrained air, other positive measures, such as adequate concrete cover over steel and rapid drainage of surface water may be essential in structures exposed to weather.

Freeze-thaw damage, which manifests itself by scaling of the surface, is magnified when deicing chemicals are used. Deicers can be applied indirectly in various ways, such as by drippings from the underside of vehicles. Table 1.3.2 (Table 4.1.1, ACI 318-89) provides the required air content for both severe and moderate exposure conditions, for various maximum aggregate sizes. Severe exposure is defined as a climate where the concrete may be in almost continuous contact with moisture prior to freezing, or where deicing salts come in contact with the concrete. Salt-laden air, as found in coastal areas, is also considered a severe exposure. A moderate exposure is one where deicing salts are not used, or where the concrete will only occasionally be exposed to moisture prior to freezing.

Table 1.3.2 Total air content for frost-resistant concrete.

Nominal maximum aggregate size,* in.	Air content, percent	
	Severe exposure	Moderate exposure
⅜	7½	6
½	7	5½
¾	6	5
1	6	4½
1½	5½	4½
2†	5	4
3†	4½	3½

*See ASTM C33 for tolerances on oversize for various nominal maximum size designations.

†These air contents apply to total mix, as for the preceding aggregate sizes. When testing these concretes, however, aggregate larger than 1½ in. is removed by handpicking or sieving and air content is determined on the minus 1½ in. fraction of mix. (Tolerance on air content as delivered applies to this value). Air content of total mix is computed from value determined on the minus 1½ in. fraction.

Table 1.3.3 (Table 4.1.2, ACI 318-89) provides the maximum permitted water-cement ratio (or, for lightweight concrete, minimum f'_c) for concrete which is exposed to an aggressive environment. ACI 318-89 also requires that the minimum cement content of concrete mixtures exposed to deicing chemicals be not less than 520 lb per cubic yard. In many plants, a water-cement ratio of 0.35 is frequently used for added durability of precast, prestressed concrete products.

Table 1.3.3 Requirements for special exposure conditions.

Exposure condition	Maximum water-cement ratio, normal weight aggregate concrete	Minimum f'_c lightweight aggregate concrete
Concrete intended to have low permeability when exposed to water	0.50	3750
Concrete exposed to freezing and thawing in a moist condition	0.45	4250
For corrosion protection for reinforced concrete exposed to deicing salts, brackish water, seawater, or spray from these sources	0.40*	4750*

*If minimum concrete cover, required by ACI 318-89, Sect. 7.7, is increased by 0.5 in., water-cement ratio may be increased to 0.45 for normal weight concrete, or f'_c reduced to 4250 for lightweight concrete.

1.3.4.2 Protection of reinforcement

Reinforcing steel is protected from corrosion by embedment in concrete. A protective iron oxide film forms on the surface of the bar, wire or strand as a result of the high alkalinity of the cement paste. As long as the high alkalinity is maintained, the film is effective in preventing corrosion.

The protective high alkalinity of the cement paste may be lost in the presence of oxygen, moisture, and chlorides. Chlorides may be found in concrete aggregates, water, cementitious materials or admixtures, and hence may be present in the concrete when cast. Concrete of low permeability and of sufficient cover over the steel will usually provide the necessary protection against chloride penetration due to the presence of deicer salts.

Moisture and oxygen alone can cause corrosion. Cracking may allow oxygen and moisture to reach the embedded steel, resulting in conditions where rusting of the steel and staining of the surface may rapidly occur. Prestressing can control cracking; in non-prestressed members, the choice of a sufficiently low stress level in the steel under permanent loads can limit the width of cracks and the intrusion of water, thus maintaining the integrity of the reinforcement.

Calcium chloride as an admixture should not be used in prestressed concrete. Consideration should be given to the chloride ion content in hardened concrete, contributed by the ingredients of the mix. Table 1.3.4 (Table 4.3.1, ACI 318-89) indicates the maximum chloride ion content. These limits are intended for use with uncoated reinforcement.

Table 1.3.4 Maximum chloride ion content for corrosion protection.

Type of member	Maximum water-soluble chloride ion (Cl⁻) in concrete, percent by weight of cement
Prestressed concrete	0.06
Reinforced concrete exposed to chloride in service	0.15
Reinforced concrete that will be dry or protected from moisture in service	1.00
Other reinforced concrete construction	0.30

In order to provide corrosion protection to reinforcement, concrete cover should conform to ACI 318-89, Section 7.7, as listed in Table 1.3.5. Concrete cover is the minimum clear distance from the reinforcement to the face of the concrete. For exposed aggregate surfaces, the concrete cover is not measured from the original surface; instead, the depth of the mortar removed between the pieces of coarse aggregate (depth of reveal) should be subtracted. Attention must also be given to scoring, false joints, and drips, as these reduce the cover. As noted above, high quality concrete provides adequate corrosion protection for reinforcement for most conditions. Even in moderate to severe aggressive environments, concrete can provide adequate protection with proper attention to mix design, steel stress level and the extent of cracking under service loads, and the depth of concrete cover. Only when these protection measures are not feasible, it may be necessary to consider other ways of protecting reinforcement, such as galvanizing or epoxy coating. These are described below.

Galvanized reinforcement. Except for exposed connections, there is rarely any technical need for galvanized reinforcement. With proper detailing and specifications, galvanizing may be superfluous. However, in architectural precast concrete, some manufacturers prefer galvanized welded wire fabric, because corrosion will not generally occur during prolonged storage of the mesh.

Galvanizing is not recommended for members subjected to chlorides, because a deleterious chemical reaction can take place when the concrete is damp and chlorides are present. Therefore, the benefit obtained by galvanizing is questionable for members subjected to deicing salts and similar exposures.

Galvanized welded wire fabric is usually available as a stock item in selected sizes. Individual wires are galvanized before they are welded together to form the fabric; zinc at each wire intersection is burned off during welding, but the resulting black spots have not caused noticeable problems. After welding, the fabric is shipped without further treatment. There is no current ASTM specification for galvanized welded wire fabric.

Table 1.3.5 Minimum cover requirements for precast and prestressed concrete.[1]

Condition	Minimum cover
Exposed to earth or weather[2] • Wall panels • Other members	#11 and smaller— ¾ in. #6 through #11— 1½ in. #5, W31 or D31 wire, and smaller—1¼ in.
Not exposed to earth or weather • Wall panels, slabs and joists • Beams and columns: Main steel Ties, stirrups or spirals	#11 and smaller—⅝ in. Diameter of bar, but not less than ⅝ in. and need not exceed 1½ in. All sizes— ⅜ in.

1. Manufactured under plant control conditions.
2. Increase cover by 50% if tensile stress of prestressed members exceeds $6\sqrt{f'_c}$.

When galvanized reinforcement is used in concrete, it should not be coupled directly to ungalvanized steel reinforcement, copper, or other dissimilar materials. Polyethylene and similar tapes can be used to provide local insulation between dissimilar materials. Galvanized reinforcement should be fastened with ties of soft stainless steel, or zinc coated or nonmetallic coated tie wire.

Epoxy coated reinforcement. Epoxy coated reinforcing bars and welded wire fabric are also available for use in members which require special corrosion protection. Epoxy coated reinforcing bars should conform to ASTM A775, and epoxy coated welded wire fabric should conform to ASTM A884. The epoxy provides excellent protection from corrosion if the bar is uniformly coated. Bars generally are coated when straight; subsequent bending should have no adverse effect on the integrity of the coating. It the coating is removed or damaged, the reinforcement should be touched up with commercially available epoxy compounds. Epoxy coating reduces bond strength; reference should be made to Section 12.2.4.3 of ACI 318-89 and Chapter 11 for the required increase in development length. Similarly, the requirements for crack control may need to be modified.

Supplemental items must also be protected to retain the full advantage of protecting the main reinforcement. For example, bar supports should be solid plastic and bar ties should be nylon-, epoxy-, or plastic-coated wire, rather than black wire. Epoxy coated reinforcing bars should be handled with nylon slings.

Epoxy coated strand. Epoxy coated strand is a relatively new product proposed for use in unusually aggressive environments. It is covered by ASTM A882, "Standard Specification for Epoxy-Coated Seven-Wire Prestressing Steel Strand."

In order for it to be used as bonded strand, the epoxy coating is impregnated with a grit to assure development of bond; without grit, the epoxy coated strand has virtually no bond strength. With adequate density and a proper distribution of the grit, the bond strength of the epoxy coated strand is comparable to that of uncoated strand.*

The behavior of epoxy coated strand at elevated temperatures is of concern due to softening of the epoxy. Pull-out tests[23] show that there is a progressive reduction in bond strength initiating at about 120°F with a virtually complete loss of bond occurring at about 200°F. This behavior necessitates a careful monitoring of concrete temperature at transfer of prestress.

Because of the uncertainties in properties noted above, particularly the behavior under elevated temperatures, it is recommended that epoxy coated strand not be used for pretensioned, prestressed concrete products.

1.3.4.3 Sulphate attack

Chemical attack of concrete materials may occur from sulphates found in ground water, soils, processing liquids, or sewage. The proper choice of cement is particularly effective in resisting sulphate attack. ACI 318-89, Section 4.2.1, and the references contained therein, provides detailed guidance.

---

*Tests have shown the transfer and the development lengths of the epoxy coated (with grit) strand to be somewhat shorter than the corresponding lengths for the uncoated strand. Also, there are differences in some other properties of the two types of strand. For example, the relaxation loss of epoxy coated strand is higher (about twice) than that of uncoated, low-relaxation strand. Assessment of the differences in various properties and their impact on design is currently under study by the PCI Ad-Hoc Committee on Epoxy Coated Strand. This committee is developing a report "Guidelines for Design and Installation of Epoxy Coated Strand in Prestressed Concrete Structures."

1.4 References

1. "Reflections on the Beginnings of Prestressed Concrete in America," JR-H-81, Precast/Prestressed Concrete Institute, Chicago, IL, 1981.

2. "Manual for Quality Control for Plants and Production of Precast and Prestressed Concrete Products," Third Edition, MNL-116-85, Precast/Prestressed Concrete Institute, Chicago, IL, 1985.

3. "Manual for Quality Control for Plants and Production of Architectural Precast Concrete Products," MNL-117-77, Precast/Prestressed Concrete Institute, Chicago, IL, 1977.

4. Pfeifer, Donald W., Landgren, J.R. and Perenchio, William, "Concrete, Chlorides, Cover and Corrosion," *PCI Journal,* V. 31, No. 4, July-August 1986.

5. "Parking Structures: Recommended Practice for Design and Construction," MNL-129-88, Precast/Prestressed Concrete Institute, Chicago, IL, 1988.

6. "Architectural Precast Concrete Cladding – Its Contribution to Lateral Resistance of Buildings," *Proceedings,* SP-CP, Precast/Prestressed Concrete Institute, Chicago, IL, 1990.

7. "Architectural Precast Concrete," Second Edition, MNL-122-89, Precast/Prestressed Concrete Institute, Chicago, IL, 1989.

8. "Recommended Practice for Glass Fiber Reinforced Concrete Panels," MNL 128-87, Precast/Prestressed Concrete Institute, Chicago, IL, 1987.

9. Abdel-Karim, A. M. and Tadros, Maher K., "Stretched-Out Precast Concrete I-Girder Bridge Spans," *Concrete International,* V. 13, No. 9, September 1991.

10. ABAM Engineers, Inc. "Precast Prestressed Concrete Horizontally Curved Bridge Beams," *PCI Journal,* V. 33, No. 5, September-October 1988.

11. Ross Bryan Associates, Inc., "Recommended Practice for Precast Prestressed Concrete Composite Bridge Deck Panels," *PCI Journal,* V. 33, No. 2, March-April 1988.

12. ACI Committee 363, "State-of-the-Art Report on High-Strength Concrete," ACI 363R-84, *Journal of the American Concrete Institute,* V. 81, No. 4, July-August 1984.

13. Ahmad, Shuaib H. and Shah, S.P., "Structural Properties of High Strength Concrete and Its Implications for Precast Prestressed Concrete." *PCI Journal,* V. 30, No. 6, November-December 1985.

14. Pfeifer, D.W., Marusin, Stella and Landgren, J.R., "Energy-Efficient Accelerated Curing of Concrete," *Technical Report,* TR-1-82, Precast/Prestressed Concrete Institute, Chicago, IL, 1982.

15. "Building Code Requirements for Reinforced Concrete," ACI 318-89, and "Commentary," ACI 318R-89, American Concrete Institute, Detroit, MI, 1989.

16. "Concrete Manual – A Manual for the Control of Concrete Construction," 8th Edition, revised, U.S. Department of Interior, Bureau of Reclamation, Denver, CO, 1981.

17. "Cementitious Grouts and Grouting," EB 111T, Portland Cement Association, Skokie, IL, 1990.

18. "Corps of Engineers Specification for Non-Shrink Grout," CRD-C588-78A, U.S. Army Corps of Engineers, 1978.

19. "Use of Epoxy Compounds with Concrete," ACI 503R-80, *ACI Manual of Concrete Practice*, Part 5, American Concrete Institute, Detroit, MI.

20. "Post-Tensioning Manual," Fifth Edition, Post-Tensioning Institute, Phoenix, AZ, 1990.

21. "Manual of Standard Practice," 25th Edition, Concrete Reinforcing Steel Institute, Schaumburg, IL, 1990.

22. "Structural Welded Wire Fabric Manual of Standard Practice," 4th Edition, Wire Reinforcement Institute, McLean, VA 1992.

23. LeClaire, Philip J. and Shaikh, A. Fattah, "Effect of Temperature on Bond Strength of Epoxy-Coated Prestressing Strand," *PCI Journal* (scheduled for publication in 1992).

CHAPTER 2
PRODUCT INFORMATION AND CAPABILITY

	Page No.
2.1 General	2-2
2.1.1 Notation	2-2
2.1.2 Introduction	2-2
2.2 Explanation of Load Tables	2-3
2.2.1 Safe Superimposed Load	2-3
2.2.2 Limiting Criteria	2-3
2.2.3 Estimated Cambers	2-4
2.2.4 Design Parameters	2-4
2.2.5 Concrete Strength and Unit Weights	2-4
2.2.6 Prestressing Strands	2-4
2.2.7 Losses	2-5
2.2.8 Strand Placement	2-5
2.2.9 Columns and Load Bearing Wall Panels	2-5
2.2.10 Piles	2-5
2.2.11 References	2-6
2.3 Stemmed Deck Members	
Double tee load tables	2-7
2.4 Flat Deck Members	
Hollow-core slab load tables	2-27
Hollow-core slab section properties	2-33
Solid slab load tables	2-36
2.5 Beam Load Tables	
Rectangular beams	2-42
L-shaped beams	2-43
Inverted tee beams	2-45
2.6 Columns and Load Bearing Wall Panels	
Precast, prestressed columns	2-48
Precast, reinforced columns	2-50
Double tee wall panels	2-52
Hollow-core wall panels	2-53
Solid wall panels	2-54
2.7 Piles	
Piles	2-56
Sheet piles	2-57

PRODUCT INFORMATION AND CAPABILITY

2.1 General

2.1.1 Notation

A	=	cross-sectional area
A_g	=	gross cross-sectional area
b_w	=	web width
D	=	unfactored dead loads
E_c	=	modulus of elasticity of concrete
e_c	=	eccentricity of prestress force from the centroid of the section at the center of the span
e_e	=	eccentricity of prestress force from the centroid of the section at the end of the span
f'_c	=	specified compressive strength of concrete
f'_{ci}	=	compressive strength of concrete at time of initial prestress
f_{pu}	=	ultimate strength of prestressing steel
f_{se}	=	effective stress in prestressing steel after losses
h	=	overall depth of member
I	=	moment of inertia
L	=	unfactored live loads
ℓ	=	span
M_n	=	nominal moment strength of a member
M_{nb}	=	nominal moment strength under balanced conditions
M_o	=	nominal moment strength of a compression member with zero axial load
M_u	=	applied factored moment at section
P_n	=	nominal axial load strength of a compression member at given eccentricity
P_{nb}	=	nominal axial load strength under balanced conditions
P_o	=	nominal axial load strength of a compression member with zero eccentricity
P_u	=	factored axial load
t	=	thickness
V_{ci}	=	nominal shear strength provided by concrete when diagonal cracking results from combined shear and moment
V_{cw}	=	nominal shear strength provided by concrete when diagonal cracking results from excessive principal tensile stress in the web
V_u	=	factored shear force
V/S	=	volume-surface ratio
y_b	=	distance from bottom fiber to center of gravity of section
y_t	=	distance from top fiber to center of gravity of section
Z	=	section modulus
Z_b	=	section modulus with respect to the bottom fiber of a cross section
Z_t	=	section modulus with respect to the top fiber of a cross section
δ	=	moment magnification factor
ϕ	=	strength reduction factor

2.1.2 Introduction

This chapter presents data on the shapes that are standard in the precast, prestressed concrete industry. Other shapes and standard shapes with depth and width variations are also available in many areas of the country. Designers should contact the local manufacturers in the geographic area of the proposed structure or project to determine the properties and dimensions of products available. This section, plus the design methods and aids provided in other chapters of this Handbook, should enable the designer to carry out safe and economical designs.

2.2 Explanation of Load Tables

The load tables on the following pages show dimensions, section properties and load carrying capabilities of the shapes most commonly used throughout the industry. These shapes include double tees, hollow-core and solid flat slabs, beams, girders, columns, piles and wall panels. The dimensions of the shapes shown in the tables may vary among manufacturers. The variations are usually small and the tables given here can still be used particularly for preliminary design. Manufacturers will in most cases have their own catalogs with load tables and geometric properties for the members they produce. Hollow-core slabs of different thicknesses, core sizes and shapes are available in the market under various trade names. Cross-sections and section properties of proprietary hollow-core slabs are shown on pages 2-33 through 2-35. Load tables on pages 2-27 through 2-32 are developed for non-proprietary hollow-core sections of thicknesses most commonly used in the industry.

Load tables for stemmed deck members, flat deck members and beams show the allowable uniform superimposed service load, estimated camber at the time of erection and the estimated long-term camber after the member has essentially stabilized, but before the application of superimposed live loads. For deck members, except pretopped double tees, the table at the top of the page gives the information for the member with no topping, and the table at the bottom of the page is for the same member with 2 in. of normal weight concrete topping acting compositely with the precast section. Values in the tables assume a uniform 2 in. topping over the full span length, and assume the member to be unshored at the time the topping is placed. Safe loads and cambers shown in the tables are based on the dimensions and section properties shown on the page, and will vary for members with different dimensions.

For beams, a single load table is used for several sizes of members. The values shown are based on sections containing the maximum practical number of prestressing strands, but in some cases, more strands could be used.

2.2.1 Safe Superimposed Load

The values for safe superimposed uniform service load are based on the capacity of the member as governed by the ACI Building Code limitations on flexural strength, service load flexural stresses, or, in the case of flat deck members without shear reinforcement, shear strength. A portion of the safe load shown is assumed to be dead load for the purpose of applying load factors and determining cambers and deflections. For untopped deck members, 10 psf of the capacity shown is assumed as superimposed dead load, typical for roof members. For topped deck members, 15 psf of the capacity shown is assumed as superimposed dead load, typical of floor members. The capacity shown is in addition to the weight of the topping. For beams, 50 percent of the capacity shown in the load table is assumed as dead load.

Example: For an 8DT24/88-D1 untopped, (p. 2-17) with a 52 ft span, the capacity shown is 70 psf. The member can safely carry service loads of 10 psf dead and 60 psf live.

2.2.2 Limiting Criteria

The criteria used to determine the safe superimposed load and strand placement are based on *Building Code Requirements for Reinforced Concrete ACI 318-89.* (This is referred to as "the Code," "the ACI Building Code" or "ACI 318" in this Handbook.) For design procedures, see Chapter 4 of this Handbook. A summary of the Code provisions used in the development of these load tables is as follows:

1. Capacity governed by design flexural strength:

   Load factors: $1.4D + 1.7L$

   Strength reduction factor, $\phi = 0.90$

   Calculation of design moments assumes simple spans with roller supports. If the strands are fully developed (see Sect. 4.2.3), the critical moment is assumed to be at midspan in members with straight strands, and at 0.4ℓ (ℓ = span) in products with strands depressed at midspan. (Note: The actual critical point can be determined by analysis, but will seldom vary significantly from 0.4ℓ). Flexural strength is calculated using strain compatibility as discussed in Chapter 4.

2. Capacity governed by service load stresses:

   Flexural stresses immediately after transfer of prestress:

   a) Compression: $0.6f'_{ci}$

   b) End tension: $6\sqrt{f'_{ci}}$

   c) Midspan tension: $3\sqrt{f'_{ci}}$

   Stresses at service loads, after all losses:

   a) Compression: $0.45f'_c$

   b) Tension:

   Stemmed deck members and beams: $12\sqrt{f'_c}$

   (Deflections must be determined based on bilinear moment-deflection relationships. See Sect. 4.6.3.)

Flat deck members: $6\sqrt{f'_c}$

Critical point for service load moment is assumed at midspan for members with straight strands and at 0.4ℓ for members with strands depressed at midspan, as described above.

3. Capacity governed by design shear strength:

   Load factors: $1.4D + 1.7L$

   Strength reduction factor, $\phi = 0.85$

   In flat deck members, since use of shear reinforcement is generally not feasible, the capacity may be limited by the design concrete shear strength. In this case, the safe superimposed load is obtained by equating the corresponding factored shear force, V_u, to the lesser of ϕV_{ci} and ϕV_{cw} (see Sect. 11.4.2 of ACI 318-89 and Sect. 4.3 of this Handbook).

   Stemmed deck members and beams do not have this limitation, namely the design concrete shear strength is not used as limiting criterion since shear reinforcement can be readily provided in such members. The design of shear reinforcement is illustrated in Sect. 4.3. In stemmed deck members, usually minimum or no shear reinforcement (see Sect. 11.5.5 of ACI 318-89) is required.

Note 1: End stresses are calculated 50 strand diameters from the end of the member, the theoretical point of full transfer.

Note 2: Release tension is not used as a limiting criterion for beams. Supplemental top reinforcement must be provided, designed as described in Sect. 4.2.2.2.

Note 3: For *stemmed deck members,* the lateral (flange) flexural and shear strengths and the service load stresses are not considered as limiting criteria. For heavy uniform loads and/or concentrated loads, these considerations may limit the capacity or require transverse reinforcement.

Note 4: *Flat deck members* show no values beyond a span/depth ratio of 50 for untopped members and 40 for topped members. These are suggested maximums for roof and floor members respectively, unless a detailed analysis is made.

2.2.3 Estimated Cambers

The estimated cambers shown are calculated using the multipliers given in Sect. 4.6.5 of this Handbook. *These values are estimates, and should not be used as absolute values.* Attachment of nonstructural elements, such as partitions, folding doors and architectural decoration to members subject to camber variations should be designed with adequate allowance for these variations. Calculation of topping quantities should also recognize that the values can vary with camber.

2.2.4 Design Parameters

The design of prestressed concrete is dependent on many variables. These include concrete strength at release and at 28 days; unit weight of concrete; grade, profile and placement of strand; jacking tension and other. The load tables given here are based on commonly used values; somewhat higher allowable loads or longer spans may be achieved by selecting other appropriate set of conditions. However, in these cases, consultation with the candidate producers is recommended.

2.2.5 Concrete Strength and Unit Weights

Twenty-eight day cylinder strength for concrete in the precast units is assumed to be 5000 psi unless otherwise indicated. Tables for units with composite topping are based on the topping concrete being normal weight concrete with a cylinder strength of 3000 psi.

Concrete strength at time of strand tension release is 3500 psi unless the value falls below the heavy line shown (Note: this area is also shown as shaded) in the load table, indicating that a cylinder strength greater than 3500 psi is required. No values are shown when the required release strength exceeds 4500 psi.

The concrete strengths used in the load tables are not intended to be limitations or recommendations for actual use. Some precast concrete manufacturers may choose to use higher or lower concrete strengths, resulting in slightly different load table values. For low levels of prestress, the concrete release strength is usually governed by handling stresses. For products such as columns, piles, wall panels and lightly stressed flexural members a minimum release strength of 2000 psi is realistic.

Unit weights of concrete are assumed to be 150 lb. per cu. ft. for normal weight and 115 lb. per cu. ft. for lightweight.

2.2.6 Prestressing Strands

Prestressing strands are available in diameters from ¼ in. to 0.6 in., grades 250 ksi and 270 ksi. The predominant current use is the low-relaxation type; thus, the load tables are here presented for this type of strand only. Since the relaxation loss of low-relaxation strand remains proportional up to an initial stress of 0.75f_{pu}[1], the tables have been developed assuming this value of initial stress. Stress at transfer of prestress has been assumed at 90 percent of the initial stress.

In developing the load tables, the stress-strain behavior of the strand was modeled with the equations given in Fig. 11.2.5. These equations are of the same form as those used in previous editions of this Handbook. However, changes in coefficients and the limiting values were made for compliance with ASTM specification minimum values and for a better correlation with experimental data.

Other stress-strain models are also available and could be used in place of the equations used here. One such model has been validated by comparison with a large number of tests.[4] This model and other valid stress-strain models are expected to produce allowable loads which differ only slightly from the values calculated using the model depicted in Fig. 11.2.5 and shown in the load tables in this chapter.

2.2.7 Losses

Losses were calculated in accordance with the recommendations given in Ref. 2. This procedure includes consideration of initial stress level (0.7 f_{pu} or higher), type of strand, exposure conditions and type of construction. A lower limit of 30,000 psi for low-relaxation strands was used. This lower limit is arbitrary; other designers may choose not to impose this limit. Additional information on losses is given in Sect. 4.5.

2.2.8 Strand Placement

Quantity, size and profile of strands are shown in the load tables under the column headed "Strand Pattern", for example, 88-S. The first digit indicates the total number of strands in the unit, the second digit is the diameter of the strand in 16ths of an inch (⁸⁄₁₆ equals ½ in. diameter), and the "S" indicates that the strands are straight. A "D1" indicates that the strands are depressed at midspan. Some precast producers choose to depress the strand at 2 points, which provides a somewhat higher capacity.

For *stemmed deck members* and *beams*, the eccentricities of strands at the ends and midspan are shown in the load tables. Strands have been placed so that the stress at 50 strand diameters from the end (theoretical transfer point) will not exceed those specified above.

For *flat deck members*, the load table values are based on prestressing strand centered 1½ in. from the bottom of the slab. Strand placement can vary from as low as ⅞ in. to as high as 2⅛ in. from the bottom, which will change the capacity and camber values shown. The higher strand placements give improved fire resistance ratings (see Sect. 9.3 of this Handbook for more information on fire resistance). The lower strand placement may require higher release strengths, or top tension reinforcement at the ends. The designer should contact the local supplier of flat prestressed concrete deck members for available and recommended strand placement locations.

2.2.9 Columns and Load Bearing Wall Panels

Interaction curves for selected precast, prestressed columns, precast reinforced columns and various types of commonly used wall panels are provided on pp. 2-48 to 2-54.

These interaction curves are based on strength design. Appropriate load factors must be applied to the service loads and moments before entering the charts. The effect of slenderness is considered through magnification of moment as described in Sect. 3.5.

For prestressed columns and wall panels, curves are shown for both partially developed strands, usually appropriate for end moment capacity, and fully developed strands, usually appropriate for mid-span moment capacity. This is discussed more completely in Sect. 4.7. The columns and wall panels which use reinforcing bars assume full development of the bars.

The column curves are terminated at a value of $P_u = 0.80\phi P_o$, the maximum allowable load for tied columns under ACI 318-89. Most of the wall panel curves show the lower portion of the curve only (flexure controlling). Actual design loads will rarely exceed the values shown.

The curves for double tee wall panels are for bending in the direction that causes tension in the stem. They are therefore conservative, in the range shown, when applied to bending in the opposite direction.

The curves for hollow-core wall panels are based on a generic section as shown. They can be used with small difference for all sections commonly marketed for wall panel use.

2.2.10 Piles

Allowable concentric service loads on prestressed concrete piles, based on the structural capacity of the pile alone, are shown in Table 2.7.1. The ability of the soil to carry these loads must be evaluated separately. Values for concrete strengths up to 8000 psi are shown. The availability of these concrete strengths should be checked with local manufacturers. The design of prestressed concrete piles is discussed in Sect. 4.7.6 of this Handbook.

Section properties and allowable service load bending moments for prestressed concrete sheet pile units are shown in Table 2.7.2. These units are available in some areas for use in earth retaining structures.

2.2.11 References

1. Martin, L.D. and Pellow, D.L. "Low-Relaxation Strand—Practical Applications in Precast Prestressed Concrete," *PCI Journal,* V. 28, No. 4, July-August 1983.

2. Zia, Paul, Preston, H.K., Scott, N.L. and Workman, E.B., "Estimating Prestress Losses," *Concrete International*, V. 1, No. 6, June 1979.

3. PCI Committee on Prestressed Concrete Piling, "Recommended Practice for Design, Manufacture and Installation of Prestressed Concrete Piling," *PCI Journal*, V. 22, No. 2, March-April 1977.

4. Devalapura, Ravi K. and Tadros, Maher K., "Stress-Strain Modeling of 270 ksi Low-Relaxation Prestressing Strands," *PCI Journal*, V. 37, No. 2, March-April 1992.

Strand Pattern Designation

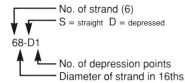

Safe loads shown include dead load of 10 psf for untopped members and 15 psf for topped members. Remainder is live load. Long-time cambers include superimposed dead load but do not include live load.

Key
- 178 — Safe superimposed service load, psf
- 0.2 — Estimated camber at erection, in.
- 0.2 — Estimated long-time camber, in.

DOUBLE TEE

8'-0" x 12"
Normal Weight Concrete

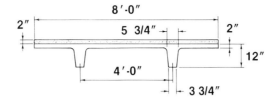

$f'_c = 5{,}000$ psi
$f_{pu} = 270{,}000$ psi

Section Properties

		Untopped		Topped	
A	=	287	in.²	—	
I	=	2,872	in.⁴	4,389	in.⁴
y_b	=	9.13	in.	10.45	in.
y_t	=	2.87	in.	3.55	in.
Z_b	=	315	in.³	420	in.³
Z_t	=	1,001	in.³	1,236	in.³
wt	=	299	plf	499	plf
		37	psf	62	psf
V/S	=	1.22	in.		

8DT12
No Topping

Table of safe superimposed service load (psf) and cambers

Strand Pattern	e_e / e_c	Span, ft.															
		12	14	16	18	20	22	24	26	28	30	32	34	36	38	40	42	44
28-S	7.13 / 7.13	178 0.2 0.2	137 0.2 0.3	108 0.2 0.3	82 0.3 0.4	61 0.3 0.4	45 0.3 0.4	34 0.3 0.3										
48-S	5.13 / 5.13			188 0.4 0.5	146 0.5 0.6	113 0.5 0.7	88 0.6 0.7	69 0.6 0.8	55 0.7 0.7	43 0.7 0.7	34 0.7 0.6						
68-S	3.13 / 3.13			162 0.4 0.5	126 0.5 0.6	99 0.5 0.6	78 0.6 0.6	62 0.6 0.6	50 0.6 0.5	40 0.5 0.4	31 0.4 0.1						
68-D1	3.13 / 6.63									94 1.4 1.6	78 1.5 1.7	65 1.6 1.7	54 1.6 1.6	45 1.7 1.5	38 1.7 1.3	31 1.6 1.0		
88-D1	1.13 / 6.38														38 1.8 1.1	32 1.7 0.6		

8DT12 + 2
2" Normal Weight Topping

Table of safe superimposed service load (psf) and cambers

Strand Pattern	e_e / e_c	Span, ft.											
		12	14	16	18	20	22	24	26	28	30	32	34
28-S	7.13 / 7.13	200 0.2 0.2	150 0.2 0.2	116 0.2 0.2	84 0.3 0.2	59 0.3 0.2	40 0.3 0.1						
48-S	5.13 / 5.13				168 0.5 0.6	127 0.5 0.5	96 0.6 0.5	73 0.6 0.4	55 0.7 0.3	41 0.7 0.1			
68-S	3.13 / 3.13					154 0.5 0.4	119 0.5 0.4	92 0.6 0.3	71 0.6 0.1	50 0.6 −0.1			
68-D1	3.13 / 6.63									100 1.4 0.9	81 1.5 0.8	65 1.6 0.6	50 1.6 0.3

Strength based on strain compatibility; bottom tension limited to $12\sqrt{f'_c}$; see pages 2-3–2-6 for explanation.
Shaded values require release strengths higher than 3500 psi.

Strand Pattern Designation

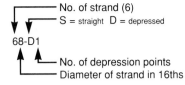

Safe loads shown include dead load of 10 psf for untopped members and 15 psf for topped members. Remainder is live load. Long-time cambers include superimposed dead load but do not include live load.

Key
- 186 — Safe superimposed service load, psf
- 0.2 — Estimated camber at erection, in.
- 0.3 — Estimated long-time camber, in.

DOUBLE TEE

8'-0" x 12"
Lightweight Concrete

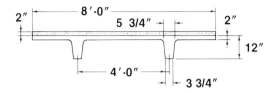

f'_c = 5,000 psi
f_{pu} = 270,000 psi

Section Properties

	Untopped	Topped
A =	287 in.²	—
I =	2,872 in.⁴	4,819 in.⁴
y_b =	9.13 in.	10.82 in.
y_t =	2.87 in.	3.18 in.
Z_b =	315 in.³	445 in.³
Z_t =	1,001 in.³	1,515 in.³
wt =	229 plf	429 plf
	29 psf	54 psf
V/S =	1.22 in.	

8LDT12
No Topping

Table of safe superimposed service load (psf) and cambers

Strand Pattern	e_e / e_c	\multicolumn{16}{c}{Span, ft.}
		12	14	16	18	20	22	24	26	28	30	32	34	36	38	40	42
28-S	7.13 / 7.13	186 / 0.2 / 0.3	144 / 0.3 / 0.4	116 / 0.4 / 0.5	89 / 0.5 / 0.6	68 / 0.5 / 0.6	53 / 0.6 / 0.7	41 / 0.6 / 0.7	31 / 0.6 / 0.6								
48-S	5.13 / 5.13			195 / 0.6 / 0.8	153 / 0.7 / 0.9	120 / 0.8 / 1.1	95 / 1.0 / 1.2	77 / 1.1 / 1.3	62 / 1.2 / 1.3	50 / 1.3 / 1.4	41 / 1.3 / 1.3	33 / 1.3 / 1.2					
68-S	3.13 / 3.13				169 / 0.6 / 0.8	133 / 0.8 / 1.0	106 / 0.9 / 1.1	86 / 1.0 / 1.1	70 / 1.0 / 1.1	57 / 1.1 / 1.1	47 / 1.1 / 1.0	39 / 1.1 / 0.8	32 / 1.0 / 0.5				
68-D1	3.13 / 6.63											72 / 2.6 / 2.9	61 / 2.8 / 2.9	52 / 3.0 / 2.9	45 / 3.1 / 2.7	38 / 3.2 / 2.4	33 / 3.2 / 2.1

8LDT12 + 2
2" Normal Weight Topping

Table of safe superimposed service load (psf) and cambers

Strand Pattern	e_e / e_c	\multicolumn{14}{c}{Span, ft.}
		12	14	16	18	20	22	24	26	28	30	32	34	36	38
28-S	7.13 / 7.13	207 / 0.2 / 0.2	157 / 0.3 / 0.3	123 / 0.4 / 0.4	91 / 0.5 / 0.4	66 / 0.5 / 0.4	47 / 0.6 / 0.3	33 / 0.6 / 0.2							
48-S	5.13 / 5.13			175 / 0.7 / 0.7	134 / 0.8 / 0.8	103 / 1.0 / 0.8	80 / 1.1 / 0.7	62 / 1.2 / 0.6	48 / 1.3 / 0.5	36 / 1.3 / 0.2					
68-S	3.13 / 3.13					162 / 0.8 / 0.7	126 / 0.9 / 0.7	99 / 1.0 / 0.6	79 / 1.0 / 0.5	62 / 1.1 / 0.3	47 / 1.1 / 0.1				
68-D1	3.13 / 6.63											72 / 2.6 / 1.3	59 / 2.8 / 1.0	46 / 3.0 / 0.6	33 / 3.1 / 0.0

Strength based on strain compatibility; bottom tension limited to $12\sqrt{f'_c}$; see pages 2-3–2-6 for explanation.
Shaded values require release strengths higher than 3500 psi.

Strand Pattern Designation

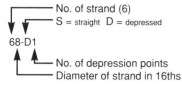

Safe loads shown include dead load of 10 psf for untopped members and 15 psf for topped members. Remainder is live load. Long-time cambers include superimposed dead load but do not include live load.

Key
168 — Safe superimposed service load, psf
0.2 — Estimated camber at erection, in.
0.2 — Estimated long-time camber, in.

DOUBLE TEE

8'-0" x 14"
Normal Weight Concrete

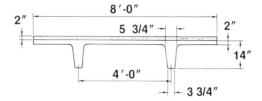

$f'_c = 5,000$ psi
$f_{pu} = 270,000$ psi

Section Properties

	Untopped	Topped
A =	306 in.²	—
I =	4,508 in.⁴	6,539 in.⁴
y_b =	10.51 in.	11.97 in.
y_t =	3.49 in.	4.03 in.
Z_b =	429 in.³	546 in.³
Z_t =	1,292 in.³	1,623 in.³
wt =	319 plf	519 plf
	40 psf	65 psf
V/S =	1.25 in.	

8DT14 — No Topping

Table of safe superimposed service load (psf) and cambers

Strand Pattern	e_e / e_c	14	16	18	20	22	24	26	28	30	32	34	36	38	40	42	44	46	48
28-S	8.51 / 8.51	168 / 0.2 / 0.2	134 / 0.2 / 0.3	103 / 0.2 / 0.3	77 / 0.3 / 0.3	58 / 0.3 / 0.3	44 / 0.3 / 0.3	33 / 0.3 / 0.3											
48-S	7.51 / 7.51				165 / 0.5 / 0.7	131 / 0.6 / 0.8	105 / 0.7 / 0.9	85 / 0.8 / 0.9	69 / 0.8 / 1.0	56 / 0.9 / 1.0	46 / 0.9 / 0.9	37 / 0.9 / 0.8	30 / 0.8 / 0.7						
68-S	4.51 / 4.51				178 / 0.5 / 0.6	142 / 0.5 / 0.7	114 / 0.6 / 0.8	93 / 0.7 / 0.8	76 / 0.7 / 0.8	62 / 0.7 / 0.8	51 / 0.7 / 0.7	41 / 0.7 / 0.6	34 / 0.6 / 0.4						
68-D1	4.51 / 8.01							145 / 1.1 / 1.4	121 / 1.2 / 1.5	101 / 1.3 / 1.6	85 / 1.4 / 1.6	72 / 1.5 / 1.7	61 / 1.6 / 1.6	51 / 1.6 / 1.6	43 / 1.6 / 1.4	36 / 1.5 / 1.2	30 / 1.4 / 0.9		
88-D1	2.51 / 7.76														61 / 2.0 / 2.0	53 / 2.0 / 1.8	45 / 2.0 / 1.6	39 / 2.0 / 1.3	33 / 1.9 / 1.0

8DT14 + 2 — 2" Normal Weight Topping

Table of safe superimposed service load (psf) and cambers

Strand Pattern	e_e / e_c	14	16	18	20	22	24	26	28	30	32	34	36	38
28-S	8.51 / 8.51	181 / 0.2 / 0.2	141 / 0.2 / 0.2	105 / 0.2 / 0.2	75 / 0.3 / 0.2	53 / 0.3 / 0.2	37 / 0.3 / 0.1							
48-S	7.51 / 7.51				179 / 0.5 / 0.5	139 / 0.6 / 0.6	109 / 0.7 / 0.6	85 / 0.8 / 0.6	67 / 0.8 / 0.5	51 / 0.9 / 0.4	39 / 0.9 / 0.2			
68-S	4.51 / 4.51					162 / 0.5 / 0.5	128 / 0.6 / 0.5	102 / 0.7 / 0.5	81 / 0.7 / 0.4	64 / 0.7 / 0.2	50 / 0.7 / 0.0			
68-D1	4.51 / 8.01							155 / 1.1 / 1.0	127 / 1.2 / 1.0	104 / 1.3 / 1.0	85 / 1.4 / 0.9	70 / 1.5 / 0.7	57 / 1.6 / 0.5	46 / 1.6 / 0.2

Strength based on strain compatibility; bottom tension limited to $12\sqrt{f'_c}$; see pages 2-3–2-6 for explanation.
Shaded values require release strengths higher than 3500 psi.

Strand Pattern Designation

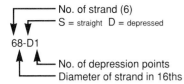

Safe loads shown include dead load of 10 psf for untopped members and 15 psf for topped members. Remainder is live load. Long-time cambers include superimposed dead load but do not include live load.

Key

176 — Safe superimposed service load, psf
0.2 — Estimated camber at erection, in.
0.3 — Estimated long-time camber, in.

DOUBLE TEE
8'-0" x 14"
Lightweight Concrete

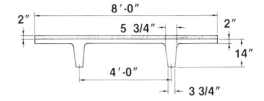

f'_c = 5,000 psi
f_{pu} = 270,000 psi

Section Properties

		Untopped	Topped
A	=	306 in.²	—
I	=	4,508 in.⁴	7,173 in.⁴
y_b	=	10.51 in.	12.40 in.
y_t	=	3.49 in.	3.60 in.
Z_b	=	429 in.³	578 in.³
Z_t	=	1,292 in.³	1,992 in.³
wt	=	244 plf	444 plf
		31 psf	56 psf
V/S	=	1.25 in.	

8LDT14
No Topping

Table of safe superimposed service load (psf) and cambers

Strand Pattern	e_e / e_c	14	16	18	20	22	24	26	28	30	32	34	36	38	40	42	44	46
28-S	8.51 / 8.51	176 / 0.2 / 0.3	142 / 0.3 / 0.4	110 / 0.4 / 0.5	85 / 0.4 / 0.5	66 / 0.5 / 0.6	52 / 0.5 / 0.6	41 / 0.5 / 0.6	32 / 0.5 / 0.5									
48-S	7.51 / 7.51				173 / 0.8 / 1.1	139 / 1.0 / 1.2	113 / 1.1 / 1.4	93 / 1.2 / 1.5	77 / 1.4 / 1.6	64 / 1.5 / 1.7	53 / 1.6 / 1.7	45 / 1.7 / 1.7	37 / 1.7 / 1.6	31 / 1.7 / 1.5				
68-S	4.51 / 4.51				186 / 0.7 / 1.0	150 / 0.9 / 1.1	122 / 1.0 / 1.2	100 / 1.1 / 1.3	83 / 1.2 / 1.4	70 / 1.3 / 1.4	58 / 1.3 / 1.4	49 / 1.4 / 1.3	41 / 1.4 / 1.2	35 / 1.4 / 1.0				
68-D1	4.51 / 8.01								109 / 2.1 / 2.6	93 / 2.3 / 2.7	80 / 2.5 / 2.8	69 / 2.7 / 2.9	59 / 2.8 / 2.9	51 / 2.9 / 2.8	44 / 3.0 / 2.7	38 / 3.1 / 2.4	33 / 3.0 / 2.1	

8LDT14 + 2
2" Normal Weight Topping

Table of safe superimposed service load (psf) and cambers

Strand Pattern	e_e / e_c	14	16	18	20	22	24	26	28	30	32	34	36	38	40	42
28-S	8.51 / 8.51	189 / 0.2 / 0.3	149 / 0.3 / 0.3	112 / 0.4 / 0.3	83 / 0.4 / 0.3	61 / 0.5 / 0.3	44 / 0.5 / 0.3	31 / 0.5 / 0.1								
48-S	7.51 / 7.51				187 / 0.8 / 0.8	147 / 1.0 / 0.9	117 / 1.1 / 1.0	93 / 1.2 / 1.0	74 / 1.4 / 1.0	59 / 1.5 / 0.9	47 / 1.6 / 0.7	36 / 1.7 / 0.5				
68-S	4.51 / 4.51					170 / 0.9 / 0.8	136 / 1.0 / 0.8	109 / 1.1 / 0.8	88 / 1.2 / 0.8	71 / 1.3 / 0.6	58 / 1.3 / 0.4	46 / 1.4 / 0.1				
68-D1	4.51 / 8.01								112 / 2.1 / 1.6	93 / 2.3 / 1.6	77 / 2.5 / 1.5	64 / 2.7 / 1.3	53 / 2.8 / 1.0	44 / 2.9 / 0.5	33 / 3.0 / 0.0	

Strength based on strain compatibility; bottom tension limited to $12\sqrt{f'_c}$; see pages 2-3–2-6 for explanation.
Shaded values require release strengths higher than 3500 psi.

Strand Pattern Designation

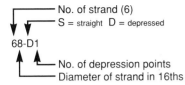

Safe loads shown include dead load of 10 psf for untopped members and 15 psf for topped members. Remainder is live load. Long-time cambers include superimposed dead load but do not include live load.

Key
- 200 — Safe superimposed service load, psf
- 0.1 — Estimated camber at erection, in.
- 0.2 — Estimated long-time camber, in.

DOUBLE TEE

8'-0" x 16"
Normal Weight Concrete

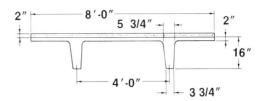

f'_c = 5,000 psi
f_{pu} = 270,000 psi

Section Properties

	Untopped		Topped	
A =	325	in.^{2}	—	
I =	6,634	in.^{4}	9,306	in.^{4}
y_b =	11.93	in.	13.52	in.
y_t =	4.07	in.	4.48	in.
Z_b =	556	in.^{3}	688	in.^{3}
Z_t =	1,630	in.^{3}	2,077	in.^{3}
wt =	339	plf	539	plf
		42	psf	67	psf
V/S =	1.29	in.		

8DT16 — No Topping

Table of safe superimposed service load (psf) and cambers

Strand Pattern	e_e / e_c	14	16	18	20	22	24	26	28	30	32	34	36	38	40	42	44	46	48	50	52	54	56
28-S	9.93 / 9.93	200 / 0.1 / 0.2	160 / 0.2 / 0.2	123 / 0.2 / 0.2	93 / 0.2 / 0.3	71 / 0.2 / 0.3	55 / 0.2 / 0.3	42 / 0.2 / 0.3	31 / 0.2 / 0.2														
48-S	8.93 / 8.93				200 / 0.4 / 0.6	159 / 0.5 / 0.7	129 / 0.6 / 0.8	105 / 0.7 / 0.8	86 / 0.7 / 0.9	70 / 0.8 / 0.9	58 / 0.8 / 0.9	47 / 0.8 / 0.9	39 / 0.8 / 0.8	31 / 0.7 / 0.6									
68-S	5.93 / 5.93					185 / 0.5 / 0.7	150 / 0.6 / 0.8	123 / 0.6 / 0.8	102 / 0.7 / 0.9	84 / 0.7 / 0.9	70 / 0.8 / 0.9	58 / 0.8 / 0.8	48 / 0.8 / 0.8	40 / 0.7 / 0.6	33 / 0.7 / 0.4								
68-D1	5.93 / 9.43							177 / 0.9 / 1.2	148 / 1.0 / 1.3	125 / 1.1 / 1.4	105 / 1.3 / 1.5	90 / 1.3 / 1.6	76 / 1.4 / 1.6	65 / 1.5 / 1.6	56 / 1.5 / 1.5	47 / 1.5 / 1.4	40 / 1.5 / 1.3	34 / 1.4 / 1.0					
88-D1	3.93 / 9.18														78 / 1.9 / 2.1	68 / 2.0 / 2.1	59 / 2.0 / 2.0	51 / 2.0 / 1.9	44 / 2.0 / 1.7	38 / 2.0 / 1.4	33 / 1.8 / 1.0		
108-D1	1.93 / 8.93																			44 / 2.2 / 1.5	39 / 2.1 / 1.2	34 / 2.0 / 0.7	

8DT16 + 2 — 2" Normal Weight Topping

Table of safe superimposed service load (psf) and cambers

Strand Pattern	e_e / e_c	16	18	20	22	24	26	28	30	32	34	36	38	40	42	44
28-S	9.93 / 9.93	167 / 0.2 / 0.2	125 / 0.2 / 0.2	91 / 0.2 / 0.2	66 / 0.2 / 0.2	47 / 0.2 / 0.2	32 / 0.2 / 0.0									
48-S	8.93 / 8.93			167 / 0.5 / 0.5	132 / 0.6 / 0.6	105 / 0.7 / 0.6	83 / 0.7 / 0.5	66 / 0.8 / 0.5	51 / 0.8 / 0.4	39 / 0.8 / 0.2						
68-S	5.93 / 5.93					164 / 0.6 / 0.6	132 / 0.6 / 0.6	107 / 0.7 / 0.5	86 / 0.7 / 0.5	69 / 0.8 / 0.4	55 / 0.8 / 0.2	44 / 0.8 / 0.0				
68-D1	5.93 / 9.43						187 / 0.9 / 0.9	154 / 1.0 / 0.9	127 / 1.1 / 1.0	106 / 1.3 / 1.0	87 / 1.3 / 0.9	72 / 1.4 / 0.8	59 / 1.5 / 0.6	49 / 1.5 / 0.4	39 / 1.5 / 0.1	
88-D1	3.93 / 9.18												75 / 1.9 / 0.9	63 / 2.0 / 0.7	51 / 2.0 / 0.3	

Strength based on strain compatibility; bottom tension limited to $12\sqrt{f'_c}$; see pages 2-3–2-6 for explanation.
Shaded values require release strengths higher than 3500 psi.

Strand Pattern Designation

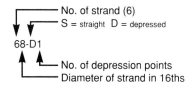

Safe loads shown include dead load of 10 psf for untopped members and 15 psf for topped members. Remainder is live load. Long-time cambers include superimposed dead load but do not include live load.

Key
- 168 — Safe superimposed service load, psf
- 0.2 — Estimated camber at erection, in.
- 0.3 — Estimated long-time camber, in.

DOUBLE TEE

8'-0" x 16"
Lightweight Concrete

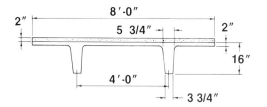

f'_c = 5,000 psi
f_{pu} = 270,000 psi

Section Properties

	Untopped	Topped
A =	325 in.²	—
I =	6,634 in.⁴	10,094 in.⁴
y_b =	11.93 in.	13.99 in.
y_t =	4.07 in.	4.01 in.
Z_b =	556 in.³	721 in.³
Z_t =	1,630 in.³	2,517 in.³
wt =	260 plf	460 plf
	33 psf	58 psf
V/S =	1.29 in.	

8LDT16 — No Topping

Table of safe superimposed service load (psf) and cambers

Strand Pattern	e_e / e_c	16	18	20	22	24	26	28	30	32	34	36	38	40	42	44	46	48	50	52	54	56
28-S	9.93 / 9.93	168 0.2 0.3	131 0.3 0.4	102 0.3 0.4	80 0.4 0.5	63 0.4 0.5	50 0.5 0.5	40 0.5 0.5	31 0.5 0.4													
48-S	8.93 / 8.93				167 0.8 1.0	137 0.9 1.2	113 1.0 1.3	94 1.2 1.4	78 1.3 1.5	66 1.4 1.6	56 1.4 1.6	47 1.5 1.6	40 1.5 1.5	33 1.5 1.4								
68-S	5.93 / 5.93				193 0.8 1.0	158 0.9 1.2	131 1.0 1.3	110 1.2 1.4	92 1.3 1.5	78 1.4 1.6	66 1.4 1.6	57 1.5 1.6	48 1.5 1.5	41 1.5 1.4	35 1.5 1.2	30 1.4 0.9						
68-D1	5.93 / 9.43						185 1.4 1.9	156 1.6 2.1	133 1.8 2.3	114 2.0 2.4	98 2.2 2.6	85 2.4 2.7	73 2.5 2.8	64 2.7 2.8	55 2.8 2.8	48 2.8 2.6	42 2.9 2.5	37 2.9 2.3	32 2.8 1.9			
88-D1	3.93 / 9.18																59 3.7 3.7	52 3.8 3.5	46 3.9 3.3	41 3.9 3.0	36 3.9 2.5	32 3.8 2.0

8LDT16 + 2 — 2" Normal Weight Topping

Table of safe superimposed service load (psf) and cambers

Strand Pattern	e_e / e_c	16	18	20	22	24	26	28	30	32	34	36	38	40	42	44	46	48
28-S	9.93 / 9.93	175 0.2 0.3	133 0.3 0.3	99 0.3 0.3	74 0.4 0.3	55 0.4 0.3	41 0.5 0.2											
48-S	8.93 / 8.93				176 0.8 0.8	140 0.9 0.9	113 1.0 0.9	91 1.2 0.9	74 1.3 0.9	59 1.4 0.8	48 1.4 0.7	38 1.5 0.5						
68-S	5.93 / 5.93					172 0.9 0.9	140 1.0 0.9	115 1.2 0.9	94 1.3 0.9	77 1.4 0.8	63 1.4 0.7	52 1.5 0.5	42 1.5 0.2					
68-D1	5.93 / 9.43						195 1.4 1.4	162 1.6 1.5	135 1.8 1.6	114 2.0 1.6	96 2.2 1.6	80 2.4 1.5	68 2.5 1.4	57 2.7 1.1	47 2.8 0.8	39 2.8 0.3		
88-D1	3.93 / 9.18																52 3.7 0.8	41 3.8 0.2

Strength based on strain compatibility; bottom tension limited to $12\sqrt{f'_c}$; see pages 2-3–2-6 for explanation.
Shaded values require release strengths higher than 3500 psi.

Strand Pattern Designation

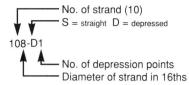

Safe loads shown include dead load of 10 psf for untopped members and 15 psf for topped members. Remainder is live load. Long-time cambers include superimposed dead load but do not include live load.

Key
- 186 — Safe superimposed service load, psf
- 0.1 — Estimated camber at erection, in.
- 0.2 — Estimated long-time camber, in.

DOUBLE TEE

8'-0" x 18"
Normal Weight Concrete

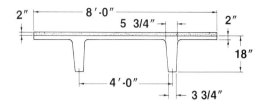

f'_c = 5,000 psi
f_{pu} = 270,000 psi

Section Properties

	Untopped	Topped
A =	344 in.²	—
I =	9,300 in.⁴	12,749 in.⁴
y_b =	13.27 in.	15.00 in.
y_t =	4.73 in.	5.00 in.
Z_b =	701 in.³	850 in.³
Z_t =	1,966 in.³	2,550 in.³
wt =	358 plf	558 plf
	45 psf	70 psf
V/S =	1.32 in.	

8DT18 — No Topping

Table of safe superimposed service load (psf) and cambers

Strand Pattern	e_e / e_c	16	18	20	22	24	26	28	30	32	34	36	38	40	42	44	46	48	50	52	54	56	58	60
28-S	11.27 / 11.27	186 / 0.1 / 0.2	144 / 0.2 / 0.2	110 / 0.2 / 0.2	85 / 0.2 / 0.2	65 / 0.2 / 0.2	51 / 0.2 / 0.2	39 / 0.2 / 0.2																
48-S	10.27 / 10.27				187 / 0.4 / 0.6	152 / 0.5 / 0.6	124 / 0.6 / 0.7	102 / 0.6 / 0.8	85 / 0.6 / 0.8	70 / 0.7 / 0.8	58 / 0.7 / 0.8	48 / 0.7 / 0.8	39 / 0.7 / 0.7	32 / 0.7 / 0.6										
68-S	7.27 / 7.27					186 / 0.5 / 0.7	153 / 0.6 / 0.8	127 / 0.7 / 0.8	107 / 0.7 / 0.9	89 / 0.8 / 0.9	75 / 0.8 / 0.9	63 / 0.8 / 0.9	53 / 0.8 / 0.8	45 / 0.8 / 0.7	37 / 0.7 / 0.6	31 / 0.6 / 0.4								
68-D1	7.27 / 10.77							175 / 0.9 / 1.2	148 / 1.0 / 1.3	126 / 1.1 / 1.4	107 / 1.2 / 1.4	92 / 1.3 / 1.5	79 / 1.3 / 1.5	68 / 1.4 / 1.5	58 / 1.4 / 1.5	50 / 1.4 / 1.4	43 / 1.3 / 1.2	36 / 1.3 / 1.0	31 / 1.2 / 0.7					
88-D1	5.27 / 10.52												109 / 1.7 / 2.0	95 / 1.8 / 2.1	83 / 1.8 / 2.1	73 / 1.9 / 2.1	64 / 1.9 / 2.0	55 / 2.0 / 1.9	48 / 2.0 / 1.7	42 / 1.9 / 1.5	36 / 1.8 / 1.2	31 / 1.6 / 0.9		
108-D1	3.27 / 10.27																		64 / 2.3 / 2.3	56 / 2.3 / 2.1	50 / 2.3 / 1.9	44 / 2.2 / 1.6	38 / 2.1 / 1.2	34 / 2.0 / 0.8

8DT18 + 2 — 2" Normal Weight Topping

Table of safe superimposed service load (psf) and cambers

Strand Pattern	e_e / e_c	16	18	20	22	24	26	28	30	32	34	36	38	40	42	44	46	48
28-S	11.27 / 11.27	193 / 0.1 / 0.1	146 / 0.2 / 0.2	107 / 0.2 / 0.2	79 / 0.2 / 0.2	58 / 0.2 / 0.1	41 / 0.2 / 0.1											
48-S	10.27 / 10.27				196 / 0.4 / 0.5	156 / 0.5 / 0.5	124 / 0.6 / 0.5	100 / 0.6 / 0.5	80 / 0.6 / 0.5	64 / 0.7 / 0.4	50 / 0.7 / 0.3	39 / 0.7 / 0.2						
68-S	7.27 / 7.27						162 / 0.6 / 0.6	133 / 0.7 / 0.6	108 / 0.7 / 0.6	89 / 0.8 / 0.5	72 / 0.8 / 0.4	58 / 0.8 / 0.3	47 / 0.8 / 0.1					
68-D1	7.27 / 10.77							181 / 0.9 / 0.9	151 / 1.0 / 0.9	126 / 1.1 / 1.0	105 / 1.2 / 0.9	88 / 1.3 / 0.9	73 / 1.3 / 0.8	61 / 1.4 / 0.6	50 / 1.4 / 0.4	41 / 1.4 / 0.2		
88-D1	5.27 / 10.52												108 / 1.7 / 1.2	92 / 1.8 / 1.1	78 / 1.8 / 1.0	66 / 1.9 / 0.8	56 / 1.9 / 0.5	47 / 2.0 / 0.2

Strength based on strain compatibility; bottom tension limited to $12\sqrt{f'_c}$; see pages 2-3–2-6 for explanation.
Shaded values require release strengths higher than 3500 psi.

Strand Pattern Designation

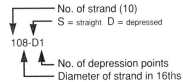

Safe loads shown include dead load of 10 psf for untopped members and 15 psf for topped members. Remainder is live load. Long-time cambers include superimposed dead load but do not include live load.

Key
194 — Safe superimposed service load, psf
0.2 — Estimated camber at erection, in.
0.3 — Estimated long-time camber, in.

DOUBLE TEE

8'-0" x 18"
Lightweight Concrete

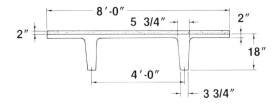

f'_c = 5,000 psi
f_{pu} = 270,000 psi

Section Properties

	Untopped	Topped
A =	344 in.²	—
I =	9,300 in.⁴	13,799 in.⁴
y_b =	13.27 in.	15.51 in.
y_t =	4.73 in.	4.49 in.
Z_b =	701 in.³	890 in.³
Z_t =	1,966 in.³	3,073 in.³
wt =	275 plf	475 plf
	34 psf	59 psf
V/S =	1.32 in.	

8LDT18 — No Topping

Table of safe superimposed service load (psf) and cambers

Strand Pattern	e_e / e_c	16	18	20	22	24	26	28	30	32	34	36	38	40	42	44	46	48	50	52	54	56	58	60	62	64
28-S	11.27 / 11.27	194 0.2 0.3	152 0.2 0.3	118 0.3 0.4	93 0.3 0.4	74 0.4 0.4	59 0.4 0.5	47 0.4 0.4	38 0.4 0.4	30 0.4 0.4																
48-S	10.27 / 10.27				196 0.7 0.9	160 0.8 1.0	133 0.9 1.1	111 1.0 1.2	93 1.1 1.3	79 1.2 1.4	67 1.2 1.4	57 1.3 1.4	48 1.3 1.4	41 1.4 1.4	35 1.4 1.3											
68-S	7.27 / 7.27					195 0.8 1.1	162 0.9 1.2	136 1.1 1.3	115 1.2 1.4	98 1.3 1.5	84 1.4 1.6	72 1.4 1.6	62 1.5 1.6	53 1.5 1.6	46 1.5 1.5	39 1.5 1.4	34 1.5 1.1									
68-D1	7.27 / 10.77							184 1.4 1.8	156 1.6 2.0	134 1.7 2.2	116 1.9 2.3	101 2.1 2.5	88 2.2 2.6	76 2.4 2.6	67 2.5 2.6	59 2.6 2.6	51 2.6 2.5	45 2.6 2.4	39 2.6 2.2	34 2.6 2.0						
88-D1	5.27 / 10.52														92 3.1 3.6	81 3.2 3.6	72 3.4 3.7	64 3.5 3.6	57 3.6 3.5	51 3.7 3.4	45 3.8 3.1	40 3.7 2.8	36 3.7 2.4	31 3.5 1.9		
108-D1	3.27 / 10.27																						47 4.5 3.6	42 4.5 3.1	38 4.4 2.5	34 4.2 1.8

8LDT18 + 2 — 2" Normal Weight Topping

Table of safe superimposed service load (psf) and cambers

Strand Pattern	e_e / e_c	16	18	20	22	24	26	28	30	32	34	36	38	40	42	44	46	48	50	52
28-S	11.27 / 11.27	201 0.2 0.2	154 0.2 0.2	116 0.3 0.3	88 0.3 0.3	66 0.4 0.3	50 0.4 0.2	37 0.4 0.1												
48-S	10.27 / 10.27				164 0.8 0.8	133 0.9 0.8	108 1.0 0.8	88 1.1 0.9	72 1.2 0.8	59 1.2 0.7	47 1.3 0.6	38 1.3 0.4								
68-S	7.27 / 7.27						171 0.9 0.9	141 1.1 1.0	117 1.2 1.0	97 1.3 1.0	81 1.4 0.9	67 1.4 0.8	55 1.5 0.6	46 1.5 0.4	37 1.5 0.1					
68-D1	7.27 / 10.77							190 1.4 1.4	159 1.6 1.5	134 1.7 1.5	114 1.9 1.6	96 2.1 1.6	82 2.2 1.5	69 2.4 1.4	59 2.5 1.1	49 2.6 0.8	41 2.6 0.5	34 2.6 0.0		
88-D1	5.27 / 10.52													87 3.1 2.0	75 3.2 1.8	65 3.4 1.5	56 3.5 1.1	48 3.6 0.7	39 3.7 0.1	

Strength based on strain compatibility; bottom tension limited to $12\sqrt{f'_c}$; see pages 2-3–2-6 for explanation.
Shaded values require release strengths higher than 3500 psi.

Strand Pattern Designation

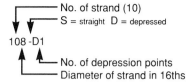

Safe loads shown include dead load of 10 psf for untopped members and 15 psf for topped members. Remainder is live load. Long-time cambers include superimposed dead load but do not include live load.

Key
- 175 — Safe superimposed service load, psf
- 0.4 — Estimated camber at erection, in.
- 0.6 — Estimated long-time camber, in.

DOUBLE TEE

8'-0" x 20"
Normal Weight Concrete

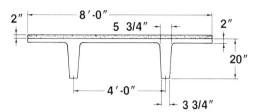

f'_c = 5,000 psi
f_{pu} = 270,000 psi

Section Properties

	Untopped	Topped
A =	363 in.²	—
I =	12,551 in.⁴	16,935 in.⁴
y_b =	14.59 in.	16.45 in.
y_t =	5.41 in.	5.55 in.
Z_b =	880 in.³	1,029 in.³
Z_t =	2,320 in.³	3,051 in.³
wt =	378 plf	578 plf
	47 psf	72 psf
V/S =	1.35 in.	

8DT20 — No Topping

Table of safe superimposed service load (psf) and cambers

Strand Pattern	e_e / e_c	24	26	28	30	32	34	36	38	40	42	44	46	48	50	52	54	56	58	60	62	64	66	68
48-S	11.59 / 11.59	175 / 0.4 / 0.6	144 / 0.5 / 0.6	119 / 0.5 / 0.7	99 / 0.6 / 0.7	82 / 0.6 / 0.7	69 / 0.6 / 0.7	57 / 0.6 / 0.7	48 / 0.6 / 0.7	39 / 0.6 / 0.6	32 / 0.6 / 0.5													
68-S	8.59 / 8.59		184 / 0.5 / 0.7	153 / 0.6 / 0.8	129 / 0.7 / 0.8	109 / 0.7 / 0.9	92 / 0.7 / 0.9	78 / 0.8 / 0.9	66 / 0.8 / 0.9	56 / 0.8 / 0.8	47 / 0.8 / 0.7	40 / 0.7 / 0.6	33 / 0.7 / 0.4											
68-D1	8.59 / 12.09			171 / 0.9 / 1.1	146 / 1.0 / 1.2	125 / 1.0 / 1.3	107 / 1.1 / 1.3	93 / 1.2 / 1.4	80 / 1.2 / 1.4	69 / 1.2 / 1.4	60 / 1.2 / 1.4	51 / 1.3 / 1.3	44 / 1.2 / 1.3	38 / 1.2 / 1.1	32 / 1.0 / 0.9									
88-D1	6.59 / 11.84						170 / 1.3 / 1.7	147 / 1.4 / 1.8	128 / 1.5 / 1.9	112 / 1.6 / 1.9	98 / 1.7 / 2.0	86 / 1.8 / 2.0	76 / 1.8 / 2.0	67 / 1.8 / 1.9	58 / 1.9 / 1.8	51 / 1.8 / 1.7	45 / 1.8 / 1.5	39 / 1.7 / 1.3	34 / 1.6 / 1.0					
108-D1	4.59 / 11.59														77 / 2.2 / 2.4	68 / 2.3 / 2.4	61 / 2.3 / 2.2	54 / 2.3 / 2.0	48 / 2.2 / 1.8	42 / 2.2 / 1.5	37 / 2.0 / 1.2	32 / 1.8 / 0.7		
128-D1	2.92 / 11.34																			54 / 2.6 / 2.1	48 / 2.5 / 1.8	43 / 2.4 / 1.4	38 / 2.3 / 0.9	34 / 2.0 / 0.4

8DT20 + 2 — 2" Normal Weight Topping

Table of safe superimposed service load (psf) and cambers

Strand Pattern	e_e / e_c	24	26	28	30	32	34	36	38	40	42	44	46	48	50	52	54
48-S	11.59 / 11.59	179 / 0.4 / 0.4	144 / 0.5 / 0.5	115 / 0.5 / 0.5	94 / 0.6 / 0.5	76 / 0.6 / 0.4	61 / 0.6 / 0.4	48 / 0.6 / 0.3	37 / 0.6 / 0.1								
68-S	8.59 / 8.59		193 / 0.5 / 0.6	158 / 0.6 / 0.6	131 / 0.7 / 0.6	108 / 0.7 / 0.6	89 / 0.7 / 0.5	73 / 0.8 / 0.5	60 / 0.8 / 0.3	49 / 0.8 / 0.2	39 / 0.8 / 0.0						
68-D1	8.59 / 12.09			173 / 0.9 / 0.8	145 / 1.0 / 0.9	122 / 1.0 / 0.9	103 / 1.1 / 0.9	87 / 1.2 / 0.8	73 / 1.2 / 0.7	60 / 1.2 / 0.6	50 / 1.3 / 0.4	41 / 1.2 / 0.2					
88-D1	6.59 / 11.84					172 / 1.3 / 1.2	148 / 1.4 / 1.3	127 / 1.5 / 1.3	109 / 1.6 / 1.2	93 / 1.7 / 1.1	80 / 1.8 / 1.0	68 / 1.8 / 0.8	58 / 1.9 / 0.6	49 / 1.9 / 0.3	41 / 1.8 / 0.0		
108-D1	4.59 / 11.59														70 / 2.2 / 0.9	60 / 2.3 / 0.5	51 / 2.3 / 0.1

Strength based on strain compatibility; bottom tension limited to $12\sqrt{f'_c}$; see pages 2-3–2-6 for explanation.

Shaded values require release strengths higher than 3500 psi.

Strand Pattern Designation

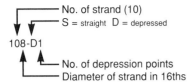

Safe loads shown include dead load of 10 psf for untopped members and 15 psf for topped members. Remainder is live load. Long-time cambers include superimposed dead load but do not include live load.

Key
- 184 — Safe superimposed service load, psf
- 0.7 — Estimated camber at erection, in.
- 0.8 — Estimated long-time camber, in.

DOUBLE TEE
8'-0" x 20"
Lightweight Concrete

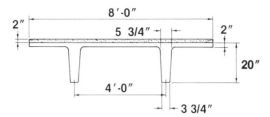

$f'_c = 5,000$ psi
$f_{pu} = 270,000$ psi

Section Properties

	Untopped	Topped
A =	363 in.²	—
I =	12,551 in.⁴	18,278 in.⁴
y_b =	14.59 in.	17.02 in.
y_t =	5.41 in.	4.98 in.
Z_b =	860 in.³	1,074 in.³
Z_t =	2,320 in.³	3,670 in.³
wt =	290 plf	490 plf
	36 psf	61 psf
V/S =	1.35 in.	

8LDT20 — No Topping

Table of safe superimposed service load (psf) and cambers

Strand Pattern	e_e / e_c	24	26	28	30	32	34	36	38	40	42	44	46	48	50	52	54	56	58	60	62	64	66	68
48-S	11.59 / 11.59	184 0.7 0.8	153 0.7 0.9	128 0.8 1.0	108 0.9 1.1	91 1.0 1.2	78 1.1 1.3	66 1.1 1.3	57 1.2 1.3	48 1.2 1.3	41 1.2 1.2	35 1.2 1.1												
68-S	8.59 / 8.59		193 0.9 1.1	162 1.0 1.2	138 1.1 1.3	118 1.2 1.4	101 1.2 1.5	87 1.3 1.6	75 1.4 1.6	65 1.5 1.6	57 1.5 1.6	49 1.5 1.5	42 1.5 1.4	37 1.5 1.2	32 1.4 1.0									
68-D1	8.59 / 12.09				180 1.4 1.8	155 1.5 1.9	134 1.6 2.0	117 1.8 2.2	102 2.0 2.3	89 2.1 2.4	78 2.2 2.4	69 2.3 2.4	61 2.3 2.4	53 2.4 2.4	47 2.4 2.3	41 2.4 2.2	36 2.4 2.1	32 2.3 1.9						
88-D1	6.59 / 11.84								137 2.4 3.0	121 2.6 3.2	107 2.8 3.3	95 2.9 3.4	85 3.1 3.5	76 3.2 3.5	68 3.4 3.5	60 3.5 3.5	54 3.5 3.3	48 3.5 3.1	43 3.5 2.8	38 3.5 2.5	34 3.4 2.1	30 3.2 1.7		
108-D1	4.59 / 11.59																	63 1.8 4.2	57 1.7 4.1	51 4.4 3.8	46 4.4 3.5	42 4.4 3.0	37 4.3 2.5	34 4.2 1.9

8LDT20 + 2 — 2" Normal Weight Topping

Table of safe superimposed service load (psf) and cambers

Strand Pattern	e_e / e_c	24	26	28	30	32	34	36	38	40	42	44	46	48	50	52	54	56	58
48-S	11.59 / 11.59	188 0.7 0.7	153 0.7 0.7	125 0.8 0.8	103 0.9 0.8	85 1.0 0.8	70 1.1 0.7	57 1.1 0.7	46 1.2 0.6	31 1.2 0.4									
68-S	8.59 / 8.59			167 1.0 0.9	140 1.1 1.0	117 1.2 1.0	98 1.2 1.0	82 1.3 0.9	69 1.4 0.8	58 1.5 0.7	48 1.5 0.5	39 1.5 0.2							
68-D1	8.59 / 12.09				182 1.4 1.4	154 1.5 1.4	131 1.6 1.4	112 1.8 1.5	96 2.0 1.5	82 2.1 1.4	70 2.2 1.3	59 2.3 1.1	50 2.3 0.8	42 2.4 0.5	35 2.4 0.1				
88-D1	6.59 / 11.84								136 2.4 2.1	118 2.6 2.1	102 2.8 2.0	89 2.9 2.0	77 3.1 1.8	67 3.2 1.6	58 3.4 1.3	50 3.5 0.9	43 3.5 0.4	36 3.5 0.2	
108-D1	4.59 / 11.59																	53 4.3 0.8	45 4.3 0.2

Strength based on strain compatibility; bottom tension limited to $12\sqrt{f'_c}$; see pages 2-3–2-6 for explanation.

Shaded values require release strengths higher than 3500 psi.

Strand Pattern Designation

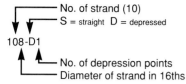

Safe loads shown include dead load of 10 psf for untopped members and 15 psf for topped members. Remainder is live load. Long-time cambers include superimposed dead load but do not include live load.

Key
- 173 — Safe superimposed service load, psf
- 0.5 — Estimated camber at erection, in.
- 0.7 — Estimated long-time camber, in.

DOUBLE TEE

8'-0" x 24"
Normal Weight Concrete

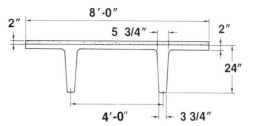

f'_c = 5,000 psi
f_{pu} = 270,000 psi

Section Properties

	Untopped	Topped
A =	401 in.²	—
I =	20,985 in.⁴	27,720 in.⁴
y_b =	17.15 in.	19.27 in.
y_t =	6.85 in.	6.73 in.
Z_b =	1,224 in.³	1,438 in.³
Z_t =	3,063 in.³	4,119 in.³
wt =	418 plf	618 plf
	52 psf	77 psf
V/S =	1.41 in.	

8DT24 — No Topping

Table of safe superimposed service load (psf) and cambers

Strand Pattern	e_e / e_c	30	32	34	36	38	40	42	44	46	48	50	52	54	56	58	60	62	64	66	68	70	72	74
68-S	11.15 / 11.15	173 / 0.5 / 0.7	147 / 0.6 / 0.8	126 / 0.6 / 0.8	108 / 0.7 / 0.8	92 / 0.7 / 0.8	79 / 0.7 / 0.8	68 / 0.7 / 0.8	58 / 0.7 / 0.8	50 / 0.7 / 0.7	43 / 0.7 / 0.7	36 / 0.6 / 0.4	30 / 0.5 / 0.2											
88-S	9.15 / 9.15		180 / 0.7 / 0.9	155 / 0.7 / 0.9	134 / 0.8 / 1.0	116 / 0.8 / 1.0	100 / 0.8 / 1.0	87 / 0.8 / 1.0	76 / 0.9 / 1.0	66 / 0.8 / 0.9	57 / 0.8 / 0.8	49 / 0.8 / 0.7	43 / 0.7 / 0.5	36 / 0.6 / 0.3	31 / 0.5 / 0.1									
88-D1	9.15 / 14.40				190 / 1.1 / 1.5	166 / 1.2 / 1.5	146 / 1.3 / 1.6	129 / 1.4 / 1.7	114 / 1.5 / 1.8	100 / 1.5 / 1.8	89 / 1.6 / 1.8	79 / 1.6 / 1.8	70 / 1.6 / 1.7	62 / 1.6 / 1.6	54 / 1.6 / 1.5	48 / 1.5 / 1.4	42 / 1.4 / 1.2	37 / 1.3 / 0.9	32 / 1.2 / 0.5					
108-D1	7.15 / 14.15							145 / 1.7 / 2.2	129 / 1.8 / 2.2	116 / 1.9 / 2.3	103 / 2.0 / 2.3	92 / 2.0 / 2.3	83 / 2.1 / 2.3	74 / 2.1 / 2.2	66 / 2.1 / 2.1	59 / 2.1 / 2.0	53 / 2.0 / 1.8	47 / 2.0 / 1.6	42 / 1.8 / 1.3	37 / 1.7 / 1.0	32 / 1.5 / 0.5			
128-D1	5.48 / 13.90															83 / 2.5 / 2.7	75 / 2.5 / 2.7	68 / 2.5 / 2.5	61 / 2.5 / 2.3	55 / 2.5 / 2.1	49 / 2.4 / 1.8	44 / 2.3 / 1.5	40 / 2.1 / 1.1	35 / 1.9 / 0.6
148-D1	4.29 / 13.65																			61 / 2.9 / 2.6	55 / 2.9 / 2.3	50 / 2.8 / 1.9	45 / 2.6 / 1.5	

8DT24 + 2 — 2" Normal Weight Topping

Table of safe superimposed service load (psf) and cambers

Strand Pattern	e_e / e_c	26	28	30	32	34	36	38	40	42	44	46	48	50	52	54	56	58	60	62	64
48-S	14.15 / 14.15	183 / 0.4 / 0.4	149 / 0.4 / 0.4	122 / 0.4 / 0.4	100 / 0.5 / 0.4	82 / 0.5 / 0.4	66 / 0.5 / 0.4	53 / 0.5 / 0.3	42 / 0.5 / 0.2	33 / 0.5 / 0.0											
68-S	11.15 / 11.15			175 / 0.5 / 0.5	147 / 0.6 / 0.6	123 / 0.6 / 0.6	103 / 0.7 / 0.5	86 / 0.7 / 0.5	72 / 0.7 / 0.4	60 / 0.7 / 0.3	49 / 0.7 / 0.2	39 / 0.7 / 0.0									
68-D1	11.15 / 14.65				184 / 0.7 / 0.8	156 / 0.8 / 0.8	133 / 0.9 / 0.8	113 / 0.9 / 0.8	96 / 1.0 / 0.7	81 / 1.0 / 0.7	69 / 1.0 / 0.6	58 / 1.0 / 0.5	48 / 1.0 / 0.3	39 / 1.0 / 0.1							
88-D1	9.15 / 14.40						190 / 1.1 / 1.1	165 / 1.2 / 1.1	143 / 1.3 / 1.2	124 / 1.4 / 1.2	107 / 1.5 / 1.2	93 / 1.5 / 1.1	80 / 1.6 / 0.9	69 / 1.6 / 0.8	59 / 1.6 / 0.6	51 / 1.6 / 0.4	43 / 1.6 / 0.1				
108-D1	7.15 / 14.15									142 / 1.7 / 1.5	124 / 1.8 / 1.5	109 / 1.9 / 1.4	96 / 2.0 / 1.3	84 / 2.0 / 1.2	74 / 2.1 / 1.0	64 / 2.1 / 0.7	56 / 2.1 / 0.5	48 / 2.1 / 0.1			
128-D1	5.48 / 13.90														74 / 2.5 / 1.0	65 / 2.5 / 0.7	57 / 2.5 / 0.3	49 / 2.5 / 0.1			

Strength based on strain compatibility; bottom tension limited to $12\sqrt{f'_c}$; see pages 2-3–2-6 for explanation.
Shaded values require release strengths higher than 3500 psi.

Strand Pattern Designation

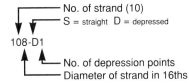

Safe loads shown include dead load of 10 psf for untopped members and 15 psf for topped members. Remainder is live load. Long-time cambers include superimposed dead load but do not include live load.

Key
- 118 — Safe superimposed service load, psf
- 1.1 — Estimated camber at erection, in.
- 1.4 — Estimated long-time camber, in.

DOUBLE TEE

8'-0" x 24"
Lightweight Concrete

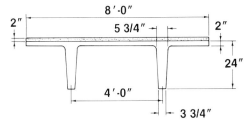

$f'_c = 5,000$ psi
$f_{pu} = 270,000$ psi

Section Properties

	Untopped		Topped	
A =	401	in.²	—	
I =	20,985	in.⁴	29,853	in.⁴
y_b =	17.15	in.	19.94	in.
y_t =	6.85	in.	6.06	in.
Z_b =	1,224	in.³	1,497	in.³
Z_t =	3,063	in.³	4,926	in.³
wt =	320	plf	520	plf
	40	psf	65	psf
V/S =	1.41	in.		

8LDT24 — No Topping

Table of safe superimposed service load (psf) and cambers

Strand Pattern	e_e / e_c	36	38	40	42	44	46	48	50	52	54	56	58	60	62	64	66	68	70	72	74	76	78	80
68-S	11.15 / 11.15	118 / 1.1 / 1.4	103 / 1.2 / 1.4	89 / 1.2 / 1.5	78 / 1.3 / 1.5	69 / 1.3 / 1.5	60 / 1.4 / 1.5	53 / 1.4 / 1.4	46 / 1.4 / 1.4	40 / 1.3 / 1.3	35 / 1.3 / 1.2	30 / 1.2 / 0.8												
88-S	9.15 / 9.15	144 / 1.2 / 1.6	126 / 1.3 / 1.6	110 / 1.4 / 1.7	97 / 1.5 / 1.8	86 / 1.6 / 1.8	76 / 1.6 / 1.8	67 / 1.6 / 1.8	59 / 1.6 / 1.7	53 / 1.6 / 1.6	47 / 1.6 / 1.5	41 / 1.6 / 1.3	36 / 1.5 / 1.0	32 / 1.4 / 0.7										
88-D1	9.15 / 14.40		176 / 1.9 / 2.4	156 / 2.1 / 2.6	139 / 2.2 / 2.8	124 / 2.4 / 2.9	111 / 2.5 / 3.0	99 / 2.7 / 3.1	89 / 2.8 / 3.2	80 / 2.9 / 3.2	72 / 3.0 / 3.2	64 / 3.0 / 3.1	58 / 3.1 / 2.9	52 / 3.1 / 2.8	47 / 3.0 / 2.6	42 / 3.0 / 2.4	38 / 2.9 / 2.1	34 / 2.8 / 1.8						
108-D1	7.15 / 14.15								113 / 3.3 / 3.9	102 / 3.4 / 4.0	93 / 3.6 / 4.1	84 / 3.7 / 4.1	76 / 3.8 / 4.1	69 / 3.9 / 4.0	63 / 4.0 / 3.9	57 / 4.1 / 3.8	52 / 4.1 / 3.5	47 / 4.0 / 3.2	43 / 3.9 / 2.8	38 / 3.8 / 2.4	35 / 3.6 / 1.9	31 / 3.4 / 1.3		
128-D1	5.48 / 13.90															65 / 4.8 / 4.7	59 / 4.9 / 4.5	54 / 4.9 / 4.2	50 / 4.9 / 3.8	45 / 4.8 / 3.4	41 / 4.7 / 2.8	38 / 4.5 / 2.2	34 / 4.3 / 1.5
148-D1	4.29 / 13.65																					46 / 5.6 / 3.8	42 / 5.5 / 3.2

8LDT24 + 2 — 2" Normal Weight Topping

Table of safe superimposed service load (psf) and cambers

Strand Pattern	e_e / e_c	26	28	30	32	34	36	38	40	42	44	46	48	50	52	54	56	58	60	62	64	66	68
48-S	14.15 / 14.15	193 / 0.6 / 0.6	159 / 0.6 / 0.6	132 / 0.7 / 0.7	110 / 0.8 / 0.7	92 / 0.8 / 0.7	76 / 0.9 / 0.6	63 / 0.9 / 0.6	52 / 1.0 / 0.5	43 / 1.0 / 0.4	34 / 1.0 / 0.2												
68-S	11.15 / 11.15			185 / 0.9 / 0.9	157 / 0.9 / 0.9	133 / 1.0 / 0.9	113 / 1.1 / 0.9	96 / 1.2 / 0.9	82 / 1.2 / 0.9	70 / 1.3 / 0.8	59 / 1.3 / 0.7	50 / 1.4 / 0.5	41 / 1.4 / 0.3	34 / 1.4 / 0.0									
68-D1	11.15 / 14.65					166 / 1.2 / 1.2	143 / 1.4 / 1.3	123 / 1.5 / 1.3	106 / 1.6 / 1.3	91 / 1.7 / 1.3	79 / 1.8 / 1.2	68 / 1.9 / 1.1	58 / 1.9 / 0.9	49 / 2.0 / 0.7	42 / 2.0 / 0.5	35 / 2.0 / 0.2							
88-D1	9.15 / 14.40							200 / 1.8 / 1.8	175 / 1.9 / 1.9	153 / 2.1 / 1.9	134 / 2.1 / 1.9	117 / 2.4 / 1.8	103 / 2.5 / 2.0	90 / 2.7 / 1.9	79 / 2.8 / 1.9	70 / 2.9 / 1.7	61 / 3.0 / 1.6	53 / 3.0 / 1.3	46 / 3.1 / 1.0	40 / 3.1 / 0.6			
108-D1	7.15 / 14.15													106 / 3.3 / 2.4	94 / 3.4 / 2.3	84 / 3.6 / 2.1	74 / 3.7 / 1.9	66 / 3.8 / 1.6	58 / 3.9 / 1.3	51 / 4.0 / 0.8	45 / 4.1 / 0.3		
128-D1	5.48 / 13.90																					53 / 4.8 / 0.7	46 / 4.9 / 0.1

Strength based on strain compatibility; bottom tension limited to $12\sqrt{f'_c}$; see pages 2-3–2-6 for explanation.
Shaded values require release strengths higher than 3500 psi.

Strand Pattern Designation

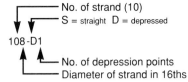

Safe loads shown include dead load of 10 psf for untopped members and 15 psf for topped members. Remainder is live load. Long-time cambers include superimposed dead load but do not include live load.

Key
186 — Safe superimposed service load, psf
1.4 — Estimated camber at erection, in.
1.8 — Estimated long-time camber, in.

DOUBLE TEE

8'-0" x 32"
Normal Weight Concrete

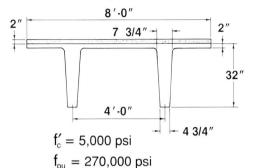

f'_c = 5,000 psi
f_{pu} = 270,000 psi

Section Properties

	Untopped		Topped	
A =	567	in.²	—	
I =	55,464	in.⁴	71,886	in.⁴
y_b =	21.21	in.	23.66	in.
y_t =	10.79	in.	10.34	in.
Z_b =	2,615	in.³	3,038	in.³
Z_t =	5,140	in.³	6,952	in.³
wt =	591	plf	791	plf
	74	psf	99	psf
V/S =	1.79	in.		

8DT32
No Topping

Table of safe superimposed service load (psf) and cambers

Strand Pattern	e_e / e_c	50	52	54	56	58	60	62	64	66	68	70	72	74	76	78	80	82	84	86	88	90	92	94
128-D1	12.04 / 17.96	186 1.4 1.8	167 1.4 1.8	151 1.5 1.8	136 1.5 1.8	123 1.5 1.8	111 1.5 1.8	100 1.5 1.7	90 1.5 1.6	82 1.5 1.6	73 1.4 1.5	66 1.4 1.3	59 1.3 1.1	53 1.1 0.8	47 1.0 0.5	42 0.8 0.2								
148-D1	9.71 / 17.71		197 1.6 2.0	179 1.7 2.1	162 1.7 2.1	147 1.8 2.2	134 1.8 2.2	121 1.8 2.1	110 1.8 2.1	100 1.8 2.0	91 1.8 1.9	83 1.7 1.8	75 1.7 1.6	68 1.6 1.4	61 1.5 1.2	55 1.3 0.9	49 1.1 0.5	44 0.9 0.1						
168-D1	8.21 / 17.46				188 1.9 2.4	171 2.0 2.5	156 2.0 2.5	142 2.1 2.5	130 2.1 2.5	119 2.2 2.4	108 2.2 2.4	99 2.2 2.3	90 2.1 2.1	82 2.0 2.0	75 1.9 1.8	68 1.8 1.5	62 1.7 1.3	56 1.5 0.9	51 1.3 0.6	46 1.1 0.1				
188-D1	6.82 / 17.21						177 2.2 2.8	162 2.3 2.8	149 2.3 2.8	136 2.4 2.8	125 2.4 2.7	115 2.4 2.7	105 2.4 2.6	96 2.4 2.4	88 2.3 2.2	81 2.3 2.0	74 2.2 1.8	67 2.0 1.5	62 1.8 1.2	56 1.6 0.8	51 1.4 0.4			
208-D1	5.71 / 16.96										130 2.6 3.0	119 2.7 3.0	110 2.7 2.9	101 2.6 2.7	93 2.6 2.5	85 2.5 2.3	78 2.4 2.0	72 2.3 1.7	66 2.1 1.4	60 1.9 1.0	55 1.7 0.5	50 1.4 0.0		
228-D1	4.80 / 16.71														105 2.9 3.0	97 2.8 2.8	89 2.8 2.5	81 2.7 2.2	74 2.5 1.9	67 2.4 1.5	62 2.2 1.1	57 2.0 0.6	52 1.7 0.1	

8DT32 + 2
2" Normal Weight Topping

Table of safe superimposed service load (psf) and cambers

Strand Pattern	e_e / e_c	44	46	48	50	52	54	56	58	60	62	64	66	68	70	72	74	76	78	80
108-D1	15.31 / 18.21	204 1.0 1.1	180 1.1 1.1	159 1.1 1.1	140 1.2 1.1	124 1.2 1.0	109 1.2 0.9	96 1.3 0.8	84 1.3 0.7	73 1.2 0.5	64 1.2 0.3	55 1.2 0.0	47 1.1							
128-D1	12.04 / 17.96			202 1.3 1.3	180 1.4 1.3	151 1.4 1.3	143 1.5 1.2	128 1.5 1.2	114 1.5 1.1	101 1.5 1.0	90 1.5 0.9	79 1.5 0.7	70 1.5 0.5	61 1.4 0.2						
148-D1	9.71 / 17.71					192 1.6 1.5	173 1.7 1.5	155 1.7 1.5	139 1.8 1.4	125 1.8 1.3	112 1.8 1.2	100 1.8 1.1	89 1.8 0.9	80 1.8 0.7	71 1.8 0.4	63 1.7 0.1				
168-D1	8.21 / 17.46							182 1.9 1.7	164 2.0 1.7	148 2.0 1.6	134 2.1 1.6	121 2.1 1.4	109 2.1 1.3	98 2.2 1.1	88 2.1 0.9	79 2.1 0.6	70 2.0 0.3			
188-D1	6.82 / 17.21									170 2.2 1.9	154 2.3 1.8	140 2.3 1.7	127 2.4 1.6	115 2.4 1.4	104 2.4 1.2	94 2.4 1.0	85 2.4 0.7	76 2.3 0.4	68 2.3 0.0	
208-D1	5.71 / 16.96												120 2.6 1.6	109 2.7 1.4	99 2.7 1.1	90 2.6 0.8	81 2.6 0.4	72 2.5 0.0		

Strength based on strain compatibility; bottom tension limited to $12\sqrt{f'_c}$; see pages 2-3–2-6 for explanation.
Shaded values require release strengths higher than 3500 psi.

PCI Design Handbook/Fourth Edition

Strand Pattern Designation

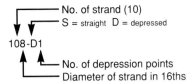

Safe loads shown include dead load of 10 psf for untopped members and 15 psf for topped members. Remainder is live load. Long-time cambers include superimposed dead load but do not include live load.

Key
150 — Safe superimposed service load, psf
2.6 — Estimated camber at erection, in.
3.2 — Estimated long-time camber, in.

DOUBLE TEE

8'-0" x 32"
Lightweight Concrete

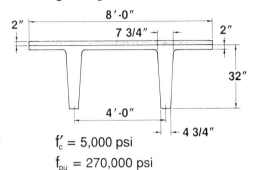

f'_c = 5,000 psi
f_{pu} = 270,000 psi

Section Properties

	Untopped	Topped
A =	567 in.²	—
I =	55,464 in.⁴	77,617 in.⁴
y_b =	21.21 in.	24.55 in.
y_t =	10.79 in.	9.45 in.
Z_b =	2,615 in.³	3,162 in.³
Z_t =	5,140 in.³	8,213 in.³
wt =	453 plf	653 plf
	57 psf	82 psf
V/S =	1.79 in.	

8LDT32 — No Topping

Table of safe superimposed service load (psf) and cambers

Strand Pattern	e_e / e_c	56	58	60	62	64	66	68	70	72	74	76	78	80	82	84	86	88	90	92	94	96	98	100
128-D1	12.04 / 17.96	150 / 2.6 / 3.2	137 / 2.7 / 3.2	125 / 2.8 / 3.3	114 / 2.9 / 3.3	105 / 2.9 / 3.2	96 / 2.9 / 3.2	88 / 2.9 / 3.1	80 / 2.9 / 3.0	73 / 2.9 / 2.9	67 / 2.9 / 2.7	61 / 2.8 / 2.5	56 / 2.7 / 2.3	51 / 2.6 / 2.0	46 / 2.4 / 1.7	42 / 2.2 / 1.3	38 / 2.0 / 0.8	34 / 1.7 / 0.2						
148-D1	9.71 / 17.71	176 / 2.9 / 3.6	161 / 3.0 / 3.7	148 / 3.1 / 3.8	136 / 3.2 / 3.8	125 / 3.3 / 3.9	114 / 3.4 / 3.9	105 / 3.5 / 3.9	97 / 3.5 / 3.8	89 / 3.5 / 3.6	82 / 3.5 / 3.5	75 / 3.5 / 3.3	69 / 3.4 / 3.0	64 / 3.3 / 2.8	58 / 3.2 / 2.5	53 / 3.1 / 2.2	49 / 2.9 / 1.8	45 / 2.7 / 1.3	41 / 2.4 / 0.8	37 / 2.1 / 0.2				
168-D1	8.21 / 17.46		185 / 3.3 / 4.1	170 / 3.4 / 4.2	157 / 3.6 / 4.3	144 / 3.7 / 4.4	133 / 3.8 / 4.5	123 / 3.9 / 4.5	113 / 4.0 / 4.5	104 / 4.0 / 4.4	97 / 4.1 / 4.3	89 / 4.1 / 4.2	82 / 4.1 / 4.0	76 / 4.1 / 3.8	70 / 4.0 / 3.5	65 / 3.9 / 3.1	60 / 3.7 / 2.7	55 / 3.6 / 2.3	51 / 3.4 / 1.9	47 / 3.1 / 1.3	43 / 2.8 / 0.8	39 / 2.5 / 0.1		
188-D1	6.82 / 17.21								129 / 4.3 / 5.0	119 / 4.4 / 5.0	110 / 4.5 / 5.0	102 / 4.6 / 4.9	95 / 4.6 / 4.7	88 / 4.6 / 4.6	82 / 4.6 / 4.4	76 / 4.6 / 4.1	70 / 4.5 / 3.7	64 / 4.4 / 3.3	59 / 4.2 / 2.8	54 / 4.0 / 2.2	50 / 3.7 / 1.7	47 / 3.4 / 1.1	44 / 3.1 / 0.4	
208-D1	5.71 / 16.96													100 / 5.1 / 5.3	92 / 5.1 / 5.1	85 / 5.1 / 4.9	79 / 5.0 / 4.6	73 / 5.0 / 4.2	67 / 4.9 / 3.8	62 / 4.7 / 3.3	57 / 4.5 / 2.8	52 / 4.3 / 2.1	48 / 4.0 / 1.4	44 / 3.7 / 0.6
228-D1	4.80 / 16.71																	81 / 5.5 / 5.1	75 / 5.5 / 4.7	69 / 5.4 / 4.3	64 / 5.2 / 3.8	59 / 5.0 / 3.2	54 / 4.8 / 2.6	50 / 4.5 / 1.8

8LDT32 + 2 — 2" Normal Weight Topping

Table of safe superimposed service load (psf) and cambers

Strand Pattern	e_e / e_c	46	48	50	52	54	56	58	60	62	64	66	68	70	72	74	76	78	80	82	84	86
108-D1	15.31 / 18.21	194 / 1.8 / 1.8	173 / 1.8 / 1.7	155 / 2.0 / 1.8	138 / 2.1 / 1.8	123 / 2.2 / 1.8	110 / 2.2 / 1.7	98 / 2.3 / 1.6	88 / 2.3 / 1.5	78 / 2.4 / 1.4	69 / 2.4 / 1.2	61 / 2.4 / 1.0	54 / 2.4 / 0.7	47 / 2.4 / 0.4	41 / 2.3 / 0.1							
128-D1	12.04 / 17.96			195 / 2.3 / 2.2	175 / 2.4 / 2.2	158 / 2.4 / 2.2	142 / 2.6 / 2.2	128 / 2.7 / 2.2	115 / 2.8 / 2.1	104 / 2.9 / 2.0	94 / 2.9 / 1.8	84 / 2.9 / 1.6	76 / 2.9 / 1.3	68 / 2.9 / 1.1	60 / 2.9 / 0.7	54 / 2.9 / 0.4						
148-D1	9.71 / 17.71					187 / 2.8 / 2.6	169 / 2.9 / 2.6	153 / 3.0 / 2.6	139 / 3.1 / 2.5	126 / 3.2 / 2.5	114 / 3.3 / 2.4	104 / 3.4 / 2.2	94 / 3.5 / 2.0	85 / 3.5 / 1.8	77 / 3.5 / 1.5	69 / 3.5 / 1.1	62 / 3.5 / 0.6	56 / 3.4 / 0.2				
168-D1	8.21 / 17.46							178 / 3.3 / 3.0	162 / 3.4 / 3.0	148 / 3.6 / 2.9	135 / 3.7 / 2.8	123 / 3.8 / 2.7	112 / 3.9 / 2.6	102 / 4.0 / 2.4	93 / 4.0 / 2.1	85 / 4.1 / 1.8	77 / 4.1 / 1.5	70 / 4.1 / 1.1	63 / 4.1 / 0.6			
188-D1	6.82 / 17.21											118 / 4.3 / 2.9	108 / 4.4 / 2.7	99 / 4.5 / 2.4	91 / 4.6 / 2.1	83 / 4.6 / 1.7	75 / 4.6 / 1.3	69 / 4.6 / 0.8				
208-D1	5.71 / 16.96																	87 / 5.1 / 1.9	80 / 5.1 / 1.5	73 / 5.1 / 0.9	65 / 5.0 / 0.3	

Strength based on strain compatibility; bottom tension limited to $12\sqrt{f'_c}$; see pages 2-3–2-6 for explanation.
Shaded values require release strengths higher than 3500 psi.

Strand Pattern Designation

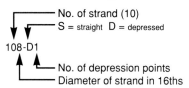

- No. of strand (10)
- S = straight D = depressed
- No. of depression points
- Diameter of strand in 16ths

Safe loads shown include dead load of 10 psf for untopped members and 15 psf for topped members. Remainder is live load. Long-time cambers include superimposed dead load but do not include live load.

Key
- 136 — Safe superimposed service load, psf
- 0.5 — Estimated camber at erection, in.
- 0.7 — Estimated long-time camber, in.

DOUBLE TEE

10'-0" x 24"
Normal Weight Concrete

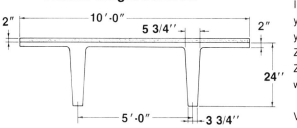

f'_c = 5,000 psi
f_{pu} = 270,000 psi

Section Properties

	Untopped	Topped
A =	449 in.²	—
I =	22,469 in.⁴	29,404 in.⁴
y_b =	17.77 in.	19.89 in.
y_t =	6.23 in.	6.11 in.
Z_b =	1,264 in.³	1,478 in.³
Z_t =	3,607 in.³	4,812 in.³
wt	468 plf	718 plf
	47 psf	72 psf
V/S =	1.35 in.	

10DT24
No Topping

Table of safe superimposed service load (psf) and cambers

Strand Pattern	e_e / e_c	30	32	34	36	38	40	42	44	46	48	50	52	54	56	58	60	62	64	66	68	70	72	74
68-S	11.77 / 11.77	136 0.5 0.7	115 0.6 0.7	97 0.6 0.8	83 0.6 0.8	71 0.7 0.8	60 0.7 0.7	51 0.7 0.7	43 0.6 0.6	37 0.6 0.6	31 0.6 0.4													
88-S	9.77 / 9.77	166 0.6 0.8	141 0.6 0.8	121 0.7 0.9	104 0.7 0.9	90 0.8 0.9	77 0.8 0.9	67 0.8 0.9	57 0.8 0.9	49 0.8 0.8	42 0.8 0.7	36 0.7 0.5	31 0.6 0.3											
88-D1	9.77 / 15.02		199 0.9 1.2	172 1.0 1.3	149 1.1 1.4	130 1.2 1.5	114 1.3 1.5	100 1.3 1.6	88 1.4 1.6	77 1.4 1.6	68 1.5 1.6	60 1.5 1.6	52 1.5 1.5	46 1.4 1.4	40 1.4 1.2	35 1.3 1.0	30 1.2 0.7							
108-D1	7.77 / 14.77						128 1.6 2.0	113 1.7 2.0	101 1.7 2.1	89 1.8 2.1	80 1.9 2.1	71 1.9 2.1	63 2.0 2.0	56 2.0 1.9	50 1.9 1.8	44 1.9 1.6	39 1.8 1.4	34 1.7 1.1	30 1.6 0.8					
128-D1	6.10 / 14.52												79 2.3 2.6	71 2.4 2.5	64 2.4 2.4	57 2.4 2.3	51 2.4 2.1	46 2.3 1.9	41 2.3 1.6	36 2.1 1.3	32 2.0 0.9			
148-D1	4.91 / 14.27																	51 2.8 2.3	46 2.7 2.0	41 2.6 1.7	37 2.5 1.3	33 2.4 0.9		

10DT24 + 2
2" Normal Weight Topping

Table of safe superimposed service load (psf) and cambers

Strand Pattern	e_e / e_c	24	26	28	30	32	34	36	38	40	42	44	46	48	50	52	54	56	58	60
48-S	14.77 / 14.77	173 0.3 0.3	139 0.3 0.3	112 0.4 0.4	91 0.4 0.3	73 0.4 0.3	58 0.5 0.3	46 0.5 0.2	35 0.5 0.1											
68-S	11.77 / 11.77		196 0.4 0.5	162 0.5 0.5	134 0.5 0.5	111 0.6 0.5	91 0.6 0.5	75 0.6 0.4	62 0.7 0.4	50 0.7 0.3	40 0.7 0.1	32 0.6 0.0								
68-D1	11.77 / 15.27			168 0.7 0.7	140 0.7 0.7	118 0.8 0.7	99 0.8 0.7	83 0.9 0.6	70 0.9 0.6	58 0.9 0.5	48 1.0 0.4	39 0.9 0.2	31 0.9 0.0							
88-D1	9.77 / 15.02				200 0.9 0.9	171 1.0 1.0	146 1.1 1.1	125 1.2 1.0	108 1.3 1.0	92 1.3 1.0	79 1.4 0.9	68 1.4 0.8	57 1.5 0.7	48 1.5 0.5	41 1.5 0.2					
108-D1	7.77 / 14.77								141 1.5 1.3	123 1.6 1.3	107 1.7 1.3	93 1.7 1.2	81 1.8 1.1	70 1.9 1.0	61 1.9 0.8	52 2.0 0.5	44 2.0 0.2			
128-D1	6.10 / 14.52																70 2.3 1.0	61 2.4 0.8	53 2.4 0.4	46 2.4 0.0

Strength based on strain compatibility; bottom tension limited to $12\sqrt{f'_c}$; see pages 2-3–2-6 for explanation.
Shaded values require release strengths higher than 3500 psi.

PCI Design Handbook/Fourth Edition

Strand Pattern Designation

```
        ┌─── No. of strand (10)
        │ ┌── S = straight   D = depressed
        ▼ ▼
       108-D1
        ▲ ▲
        │ └── No. of depression points
        └──── Diameter of strand in 16ths
```

Safe loads shown include dead load of 10 psf for untopped members and 15 psf for topped members. Remainder is live load. Long-time cambers include superimposed dead load but do not include live load.

Key

124 — Safe superimposed service load, psf
0.9 — Estimated camber at erection, in.
1.1 — Estimated long-time camber, in.

DOUBLE TEE

10'-0" x 24"
Lightweight Concrete

f'_c = 5,000 psi
f_{pu} = 270,000 psi

Section Properties

	Untopped	Topped
A =	449 in.²	—
I =	22,469 in.⁴	31,515 in.⁴
y_b =	17.77 in.	20.53 in.
y_t =	6.23 in.	5.47 in.
Z_b =	1,264 in.³	1,535 in.³
Z_t =	3,607 in.³	5,761 in.³
wt =	359 plf	609 plf
	36 psf	61 psf
V/S =	1.35 in.	

10LDT24
No Topping

Table of safe superimposed service load (psf) and cambers

Strand Pattern	e_e / e_c	32	34	36	38	40	42	44	46	48	50	52	54	56	58	60	62	64	66	68	70	72	74	76
68-S	11.77 / 11.77	124 0.9 1.1	106 1.0 1.2	92 1.1 1.3	80 1.1 1.3	69 1.2 1.4	60 1.2 1.4	52 1.3 1.4	46 1.3 1.3	40 1.3 1.2	34 1.3 1.1	30 1.2 0.9												
88-S	9.77 / 9.77	150 1.0 1.3	130 1.1 1.4	113 1.2 1.5	99 1.3 1.6	86 1.4 1.6	76 1.5 1.7	66 1.5 1.7	58 1.5 1.6	51 1.6 1.5	45 1.6 1.4	40 1.6 1.2	35 1.5 0.9	30 1.4										
88-D1	9.77 / 15.02		181 1.6 2.0	158 1.7 2.2	139 1.9 2.3	123 2.0 2.5	109 2.2 2.6	97 2.3 2.7	86 2.4 2.8	77 2.6 2.9	69 2.7 2.9	61 2.7 2.8	55 2.8 2.7	49 2.8 2.6	44 2.8 2.5	39 2.8 2.3	35 2.8 2.1	31 2.7 1.8						
108-D1	7.77 / 14.77								110 2.8 3.4	98 3.0 3.6	89 3.2 3.6	80 3.3 3.7	72 3.4 3.7	65 3.5 3.7	59 3.6 3.6	53 3.7 3.5	48 3.8 3.3	43 3.8 3.1	39 3.7 2.7	35 3.6 2.4	32 3.5 2.0			
128-D1	6.10 / 14.52															60 4.4 4.4	55 4.5 4.2	50 4.6 4.0	45 4.6 3.7	41 4.6 3.3	37 4.5 2.8	34 4.4 2.3	31 4.2 1.7	
148-D1	4.91 / 14.27																						42 5.4 3.8	38 5.3 3.2

10LDT24 + 2
2" Normal Weight Topping

Table of safe superimposed service load (psf) and cambers

Strand Pattern	e_e / e_c	24	26	28	30	32	34	36	38	40	42	44	46	48	50	52	54	56	58	60	62	64
48-S	14.77 / 14.77	182 0.5 0.5	148 0.5 0.6	121 0.6 0.6	100 0.7 0.6	82 0.7 0.6	67 0.8 0.5	55 0.8 0.5	44 0.9 0.4	35 0.9 0.3												
68-S	11.77 / 11.77			171 0.7 0.7	143 0.8 0.8	120 0.9 0.8	100 1.0 0.8	84 1.1 0.8	71 1.1 0.7	59 1.2 0.6	49 1.2 0.4	41 1.3 0.2	33 1.3									
68-D1	11.77 / 15.27				177 1.0 1.0	149 1.1 1.1	127 1.2 1.1	108 1.4 1.1	92 1.4 1.1	79 1.5 1.1	67 1.6 1.0	57 1.7 0.9	48 1.7 0.7	40 1.8 0.5	33 1.8 0.3							
88-D1	9.77 / 15.02						179 1.6 1.5	155 1.7 1.6	134 1.9 1.7	117 2.0 1.7	101 2.2 1.7	88 2.3 1.7	77 2.4 1.6	66 2.6 1.5	57 2.7 1.3	50 2.7 1.0	43 2.8 0.6	36 2.8 0.3				
108-D1	7.77 / 14.77										102 2.8 2.2	90 3.0 2.1	79 3.2 2.0	70 3.3 1.8	61 3.4 1.5	53 3.5 1.2	47 3.6 0.8	40 3.7 0.4				
128-D1	6.10 / 14.52																	48 4.4 0.7	41 4.5 0.1			

Strength based on strain compatibility; bottom tension limited to $12\sqrt{f'_c}$; see pages 2-3–2-6 for explanation.
Shaded values require release strengths higher than 3500 psi.

Strand Pattern Designation

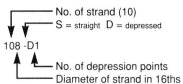

- No. of strand (10)
- S = straight D = depressed
- No. of depression points
- Diameter of strand in 16ths

Safe loads shown include dead load of 10 psf for untopped members and 15 psf for topped members. Remainder is live load. Long-time cambers include superimposed dead load but do not include live load.

Key
- 182 — Safe superimposed service load, psf
- 1.2 — Estimated camber at erection, in.
- 1.6 — Estimated long-time camber, in.

DOUBLE TEE

10'-0" x 32"
Normal Weight Concrete

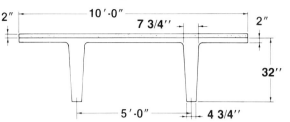

f'_c = 5,000 psi
f_{pu} = 270,000 psi

Section Properties

		Untopped	Topped
A	=	615 in.²	—
I	=	59,720 in.⁴	77,118 in.⁴
y_b	=	21.98 in.	24.54 in.
y_t	=	10.02 in.	9.46 in.
Z_b	=	2,717 in.³	3,142 in.³
Z_t	=	5,960 in.³	8,152 in.³
wt	=	641 plf	891 plf
		64 psf	89 psf
V/S	=	1.69 in.	

10DT32 — No Topping

Table of safe superimposed service load (psf) and cambers

Strand Pattern	e_e / e_c	46	48	50	52	54	56	58	60	62	64	66	68	70	72	74	76	78	80	82	84	86	88
128-D1	12.81 / 18.73	182 / 1.2 / 1.6	163 / 1.3 / 1.6	146 / 1.3 / 1.6	131 / 1.4 / 1.7	118 / 1.4 / 1.7	106 / 1.4 / 1.7	95 / 1.5 / 1.7	86 / 1.5 / 1.6	77 / 1.4 / 1.6	69 / 1.4 / 1.5	62 / 1.4 / 1.4	55 / 1.3 / 1.4	49 / 1.2 / 1.2	44 / 1.1 / 1.0	39 / 1.0 / 0.8								
148-D1	10.48 / 18.48		191 / 1.4 / 1.8	172 / 1.5 / 1.9	155 / 1.6 / 1.9	140 / 1.6 / 2.0	127 / 1.7 / 2.0	115 / 1.7 / 2.0	104 / 1.7 / 2.0	94 / 1.8 / 1.9	85 / 1.7 / 1.9	77 / 1.7 / 1.8	70 / 1.7 / 1.7	63 / 1.6 / 1.5	57 / 1.5 / 1.4	51 / 1.4 / 1.2	46 / 1.3 / 0.9	41 / 1.2 / 0.5					
168-D1	8.98 / 18.23			199 / 1.6 / 2.1	180 / 1.7 / 2.2	163 / 1.8 / 2.2	148 / 1.9 / 2.3	134 / 1.9 / 2.3	122 / 2.0 / 2.3	111 / 2.0 / 2.3	101 / 2.0 / 2.3	92 / 2.1 / 2.2	84 / 2.1 / 2.1	76 / 2.0 / 2.0	69 / 2.0 / 1.9	63 / 1.9 / 1.7	57 / 1.8 / 1.5	52 / 1.7 / 1.2	46 / 1.5 / 0.9	42 / 1.3 / 0.5			
188-D1	7.59 / 17.98							140 / 2.2 / 2.6	127 / 2.2 / 2.6	117 / 2.3 / 2.6	107 / 2.3 / 2.6	97 / 2.3 / 2.5	89 / 2.3 / 2.4	81 / 2.3 / 2.3	74 / 2.3 / 2.3	68 / 2.2 / 2.1	62 / 2.1 / 1.8	56 / 2.0 / 1.5	51 / 1.8 / 1.2	46 / 1.7 / 0.8			
208-D1	6.48 / 17.73										110 / 2.5 / 2.9	101 / 2.6 / 2.8	93 / 2.6 / 2.7	85 / 2.6 / 2.6	78 / 2.5 / 2.4	72 / 2.5 / 2.2	66 / 2.4 / 2.0	60 / 2.3 / 1.7	55 / 2.2 / 1.4	50 / 2.0 / 1.0			
228-D1	5.57 / 17.48																88 / 2.8 / 2.9	81 / 2.8 / 2.7	75 / 2.7 / 2.5	69 / 2.6 / 2.2	63 / 2.5 / 1.9	58 / 2.4 / 1.5	53 / 2.2 / 1.2

10DT32 + 2 — 2" Normal Weight Topping

Table of safe superimposed service load (psf) and cambers

Strand Pattern	e_e / e_c	42	44	46	48	50	52	54	56	58	60	62	64	66	68	70	72	74	76
108-D1	16.08 / 18.98	179 / 0.9 / 0.9	157 / 1.0 / 1.0	138 / 1.1 / 1.0	121 / 1.1 / 0.9	106 / 1.1 / 0.9	92 / 1.2 / 0.9	81 / 1.2 / 0.8	70 / 1.2 / 0.7	60 / 1.2 / 0.5	52 / 1.2 / 0.4								
128-D1	12.81 / 18.73			199 / 1.1 / 1.1	176 / 1.2 / 1.2	156 / 1.3 / 1.2	138 / 1.3 / 1.2	122 / 1.4 / 1.1	108 / 1.4 / 1.1	96 / 1.4 / 1.0	85 / 1.5 / 0.9	74 / 1.5 / 0.7	65 / 1.4 / 0.6	57 / 1.4 / 0.4					
148-D1	10.48 / 18.48					186 / 1.4 / 1.4	166 / 1.5 / 1.4	148 / 1.6 / 1.4	132 / 1.6 / 1.3	118 / 1.7 / 1.3	105 / 1.7 / 1.2	94 / 1.7 / 1.1	83 / 1.8 / 0.9	74 / 1.7 / 0.8	65 / 1.7 / 0.5	57 / 1.7 / 0.3			
168-D1	8.98 / 18.23				194 / 1.6 / 1.6	174 / 1.7 / 1.6	156 / 1.8 / 1.6	140 / 1.9 / 1.5	126 / 1.9 / 1.5	113 / 2.0 / 1.4	101 / 2.0 / 1.3	91 / 2.0 / 1.1	81 / 2.1 / 0.9	72 / 2.1 / 0.7	64 / 2.0 / 0.5				
188-D1	7.59 / 17.98								145 / 2.1 / 1.7	131 / 2.2 / 1.7	118 / 2.2 / 1.6	107 / 2.3 / 1.4	96 / 2.3 / 1.3	86 / 2.3 / 1.1	77 / 2.3 / 0.8	69 / 2.3 / 0.5	62 / 2.3 / 0.2		
208-D1	6.48 / 17.73											100 / 2.5 / 1.4	90 / 2.6 / 1.2	82 / 2.6 / 0.9	73 / 2.6 / 0.6	66 / 2.5 / 0.2			

Strength based on strain compatibility; bottom tension limited to $12\sqrt{f'_c}$; see pages 2-3–2-6 for explanation.
Shaded values require release strengths higher than 3500 psi.

Strand Pattern Designation

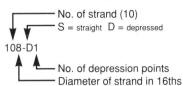

- No. of strand (10)
- S = straight   D = depressed
- No. of depression points
- Diameter of strand in 16ths

Safe loads shown include dead load of 10 psf for untopped members and 15 psf for topped members. Remainder is live load. Long-time cambers include superimposed dead load but do not include live load.

Key

- 130 — Safe superimposed service load, psf
- 2.4 — Estimated camber at erection, in.
- 2.9 — Estimated long-time camber, in.

DOUBLE TEE

10'-0" x 32"
Lightweight Concrete

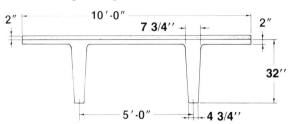

f'_c = 5,000 psi
f_{pu} = 270,000 psi

Section Properties

	Untopped	Topped
A =	615 in.²	—
I =	59,720 in.⁴	82,732 in.⁴
y_b =	21.98 in.	25.37 in.
y_t =	10.02 in.	8.63 in.
Z_b =	2,717 in.³	3,261 in.³
Z_t =	5,960 in.³	9,587 in.³
wt =	491 plf	741 plf
	49 psf	74 psf
V/S =	1.69 in.	

10LDT32
No Topping

Table of safe superimposed service load (psf) and cambers

Strand Pattern	e_e / e_c	54	56	58	60	62	64	66	68	70	72	74	76	78	80	82	84	86	88	90	92	94	96	98
128-D1	12.81 / 18.73	130 / 2.4 / 2.9	118 / 2.5 / 3.0	108 / 2.6 / 3.0	98 / 2.7 / 3.0	89 / 2.7 / 3.0	82 / 2.7 / 2.9	74 / 2.8 / 2.9	68 / 2.8 / 2.8	62 / 2.8 / 2.7	56 / 2.7 / 2.5	51 / 2.7 / 2.3	47 / 2.6 / 2.1	42 / 2.5 / 1.8	38 / 2.4 / 1.5	35 / 2.2 / 1.1	31 / 2.0 / 0.6							
148-D1	10.48 / 18.48	153 / 2.7 / 3.3	139 / 2.8 / 3.4	127 / 2.9 / 3.5	116 / 3.0 / 3.5	107 / 3.1 / 3.6	98 / 3.2 / 3.6	89 / 3.2 / 3.6	82 / 3.3 / 3.5	75 / 3.4 / 3.4	69 / 3.3 / 3.2	63 / 3.3 / 3.0	58 / 3.3 / 2.9	53 / 3.2 / 2.6	49 / 3.1 / 2.3	44 / 3.0 / 2.0	40 / 2.8 / 1.6	37 / 2.6 / 1.2	33 / 2.4 / 0.7					
168-D1	8.98 / 18.23	175 / 2.9 / 3.7	160 / 3.1 / 3.8	147 / 3.2 / 3.9	135 / 3.3 / 4.0	124 / 3.5 / 4.1	114 / 3.6 / 4.1	105 / 3.7 / 4.2	96 / 3.8 / 4.2	89 / 3.8 / 4.1	82 / 3.9 / 4.0	75 / 3.9 / 3.9	69 / 4.0 / 3.8	64 / 3.9 / 3.5	59 / 3.9 / 3.2	54 / 3.8 / 2.9	50 / 3.6 / 2.6	46 / 3.5 / 2.2	42 / 3.3 / 1.7	38 / 3.1 / 1.2	35 / 2.8 / 0.7			
188-D1	7.59 / 17.98							119 / 4.0 / 4.7	110 / 4.1 / 4.7	101 / 4.2 / 4.7	94 / 4.3 / 4.7	87 / 4.4 / 4.6	80 / 4.4 / 4.5	74 / 4.4 / 4.3	69 / 4.4 / 4.1	63 / 4.4 / 3.8	59 / 4.4 / 3.5	54 / 4.3 / 3.1	50 / 4.1 / 2.6	46 / 3.9 / 2.1	42 / 3.7 / 1.6	39 / 3.4 / 1.0	36 / 3.1 / 0.4	
208-D1	6.48 / 17.73													84 / 4.9 / 5.0	78 / 4.9 / 4.8	72 / 4.9 / 4.6	67 / 4.9 / 4.3	62 / 4.9 / 4.0	58 / 4.8 / 3.6	53 / 4.7 / 3.1	48 / 4.5 / 2.6	44 / 4.3 / 2.0	41 / 4.0 / 1.3	38 / 3.7 / 0.6
228-D1	5.57 / 17.48																	70 / 5.4 / 4.8	64 / 5.4 / 4.5	59 / 5.3 / 4.1	55 / 5.2 / 3.6	50 / 5.0 / 3.0	46 / 4.8 / 2.4	42 / 4.6 / 1.6

10LDT32 + 2
2" Normal Weight Topping

Table of safe superimposed service load (psf) and cambers

Strand Pattern	e_e / e_c	42	44	46	48	50	52	54	56	58	60	62	64	66	68	70	72	74	76	78	80	82
108-D1	16.08 / 18.98	192 / 1.5 / 1.5	169 / 1.6 / 1.5	150 / 1.7 / 1.6	133 / 1.8 / 1.6	118 / 1.9 / 1.6	105 / 2.0 / 1.6	93 / 2.1 / 1.5	82 / 2.1 / 1.4	73 / 2.2 / 1.3	64 / 2.2 / 1.2	56 / 2.2 / 1.0	49 / 2.3 / 0.8	43 / 2.3 / 0.5								
128-D1	12.81 / 18.73			188 / 2.0 / 1.9	168 / 2.1 / 2.0	150 / 2.1 / 1.9	135 / 2.3 / 2.0	121 / 2.4 / 2.0	108 / 2.5 / 1.9	97 / 2.6 / 1.8	87 / 2.7 / 1.7	77 / 2.7 / 1.5	69 / 2.8 / 1.3	61 / 2.8 / 1.1	55 / 2.8 / 0.8	48 / 2.8 / 0.5						
148-D1	10.48 / 18.48				199 / 2.3 / 2.2	178 / 2.4 / 2.3	161 / 2.5 / 2.3	145 / 2.7 / 2.3	130 / 2.8 / 2.3	118 / 2.9 / 2.3	106 / 3.0 / 2.2	96 / 3.1 / 2.1	86 / 3.2 / 1.9	77 / 3.3 / 1.7	70 / 3.3 / 1.4	62 / 3.4 / 1.1	56 / 3.3 / 0.7	50 / 3.3 / 0.3				
168-D1	8.98 / 18.23							168 / 2.9 / 2.7	152 / 3.1 / 2.7	138 / 3.2 / 2.7	125 / 3.3 / 2.6	113 / 3.5 / 2.5	103 / 3.6 / 2.4	93 / 3.7 / 2.2	84 / 3.8 / 2.0	76 / 3.8 / 1.8	69 / 3.9 / 1.5	62 / 3.9 / 1.1	56 / 4.0 / 0.7			
188-D1	7.59 / 17.98											108 / 4.0 / 2.7	99 / 4.1 / 2.5	90 / 4.2 / 2.3	82 / 4.3 / 2.0	74 / 4.4 / 1.7	67 / 4.4 / 1.3	61 / 4.4 / 0.8	55 / 4.4 / 0.3			
208-D1	6.48 / 17.43																	78 / 4.8 / 1.9	71 / 4.9 / 1.5	65 / 4.9 / 1.0	58 / 4.9 / 0.4	

Strength based on strain compatibility; bottom tension limited to $12\sqrt{f'_c}$; see pages 2-3–2-6 for explanation.

Shaded values require release strengths higher than 3500 psi.

Strand Pattern Designation

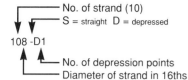

Because these units are pretopped and are typically used in parking structures, safe loads shown do not include any superimposed dead loads. Loads shown are live load. Long-time cambers do not include live load.

Key
- 196 — Safe superimposed service load, psf
- 0.4 — Estimated camber at erection, in.
- 0.5 — Estimated long-time camber, in.

PRETOPPED DOUBLE TEE

10'-0" x 26"

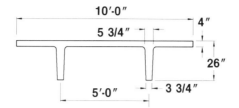

f'_c = 5,000 psi
f_{pu} = 270,000 psi

Section Properties

	Normal Weight	Lightweight
A =	689 in.^{2}	689 in.^{2}
I =	30,716 in.^{4}	30,716 in.^{4}
y_b =	20.29 in.	20.29 in.
y_t =	5.71 in.	5.71 in.
Z_b =	1,514 in.^{3}	1,514 in.^{3}
Z_t =	5,379 in.^{3}	5,379 in.^{3}
wt =	718 plf	550 plf
	72 psf	55 psf
V/S =	2.05 in.	2.05 in.

10DT26 — No Topping

Table of safe superimposed service load (psf) and cambers

Strand Pattern	e_e / e_c	26	28	30	32	34	36	38	40	42	44	46	48	50	52	54	56	58	60	62	64	66	68
68-S	14.29 / 14.29	196 / 0.4 / 0.5	161 / 0.4 / 0.5	133 / 0.4 / 0.6	109 / 0.4 / 0.6	90 / 0.5 / 0.6	74 / 0.5 / 0.6	60 / 0.5 / 0.6	49 / 0.4 / 0.6	39 / 0.4 / 0.6	30 / 0.4 / 0.5												
88-S	12.29 / 12.29			169 / 0.5 / 0.7	142 / 0.6 / 0.8	119 / 0.6 / 0.8	100 / 0.6 / 0.8	83 / 0.6 / 0.9	69 / 0.6 / 0.8	57 / 0.6 / 0.8	47 / 0.6 / 0.8	38 / 0.5 / 0.7	30 / 0.4 / 0.6										
88-D1	12.29 / 17.54				200 / 0.7 / 1.0	170 / 0.8 / 1.1	146 / 0.9 / 1.2	125 / 0.9 / 1.3	107 / 1.0 / 1.3	91 / 1.0 / 1.4	78 / 1.0 / 1.4	66 / 1.0 / 1.4	56 / 1.0 / 1.4	47 / 0.9 / 1.3	39 / 0.9 / 1.2	32 / 0.8 / 1.0							
108-D1	10.29 / 17.29					189 / 1.1 / 1.5	163 / 1.1 / 1.6	142 / 1.2 / 1.7	123 / 1.3 / 1.7	107 / 1.3 / 1.8	93 / 1.4 / 1.8	80 / 1.4 / 1.8	69 / 1.4 / 1.8	60 / 1.3 / 1.8	51 / 1.3 / 1.7	43 / 1.2 / 1.6	36 / 1.1 / 1.5						
128-D1	8.62 / 17.04								134 / 1.6 / 2.1	118 / 1.6 / 2.2	103 / 1.7 / 2.3	90 / 1.7 / 2.3	79 / 1.7 / 2.3	69 / 1.7 / 2.3	60 / 1.7 / 2.2	52 / 1.6 / 2.1	45 / 1.5 / 2.0	38 / 1.4 / 1.9	32 / 1.3 / 1.7				
148-D1	7.43 / 16.79													98 / 2.0 / 2.8	87 / 2.1 / 2.8	76 / 2.1 / 2.8	67 / 2.1 / 2.8	59 / 2.1 / 2.7	51 / 2.0 / 2.5	45 / 1.9 / 2.4	38 / 1.7 / 2.2	33 / 1.5 / 1.9	

10LDT26 — No Topping

Table of safe superimposed service load (psf) and cambers

Strand Pattern	e_e / e_c	28	30	32	34	36	38	40	42	44	46	48	50	52	54	56	58	60	62	64	66	68	70	72
68-S	14.29 / 14.29	175 / 0.6 / 0.9	146 / 0.7 / 1.0	123 / 0.7 / 1.0	104 / 0.8 / 1.1	88 / 0.8 / 1.2	74 / 0.9 / 1.2	63 / 0.9 / 1.3	53 / 0.9 / 1.3	44 / 0.9 / 1.2	36 / 0.9 / 1.2	30 / 0.8 / 1.1												
88-S	12.29 / 12.29		183 / 0.8 / 1.1	156 / 0.9 / 1.2	133 / 1.0 / 1.3	113 / 1.0 / 1.4	97 / 1.1 / 1.5	83 / 1.1 / 1.6	71 / 1.2 / 1.6	61 / 1.2 / 1.6	52 / 1.2 / 1.6	44 / 1.2 / 1.6	37 / 1.1 / 1.5	31 / 1.0 / 1.4										
88-D1	12.29 / 17.54			184 / 1.3 / 1.8	159 / 1.4 / 1.9	138 / 1.5 / 2.1	120 / 1.7 / 2.2	105 / 1.7 / 2.3	92 / 1.8 / 2.4	80 / 1.9 / 2.4	70 / 1.9 / 2.5	61 / 1.9 / 2.5	53 / 1.9 / 2.5	46 / 1.9 / 2.5	39 / 1.8 / 2.4	34 / 1.7 / 2.3								
108-D1	10.29 / 17.29				177 / 1.8 / 2.5	155 / 2.0 / 2.7	137 / 2.1 / 2.9	121 / 2.2 / 3.0	106 / 2.3 / 3.2	94 / 2.5 / 3.3	83 / 2.6 / 3.4	73 / 2.6 / 3.4	65 / 2.6 / 3.4	57 / 2.6 / 3.4	50 / 2.6 / 3.3	44 / 2.5 / 3.2	38 / 2.4 / 3.1	33 / 2.3 / 3.0						
128-D1	8.62 / 17.04							104 / 3.0 / 4.0	93 / 3.1 / 4.1	83 / 3.2 / 4.2	74 / 3.3 / 4.3	66 / 3.3 / 4.3	59 / 3.4 / 4.3	52 / 3.4 / 4.3	46 / 3.3 / 4.2	41 / 3.2 / 4.0	36 / 3.1 / 3.8	31 / 3.0 / 3.6	2.8 / 3.4					
148-D1	7.43 / 16.79												73 / 4.0 / 5.3	65 / 4.0 / 5.3	58 / 4.0 / 5.2	52 / 4.0 / 5.1	47 / 4.0 / 5.0	41 / 3.8 / 4.7	37 / 3.6 / 4.3					

Strength based on strain compatibility; bottom tension limited to $12\sqrt{f'_c}$; see pages 2-3–2-6 for explanation.
Shaded values require release strengths higher than 3500 psi.

Strand Pattern Designation

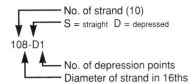

Because these units are pretopped and are typically used in parking structures, safe loads shown do not include any superimposed dead loads. Loads shown are live load. Long-time cambers do not include live load.

Key
172 — Safe superimposed service load, psf
0.9 — Estimated camber at erection, in.
1.3 — Estimated long-time camber, in.

PRETOPPED DOUBLE TEE

10'-0" x 34"

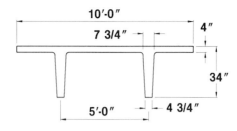

f'_c = 5,000 psi
f_{pu} = 270,000 psi

Section Properties

	Normal Weight	Lightweight
A =	855 in.²	855 in.⁴
I =	80,780 in.⁴	80,780 in.⁴
y_b =	25.07 in.	25.07 in.
y_t =	8.93 in.	8.93 in.
Z_b =	3,222 in.³	3,222 in.³
Z_t =	9,046 in.³	9,046 in.³
wt =	891 plf	683 plf
	89 psf	68 psf
V/S =	2.32 in.	2.32 in.

10DT34 — No Topping

Table of safe superimposed service load (psf) and cambers

Strand Pattern	e_e / e_c	46	48	50	52	54	56	58	60	62	64	66	68	70	72	74	76	78	80	82	84	86	88	90
128-D1	13.40 / 21.82	172 / 0.9 / 1.3	152 / 1.0 / 1.3	135 / 1.0 / 1.3	119 / 1.0 / 1.4	105 / 1.0 / 1.4	92 / 1.0 / 1.4	81 / 1.0 / 1.3	71 / 0.9 / 1.3	62 / 0.9 / 1.2	54 / 0.8 / 1.1	46 / 0.7 / 0.9	39 / 0.6 / 0.8	33 / 0.5 / 0.6										
148-D1	12.21 / 21.57		185 / 1.2 / 1.6	165 / 1.2 / 1.6	147 / 1.3 / 1.7	131 / 1.3 / 1.7	116 / 1.3 / 1.7	104 / 1.3 / 1.7	92 / 1.2 / 1.7	81 / 1.2 / 1.7	72 / 1.2 / 1.6	63 / 1.1 / 1.5	55 / 1.0 / 1.4	48 / 0.9 / 1.2	41 / 0.8 / 1.0	35 / 0.6 / 0.8								
168-D1	12.07 / 21.32			196 / 1.4 / 1.9	175 / 1.5 / 2.0	157 / 1.5 / 2.1	141 / 1.6 / 2.1	127 / 1.6 / 2.1	113 / 1.6 / 2.1	102 / 1.6 / 2.1	91 / 1.6 / 2.1	81 / 1.6 / 2.0	72 / 1.5 / 2.0	64 / 1.4 / 1.9	56 / 1.3 / 1.8	49 / 1.2 / 1.6	43 / 1.1 / 1.4	37 / 0.9 / 1.1	32 / 0.7 / 0.8					
188-D1	10.68 / 21.07					181 / 1.7 / 2.3	163 / 1.7 / 2.4	147 / 1.8 / 2.4	132 / 1.8 / 2.5	119 / 1.9 / 2.5	108 / 1.9 / 2.5	97 / 1.9 / 2.4	87 / 1.8 / 2.4	78 / 1.8 / 2.3	70 / 1.7 / 2.2	62 / 1.6 / 2.1	55 / 1.5 / 2.0	48 / 1.4 / 1.8	42 / 1.2 / 1.5	37 / 1.0 / 1.2	32 / 0.7 / 0.8			
208-D1	9.57 / 20.82						184 / 1.9 / 2.6	167 / 2.0 / 2.7	151 / 2.0 / 2.7	137 / 2.1 / 2.8	124 / 2.1 / 2.8	112 / 2.1 / 2.8	101 / 2.1 / 2.8	91 / 2.1 / 2.7	82 / 2.0 / 2.6	74 / 2.0 / 2.5	66 / 1.9 / 2.4	59 / 1.8 / 2.3	53 / 1.6 / 2.1	47 / 1.5 / 1.9	41 / 1.2 / 1.6	36 / 1.0 / 1.2	31 / 0.7 / 0.7	
228-D1	8.66 / 20.57										140 / 2.3 / 3.1	127 / 2.3 / 3.2	115 / 2.3 / 3.2	105 / 2.4 / 3.1	95 / 2.3 / 3.1	86 / 2.3 / 3.0	78 / 2.3 / 2.9	70 / 2.2 / 2.7	63 / 2.0 / 2.6	56 / 1.9 / 2.4	50 / 1.7 / 2.1	45 / 1.5 / 1.9	39 / 1.3 / 1.5	34 / 1.0 / 1.1

10LDT34 — No Topping

Table of safe superimposed service load (psf) and cambers

Strand Pattern	e_e / e_c	46	48	50	52	54	56	58	60	62	64	66	68	70	72	74	76	78	80	82	84	86	88	90
128-D1	13.40 / 21.82	189 / 1.6 / 2.1	169 / 1.6 / 2.2	152 / 1.7 / 2.3	136 / 1.8 / 2.4	122 / 1.8 / 2.4	109 / 1.9 / 2.4	98 / 1.9 / 2.5	88 / 1.9 / 2.5	79 / 1.9 / 2.5	71 / 1.9 / 2.5	63 / 1.9 / 2.4	56 / 1.8 / 2.4	50 / 1.7 / 2.3	44 / 1.6 / 2.1	39 / 1.5 / 2.0	34 / 1.3 / 1.7							
148-D1	12.21 / 21.57			182 / 2.0 / 2.7	164 / 2.1 / 2.8	148 / 2.2 / 2.9	134 / 2.3 / 3.0	121 / 2.3 / 3.1	109 / 2.3 / 3.1	99 / 2.4 / 3.1	89 / 2.4 / 3.1	80 / 2.4 / 3.1	72 / 2.4 / 3.1	65 / 2.4 / 3.0	59 / 2.3 / 2.9	52 / 2.2 / 2.8	47 / 2.1 / 2.7	42 / 1.9 / 2.5	37 / 1.8 / 2.3	32 / 1.6 / 2.0				
168-D1	12.07 / 21.32				192 / 2.4 / 3.3	174 / 2.5 / 3.4	158 / 2.7 / 3.6	144 / 2.8 / 3.7	131 / 2.9 / 3.8	119 / 2.9 / 3.9	108 / 3.0 / 4.0	98 / 3.1 / 4.0	89 / 3.1 / 4.0	81 / 3.1 / 4.0	73 / 3.0 / 3.9	67 / 3.0 / 3.8	60 / 2.9 / 3.7	54 / 2.8 / 3.5	49 / 2.7 / 3.4	44 / 2.5 / 3.2	39 / 2.4 / 3.0	35 / 2.1 / 2.7	31 / 1.9 / 2.3	
188-D1	10.68 / 21.07					198 / 2.7 / 3.8	180 / 2.9 / 3.9	164 / 3.0 / 4.1	150 / 3.1 / 4.2	136 / 3.2 / 4.4	125 / 3.3 / 4.5	114 / 3.4 / 4.5	104 / 3.5 / 4.6	95 / 3.6 / 4.6	87 / 3.6 / 4.6	79 / 3.6 / 4.6	72 / 3.5 / 4.5	66 / 3.5 / 4.3	60 / 3.4 / 4.1	54 / 3.2 / 4.0	49 / 3.1 / 3.8	44 / 2.9 / 3.5	39 / 2.7 / 3.3	35 / 2.4 / 2.9
208-D1	9.57 / 20.82										141 / 3.6 / 4.9	129 / 3.7 / 5.0	118 / 3.8 / 5.1	109 / 3.9 / 5.2	100 / 4.0 / 5.2	91 / 4.0 / 5.3	84 / 4.0 / 5.2	76 / 4.0 / 5.2	70 / 3.9 / 5.1	64 / 3.8 / 4.9	58 / 3.6 / 4.6	53 / 3.4 / 4.3	48 / 3.2 / 4.1	43 / 3.2 / 3.8
228-D1	8.66 / 20.57												112 / 4.4 / 5.8	103 / 4.4 / 5.9	95 / 4.5 / 5.9	87 / 4.5 / 5.9	80 / 4.5 / 5.8	74 / 4.5 / 5.7	67 / 4.4 / 5.6	62 / 4.4 / 5.4	56 / 4.2 / 5.1	51 / 4.0 / 4.7		

Strength based on strain compatibility; bottom tension limited to $12\sqrt{f'_c}$; see pages 2-3–2-6 for explanation.
Shaded values require release strengths higher than 3500 psi.

Strand Pattern Designation

76-S
- S = straight
- Diameter of strand in 16ths
- No. of strand (7)

Safe loads shown include dead load of 10 psf for untopped members and 15 psf for topped members. Remainder is live load. Long-time cambers include superimposed dead load but do not include live load.

Capacity of selections of other configurations are similar. For precise values, see local hollow-core manufacturer.

Key
- 306 — Safe superimposed service load, psf
- 0.2 — Estimated camber at erection, in.
- 0.2 — Estimated long-time camber, in.

HOLLOW-CORE

4'-0" x 6"
Normal Weight Concrete

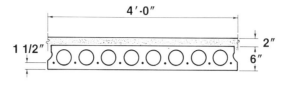

$f'_c = 5{,}000$ psi
$f'_{ci} = 3{,}500$ psi

Section Properties

		Untopped	Topped
A	=	187 in.²	—
I	=	763 in.⁴	1,640 in.⁴
y_b	=	3.00 in.	4.14 in.
y_t	=	3.00 in.	3.86 in.
Z_b	=	254 in.³	396 in.³
Z_t	=	254 in.³	425 in.³
b_w	=	16.00 in.	16.00 in.
wt	=	195 plf	295 plf
		49 psf	74 psf
V/S	=	1.73 in.	

4HC6 — No Topping

Table of safe superimposed service load (psf) and cambers

Strand Designation Code	Span, ft.																		
	12	13	14	15	16	17	18	19	20	21	22	23	24	25	26	27	28	29	30
66-S	306	257	217	184	157	135	116	100	87	75	65	56	48	42	36	30			
	0.2	0.2	0.2	0.2	0.2	0.2	0.2	0.2	0.2	0.2	0.2	0.1	0.1	0.0	−0.1	−0.2	−0.4		
	0.2	0.3	0.3	0.3	0.3	0.3	0.2	0.2	0.2	0.1	0.1	0.0	−0.2	−0.3	−0.5	−0.7	−1.0		
76-S	358	301	254	217	186	160	139	121	105	92	80	70	61	53	47	40	35		
	0.2	0.2	0.3	0.3	0.3	0.3	0.3	0.3	0.3	0.3	0.3	0.3	0.2	0.1	0.1	0.0	−0.1	−0.3	
	0.3	0.3	0.3	0.4	0.4	0.4	0.4	0.4	0.3	0.3	0.3	0.2	0.1	0.0	−0.1	−0.3	−0.5	−0.7	−1.0
96-S			384	326	279	240	208	182	159	140	123	109	97	86	76	67	60	53	46	41
			0.3	0.4	0.4	0.4	0.5	0.5	0.5	0.5	0.5	0.5	0.5	0.5	0.4	0.3	0.3	0.1	0.0	−0.1
			0.4	0.5	0.5	0.6	0.6	0.6	0.6	0.6	0.6	0.5	0.5	0.4	0.2	0.1	−0.1	−0.4	−0.6	−0.9
87-S				383	331	286	249	218	192	169	150	133	119	106	95	84	76	68	60	54
				0.5	0.5	0.6	0.6	0.7	0.7	0.7	0.7	0.8	0.8	0.7	0.7	0.7	0.6	0.5	0.4	0.3
				0.6	0.7	0.7	0.8	0.8	0.9	0.9	0.9	0.8	0.8	0.7	0.7	0.5	0.4	0.2	0.0	−0.3
97-S					364	317	277	243	214	189	168	150	134	120	107	96	87	78	70	62
					0.6	0.7	0.7	0.8	0.8	0.9	0.9	0.9	1.0	1.0	0.9	0.9	0.9	0.8	0.8	0.7
					0.8	0.9	0.9	1.0	1.0	1.1	1.1	1.1	1.1	1.0	1.0	0.9	0.8	0.6	0.4	0.2

4HC6 + 2 — 2" Normal Weight Topping

Table of safe superimposed service load (psf) and cambers

Strand Designation Code	Span, ft.																
	14	15	16	17	18	19	20	21	22	23	24	25	26	27	28	29	30
66-S	305	258	220	188	162	139	119	97	78	62	47	35					
	0.2	0.2	0.2	0.2	0.2	0.2	0.2	0.2	0.1	0.1	0.0	−0.1					
	0.2	0.2	0.2	0.1	0.1	0.0	−0.1	−0.2	−0.3	−0.5	−0.7	−0.9					
76-S	358	304	260	224	194	168	146	122	101	82	66	52	39				
	0.3	0.3	0.3	0.3	0.3	0.3	0.3	0.3	0.3	0.2	0.1	0.1	0.0				
	0.2	0.2	0.2	0.2	0.2	0.1	0.1	0.0	−0.2	−0.3	−0.5	−0.7	−0.9				
96-S		390	336	291	253	221	194	170	146	123	104	87	72	58	46	35	
		0.4	0.4	0.5	0.5	0.5	0.5	0.5	0.5	0.5	0.5	0.4	0.3	0.3	0.1	0.0	
		0.4	0.4	0.4	0.4	0.4	0.3	0.3	0.2	0.1	−0.1	−0.3	−0.5	−0.7	−1.0	−1.4	
87-S			398	346	302	265	234	206	182	158	136	117	100	85	71	59	47
			0.6	0.6	0.7	0.7	0.7	0.7	0.8	0.8	0.7	0.7	0.7	0.6	0.5	0.4	0.3
			0.5	0.6	0.6	0.6	0.5	0.5	0.4	0.4	0.2	0.1	−0.1	−0.3	−0.5	−0.8	−1.2
97-S				382	335	294	260	231	205	181	157	137	119	102	88	75	63
				0.7	0.8	0.8	0.9	0.9	0.9	1.0	1.0	0.9	0.9	0.9	0.8	0.8	0.7
				0.7	0.7	0.7	0.7	0.7	0.6	0.6	0.5	0.4	0.3	0.1	0.0	−0.2	−0.5	−0.8

Strength based on strain compatibility; bottom tension limited to $6\sqrt{f'_c}$; see pages 2-3–2-6 for explanation.

Strand Pattern Designation

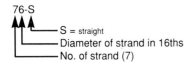

- S = straight
- Diameter of strand in 16ths
- No. of strand (7)

Safe loads shown include dead load of 10 psf for untopped members and 15 psf for topped members. Remainder is live load. Long-time cambers include superimposed dead load but do not include live load.

Capacity of selections of other configurations are similar. For precise values, see local hollow-core manufacturer.

Key
- 335 — Safe superimposed service load, psf
- 0.2 — Estimated camber at erection, in.
- 0.3 — Estimated long-time camber, in.

HOLLOW-CORE
4'-0" x 8"
Normal Weight Concrete

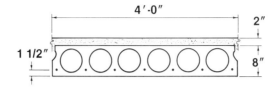

$f'_c = 5,000$ psi
$f'_{ci} = 3,500$ psi

Section Properties

	Untopped	Topped
A =	215 in.²	—
I =	1,666 in.⁴	3,071 in.⁴
y_b =	4.00 in.	5.29 in.
y_t =	4.00 in.	4.71 in.
Z_b =	416 in.³	580 in.³
Z_t =	416 in.³	652 in.³
b_w =	12.00 in.	12.00 in.
wt =	224 plf	324 plf
	56 psf	81 psf
V/S =	1.92 in.	

4HC8 — No Topping

Table of safe superimposed service load (psf) and cambers

Strand Designation Code	Span, ft.																						
	14	15	16	17	18	19	20	21	22	23	24	25	26	27	28	29	30	31	32	33	34	35	36
66-S	335	286	246	213	185	162	141	124	109	96	85	75	66	58	50	44	38	33					
	0.2	0.2	0.2	0.2	0.2	0.3	0.3	0.3	0.3	0.2	0.2	0.2	0.2	0.1	0.0	0.0	−0.1	−0.2					
	0.3	0.3	0.3	0.3	0.3	0.3	0.3	0.3	0.3	0.2	0.2	0.1	0.0	−0.1	−0.2	−0.3	−0.5	−0.7					
76-S	375	337	291	252	220	193	170	150	133	118	105	93	83	73	65	58	51	45	39	34			
	0.2	0.3	0.3	0.3	0.3	0.3	0.3	0.4	0.4	0.4	0.3	0.3	0.3	0.3	0.2	0.2	0.1	0.0	−0.1	−0.2			
	0.3	0.3	0.4	0.4	0.4	0.4	0.4	0.4	0.4	0.4	0.4	0.3	0.3	0.2	0.1	0.0	−0.1	−0.2	−0.4	−0.6	−0.8		
58-S	372	342	317	296	275	255	225	200	179	160	143	128	115	104	93	84	76	68	61	55	49	44	39
	0.3	0.3	0.4	0.4	0.5	0.5	0.5	0.5	0.6	0.6	0.6	0.6	0.6	0.5	0.5	0.5	0.4	0.3	0.2	0.1	0.0	−0.1	
	0.4	0.5	0.5	0.6	0.6	0.6	0.7	0.7	0.7	0.7	0.7	0.7	0.6	0.6	0.5	0.4	0.3	0.2	0.0	−0.2	−0.4	−0.6	−0.9
68-S		351	326	302	284	266	250	236	218	196	176	159	143	130	117	107	97	88	80	72	65	59	54
		0.4	0.5	0.5	0.6	0.6	0.7	0.7	0.7	0.8	0.8	0.8	0.8	0.8	0.8	0.8	0.8	0.7	0.7	0.6	0.5	0.4
		0.6	0.6	0.7	0.8	0.8	0.9	0.9	0.9	1.0	1.0	1.0	1.0	1.0	0.9	0.9	0.8	0.7	0.6	0.4	0.2	0.0	−0.2
78-S		360	335	311	290	272	256	242	229	215	205	188	170	154	141	128	117	106	97	89	81	74	67
		0.5	0.6	0.6	0.7	0.7	0.8	0.9	0.9	1.0	1.0	1.0	1.1	1.1	1.1	1.1	1.1	1.1	1.1	1.1	1.0	0.9	0.9
		0.7	0.8	0.8	0.9	1.0	1.0	1.1	1.2	1.2	1.3	1.3	1.3	1.3	1.3	1.3	1.2	1.2	1.1	1.0	0.9	0.7	0.5

4HC8 + 2 — 2" Normal Weight Topping

Table of safe superimposed service load (psf) and cambers

Strand Designation Code	Span, ft.																						
	16	17	18	19	20	21	22	23	24	25	26	27	28	29	30	31	32	33	34	35	36	37	38
66-S	309	267	231	201	175	153	133	117	102	89	77	67	55	44	33								
	0.2	0.2	0.2	0.3	0.3	0.3	0.3	0.2	0.2	0.2	0.2	0.1	0.0	0.0	−0.1								
	0.2	0.2	0.2	0.2	0.2	0.1	0.1	0.0	−0.1	−0.2	−0.3	−0.4	−0.6	−0.7	−0.9								
76-S	316	275	241	211	185	163	144	127	112	99	87	74	62	50	40	31							
	0.3	0.3	0.3	0.3	0.4	0.4	0.4	0.3	0.3	0.3	0.3	0.2	0.2	0.1	0.0	−0.1							
	0.3	0.3	0.3	0.3	0.2	0.2	0.2	0.1	0.0	−0.1	−0.2	−0.4	−0.5	−0.7	−0.9	−1.2							
58-S		352	317	279	248	220	196	174	156	139	124	111	98	84	71	60	50	40	32				
		0.5	0.5	0.5	0.5	0.6	0.6	0.6	0.6	0.6	0.6	0.5	0.5	0.5	0.4	0.3	0.2	0.1	0.0				
		0.5	0.5	0.5	0.5	0.5	0.4	0.4	0.3	0.3	0.2	0.1	−0.1	−0.2	−0.4	−0.6	−0.9	−1.2	−1.5				
68-S			337	316	297	268	239	215	193	173	156	141	127	114	100	87	75	64	54	45	36		
			0.6	0.7	0.7	0.7	0.8	0.8	0.8	0.8	0.8	0.8	0.8	0.8	0.8	0.7	0.7	0.6	0.5	0.4	0.2		
			0.6	0.6	0.7	0.7	0.7	0.6	0.6	0.6	0.5	0.4	0.3	0.2	0.0	−0.2	−0.4	−0.6	−0.9	−1.2	−1.6		
78-S			346	325	306	286	271	252	227	205	186	168	152	138	124	111	98	86	76	66	56	47	
			0.7	0.8	0.9	0.9	1.0	1.0	1.0	1.1	1.1	1.1	1.1	1.1	1.1	1.1	1.0	0.9	0.9	0.7	0.6	
			0.8	0.8	0.8	0.9	0.9	0.9	0.9	0.9	0.8	0.8	0.7	0.6	0.5	0.3	0.1	−0.1	−0.3	−0.6	−0.9	−1.3	

Strength based on strain compatibility; bottom tension limited to $6\sqrt{f'_c}$; see pages 2-3—2-6 for explanation.

Strand Pattern Designation

76-S
- S = straight
- Diameter of strand in 16ths
- No. of strand (7)

Safe loads shown include dead load of 10 psf for untopped members and 15 psf for topped members. Remainder is live load. Long-time cambers include superimposed dead load but do not include live load.

Capacity of selections of other configurations are similar. For precise values, see local hollow-core manufacturer.

Key
346 — Safe superimposed service load, psf
0.3 — Estimated camber at erection, in.
0.4 — Estimated long-time camber, in.

HOLLOW-CORE
4'-0" x 8"
Lightweight Concrete

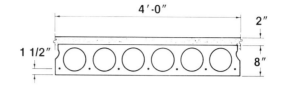

f'_c = 5,000 psi
f'_{ci} = 3,500 psi

Section Properties

	Untopped	Topped
A =	215 in.²	—
I =	1,666 in.⁴	3,529 in.⁴
y_b =	4.00 in.	5.70 in.
y_t =	4.00 in.	4.30 in.
Z_b =	416 in.³	619 in.³
Z_t =	416 in.³	821 in.³
b_w =	12.00 in.	12.00 in.
wt =	172 plf	272 plf
	43 psf	68 psf
V/S =	1.92 in.	

4LHC8 — No Topping

Table of safe superimposed service load (psf) and cambers

Strand Designation Code	Span, ft.																						
	14	15	16	17	18	19	20	21	22	23	24	25	26	27	28	29	30	31	32	33	34	35	36
66-S	346	297	257	224	196	172	152	135	120	107	95	85	76	68	61	55	49	44	39	35			
	0.3	0.3	0.4	0.4	0.4	0.5	0.5	0.5	0.5	0.5	0.5	0.5	0.5	0.4	0.4	0.3	0.3	0.2	0.1	0.0			
	0.4	0.4	0.5	0.5	0.5	0.6	0.6	0.6	0.6	0.5	0.5	0.5	0.4	0.3	0.2	0.0	-0.1	-0.3	-0.5	-0.8			
76-S		348	302	263	231	204	181	161	144	129	115	104	93	84	76	68	62	56	50	45	41	36	
		0.4	0.4	0.5	0.5	0.6	0.6	0.6	0.7	0.7	0.7	0.7	0.7	0.7	0.6	0.6	0.6	0.5	0.4	0.3	0.2	0.0	
		0.5	0.6	0.6	0.7	0.7	0.7	0.8	0.8	0.8	0.8	0.7	0.7	0.6	0.5	0.4	0.3	0.1	-0.1	-0.3	-0.6	-0.9	
58-S		350	325	304	286	265	236	211	189	170	154	139	126	114	104	95	86	79	72	66	60	55	50
		0.5	0.6	0.7	0.7	0.8	0.8	0.9	0.9	1.0	1.0	1.1	1.1	1.1	1.1	1.1	1.1	1.1	1.0	1.0	0.9	0.8	0.7
		0.7	0.8	0.9	0.9	1.0	1.1	1.1	1.2	1.2	1.2	1.2	1.2	1.2	1.1	1.1	0.9	0.8	0.7	0.5	0.2	0.0	
68-S			334	313	292	274	258	243	229	206	187	169	154	140	128	117	107	98	90	83	76	70	64
			0.7	0.8	0.9	1.0	1.1	1.1	1.2	1.3	1.3	1.4	1.5	1.5	1.5	1.6	1.6	1.6	1.6	1.6	1.5	1.5	1.4
			1.0	1.1	1.2	1.2	1.3	1.4	1.5	1.5	1.6	1.7	1.7	1.7	1.7	1.7	1.7	1.6	1.5	1.4	1.3	1.1	0.9
78-S			343	319	301	283	267	249	237	225	212	197	181	165	151	139	127	117	108	100	92	85	78
			0.9	1.0	1.1	1.2	1.3	1.4	1.5	1.6	1.7	1.7	1.8	1.9	2.0	2.0	2.1	2.1	2.1	2.2	2.2	2.1	2.1
			1.2	1.3	1.4	1.5	1.6	1.8	1.9	2.0	2.0	2.1	2.2	2.2	2.3	2.3	2.3	2.2	2.2	2.1	2.0	1.8	

4LHC8 + 2 — 2" Normal Weight Topping

Table of safe superimposed service load (psf) and cambers

Strand Designation Code	Span, ft.																						
	16	17	18	19	20	21	22	23	24	25	26	27	28	29	30	31	32	33	34	35	36	37	38
66-S	320	277	242	211	186	163	144	127	113	100	88	78	69	60	53	45							
	0.4	0.4	0.4	0.5	0.5	0.5	0.5	0.5	0.5	0.5	0.5	0.4	0.4	0.3	0.3	0.2							
	0.4	0.5	0.5	0.5	0.5	0.4	0.4	0.3	0.2	0.0	-0.1	-0.3	-0.5	-0.7	-1.0								
76-S		327	286	251	222	196	174	155	138	123	109	98	87	77	69	61	52	43					
		0.5	0.5	0.6	0.6	0.6	0.7	0.7	0.7	0.7	0.7	0.7	0.6	0.6	0.6	0.5	0.4	0.3					
		0.6	0.6	0.6	0.6	0.6	0.6	0.6	0.5	0.4	0.3	0.2	0.1	-0.1	-0.3	-0.6	-0.9	-1.2					
58-S				327	290	258	231	206	185	167	150	135	122	110	99	90	81	72	62	53	45		
				0.8	0.8	0.9	0.9	1.0	1.0	1.1	1.1	1.1	1.1	1.1	1.1	1.0	1.0	1.0	0.9	0.8	0.7		
				0.9	0.9	1.0	1.0	1.0	1.0	0.9	0.9	0.8	0.7	0.6	0.4	0.2	0.0	-0.2	-0.5	-0.9	-1.3		
68-S					323	304	278	250	225	204	184	167	151	138	125	114	103	93	83	73	64	56	48
					1.1	1.1	1.2	1.3	1.3	1.3	1.4	1.5	1.5	1.5	1.6	1.6	1.6	1.6	1.5	1.5	1.4	1.3	1.2
					1.2	1.3	1.3	1.4	1.4	1.4	1.4	1.3	1.2	1.1	1.0	0.8	0.6	0.3	0.0	-0.3	-0.7	-1.2	
78-S					332	313	297	279	263	238	216	197	179	163	149	136	125	113	102	91	81	72	64
					1.3	1.4	1.5	1.6	1.7	1.7	1.8	1.9	2.0	2.0	2.1	2.1	2.1	2.2	2.2	2.2	2.1	2.1	2.0
					1.5	1.6	1.7	1.7	1.8	1.8	1.8	1.8	1.8	1.7	1.6	1.5	1.3	1.1	0.9	0.6	0.2	-0.1	

Strength based on strain compatibility; bottom tension limited to $6\sqrt{f'_c}$; see pages 2-3–2-6 for explanation.

Strand Pattern Designation

76-S
- S = straight
- Diameter of strand in 16ths
- No. of strand (7)

Safe loads shown include dead load of 10 psf for untopped members and 15 psf for topped members. Remainder is live load. Long-time cambers include superimposed dead load but do not include live load.

Capacity of selections of other configurations are similar. For precise values, see local hollow-core manufacturer.

Key
- 239 — Safe superimposed service load, psf
- 0.3 — Estimated camber at erection, in.
- 0.4 — Estimated long-time camber, in.

HOLLOW-CORE

4'-0" x 10"
Normal Weight Concrete

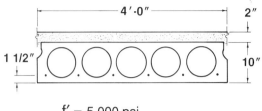

f'_c = 5,000 psi
f'_{ci} = 3,500 psi

Section Properties

	Untopped	Topped
A =	259 in.^{2}	—
I =	3,223 in.^{4}	5,328 in.^{4}
y_b =	5.00 in.	6.34 in.
y_t =	5.00 in.	5.66 in.
Z_b =	645 in.^{3}	840 in.^{3}
Z_t =	645 in.^{3}	941 in.^{3}
b_w =	10.50 in.	10.50 in.
wt =	270 plf	370 plf
	68 psf	93 psf
V/S =	2.23 in.	

4HC10
No Topping

Table of safe superimposed service load (psf) and cambers

Strand Designation Code	Span, ft.																						
	20	21	22	23	24	25	26	27	28	29	30	31	32	33	34	35	36	37	38	39	40	41	42
48-S	239 0.3 0.4	212 0.3 0.4	188 0.3 0.4	168 0.3 0.4	150 0.3 0.4	134 0.3 0.4	120 0.3 0.3	107 0.3 0.3	96 0.3 0.2	86 0.3 0.2	76 0.2 0.2	68 0.2 0.1	61 0.2 0.0	54 0.1 −0.1	48 0.1 −0.2	42 0.0 −0.3	−0.1 −0.5						
58-S	280 0.4 0.5	263 0.4 0.6	245 0.4 0.6	219 0.5 0.6	197 0.5 0.6	177 0.5 0.6	160 0.5 0.6	144 0.5 0.6	130 0.5 0.6	118 0.5 0.5	107 0.5 0.5	96 0.5 0.4	87 0.4 0.3	79 0.4 0.2	71 0.4 0.1	64 0.3 0.0	58 0.2 −0.1	52 0.2 −0.3	46 0.1 −0.5	41 0.0 −0.7	−0.1		
68-S	289 0.5 0.7	272 0.5 0.7	255 0.6 0.8	242 0.6 0.8	231 0.6 0.8	217 0.7 0.8	199 0.7 0.9	180 0.7 0.9	164 0.7 0.9	149 0.7 0.9	136 0.7 0.8	124 0.7 0.8	113 0.7 0.7	103 0.7 0.7	94 0.7 0.6	86 0.6 0.5	78 0.6 0.4	71 0.5 0.2	64 0.5 0.1	58 0.4 −0.1	53 0.3 −0.3	48 0.2 −0.6	43 0.1 −0.8
78-S	298 0.6 0.8	278 0.7 0.9	264 0.7 0.9	248 0.7 1.0	237 0.8 1.0	223 0.8 1.1	214 0.9 1.1	203 0.9 1.1	193 0.9 1.1	179 0.9 1.2	164 0.9 1.2	150 1.0 1.1	138 1.0 1.1	126 1.0 1.1	116 1.0 1.0	106 1.0 1.0	98 0.9 0.9	90 0.9 0.8	82 0.8 0.6	75 0.8 0.5	69 0.7 0.3	63 0.5 0.1	57 0.4 −0.1
88-S		287 0.8 1.0	270 0.8 1.1	257 0.9 1.2	243 0.9 1.2	229 1.0 1.3	220 1.0 1.3	209 1.1 1.4	199 1.1 1.4	189 1.2 1.4	183 1.2 1.4	174 1.2 1.5	162 1.2 1.5	149 1.2 1.5	137 1.2 1.5	126 1.2 1.4	117 1.2 1.4	107 1.2 1.3	99 1.2 1.3	91 1.1 1.2	84 1.1 1.1	78 1.0 0.9	71 0.9 0.8 0.6

4HC10 + 2
2" Normal Weight Topping

Table of safe superimposed service load (psf) and cambers

Strand Designation Code	Span, ft.																						
	20	21	22	23	24	25	26	27	28	29	30	31	32	33	34	35	36	37	38	39	40	41	42
48-S	293 0.3 0.3	258 0.3 0.3	229 0.3 0.3	203 0.3 0.2	181 0.3 0.2	161 0.3 0.2	143 0.3 0.1	127 0.3 0.1	113 0.3 0.0	101 0.2 −0.1	89 0.2 −0.2	79 0.2 −0.3	69 0.1 −0.4	60 0.1 −0.6	50 0.0 −0.8								
58-S		297 0.4 0.4	268 0.5 0.4	241 0.5 0.4	216 0.5 0.4	194 0.5 0.4	175 0.5 0.3	157 0.5 0.3	142 0.5 0.2	128 0.5 0.1	115 0.4 0.0	103 0.4 −0.1	92 0.4 −0.2	79 0.3 −0.4	68 0.2 −0.5	58 0.2 −0.7	48 0.1 −0.9						
68-S			286 0.6 0.6	272 0.6 0.6	259 0.7 0.6	244 0.7 0.6	221 0.7 0.5	200 0.7 0.5	182 0.7 0.4	165 0.7 0.4	150 0.7 0.3	136 0.7 0.2	123 0.7 0.0	109 0.6 −0.1	96 0.6 −0.3	84 0.5 −0.5	73 0.5 −0.7	63 0.4 −0.9	54 0.3				
78-S			295 0.7 0.8	278 0.8 0.8	265 0.8 0.8	250 0.9 0.8	239 0.9 0.8	226 0.9 0.8	218 0.9 0.7	201 0.9 0.7	184 1.0 0.6	168 1.0 0.5	154 1.0 0.4	138 0.9 0.3	124 0.9 0.2	111 0.9 0.0	98 0.8 −0.2	87 0.8 −0.4	77 0.7 −0.6	67 0.6 −0.9	58 0.5 −1.2	49 0.4	
88-S				287 0.9 1.0	271 1.0 1.0	259 1.0 1.0	245 1.1 1.0	232 1.1 1.0	224 1.2 1.0	213 1.2 1.0	202 1.2 0.9	193 1.2 0.9	179 1.2 0.8	163 1.2 0.7	148 1.2 0.6	134 1.2 0.5	121 1.2 0.3	110 1.1 0.1	99 1.1 −0.1	88 1.0 −0.3	78 0.9 −0.6	69	

Strength based on strain compatibility; bottom tension limited to $6\sqrt{f'_c}$; see pages 2-3–2-6 for explanation.

Strand Pattern Designation

76-S
- S = straight
- Diameter of strand in 16ths
- No. of strand (7)

Safe loads shown include dead load of 10 psf for untopped members and 15 psf for topped members. Remainder is live load. Long-time cambers include superimposed dead load but do not include live load.

Capacity of selections of other configurations are similar. For precise values, see local hollow-core manufacturer.

Key
- 127 — Safe superimposed service load, psf
- 0.3 — Estimated camber at erection, in.
- 0.4 — Estimated long-time camber, in.

HOLLOW-CORE

4'-0" x 12"
Normal Weight Concrete

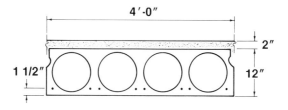

$f'_c = 5,000$ psi
$f'_{ci} = 3,500$ psi

Section Properties

	Untopped	Topped
A =	262 in.²	—
I =	4,949 in.⁴	7,811 in.⁴
y_b =	6.00 in.	7.55 in.
y_t =	6.00 in.	6.45 in.
Z_b =	825 in.³	1,035 in.³
Z_t =	825 in.³	1,211 in.³
b_w =	8.00 in.	8.00 in.
wt =	273 plf	373 plf
	68 psf	93 psf
V/S =	2.18 in.	

4HC12
No Topping

Table of safe superimposed service load (psf) and cambers

Strand Designation Code	Span, ft.																						
	28	29	30	31	32	33	34	35	36	37	38	39	40	41	42	43	44	45	46	47	48	49	50
76-S	127	115	104	94	85	76	69	62	56	50	44	39	35										
	0.3	0.3	0.3	0.3	0.3	0.2	0.2	0.1	0.1	0.0	0.0	−0.1	−0.2										
	0.4	0.3	0.3	0.3	0.2	0.1	0.1	0.0	−0.1	−0.2	−0.3	−0.5	−0.6										
58-S	173	161	147	134	122	112	102	93	85	78	71	65	59	53	48	43	39	35					
	0.5	0.5	0.5	0.5	0.5	0.5	0.5	0.5	0.4	0.4	0.3	0.3	0.2	0.1	0.1	0.0	−0.1	−0.3					
	0.6	0.6	0.6	0.6	0.6	0.5	0.5	0.4	0.4	0.3	0.2	0.1	−0.1	−0.2	−0.4	−0.6	−0.8	−1.0					
68-S	182	173	165	157	150	143	131	121	111	102	94	87	80	73	67	62	56	52	47				
	0.7	0.7	0.7	0.7	0.8	0.8	0.8	0.7	0.7	0.7	0.7	0.6	0.6	0.5	0.5	0.4	0.3	0.2	0.1				
	0.9	0.9	0.9	0.9	0.9	0.9	0.9	0.8	0.8	0.7	0.7	0.6	0.5	0.3	0.2	0.1	−0.1	−0.3	−0.5				
78-S	188	179	171	163	156	149	145	139	134	126	117	108	100	93	86	79	73	68	62	58	53	49	
	0.9	0.9	0.9	1.0	1.0	1.0	1.0	1.0	1.0	1.0	1.0	1.0	1.0	0.9	0.9	0.8	0.7	0.7	0.6	0.5	0.4	0.2	
	1.1	1.2	1.2	1.2	1.2	1.2	1.2	1.2	1.2	1.2	1.1	1.1	1.0	0.9	0.8	0.7	0.5	0.3	0.2	−0.1	−0.3	−0.5	
88-S	194	185	177	169	162	155	148	142	137	131	126	121	120	112	104	97	90	83	78	72	67	62	57
	1.0	1.1	1.1	1.2	1.2	1.2	1.3	1.3	1.3	1.3	1.3	1.3	1.3	1.3	1.3	1.3	1.2	1.2	1.1	1.0	0.9	0.8	0.6
	1.3	1.4	1.4	1.5	1.5	1.5	1.6	1.6	1.6	1.6	1.5	1.5	1.5	1.4	1.3	1.2	1.1	1.0	0.9	0.7	0.5	0.3	0.0

4HC12 + 2
2" Normal Weight Topping

Table of safe superimposed service load (psf) and cambers

Strand Designation Code	Span, ft.																						
	23	24	25	26	27	28	29	30	31	32	33	34	35	36	37	38	39	40	41	42	43	44	45
76-S	241	215	193	172	155	139	124	111	99	89	79	70	62	55	48	41							
	0.3	0.3	0.3	0.3	0.3	0.3	0.3	0.3	0.3	0.3	0.2	0.2	0.1	0.1	0.0	0.0							
	0.3	0.3	0.3	0.2	0.2	0.2	0.1	0.1	0.0	−0.1	−0.2	−0.3	−0.4	−0.5	−0.7	−0.8							
58-S	256	240	228	215	202	194	176	160	145	131	119	108	98	88	80	72	64	57	51	44			
	0.4	0.5	0.5	0.5	0.5	0.5	0.5	0.5	0.5	0.5	0.5	0.5	0.5	0.4	0.4	0.3	0.3	0.2	0.1	0.1			
	0.4	0.5	0.5	0.5	0.4	0.4	0.4	0.4	0.3	0.3	0.2	0.1	0.0	−0.1	−0.2	−0.3	−0.5	−0.7	−0.9	−1.1			
68-S	262	249	234	224	211	200	189	183	173	165	154	141	129	117	107	98	89	81	74	67	60	52	44
	0.6	0.6	0.6	0.6	0.7	0.7	0.7	0.7	0.7	0.8	0.8	0.8	0.7	0.7	0.7	0.7	0.6	0.6	0.5	0.5	0.4	0.3	0.2
	0.6	0.6	0.6	0.6	0.6	0.6	0.6	0.6	0.6	0.6	0.5	0.5	0.4	0.3	0.2	0.1	−0.1	−0.2	−0.4	−0.6	−0.8	−1.0	−1.3
78-S	271	255	243	230	217	206	195	189	179	171	163	155	148	144	134	123	113	104	95	87	80	73	65
	0.7	0.7	0.8	0.8	0.8	0.9	0.9	0.9	1.0	1.0	1.0	1.0	1.0	1.0	1.0	1.0	1.0	0.9	0.9	0.8	0.7	0.7	
	0.7	0.8	0.8	0.8	0.8	0.9	0.9	0.9	0.9	0.9	0.8	0.8	0.7	0.7	0.6	0.5	0.4	0.3	0.1	−0.1	−0.2	−0.5	−0.7
88-S	280	264	249	236	223	212	201	195	185	177	169	161	154	147	141	135	129	126	116	107	99	91	84
	0.8	0.8	0.9	0.9	1.0	1.0	1.1	1.1	1.2	1.2	1.2	1.3	1.3	1.3	1.3	1.3	1.3	1.3	1.3	1.3	1.2	1.2	1.1
	0.9	0.9	1.0	1.0	1.0	1.1	1.1	1.1	1.1	1.1	1.1	1.1	1.0	1.0	0.9	0.9	0.8	0.7	0.6	0.4	0.3	0.1	−0.1

Strength based on strain compatibility; bottom tension limited to $6\sqrt{f'_c}$; see pages 2-3–2-6 for explanation.

Strand Pattern Designation

76-S
▲▲▲
- S = straight
- Diameter of strand in 16ths
- No. of strand (7)

Safe loads shown include dead load of 10 psf for untopped members and 15 psf for topped members. Remainder is live load. Long-time cambers include superimposed dead load but do not include live load.

Capacity of selections of other configurations are similar. For precise values, see local hollow-core manufacturer.

Key
- 140 — Safe superimposed service load, psf
- 0.6 — Estimated camber at erection, in.
- 0.7 — Estimated long-time camber, in.

HOLLOW-CORE

4'-0" x 12"
Lightweight Concrete

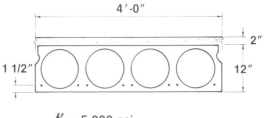

$f'_c = 5{,}000$ psi
$f'_{ci} = 3{,}500$ psi

Section Properties

		Untopped		Topped	
A	=	262	in.²	—	
I	=	4,949	in.⁴	8,800	in.⁴
y_b	=	6.00	in.	8.08	in.
y_t	=	6.00	in.	5.92	in.
Z_b	=	825	in.³	1,089	in.³
Z_t	=	825	in.³	1,486	in.³
b_w	=	8.00	in.	8.00	in.
wt	=	209	plf	309	plf
		52	psf	77	psf
V/S	=	2.18	in.		

4LHC12
No Topping

Table of safe superimposed service load (psf) and cambers

Strand Designation Code	Span, ft.																						
	28	29	30	31	32	33	34	35	36	37	38	39	40	41	42	43	44	45	46	47	48	49	50
76-S	140 0.6 0.7	128 0.6 0.7	117 0.6 0.7	107 0.6 0.7	98 0.6 0.6	90 0.6 0.6	82 0.6 0.5	75 0.5 0.5	69 0.5 0.4	63 0.5 0.3	57 0.4 0.2	52 0.4 0.2	48 0.3 0.0	43 0.2 −0.1	 −0.3								
58-S	186 0.9 1.1	174 0.9 1.1	160 0.9 1.2	147 1.0 1.2	135 1.0 1.2	125 1.0 1.2	115 1.0 1.1	106 1.0 1.1	98 1.0 1.1	91 1.0 1.0	84 1.0 0.9	78 0.9 0.8	72 0.9 0.7	66 0.8 0.6	61 0.8 0.4	57 0.7 0.3	52 0.6 0.1	48 0.5 −0.1	44 0.4 −0.4				
68-S	192 1.1 1.4	183 1.2 1.5	175 1.2 1.5	170 1.3 1.6	163 1.3 1.6	156 1.4 1.6	144 1.4 1.6	134 1.4 1.6	124 1.4 1.6	115 1.4 1.6	107 1.4 1.6	100 1.4 1.5	93 1.4 1.4	86 1.4 1.4	80 1.4 1.3	75 1.3 1.1	69 1.3 1.0	65 1.2 0.8	60 1.1 0.6	56 1.0 0.4	51 0.9 0.2	48 0.8 −0.1	
78-S	198 1.4 1.8	189 1.5 1.8	181 1.5 1.9	173 1.6 2.0	166 1.7 2.0	162 1.7 2.1	155 1.8 2.1	149 1.8 2.1	144 1.9 2.2	138 1.9 2.2	130 1.9 2.2	121 1.9 2.2	113 1.9 2.1	106 1.9 2.1	99 1.9 2.0	92 1.9 1.9	86 1.9 1.8	81 1.9 1.7	76 1.8 1.6	71 1.7 1.4	65 1.7 1.2	61 1.6 1.0	57 1.5 0.7
88-S	204 1.7 2.1	195 1.8 2.2	187 1.8 2.3	179 1.9 2.4	172 2.0 2.5	165 2.1 2.6	158 2.2 2.6	152 2.2 2.7	147 2.3 2.7	144 2.3 2.7	139 2.4 2.8	134 2.4 2.8	127 2.5 2.8	119 2.5 2.8	115 2.5 2.7	108 2.5 2.7	103 2.5 2.6	97 2.5 2.5	91 2.5 2.4	85 2.5 2.3	81 2.4 2.1	75 2.4 2.0	70 2.3 1.8

4LHC12 + 2
2" Normal Weight Topping

Table of safe superimposed service load (psf) and cambers

Strand Designation Code	Span, ft.																						
	23	24	25	26	27	28	29	30	31	32	33	34	35	36	37	38	39	40	41	42	43	44	45
76-S	254 0.5 0.5	228 0.5 0.5	206 0.6 0.5	186 0.6 0.5	168 0.6 0.5	152 0.6 0.4	137 0.6 0.4	124 0.6 0.4	112 0.6 0.3	102 0.6 0.2	92 0.6 0.1	83 0.6 0.0	75 0.5 −0.1	68 0.5 −0.2	61 0.5 −0.4	54 0.4 −0.5	49 0.4 −0.7						
58-S	266 0.7 0.7	253 0.7 0.7	238 0.8 0.8	228 0.8 0.8	216 0.9 0.8	204 0.9 0.8	189 0.9 0.8	173 1.0 0.8	158 1.0 0.8	144 1.0 0.7	132 1.0 0.7	121 1.0 0.6	111 1.0 0.6	101 1.0 0.5	93 1.0 0.4	85 1.0 0.3	77 0.9 0.1	71 0.9 0.0	64 0.8 −0.2	58 0.8 −0.4	53 0.7 −0.6	 −0.9	
68-S	275 0.9 0.9	259 0.9 1.0	244 1.0 1.0	234 1.0 1.0	222 1.0 1.1	210 1.1 1.1	203 1.1 1.1	193 1.2 1.1	183 1.2 1.1	175 1.3 1.1	167 1.3 1.1	154 1.4 1.1	142 1.4 1.0	131 1.4 1.0	120 1.4 0.9	111 1.4 0.8	102 1.4 0.7	94 1.4 0.6	87 1.4 0.4	80 1.4 0.3	73 1.3 0.1	67 1.3 −0.1 −0.3	62 1.2 −0.6
78-S	281 1.0 1.1	265 1.1 1.2	253 1.2 1.2	240 1.2 1.3	228 1.3 1.3	216 1.3 1.4	209 1.4 1.4	199 1.5 1.4	189 1.5 1.4	181 1.6 1.5	173 1.7 1.5	165 1.7 1.5	161 1.8 1.4	154 1.8 1.4	147 1.9 1.4	136 1.9 1.3	126 1.9 1.2	117 2.0 1.1	108 2.0 0.9	100 1.9 0.8	93 1.9 0.6	86 1.9 0.4	80 1.9 0.2
88-S		274 1.3 1.4	259 1.4 1.5	246 1.5 1.6	237 1.6 1.6	225 1.7 1.7	215 1.8 1.8	205 1.8 1.8	195 1.9 1.8	187 2.0 1.9	179 2.1 1.9	171 2.2 1.9	164 2.2 1.9	157 2.3 1.9	151 2.3 1.8	145 2.4 1.8	142 2.4 1.7	137 2.5 1.6	129 2.5 1.5	120 2.5 1.4	112 2.5 1.3	104 2.5 1.1	97 2.5 0.9

Strength based on strain compatibility; bottom tension limited to $6\sqrt{f'_c}$; see pages 2-3–2-6 for explanation.

HOLLOW-CORE SLABS

Fig. 2.4.1 Section Properties — normal weight concrete

Dy-Core

Trade name: Dy-Core®
Licensing Organization: Dy-Core Systems, Inc., Vancouver, British Columbia

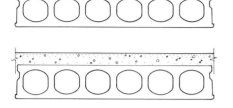

Section	Untopped				With 2" topping		
width x depth	A in.²	y_b in.	I in.⁴	wt psf	y_b in.	I in.⁴	wt psf
4'-0" x 6"	151	3.11	683	40	4.54	1,552	65
4'-0" x 8"	190	3.95	1,568	51	5.54	3,130	76
4'-0" x 10"	216	5.10	2,892	58	6.80	5,097	83
4'-0" x 12"	262	6.34	4,875	71	8.01	7,823	96
4'-0" x 15"	289	7.34	8,701	78	9.36	13,776	103

Note: All sections not available from all producers. Check availability with local manufacturers.

Fig. 2.4.2 Section Properties — normal weight concrete

Dynaspan

Trade name: Dynaspan®
Equipment Manufacturers: Dynamold Corporation, Salina, Kansas

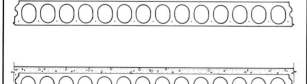

Section	Untopped				With 2" topping		
width x depth	A in.²	y_b in.	I in.⁴	wt psf	y_b in.	I in.⁴	wt psf
4'-0" x 4"	133	2.00	235	35	3.08	689	60
4'-0" x 6"	165	3.02	706	43	4.25	1,543	68
4'-0" x 8"	233	3.93	1,731	61	5.16	3,205	86
4'-0" x 10"	260	4.91	3,145	68	6.26	5,314	93
8'-0" x 6"	338	3.05	1,445	44	4.26	3,106	69
8'-0" x 8"	470	3.96	3,525	61	5.17	6,444	86
8'-0" x 10"	532	4.96	6,422	69	6.28	10,712	94
8'-0" x 12"	615	5.95	10,505	80	7.32	16,507	105

Note: All sections not available from all producers. Check availability with local manufacturers.

PCI Design Handbook/Fourth Edition

HOLLOW-CORE SLABS

Fig. 2.4.3 Section Properties — normal weight concrete — Flexicore

Trade name: Flexicore®
Licensing Organization: The Flexicore Co. Inc., Dayton,. Ohio

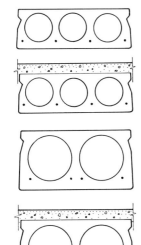

Section width x depth	Untopped				With 2" topping		
	A in.²	y_b in.	I in.⁴	wt psf	y_b in.	I in.⁴	wt psf
1'-4" x 6"	55	3.00	243	43	4.23	523	68
2'-0" x 6"	86	3.00	366	45	4.20	793	70
1'-4" x 8"	73	4.00	560	57	5.26	1,028	82
2'-0" x 8"	110	4.00	843	57	5.26	1,547	82
1'-8" x 10"	98	5.00	1,254	61	6.43	2,109	86
2'-0" x 10"	138	5.00	1,587	72	6.27	2,651	97
2'-0" x 12"	141	6.00	2,595	73	7.46	4,049	98

Note: All sections not available from all producers. Check availability with local manufacturers.

Fig. 2.4.4 Section Properties — normal weight concrete — Spancrete

Trade name: Spancrete®
Licensing Organization: Spancrete Machinery Corp., Milwaukee, Wisconsin

Trade name: Standard Spancrete®

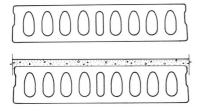

Section width x depth	Untopped				With 2" topping		
	A in.²	y_b in.	I in.⁴	wt psf	y_b in.	I in.⁴	wt psf
48" x 4"	138	2.00	238	34	3.14	739	59
48" x 6"	189	2.93	762	46	4.19	1,760	71
48" x 8"	259	3.98	1,806	63	5.22	3,443	88
48" x 10"	312	5.16	3,484	76	6.41	5,787	101
48" x 12"	355	6.28	5,784	86	7.58	8,904	111
48" x 15"	370	7.87	9,765	90	9.39	14,351	115

Trade name: Ultralight Spancrete®

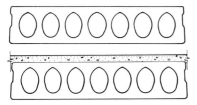

Section	A	y_b	I	wt	y_b	I	wt
48" x 8"	246	4.17	1,730	60	5.41	3,230	85
48" x 10"	277	5.22	3,178	67	6.58	5,376	92
48" x 12"	316	6.22	5,311	77	7.66	8,410	102

Note: Spancrete is also available in 40" and 60" widths. All sections are not available from all producers. Check availability with local manufacturers.

HOLLOW-CORE SLABS

Fig. 2.4.5 Section Properties — normal weight concrete   **Span Deck**

Trade name: Span Deck®
Licensing Organization: Fabcon, Incorporated, Savage, Minnesota

Section	Untopped				With 2" topping		
width x depth	A in.²	y_b in.	I in.⁴	wt* psf	y_b in.	I in.⁴	wt* psf
4'-0" x 8"	199	4.27	1,618	50-62	5.53	2,822	74-86
4'-0" x 12"	246	6.54	4,638	62-72	8.00	6,981	84-96
8'-0" x 8"	397	4.27	3,236	50-62	5.53	5,643	74-86
8'-0" x 12"	493	6.54	9,275	62-72	8.10	13,962	84-96

Weights may vary among manufacturers due to the use of light weight or heavy weight core material.

Note: All sections not available from all producers. Check availability with local manufacturers.
6" and 10" depths are of similar configuration.

Fig. 2.4.2 Section Properties — normal weight concrete   **Ultra Span**

Trade name: Ultra Span
Licensing Organization: Ultra Span Machinery Inc., Winnipeg, Manitoba, Canada

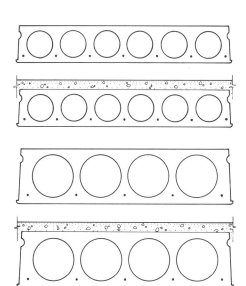

Section	Untopped				With 2" topping		
width x depth	A in.²	y_b in.	I in.⁴	wt psf	y_b in.	I in.⁴	wt psf
4'-0" x 4"	154	2.00	247	40	2.98	723	65
4'-0" x 6"	188	3.00	764	49	4.13	1,641	74
4'-0" x 8"	214	4.00	1,666	56	5.29	3,070	81
4'-0" x 10"	259	5.00	3,223	67	6.34	5,328	92
4'-0" x 12"	289	6.00	5,272	75	7.43	8,195	100

Note: All sections not available from all producers. Check availability with local manufacturers.

Strand Pattern Designation

76-S
▲▲▲
- S = straight
- Diameter of strand in 16ths
- No. of strand (7)

Safe loads shown include dead load of 10 psf for untopped members and 15 psf for topped members. Remainder is live load. Long-time cambers include superimposed dead load but do not include live load.

Key

- 196 — Safe superimposed service load, psf
- 0.1 — Estimated camber at erection, in.
- 0.1 — Estimated long-time camber, in.

SOLID FLAT SLAB

4" Thick
Normal Weight Concrete

$f'_c = 5,000$ psi
$f'_{ci} = 3,500$ psi

Section Properties

		Untopped	Topped
A	=	192 in.²	—
I	=	256 in.⁴	763 in.⁴
y_b	=	2.00 in.	2.84 in.
y_t	=	2.00 in.	3.16 in.
Z_b	=	128 in.³	269 in.³
Z_t	=	128 in.³	242 in.³
b_w	=	48.00 in.	48.00 in.
wt	=	200 plf	300 plf
		50 psf	75 psf
V/S	=	1.85 in.	

FS4
No Topping

Table of safe superimposed service load (psf) and cambers

Strand Designation Code	Span, ft.												
	10	11	12	13	14	15	16	17	18	19	20	21	22
66-S	196 0.1 0.1	165 0.1 0.1	132 0.1 0.0	105 0.0 0.0	83 0.0 −0.1	66 −0.1 −0.3	52 −0.2 −0.4	41 −0.3 −0.6	31 −0.4 −0.9				
76-S	230 0.1 0.1	190 0.1 0.1	151 0.1 0.1	122 0.1 0.0	98 0.0 −0.1	79 0.0 −0.2	63 −0.1 −0.3	50 −0.2 −0.5	40 −0.3 −0.9	30 −0.5 −1.0			
58-S	253 0.2 0.2	212 0.2 0.2	180 0.2 0.2	154 0.2 0.2	127 0.1 0.1	104 0.1 0.0	86 0.0 −0.1	70 0.0 −0.3	57 −0.1 −0.5	46 −0.3 −0.8	37 −0.4 −1.1		
68-S	300 0.2 0.3	252 0.2 0.3	214 0.2 0.3	18 0.2 0.3	152 0.2 0.2	127 0.2 0.2	105 0.2 0.1	88 0.1 −0.1	73 0.0 −0.3	60 −0.1 −0.5	50 −0.3 −0.8	40 −0.4 −1.2	32 −0.7 −1.6

FS4 + 2
2" Normal Weight Topping

Table of safe superimposed service load (psf) and cambers

Strand Designation Code	Span, ft.											
	10	11	12	13	14	15	16	17	18	19	20	21
66-S	369 0.1 0.0	296 0.1 0.0	224 0.1 0.0	167 0.0 −0.1	123 0.0 −0.2	87 −0.1 −0.3	57 −0.2 −0.5	33 −0.3 −0.7				
76-S		346 0.1 0.0	265 0.1 0.0	203 0.1 −0.1	153 0.0 −0.1	113 0.0 −0.3	80 −0.1 −0.4	53 −0.2 −0.6	31 −0.3 −0.9			
58-S		400 0.2 0.1	342 0.2 0.1	274 0.2 0.0	214 0.1 −0.1	166 0.1 −0.3	127 0.0 −0.4	95 0.0 −0.7	67 −0.1 −0.9	44 −0.3 		
68-S			335 0.2 0.1	268 0.2 0.1	213 0.2 0.0	169 0.2 −0.1	132 0.1 −0.3	101 0.0 −0.5	74 −0.1 −0.7	52 −0.3 −1.0	32 −0.4 −1.4	

Strength based on strain compatibility; bottom tension limited to $6\sqrt{f'_c}$; see pages 2-3–2-6 for explanation.

Strand Pattern Designation

76-S
- S = straight
- Diameter of strand in 16ths
- No. of strand (7)

Safe loads shown include dead load of 10 psf for untopped members and 15 psf for topped members. Remainder is live load. Long-time cambers include superimposed dead load but do not include live load.

Key
- 206 — Safe superimposed service load, psf
- 0.1 — Estimated camber at erection, in.
- 0.2 — Estimated long-time camber, in.

SOLID FLAT SLAB

4" Thick Lightweight Concrete

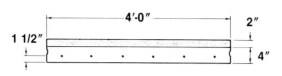

$f'_c = 5{,}000$ psi
$f'_{ci} = 3{,}500$ psi

Section Properties

	Untopped		Topped	
A =	192	in.²	—	
I =	256	in.⁴	925	in.⁴
y_b =	2.00	in.	3.10	in.
y_t =	2.00	in.	2.90	in.
Z_b =	128	in.³	298	in.³
Z_t =	128	in.³	319	in.³
b_w =	48.00	in.	48.00	in.
wt =	153	plf	253	plf
	38	psf	63	psf
V/S =	1.85	in.		

LFS4 — No Topping

Table of safe superimposed service load (psf) and cambers

Strand Designation Code	Span, ft.														
	10	11	12	13	14	15	16	17	18	19	20	21	22	23	24
66-S	206	175	143	116	95	78	64	52	42	34					
	0.1	0.1	0.1	0.1	0.1	0.0	0.0	−0.1	−0.3	−0.5					
	0.2	0.2	0.1	0.1	0.0	−0.2	−0.3	−0.5	−0.8	−1.2					
76-S	239	201	163	133	110	91	75	62	51	42	34				
	0.2	0.2	0.2	0.2	0.2	0.1	0.1	0.0	−0.1	−0.3	−0.5				
	0.2	0.2	0.2	0.2	0.1	0.0	−0.2	−0.4	−0.6	−1.0	−1.3				
58-S	263	222	190	163	136	113	95	80	68	57	48	40	33		
	0.3	0.3	0.3	0.3	0.3	0.3	0.3	0.2	0.1	0.0	−0.2	−0.4	−0.6		
	0.3	0.4	0.4	0.3	0.3	0.2	0.1	−0.1	−0.3	−0.5	−0.9	−1.3	−1.8		
68-S	309	261	223	189	158	133	112	95	81	69	59	50	42	36	30
	0.3	0.4	0.4	0.4	0.5	0.5	0.5	0.4	0.3	0.2	0.1	−0.1	−0.3	−0.5	−0.9
	0.4	0.5	0.5	0.5	0.5	0.4	0.4	0.2	0.0	−0.2	−0.5	−0.9	−1.4	−1.9	−2.6

LFS4 + 2 — 2" Normal Weight Topping

Table of safe superimposed service load (psf) and cambers

Strand Designation Code	Span, ft.										
	11	12	13	14	15	16	17	18	19	20	21
66-S	322	276	213	163	123	91	63	41			
	0.1	0.1	0.1	0.1	0.0	0.0	−0.1	−0.3			
	0.1	0.0	−0.1	−0.3	−0.5	−0.7	−1.0				
76-S	375	319	252	197	153	117	86	61	40		
	0.2	0.2	0.2	0.2	0.1	0.1	0.0	−0.1	−0.3		
	0.1	0.1	0.0	−0.1	−0.2	−0.3	−0.6	−0.8	−1.2		
58-S		351	304	258	206	164	129	100	75	54	35
		0.3	0.3	0.3	0.3	0.3	0.2	0.1	0.0	−0.2	−0.4
		0.2	0.2	0.1	0.0	−0.1	−0.3	−0.5	−0.5	−1.2	−1.7
68-S			360	310	251	204	164	131	103	79	59
			0.4	0.5	0.5	0.5	0.4	0.3	0.2	0.1	−0.1
			0.3	0.3	0.2	0.1	−0.1	−0.3	−0.6	−0.9	−1.3

Strength based on strain compatibility; bottom tension limited to $6\sqrt{f'_c}$; see pages 2-3–2-6 for explanation.

Strand Pattern Designation

76-S
- S = straight
- Diameter of strand in 16ths
- No. of strand (7)

Safe loads shown include dead load of 10 psf for untopped members and 15 psf for topped members. Remainder is live load. Long-time cambers include superimposed dead load but do not include live load.

Key
- 330 — Safe superimposed service load, psf
- 0.1 — Estimated camber at erection, in.
- 0.2 — Estimated long-time camber, in.

SOLID FLAT SLAB
6" Thick
Normal Weight Concrete

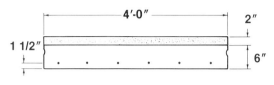

f'_c = 5,000 psi
f'_{ci} = 3,500 psi

Section Properties

	Untopped	Topped
A =	288 in.²	—
I =	864 in.⁴	1,834 in.⁴
y_b =	3.00 in.	3.82 in.
y_t =	3.00 in.	4.18 in.
Z_b =	288 in.³	480 in.³
Z_t =	288 in.³	439 in.³
b_w =	48.00 in.	48.00 in.
wt =	300 plf	400 plf
	75 psf	100 psf
V/S =	2.67 in.	

FS6 — No Topping

Table of safe superimposed service load (psf) and cambers

Strand Designation Code	Span, ft.																			
	11	12	13	14	15	16	17	18	19	20	21	22	23	24	25	26	27	28	29	30
66-S	330	284	236	195	162	135	113	94	78	65	53	43	33							
	0.1	0.1	0.1	0.1	0.1	0.1	0.1	0.1	0.0	−0.1	−0.1	−0.2	−0.3							
	0.2	0.2	0.2	0.2	0.1	0.1	0.1	0.0	−0.1	−0.2	−0.4	−0.5	−0.7							
76-S	390	336	280	233	195	164	139	117	99	83	68	55	44	35						
	0.1	0.2	0.2	0.2	0.2	0.2	0.2	0.2	0.1	0.1	0.0	0.0	−0.1	−0.2	−0.3					
	0.2	0.2	0.2	0.2	0.2	0.2	0.2	0.1	0.0	−0.1	−0.2	−0.4	−0.5	−0.8						
58-S		369	320	280	247	219	186	158	134	113	96	81	67	56	46	36				
		0.2	0.3	0.3	0.3	0.3	0.3	0.3	0.3	0.2	0.2	0.1	0.0	−0.1	−0.2	−0.3				
		0.3	0.3	0.4	0.4	0.4	0.4	0.3	0.3	0.2	0.1	0.0	−0.2	−0.4	−0.6	−0.9				
68-S			387	339	300	262	224	191	164	141	121	103	88	75	63	53	43	35		
			0.3	0.4	0.4	0.4	0.4	0.4	0.4	0.4	0.4	0.3	0.3	0.2	0.1	0.0	−0.2	−0.4		
			0.4	0.5	0.5	0.5	0.5	0.5	0.5	0.5	0.4	0.4	0.3	0.2	0.0	−0.2	−0.4	−0.7	−1.1	
78-S				396	351	303	261	225	195	168	146	126	109	94	81	69	59	49	41	33
				0.4	0.5	0.5	0.6	0.6	0.6	0.6	0.6	0.6	0.5	0.4	0.4	0.3	0.1	0.0	−0.2	−0.5
				0.6	0.7	0.7	0.7	0.7	0.7	0.7	0.7	0.6	0.5	0.4	0.2	0.0	−0.3	−0.6	−0.9	−1.4

FS6 + 2 — 2" Normal Weight Topping

Table of safe superimposed service load (psf) and cambers

Strand Designation Code	Span, ft.															
	13	14	15	16	17	18	19	20	21	22	23	24	25	26	27	28
66-S	342	284	237	198	162	127	97	71	49							
	0.1	0.1	0.1	0.1	0.1	0.1	0.0	−0.1	−0.1							
	0.1	0.1	0.1	0.0	0.0	−0.1	−0.2	−0.3	−0.5							
76-S		336	283	239	197	158	125	96	72	51	32					
		0.2	0.2	0.2	0.2	0.1	0.1	0.0	0.0	−0.1	−0.2					
		0.2	0.1	0.1	0.0	0.0	−0.1	−0.2	−0.4	−0.5	−0.7					
58-S			356	314	268	221	181	147	118	93	71	51	34			
			0.3	0.3	0.3	0.3	0.3	0.2	0.2	0.1	0.0	−0.1	−0.2			
			0.3	0.2	0.2	0.2	0.1	0.0	−0.1	−0.3	−0.4	−0.7	−0.9			
68-S				375	323	278	232	193	160	131	105	83	64	46	31	
				0.4	0.4	0.4	0.4	0.4	0.4	0.3	0.3	0.2	0.1	0.0	−0.2	
				0.4	0.4	0.3	0.3	0.2	0.1	0.0	−0.2	−0.3	−0.6	−0.9	−1.2	
78-S					371	323	281	239	201	169	140	115	93	73	56	40
					0.6	0.6	0.6	0.6	0.6	0.6	0.5	0.4	0.4	0.3	0.1	0.0
					0.5	0.5	0.5	0.3	0.4	0.3	0.1	0.0	−0.2	−0.5	−0.8	−1.1

Strength based on strain compatibility; bottom tension limited to $6\sqrt{f'_c}$; see pages 2-3–2-6 for explanation.

Strand Pattern Designation

76-S
- S = straight
- Diameter of strand in 16ths
- No. of strand (7)

Safe loads shown include dead load of 10 psf for untopped members and 15 psf for topped members. Remainder is live load. Long-time cambers include superimposed dead load but do not include live load.

Key
- 345 — Safe superimposed service load, psf
- 0.2 — Estimated camber at erection, in.
- 0.3 — Estimated long-time camber, in.

SOLID FLAT SLAB

6" Thick
Lightweight Concrete

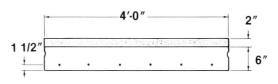

$f'_c = 5,000$ psi
$f'_{ci} = 3,500$ psi

Section Properties

		Untopped	Topped	
A	=	288 in.²	—	
I	=	864 in.⁴	2,181	in.⁴
y_b	=	3.00 in.	4.11	in.
y_t	=	3.00 in.	3.89	in.
Z_b	=	288 in.³	531	in.³
Z_t	=	288 in.³	561	in.³
b_w	=	48.00 in.	48.00	in.
wt	=	230 plf	330	plf
		58 psf	83	psf
V/S	=	2.67 in.		

LFS6 — No Topping

Table of safe superimposed service load (psf) and cambers

Strand Designation Code	Span, ft.																				
	11	12	13	14	15	16	17	18	19	20	21	22	23	24	25	26	27	28	29	30	31
66-S	345	298	250	209	177	150	127	109	93	79	68	58	49	41	34						
	0.2	0.2	0.2	0.3	0.3	0.3	0.3	0.2	0.2	0.1	0.1	0.0	−0.1	−0.3	−0.4						
	0.3	0.3	0.3	0.3	0.3	0.3	0.3	0.2	0.1	0.0	−0.1	−0.3	−0.5	−0.8	−1.1						
76-S		351	294	247	209	179	153	132	113	98	85	73	62	52	44	36					
		0.3	0.3	0.3	0.3	0.4	0.4	0.3	0.3	0.3	0.2	0.2	0.1	0.0	−0.2	−0.4					
		0.4	0.4	0.4	0.4	0.4	0.4	0.4	0.3	0.2	0.1	−0.1	−0.2	−0.5	−0.7	−1.1					
58-S			383	334	295	262	233	201	175	151	131	113	98	85	73	63	54	46	39	32	
			0.4	0.4	0.5	0.5	0.5	0.6	0.6	0.6	0.6	0.6	0.5	0.5	0.4	0.3	0.1	0.0	−0.2	−0.5	
			0.5	0.6	0.6	0.7	0.7	0.7	0.7	0.7	0.6	0.6	0.5	0.3	0.1	−0.1	−0.4	−0.7	−1.1	−1.5	
68-S				354	315	272	235	204	178	156	137	120	105	92	81	70	61	53	45	38	32
				0.6	0.6	0.7	0.8	0.8	0.8	0.8	0.8	0.8	0.8	0.8	0.7	0.6	0.4	0.3	0.1	−0.2	−0.4
				0.8	0.8	0.9	0.9	1.0	1.0	1.0	1.0	0.9	0.8	0.7	0.5	0.3	0.0	−0.4	−0.8	−1.2	−1.8
78-S					356	307	266	232	203	179	157	139	123	109	96	85	76	67	58	51	44
					0.8	0.9	0.9	1.0	1.0	1.1	1.1	1.1	1.1	1.1	1.1	1.0	0.9	0.8	0.6	0.4	0.2
					1.0	1.1	1.2	1.2	1.3	1.3	1.3	1.3	1.2	1.2	1.0	0.9	0.6	0.4	0.0	−0.4	−0.9

LFS6 + 2 — 2" Normal Weight Topping

Table of safe superimposed service load (psf) and cambers

Strand Designation Code	Span, ft.																
	14	15	16	17	18	19	20	21	22	23	24	25	26	27	28	29	30
66-S	298	251	213	181	155	132	111	86	65	47	30						
	0.3	0.3	0.3	0.3	0.2	0.2	0.1	0.1	0.0	−0.1	−0.3						
	0.2	0.2	0.2	0.1	0.0	−0.1	−0.2	−0.4	−0.6	−0.8	−1.1						
76-S	351	297	253	217	186	161	139	112	88	68	50	34					
	0.3	0.3	0.4	0.4	0.3	0.3	0.3	0.2	0.2	0.1	0.0	−0.2					
	0.3	0.3	0.3	0.2	0.2	0.1	0.0	−0.2	−0.4	−0.6	−0.8	−1.1					
58-S		371	328	283	246	214	187	163	135	110	89	70	53	38			
		0.5	0.5	0.6	0.6	0.6	0.6	0.6	0.5	0.5	0.4	0.3	0.1	0.0			
		0.5	0.5	0.5	0.4	0.4	0.3	0.2	0.1	−0.1	−0.3	−0.6	−0.9	−1.3			
68-S			390	338	294	257	226	199	175	148	124	103	83	66	51	37	
			0.7	0.8	0.8	0.8	0.8	0.8	0.8	0.8	0.8	0.7	0.6	0.4	0.3	0.1	
			0.7	0.7	0.7	0.6	0.6	0.5	0.4	0.3	0.1	−0.1	−0.4	−0.7	−1.1	−1.5	
78-S				386	337	296	261	230	204	180	155	132	111	93	77	61	47
				0.9	1.0	1.0	1.1	1.1	1.1	1.1	1.1	1.0	0.9	0.8	0.6	0.4	0.4
				0.9	0.9	0.9	0.9	0.8	0.8	0.7	0.5	0.3	0.1	−0.2	−0.5	−0.9	−1.4

Strength based on strain compatibility; bottom tension limited to $6\sqrt{f'_c}$; see pages 2-3–2-6 for explanation.

Strand Pattern Designation

76-S
▲▲▲
- S = straight
- Diameter of strand in 16ths
- No. of strand (7)

Safe loads shown include dead load of 10 psf for untopped members and 15 psf for topped members. Remainder is live load. Long-time cambers include superimposed dead load but do not include live load.

Key
- 299 — Safe superimposed service load, psf
- 0.1 — Estimated camber at erection, in.
- 0.2 — Estimated long-time camber, in.

SOLID FLAT SLAB

8" Thick
Normal Weight Concrete

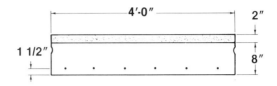

f'_c = 5,000 psi
f'_{ci} = 3,500 psi

Section Properties

		Untopped		Topped	
A	=	384	in.²	—	
I	=	2,048	in.⁴	3,630	in.⁴
y_b	=	4.00	in.	4.81	in.
y_t	=	4.00	in.	5.19	in.
Z_b	=	512	in.³	755	in.³
Z_t	=	512	in.³	699	in.³
b_w	=	48.00	in.	48.00	in.
wt	=	400	plf	500	plf
		100	psf	125	psf
V/S	=	3.43	in.		

FS8
No Topping

Table of safe superimposed service load (psf) and cambers

Strand Designation Code	Span, ft.																				
	14	15	16	17	18	19	20	21	22	23	24	25	26	27	28	29	30	31	32	33	34
66-S	299 0.1 0.2	250 0.1 0.1	210 0.1 0.1	177 0.1 0.1	149 0.1 0.1	125 0.1 0.1	105 0.1 0.0	88 0.0 0.0	73 0.0 0.0	60 0.0 −0.1	48 −0.1 −0.2	38 −0.2 −0.3		−0.2 −0.5							
76-S	357 0.1 0.2	301 0.2 0.2	255 0.2 0.2	216 0.2 0.2	184 0.2 0.2	157 0.1 0.2	134 0.1 0.1	114 0.1 0.1	97 0.1 0.1	82 0.0 0.0	68 −0.1 −0.1	57 −0.1 −0.2	46 −0.2 −0.3	37 −0.3 −0.5		−0.7					
58-S		378 0.2 0.3	338 0.2 0.3	293 0.3 0.3	252 0.3 0.3	218 0.3 0.3	189 0.3 0.3	164 0.2 0.2	142 0.2 0.2	123 0.2 0.2	107 0.1 0.1	92 0.1 0.0	78 0.0 −0.1	65 −0.1 −0.3	54 −0.2 −0.5	43 −0.3 −0.7	34 −0.4 −0.7				
68-S				358 0.3 0.5	311 0.4 0.5	271 0.4 0.5	237 0.4 0.5	207 0.4 0.5	182 0.4 0.4	159 0.4 0.4	138 0.3 0.3	120 0.3 0.3	103 0.2 0.2	88 0.2 0.0	75 0.1 −0.1	63 0.0 −0.3	53 −0.1 −0.5	43 −0.3 −0.8	34 −0.5 −1.0		
78-S					367 0.5 0.6	321 0.5 0.6	282 0.5 0.7	248 0.5 0.7	219 0.5 0.7	192 0.5 0.6	168 0.5 0.6	147 0.5 0.5	128 0.4 0.5	112 0.4 0.3	97 0.3 0.2	83 0.2 0.1	71 0.1 −0.1	60 0.0 −0.3	51 −0.1 −0.6	42 −0.3 −0.9	33 −0.5 −1.2

FS8 + 2
2" Normal Weight Topping

Table of safe superimposed service load (psf) and cambers

Strand Designation Code	Span, ft.													
	17	18	19	20	21	22	23	24	25	26	27	28	29	30
66-S	230 0.1 0.1	195 0.1 0.0	164 0.1 0.0	139 0.1 −0.1	116 0.0 −0.1	97 0.0 −0.2	73 −0.1 −0.3	52 −0.2 −0.5	34 −0.2 −0.6					
76-S	280 0.2 0.1	239 0.2 0.1	204 0.1 0.1	175 0.1 0.0	149 0.1 0.0	126 0.1 −0.1	99 0.0 −0.2	76 −0.1 −0.3	56 −0.1 −0.5	38 −0.2 −0.7				
58-S	375 0.3 0.3	324 0.3 0.2	280 0.3 0.2	243 0.3 0.2	211 0.2 0.1	183 0.2 0.1	152 0.2 0.0	124 0.1 0.0	100 0.1 −0.1	79 0.0 −0.2	59 −0.1 −0.4	42 −0.2 −0.5		−0.7
68-S		395 0.4 0.4	344 0.4 0.4	301 0.4 0.3	264 0.4 0.3	231 0.4 0.3	199 0.4 0.2	167 0.3 0.1	140 0.3 0.0	115 0.2 −0.1	94 0.2 −0.3	74 0.1 −0.4	57 0.0 −0.6	41 −0.1 −0.9

Strength based on strain compatibility; bottom tension limited to $6\sqrt{f'_c}$; see pages 2-3-2-6 for explanation.

Strand Pattern Designation

76-S
- S = straight
- Diameter of strand in 16ths
- No. of strand (7)

Safe loads shown include dead load of 10 psf for untopped members and 15 psf for topped members. Remainder is live load. Long-time cambers include superimposed dead load but do not include live load.

Key
318 — Safe superimposed service load, psf
0.2 — Estimated camber at erection, in.
0.3 — Estimated long-time camber, in.

SOLID FLAT SLAB
8" Thick
Lightweight Concrete

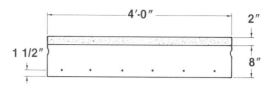

f'_c = 5,000 psi
f'_{ci} = 3,500 psi

Section Properties

	Untopped	Topped
A =	384 in.²	—
I =	2,048 in.⁴	4,234 in.⁴
y_b =	4.00 in.	5.12 in.
y_t =	4.00 in.	4.88 in.
Z_b =	512 in.³	827 in.³
Z_t =	512 in.³	868 in.³
b_w =	48.00 in.	48.00 in.
wt =	307 plf	407 plf
	77 psf	102 psf
V/S =	3.43 in.	

LFS8 — No Topping

Table of safe superimposed service load (psf) and cambers

Strand Designation Code	Span, ft.																					
	14	15	16	17	18	19	20	21	22	23	24	25	26	27	28	29	30	31	32	33	34	35
66-S	318	269	229	196	168	144	124	107	92	79	68	58	49	41	33							
	0.2	0.2	0.2	0.2	0.2	0.2	0.2	0.2	0.1	0.1	0.0	0.0	−0.1	−0.2	−0.4							
	0.3	0.3	0.3	0.3	0.3	0.2	0.2	0.1	0.1	0.0	−0.1	−0.3	−0.4	−0.6	−0.9							
76-S	376	320	274	235	203	176	153	133	116	101	88	76	66	56	48	41	34					
	0.3	0.3	0.3	0.3	0.3	0.3	0.3	0.3	0.3	0.2	0.2	0.1	0.0	−0.1	−0.2	−0.3	−0.5					
	0.3	0.4	0.4	0.4	0.4	0.4	0.4	0.4	0.3	0.3	0.2	0.1	0.0	−0.2	−0.4	−0.6	−0.8	−1.1				
58-S		398	357	312	272	238	208	183	162	143	126	111	98	87	76	67	57	49	41	34		
		0.4	0.4	0.5	0.5	0.5	0.5	0.5	0.5	0.5	0.5	0.4	0.4	0.3	0.2	0.1	0.0	−0.2	−0.3	−0.5		
		0.5	0.6	0.6	0.6	0.6	0.6	0.6	0.6	0.6	0.5	0.4	0.3	0.2	0.0	−0.2	−0.4	−0.7	−1.0	−1.4		
68-S					330	290	256	226	201	178	159	142	126	112	99	87	76	66	57	49	42	35
					0.6	0.7	0.7	0.7	0.7	0.7	0.7	0.7	0.7	0.7	0.6	0.5	0.4	0.3	0.1	0.0	−0.2	−0.5
					0.8	0.9	0.9	0.9	0.9	0.9	0.9	0.8	0.7	0.6	0.5	0.3	0.1	−0.1	−0.4	−0.7	−1.1	−1.5
78-S					386	340	301	267	237	211	188	168	151	135	120	107	95	84	74	65	57	49
					0.8	0.8	0.9	0.9	1.0	1.0	1.0	1.0	1.0	1.0	1.0	0.9	0.8	0.7	0.6	0.5	0.3	0.1
					1.0	1.1	1.1	1.2	1.2	1.2	1.2	1.2	1.2	1.1	1.0	0.9	0.7	0.5	0.3	0.0	−0.3	−0.7

LFS8 + 2 — 2" Normal Weight Topping

Table of safe superimposed service load (psf) and cambers

Strand Designation Code	Span, ft.																		
	17	18	19	20	21	22	23	24	25	26	27	28	29	30	31	32	33	34	35
66-S	250	214	184	158	136	116	100	85	72	57	41								
	0.2	0.2	0.2	0.2	0.2	0.1	0.1	0.0	0.0	−0.1	−0.2								
	0.2	0.2	0.1	0.1	0.0	−0.1	−0.2	−0.3	−0.5	−0.7	−0.9								
76-S	299	258	224	194	168	146	127	110	95	74	61	45	31						
	0.3	0.3	0.3	0.3	0.3	0.3	0.2	0.2	0.1	0.0	−0.1	−0.2	−0.3						
	0.3	0.3	0.2	0.2	0.1	0.1	0.0	−0.2	−0.3	−0.5	−0.7	−0.9	−1.2						
58-S	394	343	300	262	230	203	179	157	139	122	103	84	67	52	38				
	0.5	0.5	0.5	0.5	0.5	0.5	0.5	0.5	0.4	0.4	0.3	0.2	0.1	0.0	−0.2				
	0.5	0.5	0.5	0.4	0.4	0.4	0.3	0.2	0.1	0.0	−0.2	−0.4	−0.6	−0.9	−1.2				
68-S			364	320	283	251	222	198	176	156	139	119	100	82	67	52	40		
			0.7	0.7	0.7	0.7	0.7	0.7	0.7	0.7	0.7	0.6	0.5	0.4	0.3	0.1	0.0		
			0.6	0.6	0.6	0.6	0.6	0.5	0.4	0.3	0.2	0.0	−0.2	−0.4	−0.7	−1.0	−1.4		
78-S			374	332	295	263	235	210	188	169	151	132	113	95	79	65	51	39	
			0.9	0.9	1.0	1.0	1.0	1.0	1.0	1.0	1.0	0.9	0.8	0.7	0.6	0.5	0.3	0.1	
			0.9	0.9	0.9	0.9	0.8	0.8	0.7	0.6	0.5	0.3	0.1	−0.2	−0.4	−0.8	−1.1	−1.6	

Strength based on strain compatibility; bottom tension limited to $6\sqrt{f'_c}$; see pages 2-3–2-6 for explanation.

RECTANGULAR BEAMS

Normal Weight Concrete

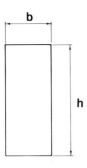

f'_c = 5,000 psi
f_{pu} = 270,000 psi
½ in. diameter
low-relaxation strand

Key
3,344 — Safe superimposed service load, plf
0.4 — Estimated camber at erection, in.
0.1 — Estimated long-time camber, in.

Section Properties							
Designation	b (in.)	h (in.)	A (in.²)	I (in.⁴)	y_b (in.)	Z (in.³)	wt (plf)
12RB16	12	16	192	4,096	8.00	512	200
12RB20	12	20	240	8,000	10.00	800	250
12RB24	12	24	288	13,824	12.00	1,152	300
12RB28	12	28	336	21,952	14.00	1,568	350
12RB32	12	32	384	32,768	16.00	2,048	400
12RB36	12	36	432	46,656	18.00	2,592	450
16RB24	16	24	384	18,432	12.00	1,536	400
16RB28	16	28	448	29,269	14.00	2,091	467
16RB32	16	32	512	43,691	16.00	2,731	533
16RB36	16	36	576	62,208	18.00	3,456	600
16RB40	16	40	640	85,333	20.00	4,267	667

1. Check local area for availability of other sizes.

2. Safe loads shown include 50% dead load and 50% live load. 800 psi top tension has been allowed, therefore additional top reinforcement is required.

3. Safe loads can be significantly increased by use of structural composite topping.

Table of safe superimposed service load (plf) and cambers

Designation	No. Strand	e	Span, ft.																
			16	18	20	22	24	26	28	30	32	34	36	38	40	42	44	46	48	50
12RB16	5	5.67	3,344 0.4 0.1	2,605 0.5 0.2	2,075 0.6 0.2	1,684 0.7 0.2	1,386 0.8 0.2	1,154 0.9 0.2	970 1.0 0.2												
12RB20	8	6.60	6,101 0.4 0.1	4,773 0.5 0.2	3,823 0.6 0.2	3,121 0.7 0.2	2,585 0.8 0.3	2,166 0.9 0.3	1,833 1.0 0.3	1,565 1.1 0.3	1,345 1.2 0.3	1,163 1.3 0.3	1,010 1.4 0.3								
12RB24	10	7.76	8,884 0.3 0.1	6,957 0.4 0.1	5,578 0.5 0.2	4,558 0.6 0.2	3,782 0.7 0.2	3,178 0.8 0.3	2,699 0.9 0.3	2,312 1.0 0.3	1,996 1.1 0.3	1,734 1.2 0.4	1,514 1.3 0.4	1,328 1.4 0.4	1,170 1.5 0.4	1,033 1.6 0.4					
12RB28	12	8.89		9,502 0.3 0.1	7,630 0.4 0.2	6,245 0.5 0.2	5,192 0.6 0.2	4,372 0.7 0.3	3,721 0.8 0.3	3,197 0.9 0.3	2,767 1.0 0.3	2,411 1.1 0.4	2,113 1.2 0.4	1,861 1.3 0.4	1,645 1.4 0.4	1,460 1.5 0.4	1,299 1.5 0.4	1,159 1.6 0.4	1,035 1.7 0.4		
12RB32	13	10.48				8,238 0.4 0.2	6,859 0.5 0.2	5,785 0.6 0.2	4,933 0.7 0.2	4,246 0.8 0.3	3,683 0.9 0.3	3,217 0.9 0.3	2,826 1.0 0.3	2,495 1.1 0.3	2,213 1.2 0.3	1,970 1.3 0.4	1,760 1.4 0.4	1,576 1.5 0.4	1,415 1.5 0.3	1,272 1.6 0.3	
12RB36	15	11.64					8,734 0.5 0.2	7,376 0.5 0.2	6,298 0.6 0.3	5,428 0.7 0.3	4,716 0.8 0.3	4,126 0.9 0.3	3,632 1.0 0.3	3,214 1.0 0.4	2,856 1.1 0.4	2,549 1.2 0.4	2,283 1.3 0.4	2,050 1.4 0.4	1,846 1.5 0.4	1,666 1.5 0.4	
16RB24	13	7.86		9,278 0.4 0.1	7,439 0.5 0.2	6,079 0.6 0.2	5,044 0.7 0.2	4,239 0.8 0.3	3,600 0.9 0.3	3,084 1.0 0.3	2,662 1.1 0.3	2,313 1.2 0.3	2,020 1.3 0.4	1,772 1.4 0.4	1,560 1.5 0.4	1,378 1.6 0.3	1,220 1.6 0.3	1,082 1.7 0.3	961 1.8 0.2		
16RB28	13	8.89			9,022 0.4 0.1	7,383 0.4 0.1	6,137 0.5 0.1	5,167 0.6 0.2	4,397 0.6 0.2	3,776 0.7 0.2	3,267 0.8 0.2	2,846 0.9 0.2	2,493 1.0 0.2	2,194 1.0 0.2	1,939 1.1 0.2	1,720 1.2 0.2	1,530 1.2 0.1	1,364 1.3 0.1	1,218 1.3 0.0	1,089 1.3 0.0	
16RB32	18	10.29				9,145 0.5 0.2	7,713 0.6 0.2	6,577 0.7 0.2	5,661 0.8 0.3	4,911 0.9 0.3	4,289 1.0 0.3	3,768 1.0 0.3	3,327 1.1 0.3	2,951 1.2 0.4	2,627 1.3 0.4	2,346 1.4 0.4	2,101 1.5 0.4	1,886 1.6 0.4	1,697 1.7 0.4		
16RB36	20	11.64					9,834 0.5 0.2	8,397 0.6 0.2	7,237 0.7 0.2	6,288 0.8 0.3	5,502 0.9 0.3	4,843 1.0 0.3	4,285 1.0 0.3	3,809 1.1 0.4	3,399 1.2 0.4	3,043 1.3 0.4	2,733 1.4 0.4	2,461 1.5 0.4	2,221 1.5 0.4		
16RB40	22	13.00						9,010 0.6 0.2	7,839 0.7 0.2	6,867 0.8 0.3	6,054 0.9 0.3	5,365 1.0 0.3	4,777 1.0 0.3	4,271 1.1 0.3	3,832 1.2 0.4	3,449 1.3 0.4	3,113 1.4 0.4	2,817 1.4 0.4			

INVERTED TEE BEAMS

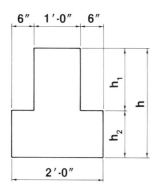

f'_c = 5,000 psi
f_{pu} = 270,000 psi

½ in. diameter
low-relaxation strand

Key
7,078 — Safe superimposed service load, plf
0.3 — Estimated camber at erection, in.
0.1 — Estimated long-time camber, in.

Normal Weight Concrete

Section Properties								
Designation	h (in.)	h_1/h_2 (in.)	A (in.²)	I (in.⁴)	y_b (in.)	Z_b (in.³)	Z_t (in.³)	wt (plf)
24IT20	20	12/8	336	10,981	8.29	1,325	938	350
24IT24	24	12/12	432	19,008	10.00	1,901	1,358	450
24IT28	28	16/12	480	30,131	11.60	2,598	1,837	500
24IT32	32	20/12	528	44,969	13.27	3,389	2,401	550
24IT36	36	24/12	576	63,936	15.00	4,262	3,045	600
24IT40	40	24/16	672	87,845	16.57	5,301	3,749	700
24IT44	44	28/16	720	116,877	18.27	6,397	4,542	750
24IT48	48	32/16	768	151,552	20.00	7,578	5,413	800
24IT52	52	36/16	816	192,275	21.76	8,836	6,358	850
24IT56	56	40/16	864	239,445	23.56	10,163	7,381	900
24IT60	60	44/16	912	293,460	25.37	11,567	8,474	950

1. Check local area for availability of other sizes.
2. Safe loads shown include 50% dead load and 50% live load. 800 psi top tension has been allowed, therefore additional top reinforcement is required.
3. Safe loads can be significantly increased by use of structural composite topping.

Table of safe superimposed service load (plf) and cambers

Designation	No. Strand	e	Span, ft.																
			16	18	20	22	24	26	28	30	32	34	36	38	40	42	44	46	48	50
24IT20	9	6.20	7,078 0.3 0.1	5,515 0.4 0.1	4,404 0.4 0.1	3,582 0.5 0.1	2,957 0.6 0.1	2,470 0.7 0.1	2,084 0.7 0.1	1,773 0.8 0.1	1,518 0.9 0.1	1,307 0.9 0.1	1,130 0.9 0.1	980 1.0 0.0	1.0 0.0					
24IT24	11	7.17		8,107 0.3 0.1	6,489 0.4 0.1	5,289 0.4 0.1	4,376 0.5 0.1	3,666 0.6 0.1	3,102 0.7 0.1	2,647 0.7 0.1	2,275 0.8 0.1	1,966 0.8 0.1	1,708 0.9 0.1	1,489 0.9 0.1	1,302 0.9 0.1	1,142 1.0 −0.1	1,002 1.0 −0.1			
24IT28	13	8.44			8,874 0.3 0.1	7,247 0.4 0.1	6,013 0.5 0.1	5,053 0.5 0.1	4,292 0.6 0.1	3,677 0.7 0.1	3,175 0.7 0.1	2,758 0.8 0.1	2,409 0.9 0.1	2,113 0.9 0.1	1,861 0.9 0.1	1,644 1.0 0.1	1,456 1.0 0.1	1,292 1.1 0.0	1,147 1.1 −0.1	1,020 1.1 −0.1
24IT32	15	9.77				9,574 0.4 0.1	7,957 0.4 0.1	6,698 0.5 0.1	5,700 0.6 0.2	4,894 0.6 0.2	4,238 0.7 0.2	3,694 0.8 0.2	3,239 0.8 0.2	2,853 0.9 0.2	2,524 1.0 0.2	2,241 1.0 0.2	1,996 1.1 0.2	1,752 1.1 0.1	1,594 1.1 0.1	1,428 1.2 0.1
24IT36	16	11.50						8,594 0.4 0.1	7,327 0.5 0.1	6,305 0.6 0.1	5,469 0.6 0.2	4,776 0.7 0.2	4,199 0.8 0.2	3,710 0.8 0.2	3,293 0.9 0.2	2,934 0.9 0.2	2,623 1.0 0.2	2,352 1.0 0.1	2,114 1.1 0.1	1,904 1.1 0.1
24IT40	19	12.02							9,061 0.5 0.1	7,802 0.5 0.1	6,775 0.6 0.2	5,926 0.6 0.2	5,214 0.7 0.2	4,611 0.8 0.2	4,097 0.8 0.2	3,654 0.9 0.2	3,271 0.9 0.2	2,936 1.0 0.2	2,642 1.0 0.2	2,383 1.1 0.2
24IT44	20	13.73								9,554 0.5 0.1	8,306 0.5 0.1	7,272 0.6 0.1	6,409 0.6 0.2	5,680 0.7 0.2	5,057 0.7 0.2	4,520 0.8 0.2	4,056 0.8 0.2	3,650 0.9 0.2	3,295 0.9 0.2	2,981 1.0 0.1
24IT48	22	15.08									9,989 0.5 0.1	8,757 0.6 0.2	7,725 0.6 0.2	6,851 0.7 0.2	6,105 0.7 0.2	5,466 0.8 0.2	4,913 0.8 0.2	4,431 0.9 0.2	4,008 0.9 0.2	3,634 1.0 0.2
24IT52	24	16.44										9,164 0.6 0.2	8,137 0.6 0.2	7,261 0.7 0.2	6,507 0.7 0.2	5,853 0.8 0.2	5,283 0.9 0.2	4,786 0.9 0.2	4,348 1.0 0.2	
24IT56	26	17.82											9,536 0.6 0.2	8,519 0.7 0.2	7,643 0.7 0.2	6,884 0.8 0.2	6,222 0.8 0.2	5,641 0.9 0.2	5,128 0.9 0.2	
24IT60	28	19.18												9,863 0.6 0.2	8,857 0.7 0.2	7,986 0.7 0.2	7,226 0.8 0.2	6,559 0.9 0.2	5,970 0.9 0.3	

INVERTED TEE BEAMS

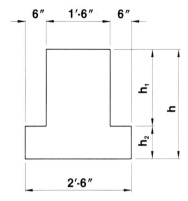

f'_c = 5,000 psi
f_{pu} = 270,000 psi

½ in. diameter
low-relaxation strand

Key
8,428 — Safe superimposed service load, plf
0.4 — Estimated camber at erection, in.
0.2 — Estimated long-time camber, in.

Normal Weight Concrete

Section Properties								
Designation	h (in.)	h_1/h_2 (in.)	A (in.²)	I (in.⁴)	y_b (in.)	Z_b (in.³)	Z_t (in.³)	wt (plf)
30IT20	20	12/8	456	15,240	8.74	1,744	1,354	475
30IT24	24	12/12	576	26,352	10.50	2,510	1,952	600
30IT28	28	16/12	648	41,824	12.22	3,423	2,650	675
30IT32	32	20/12	720	62,400	14.00	4,457	3,467	750
30IT36	36	24/12	792	88,678	15.82	5,605	4,394	825
30IT40	40	24/16	912	121,923	17.47	6,979	5,412	950
30IT44	44	28/16	984	162,161	19.27	8,415	6,557	1,025
30IT48	48	32/16	1,056	210,199	21.09	9,967	7,811	1,100
30IT52	52	36/16	1,128	266,627	22.94	11,623	9,175	1,175
30IT56	56	40/16	1,200	332,032	24.80	13,388	10,642	1,250
30IT60	60	44/16	1,272	406,997	26.68	15,255	12,215	1,325

1. Check local area for availability of other sizes.

2. Safe loads shown include 50% dead load and 50% live load. 800 psi top tension has been allowed, therefore additional top reinforcement is required.

3. Safe loads can be significantly increased by use of structural composite topping.

Table of safe superimposed service load (plf) and cambers

Designation	No. Strand	e	Span, ft.																
			18	20	22	24	26	28	30	32	34	36	38	40	42	44	46	48	50
30IT20	14	6.65	8,428 0.4 0.2	6,736 0.5 0.2	5,485 0.6 0.2	4,533 0.7 0.3	3,792 0.9 0.3	3,204 1.0 0.3	2,730 1.1 0.3	2,342 1.2 0.3	2,020 1.3 0.3	1,751 1.4 0.3	1,523 1.4 0.3	1,332 1.5 0.3	1,167 1.6 0.3	1,024 1.6 0.2			
30IT24	17	7.67		9,736 0.4 0.2	7,942 0.5 0.2	6,578 0.6 0.2	5,516 0.7 0.3	4,673 0.8 0.3	3,994 0.9 0.3	3,437 1.0 0.3	2,976 1.1 0.3	2,592 1.2 0.3	2,269 1.2 0.3	1,993 1.3 0.3	1,755 1.4 0.2	1,550 1.4 0.2	1,370 1.5 0.2	1,212 1.5 0.1	1,073 1.5 0.0
30IT28	20	9.06				9,087 0.6 0.2	7,643 0.6 0.2	6,497 0.7 0.3	5,573 0.8 0.3	4,816 0.9 0.3	4,189 1.0 0.3	3,664 1.1 0.3	3,219 1.2 0.3	2,839 1.2 0.3	2,513 1.3 0.3	2,334 1.4 0.3	1,990 1.4 0.3	1,776 1.5 0.3	1,588 1.5 0.2
30IT32	23	10.50							8,647 0.7 0.2	7,436 0.7 0.3	6,445 0.8 0.3	5,623 0.9 0.3	4,935 1.0 0.4	4,352 1.1 0.4	3,855 1.2 0.4	3,426 1.2 0.4	3,055 1.3 0.4	2,732 1.4 0.4	2,448 1.5 0.4	2,201 1.5 0.3
30IT36	24	12.32								9,492 0.7 0.2	8,243 0.7 0.2	7,207 0.8 0.3	6,340 0.9 0.3	5,605 1.0 0.3	4,978 1.0 0.3	4,439 1.1 0.3	3,971 1.2 0.3	3,563 1.3 0.3	3,205 1.3 0.3	2,892 1.4 0.3
30IT40	30	12.92									9,077 0.8 0.3	7,994 0.8 0.3	7,077 0.9 0.3	6,295 1.0 0.4	5,621 1.1 0.4	5,037 1.2 0.4	4,528 1.2 0.4	4,081 1.3 0.4	3,687 1.4 0.4
30IT44	30	14.73										9,659 0.7 0.3	8,564 0.8 0.3	7,629 0.9 0.3	6,825 1.0 0.3	6,127 1.0 0.3	5,519 1.1 0.3	4,985 1.2 0.3	4,514 1.2 0.3
30IT48	33	16.17												9,222 0.8 0.3	8,262 0.9 0.3	7,431 1.0 0.3	6,705 1.0 0.3	6,068 1.1 0.3	5,506 1.2 0.3
30IT52	36	17.62													9,836 0.9 0.3	8,858 0.9 0.3	8,004 1.0 0.3	7,255 1.1 0.3	6,594 1.1 0.4
30IT56	39	19.06															9,407 1.0 0.3	8,538 1.0 0.4	7,770 1.1 0.4
30IT60	42	20.49																9,917 1.0 0.3	9,036 1.0 0.4

INVERTED TEE BEAMS

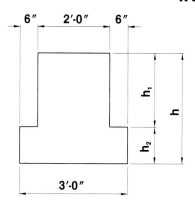

$f'_c = 5,000$ psi
$f_{pu} = 270,000$ psi

½ in. diameter
low-relaxation strand

Key
8,882 — Safe superimposed service load, plf
0.6 — Estimated camber at erection, in.
0.2 — Estimated long-time camber, in.

Normal Weight Concrete

Section Properties

Designation	h (in.)	h_1/h_2 (in.)	A (in.²)	I (in.⁴)	y_b (in.)	Z_b (in.³)	Z_t (in.³)	wt (plf)
36IT20	20	12/8	576	19,392	9.00	2,155	1,763	600
36IT24	24	12/12	720	33,523	10.80	3,104	2,540	750
36IT28	28	16/12	816	53,222	12.59	4,227	3,454	850
36IT32	32	20/12	912	79,390	14.42	5,506	4,516	950
36IT36	36	24/12	1,008	112,814	16.29	6,925	5,724	1,050
36IT40	40	24/16	1,152	155,136	18.00	8,619	7,052	1,200
36IT44	44	28/16	1,248	206,306	19.85	10,393	8,543	1,300
36IT48	48	32/16	1,344	267,410	21.71	12,317	10,172	1,400
36IT52	52	36/16	1,440	339,226	23.60	14,374	11,945	1,500

1. Check local area for availability of other sizes.
2. Safe loads shown include 50% dead load and 50% live load. 800 psi top tension has been allowed, therefore additional top reinforcement is required.
3. Safe loads can be significantly increased by use of structural composite topping.

Table of safe superimposed service load (plf) and cambers

Desig-nation	No. Strand	e	Span, ft.															
			20	22	24	26	28	30	32	34	36	38	40	42	44	46	48	50
36IT20	18	6.91	8,882 0.6 0.2	7,236 0.7 0.2	5,985 0.8 0.3	5,011 0.9 0.3	4,238 1.0 0.3	3,614 1.1 0.4	3,104 1.2 0.4	2,681 1.3 0.4	2,327 1.4 0.4	2,027 1.5 0.4	1,771 1.6 0.4	1,553 1.7 0.3	1,366 1.7 0.3	1,202 1.8 0.2	1,059 1.8 0.1	
36IT24	22	7.97			8,676 0.6 0.2	7,282 0.7 0.3	6,175 0.8 0.3	5,283 0.9 0.3	4,552 1.0 0.3	3,947 1.1 0.3	3,439 1.2 0.3	3,010 1.3 0.3	2,644 1.4 0.3	2,333 1.5 0.3	2,064 1.5 0.3	1,829 1.6 0.3	1,623 1.6 0.2	1,441 1.6 0.1
36IT28	26	9.43					8,615 0.8 0.3	7,395 0.9 0.3	6,397 1.0 0.3	5,569 1.0 0.4	4,876 1.1 0.4	4,289 1.2 0.4	3,788 1.3 0.4	3,357 1.4 0.4	2,983 1.5 0.4	2,657 1.6 0.4	2,374 1.6 0.4	2,126 1.7 0.3
36IT32	30	10.92						9,874 0.8 0.3	8,563 0.9 0.3	7,477 1.0 0.4	6,567 1.1 0.4	5,796 1.1 0.4	5,139 1.2 0.4	4,572 1.3 0.4	4,082 1.4 0.4	3,654 1.5 0.5	3,278 1.6 0.5	2,947 1.6 0.4
36IT36	32	12.79								9,640 0.9 0.3	8,485 1.0 0.3	7,508 1.0 0.4	6,674 1.1 0.4	5,955 1.2 0.4	5,333 1.3 0.4	4,790 1.4 0.4	4,314 1.5 0.4	3,893 1.5 0.4
36IT40	38	13.45									9,322 1.0 0.4	8,296 1.0 0.4	7,413 1.1 0.4	6,648 1.2 0.4	5,980 1.3 0.4	5,394 1.4 0.4	4,877 1.4 0.4	
36IT44	40	15.31											9,146 1.0 0.4	8,218 1.1 0.4	7,408 1.2 0.4	6,698 1.3 0.4	6,071 1.3 0.4	
36IT48	44	16.79													9,955 1.1 0.4	8,989 1.1 0.4	8,141 1.2 0.4	7,393 1.3 0.4
36IT52	48	18.28															9,724 1.2 0.4	8,844 1.2 0.4

PRECAST, PRESTRESSED COLUMNS

Fig. 2.6.1 Design strength interaction curves for precast, prestressed concrete columns

Criteria
1. Minimum prestress = 225 psi
2. All strand assumed 1/2 in. diameter, f_{pu} = 270 ksi
3. Curves shown for partial development of strand near member end where $f_{ps} \approx f_{se}$
4. Horizontal portion of curve is the maximum for tied columns = $0.80\phi P_o$.
5. ϕ = 0.9 for ϕP_n = 0
   0.7 for $\phi P_n \geq 0.10\, f'_c A_g$
   Varies from 0.9 to 0.7 for points between

Use of curves
1. Enter at left with applied factored axial load, P_u
2. Enter at bottom with applied magnified factored moment, δM_u
3. Intersection point must be to the left of curve indicating required concrete strength.

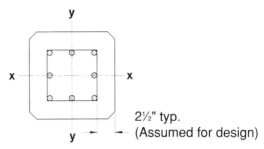

2½" typ. (Assumed for design)

Notation
ϕP_n = Design axial strength
ϕM_n = Design flexural strength
ϕP_o = Design axial strength at zero eccentricity
A_g = Gross area of the column
δ = Moment magnifier (Sect. 10.11, ACI 318-89)

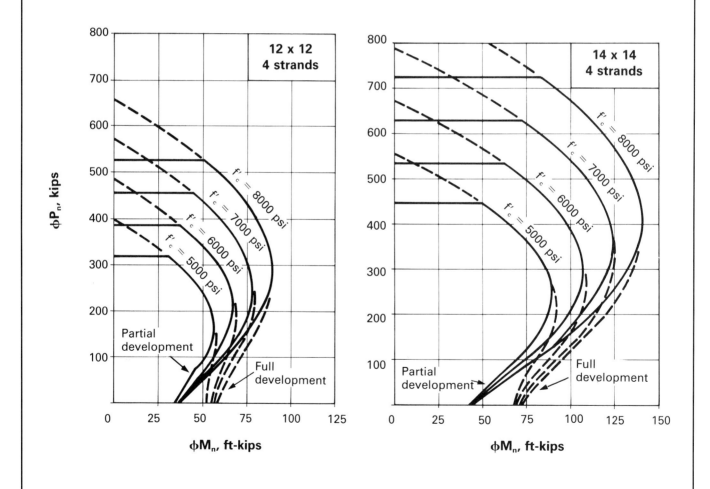

PRECAST, PRESTRESSED COLUMNS

Fig. 2.6.1 Design strength interaction curves for precast, prestressed concrete columns (continued)

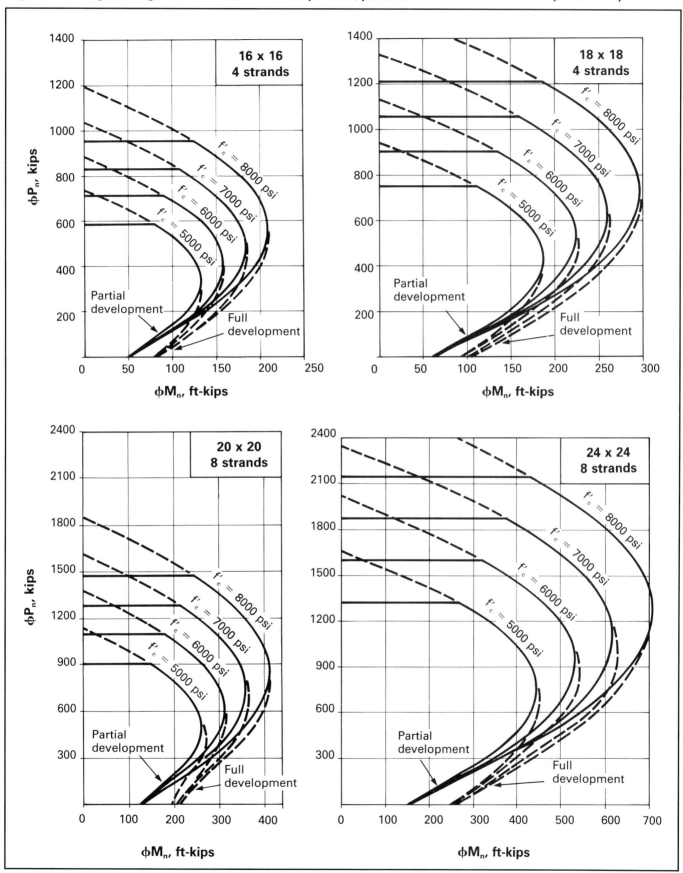

PRECAST, REINFORCED COLUMNS

Fig. 2.6.2 Design strength interaction curves for precast, reinforced concrete columns

Criteria
1. Concrete f'_c = 5,000 psi
2. Reinforcement f_y = 60,000 psi
3. Curves shown for full development of reinforcement
4. Horizontal portion of curve is the maximum for tied columns = $0.80\phi P_o$.
5. ϕ = 0.9 for $\phi P_n = 0$
   0.7 for $\phi P_n \geq 0.10 f'_c A_g$
   Varies from 0.9 to 0.7 for points between

Use of curves
1. Enter at left with applied factored axial load, P_u
2. Enter at bottom with applied magnified factored moment, δM_u
3. Intersection point must be to the left of curve indicating required reinforcement.

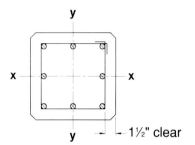

Notation
ϕP_n = Design axial strength
ϕM_n = Design flexural strength
ϕP_o = Design axial strength at zero eccentricity
A_g = Gross area of the column
δ = Moment magnifier (Sect. 10.11, ACI 318-89)

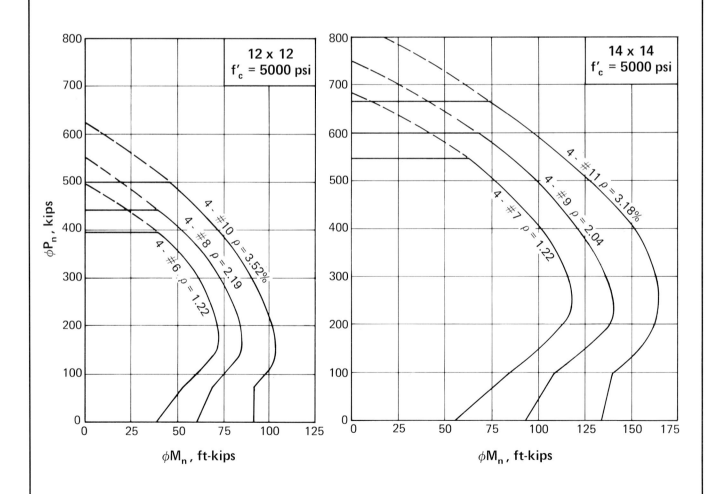

PRECAST, REINFORCED COLUMNS

Fig. 2.6.2 Design strength interaction curves for precast, prestressed concrete columns (continued)

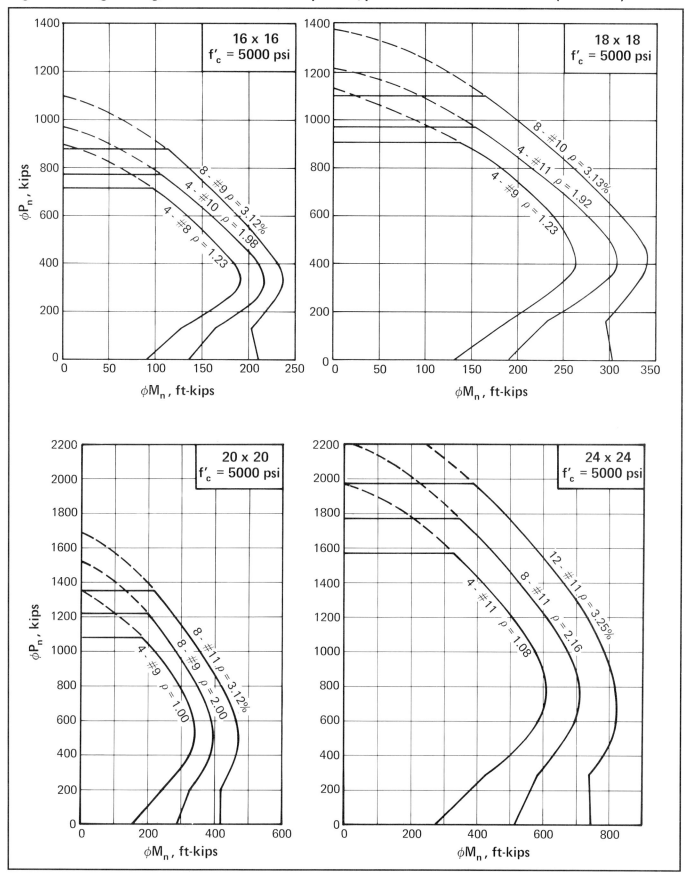

DOUBLE TEE WALL PANELS

Fig. 2.6.3 Partial interaction curve for prestressed double tee wall panels

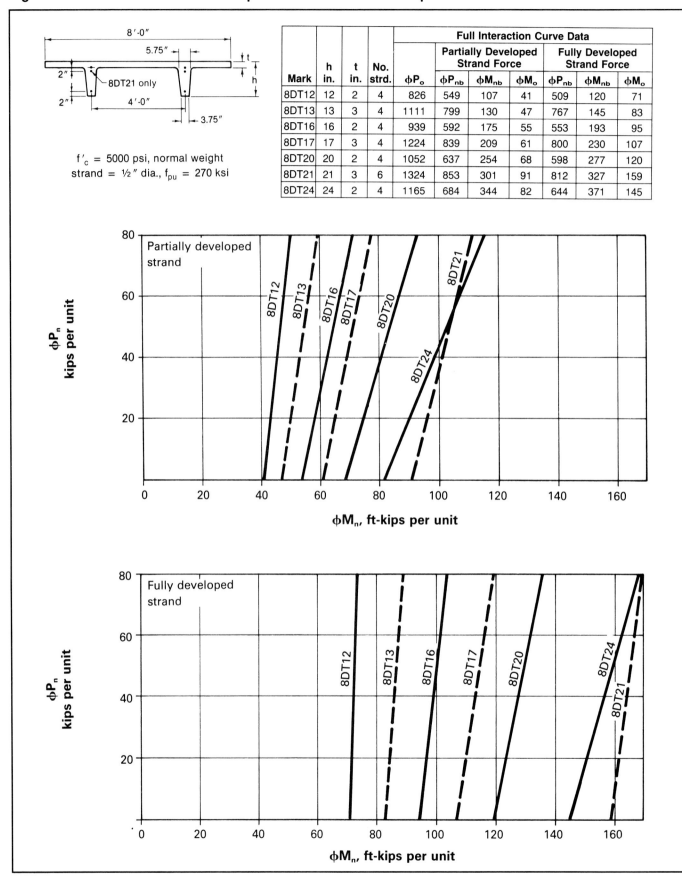

$f'_c = 5000$ psi, normal weight
strand = ½″ dia., $f_{pu} = 270$ ksi

Mark	h in.	t in.	No. strd.	ϕP_o	Partially Developed Strand Force			Fully Developed Strand Force		
					ϕP_{nb}	ϕM_{nb}	ϕM_o	ϕP_{nb}	ϕM_{nb}	ϕM_o
8DT12	12	2	4	826	549	107	41	509	120	71
8DT13	13	3	4	1111	799	130	47	767	145	83
8DT16	16	2	4	939	592	175	55	553	193	95
8DT17	17	3	4	1224	839	209	61	800	230	107
8DT20	20	2	4	1052	637	254	68	598	277	120
8DT21	21	3	6	1324	853	301	91	812	327	159
8DT24	24	2	4	1165	684	344	82	644	371	145

HOLLOW-CORE WALL PANELS

Fig. 2.6.4 Partial interaction curve for prestressed hollow-core wall panels

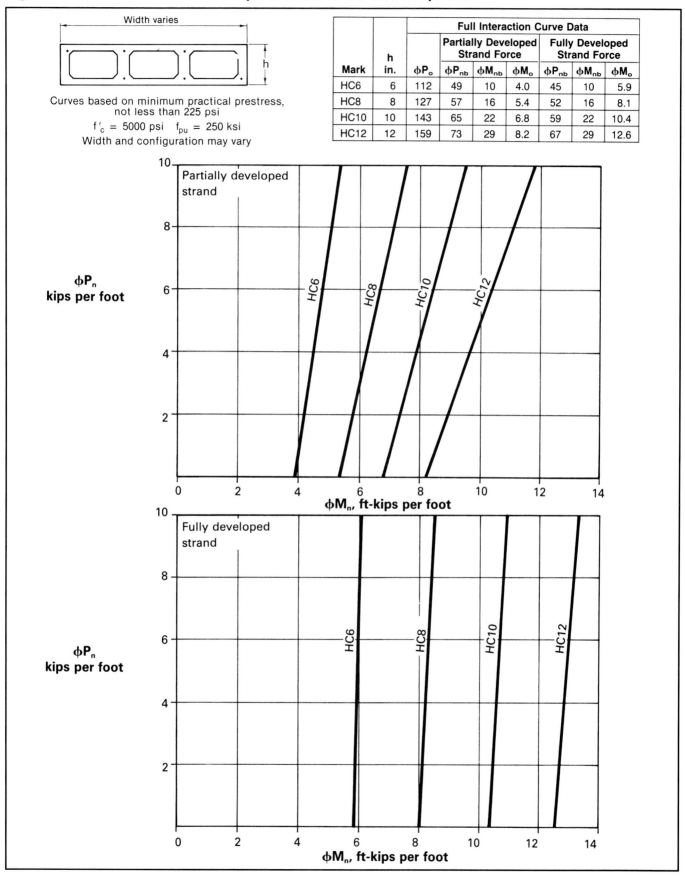

Mark	h in.	Full Interaction Curve Data						
		ϕP_o	Partially Developed Strand Force			Fully Developed Strand Force		
			ϕP_{nb}	ϕM_{nb}	ϕM_o	ϕP_{nb}	ϕM_{nb}	ϕM_o
HC6	6	112	49	10	4.0	45	10	5.9
HC8	8	127	57	16	5.4	52	16	8.1
HC10	10	143	65	22	6.8	59	22	10.4
HC12	12	159	73	29	8.2	67	29	12.6

Curves based on minimum practical prestress, not less than 225 psi

$f'_c = 5000$ psi $f_{pu} = 250$ ksi

Width and configuration may vary

PCI Design Handbook/Fourth Edition

2–53

PRECAST, PRESTRESSED SOLID WALL PANELS

Fig. 2.6.5 Partial interaction curve for prestressed solid wall panels

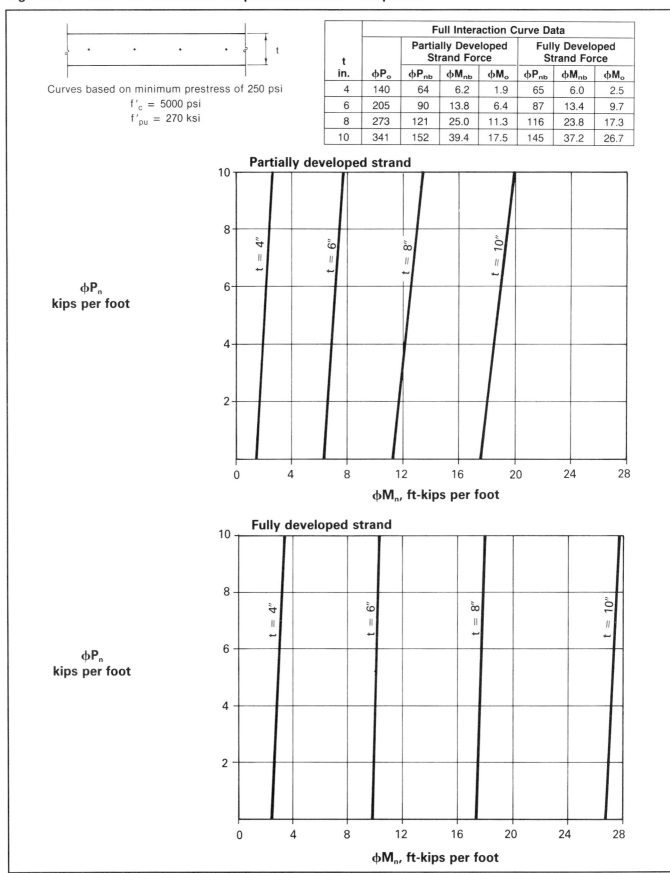

Curves based on minimum prestress of 250 psi
f'_c = 5000 psi
f'_{pu} = 270 ksi

| t in. | Full Interaction Curve Data ||||||
| | Partially Developed Strand Force ||| Fully Developed Strand Force |||
	ϕP_o	ϕP_{nb}	ϕM_{nb}	ϕM_o	ϕP_{nb}	ϕM_{nb}	ϕM_o
4	140	64	6.2	1.9	65	6.0	2.5
6	205	90	13.8	6.4	87	13.4	9.7
8	273	121	25.0	11.3	116	23.8	17.3
10	341	152	39.4	17.5	145	37.2	26.7

2–54 PCI Design Handbook/Fourth Edition

PRECAST, REINFORCED SOLID WALL PANELS

Fig. 2.6.6 Partial interaction curve for precast, reinforced concrete wall panels

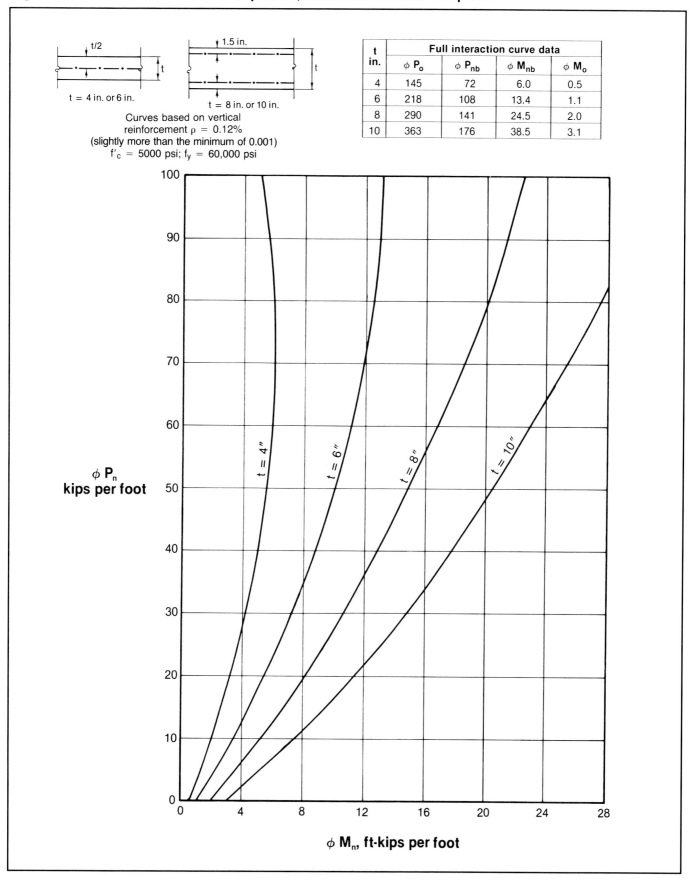

PILES

Table 2.7.1 Section properties and allowable service loads of prestressed concrete piles

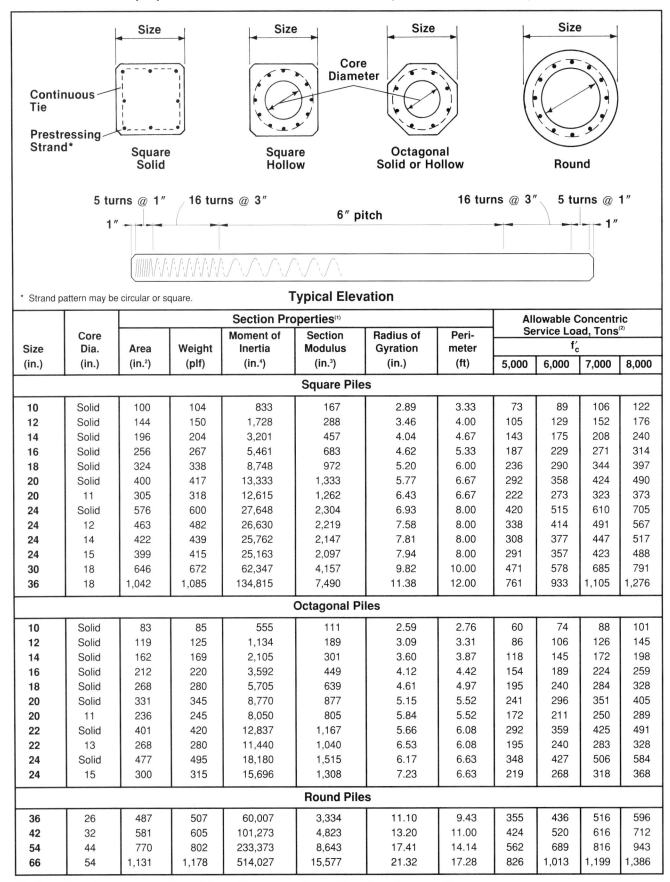

Size (in.)	Core Dia. (in.)	Area (in.²)	Weight (plf)	Moment of Inertia (in.⁴)	Section Modulus (in.³)	Radius of Gyration (in.)	Perimeter (ft)	5,000	6,000	7,000	8,000
\multicolumn{12}{c}{Square Piles}
10	Solid	100	104	833	167	2.89	3.33	73	89	106	122
12	Solid	144	150	1,728	288	3.46	4.00	105	129	152	176
14	Solid	196	204	3,201	457	4.04	4.67	143	175	208	240
16	Solid	256	267	5,461	683	4.62	5.33	187	229	271	314
18	Solid	324	338	8,748	972	5.20	6.00	236	290	344	397
20	Solid	400	417	13,333	1,333	5.77	6.67	292	358	424	490
20	11	305	318	12,615	1,262	6.43	6.67	222	273	323	373
24	Solid	576	600	27,648	2,304	6.93	8.00	420	515	610	705
24	12	463	482	26,630	2,219	7.58	8.00	338	414	491	567
24	14	422	439	25,762	2,147	7.81	8.00	308	377	447	517
24	15	399	415	25,163	2,097	7.94	8.00	291	357	423	488
30	18	646	672	62,347	4,157	9.82	10.00	471	578	685	791
36	18	1,042	1,085	134,815	7,490	11.38	12.00	761	933	1,105	1,276
\multicolumn{12}{c}{Octagonal Piles}
10	Solid	83	85	555	111	2.59	2.76	60	74	88	101
12	Solid	119	125	1,134	189	3.09	3.31	86	106	126	145
14	Solid	162	169	2,105	301	3.60	3.87	118	145	172	198
16	Solid	212	220	3,592	449	4.12	4.42	154	189	224	259
18	Solid	268	280	5,705	639	4.61	4.97	195	240	284	328
20	Solid	331	345	8,770	877	5.15	5.52	241	296	351	405
20	11	236	245	8,050	805	5.84	5.52	172	211	250	289
22	Solid	401	420	12,837	1,167	5.66	6.08	292	359	425	491
22	13	268	280	11,440	1,040	6.53	6.08	195	240	283	328
24	Solid	477	495	18,180	1,515	6.17	6.63	348	427	506	584
24	15	300	315	15,696	1,308	7.23	6.63	219	268	318	368
\multicolumn{12}{c}{Round Piles}
36	26	487	507	60,007	3,334	11.10	9.43	355	436	516	596
42	32	581	605	101,273	4,823	13.20	11.00	424	520	616	712
54	44	770	802	233,373	8,643	17.41	14.14	562	689	816	943
66	54	1,131	1,178	514,027	15,577	21.32	17.28	826	1,013	1,199	1,386

(1) Form dimensions may vary with producers, with corresponding variations in section properties.
(2) Allowable loads based on $N = A_c(0.33 f'_c - 0.27 f_{pc})$; $f_{pc} = 700$ psi. Check local producer for available concrete strengths.

SHEET PILES

Table 2.7.2 Section properties and allowable moments of prestressed sheet piles

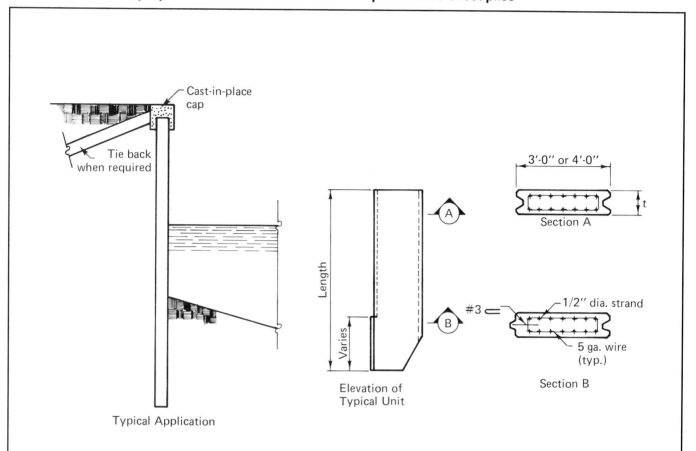

Thickness t in.	Section Properties per Foot of Width				Maximum Allowable Service Load Moment[2] ft-kips per foot	
	Area in.²	Weight[1] psf	Moment of Inertia in.⁴	Section Modulus in.³	f'_c = 5000 psi	f'_c = 6000 psi
6[3]	72	75	216	72	6.0	7.2
8[3]	96	100	512	128	10.6	12.8
10	120	125	1000	200	16.6	20.0
12	144	150	1728	288	24.0	28.8
16	192	200	4096	512	42.7	51.2
18	216	225	5832	648	54.0	64.8
20	240	250	8000	800	66.7	80.0
24	288	300	13,824	1152	96.0	115.2

(1) Normal weight concrete
(2) Based on zero tension and maximum $0.4f'_c$ compression
(3) Strand can be placed in a single layer in thin sections. Where site conditions require it, strand may be placed eccentrically.

CHAPTER 3
ANALYSIS AND DESIGN OF PRECAST, PRESTRESSED CONCRETE STRUCTURES

	Page No.
3.1 General	3-3
3.1.1 Notation	3-3
3.1.2 Introduction	3-4
3.2 Preliminary Analysis	3-5
3.2.1 Framing Dimensions	3-5
3.2.2 Span-to-Depth Ratios	3-5
3.2.3 Lateral Load Resisting Systems	3-5
3.2.4 Control of Volume Change Deformations and Restraint Forces	3-6
3.2.5 Connection Concepts	3-6
3.3 Volume Changes	3-6
3.3.1 Axial Volume Change Strains	3-6
3.3.2 Bowing	3-11
3.3.3 Expansion Joints	3-14
3.4 Component Analysis	3-15
3.4.1 Non-Bearing Wall Panels	3-15
3.4.2 Load Bearing Wall Panels	3-17
3.4.3 Non-Bearing Spandrels	3-18
3.4.4 Load Bearing Spandrels	3-18
3.4.5 Eccentrically Loaded Columns	3-20
3.5 Slenderness Effects in Columns and Wall Panels	3-21
3.5.1 Second-Order (P-Δ) Analysis	3-21
3.5.2 Moment Magnification Method	3-24
3.6 Diaphragm Design	3-28
3.6.1 Method of Analysis	3-28
3.6.2 Shear Transfer between Members	3-28
3.6.3 Chord Forces	3-28
3.7 Shear Wall Buildings	3-30
3.7.1 General	3-30
3.7.2 Rigidity of Solid Shear Walls	3-30
3.7.3 Distribution of Lateral Loads	3-31
3.7.4 Unsymmetrical Shear Walls	3-31
3.7.5 Coupled Shear Walls	3-33
3.7.6 Shear Walls with Large Openings	3-33
3.7.7 Architectural Panels as Shear Walls	3-43

Page No.

3.8 Buildings with Moment-Resisting Frames 3-43
    3.8.1   General 3-43
    3.8.2   Moment Resistance of Column Bases 3-44
    3.8.3   Fixity of Column Bases 3-48
    3.8.4   Modeling Partially Fixed Bases 3-48
    3.8.5   Volume Change Effects in Moment-Resisting Frames 3-49
            3.8.5.1   Equivalent volume change 3-49
            3.8.5.2   Calculating restraint forces 3-52
    3.8.6   Computer Models for Frame Analysis 3-53

3.9 Shear Wall-Frame Interaction 3-53

3.10 Structural Integrity 3-53
    3.10.1   Introduction 3-53
    3.10.2   Precast Concrete Structures 3-56
    3.10.3   Large Panel Bearing Wall Structures 3-57
    3.10.4   Hybrid Structures 3-57

3.11 Earthquake Analysis 3-58
    3.11.1   Notation 3-58
    3.11.2   General 3-59
    3.11.3   Building Code Requirements 3-59
    3.11.4   Design Guidelines for Wall Panels 3-60
    3.11.5   Concept of Box-Type Buildings 3-61
    3.11.6   Structural Layout and Connections 3-61
    3.11.7   Example — 1-Story Building 3-61
    3.11.8   Example — 4-Story Building 3-68
    3.11.9   Example — 3-Level Parking Structure 3-74
    3.11.10 Example — 12-Story Building 3-79
    3.11.11 Example — 23-Story Building 3-81
    3.11.12 Example — Architectural Precast Panel 3-81

3.12 References 3-86

ANALYSIS AND DESIGN OF PRECAST, PRESTRESSED CONCRETE STRUCTURES

3.1 General

3.1.1 Notation

(Note: Notation for earthquake analysis is in Sect. 3.11.1)

A	=	area (with subscripts)
A_b	=	total area of anchor bolts which are in tension
A_{ps}	=	area of prestressing steel
A_{vf}	=	area of shear-friction reinforcement
A_w	=	area of shear wall
b	=	width of a section or structure
C	=	coefficient of thermal expansion
C	=	compressive force
C_m	=	a factor relating actual moment to equivalent uniform moment
C_u	=	factored compressive force
D	=	dead load
e	=	eccentricity of axial load
E	=	modulus of elasticity (with subscripts)
E_t	=	modulus of elasticity modified for time-dependent effects
f'_c	=	concrete compressive stress
f_{pu}	=	specified tensile strength of prestressing steel
f_r	=	modulus of rupture of concrete
f_{ut}	=	factored tensile stress
f_y	=	yield strength of non-prestressed reinforcement
F	=	horizontal forces in shear wall buildings and in moment-resisting frames
F_b	=	degree of base fixity (decimal)
F_i	=	lateral force at bay i or in shear wall i
F_i	=	restraining force in multi-story columns at level i
F_u	=	factored force
F_x, F_y	=	forces in x and y directions, respectively
g	=	assumed length over which elongation of the anchor bolt takes place
h	=	column width in direction of bending
h	=	wall panel thickness or overall depth of beams or slabs
h	=	height of shear wall
h_s	=	story height
I	=	moment of inertia
I_b	=	moment of inertia of a beam
I_{bp}	=	moment of inertia of a base plate (vertical cross-section dimensions)
I_c	=	moment of inertia of a column
I_{eq}	=	approximate moment of inertia that results in a flexural deflection equal to the combined shear and flexural deflections of a wall
I_f	=	moment of inertia of the footing (plan dimensions)
I_g	=	uncracked moment of inertia
I_p	=	polar moment of inertia
k	=	effective length factor
k_b, k_f, k_m	=	coefficients used to determine forces and moments in beams and columns
k_s	=	coefficient of subgrade reaction
K	=	stiffnesses (with subscripts)
K_ℓ	=	constant used for the calculation of equivalent creep and shrinkage shortening
K_r	=	relative stiffness
K_t	=	constant used for the calculation of equivalent temperature shortening
ℓ	=	distance between wall panel supports
ℓ	=	length of span or structure
ℓ_n	=	clear span
ℓ_s	=	distance from column to center of stiffness of structure
ℓ_u	=	unbraced length
ℓ_w	=	length of weld
M	=	unfactored moment
M_c	=	factored moment to be used for design of compression member
M_j	=	moment in multi-story columns at point j
M_R	=	resisting moment
M_T	=	torsional moment
M_u	=	factored moment
M_{1b}	=	value of smaller factored end moment on compression member due to loads that result in no appreciable sidesway
M_{2b}	=	value of larger factored end moment on compression member due to loads that result in no appreciable sidesway
M_{2s}	=	value of larger factored end moment on compression member due to loads that result in appreciable sidesway
n	=	number of panels in a shear wall; number of bays in a moment-resisting frame

N	=	normal force
P	=	lateral force in wall panels to restrain bowing
P	=	applied axial load
P	=	vertical load acting at eccentricity e
P	=	lateral force applied to shear wall
P_c	=	critical buckling load
P_o	=	axial load nominal strength of a compression member with zero eccentricity
P_o	=	final prestress force in tendons
P_u	=	factored axial load
Q	=	statical moment
r	=	radius of gyration
r	=	rigidity
R_{du}	=	factored dead load reaction
t	=	thickness
T	=	tensile force
T'	=	force due to wind suction
T_u	=	factored tensile force
T_1, T_2	=	outside, inside temperature
$T_{1u}, T_{2u}, T_{3u}, T_{4u}$	=	design tie forces in transverse, peripheral, longitudinal and vertical directions, respectively
v_h	=	unit horizontal shear
v_r	=	unit shear on panel edge
v_u	=	factored unit shear
V_n	=	nominal shear strength
V_R, V_L	=	shear at right, left support
V_u	=	factored shear force
V_w	=	total wind shear
w	=	uniform load
W	=	total lateral or gravity load
x_1	=	distance from face of column to center of anchor bolts
x_2	=	distance from face of column to base plate anchorage
x, y	=	orthogonal distances of individual shear walls from center of rigidity
$\bar{x}, \bar{y}$	=	orthogonal distances locating center of rigidity of building
y_b, y_t	=	dimensions from center of gravity to opposite ends of irregular shear wall elements
Z	=	section modulus
α	=	strain gradient across thickness of wall panel
α	=	angle to direction of shear
β_d	=	dead load/total load ratio
γ	=	flexibility coefficient (with subscripts)
δ	=	moment magnifier (with subscripts)
δ	=	volume change shortening (with subscripts)
δ_e	=	equivalent volume change shortening (with subscripts)
Δ	=	total equivalent shortening or column deflection
Δ	=	thermal bow in wall panels
Δ	=	sum of flexure and shear deflections in shear walls
Δ	=	lateral deflection of a compression member
Δ_u	=	deflection due to factored loads
η	=	see Fig. 3.5.1
θ	=	see Fig. 3.5.1
θ	=	rotation of cantilever
λ	=	see Fig. 3.5.1
μ	=	static coefficient of friction
μ_e	=	effective shear-friction coefficient
ϕ	=	strength reduction factor
ϕ	=	rotation (with subscripts)
ψ	=	ratio of column to beam stiffnesses

3.1.2 Introduction

This chapter provides guidelines for the analysis and design of structures that are comprised wholly or partially of precast and/or precast, prestressed components. The primary advantages of precast concrete include:

1. Construction speed
2. Plant-fabrication quality control
3. Fire resistance and durability
4. With prestressing: greater span-depth ratios, more controllable performance, less material usage
5. Architectural precast concrete provides a wide variety of highly attractive surfaces and shapes
6. Thermal and acoustical control.

To fully realize these benefits and thereby gain the most economical and effective use of the material, the following general principles are offered:

1. Precast concrete is basically a "simple-span" material. However continuity can be, and often is, effectively achieved with properly conceived connection details.
2. Sizes and shapes of members are often limited by production, hauling and erection considerations.

3. Concrete is a massive material. This is an advantage for such matters as stability under wind loads, acoustical and vibration control and fire resistance. Also, the high dead-to-live load ratio will provide a greater safety factor against gravity overloads.

4. Maximum economy is achieved with maximum repetition, and standard sections should be used whenever possible.

5. Successful use is largely dependent on carefully conceived connection details.

6. The effects of restraint of volume changes caused by creep, shrinkage and temperature change must be considered in every structure.

7. While architectural panels are often used only as cladding, the inherent load-carrying capacity of these products should not be overlooked.

8. Prestressing improves the economy and performance of precast members, but is usually only feasible with standard shapes which are capable of being cast in "long-line" beds.

3.2 Preliminary Analysis

Maximum economy occurs when the building is laid out to take advantage of the principles discussed above. The primary considerations in preliminary analysis of the total structure are:

1. Framing dimensions
2. Span-to-depth ratios
3. Lateral load resisting systems
4. Control of volume change deformations and restraint forces
5. Connection concepts.

3.2.1 Framing Dimensions

When possible, bays should be sized to fit the module of the components selected. Standard section dimensions are shown in Chapter 2, but others may be available in a given area. Width of wall and deck units may be limited by hauling regulations.

It is often feasible to cast wall panels and columns in multi-story units, and economy is achieved with fewer pieces to handle. Length and weight limitations for hauling and erection, and stability during erection, are items to be considered in this decision.

3.2.2 Span-to-Depth Ratios

Selection of floor-to-floor dimensions should consider the practical span-to-depth ratio of the horizontal framing members, allowing adequate space for mechanical ductwork.

Typical span-to-depth ratios of flexural precast, prestressed concrete members are:

Hollow-core floor slabs	30 to 40
Hollow-core roof slabs	40 to 50
Stemmed floor slabs	25 to 35
Stemmed roof slabs	35 to 40
Beams	10 to 20

These values are intended as guidelines, not limits. The required depth of a beam or slab is influenced by the ratio of live load to total load. Where this ratio is high, deeper sections may be needed.

For non-prestressed flexural members, span-depth ratios are given in ACI 318-89, Sect. 9.5.2.1.

3.2.3 Lateral Load Resisting Systems

Often the most time consuming task in the preliminary analysis is the selection of the lateral load resisting system. Methods used to resist lateral loads, in the approximate order of economy, include:

1. *Shear walls* These can be of precast concrete, cast-in-place concrete, or masonry. These are discussed in more detail in Sect. 3.7. When architectural or structural precast members are used for the exterior cladding, they can often be used as shear walls.

2. *Cantilevered columns or wall panels* This is usually only feasible in low-rise buildings. Base fixity can be attained through a moment couple between the footing and ground-floor slab, or by fixing the column to the footing. In the latter case, a detailed analysis of the footing rotation can be made as described in Sect. 3.8.2.

3. *Steel or concrete X-bracing* This system has been used effectively in mid-rise buildings. A related resistance system occurs naturally in parking structures with sloped decks in the direction of traffic flow.

4. *Moment-resisting frames* Building function may dictate the use of moment-resisting frames; in this case forces resulting from the restraint of volume changes must be considered. It is sometimes feasible to provide a moment connection at only one end of a member, or a connection that will resist moments with lateral forces in one direction but not in the other, in order to reduce the buildup of restraint forces. To reduce the number of moment frames required, a combined shear wall-frame system may be used. Moment-resisting frames are discussed in more detail in Sect. 3.8.

All of the above systems depend on distribution of lateral loads through diaphragm action of the roof and floor systems (see Sect. 3.6).

3.2.4 Control of Volume Change Deformations and Restraint Forces

Volume changes of concrete are those resulting from creep, shrinkage, and temperature change. Creep and shrinkage cause a shortening of the member, so the critical combination is creep, shrinkage, and temperature drop.

It is important to arrange connections so that the effect of volume restraint is minimized. Sect. 3.3 provides data and guidelines for estimating the amount of shortening which may take place. Neglecting the effect of connection deformation will produce unrealistically high computed restraint forces. Sect. 3.8.5 discusses the method of estimating the force which may develop from restraint.

Problems caused by volume change movements have appeared when relatively long members were welded to their supports at both ends. When such members are connected only at the top, experience has shown that volume changes are adequately accommodated. An unyielding top connection may attract negative moments if compression resistance is encountered at the bottom. This may be difficult to accommodate.

Connections using cast-in-place concrete have exhibited few volume change problems. This is probably because microcracking and creep in the cast-in place portion effectively relieve the restraint.

Long buildings may require full height expansion joints. This is discussed in Sect. 3.3.3.

3.2.5 Connection Concepts

The types of connections to be used should be determined during the preliminary analysis, as this may have an effect on the component dimensions, the overall structural behavior, as discussed above, and on the erection procedure. Chapter 6 and some PCI publications[13,15] are devoted entirely to connections.

3.3 Volume Changes

Creep, shrinkage and temperature change, and the forces caused by restraining these strains, affect connections, service load behavior and ultimate capacity of precast, prestressed structures. Consequently, these strains and forces must be considered in the design.

Vertical members, such as load bearing wall panels, are also subject to volume change strains. The approximate magnitude can be calculated using Tables 3.3.1 through 3.3.5, adding the dead load stress to the prestress. The effects will only be significant in high rise buildings, and then only differential movements between elements will significantly affect performance of the structure. This can occur, for example, at the corner of a building where load bearing and non-load bearing panels meet.

3.3.1 Axial Volume Change Strains

Tables 3.3.1 through 3.3.5 and Figs. 3.3.1 and 3.3.2 provide the data needed to determine volume change strains.[1,2] These values can be reduced in rigid frames (see Sect. 3.8.5).

Example 3.3.1 Calculation of volume change shortening

Given:

Heated structure in Denver, Colorado

Normal weight concrete beam—12RB28

8 – ½ in. diameter, 270K, stress-relieved strands

Initial tension = $0.70f_{pu}$

Assume initial prestress loss = 10%

Release strength = 4500 psi (accelerated cure)

Length = 24 ft

Problem:

Determine the actual shortening that can be anticipated from:

a. Casting to erection at 60 days

b. Erection to the end of service life.

Solution:

From Figs. 3.3.1 and 3.3.2:

Design temperature change = 70°F

Average ambient relative humidity = 55%

Prestress force:

A_{ps} = 8(0.153) = 1.224 in.^{2}

P_o = 1.224(270)(0.70)(0.90)

= 208.2 kips

P_o/A = 208.2(1000)/(12 x 28)

= 620 psi

Volume/surface ratio

= (12 x 28)/[(2 x 12) + (2 x 28)]

= 4.2 in.

a. *At 60 days:*

From Table 3.3.1:

Creep strain = 169 x 10^{-6} in./in.

Shrinkage strain = 266 x 10^{-6} in./in.

Fig. 3.3.1 Maximum seasonal climatic temperature change, deg F

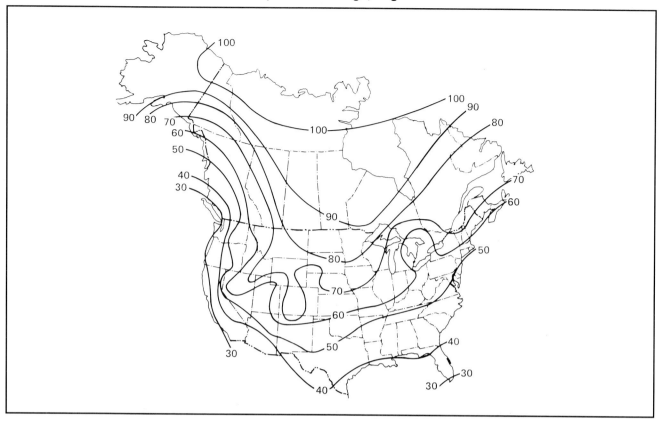

Fig. 3.3.2 Annual average ambient relative humidity, percent

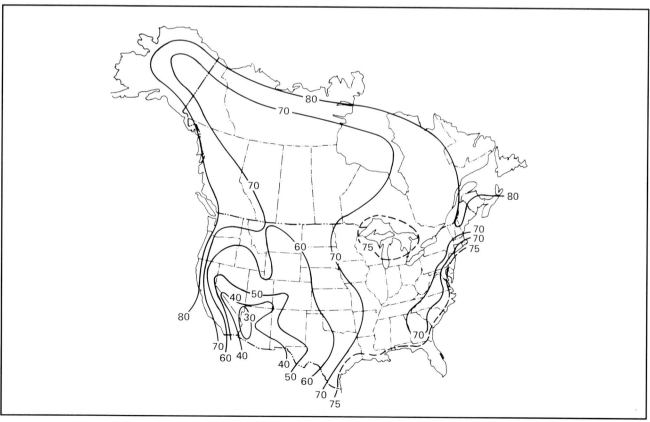

Table 3.3.1 Creep and shrinkage strains (millionths)

	Concrete Release Strength = 3500 psi Average Prestress = 600 psi Relative Humidity = 70% Volume/Surface Ratio = 1.5 in.			
Time, days	Creep		Shrinkage	
	Normal weight	Lightweight	Moist cure	Accelerated cure
1	29	43	16	9
3	51	76	44	26
5	65	97	70	43
7	76	114	93	58
9	86	127	115	72
10	90	133	124	78
20	118	176	204	136
30	137	204	258	180
40	150	224	299	215
50	161	239	329	243
60	169	252	354	266
70	177	263	373	286
80	183	272	390	302
90	188	280	403	317
100	193	287	415	329
200	222	331	477	400
1 Yr	244	363	511	443
3 Yr	273	407	543	486
5 Yr	283	422	549	495
Final	315	468	560	510

Table 3.3.2 Correction factors for prestress and concrete strength (creep only)

Ave. P/A (psi)	Release Strength, f_{ci} (psi)						
	2500	3000	3500	4000	4500	5000	6000
0	0.00	0.00	0.00	0.00	0.00	0.00	0.00
200	0.39	0.36	0.33	0.31	0.29	0.28	0.25
400	0.79	0.72	0.67	0.62	0.59	0.56	0.51
600	1.18	1.08	1.00	0.94	0.88	0.84	0.76
800	1.58	1.44	1.33	1.25	1.18	1.12	1.02
1000	1.97	1.80	1.67	1.56	1.47	1.39	1.27
1200	2.37	2.16	2.00	1.87	1.76	1.67	1.53
1400	2.76	2.52	2.33	2.18	2.06	1.95	1.78
1600		2.88	2.67	2.49	2.35	2.23	2.04
1800		3.24	3.00	2.81	2.65	2.51	2.29
2000			3.33	3.12	2.94	2.79	2.55
2200				3.43	3.23	3.07	2.80
2400				3.74	3.53	3.35	3.06
2600					3.82	3.63	3.31
2800						3.90	3.56
3000						4.18	3.82

Table 3.3.3 Correction factors for relative humidity

Ave. ambient R.H. (from Fig. 3.3.2)	Creep	Shrinkage
40	1.25	1.43
50	1.17	1.29
60	1.08	1.14
70	1.00	1.00
80	0.92	0.86
90	0.83	0.43
100	0.75	0.00

Table 3.3.4 Correction factors for volume/surface ratio

Time, days	Creep V/S						Shrinkage V/S					
	1	2	3	4	5	6	1	2	3	4	5	6
1	1.30	0.78	0.49	0.32	0.21	0.15	1.25	0.80	0.50	0.31	0.19	0.11
3	1.29	0.78	0.50	0.33	0.22	0.15	1.24	0.80	0.51	0.31	0.19	0.11
5	1.28	0.79	0.51	0.33	0.23	0.16	1.23	0.81	0.52	0.32	0.20	0.12
7	1.28	0.79	0.51	0.34	0.23	0.16	1.23	0.81	0.52	0.33	0.20	0.12
9	1.27	0.80	0.52	0.35	0.24	0.17	1.22	0.82	0.53	0.34	0.21	0.12
10	1.26	0.80	0.52	0.35	0.24	0.17	1.21	0.82	0.53	0.34	0.21	0.13
20	1.23	0.82	0.56	0.39	0.27	0.19	1.19	0.84	0.57	0.37	0.23	0.14
30	1.21	0.83	0.58	0.41	0.30	0.21	1.17	0.85	0.59	0.40	0.26	0.16
40	1.20	0.84	0.60	0.44	0.32	0.23	1.15	0.86	0.62	0.42	0.28	0.17
50	1.19	0.85	0.62	0.46	0.34	0.25	1.14	0.87	0.63	0.44	0.29	0.19
60	1.18	0.86	0.64	0.48	0.36	0.26	1.13	0.88	0.65	0.46	0.31	0.20
70	1.17	0.86	0.65	0.49	0.37	0.28	1.12	0.88	0.66	0.48	0.32	0.21
80	1.16	0.87	0.66	0.51	0.39	0.29	1.12	0.89	0.67	0.49	0.34	0.22
90	1.16	0.87	0.67	0.52	0.40	0.31	1.11	0.89	0.68	0.50	0.35	0.23
100	1.15	0.87	0.68	0.53	0.42	0.32	1.11	0.89	0.69	0.51	0.36	0.24
200	1.13	0.90	0.74	0.61	0.51	0.42	1.08	0.92	0.75	0.59	0.44	0.31
1 Yr	1.11	0.91	0.77	0.67	0.58	0.50	1.07	0.93	0.79	0.64	0.50	0.38
3 Yr	1.10	0.92	0.81	0.73	0.67	0.62	1.06	0.94	0.82	0.71	0.59	0.47
5 Yr	1.10	0.92	0.82	0.75	0.70	0.66	1.06	0.94	0.83	0.72	0.61	0.49
Final	1.09	0.93	0.83	0.77	0.74	0.72	1.05	0.95	0.85	0.75	0.64	0.54

Table 3.3.5 Design temperature strains* (millionths)

Temperature zone (from Fig. 3.3.1)	Normal weight		Lightweight	
	Heated	Unheated	Heated	Unheated
10	30	45	25	38
20	60	90	50	75
30	90	135	75	113
40	120	180	100	150
50	150	225	125	188
60	180	270	150	225
70	210	315	175	263
80	240	360	200	300
90	270	405	225	338
100	300	450	250	375

*Based on accepted coefficients of thermal expansion, reduced to account for thermal lag.
(See referenced committee report,[2] *PCI Journal,* September-October 1977)

Table 3.3.6 Volume change strains for typical building elements (millionths)

Temp. zone (from map)	Prestressed members (P/A = 600 psi)									
	Normal weight concrete					Lightweight concrete				
	Ave. R.H. (from map)					Ave. R.H. (from map)				
	40	50	60	70	80	40	50	60	70	80
Heated buildings										
0	548	501	454	407	360	581	532	483	434	385
10	578	531	484	437	390	606	557	508	459	410
20	608	561	514	467	420	631	582	533	484	435
30	638	591	544	497	450	656	607	558	509	460
40	668	621	574	527	480	681	632	583	534	485
50	698	651	604	557	510	706	657	608	559	510
60	728	681	634	587	540	731	682	633	584	535
70	758	711	664	617	570	756	707	658	609	560
80	788	741	694	647	600	781	732	683	634	585
90	818	771	724	677	630	806	757	708	659	610
100	848	801	754	707	660	831	782	733	684	635
Unheated structures										
0	548	501	454	407	360	581	532	483	434	385
10	593	546	499	452	405	619	570	521	472	423
20	638	591	544	497	450	656	607	558	509	460
30	683	636	589	542	495	694	645	596	547	498
40	728	681	634	587	540	731	682	633	584	535
50	773	726	679	632	585	769	720	671	622	573
60	818	771	724	677	630	806	757	708	659	610
70	863	816	769	722	675	844	795	746	697	648
80	908	861	814	767	720	881	832	783	734	685
90	953	906	859	812	765	919	870	821	772	723
100	998	951	904	857	810	956	907	858	809	760

Table 3.3.7 Volume change strains for typical building elements (millionths)

Temp. zone (from map)	Non-prestressed members									
	Normal weight concrete					Lightweight concrete				
	Ave. R.H. (from map)					Ave. R.H. (from map)				
	40	50	60	70	80	40	50	60	70	80
Heated buildings										
0	294	265	235	206	177	294	265	235	206	177
10	324	295	265	236	207	319	290	260	231	202
20	354	325	295	266	237	344	315	285	256	227
30	384	355	325	296	267	369	340	310	281	252
40	414	385	355	326	297	394	365	335	306	277
50	444	415	385	356	327	419	390	360	331	302
60	474	445	415	386	357	444	415	385	356	327
70	504	475	445	416	387	469	440	410	381	352
80	534	505	475	446	417	494	465	435	406	377
90	564	535	505	476	447	519	490	460	431	402
100	594	565	535	506	477	544	515	485	456	427
Unheated structures										
0	294	265	235	206	177	294	265	235	206	177
10	339	310	280	251	222	332	302	273	244	214
20	384	355	325	296	267	369	340	310	281	252
30	429	400	370	341	312	407	377	348	319	289
40	474	445	415	386	357	444	415	385	356	327
50	519	490	460	431	402	482	452	423	394	364
60	564	535	505	476	447	519	490	460	431	402
70	609	580	550	521	492	557	527	498	469	439
80	654	625	595	566	537	594	565	535	506	477
90	699	670	640	611	582	632	602	573	544	514
100	744	715	685	656	627	669	640	610	581	552

From Table 3.3.2:
Creep correction factor
= 0.88 + (20/200)(1.18 − 0.88) = 0.91

From Table 3.3.3:
Creep correction
= 1.17 − 0.5(1.17 − 1.08) = 1.13
Shrinkage correction
= 1.29 − 0.5(1.29 − 1.14) = 1.22

From Table 3.3.4:
Creep correction
= 0.48 − 0.2(0.48 − 0.36) = 0.46
Shrinkage correction
= 0.46 − 0.2(0.46 − 0.31) = 0.43

(Note: Temperature shortening is not significant for this calculation.)

Total strain:
Creep = $169 \times 10^{-6}(0.91)(1.13)(0.46)$
= 80×10^{-6} in./in.
Shrinkage = $266 \times 10^{-6}(1.22)(0.43)$
= 140×10^{-6} in./in.
Total strain = 220×10^{-6} in./in.
Total shortening = $220 \times 10^{-6}(24)(12)$
= 0.06 in.

b. *At final:*
From Table 3.3.1:
Creep strain = 315×10^{-6} in./in.
Shrinkage strain = 510×10^{-6} in./in.

Factors from Tables 3.3.2 and 3.3.3 same as for 60 days.

From Table 3.3.4:
Creep correction
= 0.77 − 0.2(0.77 − 0.74) = 0.76
Shrinkage correction
= 0.75 − 0.2(0.75 − 0.64) = 0.73

From Table 3.3.5:
Temperature strain = 210×10^{-6} in./in.

Total creep and shrinkage strain:
Creep = $315 \times 10^{-6}(0.91)(1.13)(0.76)$
= 246×10^{-6} in./in.
Shrinkage = $510 \times 10^{-6}(1.22)(0.73)$
= 454×10^{-6} in./in.

Total = 700×10^{-6} in./in.

Difference from 60 days to final
= 700 − 220
= 480×10^{-6} in./in.

Total strain = 480 + 210
= 690×10^{-6} in./in.

Total shortening
= $690 \times 10^{-6}(24)(12)$
= 0.20 in.

For the effects of shortening in frame structures, see Sect. 3.8.5.

The behavior of actual structures indicates that reasonable estimates of volume change characteristics are satisfactory for the design of most structures even though test data relating volume changes to the variables shown in Tables 3.3.1 through 3.3.5 exhibit considerable scatter. Therefore, it is possible to reduce the variables and use approximate values as shown in Tables 3.3.6 and 3.3.7.

Example 3.3.2 Determine volume change shortening by Tables 3.3.6 and 3.3.7

Given:
Same as Example 3.3.1.

Solution:
For prestressed, normal weight concrete, in a heated building, use Table 3.3.6.

For 55% relative humidity and 70°F temperature, interpolating from Table 3.3.6:

Actual strain = 687×10^{-6} in./in.

This result compares with the value of 690×10^{-6} calculated from Tables 3.3.1 through 3.3.5.

3.3.2 Bowing

A strain gradient between the inside and outside of a wall panel, or between the top and underside of an uninsulated roof member, can cause the member to bow. The theoretical magnitude of bowing (see Fig. 3.3.3) can be determined by:

$$\Delta = \alpha \frac{\ell^2}{8h} \quad \text{(Eq. 3.3.1)}$$

where:

α = strain gradient across panel thickness
ℓ = distance between supports
h = member thickness

For a temperature difference between inside and outside of a panel

$$\alpha = C(T_1 - T_2) \quad \text{(Eq. 3.3.2)}$$

where:

C = coefficient of thermal expansion
T_2, T_1 = inside and outside temperature

Limited records of temperature measurements indicate that in open structures, such as the roofs of parking decks, the temperature differential $(T_1 - T_2)$ seldom exceeds 30 to 40°F. In an insulated sandwich wall panel, the theoretical difference can be higher, but this is tempered by "thermal lag" due to the mass of the concrete (see Sect. 9.1).

Fig. 3.3.3 Thermal bow of wall panel

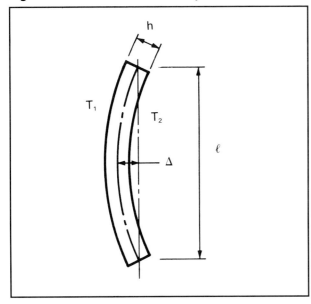

Moisture differences between the inside and outside of an enclosed building can also cause bowing, however, calculation is much less precise and involves more variables. The exterior layer of the concrete panel absorbs moisture from the atmosphere and periodic precipitation, while the interior layer is relatively dry, especially when the building is heated. This causes the inside layer to shrink more than the outside, causing an outward bow. This would tend to balance the theoretical inward thermal bowing in cold weather, which is believed to explain the observation that "wall panels always bow out".

Example 3.3.3 Thermal bow in a wall panel

Given:

A 20 ft high, 6 in. thick wall panel as shown below. Assume a coefficient of thermal expansion $C = 6 \times 10^{-6}$ in./in./°F; a temperature differential $T_1 - T_2 = 35°F$; $E_c = 4300$ ksi.

Problem:

Determine the potential thermal bow, Δ_1, the force, P, required at mid-height to restrain the bowing, the stress in the panel caused by the restraint, and the residual bow, Δ_2.

Solution:

From Eqs. 3.3.1 and 3.3.2:

$$\Delta_1 = \frac{(6 \times 10^{-6})(35)(20 \times 12)^2}{8(6)}$$

$= 0.25$ in.

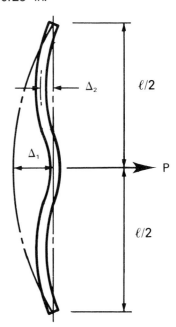

From Table 3.3.8:

$E_t = 0.75(4300) = 3225$ ksi
$I = (bh^3)/12 = [12(6)^3]/12$
$= 216$ in.^{4}/ft.

From Table 3.3.8, Case 1:

$P = 48E_t I\Delta/\ell^3$
$= [48(3225)(216)(0.25)]/(20 \times 12)^3$
$= 0.605$ k/ft width
$M = P\ell/4 = [0.605(20)]/4$
$= 3.02$ ft-kips/ft width

Panel stress $= My/I$
$= [3.02(12,000)(3)]/216$
$= 503$ psi

The residual bow can be calculated by adjusting the equation in Table 3.3.8, Case 5, to read:

$\Delta = M\ell^2/16E_t I$; and substituting
$\ell/2$ for ℓ

$\Delta_2 = [(3.02 \times 12)(10 \times 12)^2]/[16(3225)(216)]$
$= 0.05$ in.

While the magnitude of bowing is usually not very significant, in the case of wall panels it may cause unacceptable separation at the corners (see Fig. 3.3.4), and possible damage to joint sealants. It may therefore be desirable to restrain bowing with one or more connectors between panels. Table 3.3.8 gives equations for calculating the required restraint and the moments this would cause in the panel.

Similarly, differential temperature can cause upward bowing in roof members, especially in open structures such as parking decks. If these members are restrained from rotations at the ends, positive moments (bottom tension) can develop at the support, as shown in Cases 4 and 5, Table 3.3.8. The bottom tension can cause severe cracking, but once the cracks occur, the tension is relieved. Examination of the equations shows that the thermal induced positive moments are independent of the span length. (For example, substitute Eq. 3.3.1 for Δ in Table 3.3.8, Case 4). Note from Table 3.3.8 that if only one end is restrained, as is sometimes done to relieve axial volume change force, the restraint moment is doubled. Also note that, since thermal bow occurs with daily temperature changes, the cyclical effects could magnify the potential damage.

Fig 3.3.4 Corner separation due to thermal bow

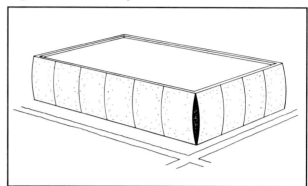

Table 3.3.8 Forces required to restrain bowing

Intermediate restraint (ends free to rotate)	End restraint
Case 1: single restraint at midspan $P = \dfrac{48 E_t I \Delta}{\ell^3}$ Moment in panel $= \dfrac{P\ell}{4}$	**Case 4: both ends restrained** $M = \dfrac{8 E_t I \Delta}{\ell^2}$
Case 2: two restraint points $P = \dfrac{24 E_t I \Delta}{3a\ell^2 - 4a^3}$ Moment in panel $= Pa$	**Case 5: one end restrained** $M = \dfrac{16 E_t I \Delta}{\ell^2}$
Case 3 – Three or more restraint points (Approx. uniform continuous restraint) $\Sigma P = w\ell = \dfrac{77 E_t I \Delta}{\ell^3}$ Moment in panel $= \dfrac{w\ell^2}{8} = \Sigma P \left(\dfrac{\ell}{8}\right)$	For daily temperature change, use $E_t = 0.75 E_c$ For seasonal changes, use $E_t = 0.50 E_c$

Example 3.3.4 Thermal bow in a roof member

Given:

A 30IT24 inverted tee beam supporting the double tees of the upper level of a parking deck, as shown in Fig. 3.3.5, is welded at each end at the bearing to the column, and subject to a thermal gradient of 35°F. Coefficient of thermal expansion $C = 6 \times 10^{-6}$ in./in./°F; $T_1 - T_2 = 35°F$; $E_c = 4300$ ksi; for the composite section $I = 51,725$ in.[4]

Problem:

Find the tensile force developed at the support.

Solution:

From Eqs. 3.3.1 and 3.2.2:

$$\Delta = \frac{(6 \times 10^{-6})(35)(24 \times 12)^2}{8(27)} = 0.081 \text{ in.}$$

From Table 3.3.8, Case 4:

$E_t = 0.75(4300) = 3225$ ksi

$$M = \frac{8(3225)(51,725)(0.081)}{(24 \times 12)^2}$$

$$= 1303 \text{ in.-kips}$$

$$T = \frac{1303}{(24 + 1.5)} = 51.1 \text{ kips}$$

This example illustrates the very high forces that can occur when roof members are welded at the bearings, and why cracking often occurs. The force calculated is an upper bound value, and does not consider the relieving effects of connection extension, microcracking and column flexibility. Nevertheless, an alternative method which does not employ welding at the bearings is strongly recommended.

3.3.3 Expansion Joints

Joints are placed in structures to limit the magnitude of forces which result from volume change deformations (temperature changes, shrinkage and creep), and to permit movements (volume change, seismic) of structural elements. If the forces generated by temperature rise are significantly greater than shrinkage and creep forces, a true "expansion joint" is needed. However, in concrete structures, joints which provide for expansion are seldom required. Instead, joints that permit contraction of the structure are needed to relieve the strains caused by temperature drop, creep and shrinkage, which are additive. Such joints are properly called contraction or control joints but are commonly referred to as expansion joints.

It is desirable to have as few expansion joints as possible. The purpose of this section is to present guidelines for determining the spacing and width of expansion joints.

Spacing of expansion joints

There is a divergence of opinion concerning the spacing of expansion joints. Typical practice in concrete structures, prestressed or non-prestressed, is to locate expansion joints at distances between 150 and 300 ft. However, concrete buildings exceeding these limits have performed well without expansion joints. Recommended joint spacings for precast concrete buildings are generally based on experience. Evaluation of joint spacing should consider the types of connections used, the column stiffnesses in simple span structures, the relative stiffness between beams and columns in framed structures, location of lateral load resisting elements and the weather exposure conditions. Non-heated structures, such as parking garages, are subjected to greater temperature changes than occupied structures, so shorter distances between expansion joints may be warranted.

Sects. 3.3 and 3.8 present methods for analyzing the potential movement of framed structures, and the effect of restraint of movement on the connections and structural frame. This information along with the connection design methods in Chapter 6 can aid in determining spacing of expansion joints.

Fig. 3.3.6 shows joint spacing as recommended by the Federal Construction Council, and is adapted from Expansion Joints in Buildings, Technical Report No. 65, prepared by the Standing Committee on Structural Engineering of the Federal Construction Council, Building Research Advisory Board, Division of Engineering, National Research Council, National Academy of Sciences, 1974. Note that the spacings obtained from the graph in Fig. 3.3.6 should be modified for various conditions as shown in the notes below the graph. Values for the design temperature change can be obtained from Fig. 3.3.1.

When expansion joints are required in non-rectangular structures, they should be located at places where the plan or elevation dimensions change radically.

Width of expansion joints

The width of the joint can be calculated theoretically using a coefficient of expansion of 6×10^{-6} in./in./°F for normal weight and 5×10^{-6} in./in./°F for sand-lightweight concrete. The Federal Construction Council report, referenced above, recommends a minimum width of 1 in. However, since the primary problem in concrete buildings is contraction rather than expansion, joints that are too wide may result in problems with reduced bearing or loss of filler material. The joint width should consider the ambient temperature at time of erection. Seismic codes also stipulate joint widths to accommodate earthquake movement.

Fig. 3.3.5 Inverted tee beam for Example 3.3.4

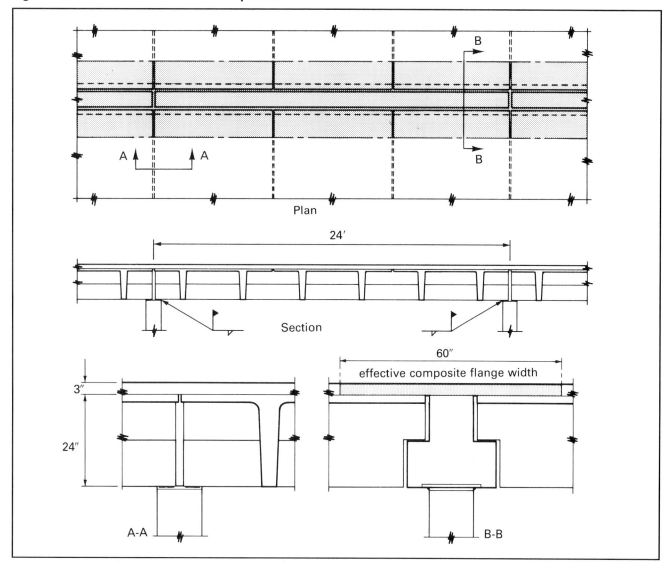

3.4 Component Analysis

The design of components is covered in Chapter 4. This section is intended to assist in establishing design parameters for some components, and to discuss the effects of non-frame components on the frame.

3.4.1 Non-Bearing Wall Panels

Non-bearing panels are designed to resist wind, seismic forces generated from self weight, and forces required to transfer the weight of panel to the support. It is rare that these externally applied loads will produce the maximum stresses; the forces imposed during manufacturing and erection will usually govern the design, except for the connections.

Deformations

The relationship of the deformations of the panel and the supporting structure must be evaluated, and care taken to prevent unintended restraints from imposing additional loads. Such deformation may be caused by the weight of the panel, volume changes of concrete frames, and rotation of spandrel beams. To prevent imposing loads on the panel, the connections must be designed and installed to permit these deformations to freely occur.

For example, the tendency for the panels to follow the beam shown in Fig. 3.4.1 may cause restraining forces to develop at panel joints. The connections should be designed to allow the supporting beam to deflect, but the beam should be stiff enough that panel joint widths are maintained within the specified tolerances. These effects can sometimes be controlled by adjusting the erection sequence.

The most prevalent cause of panel deformation is bowing due to thermal gradients. If supported in a manner that will permit bowing, the panel will not be subjected to stress. However, if the panel is re-

Fig. 3.3.6 Maximum building length without use of expansion joints

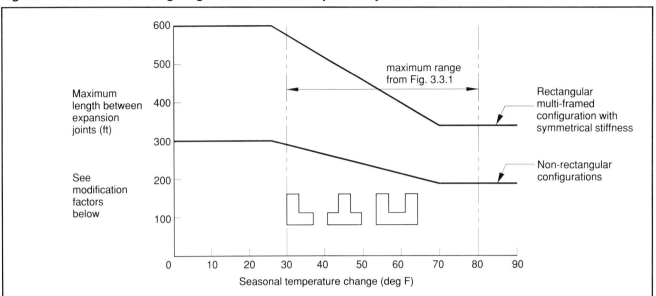

These curves are directly applicable to buildings of beam-and-column construction, hinged at the base, and with heated interiors. When other conditions prevail, the following rules are applicable:

(a) If the building will be heated only and will have hinged-column bases, use the allowable length as specified;

(b) If the building will be air conditioned as well as heated, increase the allowable length by 15 percent (provided the environmental control system will run continuously);

(c) If the building will be unheated, decrease the allowable length by 33 percent;

(d) If the building will have fixed-column bases, decrease the allowable length by 15 percent;

(e) If the building will have substantially greater stiffness against lateral displacement at one end of the plan dimension, decrease the allowable length by 25 percent.

When more than one of these design conditions prevail in a building, the percentile factor to be applied should be the algebraic sum of the adjustment factors of all the various applicable conditions.

Source: "Expansion Joints in Buildings," Technical Report No. 65, National Research Council, National Academy of Sciences, 1974.

strained laterally between supports, stresses in the panel and forces in the structure will occur. This is discussed in Sect. 3.3.2.

Non-bearing panels should be designed and installed so that they do not restrain frames from lateral translation. If such restraint occurs, the panels may tend to act as shear walls and become overstressed (Fig. 3.4.2). Panels which are installed on a frame should be connected in a manner to allow frame distortion. In some cases, especially in high seismic regions, special connections which can accommodate inter-story drift may be required.

The shortening of concrete columns from elastic and plastic deformation should be considered in very tall structures. At intermediate levels, the differential shortening between two adjacent floors will be negligible, and the panel will follow the frame movement. At the lowest level, if the panel is rigidly supported at the base (such as foundation or transfer girder), the accumulated shortening of the structure above may induce unintended loading on the panel. In such cases, the panel connection should be designed to permit the

Fig. 3.4.1 Deformation of panels on flexible beam

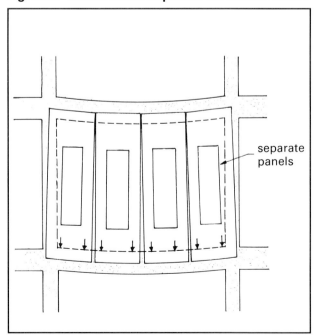

Fig. 3.4.2 Panel forces induced by frame distortion

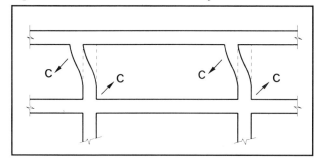

Fig. 3.4.3 Shearing wind on ribbed panels

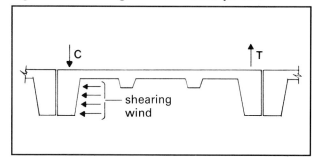

Fig. 3.4.4 Forces on a panel subjected to wind suction

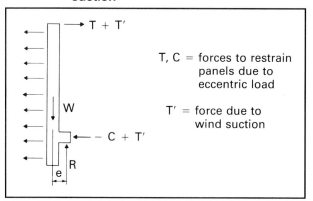

calculated deformation.

A similar situation will result when two adjacent columns have significantly different loads. For example, the corner column of a structure will usually be subjected to a smaller load than the adjacent columns. If both columns are the same size (as is often the situation for architectural reasons) and reinforced approximately the same, they will undergo different shortening.

Wind load

Building codes specify the wind pressure for which a building is to be designed, including magnified loads for localized portions due to gusting or funnel effects produced by adjacent structures.

The lateral deflection of thin panels when subjected to wind should be determined, particularly if they are attached to, or include, windows. Panels with deep protruding ribs may require analysis for shearing winds, as indicated in Fig. 3.4.3, and the connections designed for the twist produced. Although the design of the panel itself will generally not be critical for wind, the connections may be. This is particularly true for the tension connection which resists wind suction along with eccentric gravity loads, as indicated in Fig. 3.4.4.

Panels with openings

Non-bearing panels which contain openings may develop stress concentrations at these openings, resulting from unintended loading or restrained bowing. Hairline cracks radiating from the corners can result (Fig. 3.4.5). While these stress concentrations may be partially resisted by reinforcement, the designer should consider methods of eliminating imposed restraints. Areas of abrupt change in cross-section should be reinforced, and should be rounded or chamfered.

Loads from adjacent floors can be imposed on non-load bearing panels by methods of joinery, and these loads can cause excessive stresses at the "beam" portion of an opening (Fig. 3.4.6). This can be prevented by locating connections away from critical sections.

Unless a method of preventing load transfer can be developed and permanently maintained, the "beam" should be designed for some loads from the floor. The magnitude of such loads requires engineering judgment.

3.4.2 Load Bearing Wall Panels

Most of the items in the previous section must also be considered in the analysis of load bearing wall panels. Panels may be designed to span horizontally between columns, or vertically. When spanning horizontally, they are designed as beams, or if they have

Fig. 3.4.5 Corner cracking due to restrained bowing

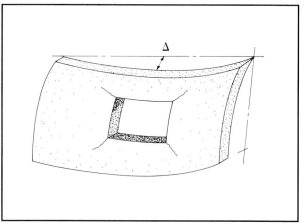

Fig. 3.4.6 Unanticipated loading on a non-load bearing panel

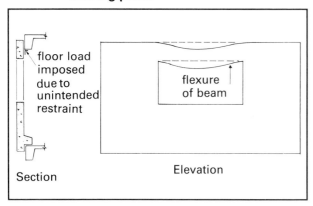

Fig. 3.4.7 Horizontal and vertical rib panels

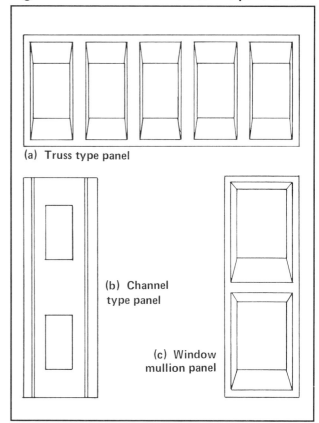

frequent, regularly spaced window openings, as shown in Fig. 3.4.7(a), as Vierendeel trusses. When so designed there must be a space or joint horizontally between panels, to insure that they will not transfer loads to panels below.

When the panels are placed vertically, they are usually designed as columns, and slenderness, as described in Sect. 3.5, should be considered. If a large portion of the panel is window opening, as in Fig. 3.4.7(c), it may be necessary to analyze it as a rigid frame.

Fig. 3.4.7 shows architectural wall panels, generally used with relatively short vertical spans, although they will sometimes span continuously over two or more floors. They are usually custom-made for each project, and reinforced with mild steel. Standard flat, hollow-core and stemmed members are also used as wall panels, and are frequently prestressed.

Dimensions of architectural panels are usually selected based on a desired appearance. When these panels are also used to carry loads, or act as shear walls, it is obviously important to have some engineering input early in the preliminary stages of the project.

3.4.3 Non-Bearing Spandrels

These are precast elements which are less than story height, made up either as a series of individual units or as one unit extending between columns. Support for spandrel weight may be the floor or the column, and stability against eccentric loading is achieved by connections to the underside of the floor or to the column (see Fig. 3.4.8).

Spandrels are usually part of a window wall, so consideration should be given to the effect of deflections and rotations of the spandrel on the window. Deformation calculations should be based on gross concrete section since the stresses will generally be less than those which cause cracking. For elements which extend in one piece between columns, it is preferable that the connections which provide vertical support be located close to the ends. This arrangement will minimize interaction and load transfer between floor and spandrel.

Consideration should also be given to spandrels which are supported at the ends of long cantilevers. The designer must determine the effect of deflection and rotation of the support, including the effects of creep, and arrange the details of all attachments to accommodate this condition (Fig. 3.4.9). A particularly critical condition can occur at corners of buildings, especially when there is a cantilever on both faces.

3.4.4 Load Bearing Spandrels

Load bearing spandrels are panels which support floor or roof loads. Except for the magnitude and location of these additional loads, the design is the same as for non-bearing spandrels.

Load bearing spandrels support structural loads which are usually applied eccentrically with respect to the support. A typical arrangement of spandrel and supported floor is shown in Fig. 3.4.10

Torsion due to eccentricity must be resisted by the spandrel or by a horizontal couple developed in the floor construction (see Fig. 3.4.10b and c). In order to take torsion in the floor construction, the details must provide for a compressive force transfer at the top of the floor, and a tensile force transfer at the bearing of the precast floor element. The load path of these floor forces must be followed through the structure, and considered in the design of other members in the building. Even when torsion is resisted in this manner in the completed structure, twisting on the spandrel prior to completion must be considered.

If torsion cannot be accommodated by floor connections, the spandrel panel should be designed for induced stresses. Torsion design is illustrated in Chapter 4.

Example 3.4.1 Spandrel panel rotation

Given:

The spandrel panel shown.

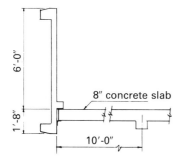

Panel weight = 417 plf

Normal weight concrete slab

E = 4000 ksi

Assume creep reduces effective E to 2000 ksi

Superimposed dead load = 10 psf

Problem:

Determine rotation and displacement at top of spandrel.

Solution:

Rotation of cantilever due to slab weight plus superimposed loads (neglecting support rotation):

w = 8(150)/12 + 10 = 110 psf

I = $bh^3/12$ = $12(8)^3/12$ = 512 in.^{4}

$$\theta = \frac{w\ell^3}{6EI} = \frac{(0.110/12)(10 \times 12)^3}{6(2000)(512)}$$

= 0.00258 radians

Rotation of cantilever due to weight of spandrel:

$$\theta = \frac{W\ell^2}{2EI} = \frac{(0.417)(10 \times 12)^2}{2(2000)(512)}$$

= 0.00293 radians

Total rotation of cantilever

= 0.00258 + 0.00293 = 0.00551 radians

Displacement of top of spandrel

= (0.00551)(72 + 8/2) = 0.42 in.

Fig. 3.4.8 Typical spandrel connections

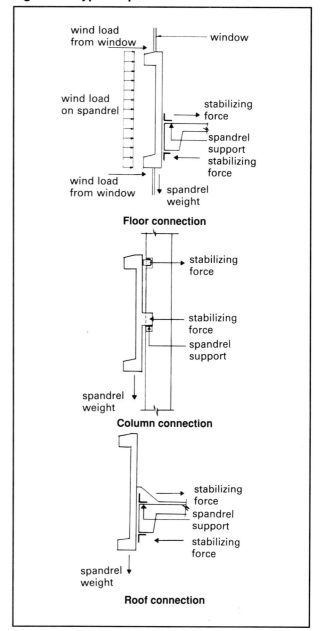

Fig. 3.4.9 Effect of cantilever supports

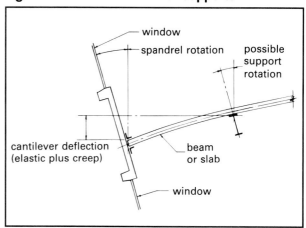

Fig. 3.4.10 Load bearing spandrel

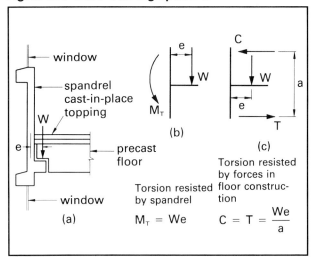

3.4.5 Eccentrically Loaded Columns

Many precast concrete structures utilize multi-story columns with simple-span beams resting on haunches. Fig. 3.4.11 and Table 3.4.1 are provided as aids for determining the various combinations of load and moment that can occur with such columns.

The following conditions and limitations apply to Fig. 3.4.11 and Table 3.4.1:

1. The coefficients are only valid for columns braced against sidesway.

2. For partially fixed column bases (see Sect.3.8.3), a straight line interpolation between the coefficients for pinned and fixed bases can be used with small error.

3. For taller columns, the coefficients for the 4-story columns can be used with small error.

4. The coefficients in the "Σ Max" line will yield the maximum required restraining force, F_i, and column moments caused by loads (equal at each level) which can occur on either side of the column, for example, live loads on interior columns. The maximum force will not necessarily occur with the same loading pattern that causes the maximum moment.

5. The coefficients in the "Σ One Side" line will yield the maximum moments which can occur if the column is loaded on only one side, such as the end column in a bay.

Example 3.4.2 Use of Fig. 3.4.11 and Table 3.4.1

Problem:

Using Table 3.4.1, determine the maximum restraining force and moment in the lowest story of a 3-story frame for:

a. An interior column in a multi-bay frame
b. An exterior column

Given:

Beam reactions to column haunch at each level:
D.L. = 50 kips
L.L. = 20 kips
Eccentricity e = 14 in.
Story height h_s = 16 ft
Column base is determined to be 65% fixed.

Solution:

Factored loads: D.L. = 1.4(50) = 70.0 kips
L.L. = 1.7(20) = 34.0 kips
104.0 kips

a. For the interior column, the dead load reaction would be the same on either side, thus, no moment results. The live load could occur on any one side at any floor, hence, use of the coefficients in the "Σ Max" line:

$P_u e$ = 34.0(14) = 476 in.-kips = 39.7 ft-kips

To determine the maximum moment at point B:
For a pinned base: k_m = 0.67
For a fixed base: k_m = 0.77
For 65% fixed: k_m = 0.67 + 0.65(0.77 − 0.67)
= 0.74
M_u = $k_m P_u e$ = 0.74(39.7) = 29.4 ft-kips

Maximum restraining force at level 2:

F_u = $k_f P_u e / h_s$
k_f = 1.40 + 0.65(1.62 − 1.40) = 1.54
F_u = 1.54(39.7)/16 = 3.82 kips (tension or compression)

b. For the exterior column, the total load is eccentric on the same side of the column, hence use the coefficients in the "Σ One Side" line:

$P_u e$ = 104.0(14) = 1456 in.-kips = 121.3 ft-kips

To determine the maximum moment at point B:
For a pinned base, k_m = 0.40
For a fixed base, k_m = 0.46
For 65% fixed, k_m = 0.40 + 0.65(0.46 − 0.40)
= 0.44
M_u = $k_m P_u e$ = 0.44(121.3) = 53.4 ft-kips

Fig. 3.4.11 Use of Table 3.4.1

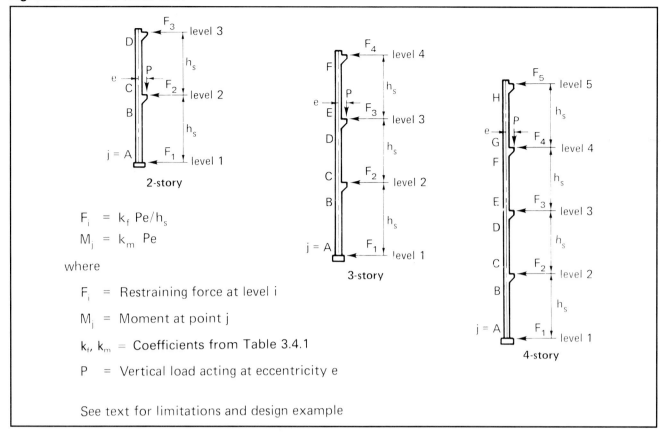

$F_i = k_f Pe/h_s$

$M_j = k_m Pe$

where

F_i = Restraining force at level i

M_j = Moment at point j

k_f, k_m = Coefficients from Table 3.4.1

P = Vertical load acting at eccentricity e

See text for limitations and design example

Maximum restraining force at level 2:

$F_u = k_f P_u e/h_s$

$k_f = -0.60 - 0.65(-0.60 + 0.22) = -0.35$

$F_u = -0.35(121.3)/16$

= −2.65 kips (tension)

3.5 Slenderness Effects in Columns and Wall Panels

The term "slenderness effects," can be described as the moments in a member produced when the line of action of the axial force is not coincident with the displaced centroid of the member. These moments, which are not accounted for in the primary analysis, are thus termed "secondary moments." These secondary moments arise from changes in the geometry of the structure, and may be caused by one or more of the following:

1. Relative displacement of the ends of the member due to:
   a. Lateral or unbalanced vertical loads in an unbraced frame, usually labeled "translation" or "sidesway."
   b. Manufacturing and erection tolerances.
2. Deflections away from the end of the member due to:
   a. End moment due to eccentricity of the axial load.
   b. End moments due to frame action—continuity, fixity or partial fixity of the ends.
   c. Applied lateral loads, such as wind.
   d. Thermal bowing from differential temperature (see Sect. 3.3.2).
   e. Manufacturing tolerances.
   f. Camber due to prestressing.

These secondary effects can be considered in the design of the member by either using the Moment Magnification Method, described in Sect. 3.5.2, or by a direct iterative analysis usually termed "second-order" or "P-Δ" analysis.

3.5.1 Second-Order (P-Δ) Analysis

The usual procedure for this analysis is to perform an elastic type analysis using factored loads. Out-of-plumbness (items 1b and 2e above) are initially assumed based on experience and/or specified tolerances. The thermal bowing effect is usually neglected in columns, but may be significant in exterior wall panels (see Sect. 3.3.2).

At each iteration, the lateral deflection is calculated, and the moments caused by the axial load acting

Table 3.4.1 Coefficients k_f and k_m for determining moments and restraining forces on eccentrically loaded columns braced against sidesway

See Fig. 3.4.11 for explanation of terms

+ Indicates clockwise moments on the columns and compression in the restraining beam

No. of stories	Base Fixity	P acting at level	k_f at level 1	2	3	4	5	k_m at point A	B	C	D	E	F	G	H
2	PINNED	3	+0.25	−1.50	+1.25			0	−0.25	+0.25	+1.0				
		2	−0.50	0	+0.50			0	+0.50	+0.50	0				
		Σ Max	±0.75	±1.50	±1.75			0	±0.75	±0.75	±1.0				
		Σ One Side	−0.25	−1.50	+1.75			0	+0.25	+0.75	+1.0				
	FIXED	3	+0.43	−1.72	+1.29			−0.14	−0.29	+0.29	+1.0				
		2	−0.86	+0.43	+0.43			+0.29	+0.57	+0.43	0				
		Σ Max	±1.29	±2.15	±1.72			±0.43	±0.86	±0.72	±1.0				
		Σ One Side	−0.43	−1.29	+1.72			+0.15	+0.28	+0.72	+1.0				
3	PINNED	4	−0.07	+0.40	−1.60	+1.27		0	+0.07	−0.07	−0.27	+0.27	+1.0		
		3	+0.13	−0.80	+0.20	+0.47		0	−0.13	+0.13	+0.53	+0.47	0		
		2	−0.47	−0.20	+0.80	−0.13		0	+0.47	+0.53	+0.13	−0.13	0		
		Σ Max	±0.67	±1.40	±2.60	±1.87		0	±0.67	±0.73	±0.93	±0.87	±1.0		
		Σ One Side	−0.41	−0.60	−0.60	+1.61		0	+0.40	+0.60	+0.40	+0.60	+1.0		
	FIXED	4	−0.12	+0.47	−1.62	+1.27		+0.04	+0.08	−0.08	−0.27	+0.27	+1.0		
		3	+0.23	−0.92	+0.23	+0.46		−0.08	−0.15	+0.15	+0.54	+0.46	0		
		2	−0.81	+0.23	+0.70	−0.12		+0.27	+0.54	+0.46	+0.12	−0.12	0		
		Σ Max	±1.16	±1.62	±2.55	±1.85		±0.38	±0.77	±0.69	±0.92	±0.85	±1.0		
		Σ One Side	−0.70	−0.22	−0.69	+1.61		+0.23	+0.46	+0.54	+0.38	+0.62	+1.0		
4	PINNED	5	+0.02	−0.11	+0.43	−1.61	+1.27	0	−0.02	+0.02	+0.07	−0.07	−0.27	+0.27	+1.0
		4	−0.04	+0.22	−0.86	+0.22	+0.46	0	+0.04	−0.04	−0.14	+0.14	+0.54	+0.46	0
		3	+0.13	−0.75	0	+0.75	−0.12	0	−0.13	+0.13	+0.50	+0.50	+0.12	−0.12	0
		2	−0.46	−0.22	+0.86	−0.22	+0.04	0	+0.46	+0.54	+0.14	−0.14	−0.04	+0.04	0
		Σ Max	±0.65	±1.30	±2.15	±2.80	±1.89	0	±0.64	±0.72	±0.86	±0.86	±0.97	±0.89	±1.0
		Σ One Side	−0.35	−0.86	+0.43	−0.86	+1.65	0	+0.35	+0.65	+0.57	+0.43	+0.35	+0.65	+1.0
	FIXED	5	+0.03	−0.12	+0.43	−1.61	+1.27	−0.01	−0.02	+0.02	+0.07	−0.07	−0.27	+0.27	+1.0
		4	−0.06	+0.25	−0.87	+0.22	+0.46	+0.02	+0.04	−0.04	−0.14	+0.14	+0.54	+0.46	0
		3	+0.22	−0.87	+0.03	+0.74	−0.12	−0.07	−0.14	+0.14	+0.51	+0.50	+0.12	−0.12	0
		2	−0.80	+0.21	+0.74	−0.18	+0.03	+0.27	+0.54	+0.46	+0.12	−0.12	−0.03	+0.03	0
		Σ Max	±1.11	±1.45	±2.07	±2.75	±1.88	±0.37	±0.74	±0.67	±0.84	±0.83	±0.96	±0.88	±1.0
		Σ One Side	−0.61	−0.53	+0.33	−0.83	+1.64	+0.21	+0.41	+0.59	+0.56	+0.44	+0.36	+0.64	+1.0

at that deflection are accumulated. After three or four iterations, the increase in deflection should be negligible (convergence). If it is not, the member may be approaching stability failure, and the section dimensions should be re-evaluated.

If the calculated moments indicate that cracking will occur, this needs to be taken into account in the deflection calculations. This may involve iterations within iterations, greatly complicating the procedure, although approximations of cracked section properties are usually satisfactory.

Effects of creep should also be included. The most common method is to divide the stiffness (EI) by the factor $1 + \beta_d$ as specified in the ACI moment magnification method.

In unbraced continuous frames, the joint translations effect and the frame response are interdependent. Thus, a practical analysis, especially when potential cracking is a parameter, usually will require the use of computers.

A good review of second-order analysis, along with an extensive bibliography and an outline of a complete program, is contained in Ref. 3.

An example of P-Δ calculations for a simple yet frequently encountered problem is shown in Example 3.5.1.

Example 3.5.1 Second-order analysis of an uncracked member

Given:

A 6 in. thick, 8 ft wide prestressed wall panel as shown.

Loading assumptions are as follows:

1. Axial load eccentricity = 1 in.
2. Midspan bowing due to temperature = 0.6 in. outward.
3. Wind load = 10 psf.
4. Braced frame — no joint translation.
5. Joints assumed pinned top and bottom.

Concrete: f'_c = 5000 psi
E_c = 4300 ksi

D.L. = 20 kips
L.L. = 10 kips
30'-0" = 360"
6"

Problem:

Determine if standard panel (Fig. 2.6.5) is adequate.

Solution:

Case 1 — No wind

P_u = 1.4D + 1.7L = 1.4(20) + 1.7(10)
    = 45 kips

I_g = $bh^3/12$ = $96(6)^3/12$ = 1728 in.^{4}

ϕ = 0.7 at P_u = $0.1 f'_c A_g$ = 288 kips
  = 0.9 at P_u = 0

ϕ = 0.9 − 0.2(45/288) = 0.87
β_d = 28/45 = 0.62

Note: The ϕ-factor is used to allow for unplanned deviations in the material dimensions and properties. The β_d factor is used to account for sustained load effects. Thus, both factors affect the stiffness, EI. It is reasonable to use:

$$EI = \frac{\phi E_c I_g}{1+\beta_d} = \frac{0.87(4300)(1728)}{1.62}$$

$$= 3.99 \times 10^6 \text{ k-in.}^2$$

Moments on panel:

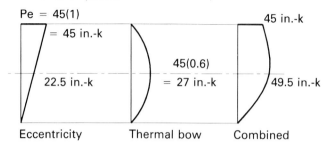

Eccentricity: Pe = 45(1) = 45 in.-k, 22.5 in.-k
Thermal bow: 45(0.6) = 27 in.-k
Combined: 45 in.-k, 49.5 in.-k

Deflection at midspan due to Pe:

$$\Delta = \frac{Pe\ell^2}{16EI} = \frac{45(1)(360)^2}{16(3.99 \times 10^6)} = 0.09$$

Total midspan deflection including thermal:
 = 0.6 + 0.09 = 0.69 in.

Deflection due to P-Δ moment at midspan:

$$\Delta = \frac{Pe\ell^2}{8EI} = \frac{45e(360)^2}{8(3.99 \times 10^6)}$$

$$= 0.183e$$

First iteration:
  Δ = 0.183(0.69) = 0.13 in.

Second iteration:
  e = 0.69 + 0.13 = 0.82 in.
  Δ = 0.183(0.82) = 0.15 in.

Third iteration:
  e = 0.69 + 0.15 = 0.84 in.
  Δ = 0.183(0.84) = 0.15 in. (convergence)

M_u at midheight = 22.5 + 45(0.84)
           = 60.3 in.-kips

Check for cracking at midheight:
  My/I = 60.3(3)(1000)/1728 = 105 psi
  ½ Panel weight: 75(15)/6(12) = −16 psi
  Prestress   = −250 psi
  Net stress = −161 psi (compression)

Therefore, the analysis is valid.

Using interaction curve p. 2-54:
P_u = 45/8 = 5.63 kips/ft
M_u = 60.3/(12 x 8) = 0.63 ft-kips/ft

Point is below curve, OK.

Case 2— Include wind

Axial load without wind is:
P_u = 0.75(1.4D + 1.7L + 1.7W)
  = 0.75[1.4(20) + 1.7(10)] = 34 kips

Deflection due to Pe (see Case 1) = 34/45(0.09)
  = 0.07 in.

Additive wind load would be suction = 10 psf
w_u = 10 x 8 x 0.75 x 1.7 = 102 lb/ft

When considering wind, $\beta_d = 0$ and $\phi = 0.9$ (bending)

$$EI = \frac{\phi E_c I_g}{1+\beta_d} = \frac{0.9(4300)(1728)}{1.0}$$

= 6.69 x 10^6 k-in.^{2}

Deflection due to wind:

$$\frac{5w_u \ell^4}{384\, EI} = \frac{5\left(\frac{0.102}{12}\right)(360)^4}{384(6.69 \times 10^6)} = 0.28 \text{ in.}$$

Total initial midspan bow including eccentricity, thermal and wind:
Δ = 0.60 + 0.07 + 0.28 = 0.95 in.

Deflection due to P-Δ moment at midspan:

$$\Delta = \frac{Pe\ell^2}{8EI} = \frac{34e\,(360)^2}{8(3.99 \times 10^6)} = 0.138e$$

First iteration:
Δ = 0.138(0.95) = 0.13 in.

Second iteration:
e = 0.95 + 0.13 = 1.08 in.
Δ = 0.138(1.08) = 0.15 in.

Third iteration:
e = 0.95 + 0.15 = 1.10 in.
Δ = 0.138(1.10) = 0.15 in. (convergence)

$$M_u = \frac{34(1)}{2} + 34(1.10) + \frac{\left(\frac{0.102}{12}\right)(360)^2}{8}$$

= 192 in.-kips

My/I = 192(3)(1000)/1728 = 333 psi
Panel weight = –16 psi
Prestress = –250 psi
Net stress = 67 psi (tension)
$f_r = 7.5\sqrt{f'_c}$ = 530 psi > 67 psi

Therefore, the analysis is valid.
P_u = 34/8 = 4.2 kips/ft
M_u = 192/(12 x 8) = 2.0 ft-kips/ft

By comparison with Fig. 2.6.5, OK.

3.5.2 Moment Magnification Method

This is an approximate method described in Sect. 10.11 of ACI 318-89, and is applicable to precast and prestressed members. The ACI equations are repeated here for convenience:

$$M_c = \delta_b M_{2b} + \delta_s M_{2s} \quad \text{(Eq. 3.5.1)}$$

where:

$$\delta_b = \frac{C_m}{1 - \frac{P_u}{\phi P_c}} \geq 1.0 \quad \text{(Eq. 3.5.2)}$$

$$\delta_s = \frac{1}{1 - \frac{\Sigma P_u}{\phi \Sigma P_c}} \geq 1.0 \quad \text{(Eq. 3.5.3)}$$

$$P_c = \frac{\pi^2 EI}{(k\ell_u)^2} \quad \text{(Eq. 3.5.4)}$$

$$C_m = 0.6 + 0.4 \frac{M_{1b}}{M_{2b}} \geq 0.4 \quad \text{(Eq. 3.5.5)}$$

The Code suggests Eqs. 3.5.6 and 3.5.7 for the value of EI in Eq. 3.5.4. They were developed for reinforced concrete columns with at least 1% reinforcement, and may be used for precast columns which meet that criterion.

$$EI = \frac{(E_c I_g / 5) + E_s I_{se}}{1 + \beta_d} \quad \text{(Eq. 3.5.6)}$$

or conservatively

$$EI = \frac{E_c I_g / 2.5}{1 + \beta_d} \quad \text{(Eq. 3.5.7)}$$

Since most prestressed compression members and precast load bearing wall panels have much less than 1% steel, modified equations for EI are recommended. Fig 3.5.1 presents such equations based on "curve fitting" with theoretically exact procedures, and agrees reasonably well for $k\ell_u/r$ values up to 150. Other modifications have also been used successfully. These approximations are necessarily

Fig. 3.5.1 Coefficients, λ, for modified EI

$$EI = \frac{E_c I_g / \lambda}{1 + \beta_d} \quad \text{(for } P_c \text{ equation)}$$

where:

$$\eta = 2.5 + \frac{1.6}{P_u / P_o}$$

$6 \leq \eta \leq 70$

$\lambda = \eta\theta$

And θ is given below

(a) Compression flange

$$\theta = \frac{35}{k\ell_u / r} - 0.09$$

(b) No compression flange

$$\theta = \frac{27}{k\ell_u / r} - 0.05$$

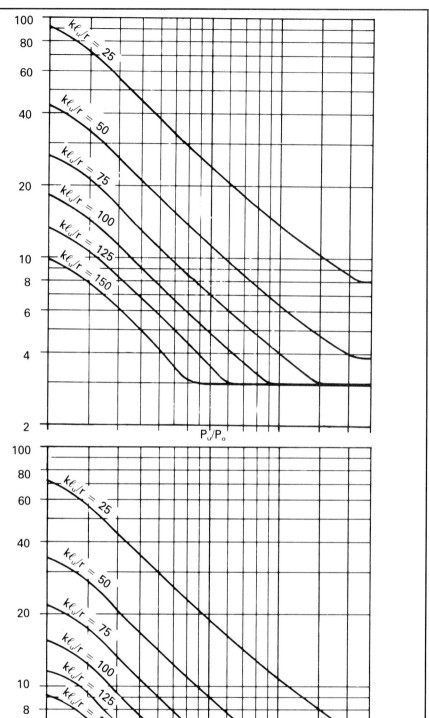

Fig. 3.5.2 Alignment charts for determining effective length factors

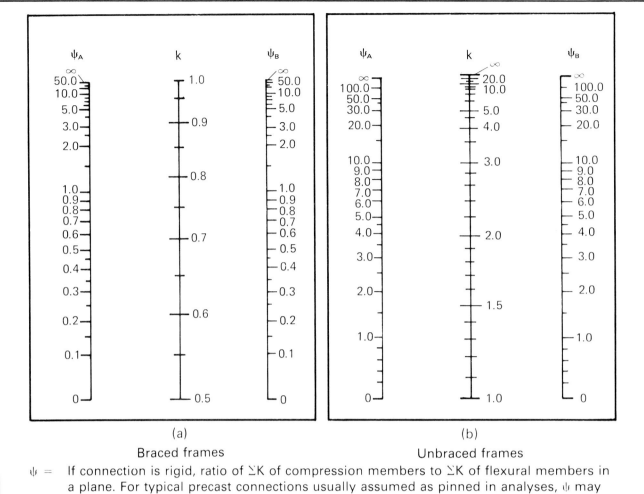

(a) Braced frames

(b) Unbraced frames

ψ = If connection is rigid, ratio of ΣK of compression members to ΣK of flexural members in a plane. For typical precast connections usually assumed as pinned in analyses, ψ may be taken as 10. Lower values may be used if justified by analysis. For example, ψ at footings may be equal to K_c/K_b as calculated in Sect. 3.8.3.

k = effective length factor

conservative, so the second-order analysis described in Sect. 3.5.1 is preferred for major structures. In Eq. 3.5.1, the subscript "s" is used to denote loads and moments which contribute to sidesway in a frame, and the subscript "b" those which do not. In a braced frame, the second term becomes zero. Example 3.5.2 illustrates the use of this equation.

In Eq. 3.5.4, the value of k can be determined from the Jackson-Moreland alignment charts, Fig. 3.5.2. Since most precast members are used in braced frames, and the connections are not designed to transfer moment to horizontal members, a k of 1 is usually used. ψ_a and ψ_b refer to the ends of the compression member.

Example 3.5.2 Slenderness effects using moment magnification

Given:

The structure shown (page 3-27) is the interior portion of a long building that is isolated from the remaining structure by expansion joints, creating an unbraced frame. A frame analysis of the structure yields the following data:

Each wall panel:
   Axial load: D = 14.4 kips, L = 7.2 kips, W = 0
   Top moment: D = 9.6 ft-kips, L = 4.8 ft-kips, W = 0
   Bott. moment: D = 4.2 ft-kips, L = 2.1 ft-kips, W = 17.0 ft-kips (flange compression)

Each column:
   Axial load: D = 115.2 kips, L = 57.6 kips, W = 0
   Top moment: D = L = W = 0 (pinned)
   Bott. moment: D = L = 0, W = 3.0 ft-kips

Wall panel properties:
  $A = 401$ in.^{2}; $I = 20{,}985$ in.^{4}; $r = 7.23$ in.
  P_o (see p. 2-52) $= 1165/\phi = 1664$ kips
  $f'_c = 5000$ psi, $E_c = 4300$ ksi

Column properties:
  $A = 576$ in.^{2}; $I = 27{,}648$ in.^{4}; $r = 7.2$ in.
  P_o (see p. 2-49) $= 1650/0.7 = 2357$ kips
  $f'_c = 5000$ psi, $E_c = 4300$ ksi

Problem:

Find magnified moments for wall panels and columns.

Solution:

Wall panel:

Case 1 (Dead + Live)
  $P_u = 1.4(14.4) + 1.7(7.2) = 32.4$ kips
  $M_2 = 1.4(9.6) + 1.7(4.8) = 21.6$ ft-kips
  $M_1 = 1.4(4.2) + 1.7(2.1) = 9.5$ ft-kips

In this case, the larger moment M_2 occurs at the top.

$\beta_d = 1.4(14.4)/32.4 = 0.62$

When gravity loads are unsymmetrical, frame moments can be very small, but a sidesway case must be checked for overall stability. Since in this structure the axial loads are symmetrical and do not contribute to sidesway, the value of k can be taken as 1 and $\delta_s M_{2s} = 0$.

$k\ell_u/r = 16(12)/7.23 = 26.6 > 22$
$P_u/P_o = 32.4/1664 = 0.02$
From Fig. 3.5.1(a): $\lambda = 86$

$$EI = \frac{E_c I_g / \lambda}{1 + \beta_d}$$

$$= \frac{4300(20{,}985)/86}{1 + 0.62}$$

$= 647{,}685$ kips-in.^{2}

$C_m = 0.6 + 0.4(9.5/21.6) = 0.78$

$$P_c = \frac{\pi^2 EI}{(k\ell_u)^2} = \frac{\pi^2(647{,}685)}{(16 \times 12)^2}$$

$= 173$ kips

$\phi = 0.7$ at $P_u = 0.1 f'_c A_g = 200$ kips
   $= 0.9$ at $P_u = 0$
$\phi = 0.9 - 0.2(32.4/200) = 0.87$

$$\delta_b = \frac{C_m}{1 - \dfrac{P_u}{\phi P_c}} = \frac{0.78}{1 - \dfrac{32.4}{0.87(173)}}$$

$= 0.99$, use 1.0

$M_c = \delta_b M_{2b} = 1.0(21.6) = 21.6$ ft-kips

Case 2 (Dead + Live + Wind)

Since wind loads contribute to sidesway:
  $P_u = 0.75(32.4) = 24.3$ kips
  $M_{2b} = 0.75(9.5) = 7.1$ ft-kips
  $M_{2s} = 0.75(1.7)(17.0) = 21.7$ ft-kips

In this case, the larger moment M_2 occurs at the bottom.

$\beta_{db} = 0.62$ as before; $\beta_{ds} = 0$

Assume ψ_A for use in Jackson-Moreland alignment charts is approximately the ratio of ℓ_u to length below floor $= 16/3 = 5.3$; $\psi_B = 10$ (max.)

$k = 2.6$; $k\ell_u = 2.6(16)(12) = 499.2$ in.

Using Fig. 3.5.1(a):
$P_u/P_o = 24.3/1664 = 0.015$, $k\ell_u/r = 69.0$
$\lambda = 29$

$$(EI)_b = \frac{4300(20{,}985)/29}{1.62}$$

$= 1.92 \times 10^6$ kips-in.^{2}

$(C_m)_b = 0.78$

$$(P_c)_b = \frac{\pi^2(1.92 \times 10^6)}{(16 \times 12)^2} = 514 \text{ kips}$$

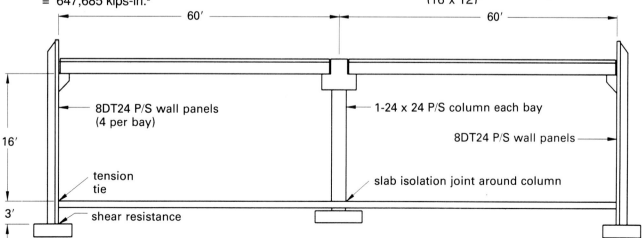

$\phi = 0.9 - 0.2(24.3/200) = 0.88$

$\delta_b = \dfrac{0.78}{1 - \dfrac{24.3}{0.88(514)}} = 0.82$, use 1.0

$(EI)_s = \dfrac{4300(20{,}985)/29}{1.0}$

$\qquad = 3.11 \times 10^6$ kip-in.²

$C_m = 1.0$

$(P_c)_s = \dfrac{\pi^2(3.11 \times 10^6)}{(499.2)^2} = 123$ kips

Column (Case 2 continued):

Column $P_u = 0.75[1.4(115.2) + 1.7(57.6)]$
$\qquad\quad = 194.4$ kips

From analysis of column/base relationship (see Sect. 3.8.3), $K_c/K_b = 1.0$

Using alignment charts: $\psi_A = 1$; $\psi_B = 10$

$k = 1.9$
$k\ell_u = 1.9(16)(12) = 364.8$ in.
$k\ell_u/r = 364.8/7.2 = 51$
$P_u/P_o = 194.4/2357 = 0.08$

From Fig. 3.5.1(b): $\lambda = 13$

$(EI)_s = \dfrac{4300(27{,}648)/13}{1.0}$

$\qquad = 9.15 \times 10^6$ kip-in.²

$(P_c)_s = \dfrac{\pi^2(9.15 \times 10^6)}{(364.8)^2} = 678$ kips

$\Sigma P_u = 8(24.3) + 194.4 = 389$ kips
$\Sigma(P_c)_s = 8(123) + 678 = 1662$ kips
$0.1 f'_c A_g = 0.1(5)[8(401) + 576] = 1892$ kips
$\phi = 0.9 - 0.2(389/1892) = 0.86$

$\delta_s = \dfrac{1.0}{1 - \dfrac{389}{0.86(1662)}} = 1.37$

Wall panel $M_c = \delta_b M_{2b} + \delta_s M_{2s}$
$\qquad\qquad = 1.0(7.1) + 1.37(21.7)$
$\qquad\qquad = 36.8$ ft-kips

Column $M_{2b} = 0$
$M_{2s} = 0.75(1.7)(3.0) = 3.82$ ft-kips
$M_c = 1.37(3.82) = 5.23$ ft-kips

Use interaction diagrams in Chapter 2 to complete solution.

3.6 Diaphragm Design

Horizontal loads from wind or earthquake are usually transmitted to shear walls or moment-resisting frames through the roof and floors acting as horizontal diaphragms.

3.6.1 Method of Analysis

The diaphragm is analyzed by considering the roof or floor as a deep horizontal beam, analogous to a plate girder or I-beam. The shear walls or structural frames are the supports for this beam. Thus, the lateral loads are transmitted to these supports as reactions. As in a beam, tension and compression are induced in the chords or "flanges" of the analogous beam as shown in Fig. 3.6.1.

When precast concrete members which span parallel to the supporting shear walls or frames are used for the diaphragm, the shear in the analogous beam must be transferred between adjacent members and also to the supporting elements. The "web" shear must also be transferred to the chord elements. Thus, the design of a diaphragm is essentially a connection design problem.

3.6.2 Shear Transfer between Members

In floors or roofs without composite topping, the shear transfer between members is usually accomplished by weld plates or grout keys, depending on the member.

Weld plates may be analyzed as illustrated in Fig. 3.6.2. In addition to the hardware details shown, many others are used by precast concrete manufacturers.

For members connected by grout keys, a conservative value of 80 psi can be used for the design strength of the grouted key. If necessary, reinforcement placed as shown in Fig. 3.6.3 can be used to transfer the shear. This steel is designed by the shear-friction principles discussed in Chapter 6.

In floors or roofs with composite topping, the topping itself can act as the diaphragm, if it is adequately reinforced. Reinforcement requirements can be determined by shear-friction analysis.

Connections between members often serve functions in addition to the transfer of shear for lateral loads. For example, weld plates in flanged members are often used to adjust differential camber. Grout keys may be called upon to distribute concentrated loads.

Connections which transfer shear from the diaphragm to the shear walls or moment-resisting frames are analyzed in the same manner as the connection between members.

3.6.3 Chord Forces

Chord forces are calculated as shown in Fig. 3.6.1. For roofs with intermediate supports as shown, the shear stress is carried across the beam with weld plates or bars in grout keys as shown in Section A-A. Bars are designed by shear-friction. Stresses are usually quite low, and only as many bars or weld plates as required should be used.

Fig. 3.6.1 Analogous beam design of a diaphragm

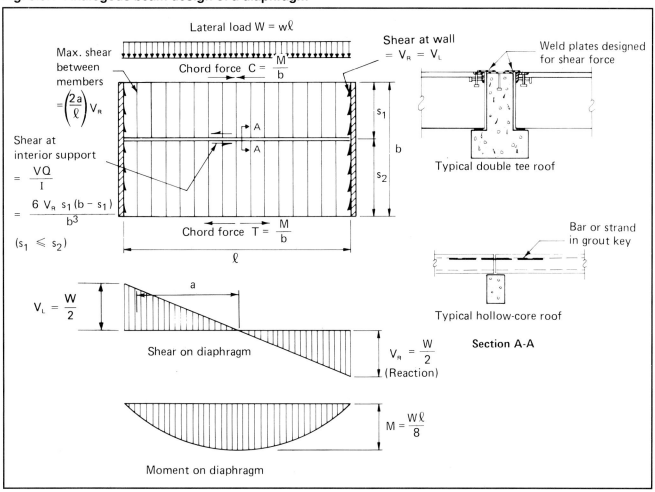

Fig. 3.6.2 Typical flange weld plate details

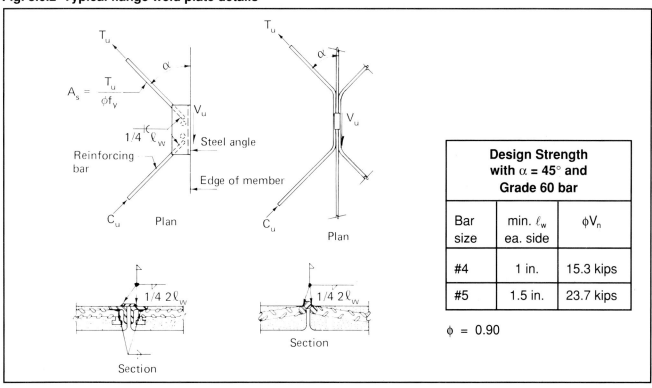

Design Strength with $\alpha = 45°$ and Grade 60 bar		
Bar size	min. ℓ_w ea. side	ϕV_n
#4	1 in.	15.3 kips
#5	1.5 in.	23.7 kips

$\phi = 0.90$

PCI Design Handbook/Fourth Edition

Fig. 3.6.3 Use of perimeter reinforcement as shear-friction steel

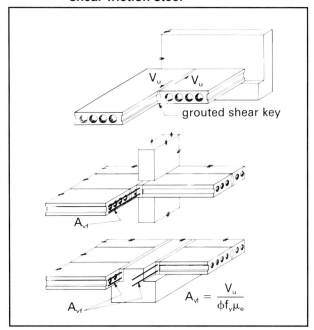

In flanged deck members the chord tension at the perimeter of the building is usually transferred between members by the same type of connection used for shear transfer (Fig. 3.6.2).

In all buildings, a minimum amount of perimeter reinforcement is required to satisfy structural integrity (see Sect. 3.10)[7]. These minimum requirements may be more than enough to resist the chord tension.

3.7 Shear Wall Buildings

3.7.1 General

In most precast, prestressed concrete buildings, it is desirable to resist lateral loads with shear walls of precast or cast-in-place concrete, or unit masonry. Shear walls are usually the exterior wall system, interior walls, or walls of elevator, stairway, and mechanical shafts or cores. The transfer of load from horizontal diaphragm to shear walls, or walls of elevator and stairway cores or mechanical shafts, can be achieved either through connections or by direct bearing.

Shear walls act as vertical cantilever beams which transfer lateral forces from the superstructure to the foundation. Most structures contain a number of walls which resist lateral load in orthogonal directions. The portion of the total lateral force which each wall resists depends on the bending and shear resistance of the wall, the participation of the floor, and the characteristics of the foundation. It is common practice to assume that floors act as rigid elements for loads in the plane of the floor, and that the deformations of the footings and soil can be neglected. Thus, for most structures, lateral load distribution is based only on the properties of the walls.

If the floor is considered to be a rigid body, it will translate in a direction parallel to the applied load an amount related to the flexural and shear rigidity of the participating shear walls (Fig. 3.7.1). If the center of rigidity is not coincident with the line of action of the applied loads, the floor will tend to rotate about the center of rigidity, introducing additional forces (Fig. 3.7.1b). Therefore, the load on each shear wall will be determined by combining the effects produced by rigid body translation and rotation.

A shear wall need not consist of a single element. It can be composed of independent units such as double tee, hollow-core, or architectural precast wall panels. If such units have adequate shear ties between them, they can be designed to act as a single unit, greatly increasing their shear resistance. Connecting such units can, however, result in a buildup of volume change forces, so it is usually desirable to connect only as many units as necessary to resist the overturning moment. Connecting the required units near mid-length of the wall will minimize the volume change restraint forces.

Connection of rectangular wall units to form "T" or "L" shaped walls increases their flexural rigidity, but has little effect on shear rigidity. The effective flange width that can be assumed for such walls is illustrated in Fig. 3.7.2. The designer should evaluate whether such connections are worth the extra cost. In some structures, such as Example 3.7.3, it may be desirable to provide shear connections between non-load bearing and load bearing shear walls in order to increase the dead load resistance to moments caused by lateral loads.

3.7.2 Rigidity of Solid Shear Walls

In order to determine the distribution of lateral loads, the relative rigidity of all shear walls must be established. Rigidity is defined as:

$$r = 1/\Delta \quad \text{(Eq. 3.7.1)}$$

where:

Δ = sum of flexural and shear deflections.

For a structure with rectangular shear walls of the same material, flexural deflections can be neglected when the wall height-to-length ratio is less than about 0.3. The rigidity of the element is then directly proportional to its cross-sectional area. When the wall height-to-length ratio is greater than about 3.0, shear deflections can be neglected, and the rigidity is proportional to the moment of inertia (plan dimensions). When the height-to-length ratio is between 0.3 and 3.0, an equivalent moment of inertia, I_{eq}, can be de-

Fig. 3.7.1 Translation and rotation of rigid floors

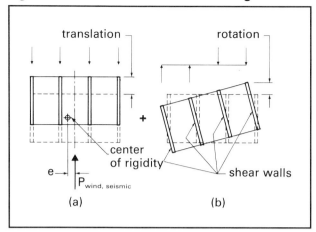

Fig. 3.7.2 Effective width of walls perpendicular to shear walls[7]

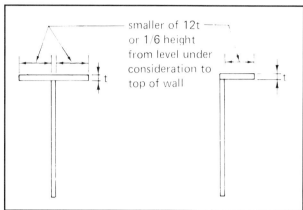

rived for simplifying the calculation of wall rigidity. I_{eq} is an approximation of the moment of inertia that would result in a flexural deflection equal to the combined flexural and shear deflections of the wall. Table 3.7.1 compares the deflections and I_{eq} for several load and restraint conditions.

Connecting or coupling of shear walls and large openings in walls also affect stiffness, as discussed in Sects. 3.7.5 and 3.7.6.

3.7.3 Distribution of Lateral Loads

Lateral loads are distributed to each shear wall in proportion to its rigidity. It is usually considered sufficient to design for lateral loads in only two orthogonal directions.

When the shear walls are symmetrical with respect to the center of load application, the force resisted by any shear wall is:

$$F_i = (r_i/\Sigma r)W \qquad \text{(Eq. 3.7.2)}$$

where:

F_i = the force resisted by an individual shear wall, i

r_i = the rigidity of wall i

Σr = sum of rigidities of all shear walls

W = total lateral load

3.7.4 Unsymmetrical Shear Walls

Structures which have shear walls placed unsymmetrically with respect to the center of the lateral load should take the torsional effect into account. Typical examples are shown in Fig. 3.7.3. For wind loading on most structures, a simplified method of determining the torsional resistance may be used in lieu of more exact design. The method is similar to the design of bolt groups in steel connections, and is illustrated in the following example.

Table 3.7.1 Shear wall deflections

Case	Deflection due to		Equivalent Moment of Inertia, I_{eq}	
	Flexure	Shear	Single story	Multi-story
	$\dfrac{Ph^3}{3EI}$	$\dfrac{2.78Ph}{A_wE}$	$\dfrac{I}{1 + \dfrac{8.34I}{A_wh^2}}$	$\dfrac{I}{1 + \dfrac{13.4I}{A_wh^2}}$
	$\dfrac{Wh^3}{8EI}$	$\dfrac{1.39Wh}{A_wE}$	—	$\dfrac{I}{1 + \dfrac{23.6I}{A_wh^2}}$
	$\dfrac{Ph^3}{12EI}$	$\dfrac{2.78Ph}{A_wE}$	$\dfrac{I}{1 + \dfrac{33.4I}{A_wh^2}}$	—

Example 3.7.1 Design of unsymmetrical shear walls

Given:

The structure of Fig. 3.7.3c. All walls are 8 ft high and 8 in. thick.

Problem:

Determine the shear in each wall, assuming the floors and roof are rigid diaphragms. Walls D and E are not connected to Wall B.

Solution:

Maximum height-to-length ratio of north-south walls = 8/30 < 0.3. Thus, for distribution of the direct wind shear, neglect flexural stiffness. Since walls are the same thickness and material, distribute in proportion to length.

Total lateral load, W = 0.20 × 200 = 40 kips

Determine center of rigidity:

$$\bar{x} = \frac{40(75) + 30(140) + 40(180)}{40 + 30 + 40}$$

= 130.9 ft from left

$\bar{y}$ = center of building, since walls D and E are placed symmetrically about the center of the building in the north-south direction.

Torsional moment, M_T = 40(130.9 − 200/2)
= 1236 ft-kips

Fig. 3.7.3 Unsymmetrical shear walls

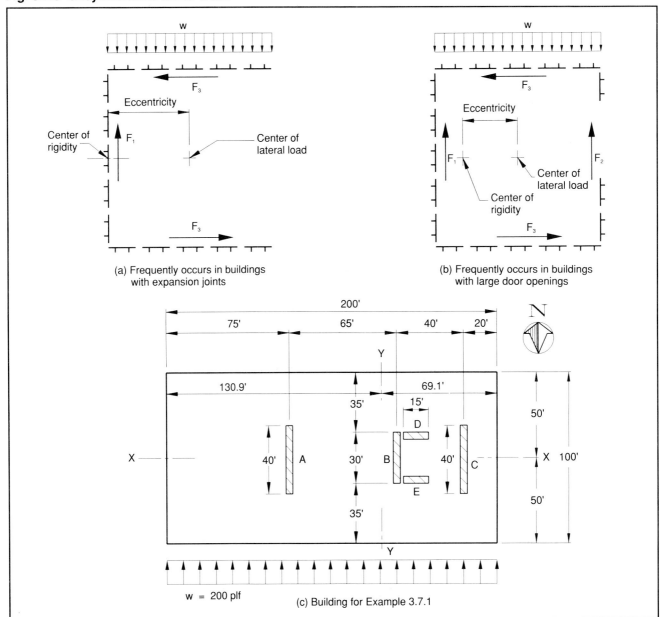

(a) Frequently occurs in buildings with expansion joints

(b) Frequently occurs in buildings with large door openings

(c) Building for Example 3.7.1

Determine the polar moment of inertia of the shear wall group about the center of rigidity:

$I_p = I_{xx} + I_{yy}$

$I_{xx} = \Sigma \ell y^2$ of east-west walls
$= 2(15)(15)^2 = 6750 \text{ ft}^3$

$I_{yy} = \Sigma \ell x^2$ of north-south walls
$= 40(130.9 - 75)^2 + 30(140 - 130.9)^2$
$\quad + 40(180 - 130.9)^2$
$= 223,909 \text{ ft}^3$

$I_p = 6750 + 223,909 = 230,659 \text{ ft}^3$

Shear in north-south walls $= \dfrac{W\ell}{\Sigma \ell} + \dfrac{M_T x \ell}{I_p}$

Wall A $= \dfrac{40(40)}{110} + \dfrac{1236(130.9 - 75)(40)}{230,659}$
$= 14.5 + 12.0 = 26.5 \text{ kips}$

Wall B $= \dfrac{40(30)}{110} + \dfrac{1236(-9.1)(30)}{230,659}$
$= 10.9 - 1.5 = 9.4 \text{ kips}$

Wall C $= \dfrac{40(40)}{110} + \dfrac{1236(-49.1)(40)}{230,659}$
$= 14.5 - 10.5 = 4.0 \text{ kips}$

Shear in east-west walls $= \dfrac{M_T y \ell}{I_p}$

$= \dfrac{1236(15)(15)}{230,659} = 1.2 \text{ kips}$

3.7.5 Coupled Shear Walls

Fig. 3.7.4 shows two examples of coupled shear walls. The effect of coupling is to increase the stiffness by transfer of shear through the coupling beam. The wall curvatures are altered from that of a cantilever because of the frame action developed. Fig. 3.7.5 shows how the deflected shapes differ in response to lateral loads.

Several approaches may be used to analyze the response of coupled shear walls. A simple approach is to ignore the coupling effect by considering the walls as independent cantilevers. This method results in a conservative wall design. However, if the coupling beam is rigidly connected, significant shears and moments will occur in the beam that may cause unsightly and possibly dangerous cracking. To avoid the problem, the beam-to-panel connection can be detailed for little or no rigidity, or the beam can be designed to resist the actual shears and moments.

Finite element analysis may be used to determine the distribution of shears and moments within a coupled shear wall. The accuracy (and cost) of such an analysis is a function of the element size used. This method is usually reserved for very complex structures.

A "plane frame" computer analysis will be sufficiently accurate for the great majority of structures. In modeling the coupled shear wall as a frame, the member dimensions must be considered, as a centerline analysis may yield inaccurate results. A suggested model is shown in Fig. 3.7.6(a).

Either a finite element or frame analysis may be used to determine the deflection of a coupled shear wall, and hence its equivalent moment of inertia. This may then be used to determine distribution of shears in a building which contains both solid and coupled shear walls. Some frame analysis programs do not calculate shear deformations; if significant, shear deformations may have to be calculated separately.

3.7.6 Shear Walls with Large Openings

Window panels and other wall panels with large openings may also be analyzed with a plane frame computer program. Fig. 3.7.6(b) shows suggested models. Where length-to-depth ratios for vertical and horizontal segments are similar, a frame model based on segment centerlines will be reasonably ac-

Fig. 3.7.4 Coupled shear walls

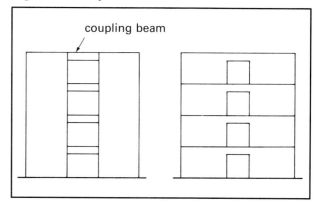

Fig. 3.7.5 Response to lateral loads

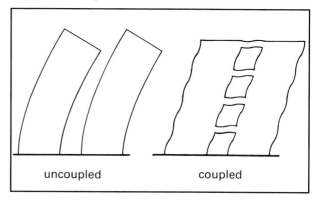

curate. Otherwise, an analysis similar to that described for coupled shear walls may be used.

As with coupled shear walls, the deflections yielded by the computer analysis may be used to determine equivalent stiffness for determining lateral load distribution. Shear deflections, if significant, may have to be hand calculated and added to the flexural stiffnesses from the frame analysis.

In very tall structures, vertical shear and axial deformations influence the rigidity of panels with large openings, so a more rigorous analysis may be required.

Example 3.7.2 One-story building

Given:

The wind load analysis and design of a typical one-story industrial building are illustrated by the structure shown in Fig. 3.7.7. 8-ft wide double tees are used for both the roof and walls. The local building code specifies that a wind load of 25 psf be used for buildings of this height.

Solution:

1. Calculate forces, reactions, shears and moments:

   Total wind force to roof:

   W = [25 (160)(18/2 + 2.5)]/1000 = 46 kips

   $V_L = V_R$ = 23 kips

   Diaphragm moment $= \dfrac{W\ell}{8} = \dfrac{46(160)}{8}$

   = 920 ft-kips

2. Check sliding resistance of the shear wall:

   Determine dead load on the footing:

   8DT12 wall = 37(23.5)(120) = 104,340 lb

   12" x 18" footing
   = 1(1.5)(150)(120)   = 27,000 lb

   Assume 2 ft backfill
   = (100)(1.5)(120)(2)   = 36,000 lb
   Total   = 167,340 lb

   Assume coefficient of friction against granular soil, μ_s = 0.5

   Sliding resistance = μ_sN = 0.5(167.34)
                                  = 83.67 kips

   Factor of safety   = 83.67/23 = 3.64 OK

   (Note: A factor of safety of 1.5 is specified by some building codes.)

3. Check overturning resistance:

   Applied overturning moment = 23(4 + 18)
                              = 506.0 ft-kips

   Resistance to overturning:

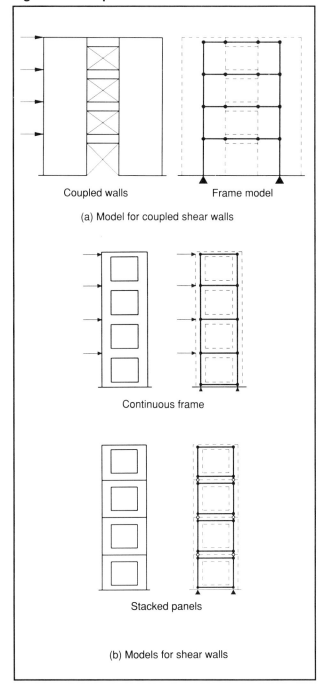

Fig. 3.7.6 Computer models

Coupled walls   Frame model

(a) Model for coupled shear walls

Continuous frame

Stacked panels

(b) Models for shear walls

Assume axis of rotation at leeward edge of the building.

(Note: Some engineers prefer to use the more conservative assumption of an axis at b/5, b/4 or b/3 from the leeward edge, depending on the foundation conditions.)

Resisting moment = 167.34 (120/2)
                 = 10,040 ft-kips

Factor of safety = 10,040/506
                 = 19.8 > 1.5 OK

Fig. 3.7.7 Example 3.7.2

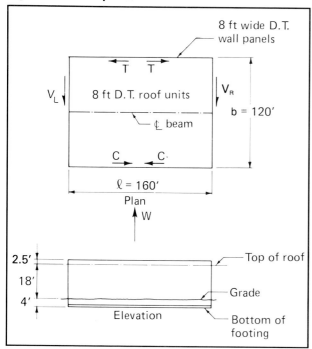

4. Analyze connections:
   a. Shear ties in double tee roof joint:
      (Maximum load at next to last joint)
      Applied shear = [(80 − 8)/80](23)
      = 20.7 kips
      Load factor by ACI 318-89 = 1.3
      Connection load factor (see Sect. 6.3)
      choose 1.2
      V_u = 20.7(1.3)(1.2) = 32.3 kips
      v_u = 32.3/120 = 0.269 kips/ft
      Use #4 ties as shown in Fig. 3.6.2
      ϕV_n = 15.3 kips
      Required spacing = 15.3/0.269 = 56.9 ft

      (Note: Most engineers and precasters prefer a maximum connection spacing of about 8 to 15 ft.)

   b. Shear ties at the shear walls:
      V_u = 23(1.3)(1.2) = 35.88 kips
      v_u = 35.88/120 = 0.299 kips/ft

A connection as shown in Fig. 3.6.2 is designed similar to the shear tie between tees. This would require a spacing of 15.3/0.299 = 51.2 ft. In order to distribute the load to the wall panels, at least one connection per panel is required. From Fig. 3.7.8 it is apparent that these connections should occur at the tee stems. Thus a spacing of 4 ft or 8 ft would be used in this case.

Other types of connections using headed studs are commonly used for this application. Design of studs is shown in Chapter 6.

In some cases, the designer may find it necessary to provide a connection that permits vertical movement of the roof member. This is illustrated in Fig. 3.7.8(b).

c. Chord force (see Fig. 3.7.9)
   T = C = M/b = 920/120
   = 7.67 kips
   T_u = 1.3(7.67) = 9.97 kips

This force can be transmitted between members by ties at the roof tees, wall panels or a combination, as illustrated in Fig. 3.7.9. The force through the member flanges can be transmitted by the flange mesh. These ties and transmission of forces will typically provide the tie requirements given in Sect. 3.10.

d. Wall panel connections:
   This shear wall may be designed to act as a series of independent units, without ties between the panels. The shear force is assumed to be distributed equally among the wall panels (see Fig. 3.7.10).
   n = 120/8 = 15 panels
   V = V_R/n = 23/15 = 1.53 kips
   D = 37(8)(23.5) = 6956 lb = 6.96 kips
   Design base connection for 1.3W − 0.9D
   T_u = [1.3(1.53)(21) − 0.9(6.96)(2)]/4
   = 7.31 kips tension

As an alternative, the shear walls may be designed with two or more panels connected together. Two analysis options are available. Example 3.11.7 illustrates an analysis where tension and compression compensate one another with simple shear connections across the vertical joints. In that system, tie-down connections may be required at the end panels of a string of interconnected panels. The second method is to connect a number of panels with rigid connections so as to have the connected panels act as a monolithic unit. By engaging more weight, tie-down forces can be reduced or eliminated. Volume change restraint must be considered when determining the number of panels to be interconnected.

Analysis of a rigidly connected panel group is dependent upon a number of factors. Connection stiffness determines whether the group will act monolithically or act as analyzed in Example 3.11.7. The aspect ratio of the rigidly connected group will affect the stress distribution at the base. While a panel group with a high height-to-length ratio may act nearly as a cantilever beam, a group with a medium to low ratio of height to length will act as a deep beam and ex-

Fig. 3.7.8 Connection of roof tee to wall

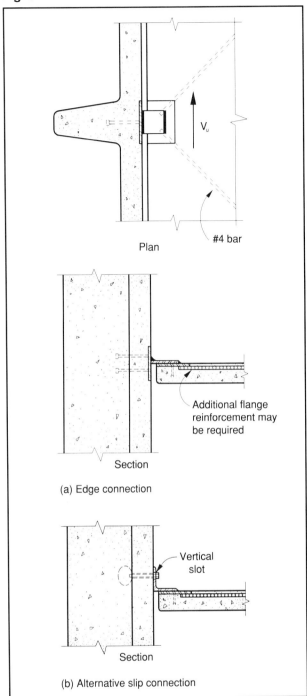

(a) Edge connection

(b) Alternative slip connection

Fig. 3.7.9 Chord forces

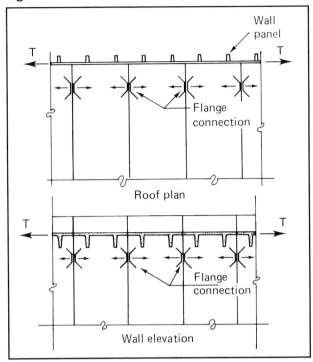

Fig. 3.7.10 Panels acting as individual units in a shear wall

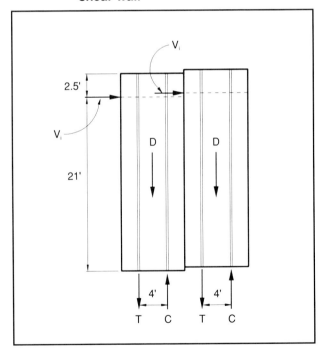

hibit non-linear stress distributions. Determination of tie-down forces, vertical joint forces, and base shear distribution will be a function of the aspect ratio. The shear and flexural stiffness of the individual panels will have a similar effect on force distribution.

A simplified analysis is presented for this problem. The analysis is fairly accurate for the aspect ratio used, and assumes highly rigid panel-to-panel connections. However, extrapolation to other aspect ratios is not recommended.

Factored shear = (1.3)(1.53) = 2 kips per panel
Factored weight = (0.9)(6.96) = 6.26 kips per panel

From Fig. 3.7.11(a), try connecting two panel units:

For flexural integrity:
$M_u = (2)(2)(21) = 84$ ft.-kips
$\phi M_n = 0.9[2(6.26)(6) + T_u(12)] = 84$ ft.-kips
$T_u = 1.52$ kips; a nominal connection will satisfy.

For equlibrium:
$T_u = [(84 - 2(6.26)(6)]/12 = 0.74$ kips

From Fig. 3.7.11(b), examine joint forces using individual free body diagrams. On the tension side, assume the tie-down connection is concentrated at one stem:
ΣF_y: $V_{u1} = 6.26 + 0.74 = 7$ kips
$\Sigma M = 2(21) - 6.26(4) - 0.74(6) + 17.5 C_u = 0$
$C_u = 0.72$ kips
ΣF_x: $V_{u2} = 2 - 0.72 = 1.28$ kips

On the compression side, from Fig. 3.7.11(c):
$\Sigma F_y = 7 + 6.26 - C_u = 0$; thus $C_u = 13.26$ kips
$\Sigma F_x = 2 + 0.72 - V_{u2} = 0$; thus $V_{u2} = 2.72$ kips
$\Sigma M = 2(21) + 0.72(17.5) - 7(6) - 6.26(2) = 0$ OK

Fig. 3.7.11 Rigidly connected panel group

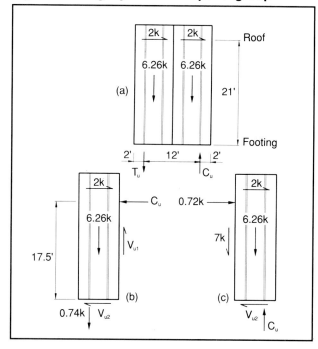

Example 3.7.3 Four-story building

Given:

The wind load analysis and design of a typical four-story residential building are illustrated by the structure shown in Fig. 3.7.12. 8-in. deep hollow-core units are used for the floors and roof, and 8-in. thick precast concrete walls are used for all walls shown. Unfactored loads are given as follows:

Gravity loads:	L.L.	D.L.
Roof	30	
Roofing, mechanical, etc.		10
Hollow-core slabs		64
	30 psf	74 psf
Typical floor		
Living areas	40	
Corridors	100	
Partitions		10
Hollow-core slabs		64
		74 psf
Walls		100 psf
Stairs	100	130 psf

Wind loads:
0 to 30 ft above grade = 25 psf
30 to 34 ft 8 in. above grade = 30 psf

Solution:

1. For wind in the transverse (east-west) direction, normal practice for this structure would be to conservatively neglect the resistance provided by the stair, elevator and longitudinal walls. Thus, two 27-ft long interior bearing walls can be assumed to resist the wind on one 26-ft bay. The wind and gravity loads on the wall are shown in Fig. 3.7.13.

Concentrated loads from the corridor lintels can be assumed to be distributed as shown in Fig. 3.7.13. In this example, these loads are conservatively neglected to simplify the calculations.

Check overturning of shear wall:
D.L. resisting moment about toe of wall
= $27(27/2)[1.92 + 3(2.72) + 0.8]$
= 3966 ft-kips
Factor of safety = 3966/205.1
= 19.3 > 1.5 OK

Check for tension using factored loads:
Dead weight on wall
$P = [1.92 + 3(2.72) + 0.8](27)$
= 293.8 kips

Maximum moment at foundation
$f_{ut} = \dfrac{1.3M}{(\ell^2/6)} - \dfrac{0.9(293.8)}{27}$

$= \dfrac{1.3(205.1)}{(27^2/6)} - \dfrac{0.9(293.8)}{27}$

= −7.60 klf (compression)
= 205.1 ft-kips

No tension connections are required between panels and the foundation. Thus, the building is stable under wind loads in east-west direction.

Fig. 3.7.12 Example 3.7.3—Four-story building design

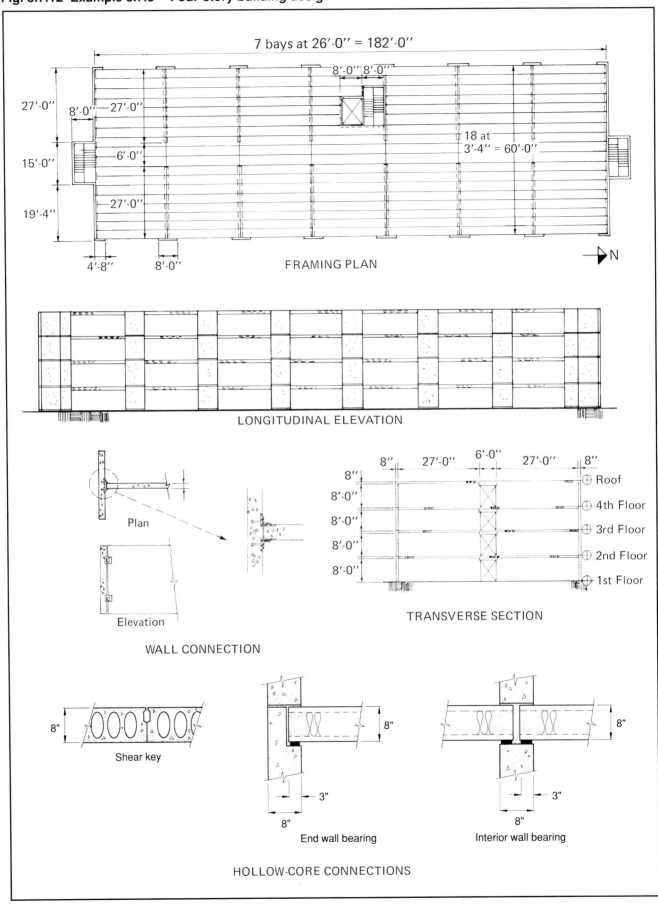

Fig. 3.7.13 Loads to transverse walls—four-story design example

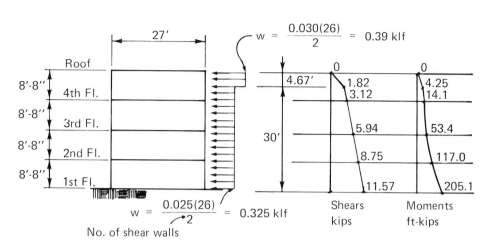

EAST-WEST LATERAL WIND LOADING

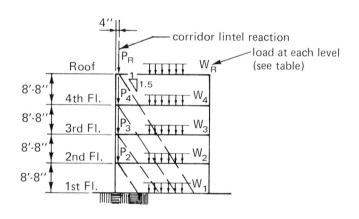

GRAVITY LOADS ON BEARING WALL

Summary of gravity loads

Load Mark	Tributary Area	Unit loads, psf		Wall weight, klf	Total unfactored loads		
		L.L.	D.L.		L.L.	D.L.	T.L.
P_R	78 sq. ft.	30	74	—	2.3 kips	5.8 kips	8.1 kips
P_4	78 sq. ft.	100	64	—	7.8 kips	5.0 kips	12.8 kips
P_3	78 sq. ft.	100	64	—	7.8 kips	5.0 kips	12.8 kips
P_2	78 sq. ft.	100	64	—	7.8 kips	5.0 kips	12.8 kips
W_R	26 lin. ft.	30	74	—	0.78 klf	1.92 klf	2.70 klf
W_4	26 lin. ft.	16*	74	0.8	0.42 klf	2.72 klf	3.14 klf
W_3	26 lin. ft.	16*	74	0.8	0.42 klf	2.72 klf	3.14 klf
W_2	26 lin. ft.	16*	74	0.8	0.42 klf	2.72 klf	3.14 klf
W_1	N/A	—	—	0.8	0	0.80 klf	0.80 klf

*Includes live load reduction allowed by codes

(Note: Structural integrity considerations may dictate the use of minimum vertical ties. See Sect. 3.10.)

2. For wind in the longitudinal (north-south) direction, the shear walls will be connected to the load bearing walls. The assumed resisting elements are shown in Fig. 3.7.14; a summary of the properties is shown in Table 3.7.2. Sample calculations of these properties are given below for element A.

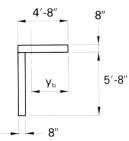

Effective width of perpendicular wall (see Fig. 3.7.2) is the smaller of:

12t = 12(8) = 96 in., or
1/6(34.67 x 12) = 69.3 in. Use 5 ft 8 in.

$$\text{Area of web} = 4.67 \times 0.67 = 3.11 \text{ ft}^2$$
$$\text{Area of flange} = 5.67 \times 0.67 = \underline{3.78}$$
$$6.89 \text{ ft}^2$$

$$y_b = \frac{3.11(4.67/2) + 3.78(4.67 - 0.33)}{6.89}$$

$$= 3.43 \text{ ft}$$

$$y_t = 4.67 - 3.43 = 1.24 \text{ ft}$$

$$I = \frac{0.67(4.67)^3}{12} + 3.11(3.43 - 2.33)^2 + 3.78(1.24 - 0.33)^2$$

$$= 12.58 \text{ ft}^4$$

Equivalent stiffness is calculated using the Case 1 multi-story formula from Table 3.7.1.

$$I_{eq} = \frac{I}{1 + \frac{13.4 I}{A_w h_s^2}}$$

$$= \frac{12.58}{1 + \frac{13.4(12.58)}{3.11(8.67)^2}} = 7.31 \text{ ft}^4$$

I_{eq} is essentially a relative stiffness:

$$K_r = 1/\Delta; \quad \Delta = \frac{Ph^3}{3EI_{eq}}$$

$$K_r = \frac{3EI_{eq}}{Ph^3}$$

Since 3, E, P, and h are all constants when comparing stiffnesses, $K_r = I_{eq}$.

Distribution of load to element A based on its relative stiffness is (see Table 3.7.2):

$$\frac{I_{eq}}{\Sigma n I_{eq}} = \frac{7.31(100)}{368.54} = 1.98\%$$

The shears and moments in the north-south direction are shown in Fig. 3.7.15, and the distributions are shown in Table 3.7.3.

To check overturning, consider element B at the first floor:

From Fig. 3.7.13 the dead load on the 6'-4" portion of element B:

= 1.92 + 3(2.72) + 0.8 = 10.88 kips/ft

The dead load on the 8'-0" portion of element B is the weight of the wall:

= 34.67 x 0.1 = 3.47 kips/ft

The resisting moment is then:

M_R = 10.88(5.67)(4) + 3.47(8)(4)

= 358 ft-kips x 11 elements

= 3938 ft-kips

Factor of safety = 3938/966.9 = 4.1 > 1.5 OK

(Note: This conservatively neglects the contribution of the other elements.)

To check for tension, also consider element B:

Total dead weight on the wall

= 10.88(5.67) + 3.47(8) = 89.45 kips

Total wall area

= (8.0 + 5.67)0.67 = 9.16 ft²

M = 38.5 ft-kips (see Table 3.7.3)

$$f_{ut} = \frac{1.3M(d/2)}{I} - \frac{0.9P}{A}$$

$$= \frac{1.3(38.5)(4.0)}{28.7} - \frac{0.9(89.45)}{9.16}$$

$$= -1.81 \text{ ksf (compression)}$$

No net uplift exists between panels and the foundation. The building is stable under wind loads in the north-south direction.

The connections required to assure that the elements will act in a composite manner as assumed can be designed by considering element A. The unit stress at the interface is determined using the classic equation for horizontal shear:

$$v_h = \frac{VQ}{I}$$

Q = 5.67(0.67)(1.24 − 0.33) = 3.46 ft³

$$v_h = \frac{1.08(3.46)}{12.58} = 0.297 \text{ kips/ft}$$

Total shear = 0.297(8.0) = 2.37 kips

Fig. 3.7.14 Wind resisting elements for north-south wind

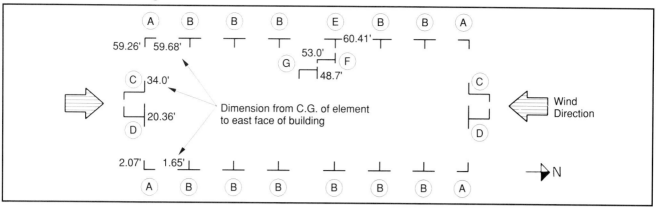

Table 3.7.2 Properties of resisting elements for wind in longitudinal direction

Element		A_w	I	y_b	I_{eq}	No. of elements	nI_{eq}	$\dfrac{I_{eq}}{\Sigma nI_{eq}}(100)$	$\Sigma \bar{y}$	$I_{eq}(\Sigma \bar{y})$
⌐ 4'-8" / 6'-4"	Ⓐ	3.11	12.6	3.43	7.31	4	29.24	1.98	123	899
⊥ 8'-0" / 6'-4"	Ⓑ	5.36	28.7	4.0	14.68	11	161.5	3.98	308	4521
8'-8" / 6'-4" / 6'-4"	Ⓒ	5.81	158.1	4.34	27.02	2	54.04	7.33	68	1837
6'-4" / 12'-0" / 8'-8"	Ⓓ	5.81	205.6	3.45	28.13	2	56.26	7.63	41	1153
8'-0" / 4'-0"	Ⓔ	5.36	29.0	4.0	14.76	1	14.76	4.01	60	886
8'-8" / 2'-0" / 9'-8"	Ⓕ	5.81	114.1	2.72	25.35	1	25.35	6.88	53	1344
8'-8" / 6'-4" / 7'-8"	Ⓖ	5.81	171.6	4.09	27.39	1	27.39	7.43	49	1342

$\Sigma nI_{eq} = 368.54$ $\Sigma = 11{,}982$
Center of rigidity = 11,982/368.54 = 32.51 ft from east
Note: The north-south wind load is slightly eccentric by 32.51 − 61.33/2 = 1.85 ft.
Torsion due to this eccentricity is neglected in calculating shears and moments in Table 3.7.3.

Table 3.7.3 Distribution of wind shears and moments (north-south direction)

Element	% Dist.	4th floor Shear 14.71 kips	4th floor Moment 66.7 ft-kips	3rd floor Shear 27.98 kips	3rd floor Moment 251.7 ft-kips	2nd floor Shear 41.24 kips	2nd floor Moment 551.8 ft-kips	1st floor Shear 54.51 kips	1st floor Moment 966.9 ft-kips
A	1.98	0.29	1.32	0.55	4.98	0.82	10.9	1.08	19.1
B	3.98	0.59	2.65	1.11	10.0	1.64	22.0	2.17	38.5
C	7.33	1.08	4.89	2.05	18.4	3.02	40.4	4.00	70.9
D	7.63	1.12	5.09	2.13	19.2	3.15	42.1	4.16	73.8
E	4.01	0.59	2.67	1.12	10.1	1.65	22.1	2.19	38.8
F	6.88	1.01	4.59	1.93	17.3	2.84	38.0	3.75	66.5
G	7.43	1.09	4.96	2.08	18.7	3.06	41.0	4.05	71.8

The relative stiffness and percent distribution for the elements in this table are assumed the same for all stories. The exact values may be slightly different for each story because the values change due to reduced flange width (see Fig. 3.7.2).

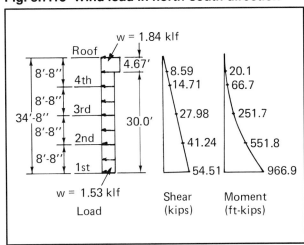

Fig. 3.7.15 Wind load in north-south direction

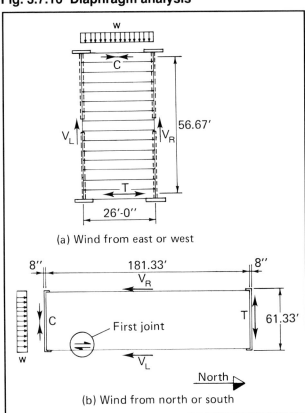

Fig. 3.7.16 Diaphragm analysis

Connections similar to those shown in Fig. 3.7.12 can be designed using the principles outlined in Chapter 6.

Design of floor diaphragm:

Analysis procedures for the floor diaphragm are described in Sect. 3.6. For this example refer to Fig. 3.7.16.

The factored wind load for a typical floor is:

$w_u = 1.3(25)(8.67) = 282$ plf

For wind from the east or west:

$V_{Ru} = \dfrac{0.282(26)}{2} = 3.67$ kips

$C_u = T_u = \dfrac{M_u}{\ell} = \dfrac{0.282(26)^2}{8(56.67)}$

$= 0.42$ kips

The reaction V_{Ru} is to be transferred to the shear wall by connections.

The chord tension, T_u, is resisted by the tensile strength of the floor slab. The grout key between slabs must also resist approximately the same force.

Assume area of exterior slab = 218 in.^{2}

Grout key = 3 in. deep

Concrete f'_c = 5000 psi

Use a resisting tensile strength of:

$3\lambda \sqrt{f'_c}$ = 212 psi

Grout key resisting strength (see Sect. 3.6.2)
= 80 psi

Resisting tensile strength of slab
= 218(0.212) = 46.2 kips > 0.42 OK

Resisting strength of grout key
= 26(12)(3)(0.080)
= 74.9 kips > 0.42 OK

For wind from the north or south:

$V_{Ru} = \dfrac{0.282(61.33)}{2}$ = 8.65 kips

Resisting force in the first joint
= 181.33(12)(3)(0.080) = 522 kips OK

$C_u = T_u = \dfrac{0.282(61.33)^2}{8(181.33)}$

= 0.73 kips

Resisting force at end wall bearing
= 8(3)(0.212) = 5.1 kips > 0.73 OK

The reaction V_{Ru} needs to be transferred to the shear wall by connections.

In this example, only the resistance to wind loading was analyzed. Any other required loading (and the requirements of structural integrity) must be reviewed for a complete analysis.

3.7.7 Architectural Panels as Shear Walls

In many structures it is economical to take advantage of the strength and rigidity of exterior panels, and design them to serve as the lateral load resisting system. The effectiveness of such a system is largely dependent on the panel-to-panel connections.

Fig. 3.7.17 illustrates the foundation reaction distributions of exterior architectural precast shear wall systems under the action of lateral load, with and without connections between the shear walls and the windward or leeward walls. The structure with corner connections is structurally more efficient for resisting lateral loads.

The lateral load resisting system shown in Fig. 3.7.17(b) is frequently labeled a "tube". However, because the components and the connections are not perfectly rigid, full tube behavior does not develop. Fig. 3.7.18 illustrates the difference. The peaking of the foundation reaction at the corner results from shear lag, which limits the effective width of the "flange". Accurate evaluation of shear lag is difficult, but analytical and research studies have indicated the following limitations on the effective flange are sufficiently accurate for most structures:

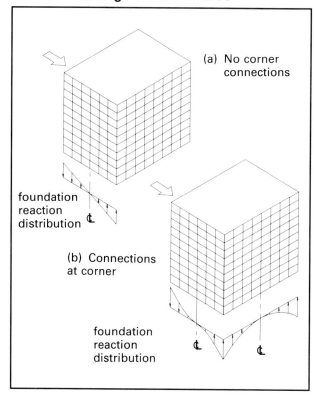

Fig. 3.7.17 Foundation reaction distributions resulting from lateral loads

(a) No corner connections

(b) Connections at corner

1. One-half the length of the shear wall
2. One-third the length of the windward or leeward wall
3. One-tenth the height of the building
4. Six times the thickness of the "flange" wall
5. Distance to nearest major opening
6. One-half the distance to the nearest shear wall.

3.8 Buildings With Moment-Resisting Frames

3.8.1 General

Precast, prestressed concrete beams and deck members are usually most economical when they can be designed and connected into a structure as simple-span members. This is because:

1. Positive moment-resisting capacity is much easier and less expensive to attain with pretensioned members than negative moment capacity at supports.
2. Connections which achieve continuity at the supports are usually complex and costly.
3. The restraint to volume changes that occurs in rigid connections may cause serious cracking and unsatisfactory performance or, in extreme cases, even structural failure.

Fig. 3.7.18 Influence of shear lag on tube behavior

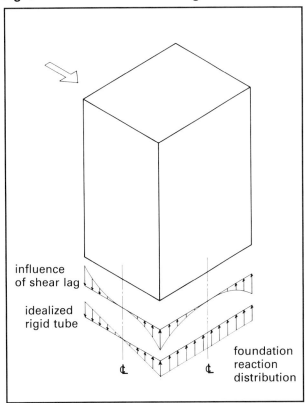

Therefore, it is most desirable to design precast, prestressed concrete structures with connections which allow lateral movement and rotation, and to design the structure to achieve lateral stability through the use of floor and roof diaphragms and shear walls.

However, in some structures, adequate shear walls interfere with the function of the building, or are more expensive than alternate solutions. In these cases, the lateral stability of the structure depends on the moment-resisting capacity of either the column bases, a beam-column frame, or both.

When moment connections between beams and columns are required to resist lateral loads, it is desirable to make the moment connection after most of the dead loads have been applied. This requires careful detailing, specification of the construction process, and inspection. If such details are possible, the moment connections need only resist the negative moments from live load, lateral loads and volume changes, and will then be less costly.

3.8.2 Moment Resistance of Column Bases

Single-story and some low-rise buildings without shear walls may depend on the fixity of the column base to resist lateral loads. The ability of a spread footing to resist moments caused by lateral loads is dependent on the rotational characteristics of the base. The total rotation of the column base is a function of rotation between the footing and soil, bending in the base plate, and elongation of the anchor bolts, as shown in Fig. 3.8.1.

The total rotation of the base is:

$$\phi_b = \phi_f + \phi_{bp} + \phi_{ab} \quad \text{(Eq. 3.8.1)}$$

If the axial load is large enough so that there is no tension in the anchor bolts, ϕ_{bp} and ϕ_{ab} are zero, and:

$$\phi_b = \phi_f \quad \text{(Eq. 3.8.2)}$$

Rotational characteristics can be expressed in terms of flexibility or stiffness coefficients:

$$\phi = \gamma M = M/K \quad \text{(Eq. 3.8.3)}$$

where:

M = applied moment = Pe

e = eccentricity of the applied load, P

γ = flexibility coefficient

K = stiffness coefficient = $1/\gamma$

If bending of the base plate and strain in the anchor bolts are assumed as shown in Fig. 3.8.1, the flexibility coefficients for the base can be derived, and the total rotation of the base becomes:

$$\phi_b = M(\gamma_f + \gamma_{bp} + \gamma_{ab})$$
$$= Pe(\gamma_f + \gamma_{bp} + \gamma_{ab}) \quad \text{(Eq. 3.8.4)}$$

$$\gamma_f = \frac{1}{k_s I_f} \quad \text{(Eq. 3.8.5)}$$

Fig. 3.8.1 Assumptions used in derivation of rotational coefficients for column bases

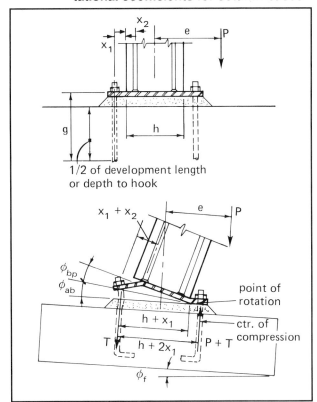

$$\gamma_{bp} = \frac{(x_1 + x_2)^3 [2e/(h + 2x_1) - 1]}{6eE_s I_{bp}(h + x_1)} \geq 0 \quad \text{(Eq. 3.8.6)}$$

$$\gamma_{ab} = \frac{g[2e/(h + 2x_1) - 1]}{2eA_b E_s (h + x_1)} \geq 0 \quad \text{(Eq. 3.8.7)}$$

where:

$\gamma_f, \gamma_{bp}, \gamma_{ab}$ = flexibility coefficients of the footing/soil interaction, the base plate and the anchor bolts, respectively

k_s = coefficient of subgrade reaction from Fig. 3.8.2

I_f = moment of inertia of the footing (plan dimensions)

E_s = modulus of elasticity of steel

I_{bp} = moment of inertia of the base plate (vertical cross-section dimensions)

A_b = total area of anchor bolts which are in tension

h = width of the column in the direction of bending

x_1 = distance from face of column to the center of the anchor bolts, positive when anchor bolts are outside the column, and negative when anchor bolts are inside the column

x_2 = distance from the face of the column to base plate anchorage

g = assumed length over which elongation of the anchor bolt takes place = ½ of development length + projection for anchor bolts made from reinforcing bars, or the length to the hook + projection for smooth anchor bolts (see Fig. 3.8.1)

Rotation of the base may cause an additional eccentricity of the loads on the columns, causing moments which must be added to the moments induced by the lateral loads.

Note that in Eqs. 3.8.6 and 3.8.7, if the eccentricity, e, is less than $h/2 + x_1$ (inside the center of compression), γ_{bp} and γ_{ab} are less than zero, meaning that there is no rotation between the column and the footing, and only the rotation from soil deformation (Eq. 3.8.5) need be considered.

Values of Eq. 3.8.5 through 3.8.7 are tabulated for typical cases in Tables 3.8.1 and 3.8.2.

Example 3.8.1 Stability analysis of an unbraced frame

Given:

The column shown in Fig. 3.8.3
Soil bearing capacity = 5000 psf
P = 80 kips dead load, 30 kips live load
W = 2 kips wind load

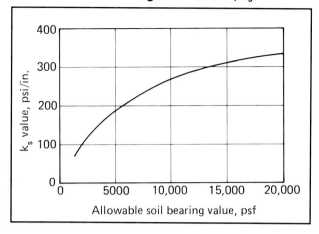

Fig. 3.8.2 Approximate relationship between allowable soil bearing value and coefficient of subgrade reaction, k_s

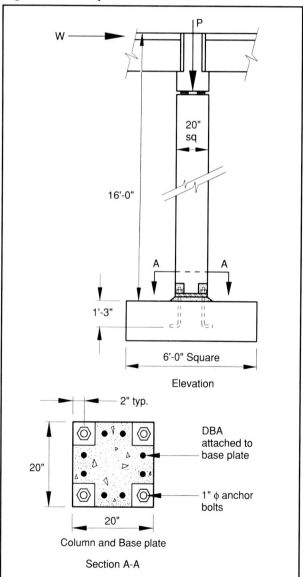

Fig. 3.8.3 Examples 3.8.1 and 3.8.2*

*These examples assume that the section properties of the column extend to the top of the roof.

Table 3.8.1 Flexibility coefficients for footing/soil interaction

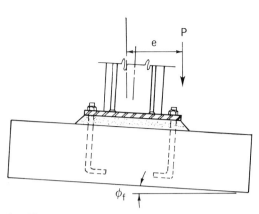

Flexibility of base $= \gamma_b = \gamma_f + \gamma_{ab} + \gamma_{bp}$
Rotation of base $= \phi_b = \gamma_b Pe$
Stiffness of base $= K_b = 1/\gamma_b$
Fixity of base $= K_b/(K_c + K_b)$

K_c = Column stiffness = $4E_c I_c/h_s$
E_c = Modulus of elasticity of column concrete, psi
I_c = Moment of inertia of column, in.^{4}
h_s = Story height, in.

γ_f, 1/in.-lb × 10^{-10} for square footings

Footing Size (ft)	k_s				
	100	150	200	250	300
2.0 X 2.0	3616.9	2411.3	1808.4	1446.8	1205.6
2.5 X 2.5	1481.5	987.7	740.7	592.6	493.8
3.0 X 3.0	714.4	476.3	357.2	285.8	238.1
3.5 X 3.5	385.6	257.1	192.8	154.3	128.5
4.0 X 4.0	226.1	150.7	113.0	90.4	75.4
4.5 X 4.5	141.1	94.1	70.6	56.5	47.0
5.0 X 5.0	92.6	61.7	46.3	37.0	30.9
5.5 X 5.5	63.2	42.2	31.6	25.3	21.1
6.0 X 6.0	44.7	29.8	22.3	17.9	14.9
6.5 X 6.5	32.4	21.6	16.2	13.0	10.8
7.0 X 7.0	24.1	16.1	12.1	9.6	8.0
7.5 X 7.5	18.3	12.2	9.1	7.3	6.1
8.0 X 8.0	14.1	9.4	7.1	5.7	4.7
9.0 X 9.0	8.8	5.9	4.4	3.5	2.9
10.0 X 10.0	5.8	3.9	2.9	2.3	1.9
11.0 X 11.0	4.0	2.6	2.0	1.6	1.3
12.0 X 12.0	2.8	1.9	1.4	1.1	0.9

Problem:

Determine the column design loads and moments for stability as an unbraced frame.

Solution:

ACI 318-89 requires that the column be designed for the following conditions:

1. 1.4D + 1.7L
2. 0.75(1.4D + 1.7L + 1.7W)
3. 0.9D + 1.3W

The maximum eccentricity would occur when 3 is applied. Moment at base of column

$= 2(16) = 32$ ft-kips $= 384$ in.-kips
$0.9D = 0.9(80) = 72$ kips
$1.3W = 1.3(384) = 499.2$ in.-kips

Eccentricity due to wind load

$$e = \frac{M_u}{P_u} = \frac{499.2}{72} = 6.93 \text{ in.}$$

To determine the moments caused by base rotation, an iterative procedure is required.

Estimate eccentricity due to rotation = 0.25 in.
$e = 6.93 + 0.25 = 7.18$ in.

Check rotation between column and footing:
$h/2 + x_1 = 20/2 + (-2) = 8$ in. > 7.18

thus, there is no tension in the anchor bolts and no rotation between the column and footing.

$I_f = (6 \times 12)^4/12 = 2.24 \times 10^6$ in.^{4}

From Fig. 3.8.2: $k_s \approx 200$ psi/in.

$\gamma_f = 1/k_s I_f = 1/[200(2.24 \times 10^6)]$
$= 22.3 \times 10^{-10}$

Table 3.8.2 Flexibility coefficients for anchor bolts and base plates

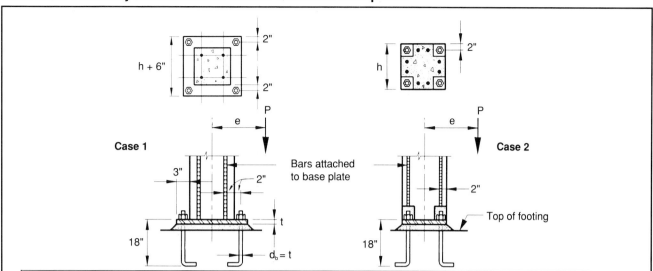

$\gamma_{ab} + \gamma_{bp}$, 1/in.-lb × 10⁻¹⁰ for typical details (see Eq. 3.8.6 and 3.8.7)

Column size, h (in)	e (in)	Case 1: Exterior anchor bolts				Case 2: Interior anchor bolts			
		Base plate thickness & anc. bolt diameter				Base plate thickness & anc. bolt diameter			
		.75	1.00	1.25	1.50	.75	1.00	1.25	1.50
12 × 12	4	.0	.0	.0	.0	.0	.0	.0	.0
	6	.0	.0	.0	.0	29.3	16.5	10.5	7.3
	8	.0	.0	.0	.0	43.9	24.7	15.8	11.0
	10	16.6	7.9	4.5	2.9	52.7	29.6	19.0	13.2
	12	27.7	13.2	7.5	4.8	58.5	32.9	21.1	14.6
	14	35.7	16.9	9.6	6.1	62.7	35.3	22.6	15.7
	16	41.6	19.8	11.2	7.2	65.8	37.0	23.7	16.5
	18	46.2	22.0	12.5	8.0	68.3	38.4	24.6	17.1
16 × 16	6	.0	.0	.0	.0	.0	.0	.0	.0
	8	.0	.0	.0	.0	10.5	5.9	3.8	2.6
	10	.0	.0	.0	.0	16.7	9.4	6.0	4.2
	12	7.7	3.7	2.1	1.4	20.9	11.8	7.5	5.2
	14	13.1	6.3	3.6	2.3	23.9	13.4	8.6	6.0
	16	17.2	8.3	4.8	3.1	26.1	14.7	9.4	6.5
	18	20.4	9.8	5.7	3.6	27.9	15.7	10.0	7.0
	20	23.0	11.1	6.4	4.1	29.3	16.5	10.5	7.3
20 × 20	8	.0	.0	.0	.0	.0	.0	.0	.0
	10	.0	.0	.0	.0	4.9	2.7	1.8	1.2
	12	.0	.0	.0	.0	8.1	4.6	2.9	2.0
	14	4.1	2.0	1.2	.7	10.5	5.9	3.8	2.6
	16	7.1	3.5	2.0	1.3	12.2	6.9	4.4	3.0
	18	9.5	4.6	2.7	1.7	13.5	7.6	4.9	3.4
	20	11.4	5.6	3.2	2.1	14.6	8.2	5.3	3.7
	22	13.0	6.3	3.7	2.4	15.5	8.7	5.6	3.9
24 × 24	10	.0	.0	.0	.0	.0	.0	.0	.0
	12	.0	.0	.0	.0	2.7	1.5	1.0	.7
	14	.0	.0	.0	.0	4.6	2.6	1.6	1.1
	16	2.4	1.2	.7	.5	6.0	3.4	2.2	1.5
	18	4.3	2.1	1.2	.8	7.1	4.0	2.6	1.8
	20	5.8	2.8	1.7	1.1	8.0	4.5	2.9	2.0
	22	7.0	3.4	2.0	1.3	8.7	4.9	3.1	2.2
	24	8.0	3.9	2.3	1.5	9.3	5.2	3.4	2.3

(Note: This could also be read from Table 3.8.1)

$M_u = 72(7.18) = 517$ in.-kips

$\phi_b = \gamma_f M_u = (22.3 \times 10^{-10})(517 \times 10^3)$
$= 0.00115$ radians

Eccentricity caused by rotation:

$\phi_b h_s = 0.00115(16 \times 12) = 0.22$ in. ≈ 0.25

no further trial is required

Design requirements for $0.9D + 1.3W$:

$P_u = 72$ kips

$M_u = 517$ in.-kips $= 43.1$ ft-kips

Check for $0.75(1.4D + 1.7L + 1.7W)$:

$P_u = 0.75(1.4D + 1.7L)$
$= 0.75[1.4(80) + 1.7(30)]$
$= 122.3$ kips

$M_u = 0.75(1.7W) = 0.75[1.7(384)]$
$= 489.6$ in.-kips

$e = \dfrac{489.6}{122.3} = 4.0$ in.

Estimate eccentricity due to rotation $= 0.22$ in.

$M_u = 122.3(4.22) = 516.1$ in.-kips

$\phi_b = \gamma_f M_u = (22.3 \times 10^{-10})(516.1 \times 10^3)$
$= 0.00115$ radians

$\phi_b h_s = 0.00115(16 \times 12) = 0.22$ in. OK

Design requirements for $0.75[1.4D + 1.7L + 1.7W]$:

$P_u = 122.3$ kips

$M_u = 516.1$ in.-kips $= 43.0$ ft-kips

Section 10.11.5.5 (ACI 318-89) also requires that the moment caused by a minimum eccentricity of $0.6 + 0.03h$ be considered when designing for $1.4D + 1.7L$.

$P_u = 1.4D + 1.7L = 1.4(80) + 1.7(30)$
$= 163$ kips

$e = 0.6 + 0.03h = 0.6 + 0.03(20)$
$= 1.2$ in.

Estimate eccentricity due to rotation $= 0.1$ in.

$M_u = P_u e = 163(1.2 + 0.1)$
$= 211.9$ in.-kips

$\phi_b = (22.3 \times 10^{-10})(211.9 \times 10^3)$
$= 0.000473$ radians

$\phi_b h_s = 0.000473(16 \times 12) = 0.09$ in.
≈ 0.1 in. OK

Design requirements for $1.4D + 1.7L$:

$P_u = 163$ kips

$M_u = 212$ in.-kips $= 17.7$ ft-kips

3.8.3 Fixity of Column Bases

The degree of fixity of a column base is the ratio of the rotational stiffness of the base to the sum of the rotational stiffnesses of the column plus the base:

$$F_b = \dfrac{K_b}{K_b + K_c} \quad \text{(Eq. 3.8.8)}$$

where:

F_b = degree of base fixity, expressed as a decimal
$K_b = 1/\gamma_b$
$K_c = \dfrac{4E_c I_c}{h_s}$
E_c = modulus of elasticity of the column concrete
I_c = moment of inertia of the column
h_s = column height

Example 3.8.2 Calculation of degree of fixity

Determine the degree of fixity of the column base in Example 3.8.1; $E_c = 4300$ ksi:

$K_b = 1/y_b = 1/(22.3 \times 10^{-10}) = 4.48 \times 10^8$

$I_c = 20^4/12 = 13,333$

$K_c = \dfrac{4(4.3 \times 10^6)(13,333)}{16 \times 12}$

$= 11.94 \times 10^8$

$F_b = \dfrac{4.48}{4.48 + 11.94} = 0.27$

3.8.4 Modeling Partially Fixed Bases

Contemporary computer programs permit the direct modeling of various degrees of base fixity by the use of spring options. A simple way to model base fixity is to incorporate an imaginary column below the actual column base. If the bottom of the imaginary column is modeled as pinned, then the expression for its rotational stiffness is:

$$K_{ci} = 3E_{ci}I_{ci}/h_{ci} \quad \text{(Eq. 3.8.9)}$$

where the subscript "ci" denotes the properties of the imaginary column (Fig. 3.8.4), and $K_{ci} = K_b$ as calculated in Sect. 3.8.2. Or, the degree of base fixity, F_b, can be determined or estimated, and K_b calculated from Eq. 3.8.8.

For the computer model, either I_{ci} or h_{ci} may be varied for different values of K_{ci}, with the other terms left constant for a given problem. It is usually preferable to use $E_{ci} = E_c$. For the assumptions of Fig. 3.8.4:

$$h_{ci} = 3E_{ci}I_{ci}/K_{ci} \qquad \text{(Eq. 3.8.9a)}$$

or

$$I_{ci} = K_{ci}h_{ci}/(3E_{ci}) \qquad \text{(Eq. 3.8.9b)}$$

Example 3.8.3 Imaginary column for computer model

Determine the length of an imaginary column to model the fixity of the column base of Examples 3.8.1 and 3.8.2; $f'_c = 5000$ psi, $E_c = 4300$ ksi.

Assume:

$$E_{ci} = E_c = 4.3 \times 10^6 \text{ psi}$$
$$I_{ci} = I_c = 13{,}333 \text{ in.}^4$$
$$K_{ci} = K_b = 4.48 \times 10^8$$
$$h_{ci} = \frac{3E_{ci}I_{ci}}{K_{ci}} = \frac{3(4.3 \times 10^6)(13{,}333)}{4.48 \times 10^8}$$
$$= 383 \text{ in.} = 31.9 \text{ ft}$$

Fig. 3.8.4 Model for partially fixed column base

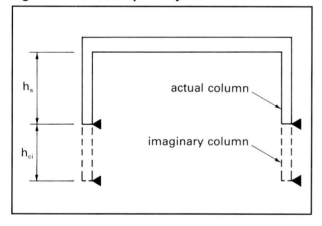

3.8.5 Volume Change Effects in Moment-Resisting Frames

The restraint of volume changes in moment-resisting frames causes tension in the girders and deflections and moments in the columns. The magnitude of these tensions, moments and deflections is dependent on the distance from the center of stiffness of the frame.

The center of stiffness is that point of a frame, which is subject to a uniform unit shortening, at which no lateral movement will occur. For frames which are symmetrical with respect to bay sizes, story heights and member stiffnesses, the center of stiffness is located at the midpoint of the frame, as shown in Fig. 3.8.5.

Tensions in girders are maximum in the bay nearest the center of stiffness. Deflections and moments in columns are maximum furthest from the center of stiffness. Thus in Fig. 3.8.5:

$$F_1 < F_2 < F_3$$
$$\Delta_1 > \Delta_2 > \Delta_3$$
$$M_1 > M_2 > M_3$$

The degree of fixity of the column base as described in Sect. 3.8.3 has a great effect on the magnitude of the forces and moments caused by volume change restraint. An assumption of a fully fixed base in the analysis of the structure may result in significant overestimation of the restraint forces, whereas assuming a pinned base may have the opposite effect. The degree of fixity used in the volume change analysis should be consistent with that used in the analysis of the column for other loadings, and the determination of slenderness effects. The horizontal shear force at the foundation may result in a lateral displacement of the foundation; this will reduce the volumetric restraint force.

3.8.5.1 Equivalent volume change

If a horizontal framing member is connected at the ends such that the volume change shortening is restrained, a tensile force is built up in the member and transmitted to the supporting elements. However, since the shortening takes place gradually over a period of time, the effect of the shortening on the shears and moment of the support is lessened because of creep and microcracking of the member and its support.

For ease of design, the volume change shortenings can be treated in the same manner as short term elastic deformations by using a concept of "equivalent" shortening.

Thus, the following relations can be assumed:

$$\delta_{ec} = \delta_c/K_\ell \qquad \text{(Eq. 3.8.10)}$$
$$\delta_{es} = \delta_s/K_\ell \qquad \text{(Eq. 3.8.11)}$$

where:

δ_{ec}, δ_{es} = equivalent creep and shrinkage shortenings, respectively

δ_c, δ_s = calculated creep and shrinkage shortenings, respectively

K_ℓ = a constant for design purposes which varies from 4 to 6

The value of K_ℓ will be near the lower end of the range when the members are heavily reinforced, and near the upper end when they are lightly reinforced. For most common structures, a value of $K_\ell = 5$ is sufficiently conservative.

Fig. 3.8.5 Effect of volume change restraints in building frames

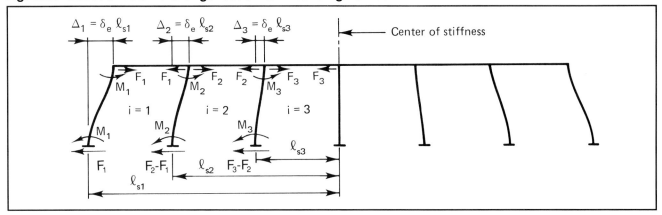

Shortening due to temperature change* will be similarly modified. However, the maximum temperature change will usually occur over a much shorter time, probably within 60 to 90 days. Thus,

$$\delta_{et} = \delta_t/K_t \quad \text{(Eq. 3.8.12)}$$

where:

δ_{et}, δ_t = the equivalent and calculated temperature shortening, respectively

K_t = a constant; recommended value = 1.5

The total equivalent shortening to be used for design is:

$$\Delta = \delta_{ec} + \delta_{es} + \delta_{et}$$
$$= \frac{\delta_c + \delta_s}{K_\ell} + \frac{\delta_t}{K_t} \quad \text{(Eq. 3.8.13)}$$

When the equivalent shortening is used in frame analysis for determining shears and moments in the supporting elements, the actual modulus of elasticity of the members is used, rather than a reduced modulus as used in other methods.

Tables 3.8.3 and 3.8.4 provide equivalent volume change strains for typical building frames.

Example 3.8.4 Calculation of column moment caused by volume change shortening of a beam

Given:

The beam of Example 3.3.1 is supported and attached to two 16 x 16-in. columns as shown in the sketch. Use $E_c = 4.3 \times 10^6$ psi and f'_c (col.) = 5000 psi.

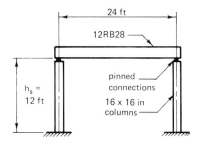

Problem:

Determine the horizontal force at the top of the column and the moment at the base of the column caused by volume change shortening of the beam.

Solution:

$I_c = bh^3/12 = 16^4/12 = 5461$ in.^{4}

From Example 3.3.1:

Total volume change shortening from erection to final is 0.20 in., or 0.10 in. each end.

Calculate the equivalent shortening:

$$\Delta = \frac{\delta_c + \delta_s}{K_\ell} + \frac{\delta_t}{K_t}$$

$$= \left[\frac{246-80+454-140}{5} + \frac{210}{1.5}\right] \times (10^{-6})(24)(12)$$

$$= 236 \times 10^{-6}(288) = 0.07 \text{ in.}$$

$$\frac{\Delta}{2} = \frac{0.070}{2} = 0.035 \text{ in. each end}$$

$$\frac{\Delta}{2} = \frac{Fh_s^3}{3E_cI_c}$$

$$F = \frac{3E_cI_c\left(\frac{\Delta}{2}\right)}{h_s^3}$$

$$= \frac{3(4.3 \times 10^6)(5461)(0.035)}{(12 \times 12)^3} = 826 \text{ lb}$$

$$M = Fh_s = 826(144)$$
$$= 118,944 \text{ in.-lb} = 9.91 \text{ ft-kips}$$

*Temperature change is, of course, a reversible effect; increases cause expansion and are important in design and location of expansion joints (see Sect. 3.3.3). Temperature differentials in roof and wall elements should also be considered.

Table 3.8.3 Equivalent volume change strains for typical continuous building frames (millionths)

Temp. zone (from map)	Prestressed members (P/A = 600 psi)									
	Normal weight concrete					Lightweight concrete				
	Ave. R.H. (from map)					Ave. R.H. (from map)				
	40	50	60	70	80	40	50	60	70	80
Heated buildings										
0	110	100	91	81	72	116	106	97	87	77
10	130	120	111	101	92	133	123	113	104	94
20	150	140	131	121	112	150	140	130	120	110
30	170	160	151	141	132	166	156	147	137	127
40	190	180	171	161	152	183	173	163	154	144
50	210	200	191	181	172	200	190	180	170	160
60	230	220	211	201	192	216	206	197	187	177
70	250	240	231	221	212	233	223	213	204	194
80	270	260	251	241	232	250	240	230	220	210
90	290	280	271	261	252	266	256	247	237	227
100	310	300	291	281	272	283	273	263	254	244
Unheated structures										
0	110	100	91	81	72	116	106	97	87	77
10	140	130	121	111	102	141	131	122	112	102
20	170	160	151	141	132	166	156	147	137	127
30	200	190	181	171	162	191	181	172	162	152
40	230	220	211	201	192	216	206	197	187	177
50	260	250	241	231	222	241	231	222	212	202
60	290	280	271	261	252	266	256	247	237	227
70	320	310	301	291	282	291	281	272	262	252
80	350	340	331	321	312	316	306	297	287	277
90	380	370	361	351	342	341	331	322	312	302
100	410	400	391	381	372	366	356	347	337	327

Table 3.8.4 Equivalent volume change strains for typical continuous building frames (millionths)

Temp. zone (from map)	Non-prestressed members									
	Normal weight concrete					Lightweight concrete				
	Ave. R.H. (from map)					Ave. R.H. (from map)				
	40	50	60	70	80	40	50	60	70	80
Heated buildings										
0	59	53	47	41	35	59	53	47	41	35
10	79	73	67	61	55	76	70	64	58	52
20	99	93	87	81	75	92	86	80	75	69
30	119	113	107	101	95	109	103	97	91	85
40	139	133	127	121	115	126	120	114	108	102
50	159	153	147	141	135	142	136	130	125	119
60	179	173	167	161	155	159	153	147	141	135
70	199	193	187	181	175	176	170	164	158	152
80	219	213	207	201	195	192	186	180	175	169
90	239	233	227	221	215	209	203	197	191	185
100	259	253	247	241	235	226	220	214	208	202
Unheated structures										
0	59	53	47	41	35	59	53	47	41	35
10	89	83	77	71	65	84	78	72	66	60
20	119	113	107	101	95	109	103	97	91	85
30	149	143	137	131	125	134	128	122	116	110
40	179	173	167	161	155	159	153	147	141	135
50	209	203	197	191	185	184	178	172	166	160
60	239	233	227	221	215	209	203	197	191	185
70	269	263	257	251	245	234	228	222	216	210
80	299	293	287	281	275	259	253	247	241	235
90	329	323	317	311	305	284	278	272	266	260
100	359	353	347	341	335	309	303	297	291	285

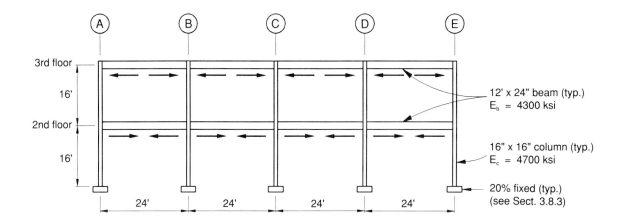

3.8.5.2 Calculating restraint forces

Most "plane frame" computer analysis programs allow the input of shortening strains of members from volume changes. The equivalent strains as described in Sect. 3.8.5.1 can be input directly into such programs.

For frames that are approximately symmetrical, the coefficients from Tables 3.8.5 and 3.8.6 can be used with small error. The notation for these tables is described in Fig. 3.8.6.

Example 3.8.5 Volume change restraint forces

Given:

The 4-bay, 2-story frame shown above
Beam modulus of elasticity = E_b = 4300 ksi
Column modulus of elasticity = E_c = 4700 ksi
Column bases 20% fixed (see Sect. 3.8.3)
Design R.H. = 70%
Design temperature change = 70°F

Problem:

Determine the maximum tension in the beams and the maximum moment in the columns caused by volume change restraint.

Solution:

1. Determine relative stiffness between columns and beams:

   I_b = 12(24)³/12 = 13,824 in.⁴

   $E_b I_b / \ell$ = 4300(13,824)/(24 × 12)
   = 206,400

   I_c = 16(16)³/12 = 5461 in.⁴

   $E_c I_c / h_s$ = 4700(5461)/(16 × 12)
   = 133,681

Table 3.8.5 Build-up of restraint forces in beams (k_b)

Total number of bays (n)	Number of bays from end (i)							
	1	2	3	4	5	6	7	8
2	1.00							
3	1.00	4.00						
4	1.00	3.00						
5	1.00	2.67	9.00					
6	1.00	2.50	6.00					
7	1.00	2.40	5.00	16.00				
8	1.00	2.33	4.50	10.00				
9	1.00	2.29	4.20	8.00	25.00			
10	1.00	2.25	4.00	7.00	15.00			
11	1.00	2.22	3.86	6.40	11.67	36.00		
12	1.00	2.20	3.75	6.00	10.00	21.00		
13	1.00	2.18	3.67	5.71	9.00	16.00	49.00	
14	1.00	2.17	3.60	5.50	8.33	13.50	28.00	
15	1.00	2.15	3.55	5.33	7.86	12.00	21.00	64.00
16	1.00	2.14	3.50	5.20	7.50	11.00	17.50	36.00

$$K_r = \frac{E_b I_b / \ell}{E_c I_c / h_s} = \frac{206{,}400}{133{,}681} = 1.5$$

2. Determine deflections:

   From Table 3.8.3: $\delta_e = 221 \times 10^{-6}$ in./in.

   $\Delta_B = \delta_e \ell = 0.000221(24)(12) = 0.064$ in.

   $\Delta_A = \delta_e(2\ell) = 0.128$ in.

3. Determine maximum beam tension:

   Maximum tension is nearest the center of stiffness, i.e., beams BC and CD, 2nd floor.

   From Table 3.8.5:

   For n = 4 and i = 2; $k_b = 3.00$

   From Table 3.8.6:

   For $K_r = 1.0$, fixed base; $k_f = 11.2$

   For $K_r = 2.0$, fixed base; $k_f = 11.6$

   Therefore for $K_r = 1.5$; $k_f = 11.4$

   For pinned base, $k_f = 3.4$

   (for $K_r = 1.0$ and 2.0)

   For 20% fixed:

   $k_f = 3.4 + 0.20(11.4 - 3.4) = 5.0$

   $F_2 = k_f k_b \Delta_i E_c I_c / h_s^3$

   $= \dfrac{5.0(3.0)(0.064)(4700)(5461)}{(16 \times 12)^3}$

   $= 3.48$ kips

4. Determine maximum column moments:

   For base moment, M_1:

   From Table 3.8.6, by interpolation similar to above:

   k_m (fixed) = (4.9 + 5.2)/2 = 5.05

   k_m (pinned) = 0

   k_m (20% fixed) = 0 + 0.20(5.05) = 1.0

   $M_1 = k_m \Delta_i E_c I_c / h_s^2$

   $= 1.0(0.128)(4700)(5461)/(16 \times 12)^2$

   $= 89.1$ in.-kips

   For second floor moment, M_{2L}:

   k_m (fixed) = (3.9 + 4.5)/2 = 4.2

   k_m (pinned) = (2.1 + 2.4)/2 = 2.25

   k_m (20% fixed) = 2.25 + 0.20(4.20 - 2.25)

   $= 2.64$

   $M_{2L} = \dfrac{2.64(0.128)(4700)(5461)}{(16 \times 12)^2}$

   $= 235$ in.-kips

3.8.6 Computer Models for Frame Analysis

When precast frames are modeled as "sticks", as is usually done with steel frames, the results are often very misleading. For example, the structure as modeled in Fig. 3.8.7(a) will indicate more flexibility than is actually true. Lateral drift will be overestimated, and the moments caused by axial shortening may be underestimated. Fig. 3.8.7(b) shows a suggested model which will better estimate the true condition.

3.9 Shear Wall-Frame Interaction

Rigid frames and shear walls exhibit different responses to lateral loads, which may be important, especially in high-rise structures. This difference is illustrated in Fig. 3.9.1.

A frame bends predominantly in a shear mode as shown in Fig. 3.9.1(a), while a shear wall deflects predominantly in a cantilever bending mode, Fig. 3.9.1(b). Elevator shafts, stairwells, and concrete walls normally exhibit this behavior.

It is not always easy to differentiate between modes of deformation. For example, a shear wall weakened by a row, or rows of openings may tend to act like a frame, and an infilled frame will tend to deflect in a bending mode. Also, shear deformation of a shear wall can be more important than bending deformation if the height-to-length ratio is small, as discussed in Sect. 3.7.2.

If all vertical elements of a structure exhibit the same behavior under load, that is, if they are all frames or all shear walls, the load can be distributed to the units in proportion to their stiffnesses (Sect. 3.7.3). However, because of the difference in bending modes, the load distribution in structures with both frames and shear walls is considerably more complex. References 8 through 12 address this problem in more detail.

3.10 Structural Integrity

3.10.1 Introduction

ACI 318-89, Sect. 7.13, requires that, in the detailing of reinforcement and connections, members of a structure be effectively tied together to improve integrity of the overall structure. For precast concrete construction this is to be achieved by providing tension ties in the transverse, longitudinal, and vertical directions, and around the perimeter of the structure. The commentary for Sect. 7.13 emphasizes that the overall integrity of a structure can be substantially enhanced by nominal changes of reinforcement. In the event of damage to a beam, for example, it is important that displacement of its support member be

Table 3.8.6 Coefficients k_f and k_m for forces and moments caused by volume change restraint forces (see Fig. 3.8.6 for notation)

No. of Stories	$K_r = \dfrac{\Sigma E_b I_b / \ell}{\Sigma E_c I_c / h_s}$	Base Fixity	Values of k_f				Values of k_m					
			F_1	F_2	F_3	F_4	Base M_1	2nd floor		3rd floor		4th
								M_{2L}	M_{2U}	M_{3L}	M_{3U}	M_4
1	0	Fixed	3.0	3.0			3.0	0				
		Pinned	0	0			0	0				
	0.5	Fixed	6.0	6.0			4.0	2.0				
		Pinned	1.2	1.2			0	1.2				
	1.0	Fixed	7.5	7.5			4.5	3.0				
		Pinned	1.7	1.7			0	1.7				
	2.0	Fixed	9.0	9.0			5.0	4.0				
		Pinned	2.2	2.2			0	2.2				
	4.0 or more	Fixed	10.1	10.1			5.4	4.7				
		Pinned	2.5	2.5			0	2.5				
2	0	Fixed	6.8	9.4	2.6		4.3	2.6	2.6	0		
		Pinned	0	3.0	1.5		0	1.5	1.5	0		
	0.5	Fixed	8.1	10.7	2.6		4.7	3.4	2.1	0.4		
		Pinned	1.9	3.4	1.4		0	1.9	1.2	0.2		
	1.0	Fixed	8.9	11.2	2.3		4.9	3.9	1.8	0.5		
		Pinned	2.1	3.4	1.3		0	2.1	1.0	0.3		
	2.0	Fixed	9.7	11.6	1.9		5.2	4.5	1.4	0.5		
		Pinned	2.4	3.4	1.0		0	2.4	0.8	0.3		
	4.0 or more	Fixed	10.4	11.9	1.4		5.5	5.0	1.0	0.4		
		Pinned	2.6	3.4	0.8		0	2.6	0.5	0.2		
3 or more	0	Fixed	7.1	10.6	4.1	0.7	4.4	2.8	2.8	0.7	0.7	0
		Pinned	1.6	3.6	2.4	0.4	0	1.6	1.6	0.4	0.4	0
	0.5	Fixed	8.2	11.1	3.5	0.5	4.7	3.5	2.2	0.7	0.4	0.09
		Pinned	1.9	3.6	1.9	0.3	0	1.9	1.2	0.4	0.2	0.05
	1.0	Fixed	8.9	11.4	2.9	0.4	5.0	3.9	1.9	0.7	0.3	0.09
		Pinned	2.2	3.5	1.6	0.2	0	2.2	1.0	0.4	0.2	0.05
	2.0	Fixed	9.7	11.7	2.2	0.2	5.2	4.7	1.4	0.6	0.2	0.06
		Pinned	2.4	3.5	1.2	0.1	0	2.4	0.8	0.3	0.1	0.03
	4.0 or more	Fixed	10.4	11.9	1.5	0.04	5.5	5.0	1.0	0.5	0.04	0.01
		Pinned	2.6	3.4	0.8	0.02	0	2.6	0.5	0.2	0.02	0.00

Fig. 3.8.6 Use of Table 3.8.6

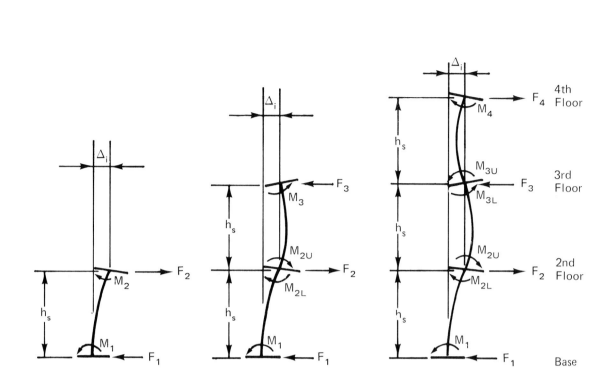

$$\Delta_i = \delta_e \ell_s$$

$$F_i = k_f k_b \Delta_i E_c I_c / h_s^3$$

$$M_i = k_m \Delta_i E_c I_c / h_s^2$$

where: δ_e = equivalent unit strain (see Sect. 3.3)

ℓ_s = distance from column to center of stiffness

F_i = F_1, F_2, etc., as shown above

k_f, k_m = coefficients from Table 3.8.6

k_b = $i\left(\dfrac{n+1-i}{n+2-2i}\right)$ (or from Table 3.8.5)

n = no. of bays

i = as shown in Fig. 3.8.5

E_c = modulus of elasticity of the column concrete

I_c = moment of inertia of the column

Fig. 3.8.7 Computer models

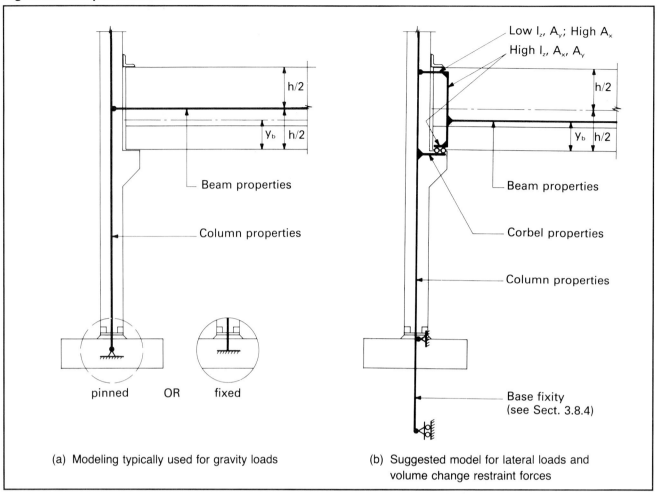

(a) Modeling typically used for gravity loads

(b) Suggested model for lateral loads and volume change restraint forces

minimized, so that other members will not be affected; for this reason, the commentary indicates that connection details which rely solely on friction caused by gravity loads are not permitted. Connections shall be arranged to minimize the potential for cracking due to restraint of volume changes.

3.10.2 Precast Concrete Structures

For typical precast concrete structures, longitudinal and transverse tie requirements are achieved by connecting members into a load path to the lateral load resisting system. The load path in the lateral load resisting system shall be continuous to the foundation.

Any individual member may be connected into this load path by alternative methods. For example, a load bearing spandrel could be connected to a diaphragm (part of the lateral load resisting system). Structural integrity could be achieved by connecting the spandrel into all or a portion of the deck members forming the diaphragm. Alternatively, the spandrel could be connected only to its supporting columns, which in turn must then be connected to the diaphragm.

Vertical tension tie requirements are achieved by providing connections at horizontal joints of vertical members.

Fig. 3.9.1 Deformation modes

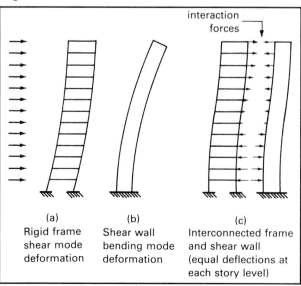

(a) Rigid frame shear mode deformation

(b) Shear wall bending mode deformation

(c) Interconnected frame and shear wall (equal deflections at each story level)

For precast concrete structures, the following provisions will satisfy the requirements of ACI 318-89, Sect. 7.13.3:

1. All members shall be connected to members which are part of the lateral load resisting system.

2. The lateral load resisting system shall be continuous to the foundation.

3. Brittle failure of tie connections shall be precluded by designing and detailing them so that the failure mode is by yielding of steel.

4. A diaphragm shall be complete with tension ties around its perimeter and around openings which significantly interrupt diaphragm action, and with connections to load and unload the diaphragm.

5. Column splices and column base connections shall have a tensile design strength not less than $200A_g$ in pounds, where A_g is the gross area of the column. For a compression member with a larger cross-section than required by consideration of loading, a reduced effective area, A_g, not less than one-half the total area, may be used.

6. Precast walls which are essential vertical structural elements, including shear walls, shall be vertically connected by a minimum of two connections per panel. Each connection shall have a nominal tensile capacity of not less than 10 kips.

3.10.3 Large Panel Bearing Wall Structures

Large panel bearing wall structures are a special category of precast concrete structures, with respect to structural integrity. Large panel structures are typically constructed with precast walls having a horizontal dimension greater than the vertical dimension, which is generally the height of one story. The panels are stacked for the height of the building and support the floor and roof decks. Criteria have been developed[7] for redundant load paths in such buildings three stories or more in height. Fig. 3.10.1 illustrates the tie forces required to achieve these redundant load paths. Use of the following forces is recommended for large panel bearing wall structures. It is not intended that these forces replace an analysis of the actual design forces required in the structure; these forces are not additive to the actual design forces.

T_{1u} = Design force equal to 1500 lb per lin ft of floor or roof span. Ties may be encased in the floor units or in a topping, or may be concentrated at the wall.

T_{2u} = Peripheral design force sufficient to develop diaphragm action, but not less than 16,000 lb, located within the depth of the floor of roof slab. Ties may be reinforcing steel or prestressing strand in a grout joint or in an edge beam; reinforced spandrels or walls anchored to the floor or roof may also be considered.

T_{3u} = Design force equal to 2.5% of the service load on the bearing wall, but not less than 1500 lb per lin ft of wall. Ties should be spaced not greater than 8 ft on centers. They may project from the precast element or be embedded in grout joints, with sufficient length and cover to develop the specified design force. At end walls, project wall reinforcement into the floor or provide a mechanical anchorage between floor and wall.

T_{4u} = Design force of 3000 lb per lin ft of wall in all bearing walls, with a minimum of two ties per wall. These ties should be continuous from foundation to the roof.

3.10.4 Hybrid Structures

The provisions of ACI 318-89 relate to concrete buildings only. Those connections which interface precast concrete components with other structural materials (e.g., masonry walls, steel or wood roofs) should provide the same load paths and follow the design philosophy described above. Since the pre-

Fig. 3.10.1 Recommended tie forces in precast concrete bearing wall buildings

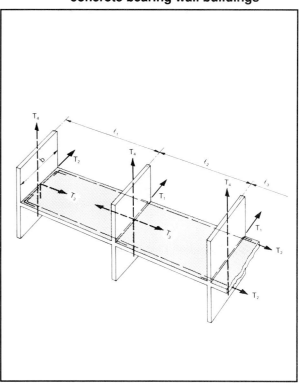

cast concrete supplier rarely has control over such other materials, the Engineer of Record must design and detail the connections to satisfy structural integrity and include them in the contract drawings.

3.11 Earthquake Analysis

3.11.1 Notation

A_c = $\Sigma A_e [0.2 + (D_e/h_n)^2]$

A_e = minimum cross-sectional shear area in any horizontal plane in the first story of a shear wall (UBC-88)

A_ℓ = cross-sectional area in linear measure

A_s = area of non-prestressed reinforcement

A_v = coefficient representing effective peak velocity-related acceleration (SBC-88)

A_{vf} = area of shear-friction reinforcement

b = width of panel

C = coefficient for base shear (UBC-88)(SBC-88); total compressive force

C_p = coefficient for horizontal force (UBC-88)

C_o = compressive chord force

C_1 = overturning couple force

C_u = factored compressive force

D_e = length of shear wall in the first story in the direction parallel to the applied force (UBC-88)

D_s = longest dimension of a shear wall or braced frame in the direction parallel to the applied force (SBC-88)

d = dimension of building; distance from extreme compression fiber to centroid of tension reinforcement

F_i, F_n, F_x = lateral forces applied to level i, n, or x, respectively (SBC-88)

F_p = lateral force on the part of the structure and in the direction under consideration (UBC-88) (SBC-88)

F_{px} = force on floor diaphragms and collectors (SBC-88)

F_t = that portion of V considered concentrated at the top of the structure, level n (UBC-88), or in addition to F_n (SBC-88)

F_x = forces in x direction; force at level x

F_y = forces in y direction

f'_c = concrete compressive strength

f_s = stress in steel

f_y = yield strength of non-prestressed reinforcement

H = horizontal force needed to overcome friction

h = height of member

h_i, h_n, h_x = height above base level to level i, n, or x, respectively (UBC-88)(SBC-88)

I = occupancy importance factor (UBC-88)(SBC-88)

K = coefficient relating to type of construction (SBC-88)

ℓ = length of building or member

M = moment

M_1 = overturning moment

M_R = overturning moment resistance

N = force normal to friction plane

n = uppermost level in the structure (UBC-88) (SBC-88)

$P_{1,2,3,4}$ = forces

R_o, R_1 = reactions

R_s = resistance to sliding

R_w = coefficient relating to type of construction (UBC-88)

S, S_1, S_2 = soil factor (SBC-88)

s = spacing of weld clips

T = fundamental period of vibration of the building in the direction under consideration (UBC-88) (SBC-88); total tensile force

T_o = tensile chord force

T_1 = overturning couple force

T_u = factored tensile force

V = total lateral load or shear at the base (UBC-88) (SBC-88); shear force

V_{Ru} = design shear strength

V_u = factored shear force

v = unit shear stress

$v_{0,1,2,3}$ = unit shear stresses

v_{ru} = design unit shear stress

v_u = factored shear stress

W = total dead load of building (UBC-88); total dead load of buildings and structures with modification (SBC-88)

W_p = total weight of a part or portion of a structure (UBC-88) (SBC-88)

w_{px} = weight of diaphragms and collectors and tributary elements at level x plus allowable portions of live load (SBC-88)

w_i, w_x = that portion of W which is located at or is assigned to level i or x, respectively (SBC-88)

x = level which is under design consideration (SBC-88)

Z = coefficient dependent upon the seismic zone (UBC-88); coefficient dependent upon the effective peak velocity-related acceleration, A_v (SBC-88)

Z_ℓ = section modulus in linear measure

ϕ = capacity reduction factor (ACI-318-89)

μ = shear-friction coefficient

μ_s = static coefficient of friction

3.11.2 General

Earthquakes generate horizontal and vertical ground movement. When the earthquake passes beneath a structure, the foundation will tend to move with the ground, while the superstructure will tend to remain in its original position. The lag between foundation and superstructure movement will cause distortions and develop forces in the structure. As the ground moves, changing distortions and forces are produced throughout the height of the structure.

Precast concrete structures are jointed construction, and individual elements are connected at these joints using a variety of methods. These connections may be embedded steel shapes, such as flat bars and angles, with headed stud or reinforcing bar anchorages. The steel embedments are field bolted or welded. This connection type is referred to as a "dry" connection. Another type of connection, referred to as a "wet" connection, consists of protruding reinforcing bars which may be mechanically coupled or lap spliced. The joint between members is then completed with cast-in-place concrete. The "wet joint" method of assembling precast members is particularly appropriate in seismic Zones 3 and 4. For zones of lower seismicity, dry connections are generally used.

It is imperative that lateral load paths and resisting elements are clearly defined. Where significant movement between adjacent elements is anticipated, ductile connections must be provided.

The current philosophy for the design of earthquake resistant structures permits minor damage for moderate earthquakes, and accepts major damage for severe earthquakes, provided that complete collapse is prevented. Seismic performance can be improved by setting limitations on structural deflections. The design details often require large, inelastic deformations to occur in order to absorb the inertia forces. This is achieved by providing member and connection ductility. While this ductility prevents total collapse, the resultant distortions may lead to significant damage to mechanical, electrical, and architectural elements.

To limit damage, three paths are open to the designer. First, the elements may be uncoupled from the structural system, so that these elements are not forced to undergo as much deformation as the supporting structure. Second, the deflections of the supports could be reduced in order to minimize deformations of the architectural elements. Third, the connection between individual elements and the supporting frame could be designed to sustain large deformations and rotations without failure. Generally, the first or third approach is adopted for non-structural architectural wall panels (see Sect. 3.11.12).

Buildings may be designed as either flexible or stiff. Flexible structures will develop large deflections and small inertial forces; conversely, stiff structures will develop large inertial forces but small deflections. Either type may be designed to be safe against total collapse. However, experience demonstrates that a stiff structure, properly designed to account for the large inertia forces, will incur significantly less damage to architectural, mechanical, and electrical elements.

Since ground motion is random in direction, a structure which is shaped so as to be equally resistant in any direction is the optimal solution. Furthermore, closed sections (i.e., boxes or tubes) have demonstrated markedly better behavior when compared with open sections, because (1) closed sections provide a high degree of torsional resistance, and (2) the higher axial stresses and resultant deformations in the exterior columns provide significant energy absorption.

Load tests of prestressed concrete members have consistently shown that large deflections occur as the design strength is approached. Because of prestressing, the transition from linear to nonlinear response is gradual and smooth. Cyclic load tests have shown that prestressed concrete beams can undergo several cycles of intense load reversals and still maintain their design strength.

3.11.3 Building Code Requirements

All major building codes include seismic provisions. In this section, examples are presented using the Uniform Building Code (UBC) and the Standard

Building Code (SBC). The following discussion is based on the provisions of UBC-88.

The response of a structure to the ground motion of an earthquake depends on the structural system with its damping characteristics, and on the distribution of its mass. With mathematical idealization, a designer can determine the probable response of the structure to an imposed earthquake. UBC-88 requires a dynamic analysis for structures which have highly irregular shapes or framing systems and allows it for other structures. However, most buildings have structural systems and shapes which are more or less regular, and many designers use the equivalent static load method for these structures.

In its simplest form, UBC-88 requires that a total base shear, V, be applied to the building in any horizontal direction, where

$$V = \frac{ZICW}{R_w} \quad \text{(Eq. 3.11.1)}$$

in which

Z is based on the expected earthquake intensity,

I is based on the type of occupancy anticipated,

R_w is based on the type of framing,

C is based on the flexibility of the structure and the site soil condition, and

W is the dead load of the structure, plus portions of the live load for some occupancies.

This total shear is divided among the story levels, with the upper stories being assigned more of the horizontal load than the lower stories. This is the method of equivalent static loads.

Also, UBC-88 requires that parts of the structure such as roofs, floors, walls, and their connections be designed locally for either the distributed base shear or the lateral forces determined by the expression:

$$F_p = ZIC_pW_p \quad \text{(Eq. 3.11.2)}$$

in which the subscript "p" denotes the effects of the building part. For certain parts of buildings, F_p requirement is sometimes more severe than the distributed base shear.

The occupancy importance factor, I, has the effect of making structures that would be essential during an earthquake disaster (e.g., hospitals, fire stations, etc.), or those that could house large numbers of people, less likely to be severely damaged.

The values C, C_p and R_w may be determined in accordance with UBC-88, Chapter 23. The value of C need not exceed 2.75, and may be used for any structure without regard to soil type or structure period. The value of R_w is 6 for concrete bearing wall systems. For shear walls which are not load bearing, R_w is 8.

3.11.4 Design Guidelines for Wall Panels

The sections that follow deal primarily with totally precast concrete structures. In addition, architectural wall panels, whether connected to precast concrete or other materials, require special considerations:

1. Exterior walls perforated for windows will act somewhere between a solid wall and a flexible frame. For tall buildings, this will result in a non-linear distribution of forces, due to shear lag (see Sect. 3.7.7). When similar to a flexible frame, they must be designed as moment-resisting frames to resist seismic loads.

2. Portions of walls with opening can be subjected to significant axial loads. These portions may require reinforcement with closely spaced ties, in accordance with ACI 318-89, Chapter 21.

3. Connected walls may act as coupled walls (see Sect. 3.7.5).

4. Walls will be subjected to lateral loads perpendicular to the plane of the wall (wind, seismic) in combination with loads in the plane.

5. Design should consider the eccentricities produced by story drift. These are combined with the eccentricities due to manufacturing and erection tolerances. Drift is defined as the relative movement of one story with respect to the stories immediately above or below the level under consideration. Between points of connection, non-load bearing panels should be separated from the building frame to avoid contact under seismic action. In the immediate area of connections, the panel will tend to distort the same amount as the supporting frame. Internal stresses induced due to a statically indeterminate support system should be checked. Even in a statically determinate panel there may be some built-in restraint at the connections, so that some allowance for internal stresses should be considered.

6. Under severe earthquake, large deflections may be anticipated. The investigation of individual walls and of the entire structure should include the consideration of deflection (P-Δ effect).

7. To account for accidental torsion, the mass at each level shall be considered displaced from the calculated center of mass an amount equal to 5 percent of the building dimension perpendicular to the direction of force.

8. Seismic induced forces are reversible. This is particularly important at joints.

9. The best energy absorbing members are those with high moment-rotation capabilities. The energy absorbing capacity of a flexural member is measured by the area under the moment-rotation curve. Correctly reinforced, concrete can

exhibit high ductility. Refer to ACI 318-89, Chapter 21, for proper methods of reinforcing to achieve ductility.

10. Joints represent discontinuities, and may be the location of stress concentrations. Reinforcement or mechanical anchorage must be provided through the joint to fully transmit the horizontal shear and flexure developed during seismic activity. See Chapter 6 for a discussion on connections. In zones of high seismicity, cast-in-place reinforced concrete in combination with precast concrete has proved successful in economically transferring seismic forces (see Sect. 7.3).

11. Where possible, make panel connections to the supporting structure statically determinate, in order to permit a more accurate determination of force distribution.

12. Choose the number and location of connections to permit movements in the plane of the panel to accommodate story drift and volume changes.

13. Locate connections to minimize torsional moments on supporting spandrel beams, particularly if the beams are structural steel.

14. Provide separation between the panel and the building frame to prevent contact during an earthquake.

3.11.5 Concept of Box-Type Buildings

A box-type building consists of roof and floor diaphragms, and shear walls. When these components are appropriately connected they form a structure which is very resistant to lateral loads.

Since an earthquake is a ground motion reacting with the inertia of the building and its parts, the equivalent static loads are applied at the centroids of the parts. The internal forces that link the applied loads and the ground reaction follow the stiffest paths consistent with equilibrium and compatibility of deflection. With box-type buildings of moderate height-to-width ratio, the stiffnesses of diaphragms and walls in their own planes greatly exceed other resistances. The load paths are then along diaphragms and walls rather than through moment-resisting frames.

In multi-story, box-type buildings, the equivalent static loads find resistances in the several diaphragms and walls, and downward to the footings. The designer should try to arrange the path of resistance to be direct.

A diaphragm made up of precast elements requires that it be strong enough in shear and moment. When walls have numerous openings, the designer must judge if the wall should be considered a shear wall or a moment-resisting frame.

3.11.6 Structural Layout and Connections

A box-type structure may have a large number of precast concrete elements that are assembled into walls, floors, roof, and occasionally frames. Proper connections between the many pieces create the diaphragms and shear walls, and the connections between these, in turn, create the box-type structure. In the seismic design of a box-type structure, there are two fundamental and different requirements of the two groups of connections:

1. One group of connections that transmits forces between elements within a horizontal diaphragm or a shear wall, and

2. Another group of connections that transmits forces between a horizontal diaphragm and a shear wall.

In seismic design, forces must be positively transmitted. Load paths must be as direct as possible. Anchors should be attached to or hooked around reinforcing bars or otherwise terminated so as to effectively transfer forces to the bars. Reinforcement in the vicinity of the anchors should be designed to distribute the forces so as to preclude local failure. Concrete dimensions must be ample, so that the hardware of the connection is confined, and the connection thus can transmit accidental forces that are normal to the usual plane of the load path. Finally, a connection should be such that, if it were to yield, it will do so in a ductile manner, i.e. without loss of load carrying capacity when the concrete cracks.

3.11.7 Example—1-Story Building

General

By taking advantage of walls already present, one-story buildings usually can be designed to resist lateral loads (wind or earthquake) by shear wall and diaphragm action. If a shear wall and diaphragm con-

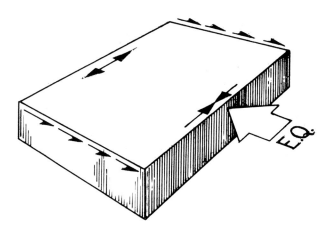

cept is feasible, it is generally the most economical concept. This section of the Handbook is intended to assist the designer with the shear wall/diaphragm concept for precast, prestressed concrete buildings.

To show a fairly complete design, the simple one-story example building in Fig. 3.11.1 will be illustrated. It is 128 ft x 160 ft in plan, and has 16-ft clear height inside. It is entirely precast above the floor, using 16-in. double tees for the walls and 24-in. double tees for the roof. The double tees are 8 ft wide, with stems 4 ft on centers. The flanges on the roof tees are 2 in. thick, and on the wall tees, 4 in. thick. The weight of roofing and mechanical equipment is 10 psf.

Because all loads must funnel through the connections, gravity and lateral loads must be considered together. Thus, this example shows both gravity and lateral load connections. The example emphasizes both free bodies and the concept of load path.

Load analysis

Earthquakes impose lateral and vertical ground motions upon a structure. The structure responds to these motions with its own deflections. These deflections are accompanied by corresponding strains and the resulting stresses. However, most designers are used to thinking of stress as caused by load rather than as caused by deflection. Consequently, all the common methods of design use a set of static lateral loads intended to be equivalent to (i.e. produce the same stresses as) the real, dynamic loads caused by deflection. One such method is used in UBC-88. The example building is analyzed here for N-S earthquake only. In a real design situation, it would also have to be analyzed for E-W earthquake.

The example building resists lateral load by diaphragm and shear wall action. Inertia loads (mass x acceleration) are delivered to the roof diaphragm. The diaphragm acts like a plate girder laid flat, spanning between the shear walls. The diaphragm may be taken as the interconnected flanges of the precast roof elements. In determining the equivalent static loads on this diaphragm, the mass tributary to the diaphragm must be determined. This is done in terms of dead weight, W, as follows:

N and S walls:

half ht + parapet	= 11.5 ft
total length = (2)(160)	= 320 ft
weight = 11.5(320)(0.067)	= 247 kips

Roof (including 10 psf dead load):

| weight = 128(160)(0.062) | = 1270 kips |
| Total W | = 1517 kips |

The equivalent lateral load V is computed by multiplying W by an acceleration. The acceleration is determined from ZIC/R_w. The Z-factor, denoting geographical zones of equal probability of serious earthquake, is here taken as Zone 3 value of 0.30. The occupancy importance factor, I, is assumed as 1.0 and the R_w factor, used to indicate the performance different framing systems have shown in actual earthquakes, is taken as 6.

The value of C is calculated as follows:

$$C = \frac{1.25S}{T^{2/3}} \leq 2.75$$

where:

$T = C_T(h_n)^{2/3}$

$C_T = 0.1\sqrt{A_c} \geq 0.020$

Assume a stiff soil of depth > 200 ft.

$S = S_2 = 1.2$

Calculate C_T in North-South direction:

There are 2 shear walls, each 128 ft long and 18 ft high (to roof) and 4 in. thick. Thus,

$A_e = (128)(0.33) = 42.67$ sq ft

$D_s/h_n = 128/18 = 7.11 > 0.9$; use 0.9

$A_c = (2)(42.67)(0.2 + 0.9^2) = 86.19$ sq ft

$C_T = 0.1/\sqrt{86.19} = 0.011 < 0.02$; use 0.02

$T = 0.02(18)^{2/3} = 0.137$ seconds

$C = 1.25(1.2)/(0.137)^{2/3} = 5.6 > 2.75$; use 2.75

The coefficient used in design is thus:

$ZIC/R_w = 0.30(1)(2.75)/6 = 0.14$

Shear on the diaphragm due to earthquake is:

$V = (ZIC/R_w)W = 0.14(1517) = 212$ kips

In comparison, the shear caused by a 20 psf wind load is:

$V = 11.5(160)(0.020) = 37$ kips < 212 kips

Thus the building is governed by earthquake.

A word of caution is in order. The base shear due to earthquake as calculated above is a "service" load. Accelerations in real earthquakes may be many times the ZIC/R_w shown above. This will cause elements of the structure to be strained to first yield point and beyond in a severe earthquake. The structure must have deformation capability to absorb overloads of short duration without failure.

To guard against displacement of mass from actual location UBC-88 requires that a minimum 5 percent eccentricity be assumed with respect to the center of mass as shown in Fig. 3.11.2. With this in

Fig. 3.11.1 1-story example building

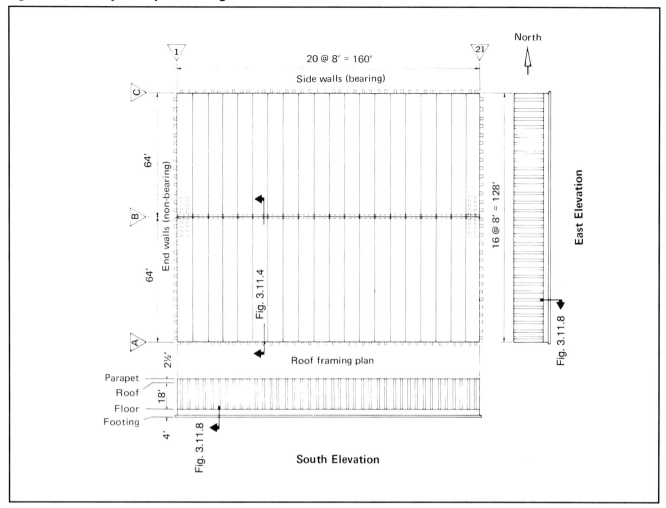

mind, the forces internal to the roof diaphragm are:

Max. shear reaction,

$R_o = (88/160)V = 0.55(212) = 117$ kips

Max. shear intensity,

$v_o = 117/128 = 0.91$ klf

Max. bending moment,

$M_o = V\ell/8 = 212(160/8) = 4240$ ft-kips

Max. chord forces,

$C_o = T_o = M_o/d = 4240/128$ ft-kips $= 33$ kips

The shear forces are analogous to those in the web of a plate girder. The chord forces are analogous to those in the flanges of a plate girder.

Considering the roof as a free body, Fig. 3.11.2 shows that equilibrium is maintained by the reactions R_o from the tops of the shear wall as a free body. Fig. 3.11.2 shows that the wall can be in equilibrium only if sufficient sliding resistance and overturning resistance are provided. The forces acting which must be resisted by the shear wall are:

Sliding force,

$R_o = 117$ kips $= R_1$

Overturning moment,

$M_1 = R_o h = 117(22) = 2574$ ft-kips

The weight of the end wall, tributary roof, floor, backfill, etc., is about $N = 400$ kips. The available resistance to overturning, then, is:

$M_R = N(d/2) = 400(128/2) = 25,600$ ft-kips

The sliding force is resisted by friction at the bottom of the wall footing. Assume a granular soil, the coefficient of sliding friction is about 0.5.

$\mu_s N = 0.5(400) = 200$ kips

The factors of safety (load factors) for overturning and sliding are seen to be sufficient:

$M_R/M_1 = 25,600/2574 = 9.95$

$R_s/R_1 = 200/117 = 1.71$

The designer must be careful to follow the loads all the way down into the ground. This is the "load-path" concept.

Fig. 3.11.2 Forces acting on roof and walls

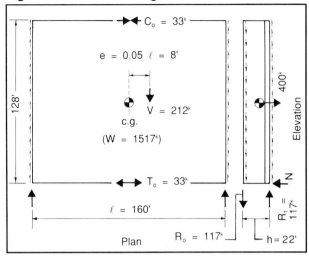

Strength analysis—roof

In this example, strengths of concrete components are analyzed using the load factors of ACI 318-89, and strength of structural steel parts by working stress methods. Reinforcing bars are ASTM A615, Grade 60. (See Chapter 6 for guides in welding reinforcing bars.)

Following the load path, the diaphragm is first analyzed for shear. The applied design shear in the double tee flanges is:

$$v_u = 1.4 v_o = 1.4(0.91) = 1.28 \text{ klf}$$

(The load factor, 1.4, is derived from Sects. 9.2.2 and 9.2.3 of ACI 318-89. From Eq. 9-2, $U = 0.75(1.7 \times 1.1E) = 1.4E$, and from Eq. 9-3, $U = 1.3 \times 1.1E \approx 1.4E$. Also, UBC-88 stipulates a load factor of 1.4.)

The design shear strength of the reinforced concrete in the double tee flanges, using a ϕ factor of 0.6 per UBC-88, and using a minimum ρ_n of 0.0025, is:

$$v_{ru} = \phi(A_{cv})(2\sqrt{f'_c} + \rho_n f_y)$$
$$= 0.6(12)(2)[2\sqrt{6000} + 0.0025(60,000)]$$
$$v_{ru} = 4.39 \text{ kips per lin ft} > 1.28$$

This is greater than required, therefore, satisfactory.

The double tee flanges must be connected at their edges to each other and to the end shear walls. This is analogous to providing shear strength along vertical joints in the web of a plate girder, and is done by weld clips as shown in Fig. 3.11.3(a), (b) and (c). This clip is analyzed by truss analogy, illustrated in Fig. 3.11.3(d). The design forces in the bars, and their resultant along the double tee edge, are:

$$C_u = T_u = \phi A_s f_y = 0.9(0.31)(60)$$
$$= 16.7 \text{ kips}$$
$$V_{Ru} = (C_u + T_u) \cos 45° = (16.7 + 16.7)(0.707)$$
$$= 23.6 \text{ kips}$$

Hence, at the junction with the end walls, the spacing between clips (as limited by horizontal or diaphragm shear) must be no more than:

$$s = V_{Ru}/v_u = 23.6/1.28 = 18.4 \text{ ft, but use 10 ft}$$

Ratioing by distance from the center of the roof, the shear between the first and second double tee and the corresponding spacing between clips, is:

$$v_u = (72/80)(1.28) = 1.15 \text{ klf}$$
$$s = 23.6/1.15 = 20.5 \text{ ft, but use maximum spacing of 10 ft}$$

Ten feet is a reasonable maximum spacing. However, in an earthquake, the double tee next to the end wall is subject to vertical bouncing. This could destroy the essential connection of diaphragm to shear wall, unless the connection is strengthened sufficiently to yield the double tee flange as a cantilever in bending where it joins the first interior stem. The flanges of the double tees used here are reinforced with 4 x 4-W2.1 x W2.1 welded wire fabric ($A_s = 0.062$ in.^{2}/ft), ½ in. clear to top. If standard weld clips are placed at 4 ft on centers (2 per wall panel), they will easily force uniform yield in the flange. Hence, the clips are spaced:

$$s = 10 \text{ ft on centers typically, and}$$
$$s = 4 \text{ ft on centers at end walls}$$

The double tee flanges must also be connected to the north and south walls, in order to transfer the "VQ/Ib" type web shears to the chords. This is analogous to the connection between web and flange of a plate girder. This same connection must also function as a tension tie, by holding the wall panels onto the roof against wind and earthquake, and by holding the building together against shrinkage and related forces.

The connection which holds the wall panel to the roof must be designed for an earthquake force, F_p, which is obtained by multiplying the tributary weight of this part of the building, W_p, by an acceleration, ZIC_p. Z and I are the same zone and occupancy factors as before. C_p is a factor used to correlate for past good and bad experience in earthquakes. The UBC-88 requires a C_p of 0.75 for connections of wall panels, but also requires that the fasteners, such as bolts, inserts, welds, dowels, etc., be designed for four times the load determined from the above formula. Thus, for most parts of the connection, design acceleration is:

$$ZIC_p = 0.30(1.0)(0.75)(4) = 0.90$$

Fig. 3.11.3 Connections between flanges of roof and double tees

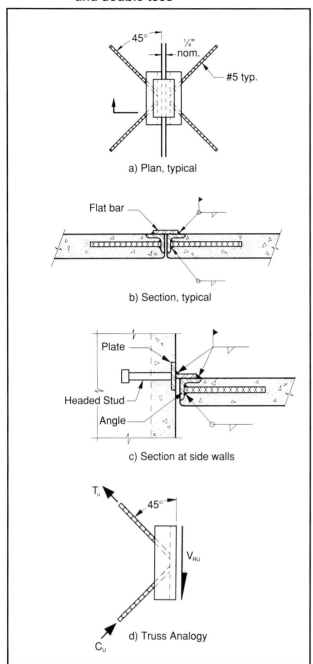

a) Plan, typical

b) Section, typical

c) Section at side walls

d) Truss Analogy

Fig. 3.11.4 Connections between roof double tees at (a) side walls and (b) ledger beam

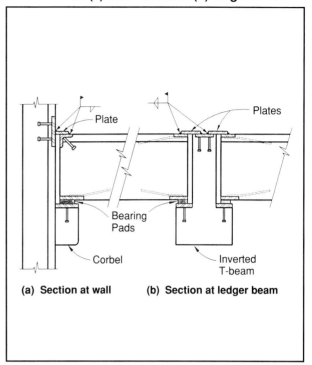

(a) Section at wall (b) Section at ledger beam

load end rotation of the simple span roof tees. This is done by welding the roof and wall tees together at the top of the roof tee, but not at the bearing (Fig 3.11.4). The bearing will then slide under live load end rotation, and friction at the bearing produces tension in the top connection.

In the case illustrated, bearing pads are used; the deformation of the pads will minimize the horizontal restraint at the bearing, and the connection design will be governed by the earthquake force. The bar anchorage is sized thus:

$T_u = (1.4)F_p = 1.4(2.78) = 3.89$ kips

$A_s = T_u/\phi f_y = 3.89/(0.9 \times 60) = 0.072$ in.²

Use 2 – #3 bars ($A_s = 0.22$ in.²)

The flange, or "chord" reinforcement is sized for tension as follows:

$T_u = C_u = 1.4T_o = 1.4(33) = 46$ kips

$A_s = T_u/\phi f_y = 46/(0.9 \times 60) = 0.86$ in.²

Use 2 – #6 bars ($A_s = 0.88$ in.²)

These bars are located as shown in Fig. 3.11.5. The cross-sectional area of the plates and the welds should be sufficient to ensure that yielding in the bars occurs first.

The diaphragm must be continuous in shear over the centerline ledger beam (see Fig. 3.11.4b). For the design of this connection, refer to Sect. 3.6 and Fig.

Tributary weight of an 8-ft wide wall panel is:

W_p (half ht + parapet) = (11.5)(8)(0.067)
= 6.16 kips

The corresponding lateral earthquake force is:

$F_p = ZIC_pW_p = 0.90(6.16) = 5.55$ kips
= 2.78 kips per double tee stem

As this is equivalent to a wind force of 60 psf, wind is no problem. The connection must transmit diaphragm shear and at the same time allow for the live

Fig. 3.11.5 Flange or chord reinforcement

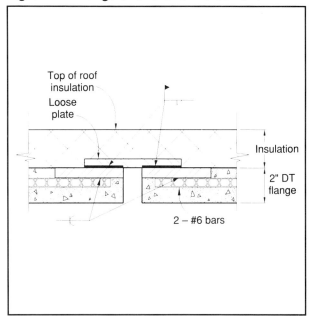

3.6.1. To provide ductility, the weld should be designed for 1.33 times the yield strength of the connection, and the headed studs made long enough to preclude a premature shear cone failure. Reinforcing bars may be substituted for the headed studs.

This completes the strength analysis of the diaphragm for N-S earthquake. For E-W earthquake, chords would also be needed in the east and west walls.

Strength analysis—walls

Following the load path, the shear walls are analyzed next. For simplicity, it is assumed the walls have no openings. Thus, there are interior and exterior (corner) wall panels, as shown in Fig. 3.11.6. Connections are made across the vertical panel joints to take advantage of the fact that compensating forces are generated in the panels. Considering an interior panel:

$V(h) = V_1(b) + W(b/2 - a)$

$V_1 = [V(h) - W(b/2 - a)]/b$

For vertical equilibrium:

$C = W$

Since this force system can exist for all interior panels, edge shears will balance to zero, when all panels have the same dimensions and weight. The only requirement for the connections is a transfer of vertical shear. Therefore, longitudinally sliding connections can be used if volume restraint is of concern. At the exterior panels, the edge shear V_1 from an exterior panel will be applied at one edge only. Because tension and compression base connections are not located at the panel edges, equilibrium may have to be satisfied with tension and compression connections to the foundation.

At the tension side exterior panel, equilibrium can be determined by summing moments about the compression force:

$T = [V(h) - W(b/2 - a) - V_1(a)]/d = V_1(b - a)/d$

$C = T + W - V_1$

At the compression side exterior panel, equilibrium can also be determined by summing moments about the compression force:

$T = [V(h) - W(b/2 - a) - V_1(b - a)]/d = V_1(a)/d$

$C = T + W + V_1$

For this example, stemmed panels are used for the walls. Connections are most conveniently located at the stems, so the pertinent dimensions are:

$b = 8$ ft, $a = 2$ ft, and $d = 4$ ft

Working with factored loads:

$V_u = 1.4[(0.91)(8)] = 10.19$ kips per panel

$W_u = 0.9(8)(20.5)(0.067) = 9.9$ kips

Fig. 3.11.6 Forces acting on wall panels

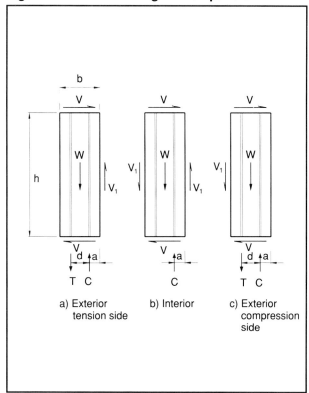

a) Exterior tension side b) Interior c) Exterior compression side

For the typical interior panel:
$V_{1u} = [10.19(18) - 9.9(8/2 - 2)]/8 = 20.45$ kips
$C_u = 9.9$ kips;
$V_u = 10.19$ kips

Weld clips similar to Fig. 3.11.3 will be used in this example at the vertical joints. The number required is 20.45 kips/23.6 kips per connection = 0.87; use 2.

The shear transferred to the foundation might be accomplished with weld plates, dowels, splice sleeves, or similar methods.

The tension side exterior panel will require a tie-down connection in this example.

$T_u = 20.45(8 - 2)/4 = 30.68$ kips
$C_u = 30.68 + 9.9 - 20.45 = 20.13$ kips
$V_u = 10.19$ kips

The compression side exterior panel will also require a tie-down connection.

$T_u = 20.45 (2)/4 = 10.23$ kips
$C_u = 10.23 + 9.9 + 20.45 = 40.58$ kips
$V_u = 12.32$ kips

The controlling tie-down force is $T_u = 30.68$ kips

The reinforcement required to resist T_u is determined as follows:

$$A_s = \frac{T_u}{\phi f_y} = \frac{30.68}{(0.90)(60)} = 0.57 \text{ in.}^2 \text{; use } 2-6\# \text{ bars}$$

Weld of No. 6 bar to plate, to ensure yielding of bar (Table 6.20.3):

$\ell_w = 1.33(3.0) = 4$ in.

Connector plate area required:

$30.68/(0.9)(36) = 0.95$ in.²; use 4 in. long plate

Connector plate thickness required to ensure yielding of the 2 – #6 bars:

$t = (1.33)(60)(0.88)/(4)(0.9)(36) = 0.54$ in.

Therefore, use connector plate 4" × ⅝" × 6," with 45° bevel and ⅛" × ½" × 4" backing plate.

Plate in foundation: use 4 – #4 bars (Grade 60) to anchor the plate:

$A_s = (4)(0.20) = 0.8$ in.² OK

Determine plate thickness:

$(M_u)_{pl} = \frac{T_u \ell}{A} = (30.68)\left(\frac{2}{4}\right) = 15.34$ in.-kips

Fig. 3.11.7 Wall double tee connections

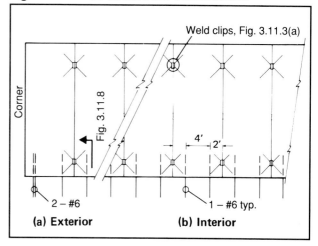

(a) Exterior (b) Interior

Fig. 3.11.8 Wall-footing connections

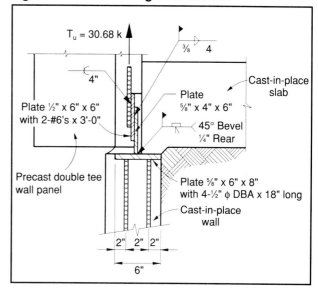

where ℓ = spacing of anchors

$Z_{req} = 18.93/(0.9 \times 36) = 0.58$ in.³

$t_{req} \geq \sqrt{\frac{(4)(0.58)}{(8)}} = 0.54$ in.

Use plate ⅝" × 6" × 8," with 4 – ½ in. diameter × 18 in. DBA bars.

Details for wall panel flange connections and footing connections are shown in Figs. 3.11.7 and 3.11.8.

Check the wall panel reinforcing to determine if two curtains of steel are required, (ACI 318-89, Sect. 21.5.2.2):

$2A_{cv}\sqrt{f'_c} = 2(12)(4)\sqrt{6000} = 7.44$ klf

This is much in excess of the actual factored shear. Two curtains are not required.

Check to see if a boundary member is required (ACI 318-89, Sect. 21.5.3.1):

Gross section, neglecting legs:
A = (4/12)(128) = 42.67 ft²
S = (4/12)(128²/6) = 910.2 ft³
Roof lateral load = 117 kips
Wall lateral load = 10.99 x 16 x 0.14 = 24.6 kips
≈ 25 kips
P_u = 1.4(W_p)(16) = 1.4(10.99)(16) = 246 kips
M_u = 1.4[(117)(18) + 25(20.5/2)] = 3307 ft-kips
f = P/A + M/S = 246/42.67 + 3307/910.2
= 9.40 ksf, or 65 psi
0.2 f'_c = 1200 psi, > 65 psi
Thus boundary elements are not required.

3.11.8 Example— 4-Story Building

General

The following is an example analysis of a four-story building in which the structural system resisting lateral forces is of the box type, that is, floors and roof are horizontal diaphragms, and exterior walls are shear walls. The building is rectangular in plan, and dimensions and layout are shown in Figs. 3.11.9 and 3.11.10. Floors and roofs are made up of 8 ft wide precast, prestressed concrete double tees, which are supported by exterior bearing walls and a central interior frame of precast, reinforced concrete beams and columns. There are smaller frames at the two elevator-stairway shafts, one at each end of the building. The interior frames are made up of full height columns and short beams. The exterior walls are 8 ft wide precast, prestressed concrete double tees placed on end, and are continuous from top of footing to top of parapet. Seismic code is UBC-88.

Approximations and partial designs

In the example, several approximations are made, but they are all on the conservative side. The designs are not complete in all respects, but in the case of several similar parts the more critical part is analyzed to illustrate procedure. Some items, such as number or spacing of connections, are in some cases determined by judgment.

Weights

The unit dead loads are:
Roof: 100 psf
Floors: 110 psf
Walls: 90 psf max., 67 psf average

The dead load assigned to each level of the building is the weight of the roof or floor plus the walls from midstory to midstory (only walls for first level).

This gives values of w_x as listed in Table 3.11.1 (column 3).

Equivalent static loads

The building is located in a seismic Zone 2A:

Z = 0.15
The structural system is bearing walls:
R_w = 6
The site coefficient is (say):
S_2 = 1.20
The fundamental period is approximately:
T = $C_T(h_n)^{3/4}$ = 0.02(50)$^{3/4}$ = 0.376 sec.
W = 3823 kips, as shown in Table 3.11.1
C = 1.25S/T$^{2/3}$ = 1.25 x 1.20/0.376$^{2/3}$
= 2.88 > 2.75
The base shear is then:
V = ZICW/R_w = (0.15 x 1.0 x 2.75 x (3823)/6
= 263 kips

The base shear is distributed over the height of the building. In this example, the extra lateral force at the top level is F_t = 0, since T < 0.7 sec. The distribution to the several levels is then:

$$F_x = \frac{(V - F_t)w_x h_x}{\sum_{i=1}^{n} w_i h_i} = \frac{263 w_x h_x}{120,300} \quad \text{(Eq. 3.11.3)}$$

which gives the values of F_x listed in Table 3.11.1 (Column 4). These lateral loads are used for design of the building as a whole, like shear in and overturning of walls. For floor and roof diaphragms the following equations are used:

$$F_{px} = \frac{F_t + \sum_{i=x}^{n} F_i}{\sum_{i=x}^{n} w_i} w_{px} \quad \text{(Eq. 3.11.4)}$$

0.35 ZIw_{px} < F_{px} < 0.75 ZIw_{px}

The resulting values of F_{px} are listed in Table 3.11.1 (Column 8). Note that these loads include the effects of the weights of walls that are parallel to the direction of motion, resulting in a conservative design.

Table 3.11.1

(1) Level	(2) h_x, ft	(3)* w_x, kips	(4) F_x, kips	(5)	(6)	(7) (6)/(5)	(8) F_{px}
Roof	50	831	90.8	831	90.8	0.1093	91.0
4th	39	926	79.0*	1757	169.8	0.0966	89.7
3rd	28	926	56.7*	2683	226.5	0.0844	78.3
2nd	17	962	35.8	3645	262.3	0.0720	69.3
1st	2	178	0.8	—			
Totals		3823	263.1				

*w_{px} is assumed = w_x

Fig. 3.11.9 Four-story example building—elevations

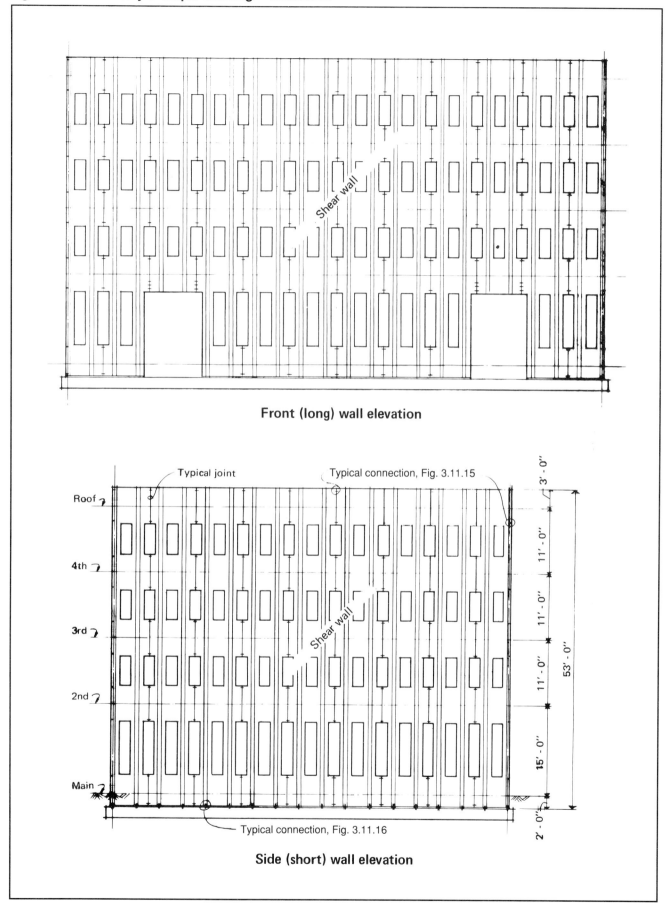

Fig. 3.11.10 Four-story example building—typical floor plan

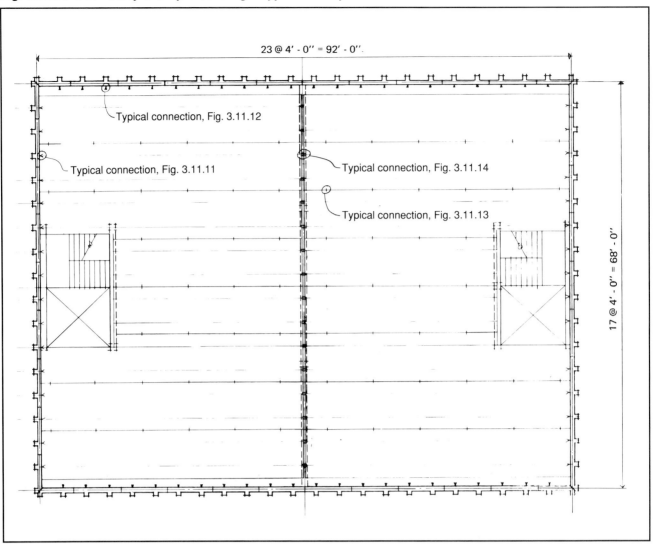

Chord forces in diaphragm

The roof is the most heavily loaded diaphragm, with a total lateral load of 91 kips. With the diaphragm spanning in the long direction, the moment $T(d) = 1/8 W\ell$ and the chord force

$$T = \frac{w\ell}{8d} = \frac{91(92)}{8(68-2)} = 15.8 \text{ kips}$$

Design of the chord reinforcement:

$$A_s = T_u/\phi f_y = 1.4(15.8)/(0.9)(60) = 0.41 \text{ in.}^2$$

Use 2 – #5 bars placed as shown in Figs. 3.11.11 and 3.11.12.

Requirements at lower levels and for the diaphragm spanning in the short direction are even less, but, for simplicity, the same diaphragm chord reinforcement is used throughout.

Fig. 3.11.11 Typical connection between diaphragm and wall at end of double tee

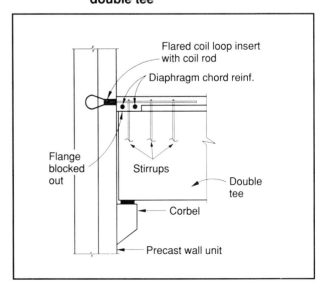

Fig. 3.11.12 Typical connection between diaphragm and wall at side of double tee

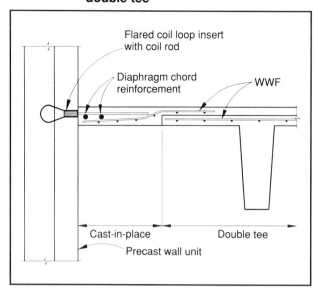

Shear stresses in diaphragm

This example building is assumed to be symmetrical, and the centers of mass and rigidity coincide. It is then necessary to include the effect of an arbitrary torsion, that is, the lateral load is offset from the center of rigidity by 5 percent of the maximum dimension of the diaphragm.

The torsional moment is resisted by all four exterior walls, depending on rigidities and distances from the center or rigidity. However, it is conservative to assume that the torsional moment is resisted by the two walls that are parallel to the direction of motion. For earthquake in the transverse direction, the maximum total shear to one wall is:

$R_o = 0.55(F_x \text{ or } F_{px}) = 0.55(91) = 50.1 \text{ kips}$

Deducting the length of the service shafts, the maximum unit shear is:

$v_o = R_o/\ell = 50.1/(68 - 20) = 1.04 \text{ klf}$

For earthquake in the longitudinal direction, the maximum unit shear to the long walls is less. For simplicity in production of elements and erection at the site, the 1.04 klf is used throughout all diaphragms and for their connections to the walls.

Typical connections for interior of diaphragm

A typical connection transmitting shear between double tees is illustrated in Fig. 3.11.13. In each edge there is a small angle anchored with two bars at 45°. The shear strength is calculated from force components in the anchor bars, as in the previous example:

Fig. 3.11.13 Typical diaphragm connections between double tees

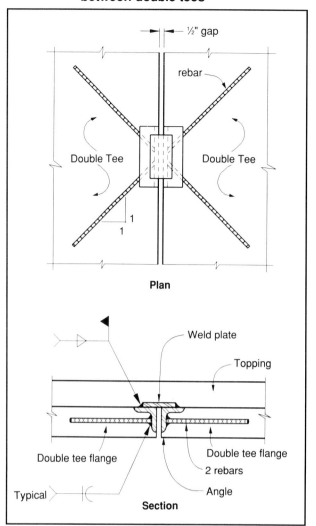

Fig. 3.11.14 Typical end connection between double tees

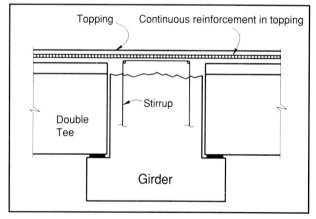

$V_{Ru} = 23.6 \text{ kips}$

To equalize cambers and deflections in neighboring elements, a maximum spacing should not exceed about 8 ft. This provides a resisting unit shear of:

$v_u = 23.6/8 = 2.95$ klf

which compares to the requirement of:

$1.4(1.04) = 1.46$ klf

The cast-in-place strip along the longitudinal wall, Fig. 3.11.12, is reinforced with 8 x 4-W2.9 x W2.9 WWF. By shear-friction, this is good for:

$V_{Ru} = \phi A_s f_y \mu = 0.85(0.087)(60)(1.0)$
$= 4.43$ klf

which is several times that required.

Across the center of the building, the double tee elements are supported by a beam, and they meet end to end. It is desirable to make the shear path stay at the level of the flange, that is, making the path as direct as feasible. The easiest way to achieve this is by use of continuous reinforcement in the topping as shown in Fig. 3.11.14. Even so, it may be desirable to use some of the hardware type connections to make the building resistant to lateral loads during construction.

Connection of diaphragm to wall

First consider the connections between the ends of the double tees and the short wall. The ribs of the double tees and of the wall elements are aligned, and the vertical loads are carried by corbels projecting from the interior wall surface. Two requirements must be satisfied: a) horizontal shear between diaphragm and wall, and b) direct horizontal tension caused by the tendency of the wall to fall away from the building. The two requirements will be superimposed because earthquake motion may occur in any horizontal direction, and is not limited to the principal axes of the building.

It is convenient to space the connections 4 ft apart so that they will occur at wall ribs. At each connection the forces are:

a) Horizontal shear, $V = 1.04(4) = 4.16$ kips

This can be handled by shear-friction. Then the tensile capacity must be:

$V_u/\mu = 1.4 \times 4.16/1.0 = 5.8$ kips

b) Direct horizontal tension is given by:

$F_p = ZIC_p W_p$

UBC-88, Table 23-P, gives $C_p = 0.75$. For the weight of wall contributing to one connection, a unit weight of 90 psf over an area of 4 x 14 = 56 ft² will be used, so that:

$W_p = 0.090(56) = 5.04$ kips

$F_p = 0.15(1.0)(0.75)(5.04) = 0.57$ kips

The minimum anchorage between walls and floors, per UBC-88 Sect 2310, is 200 lb per ft. Thus

$F_p = 0.2(4) = 0.80$ kips

Design tensile strength is $1.4(0.80) = 1.12$ kips

The connection must also satisfy the requirements of structural integrity; refer to Sect. 3.10.

A choice must be made of the type and size of connection. The coil-loop insert with continuous threaded coil rod is a simple and useful type, and there are several proprietary makes. The installation is illustrated in Fig. 3.11.11. A note of caution is in order: the coil-loop insert and the coil rod should each be amply anchored in their respective parts so that failure by yielding in the coil rod is ensured.

A continuous tie is required, per UBC-88, Sect. 2312. The connections between the edge of the double tee and the long wall are of the same kind, and are calculated in the same way. The primary difference is caused by the cast-in-place strip of diaphragm adjacent to the wall. The threaded coil rods extend into this strip and overlap the welded wire fabric that projects out of the edge of the double tee flange (see Fig. 3.11.12).

Special situation at service shafts

From the typical floor plan, Fig. 3.11.10, it is seen that several wall ribs are not tied to the floor, and the wall, because of vertical joints, is not suited to span horizontally past the shaft openings. This situation is readily alleviated with a strongback type beam placed horizontally at each floor level, and spanning the shaft opening. The adjacent connections at each side of the shaft will have to be stronger than the standard connection designed in the preceding section.

Shear wall

The exterior walls of the building provide the horizontal reaction for the diaphragms. These forces are in the plane of the wall and become horizontal shears, therefore the term shear wall. The end walls in the example building do double duty in that they are also bearing walls.

The design of the individual wall elements can be important, but will not be covered here. For the example, next examine the shear in the joints, and overall stability. The equivalent static loads, which are assumed to act horizontally at each floor level, cause a tendency for the wall to overturn. Gravity loads and footing will stabilize the wall and prevent overturning, but internal shear and direct stresses will develop. In this case, no advantage will be taken of the slightly reduced overturning permitted by UBC-88, but the full value of the cantilever moment will be used. Results of the calculations are shown in Table 3.11.2 for one wall.

Shear in shear wall

The maximum shear in the short end wall due to the equivalent static loads and the arbitrary 5 percent eccentricity is

V = 0.55(263) = 145 kips

v = 145/68 = 2.13 klf

v_u = 0.85(2.13) = 1.81 klf

This unit shear occurs near the bottom of the wall, and is both horizontal and vertical.

A typical connection for the vertical joints between wall elements is shown in Fig. 3.11.15. In each edge there is a small angle anchored with two No. 5 bars. With 6-in. thick walls, it is possible to drypack the joints effectively. Apply the shear-friction concept, with the anchor bars placed at 90° to the joint. The shear capacity of one connection is:

V_{Ru} = $\phi A_s f_y \mu$ = 0.85(2)(0.31)(60)(1.0)

= 31.6 kips

Near the bottom of the wall, the average vertical spacing should not be more than 31.6/1.81 = 17.4 ft, however, use three connections in each story. Since windows cross the joints, be sure to allow for this when locating the connections.

Moment in shear wall

The maximum moment in the short end wall, with allowance for the arbitrary 5 percent eccentricity, is:

M = 4911 x 1.05 = 5157 ft-kips

The maximum gravity load is:

W of one end wall, 68(53)(0.067)	= 241
W of ¼ of roof, 68(92/4)(0.100)	= 156
W of ¼ of floors, 68(92/4)(0.110)(3)	= 516
Total W	= 913 kips

With wall ribs neglected, section properties in linear measure are, area A_ℓ = 68 ft, and section modulus Z_ℓ = (1/6)(68)² = 771 ft².

Fiber stresses at bottom of wall are:

$$f = \frac{W}{A} \pm \frac{M}{Z} = \frac{913}{68} \pm \frac{5157}{771} = 20.1 \text{ klf and } 6.7 \text{ klf}$$

These values are a good measure of the stresses in the gross wall section, and can also be used to calculate the necessary width of wall footing. Additional refinement of stress calculations are necessary at window openings.

Table 3.11.2 Shear and moments by level

Level	V_x, kips	M_x, ft-kips
Roof	45.4	0
4th	84.9	499
3rd	113.3	1433
2nd	131.2	2680
1st	131.6	4648
Top of footing		4911

Fig. 3.11.15 Typical connections, wall to wall

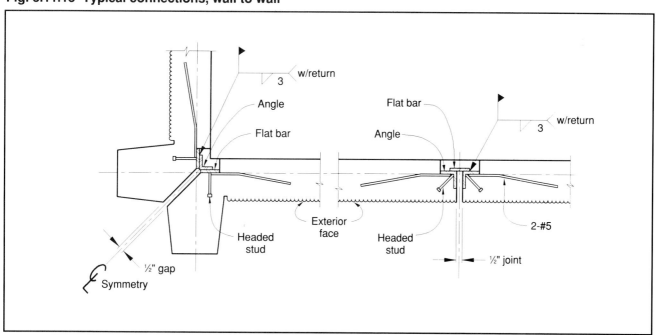

Fig. 3.11.16 Typical connection of wall to footing

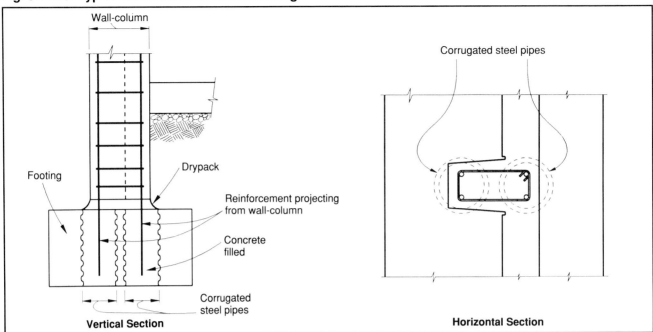

Other shear walls

The long shear walls, which provide the resistance to earthquake in the longitudinal direction of the building, are less critical than the short walls, and they will not be discussed. Details of design will be the same as for the short walls.

There is a special situation at the building's entrance doors. The wall panel above the doorway is supported by connections to neighboring units. The necessary strength can be provided by two additional connections in each joint above the doorway.

Connections of walls to footings

The footings can be proportioned on the basis of reactions from the walls, as indicated above. Each wall element can be connected to the footing by projecting the main reinforcement of the ribs downward into blockouts in the footings, and filling the blockouts with concrete. The space between wall elements and footing must be drypacked as shown in Fig. 3.11.16.

3.11.9 Example — 3-Level Parking Structure

General

Over the years precast, prestressed concrete parking structures have proven to be a reliable, economical means of providing attractive and secure parking for the public. Since prestressed concrete is capable of long spans which accommodate ease of traffic circulation, a structure with minimum structural elements for vertical load support within the area of parking results. The resulting openness requires the designer to consider, early in the planning, the method of providing for lateral load resistance. As is demonstrated in this design example, sufficient shear walls can be provided in a manner so as to not intrude upon the functionality of the design. A detailed presentation of all aspects of precast concrete parking structures may be found in Ref. 16.

In this example, the applicable building code is the Standard Building Code, 1988. It should be noted that the requirements of the BOCA National Building Code, 1990, are similar. This structure is in seismic Zone 2, and the design wind speed is 100 mph. A conceptual determination of the lateral load resisting system is first given, followed by a detailed analysis of key elements and their connections.

The structure is shown in Fig. 3.11.17.

Lateral load resisting system — concept

The functional design has resulted in a structure with three, 60-ft wide bays in one direction, and a 264-ft length in the other direction. Three elevated levels, with 10 ft 6 in. floor-to-floor height are needed to provide the required number of parking spaces. It has been determined that the structure is to be open on all four sides.

Since the structure must resist Zone 2 seismic forces and the forces generated by a 100-mph wind, as well as the vertical loads, the magnitude of generated forces must first be determined in order to determine which combination of loads will control. By arranging support elements in a manner as to resist both vertical and lateral loads, the goal of maintaining an open structure will be achieved.

Fig. 3.11.17 Layout of parking structure

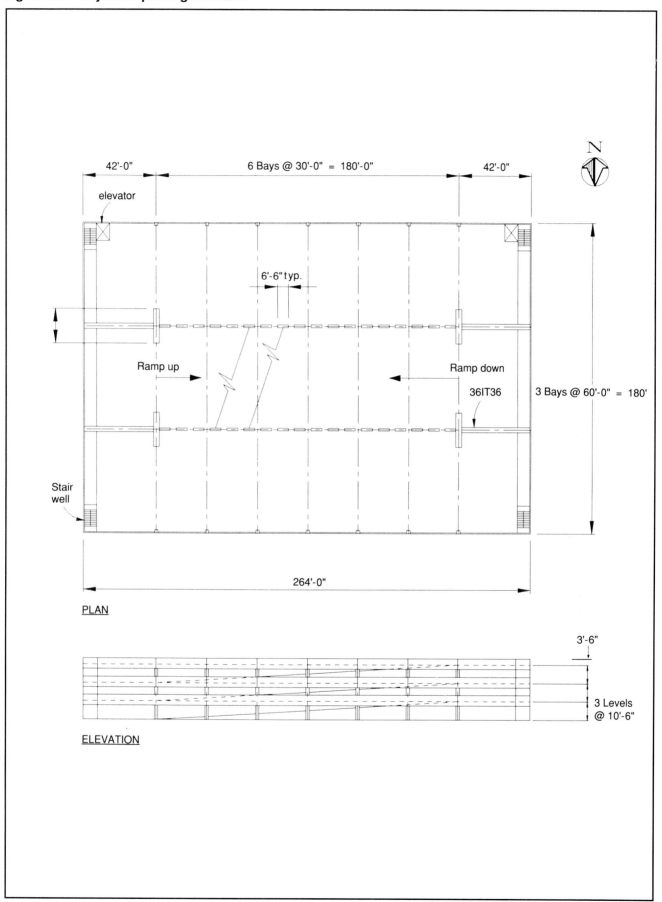

Load analysis

Obtain an approximate magnitude of the applied loads. For gravity loads, 26 in. deep, 10 ft wide pretopped double tees will be used. The total weight of double tees, beams, columns, and curbs will be taken as 110 psf. The code specified live load is 50 psf. It is determined that for this magnitude of loading, 30-ft bays with 24 in. square columns, and a 36IT36 girder in the end bays, will support vertical loads.

For lateral loads, a check is made to determine whether seismic or wind will govern the design. The basic wind pressure is 30 psf, and, with a 1.2 factor for combination of direct and suction forces, the total wind force in the north-south direction is (0.030)(1.2)(35)(264) = 333 kips. Similarly, the total wind force in the east-west direction is 227 kips. The code seismic force is determined from the equation V = ZIKCW, where W per level is 110 psf dead load plus 25% of the 50 psf live load (use group S). Therefore, W per level is (0.1225)(180)(264) = 5821 kips. Since shear walls will be provided as the lateral load resisting system, K = 1.33. I = 1.0, and CS will be assumed as the maximum value, 0.14; for Zone 2 with A_v = 0.15, Z = 0.375. Therefore, V = (0.375) x (1.33)(1.0)(0.14)W = 0.07W. Consequently, for this 3-level structure, the total seismic shear in the north-south and east-west directions is equal to (0.07) x (3)(5821) = 1222 kips. Seismic forces govern in both the north-south and east-west directions.

Substantial shear resisting elements are required. Load bearing shear walls are chosen, primarily because the vertical gravity load will help resist the overturning moments due to applied lateral loads. While the corner stairwells and elevator shafts could be used as part of the lateral load resisting system, this may result in high forces due to restraint of volumetric deformations; consequently, it is decided that the corners will be isolated from the main structure. Alternatively, it might have been decided to use these corner elements, and provide connections which are flexible in the direction of volumetric restraint.

A preliminary distribution of seismic shears is next made. Since equivalent static loads approximate an inverted triangle, the first level is assigned a unit shear of 1, the second level a unit shear of 2, and the third level a unit shear of 3, as shown in Fig. 3.11.18.

For the north-south load resisting system, try an 8 in. thick load bearing shear wall, located at each end of the ramp. These walls support the 36IT36 girder, and may be as long as 30 ft without interfering with the traffic flow; a 20-ft length is used as a first try. Fig. 3.11.19 illustrates the arrangement and loading. Since there are 4 walls, the total shear per wall is 1222/4 = 306 kips, and the total overturning moment is 30,000/4 = 7500 ft-kips. The wall supports its own weight plus a superimposed dead load of (21)(60)(3)(0.110), or a total of 480 kips. Assuming an 18 ft. moment arm, and using ACI 318-89 load factor 0.9D ± 1.4E, the net factored uplift force will be [(1.4)(7500) − (0.9)(480)(10)]/18 = 343 kips. This can be satisfied by 343/(0.9)(60) = 6.35 in.², or 5 − No. 10 bars which will be located so as to be centered 2 ft from each end of the wall. The force transfer between precast shear wall and the foundation can be accomplished by reinforcing bars with splice sleeves to develop full 125 percent tension splices. Alternatively, post-tensioning bars could be chosen. The preliminary analysis is completed by examining the capacity of the foundation system to transfer this force to the supporting ground; this analysis is not shown here.

For resistance in the east-west direction, 36 individual load bearing walls located along the length of the interior ramped bay will be used. These 8-in. thick walls are spaced 10 ft on centers, supporting one 60-ft double tee each side of the wall, as shown in Fig. 3.11.20. Each wall is 6'-6" wide to accommodate the 5-ft stem spacing of the double tees, and to allow for visibility between the wall units. The total shear per wall is 1222/36 = 34 kips, and the total overturning moment per wall is 30,000/36 = 833 ft-kips. Each unit supports its own weight plus superimposed dead load of (60)(10)(0.110)(3), or a total of 218 kips. The approximate new uplift per wall will be [1.4(833) − 0.9(218)(3.25)]/5 = 106 kips. This can be satisfied by 106/(0.9)(60) = 1.96 in², or 2 − No. 9 bars centered 18 in. from each end of each wall.

In the final design, a more accurate determination of forces, including the effect of torsional eccentricity, will be prepared. For a structure of this configuration, the final forces will be within about 10 percent of those calculated here.

Diaphragm—general

The floors act as a diaphragm to transfer the lateral loads to the shear walls. Prior to analyzing the diaphragms, it is necessary to determine if a concentrated seismic force need be applied to the top level. As specified in the code, a concentrated force F_t must be applied to the top level when the structural period T is greater than 0.7 seconds. Using D_s = 6.5 ft in the east-west direction, and 20 ft in the north-south direction, and h_n = 31.5 for both directions:

$$T_{EW} = \frac{(0.05)(31.5)}{\sqrt{6.5}} = 0.62 \text{ sec}$$

$$T_{NS} = \frac{(0.05)(31.5)}{\sqrt{20}} = 0.35 \text{ sec}$$

After calculating story shears, the individual shear force on each diaphragm is calculated. The diaphragm shear is not the same as the story shear. As stated in the Code, "floor and roof diaphragms and collectors shall be designed to resist forces determine in accordance with the formula:

Fig. 3.11.18 Summary of vertical distribution of seismic shears

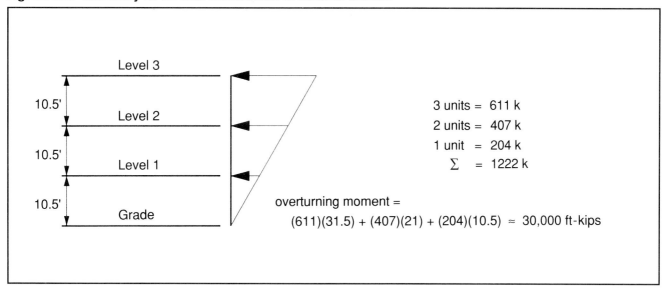

Fig. 3.11.19 Conceptual sketch of typical N-S shear wall

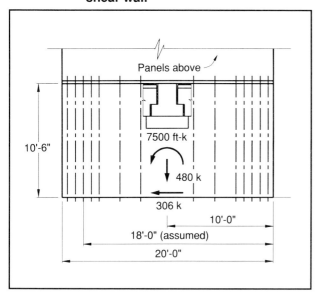

Fig. 3.11.20 Conceptual sketch of typical E-W shear wall

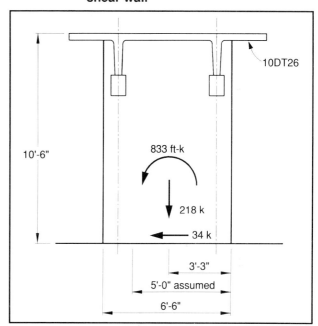

Table 3.11.3

(1)	(2)	(3)	(4)[1]	(5)	(6)	(7)	(8)[2]	(9)
Level	x	h_x, kips	w_x, kips	F_x, kips	Σw_i, kips	ΣF_i, kips	(7)/(6)	F_{px}, kips (8)(4)
3	3	31.5	5821	611	5821	611	0.105	611
2	2	21.0	5821	407	11,642	1018	0.087	506
1	1	10.5	5821	204	17,463	1222	0.070	407
Totals			17,463	1222				

Notes: 1. w_{px} is assumed equal to w_x
2. $0.3ZI = 0.1125$ and $0.14ZI = 0.0525$ as upper and lower bounds on column 8.

$$F_{px} = \frac{\sum_{i=x}^{n} F_i}{\sum_{i=x}^{n} w_i} w_{px} \qquad \text{(Eq. 3.11.5)}$$

The force F_{px} need not exceed $0.30Zlw_{px}$, and shall not be less than $0.14Zlw_{px}$.

These limits and forces are calculated and shown in Table 3.11.3.

North-south seismic forces on diaphragm

The diaphragm is modeled as shown in Fig. 3.11.21. Since the diaphragm forces on Level 3 are the greatest, only this level will be analyzed and designed. To simplify, the diaphragm will be divided into 3 separate analyses. One is the flat area, which is further divided by a line along the center of the ramp, and the other is the ramp area.

$w_1 = [(60)(264)/(264)(180)] \times [(611)/(264)] = 0.77$ klf

$w_2 = [(30)(42)/(264)(180)] \times [(611)/(42)] = 0.39$ klf

$w_3 = [(60)(180)/(264)(180)] \times [(611)/(180)] = 0.77$ klf

Because the overhanging cantilevers will reduce the stresses in the level area of the ramp diaphragm, the ramp diaphragm will be analyzed and the results used for the level area diaphragm as well. However, the negative moment for the overhangs will be determined.

Summary of north-south diaphragm analysis

$+M_u = (1.4)(0.77)(180)^2/8 = 4366$ ft-kips

$-M_u = (1.4)(0.77 + 0.39)(42^2/2) = 1432$ ft-kips

$V_u = (1.4)(0.77)(180)/2 = 97$ kips

$R_1 = (264/2)(0.77) + (0.39)(42) = 118$ kips

$R_2 = (180/4)(0.77) = 35$ kips

$R_{tot} = R_1 + R_2 = 118 + 35 = 153$ kips

R_u per wall $= (1.4)(153) = 214$ kips

Diaphragm—Moment Design

Assuming a 58-ft moment arm, $T_{3u} = (4366)/58 = 75.3$ kips. This tensile force may be resisted by reinforcing bars placed into field applied concrete topping or curbs located at each end of the double tees, or by reinforcing steel shop welded to plates cast in the edges of the double tee flanges. These plates would be connected together in the field across the joint using splice plates and welds. Various suggestions for connections are provided in Ref. 15.

$A_s = T_{3u}/\phi f_y = (75.3)/(0.9)(60) = 1.39$ in.^{2}
use 3 – #6 bars

$A_{pl} = 1.39 (60/36) = 2.32$ in.^{2}
use a plate 6" x ⅜" x 0'-9"

The required length of a 5/16 in. weld, using a factor 1.33 to assure yielding of the plate or reinforcement first is:

$\ell_w = 1.33(60)(1.39)[1.67(21)(0.70)(5/16)]$
$= 14.35$ in.; use $\ell_w = 15$ in.

The arrangement of reinforcement is as shown:

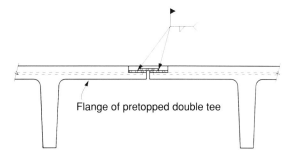

Flange of pretopped double tee

Shear Design

$V_u = 97/60 = 1.62$ kips per ft. If flange connectors are provided 5 ft on centers, V_u per connector $= 1.62 \times 5 = 8.10$ kips.

The arrangement of flange connectors to provide this resistance is shown at the top of the next column.

Fig. 3.11.21 North-south diaphragm analysis

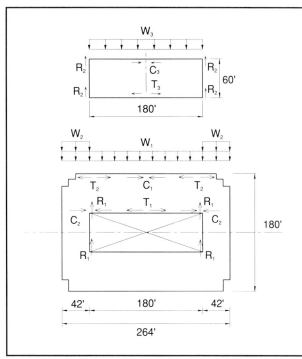

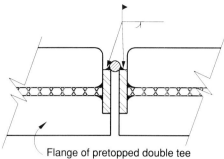
Flange of pretopped double tee

Conclusion

The preliminary analysis indicates that the presumed sizes and arrangement of load resisting elements is reasonable. Refinements as may be architecturally required are then made, and the final analysis performed.

3.11.10 Example—12-Story Building

Box-type buildings can be built to a height of 160 ft, according to UBC-88. For buildings approaching that height, the structural layout and connection details may differ from those of the example, Sect. 3.11.8, but the design procedure is the same. The base shear, its distribution as equivalent static loads over the height of the building, and the load paths down to the foundations, are calculated as illustrated in Sect. 3.11.8.

Fig. 3.11.22 shows a 12-story building which is square in plan. Wall units are 12 ft high and 30 ft or 15 ft in width. The "running bond" pattern illustrated is not an essential feature of seismic design, but it does help tie the building together, thus reducing the likelihood of progressive collapse in case of accidental overloads.

Floors consist of precast, prestressed hollow-core slabs, 12 in. deep and 4 ft wide. They bear on continuous corbels on the walls and on interior beams.

Buildings of this type are generally erected one story at a time, and location of joints, temporary bracing, and connection details must be correlated with the erection sequence. Fig. 3.11.23 indicates the erection cycle. The walls of the story below have been erected, braced, and fully connected. The hollow-core slabs are then erected, bearing on wall corbels and interior beams. The wall panels of the next story are then erected; these have dowels projecting downward into sleeves filled with grout in the wall panels below (Fig. 3.11.24a). Hardware in the vertical joints for welded connections is shown in Fig. 3.11.24(b). The next step is to place continuous welded wire fabric for the topping, and finally place the topping. The erection cycle can be repeated as soon as the grout and concrete have adequate strength, and the welding of the connections between the wall panels has been completed.

The connections shown in Fig. 3.11.24 are illustrations of methods of providing the diaphragms and con-

Fig. 3.11.22 12-story building—elevation

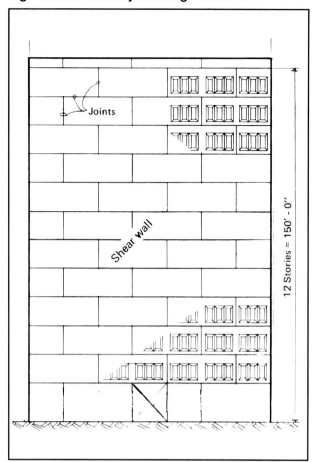

Fig. 3.11.23 12-story building—isometric view of assembly

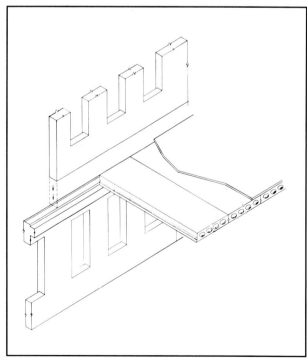

Fig. 3.11.24 Typical connections for 12-story example building—seismic loads

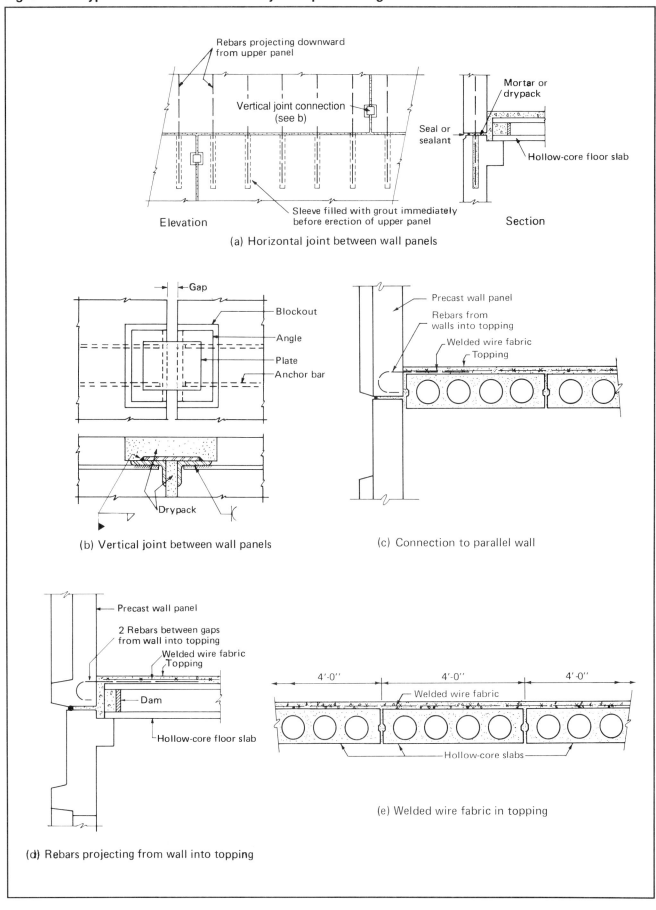

tinuous load paths discussed in the previous examples. Many other details have been used successfully on similar structures in high-intensity earthquake areas.

3.11.11 Example — 23-Story Building

Figs. 3.11.25 and 3.11.26 show a 23-story building 300 ft high. The building was designed for earthquake resistance with a ductile moment-resisting space frame.

All floors consist of 10-ft wide single-tee units that were precast and prestressed. Adjacent tees were connected by means of hardware in the flanges, but these connections were not considered as diaphragm connections. The tees have a structural concrete topping reinforced with welded wire fabric. The diaphragm is provided by the topping rather than by the flanges of the tees.

The core of the building has no structural walls or frames that contribute measurably to lateral stability of the building. Each single tee is supported at the core by a cast-in-place column, but the connection is immaterial for earthquake resistance. The core has additional columns and partial slabs, but in effect, the core creates a large square hole in the floor diaphragm. Reinforcement must be provided in the diaphragm around the core in the same manner as around any large hole in a slab.

The exterior walls of the building are the moment-resisting frames. They consist of reinforced concrete columns and spandrel beams. For these frames to be ductile, they must have closely spaced spirals in the columns and ties in the spandrels. At the intersections, the congestion of reinforcement is so great that it is not feasible to have the tees align with the columns. Instead they are located midway between columns, but again the connection of the tee stem to the spandrel is immaterial to the earthquake resistance. The vital connection here is between the flange and the spandrel. This was accomplished by reinforcement projecting out of the end of the tee flange, which is later embedded in the cast-in-place concrete of the spandrel.

3.11.12 Example — Architectural Precast Panel

General

The code requires that precast or prefabricated non-bearing, non-shear wall panels or similar elements which are attached to or enclose the exterior shall be designed to resist the inertia[6] forces and shall accommodate movements of the structure resulting from lateral forces or temperature changes. Panels typically have two rigid load bearing connections with volume change relief provided only by the ductility of the connections, and two or more tie-back connec-

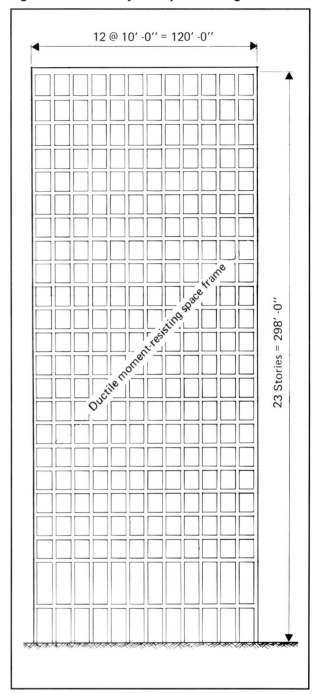

Fig. 3.11.25 23-story example building—elevation

tions with full freedom of movement in the plane of the panel.

Connections and joints between panels should be designed to accommodate the movement of the structure under seismic action. Connections which permit movement in the plane of the panel for story drift by bending of steel, sliding connections using slotted or oversized holes, or other similar methods are also permissible. Story drift, which is the relative movement of one floor with respect to the adjacent floor, must be considered when determining panel

Fig. 3.11.26 23-story example building— typical floor plan

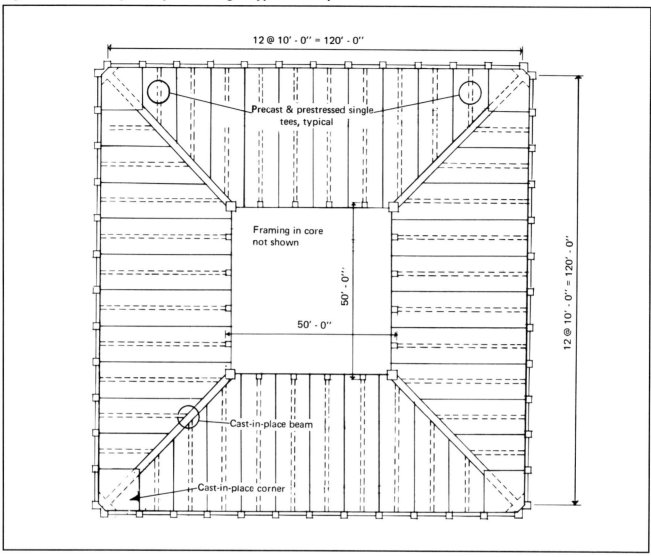

joint locations as well as connection locations and types. Between points of connections, non-load bearing panels should be separated from the building frame to avoid contact under seismic action.

For seismic forces, UBC-88 requires that the body of a connector be designed for a force equal to 1.33 times the inertia force, and that all fasteners (i.e., bolts) be designed for a force equal to 4 times the inertia force. Anchorage to concrete is required to engage reinforcing steel in a manner so as to distribute forces to the concrete and/or reinforcement thus averting sudden or localized failure. Since the force distribution philosophy is critical to design in high seismic zones, many designers specify confining hoop steel, or the use of long deformed anchor bars or reinforcement, as opposed to short headed studs or inserts. When studs or inserts are used, and located near panel edges, it is recommended that they be enclosed in sufficient reinforcing steel to carry loads

Fig. 3.11.27 Architectural panel section— example 3.11.12

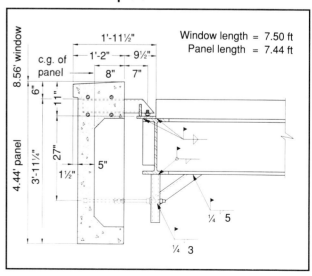

back into the panel, thus averting a sudden tensile failure mode in the concrete.

The following example illustrates the design of a non-load bearing architectural panel in a Zone 4 seismic zone; the design follows the requirements of UBC-88.

Material

Concrete: f'_c = 5000 psi, normal weight

Reinforcing steel: A615, Gr. 60

Structural steel: Plates and shapes A-36

f_y = 36 ksi

Tubes A-500 Gr. B

f_y = 46 ksi

Bolts A-307

Welds AWS E70

Loads

Seismic $F_p = ZIC_pW_p$ $Z = 0.4, I = 1$

For panel design: C_p = 0.75 (Sect. 2312, Table 23-P)

For body of connection: C_p = 1.00 (Sect. 2312(h)-2D)

For fasteners: C_p = 3.00 (Sect. 2112(h)2D)

Wind

Direct = 31 psf

Suction = 64 psf

Window weight = 10 psf

Story drift = 0.015 (story height)
= 0.015 (13 × 12) = 2.34 in.

Section Properties

A = 351 in.²

I_x = 117,379 in.⁴ I_y = 4663 in.⁴

y_t = 23.32 in. y_ℓ = 5.97 in.

y_b = 29.93 in. y_r = 8.03 in.

Gravity Loads

Panel weight = 351(150/144)(7.44) = 2720 lb

Window = 10(8.56)(7.5) = 642 lb

Total = 3362 lb

Vertical load on bearing connections
= 3362/2 = 1681 lb

Lateral load on tieback connections (e = 15 in.)
= 1681(15)/27 = 934 lb

Wind Loads

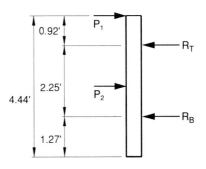

Assume ¼ of the total wind load on the window as a concentrated load at the top of the precast panel.

$P_1 = \dfrac{8.56(7.5)}{4}w = 16.05w$

$P_2 = (7.44)(4.44)w = 33.03w$

$R_T = \dfrac{16.05(2.25 + 0.92)w + 33.03(2.22 - 1.27)w}{2.25}$

= 36.6w

$R_B = \dfrac{33.03(2.22 - 0.92)w - 16.05(0.92)w}{2.25}$

= 12.5w

Wind load per connection:

R_T = 36.6(31)/2 = 567 lb (compression)

or

R_T = 36.6(−64)/2 = −1171 lb (tension)

R_b = 12.5(31)/2 = 194 lb (compression)

or

R_b = 12.5(−64)/2 = −400 lb (tension)

Seismic Acting Perpendicular to Panel

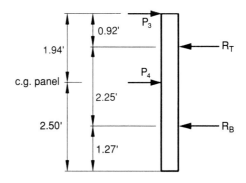

Lateral load = ZIC_pW_p
= $(0.4)(1.0)C_pW_p$
= $0.4C_pW_p$

For the window: $W_p = 642$ lb
For the precast panel: $W_p = 2720$ lb
Thus:
$$R_T = \frac{0.4 C_p}{2.25}[2720(2.50-1.27) + 642(0.92+2.25)]$$
$$= 957 \, C_p$$
$$R_B = \frac{0.4 C_p}{2.25}[2720(1.94-0.92) - 642(0.92)]$$
$$= 388 \, C_p$$

Seismic loads per connection:

Panel:	$R_T = 0.75(957)/2$	$=$	359 lb
	$R_B = 0.75(388)/2$	$=$	146 lb
Body:	$R_T = 1.0(957)/2$	$=$	478 lb
	$R_B = 1.0(388)/2$	$=$	194 lb
Fasteners:	$R_T = 3.0(957)/2$	$=$	1435 lb
	$R_B = 3.0(388)/2$	$=$	583 lb

Seismic Acting Parallel to Panel

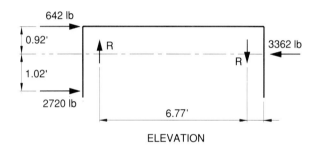
ELEVATION

Seismic load per connection:
$$R = \frac{(2720)(0.4)(1.02) - (642)(0.4)(0.92)}{6.67} C_p = 129 C_p$$

Panel:	$R = 0.75(129) =$	97 lb
Body:	$R = 1.0(129) =$	129 lb
Fasteners:	$R = 3.0(129) =$	387 lb

Bearing Connection

Refer to Chapter 6 for design procedure

DL + E: $V_u = 1.4(1681 + 129)$
$= 2534$ lb

Assume a 4 in. x 4 in. structural tube (see Sect. 6.12)

$\ell_e = 11.0$ in., $e = 7 + 11.0/2 = 12.5$ in.,
$b = 4$ in., $f'_c = 5000$ psi

Concrete:
$$V_c = \frac{0.85 f'_c b \ell_e}{1 + 3.6 e/\ell_e}$$
$$V_c = \frac{0.85(5000)(4)(11.0)}{1 + \frac{(3.6)(12.5)}{11.0}} = 36{,}739 \text{ lb}$$
> 2534 lb OK

Steel section:
$$e = a + \frac{V_u}{0.85 f'_c b} = 7 + \frac{2534}{(0.85 \times 5000 \times 4)}$$
$$= 7.15 \text{ in.}$$
$M = 1681 \times 7.15 = 12{,}019$ in.-lb
Allowable stress $= 0.60 f_y = 27{,}600$ psi
$$S_{req} = \frac{12{,}019}{27{,}600} = 0.44 \text{ in.}^3$$

Use ST 4" x 4" x 3/16": $S = 3.30$ in.^{3} > 0.44 in.^{3}
Bolt: use 3/4 in. (Table 6.20.9)
Allowable load $= 8800$ lb > 1681 lb

Plate:

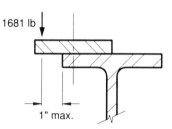

$M = 1681 \times 1 = 1681$ in.-lb
Try 3/4" x 5" plate; $S = 5(0.75)^2/6 = 0.47$ in.^{3}
Allowable stress $= 0.66 f_y = 23{,}760$ psi
$f_b = 1681/0.47 = 3570$ psi $< 23{,}760$ psi
Welds: try 3 in. length; $S = 2 \times 3^2/6 = 3.0$ in.^{3}
$f_w = 1681/3 = 566$ lb per in. Use 1/4 x 3 in. weld.
Allowable (Table 6.20.2) $= 3.71$ kips/in.

Tie-Back at Top (See plan view sketch pg. 3-85)

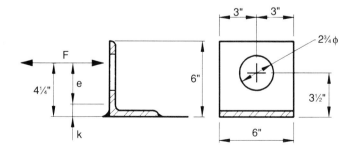

Load per connection (Tension)

Panel: DL = 934 lb

DL + W = 934 + 1171 = 2105 lb

Body: DL + E = 934 + 478 = 1412 lb

Fasteners: DL + E = 934 + 1435 = 2369 lb

Load per connection (Compression)

DL + W = −934 + 567 = −367 lb (tension)

DL + E = −934 + 1435 = 501 lb

Check 6" x 3 ½" x ½" x 0'-6" long angle; k = 1 in.

M = 2105(4.25 − 1) = 6841 in.-lb

Allowable overstress for wind = 1.33

$$f = \frac{6 \times 6841}{1.33 \times 6 \times 0.5^2} = 20{,}575 \text{ psi} < 23{,}760 \text{ psi}$$

Bottom Tie-Back

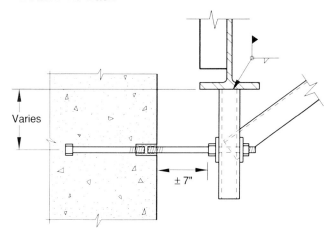

Compression is critical

DL = 934 lb

DL + W = 934 + 194 = 1128 lb

DL + E (Body) = 934 + 194 = 1128 lb

DL + E (Fastener) = 934 + 583 = 1517 lb

¾" Bolt:

Compression ℓ/r = 7/0.188 = 37.2; thus

F_a = 19.4 ksi (AISC Manual)

P = 0.44 x 19.4 = 8.5 kips > 1.517 OK

Check torsional stress in tube where kicker angle is attached.

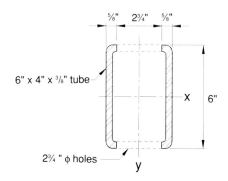

$$I_x = 29.7 - \frac{2.75(0.375)(5.625)^2}{2} = 13.4 \text{ in.}^4$$

$$I_y = 15.6 - \frac{2(0.375)(2.75)^2}{12} = 15.1 \text{ in.}^4$$

I_p = 28.5 in.^{4}

With allowable overstress of 1.33 for wind or earthquake, DL only is more critical for body of connection.

Torsion = 934(3) = 2802 in.-lb

$$f = 2802\sqrt{2^2 + 3^2} / 28.5 = 355 \text{ psi} < 27{,}600$$

Seismic Parallel to Panel

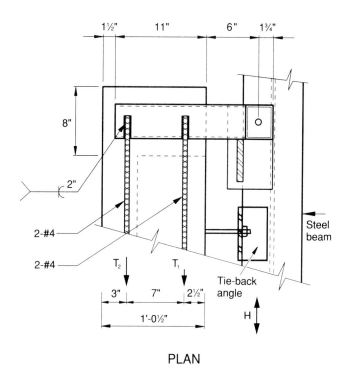

PLAN

H = C_pZ(DL) = 3.0(0.4)(3362) = 4034 lb

Bar: try 2 − #4 top and bottom of tube (assume T_2 = 0)

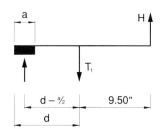

$$a = \frac{A_s f_y}{0.85 f'_c b} = \frac{0.4 \times 60{,}000}{0.85 \times 5000 \times 4} = 1.41 \text{ in.}$$

$d - a/2 = 8.5 - 1.41/2 = 7.80$ in.

$$T_1 = \frac{(9.50 + 7.80)}{7.80}(4034) = 8947 \text{ lb}$$

$$T_{allow} = \frac{\phi f_y A_s}{\text{load factor}} = \frac{(0.9)(60{,}000)(0.4)}{(1.4)}$$
$$= 15{,}430 \text{ lb} \quad \text{OK}$$

Weld of tube to building beam:

PL 4" x 3/8" x 0'-5"

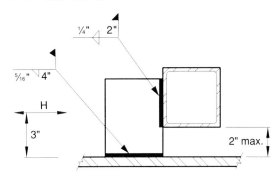

Weld to tube:

Use $\ell_w = 2$ in.; $t_w = 1/4$ in.; weld both sides

f_w (allow.) = 3.71 kips/in. (Table 6.20.2)

$F_w = 3.71 \times 2 \times 2 = 14.84$ kips > 4.03 kips

Weld to beam:

$M = 4034 \times 3 = 12{,}102$ in.-lb

Use $\ell_w = 4$ in.; $t_w = 5/16$ in.; weld both sides

$$S = \frac{d^2}{3} \quad \text{(Fig. 6.5.12)}$$

$$S = \frac{4^2}{3} = 5.3 \text{ in.}^3$$

$$f_w \text{ (req'd.)} = \frac{4034}{4 \times 2} + \frac{12{,}102}{5.3} = 2.79 \text{ kips/in.}$$

f_w (allow.) = 4.64 kips/in. OK (Table 6.20.2)

3.12 References

1. Branson, Dan E., "Deformation of Concrete Structures," McGraw-Hill International Book Co., New York, NY, 1977.

2. PCI Committee on Design Handbook, "Volume Changes in Precast, Prestressed Concrete Structures", *PCI Journal*, V. 22, No. 5, September-October 1977.

3. Nathan, Noel D., "Rational Analysis and Design of Prestressed Concrete Beam Columns and Wall Panels," *PCI Journal*, V. 30, No. 3, May-June 1985.

4. PCI Committee on Prestressed Concrete Columns, "Recommended Practice for the Design of Prestressed Concrete Columns and Walls", *PCI Journal*, V. 33, No. 4, July-August 1988.

5. "Notes on ACI 318-89, Building Code Requirements for Reinforced Concrete with Design Applications," EB070D, Portland Cement Association, Skokie, IL, 1990.

6. "Commentary on Building Code Requirements for Reinforced Concrete", ACI 318R-89, American Concrete Institute, Detroit, MI, 1989.

7. Speyer, Irwin J., "Considerations for the Design of Precast Concrete Bearing Wall Buildings to Withstand Abnormal Loads", *PCI Journal*, V. 21, No. 2, March-April 1976.

8. "Response of Multistory Concrete Structures to Lateral Forces", Special Publication SP-36, American Concrete Institute, Detroit, MI, 1973.

9. ACI Committee 442, "Response of Buildings to Lateral Forces", *Journal of the American Concrete Institute,* V. 68, No. 2, February 1971.

10. "Design of Combined Frames and Shear Walls", Advanced Engineering Bulletin No. 14, Portland Cement Association, Skokie, IL, 1965.

11. Fintel, Mark (Editor), "Handbook of Concrete Engineering", 2nd Edition, Van Nostrand Reinhold Company, New York, NY, 1985.

12. MacLeod, I.A., "Shearwall-Frame Interaction, A Design Aid," EB066D, Portland Cement Association, Skokie, IL, April 1970.

13. Martin, L.D. and Korkosz, W.J., "Connections for Precast Prestressed Concrete Buildings, Including Earthquake Resistance," Technical Report No. 2, Precast/Prestressed Concrete Institute, Chicago, IL, 1982.

14. Clough, D.P., "Design of Connections for Precast Prestressed Concrete Buildings for the Effects of Earthquake," Technical Report No. 5, Precast/Prestressed Concrete Institute, Chicago, IL, 1986.

15. "Design and Typical Details of Connections for Precast and Prestressed Concrete," Second Edition, MNL-123-88, Precast/Prestressed Concrete Institute, Chicago, IL, 1988.

16. "Parking Structures: Recommended Practice for Design and Construction," MNL-129-88, Precast/Prestressed Concrete Institute, Chicago, IL, 1988.

CHAPTER 4
DESIGN OF PRECAST AND PRESTRESSED CONCRETE COMPONENTS

		Page No.
4.1	General	4-3
	4.1.1 Notation	4-3
	4.1.2 Introduction	4-6
4.2	Flexure	4-6
	4.2.1 Strength Design	4-6
	4.2.2 Service Load Design	4-16
	4.2.2.1 Non-prestressed element design	4-16
	4.2.2.2 Prestressed element design	4-18
	4.2.3 Prestress Transfer and Strand Development	4-22
4.3	Shear	4-22
	4.3.1 Shear Resistance of Non-Prestressed Concrete	4-23
	4.3.2 Shear Resistance of Prestressed Concrete Members	4-23
	4.3.3 Design Using Design Aids	4-25
	4.3.4 Shear Reinforcement	4-27
	4.3.5 Horizontal Shear Transfer in Composite Members	4-27
4.4	Torsion	4-30
	4.4.1 Design for Shear and Torsion — Non-Prestressed Members	4-30
	4.4.2 Design for Shear and Torsion — Prestressed Members	4-32
4.5	Loss of Prestress	4-36
	4.5.1 Sources of Stress Loss	4-37
	4.5.2 Range of Values for Total Loss	4-37
	4.5.3 Estimating Prestress Loss	4-37
	4.5.4 Critical Locations	4-38
4.6	Camber and Deflection	4-40
	4.6.1 Initial Camber	4-40
	4.6.2 Elastic Deflections	4-42
	4.6.3 Bilinear Behavior	4-42
	4.6.4 Effective Moment of Inertia	4-43
	4.6.5 Long-Time Camber/Deflection	4-44

4.7 Compression Members 4-45
    4.7.1   Strength Design of Precast Concrete Compression Members 4-45
    4.7.2   Slenderness Effects 4-50
    4.7.3   Service Load Stresses 4-50
    4.7.4   Effective Width of Wall Panels 4-50
    4.7.5   Varying Section Properties of Compression Members 4-50
    4.7.6   Piles 4-52
        4.7.6.1   General 4-52
        4.7.6.2   Strengths under direct loads 4-52
        4.7.6.3   Moment resisting capacities 4-52
        4.7.6.4   Combined moment and direct load 4-53

4.8 Special Considerations 4-55
    4.8.1   Load Distribution 4-55
    4.8.2   Effect of Openings 4-57
    4.8.3   Continuity 4-57
    4.8.4   Cantilevers 4-57

4.9 References 4-57

4.10 Design Aids 4-59
    4.10.1   Flexure 4-59
    4.10.2   Shear 4-62
    4.10.3   Torsion 4-67
    4.10.4   Camber and Deflection 4-72

DESIGN OF PRECAST AND PRESTRESSED CONCRETE COMPONENTS

4.1 General

4.1.1 Notation

A = cross-sectional area

A = average effective area around one reinforcing bar

A_{comp} = cross-sectional area of the equivalent rectangular stress block

A_{cr} = area of crack interface

A_{cs} = area of horizontal shear ties

A_g = gross area of concrete cross section

A_ℓ = total area of longitudinal reinforcement to resist torsion

A_{ps}, A'_{ps} = area of prestressed reinforcement

A_{pw} = portion of prestressed reinforcement to develop web strength

A_s = area of non-prestressed tension reinforcement

A'_s = area of non-prestressed compression reinforcement

A_t = area of one leg of closed stirrup

A_{top} = effective area of cast-in-place composite topping

A_v = area of shear reinforcement

a = depth of equivalent rectangular stress block

b = width of compression or tension face of member

b_v = width of interface surface in a composite member (Sect. 4.3.5)

b_w = web width

C = coefficient as defined in section used (with subscripts)

C = compressive force

C_c = compressive force capacity of composite topping

C_t = coefficient used in torsion design

CR = creep of concrete

c = distance from extreme compression fiber to neutral axis

d, d_s = distance from extreme compression fiber to centroid of non-prestressed tension reinforcement

d', d'_s = distance from extreme compression fiber to centroid of non-prestressed compression reinforcement

d_b = nominal diameter of reinforcing bar or prestressing strand

d_c = concrete cover to the center of reinforcement closest to the tension face

d_p, d'_p = distance from extreme compression fiber to centroid of prestressed reinforcement

E = modulus of elasticity

E_c = modulus of elasticity of concrete

E_{ci} = modulus of elasticity of concrete at time of initial prestress

ES = elastic shortening

E_{ps} = modulus of elasticity of prestressed reinforcement

E_s = modulus of elasticity of steel

e = eccentricity of design load or prestressing force parallel to axis measured from the centroid of the section

e' = distance between c.g of strand at end and c.g of strand at lowest point = $e_c - e_e$

e_c = eccentricity of prestressing force from the centroid of the section at the center of the span

e_e = eccentricity of prestressing force from the centroid of the section at the end of the span

F = force as defined in section used (with subscripts)

F_h = horizontal shear force

f_b = stress in the bottom fiber of the cross section

f'_c = specified compressive strength of concrete

f'_{cc} = specified compressive strength of composite topping

f_{cds} = concrete stress at center of gravity of prestressing force due to all permanent (dead) loads not used in computing f_{cir}

f'_{ci} = compressive strength of concrete at time of initial prestress

f_{cir} = concrete stress at center of gravity of prestressing force immediately after transfer

f_{ct}	=	splitting tensile strength of lightweight concrete
f_d	=	stress due to service dead load
f_e	=	total load stress in excess of f_r
f_ℓ	=	stress due to service live load
f_{pc}	=	compressive stress in concrete at centroid of cross section due to prestress (after allowance for all prestress losses)
f_{pe}	=	compressive stress in concrete due to effective prestress forces only (after allowance for all prestress losses) at extreme fiber of section where tensile stress is caused by externally applied loads
f_{pi}	=	compressive stress in concrete due to initial prestress force
f_{ps}, f'_{ps}	=	stress in prestressed reinforcement at nominal strength of member
f_{pu}	=	ultimate strength of prestressing steel
f_{py}	=	specified yield strength of prestressing steel
f_r	=	modulus of rupture of concrete
f'_r	=	allowable flexural tension, computed using gross concrete section
f_s, f'_s	=	stress in non-prestressed reinforcement
f_{se}	=	effective stress in prestressing steel after losses
f_t	=	stress in the top fiber of a cross section
f_t	=	tensile stress at extreme tension fiber
$f_{t\ell}$	=	final calculated stress in the member
f_y, f'_y	=	specified yield strength of non-prestressed reinforcement
h	=	wall thickness of a box section; overall depth of a compression member
h	=	unsupported length of a pile
h_f	=	depth of flange
h_1	=	distance from centroid of tensile reinforcement to neutral axis
h_2	=	distance from extreme tension fiber to neutral axis
I	=	moment of inertia
I_{cr}	=	moment of inertia of cracked section transformed to concrete
I_e	=	effective moment of inertia for computation of deflection
I_g	=	moment of inertia of gross section
J	=	coefficient as defined in Sect. 4.5.3
j, k	=	factors used in service load design (Sect. 4.2.2)
K	=	coefficient as defined in Sect. 4.5.3 (with subscripts)
K_t	=	coefficient used in torsion design
K_u	=	$[M_u/\phi - A'_s f'_y (d - d')]/f'_c bd^2$
K'_u	=	a coefficient = $\phi M_n (12{,}000)/bd_p^2$
ℓ	=	span length
ℓ_d	=	development length
ℓ_{vh}	=	horizontal shear length as defined in Fig. 4.3.5
M	=	service load moment
M'	=	allowable moment
M_a	=	total moment at the section
M_{cr}	=	cracking moment
M_d	=	moment due to service dead load (unfactored)
M_g	=	moment due to weight of member (unfactored)
M_ℓ	=	moment due to service live load (unfactored)
M_{max}	=	maximum factored moment at section due to externally applied loads
M_n	=	nominal moment strength of a section
M_{nb}	=	nominal moment strength under balanced conditions
M_o	=	nominal moment strength of a compression member with zero axial load
M_{sd}	=	moment due to superimposed dead load (unfactored)
M_{top}	=	moment due to topping (unfactored)
M_u	=	applied factored moment at a section
N	=	unfactored axial load
N'	=	allowable axial load
n	=	modular ratio = E_s/E_c
n	=	number of reinforcing bars
P	=	prestress force after losses
P_i	=	initial prestress force
P_n	=	axial load nominal strength of a compression member at given eccentricity
P_{nb}	=	axial load nominal strength under balanced conditions
P_o	=	prestress force at transfer
P_o	=	axial load nominal strength of a compression member with zero eccentricity

P_u	=	factored axial load
RE	=	relaxation of tendons
R.H.	=	average ambient relative humidity
R_t, R_v	=	coefficients used in torsion design
r	=	radius of gyration
SH	=	shrinkage of concrete
s	=	shear or torsion reinforcement spacing in a direction parallel to the longitudinal reinforcement
T	=	tensile force
T_c, T_c'	=	nominal torsional moment strength provided by concrete under combined shear and torsion and under pure torsion, respectively
TL	=	total prestress loss
T_n	=	nominal torsional moment strength
T_s	=	nominal torsional moment strength provided by torsion reinforcement
T_u	=	factored torsional moment on a section
$T_{u(max)}$	=	upper limit of factored torsion for design
$T_{u(min)}$	=	torsion level below which its effect may be neglected
t	=	thickness (used for various parts of members with subscripts)
V_c	=	nominal shear strength provided by the concrete
V_c, V_c'	=	nominal shear strength provided by concrete under combined shear and torsion and under pure shear, respectively
V_{ci}	=	nominal shear strength provided by concrete when diagonal cracking is the result of combined shear and moment
V_{cw}	=	nominal shear strength provided by concrete when diagonal cracking is the result of excessive principal tensile stress in the web
V_d	=	dead load shear (unfactored)
V_i	=	factored shear force at section due to externally applied loads occurring simultaneously with M_{max}
V_ℓ	=	live load shear (unfactored)
V_n	=	nominal shear strength
V_p	=	vertical component of the effective prestress force at the section considered
V_s	=	nominal shear strength provided by the shear reinforcement
V/S	=	volume-surface ratio
V_u	=	factored shear force at section
$V_{u(max)}$	=	upper limit of factored shear for design
w	=	maximum crack width at extreme tension fiber
w	=	unfactored load per unit length of beam or per unit area of slab
w_d	=	unfactored dead load per unit length
w_ℓ	=	unfactored live load per unit length
w_{sd}	=	dead load due to superimposed loading
$w_{t\ell}$	=	unfactored total load per unit length = $w_d + w_\ell$
w_u	=	factored total load per unit length or area
x	=	distance from support to point being investigated
x, x_1	=	shorter side of component rectangle and closed stirrup, respectively (torsion design)
y, y_1	=	longer side of component rectangle and closed stirrup, respectively (torsion design)
y'	=	distance from top to c.g. of A_{comp}
y_b	=	distance from bottom fiber to center of gravity of the section
y_s	=	distance from centroid of prestressed reinforcement to bottom fiber
y_t	=	distance from top fiber to center of gravity of the section
y_t	=	distance from centroid of gross section to extreme fiber in tension
Z	=	section modulus
Z_b	=	section modulus with respect to the bottom fiber of a cross section
Z_t	=	section modulus with respect to the top fiber of a cross section
z	=	quantity limiting distribution of flexural reinforcement (Sect. 4.2.2.1)
α	=	distance from end of member to strand depression point (Fig. 4.10.14)
$\alpha_t, \beta_t, \gamma_t$	=	coefficients used in torsion design
β_1	=	factor defined in Sect 4.2.1
γ_p	=	factor for type of prestressing tendon (see ACI Code Sect. 18.0 for values)
Δ	=	deflection (with subscripts)
ε_{cu}	=	ultimate concrete strain
$\varepsilon_{ps}, \varepsilon_{ps}'$	=	strain in prestressing steel corresponding to f_{ps}, f_{ps}'
ε_s	=	strain in non-prestressed tension reinforcement

ε'_s = strain in non-prestressed compression reinforcement

ε_{sa} = strain in prestressing steel caused by external loads = $\varepsilon_{ps} - \varepsilon_{se}$

ε_{se} = strain in prestressing steel after losses

ε_y = maximum allowable strain in non-prestressed reinforcement

λ = a conversion factor for lightweight concrete

λ = multiplier applied to initial deflection

ξ = time-dependent factor for sustained loads

μ = shear-friction coefficient

μ_e = effective shear-friction coefficient

ρ = A_s/bd = ratio of non-prestressed tension reinforcement

ρ' = A'_s/bd = ratio of non-prestressed compression reinforcement

ρ_p = A_{ps}/bd_p = ratio of prestressed reinforcement

ρ_w = $A_s/b_w d$ = ratio of non-prestressed tension reinforcement based on web width

ρ_{bal} = non-prestressed reinforcement ratio producing balanced strain conditions

ρ_{max} = maximum reinforcement ratio for non-prestressed members

ρ_{min} = minimum ratio of non-prestressed reinforcement

ϕ = strength reduction factor

ω = $\rho f_y/f'_c$

ω' = $\rho' f_y/f'_c$

ω_p = $\rho_p f_{ps}/f'_c$

ω_{pu} = $\rho_p f_{pu}/f'_c$

$\omega_w, \omega_{pw}, \omega'_w$ = reinforcement indices for flanged sections computed as for ω, ω_p, and ω' except that b shall be the web width, and reinforcement area shall be that required to develop compressive strength of web only

4.1.2 Introduction

This chapter of the Handbook provides a summary of theory and procedures used in the design of precast and prestressed concrete structures. Designs are based on the provisions of the ACI Building Code[1] (referred to as "the Code" or "ACI 318" in the Handbook).

Two different phases must be considered in designing precast concrete elements: (1) the manufacturing through erection phase and (2) the in-service conditions. The designer is referred to Chapter 5 for the first phase. This chapter is concerned with the in-service conditions.

The load tables in Chapter 2 will generally not provide all the design data necessary. In most cases, the engineer will select a standard section, with the detailed design calculations furnished by the staff or consulting engineer of the prestressed concrete producer. The engineer of record should verify that the section selected is capable of satisfying both strength and performance criteria for the use intended. In cases where the engineer of record undertakes the complete design responsibility, consultation with producers in the area will ensure compatibility of design with production and will result in optimum quality and economy.

4.2 Flexure

Design for flexure in accordance with the Code requires that precast and prestressed concrete members be checked for both design strength and service load.

4.2.1 Strength Design

Strength design is based on solution of the equations of equilibrium, normally using the rectangular stress block in accordance with Sect. 10.2.7 of the Code (see Fig.4.2.1). The stress in the prestressed steel at nominal strength, f_{ps}, can be determined by strain compatibility[2] or by the approximate equation given in the Code (Eq. 18-3). In carrying out strain compatibility analysis, the manufacturer's stress-strain relations may be used.

Fig. 4.2.1 Nominal flexural resistance

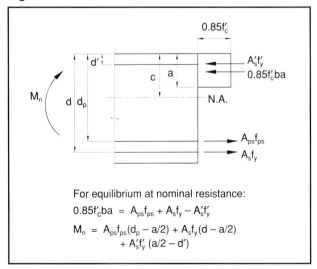

For equilibrium at nominal resistance:
$0.85f'_c ba = A_{ps}f_{ps} + A_s f_y - A'_s f'_y$
$M_n = A_{ps}f_{ps}(d_p - a/2) + A_s f_y(d - a/2) + A'_s f'_y(a/2 - d')$

Alternatively, the idealized stress-strain equations given in Design Aid 11.2.5 may be used. For elements with compression reinforcement, the nominal strength can be calculated by assuming that the compression reinforcement yields. This assumption should be subsequently verified from the strain diagram. The designer will normally choose a section and reinforcement and then determine if it meets the basic design strength requirement:

$$\phi M_n \geq M_u$$

A flow chart illustrating the calculation of the nominal strength of flexural elements is given in Fig. 4.2.2.

Depth of stress block

The depth "a" of the rectangular stress block is related to the depth to the neutral axis "c" by the equation:

$$a = \beta_1 c$$

where:

β_1	f'_c, psi
0.85	3000
0.85	4000
0.80	5000
0.75	6000
0.70	7000
0.65	8000 and higher

Flanged elements

The equations for nominal strength given in Fig. 4.2.1 apply to rectangular cross sections and flanged sections in which the stress block lies entirely within the depth of the flange h_f. The depth of the stress block "a", is obtained from the first equation of equilibrium in Fig. 4.2.1:

$$a = \frac{A_{ps} f_{ps} + A_s f_y - A'_s f'_y}{0.85 f'_c b} \quad \text{(Eq. 4.2.1)}$$

If $a > h_f$, and compression reinforcement is present, use of strain compatibility or reasonable approximations for it may be made to find the nominal strength. If compression reinforcement is not present the nominal strength can be found using the Code equations shown in Fig. 4.2.2 or by strain compatibility.

Limitations on reinforcement

For non-prestressed flexural elements, except slabs of uniform thickness, the minimum reinforcement ratio, ρ_{min} is:

$$\rho_{min} = \frac{200}{f_y} \quad \text{(Eq. 4.2.2)}$$

unless the area of reinforcement provided is ⅓ greater than that required by analysis. For flanged sections, ρ is based upon the width of the web. For slabs, the minimum flexural reinforcement is that amount required for shrinkage and temperature reinforcement.

The maximum reinforcement ratio for non-prestressed elements is limited to 0.75 times the balanced reinforcement ratio:

$$\rho_{max} = 0.75 \, \rho_{bal} \quad \text{(Eq. 4.2.3)}$$

where:

$$\rho_{bal} = \frac{0.85 \beta_1 f'_c}{f_y} \left(\frac{87,000}{87,000 + f_y} \right)$$

Substituting the equation for ω yields:

$$\omega_{max} = \frac{\rho_{max} f_y}{f'_c}$$

$$= 0.64 \, \beta_1 \left(\frac{87,000}{87,000 + f_y} \right) \quad \text{(Eq. 4.2.4)}$$

For prestressed elements, the Code requires that the total prestressed and non-prestressed reinforcement be adequate to develop a factored load at least 1.2 times the cracking load, except for flexural members with shear and flexural strength at least twice that required by analysis. The cracking strength is based on a modulus of rupture of:

$$f_r = 7.5 \sqrt{f'_c} \quad \text{(Eq. 4.2.5)}$$

No upper limit is placed on the reinforcement for prestressed elements. However, when either:

$\omega_p, [\omega_p + d/d_p(\omega - \omega')]$

or $[\omega_{pw} + d/d_p (\omega_w - \omega_w')] > 0.36 \beta_1$

or alternatively:

$0.85 a/d_p > 0.36 \beta_1$

the nominal strength, as shown in Fig. 4.2.2, is calculated based on the compression force of the moment couple.

Critical section

For simply supported, uniformly loaded, prismatic non-prestressed elements, the critical section for flexural design will occur at midspan. For uniformly loaded prestressed elements, in order to reduce the end stresses at release some strands are often depressed near midspan, or debonded for a length near the ends. For strands with a single point depression, the critical section can usually be assumed at 0.4ℓ. For straight strands, the critical section will be at midspan, but if some strands are debonded near the end, an additional critical section may occur near the end of the debonded length as shown in Fig. 4.2.3.

The presence of concentrated loads or non-prestressed reinforcement may further complicate the location of the critical section. In such cases, compu-

Fig. 4.2.2 Flow chart for nominal strength calculations for flexure

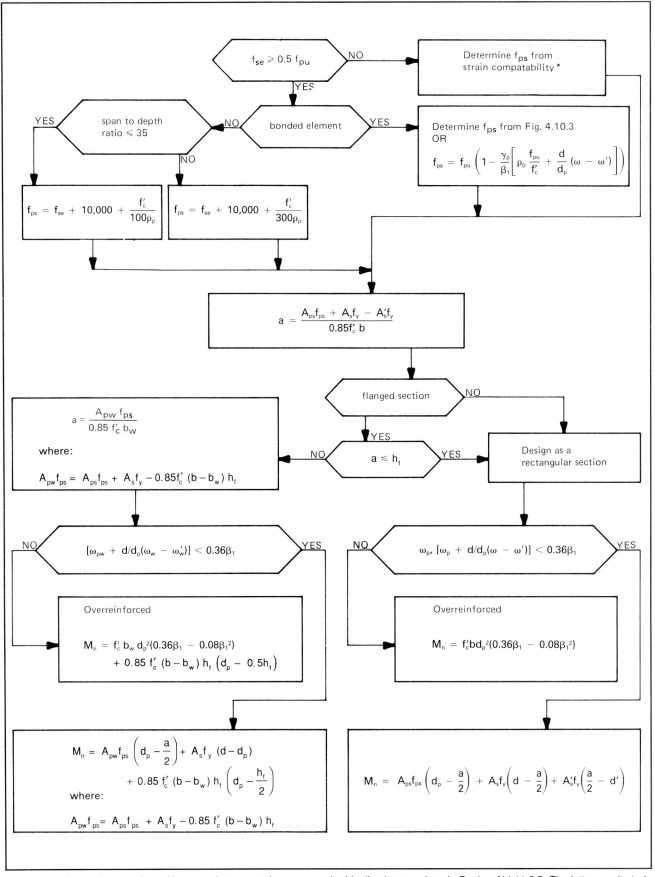

* This analysis may be based on either actual stress-strain curves or the idealized curve given in Design Aid 11.2.5. The latter results in f_{ps} values given in Fig. 4.10.3.

Fig. 4.2.3 Critical section for flexural design

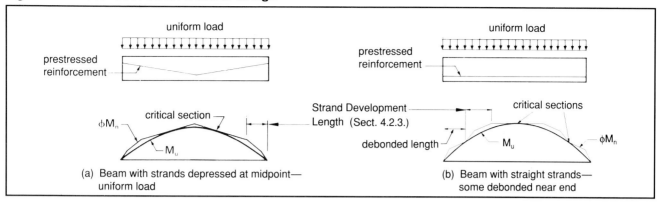

(a) Beam with strands depressed at midpoint—uniform load

(b) Beam with straight strands—some debonded near end

ter programs with the capability of checking the capacity at short intervals along the member length can be used to expedite analysis.

Analysis using Code equations

Fig. 4.2.2 essentially outlines the design procedures using the Code equations for prestressed and partially prestressed members. If compression reinforcement is present, the Code requires certain checks to assure that the stress in the compression reinforcement is at its yield strength. In computing f_{ps}, if any compression reinforcement is taken into account, the term

$$[\, \rho_p \frac{f_{pu}}{f'_c} + \frac{d}{d_p}(\omega - \omega') \,]$$

shall be taken not less than 0.17 and d' shall be no greater than $0.15 d_p$. Alternatively, the yielding of the compression reinforcement is ensured if Eq. 4.2.6 is satisfied.

$$\frac{A_{ps} f_{ps} + A_s f_y - A'_s f'_y}{bd}$$

$$\geq 0.85 \, \beta_1 \, f'_c \, \frac{d'}{d} \left(\frac{87,000}{87,000 - f_y} \right) \quad \text{(Eq. 4.2.6)}$$

Analysis using strain compatibility

Strain compatibility is recognized as being the more accurate alternative method to the Code equations. The procedure consists of assuming the location of the neutral axis, computing the strains in the prestressed and non-prestressed reinforcement, and establishing the depth of the stress block. Knowing the stress-strain relationship for the reinforcement, and assuming that the maximum strain in the concrete is 0.003, the forces in the reinforcement and in the concrete are determined and the sum of compression and tension forces is computed. If necessary, the neutral axis location is moved on a trial and error basis until the sum of forces is zero. The moment of these forces is then computed to obtain the nominal strength of the section.

Design aids

Figs. 4.10.1 through 4.10.3 are provided to assist in the strength design of flexural members. Fig. 4.10.1 can be used for members with prestressed or non-prestressed reinforcement, or combinations (partial prestressing). Note that to use this aid, it is necessary to determine f_{ps} from some other source, such as Eq. 18-3 of ACI 318-89 or Fig. 4.10.3.

Fig. 4.10.2 is for use only with fully prestressed members. The value of f_{ps} is determined by strain compatibility in a manner similar to that used in Fig. 4.10.3. The reduction factor, ϕ, is included in the values of K'_u. The following examples illustrate the use of these design aids.

Example 4.2.1 Use of Fig. 4.10.1 for determination of non-prestressed reinforcement

Given:

The ledger beam shown.

Applied factored moment, M_u = 1460 ft-kips

f'_c = 5000 psi, normal weight concrete

d = 72 in. f_y = 60 ksi

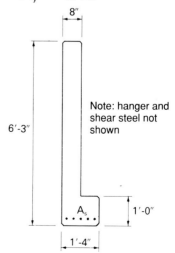

Note: hanger and shear steel not shown

Problem:

Find the amount of mild steel reinforcement, A_s, required.

Solution:

Referring to Fig. 4.10.1, $A_{ps} = 0$; $A'_s = 0$

$$K_u = \frac{M_u / \phi}{f'_c bd^2} = \frac{(1460 \times 12{,}000)/0.9}{5000(8)(72)^2} = 0.0939$$

Required $\bar{\omega} = 0.10 < \omega_{max} = 0.302$

$$A_s = \frac{\bar{\omega} bd f'_c}{f_y} = \frac{0.10(8)(72)(5)}{60}$$

$$= 4.80 \text{ sq. in.}$$

Use 5 – #9; $A_s = 5$ sq in.

$\rho = A_s / bd = 5/(8)(72) = 0.0087$

$\rho_{min} = 200/f_y = 200/60{,}000$

$= 0.0033 < 0.0087$ OK

Note: Detailing of reinforcement must provide for adequate crack control (see Sect. 4.2.2.1 and ACI 318-89, Sect. 10.6)

Example 4.2.2 Use of Fig. 4.10.2 for determination of prestressing steel requirements—bonded strand

Given:

PCI standard rectangular beam 16RB24

Applied factored moment, M_u = 600 ft-kips

f'_c = 6000 psi normal weight concrete

f_{pu} = 270 ksi, low-relaxation strand

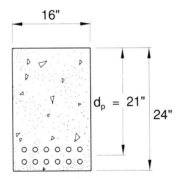

Problem:

Find the required amount of prestressing steel.

Solution:

Referring to Fig. 4.10.2:
$$M_u \leq \phi M_n = K'_u bd_p^2 / 12{,}000$$

Required $K'_u = \dfrac{M_u(12{,}000)}{bd_p^2} = \dfrac{600(12{,}000)}{16(21)^2}$

$= 1020$

for $\omega_{pu} = 0.22$, $K'_u = 1005$

for $\omega_{pu} = 0.23$, $K'_u = 1041$

therefore

$$\omega_{pu} = 0.22 + \frac{1020 - 1005}{1041 - 1005}(0.01) = 0.224$$

$$A_{ps} = \frac{\omega_{pu} bd_p f'_c}{f_{pu}} = \frac{0.224(16)(21)(6)}{270}$$

$= 1.67$ sq in.

Use 12 – ½ in. diameter strands; $A_{ps} = 1.84$ sq in.

Example 4.2.3 Use of Fig. 4.10.3 — values of f_{ps} by stress-strain relationship— bonded strand

Given:

3'-4" x 8" hollow-core slab

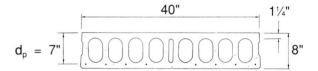

Concrete:

f'_c = 5000 psi normal weight concrete

Prestressing steel:

8 – ⅜ in. diameter 270K low-relaxation strand

A_{ps} = 8(0.085) = 0.68 sq in.

Section properties:

A = 218 in.²

Z_b = 381 in.³

y_b = 3.98 in.

Problem:

Find design flexural strength, ϕM_n

Solution:

Determine $C\omega_{pu}$ for the section:

$$C\omega_{pu} = C\frac{A_{ps} f_{pu}}{bd_p f'_c} + \frac{d}{d_p}(\omega - \omega')$$

since $\omega = \omega' = 0$

$$C\omega_{pu} = \frac{1.06(0.68)(270)}{(40)(7)(5)} = 0.139$$

Entering Fig. 4.10.3 with this parameter and an assumed effective stress, f_{se} = 150 ksi gives a value of:

f_{ps} = 266 ksi

Determine the flexural strength:

$\phi M_n = \phi[A_{ps} f_{ps}(d_p - a/2)]$

$a = A_{ps}f_{ps} / (0.85f'_c b)$

$a = \dfrac{0.68(266)}{0.85(5)(40)} = 1.064$ in.

$\phi M_n = 0.9[0.68(266)(7 - 1.064/2)]$

$= 1053$ in-kips $= 87.7$ ft-kips

Check the ductility requirement, $\phi M_n > 1.2 M_{cr}$

$P = f_{se}A_{ps} = 150(0.68)$
$= 102$ kips

$1.2M_{cr} = 1.2(P/A + Pe/Z_b + 7.5\sqrt{f'_c})Z_b$

$= 1.2\left(\dfrac{102}{218} + \dfrac{102(2.98)}{381}\right.$

$\left. + \dfrac{7.5\sqrt{5000}}{1000}\right)381$

$= 823$ in.-kips $= 68.6$ ft-kips

< 87.7 ft-kips OK

Example 4.2.4 Use of Fig. 4.10.3 and Eq. 18-3 (ACI 318-89) for partial prestressed member

Given:

PCI standard double tee
8DT24 + 2

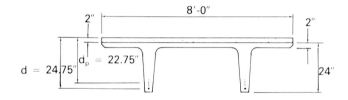

Concrete:

Precast: $f'_c = 5000$ psi

Topping: $f'_c = 3000$ psi, normal weight

Reinforcement:

12 – ½ in. diameter 270K low-relaxation strands (6 each stem)

$A_{ps} = 12(0.153) = 1.84$ sq in.

A_s (2 – #6) = 0.88 sq in.

Problem:

Find the design flexural strength of the composite section by the stress-strain relationship, Fig. 4.10.3, and compare using Code Eq. 18-3.

Solution:

Assume $f_{se} = 150$ ksi

$C\omega_{pu} = C\dfrac{A_{ps}f_{pu}}{bd_p f'_c} + \dfrac{d}{d_p}(\omega - \omega')$

$\omega = A_s f_y / bdf'_c$

$= \dfrac{0.88(60)}{96(24.75)(3)} = 0.0074$

$C\omega_{pu} = \dfrac{1.00(1.84)(270)}{96(22.75)(3)} + \dfrac{24.75}{22.75}(0.0074)$

$= 0.084$

From Fig. 4.10.3,

$f_{ps} = 268$ ksi

$a = \dfrac{1.84(268) + 0.88(60)}{0.85(3)(96)} = 2.23$ in.*

$M_n = 1.84(268)[22.75 - (2.23/2)]$
$+ 0.88(60)[(24.75 - (2.23/2)]$

$= 11,917$ in.-kips $= 993$ ft-kips

$\phi M_n = 0.9(993) = 894$ ft-kips

Find f_{ps} using Eq. 18-3 (ACI 318-89)

$f_{ps} = f_{pu}\left(1 - \dfrac{\gamma_p}{\beta_1}\left[\rho_p \dfrac{f_{pu}}{f'_c} + \dfrac{d}{d_p}(\omega - \omega')\right]\right)$

$\omega' = 0$ in this example

$f_{ps} = 270\left(1 - \dfrac{0.28}{0.85}\left[\dfrac{1.84(270)}{96(22.75)(3)}\right.\right.$

$\left.\left. + \dfrac{24.75}{22.75}(0.0074)\right]\right)$

$= 263$ ksi (vs. 268 ksi from Fig. 4.10.3)

$\phi M_n = 976$ ft-kips

Example 4.2.5 Flexural strength of double tee flange in transverse direction

For flanged sections, in addition to providing adequate flexural strength in the longitudinal direction, the flanges must be designed for bending in the transverse direction. This example illustrates a design for both uniformly distributed loads (Part A) and concentrated loads (Part B).

---

*Since $a = 2.23$ in. > 2.0 in. — the topping thickness, a more exact analysis requires revised calculation to account for the higher strength concrete of the double tee flange. However, such a refinement is expected to produce negligible difference in results. In this case, revised calculation yields $a = 2.16$ in. versus 2.23 in. and $\phi M_n = 896$ ft-kips versus 894 ft-kips.

Part A

Given:

PCI standard double tee of Example 4.2.4. Topping is reinforced with 6 x 6-W1.4 x W1.4 WWF, and the flange is reinforced with 6 x 6-W4 x W4 WWF.

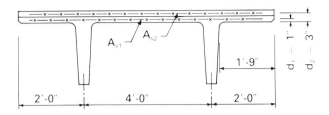

f'_c (topping) = 3000 psi; f'_c (precast) = 5000 psi
f_y = 60,000 psi

Problem:

Find the uniform live load which the flange can support.

Solution:

A_{s1} (WWF in precast) = 0.080 sq in./ft

A_{s2} (WWF in topping) = 0.028 sq in./ft

The cantilevered flange controls the design since the negative moment over the stem reduces the positive moment between the stems. Construct strain diagram:

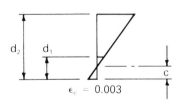

Try c = ¼ in.

$$\frac{\varepsilon_s + 0.003}{0.003} = \frac{d}{c}$$

Therefore:

$$\varepsilon_{s1} = \frac{0.003\, d_1}{c} - 0.003$$

$$\varepsilon_{s2} = \frac{0.003\, d_2}{c} - 0.003$$

$$\varepsilon_{s1} = \frac{0.003 \times 1}{0.25} - 0.003 = 0.009$$

$$\varepsilon_{s2} = \frac{0.003 \times 3}{0.25} - 0.003 = 0.033$$

$$\varepsilon_y = \frac{f_y}{E_s} = \frac{60}{29,000} = 0.0021 < 0.009$$

Therefore, reinforcement yields, and

$f_s = f_y$ = 60 ksi

$a = \beta_1 c = 0.80(0.25) = 0.20$ in.

$C = 0.85 f'_c ba = 0.85(5)(12)(0.20)$
= 10.2 kips/ft

$T_1 = A_{s1} f_y = 0.08(60) = 4.80$ kips/ft

$T_2 = A_{s2} f_y = 0.028(60) = 1.68$ kips/ft

$T_1 + T_2 = 6.48 < 10.2$ kips/ft

Therefore, try

$a = (T_1 + T_2)/0.85 f'_c b$

= 6.48/0.85(5)(12)

= 0.13 in.

c = 0.13/0.80 = 0.16

Check:

$$\varepsilon_{s1} = \frac{0.003 \times 1}{0.16} - 0.003$$

= 0.016 > 0.0021

Therefore, the reinforcement yields, and the analysis is valid.

$\phi M_n = \phi[T_1(d_1 - a/2) + T_2(d_2 - a/2)]$

= 0.9[4.80(1 - 0.13/2)
+ 1.68(3 - 0.13/2)]

= 8.48 in.-kips/ft = 707 ft-lb/ft

Check ρ_{min} and ρ_{max}:

$d = (0.028 \times 3 + 0.080 \times 1)/(0.028 + 0.080)$

= 1.52 in.

$$\rho = \frac{0.028 + 0.080}{12 \times 1.52} = 0.0059$$

$\rho_{min} = 200/60,000 = 0.0033 < 0.0059$ OK

$\rho_{max} = 0.75 \rho_{bal}$

$$= 0.75(0.85\beta_1)\left(\frac{f'_c}{f_y}\right)\left(\frac{87,000}{87,000 + f_y}\right)$$

$$= 0.75(0.85)(0.80)\left(\frac{5000}{60,000}\right)\left(\frac{87}{87+60}\right)$$

= 0.025 > 0.0059 OK

Calculate allowable load:

w_d (flange self weight) = 50 psf

$$M_d = \frac{w_d \ell^2}{2} = \frac{50(1.4)(1.75)^2}{2} = 107.2 \text{ ft-lb/ft}$$

$$M_\ell = 707 - 107.2 = 599.8 \text{ ft-lb/ft}$$

$$w_\ell = \frac{599.8(2)}{(1.75)^2(1.7)} = 230 \text{ psf}$$

Part B

Given:

Pretopped double tee 10DT34 (see Chapter 2)

$f'_c = 5000$ psi; $f_y = 65$ ksi (WWF)

Problem:

Design the flange for bending in the transverse direction for a concentrated live load of 2 kips (typical for parking structures) as shown.

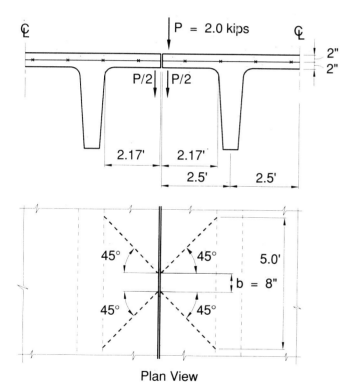

Plan View

Solution:

The following assumptions related to distribution of the concentrated load are typical:

(a) Because of the flange-to-flange connection (usually spaced 4 to 5 ft center to center), the 2 kip load is distributed to two adjacent double tees (1 kip per double tee).

(b) The load is considered applied over an area of about 20 sq in. with the dimension b equal to 6 to 10 in.

(c) An angle of 45° is typically used for distribution of the concentrated load in each double tee flange.

Note: For this example, a 45° angle and the dimension b equal to 8 in. results in a distribution width at face of stem of 5 ft as shown.

Calculate factored moment per foot width:

ω_d (self weight of flange) = $\frac{4}{12}(150)$ = 50 psf

$$M_d \text{ (factored)} = 1.4(50)(2.17)\left(\frac{2.17}{2}\right)\left(\frac{12}{1000}\right)$$

$$= 1.98 \text{ in.-kips/ft}$$

$$M_\ell \text{ (factored)} = \frac{1.7(1)(2.17)(12)}{(5)}$$

$$= 8.85 \text{ in.-kips/ft}$$

$$M_u = 1.98 + 8.85 = 10.83 \text{ in.-kips/ft}$$

Calculate design moment strength with trial wire mesh reinforcement that has W4 wire at 4 in. on centers.

$A_s = 0.12 \text{ in.}^2/\text{ft}$

$$a = \frac{A_s f_y}{0.85 f'_c b} = \frac{0.12(65)}{0.85(5)(12)} = 0.15 \text{ in.}$$

$$\phi M_n = \phi A_s f_y \left(d - \frac{a}{2}\right)$$

$$= 0.9(0.12)(65)\left(2 - \frac{0.15}{2}\right)$$

$$= 13.5 \text{ in.-kips/ft} > 10.83 \text{ OK}$$

Check ACI 318-89, Sects. 10.5.3 and 7.12 for required minimum reinforcement:

A_s (shrinkage and temperature)

$$= 0.0018(12 \times 4)\left(\frac{60}{65}\right)$$

$$= 0.080 \text{ in.}^2/\text{ft} < 0.12 \quad \text{OK}$$

Use 12 x 4 – W2.0 x W4.0 (one layer)

(Note: Since most double tees meet the requirements of ACI 318-89, Sect. 7.12.3, shrinkage and temperature reinforcement in the longitudinal direction typically is not required. In such cases, only a nominal amount is used to facilitate shipping and handling of the fabric. Note that the Wire Reinforcement Institute requires that the longitudinal wire must have an area at least equal to 0.4 times the transverse wire area.)

Example 4.2.6 Design of a partially prestressed flanged section using strain compatibility

Given:
Inverted tee beam with 2 in. composite topping as shown

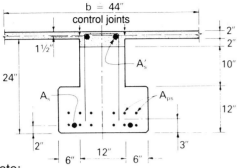

Concrete:
f'_c (precast) = 5000 psi
f'_c (topping) = 3000 psi

Reinforcement:
12 – ½ in. diameter 270K low relaxation strand
A_{ps} = 12 x 0.153 = 1.836 sq in.
E_{ps} = 28,500 ksi
A_s = 2 – #7 = 1.2 in.² E_s = 29,000 ksi
A'_s = 2 – #9 = 2.0 in.² E_s = 29,000 ksi

Problem:
Find flexural strength, ϕM_n

Solution:
Determine effective flange width, b, from Sect. 8.10.2 of the Code; overhanging width = 8 times thickness
$b = b_w + 2(8t) = 12 + 2(8)(2) = 44$ in.
$d_p = 26 - 3 = 23$ in.
$d = 26 - 2 = 24$ in.
$d' = 1½$ in.
Assume 20% loss of prestress
Strand initially tensioned to 75% of f_{pu}
$f_{se} = 0.80(0.75)(270) = 162$ ksi

$$\varepsilon_{se} = \frac{f_{se}}{E_{ps}} = \frac{162}{28,500} = 0.0057$$

Construct a strain diagram as below:

$$\frac{\varepsilon_{sa} + 0.003}{0.003} = \frac{d_p}{c}$$

$$\varepsilon_{sa} = \frac{0.003 d_p}{c} - 0.003$$

$$\varepsilon_s = \frac{0.003 d}{c} - 0.003$$

$$\varepsilon'_s = 0.003 - \frac{0.003 d'}{c}$$

$$\varepsilon_{sa} = \frac{0.003(23)}{c} - 0.003 = \frac{0.069}{c} - 0.003$$

$$\varepsilon_{ps} = \varepsilon_{sa} + \varepsilon_{se}$$
$$= \frac{0.069}{c} + 0.0027$$

$$\varepsilon_s = \frac{0.003(24)}{c} - 0.003 = \frac{0.072}{c} - 0.003$$

$$\varepsilon'_s = 0.003 - \frac{0.003(1.5)}{c} = 0.003 - \frac{0.0045}{c}$$

$$\varepsilon_y = \frac{f_y}{E_s} = \frac{60}{29,000} = 0.0021$$

Try c = 9 in.

$$\varepsilon_{ps} = \frac{0.069}{9} + 0.0027 = 0.0104$$

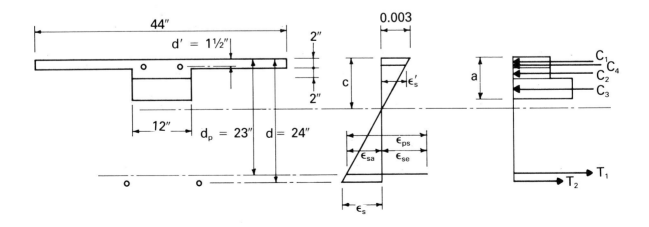

From the strand stress-strain curve equations in Design Aid 11.2.5

$$f_{ps} = 270 - \left(\frac{0.04}{0.0104 - 0.007}\right) = 258.2 \text{ ksi}$$

$$\varepsilon_s = \frac{0.072}{9} - 0.003 = 0.0050 > 0.0021$$

$$f_s = 60 \text{ ksi}$$

$$\varepsilon'_s = 0.003 - \frac{(0.0045)}{9} = 0.0025 > 0.0021$$

$$f'_s = 60 \text{ ksi}$$

$$a_3 = \beta_1 c - 4 \text{ in.} = (0.85)(9) - 4 = 3.7 \text{ in.}$$

$$C_1 = (0.85)(3)(2)(44) = 224.4 \text{ kips}$$

$$C_2 = (0.85)(3)(2)(12) = 61.2 \text{ kips}$$

$$C_1 + C_2 = 285.6 \text{ kips}$$

$$C_3 = (0.85)(5)(3.7)(12) = 188.7 \text{ kips}$$

$$C_4 = (2.0)(60) = 120 \text{ kips}$$

$$C_1 + C_2 + C_3 + C_4 = 594.3 \text{ kips}$$

$$T_1 = A_{ps}f_{ps} = (1.836)(258.2) = 474.1 \text{ kips}$$

$$T_2 = (1.2)(60) = 72 \text{ kips}$$

$$T_1 + T_2 = 546.1 \text{ kips} < 594.3 \text{ kips}$$

Try $c = 8.25$ in.

$$\varepsilon_{ps} = \frac{0.069}{8.25} + 0.0027 = 0.0111$$

$$f_{ps} = 270 - \left(\frac{0.04}{0.0111 - 0.007}\right) = 260.2 \text{ ksi}$$

$$\varepsilon_s = \frac{0.072}{8.25} - 0.003 = 0.0057 > 0.0021$$

$$f_s = 60 \text{ ksi}$$

$$\varepsilon'_s = 0.003 - \frac{(0.0045)}{8.25} = 0.0025 > 0.0021$$

$$f'_s = 60 \text{ ksi}$$

$$a_3 = (0.85)(8.25) - 4 = 3.01 \text{ in.}$$

$$C_1 + C_2 = 285.6 \text{ kips}$$

$$C_3 = (0.85)(5)(3.01)(12) = 153.5 \text{ kips}$$

$$C_4 = (2.0)(60) = 120 \text{ kips}$$

$$C_1 + C_2 + C_3 + C_4 = 559.1 \text{ kips}$$

$$T_1 = (1.836)(260.2) = 477.7 \text{ kips}$$

$$T_2 = (1.2)(60) = 72 \text{ kips}$$

$$T_1 + T_2 = 549.7 \text{ kips} \approx 559.1 \text{ kips OK}$$

Check whether the section is over-reinforced (see ACI 318-89, Sect. 18.8.1):

Determine force capacity of the overhanging flanges:

$$C_f = 0.85 f'_c (b - b_w) h_f$$

$$= (0.85)(3)(44 - 12)(2) = 163.2 \text{ kips}$$

Using average $f'_c = 3.5$ ksi, find $\omega_{pw}, \omega_w, \omega'_w$

$$\omega_{pw} = \frac{A_{ps}f_{ps} - C_f}{b_w d_p f'_c}$$

$$= \frac{(1.836)(260.2) - 163.2}{(12)(23)(3.5)} = 0.326$$

$$\omega_w = \frac{A_s f_y}{b_w d f'_c} = \frac{(1.2)(60)}{(12)(24)(3.5)} = 0.071$$

$$\omega'_w = \frac{(2.0)(60)}{(12)(24)(3.5)} = 0.119$$

$$\omega_{pw} + \frac{d}{d_p}(\omega_w - \omega'_w) < 0.36\beta_1$$

$$0.326 + \frac{24}{23}(0.071 - 0.119) < (0.36)(0.85)$$

$$0.276 < 0.306$$

The section is not over-reinforced and tension governs.

Note: Alternatively, the check $0.85a/d_p \leq 0.36\beta_1$ may be used as shown below to arrive at the same conclusion:

From last iteration, $c = 8.25$ in., thus:

$$a = \beta_1 c = 0.85(8.25) = 7.01 \text{ in.}$$

$$0.85a/d_p = 0.85(7.01)/23$$

$$= 0.259 < 0.306$$

The difference in the values from the two alternatives, namely 0.276 and 0.259 is the result of using average f'_c in the calculation of ω terms.

$$M_n = 224.4(8.25 - 1) + 61.2(8.25 - 3)$$

$$+ 153.5\left(8.25 - 4 - \frac{3.01}{2}\right) + 120(8.25 - 1.5)$$

$$+ 477.7(23 - 8.25) + 72(24 - 8.25)$$

$$= 11,359 \text{ in.-kips}$$

$$= 947 \text{ ft-kips}$$

$$\phi M_n = 0.9(947) = 852 \text{ ft-kips}$$

Notes:

(1) This example shows the exact method; approximate methods may be satisfactory in many situations.

(2) In this example, since the cast-in-place topping carries significantly more compression force than the precast member, $\beta_1 = 0.85$ corresponding to topping concrete. In other cases, where the compression is shared by the topping and precast in different proportions, a reasonable average value for β_1 may be used.

(3) In evaluating $\omega_{p\omega}$ and other similar factors, $f'_c = 3.5$ ksi is used to reflect the contribution of topping vs. precast member to the total compression force.

(4) For over-reinforced members, nominal moment strength must be calculated based on the compression portion of the internal couple. The equations for that case are given in Fig. 4.2.2.

4.2.2 Service Load Design

Precast members are checked under service load, primarily for meeting performance criteria and to control cracking.

4.2.2.1 Non-prestressed element design

Non-prestressed flexural elements are normally proportioned, and reinforcement selected, on the basis of the procedures described in Sect. 4.2.1. However, depending upon the application and exposure of the member, designers may want to control the degree of cracking. In some applications, such as architectural precast concrete panels, they may not want any discernible cracking. In other cases cracking may be permitted, but the crack width must be limited. In addition, the Code requires that the crack width be limited when the yield strength of the reinforcement exceeds 40,000 psi.

If no discernible cracking is the criteria, the flexural tensile stress level should be limited to:

$$f'_r \leq 5\lambda\sqrt{f'_c} \qquad \text{(Eq. 4.2.7)}$$

where:

f'_r = allowable flexural tension, computed using gross concrete section

f'_c = concrete strength at the time considered

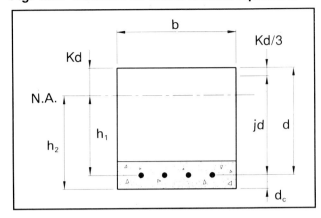

Fig. 4.2.4 Notation for crack-control equations

λ = 1.0 for normal weight concrete

= 0.85 for sand-lightweight concrete

= 0.75 for all-lightweight concrete

When tensile stress, f_t, exceeds this value, required reinforcement is determined by a Code limitation on the maximum value of the quantity z as shown in Table 4.2.1, in which z is calculated from the equation:

$$z = f_s \sqrt[3]{d_c A} \qquad \text{(Eq. 4.2.8)}$$

where:

f_s = reinforcement stress at service load, ksi

= 0.6 f_y, unless otherwise determined

d_c = concrete cover to the center of the reinforcement closest to the tension face, in.

A = average effective area around one reinforcing bar, in.²

= 2b d_c/n

b = width of tension face, in.

n = number of reinforcing bars

It is recommended that a minimum amount of reinforcement equivalent to $\rho = 0.001$ should be provided in precast concrete panels.

Table 4.2.1 Recommended maximum values of z and crack widths, w

Type	Not exposed to view		Critical appearance	
Exposure	Not exposed to weather	Exposed to weather	Not exposed to weather	Exposed to weather
Max. value of z (k/in.)	175	145	105	53
Corresponding value of w (in.)	0.016	0.013	0.010	0.005

Based on $h_2/h_1 = 1.2$ in Eq. 4.2.9

This equation is derived from the Gergely-Lutz[3] expression:

$$w = (7.6 \times 10^{-5}) \frac{h_2}{h_1} f_s \sqrt[3]{d_c A} \quad \text{(Eq. 4.2.9)}$$

where:

- w = maximum crack width at extreme tension fiber, in.
- h_1 = distance from centroid of tensile reinforcement to neutral axis, in.
- h_2 = distance from extreme tension fiber to neutral axis, in.

If values of f_s under service load conditions are required to be less than $0.6 f_y$ to satisfy crack control requirements, reinforcement should be provided equal to:

$$A_s = \frac{M}{0.9 f_s d} \quad \text{(Eq. 4.2.10)}$$

where M = service load moment.

This equation is based on working stress design principles with the assumption that $j = 0.9$ and $k = 0.3$.

Example 4.2.7 Non-prestressed panel design

Given:

A 6 in. thick architectural panel exposed to the weather, with dimensions as shown:

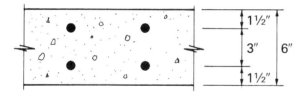

Concrete $f'_c = 5000$ psi

Service load moment, $M = 2.4$ ft-kips/ft

Problem:

Determine if reinforcement is needed to control cracking, and if so, amount required.

Solution:

For a 12 in. width:

$$f_t = \frac{M}{Z} = \frac{2.4(12)(1000)}{12(6)^2/6} = 400 \text{ psi}$$

From Eq. 4.2.7:

$$f'_r = 5(1.0)\sqrt{5000}$$
$$= 354 \text{ psi} < 400, \text{ reinforcement required}$$

For a panel with critical appearance exposed to the weather, the recommended maximum value of w from Table 4.2.1 is:

$$w = 0.005 \text{ in.}$$

Assuming $j = 0.9$ and $k = 0.3$, calculate:

- $d = 6 - 1.5 = 4.50$ in.
- $kd = (0.3)(4.5) = 1.35$ in.
- $h_1 = 4.5 - 1.35 = 3.15$ in.
- $h_2 = 6 - 1.35 = 4.65$ in.
- $\frac{h_2}{h_1} = \frac{4.65}{3.15} = 1.48$
- $d_c = 1.5$ in.

Try a bar spacing of 4 in.

$$A = 2(4)(1.5) = 12 \text{ sq in.}$$

From Eq. 4.2.9:

$$f_s = \frac{w}{(7.6 \times 10^{-5}) \frac{h_2}{h_1} \sqrt[3]{d_c A}}$$

$$= \frac{0.005}{(7.6 \times 10^{-5})(1.48)\sqrt[3]{1.5(12)}} = 16.96 \text{ ksi}$$

$$A_s = \frac{M}{0.9 f_s d}$$

$$= \frac{2.4(12)}{0.9(16.96)(4.5)} = 0.42 \text{ sq in./ft}$$

Fig. 4.2.5 Calculation of service load stresses

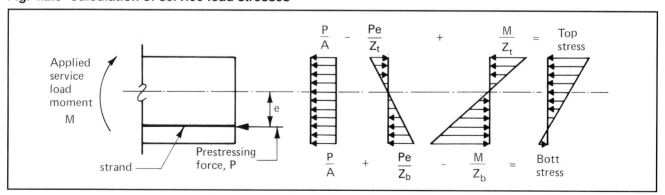

Use No. 4 at 4 in.; A_s = 0.60 sq in./ft

Note: This is an unusually high amount of reinforcement for this type of panel. Ordinarily, span would be reduced to keep the stress below f'_r.

4.2.2.2 Prestressed element design

For prestressed concrete members, the ACI Code requires that service load stresses be checked at critical points, in addition to the design strength of the member. Code limitations on the service load stresses are summarized as follows (see Code Sects. 18.4 and 18.5):

Concrete

1. At release (transfer) of prestress, before time-dependent losses:

    a. Compression $0.60 f'_{ci}$

    b. Tension (except at ends) $3\sqrt{f'_{ci}}$

    c. Tension at ends* of simply supported members $6\sqrt{f'_{ci}}$

2. Under service loads:

    a. Compression $0.45 f'_c$

    b. Tension in precompressed tensile zone when deflections are calculated based on gross section $6\sqrt{f'_c}$

    c. Tension on precompressed tensile zone when deflections are calculated based on bilinear relationships (see Sect.4.6.3) $12\sqrt{f'_c}$

Prestressing steel

a. Tension due to tendon jacking force:
$$0.85 f_{pu} \text{ or } 0.94 f_{py}$$

b. Tension immediately after prestress transfer:

    Stress-relieved strand: $0.7 f_{pu}$

    Low-relaxation strand: $0.74 f_{pu}$

It is common practice in the precast, prestressed concrete industry to follow the above recommendations with the following clarifications:

Tension in precompressed tensile zone at service loads:

    Hollow-core and solid flat slabs: $6\sqrt{f'_c}$

    Stemmed deck members and beams: $12\sqrt{f'_c}$

Initial stress in steel due to jacking forces:

    Stress-relieved strand: $0.70 f_{pu}$

    Low-relaxation strand: $0.75 f_{pu}$

These values should not be exceeded without consulting the product manufacturer.

Calculations of stresses at critical points follow classical straight line theory as illustrated in Fig. 4.2.5.

*May be considered at transfer length from end (see Sect. 4.2.3).

Load	Transfer Pt. at Release $P = P_o$		Midspan at Release $P = P_o$		Midspan at Service Load $P = P$	
	f_b	f_t	f_b	f_t	f_b	f_t
P/A	+ 525	+ 525	+ 525	+ 525	+ 479	+ 479
Pe/Z	+ 931	− 744	+ 931	− 744	+ 850	− 680
M_d/Z	− 164	+ 131	− 751	+ 600	− 751	+ 600
M_{top}/Z					− 271	+ 217
M_{sd}/Z^c					− 175	+ 84
M_ℓ/Z^c					− 437	+ 210
Stresses	+ 1292	− 88	+ 705	+ 381	− 305	+ 910
Allowable Stresses	$0.6f'_{ci}$	$6\sqrt{f'_{ci}}$	$0.6f'_{ci}$	$0.6f'_{ci}$	$6\sqrt{f'_c}$	$0.45f'_c$
	2400	− 379	2400	2400	− 465	2700
	OK	OK	OK	OK	OK	OK

Composite members

It is usually more economical to place cast-in-place composite topping without shoring the member, especially for deck members. This means that the weight of the topping and simultaneous construction live load must be carried by the precast member alone. Additional superimposed dead and live loads are carried by the composite section.

The following examples illustrate a tabular form of superimposing the stresses caused by the prestress force and the dead and live load moments.

Sign Convention

The customary sign convention used in the design of precast, prestressed concrete members for service load stresses is positive (+) for compression and negative (−) for tension. This convention is used throughout this Handbook.

Example 4.2.8 Calculation of critical stresses—straight strands

Given:

Span = 36 ft

Select 4HC12 + 2

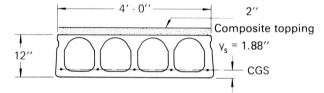

Section properties:

	Non-composite	Composite
A	= 265 sq in.	361 sq in.
I	= 4771 in.⁴	7209 in.⁴
y_b	= 6.67 in.	8.10 in.
y_t	= 5.33 in.	5.90 in.
Z_b	= 715.3 in.³	890.0 in.³
Z_t	= 895.1 in.³	1221.9 in.³
wt	= 276 plf	376 plf
e	= 4.79 in.	

Superimposed dead load = 20 psf = 80 plf
Superimposed live load = 50 psf = 200 plf

Precast concrete:

f'_c = 6000 psi
f'_{ci} = 4000 psi
E_c = 4700 ksi
Normal weight

Topping concrete:

f'_c = 4000 psi
E_c = 3800 ksi
Normal weight

Prestressing steel:

5 – ½ in. dia. 270K low-relaxation strand
A_{ps} = 5 × 0.153 = 0.765 sq in.
Straight strands

Problem:

Find critical service load stresses.

Solution:

Prestress force:

P_i = 0.765(0.75 × 270) = 155 kips
P_o (assume 10% initial loss)
= 0.90(155) = 139 kips
P (assume 18% total loss)
= 0.82(155) = 127 kips

Midspan service load moments:

M_d = 0.276 (36)²(12)/8 = 537 in. kips
M_{top} = 0.100 (36)²(12)/8 = 194 in. kips
M_{sd} = 0.080 (36)²(12)/8 = 156 in. kips
M_ℓ = 0.200 (36)²(12)/8 = 389 in. kips

Allow $6\sqrt{f'_c}$ tension at service load.

See table, page 4-18, for stresses.

Example 4.2.9 Calculation of critical stresses—single point depressed strand

Given:

Span = 70 ft
Superimposed dead load = 10 psf = 80 plf
Superimposed live load = 35 psf = 280 plf
Select 8DT24 as shown

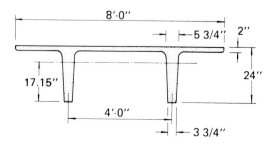

Load	Transfer Pt. at Release $P = P_o$		Midspan at Release $P = P_o$		0.4ℓ at Service Load $P = P$	
	f_b	f_t	f_b	f_t	f_b	f_t
P/A	+ 835	+ 835	+ 835	+ 835	+ 743	+ 743
Pe/Z	+ 1500	− 599	+ 3804	− 1520	+ 2985	− 1189
M_d/Z	− 290	+ 116	−2510	+1003	− 2409	+ 962
M_{sd}/Z^c					− 461	+ 184
M_ℓ/Z^c					− 1614	+ 645
Stresses	+ 2045	+ 352	+2129	+ 318	− 756	+ 1345
Allowable Stresses	$0.60f'_{ci}$	$0.60f'_{ci}$	$0.60f'_{ci}$	$0.60f'_{ci}$	$12\sqrt{f'_c}$	$0.45f'_c$
	+ 2100	+ 2100	2100	2100	− 848	2250
	OK	OK	HIGH	OK	OK	OK

Concrete:
f'_c = 5000 psi
f'_{ci} = 3500 psi
Normal weight

Prestressing steel:
12 − ½ in. dia. 270K low-relaxation strand
A_{ps} = 12 x 0.153 = 1.836 sq in.

Section properties:
A = 401 sq in.
I = 20,985 in.^{4}
y_b = 17.15 in.
y_t = 6.85 in.
Z_b = 1224 in.^{3}
Z_t = 3063 in.^{3}
wt = 418 plf = 52 psf

Eccentricities, single point depression
e_e = 5.48 in., e_c = 13.90 in.
e @ 0.4ℓ = 12.22 in.
e' = 13.90 − 5.48 = 8.42 in.

Problem:
Find critical service load stresses.

Solution:
Prestress force:
P_i = 1.836 (0.75 x 270) = 372 kips

P_o (assume 10% initial loss)
= 0.90 (372) = 335 kips
P (assume 20% total loss)
= 0.80(372) = 298 kips

Service load moments
at midspan:
M_d = 0.418(70)^2 (12)/8 = 3072 in.-kips
M_{sd} = 0.080(70)^2 (12)/8 = 588 in.-kips
M_ℓ = 0.280(70)^2 (12)/8 = 2058 in.-kips
at 0.4ℓ:
M_d = 3072(0.96) = 2949 in.-kips
M_{sd} = 588(0.96) = 564 in.-kips
M_ℓ = 2058(0.96) = 1976 in.-kips

Allow $12\sqrt{f'_c}$ final tension.

See table above for stresses.

In this example, a minimum release strength of f'_{ci} = 2129/0.6 = 3548 psi should be provided. Also deflection should be checked.

Example 4.2.10 Tensile force to be resisted by top reinforcement

Given:
Span = 24 ft
24IT26 as shown on next page

Concrete:
f'_c = 6000 psi
f'_{ci} = 4000 psi

Load	Transfer Point at Release $P = P_o$		Midspan at Release $P = P_o$	
	f_b	f_t	f_b	f_t
P/A	+ 917	+ 917	+ 917	+ 917
Pe/Z	+1401	− 1975	+1401	− 1975
M_d/Z	− 58	+ 82	− 183	+ 258
Stresses	+ 2260	− 976	+ 2135	− 800
Allowable Stresses	$0.60 f'_{ci}$	$6\sqrt{f'_{ci}}$	$0.60 f'_{ci}$	$3\sqrt{f'_{ci}}$
	2400	− 379	2400	− 190
	OK	HIGH	OK	HIGH

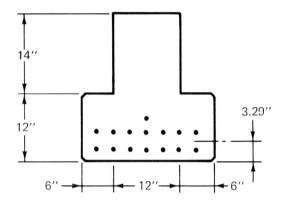

Prestressing steel:

15 − ½ in. dia. 270K low-relaxation strand

A_{ps} = 15(0.153) = 2.295 sq in.

Section properties:

A = 456 sq in.
I = 24,132 in.⁴
y_b = 10.79 in.
y_t = 15.21 in.
Z_b = 2237 in.³
Z_t = 1587 in.³
w_t = 475 plf
e = 7.5 in.

Problem:

Find critical stresses at release.

Solution:

Prestress force:

P_i = 2.295(0.75)(270) = 465 kips
P_o (assume 10% initial loss)
   = 0.90(465) = 418 kips

Moment due to member weight:

At midspan:

M_d = 0.475(24)²(12)/8 = 410 in.-kips

At 50 strand diameters (2.08 ft) transfer point:

$$M_d = \frac{wx}{2}(\ell - x)$$

$$= \frac{0.475(2.08)}{2}(24 - 2.08)(12)$$

$$= 130 \text{ in.-kips}$$

See table above for stresses.

Since the tensile stress exceeds allowable limits, reinforcement is required to resist the total tensile force, as follows:

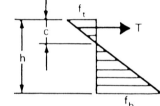

$$c = \frac{f_t}{f_t + f_b}(h) = \frac{976}{976 + 2260}(26)$$

$$= 7.84 \text{ in.}$$

$$T = \frac{c f_t b}{2} = \frac{7.84(976)(12)}{2} = 45,911 \text{ lb}$$

Similarly, the tension at midspan can be found as 34,000 lb.

The Commentary to the Code, Sect. 18.4.1 (b) and (c), recommends that reinforcement be proportioned to resist this tensile force at a stress of $0.6 f_y$, but not more than 30 ksi. Using reinforcement with f_y = 60 ksi:

0.6 (60) = 36 ksi, use 30 ksi

$$A_s \text{ (end)} = \frac{45.9}{30} = 1.53 \text{ sq in.}$$

$$A_s \text{ (midspan)} = \frac{34.0}{30} = 1.13 \text{ sq in.}$$

Top strands used as stirrup supports may also be used to carry this tensile force.

4.2.3 Prestress Transfer and Strand Development

In a pretensioned member, the prestress force is transferred to the concrete by bond. The length required to accomplish this transfer is called the "transfer length." It is given in the Commentary to the Code to be equal to $(f_{se}/3)d_b$ which for $f_{se} = 150$ ksi results in $50d_b$, where d_b is the nominal diameter of the strand.

However, the length required to develop the design strength of the strand is much longer, and is specified in the Code in Sect. 12.9.1 by the equation:

$$\ell_d = (f_{ps} - 2/3 f_{se})d_b \qquad \text{(Eq. 4.2.11)}$$

Where bonding of strand does not extend to end of member and the member is designed such that tension will occur under service loads, the development length by Eq. 4.2.11 must be doubled as specified in Code Sect. 12.9.3.

In the Commentary to the Code, the variation of strand stress along the development length is given as shown in Fig.4.2.6. For convenience, this curve may be approximated by straight lines. Also, to be consistent with the 50 diameter transfer length specified in other Code sections, the value of f_{se}, the stress which must be transferred, is assumed to be 150 ksi.

Fig.4.10.4 is a curve plotted according to the above assumptions, and can be used as a design aid as illustrated in Example 4.2.11. A more general design aid which includes all currently used strand sizes and an additional value of f_{se} is given in Chapter 11 (see Design Aid 11.2.6).

Example 4.2.11 Use of Fig. 4.10.4—Design stress for underdeveloped strand

Given:

Span = 12 ft

4HC8 as shown

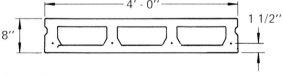

Concrete:

f'_c = 5000 psi

Normal weight

Prestressing steel:

4 – ½ in. diameter 270K strands

A_{ps} = 4 x 0.153 = 0.612 sq in.

Fig. 4.2.6 Variation of steel stress with development length

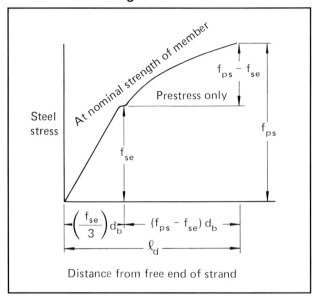

Distance from free end of strand

Solution:

If the strand is fully developed (see Fig.4.10.3):

$$C\omega_{pu} = \frac{CA_{ps}f_{pu}}{bd_p f'_c} = \frac{1.06(0.612)(270)}{48(6.5)(5)} = 0.11$$

From Fig.4.10.3 with f_{se} = 150 ksi

f_{ps} = 268 ksi

The maximum development length available is:

$\ell/2$ = 12 x 12/2 = 72 in.

From Fig 4.10.4 the maximum f_{ps} = 244 ksi

This value, rather than 268 ksi, should be used to calculate the design strength (ϕM_n) of the member at midspan.

4.3 Shear

The shear design of precast concrete members is covered in Chapter 11 of ACI 318-89. The shear resistance of precast concrete elements must meet the requirement:

$V_u \leq \phi V_n$

where $V_n = V_c + V_s$, and:

V_c = nominal shear strength of concrete

V_s = nominal shear strength of shear reinforcement

ϕ = 0.85

If V_u is less than $\phi V_c/2$, no shear reinforcement is required. However, for flat deck members (hollow-core and solid slabs), and others proven by test, no shear reinforcement is required if the factored shear force, V_u, does not exceed the design shear strength of the concrete, ϕV_c. For other members, the minimum shear reinforcement is usually adequate.

The critical section for shear and torsion is indicated in the Code to be a distance "d" from the face of the support for non-prestressed members and "h/2" for prestressed members. However, precast concrete members on which the load is not applied at the top of the member, such as L-shaped beams, the distance "d" or "h" should be measured from the point of load application to the bottom, or, conservatively, the critical section taken at the face of the support. Also, if a concentrated load is applied near a support face, the critical section must be redefined so as not to exclude its effect on shear design. See ACI Sect. 11.1.3.

4.3.1 Shear Resistance of Non-Prestressed Concrete

In the absence of torsion and axial forces, the nominal shear resistance of concrete is given by:

$$V_c = 2\sqrt{f'_c}\, b_w d \qquad \text{(Eq. 4.3.1)}$$

or if one performs a more detailed analysis:

$$V_c = \left(1.9\sqrt{f'_c} + 2500\rho_w \frac{V_u d}{M_u}\right) b_w d \leq 3.5\sqrt{f'_c}\, b_w d \qquad \text{(Eq. 4.3.2)}$$

where $\dfrac{V_u d}{M_u} \leq 1.0$

See ACI Code for members subjected to significant axial or torsional forces in addition to shear.

Example 4.3.1 Design of shear reinforcement—non-prestressed member

Given:

A spandrel beam as shown:

b = 8 in.

d = 87 in.

Span, ℓ = 30 ft.

f'_c = 5 ksi

Loading at top of member

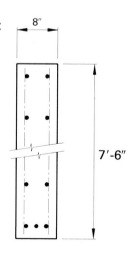

Loading x load factor	= factored load
Self wt = 0.75 x 1.4	= 1.05
Dead load = 3.21 x 1.4	= 4.49
Live load = 1.31 x 1.7	= 2.23
	w_u = 7.77 k/ft

Problem:

Determine what size welded wire fabric will satisfy the shear requirements.

Solution:

Determine V_u at a distance d from support:

$V_u = 7.77(30/2 - 87/12) = 60.2$ kips

$V_c = 2\sqrt{f'_c}\, b_w d = 2\sqrt{5000}\,(8 \times 87) = 98.4$ kips

$\dfrac{\phi V_c}{2} = \dfrac{0.85(98.4)}{2} = 41.8$ kips

$V_c > V_u > \dfrac{\phi V_c}{2}$

therefore minimum shear reinforcement is required.

ACI 318-89, Eq. 11-14 requires a minimum amount of reinforcement be provided as follows:

$A_v = 50 b_w s / f_y = 50(8)(12)/60{,}000$

$\quad = 0.080$ sq in./ft

From Design Aid 11.2.11, select a WWF that has vertical wires:

W4 @ 6 in. $A_v = 0.08$ sq in./ft

4.3.2 Shear Resistance of Prestressed Concrete Members

Shear design of prestressed concrete members is covered in ACI 318-89 by Eqs. 11-10 through 11-13, reproduced below.

Either Eq. 4.3.3 or the lesser of Eqs. 4.3.4 or 4.3.6 may be used, however, Eq. 4.3.3 is valid only if the effective prestress force is at least equal to 40% of the tensile strength of the prestressing strand. The Code places certain upper and lower limits on the use of these equations, which are shown in Fig. 4.3.1.

$$V_c = \left(0.6\sqrt{f'_c} + 700 \frac{V_u d}{M_u}\right) b_w d \qquad \text{(Eq. 4.3.3)}$$

where $\dfrac{V_u d}{M_u} \leq 1.0$

$$V_{ci} = 0.6\sqrt{f'_c}\, b_w d + V_d + \frac{V_i M_{cr}}{M_{max}} \qquad \text{(Eq. 4.3.4)}$$

$$M_{cr} = \left(\frac{I}{y_t}\right)\left(6\sqrt{f'_c} + f_{pe} - f_d\right) \qquad \text{(Eq. 4.3.5)}$$

$$V_{cw} = \left(3.5\sqrt{f'_c} + 0.3 f_{pc}\right) b_w d + V_p \qquad \text{(Eq. 4.3.6)}$$

The value of d in the term $V_u d/M_u$ in Eq. 4.3.3 is the distance from the extreme compression fiber to the centroid of the prestressed reinforcement. In all other equations, d need not be less than 0.8h.

In unusual cases, such as members which carry heavy concentrated loads, or short spans with high superimposed loads, it may be necessary to construct a shear resistance diagram (V_c) and superimpose upon that a factored shear (V_u) diagram. The procedure is illustrated in Fig. 4.3.1.

The steps for constructing the shear resistance diagram are as follows:

Fig. 4.3.1 Shear design

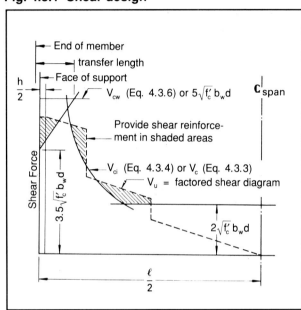

1. Draw a horizontal line at a value of $2\sqrt{f'_c} b_w d$ (Note: The Code requires that this minimum be reduced to $1.7\sqrt{f'_c} b_w d$ when the stress in the strand after all losses is less than $0.4 f_{pu}$. For precast, prestressed members the value will generally be above $0.4 f_{pu}$.)

2. Construct the curved portion of the diagram. For this, either Eq. 4.3.4 or, more conservatively, Eq. 4.3.3 may be used. Usually it is adequate to find 3 points on the curve.

3. Draw the upper limits line, V_{cw} from Eq. 4.3.6 if Eq. 4.3.4 has been used in Step 2, or $5\sqrt{f'_c} b_w d$ if Eq. 4.3.3 has been used.

4. The diagonal line at the upper left of Fig. 4.3.1 delineates the upper limit of the shear resistance diagram in the prestress transfer zone. This line starts at a value of $3.5\sqrt{f'_c} b_w d$ at the end of the member, and intersects the V_{cw} line or $5\sqrt{f'_c} b_w d$ line at transfer length from the end of the member.

Example 4.3.2 Construction of applied and resisting design shear diagrams

Given:

2HC8 with span and loadings shown

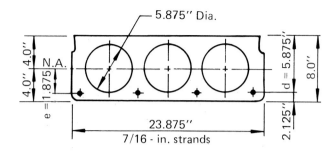

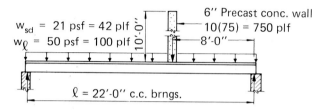

Section properties:

A = 110 sq in.

I = 843 in.⁴

y_b = 4.0 in.

b_w = 6.25 in.

d = 5.875 in.

h = 8.0 in. (0.8h = 6.4 in.)

wt = 57 psf = 114 plf

Concrete:

f'_c = 5000 psi normal weight

Solution:

1. Determine factored loads

   Uniform dead = 1.4(42 + 114)
   = 218 plf

   Uniform live = 1.7(100)
   = 170 plf

   Concentrated dead = 1.4(2 × 750)
   = 2100 lb

2. Construct shear diagram as shown in Fig. 4.3.2

3. Construct the shear resistance diagram as described in previous section.

   a. Construct line at $2\sqrt{f'_c} b_w d$ = 5.7 kips

   b. Construct V_c line by Eq. 4.3.3

$$V_c = \left(0.6\sqrt{f'_c} + 700\frac{V_u d}{M_u}\right)b_w d$$

$$= 1.70 + 28.0\frac{V_u d}{M_u}$$

where $d = 5.875$ in.

At 1, 2, and 4 ft from each end:

V_u (left) $= 5.03 - 0.39x$

M_u (left) $= \left(5.03x - \dfrac{0.39x^2}{2}\right)12$

V_u (right) $= 5.61 - 0.39x$

M_u (right) $= \left(5.61x - \dfrac{0.39x^2}{2}\right)12$

Point	x	V_u (kips)	M_u (in.-kips)	$V_u d/M_u$	V_c (kips)
1	1	4.64	58.02	0.470	14.9
2	2	4.25	111.36	0.224	8.0
3	4	3.47	204.00	0.100	4.5
4	1	5.22	64.98	0.472	14.9
5	2	4.83	125.28	0.227	8.1
6	4	4.05	231.84	0.103	4.6

c. Construct upper limit line at

= 14.1 kips

d. Construct diagonal line at transfer zone from $3.5\sqrt{f'_c} = 9.9$ kips at end of member to 14.1 kips at $50d_b = 50(7/16) = 21.9$ in. = 1.82 ft

e. Construct V_u/ϕ diagram:

$5.03/0.85 = 5.92$

$0.40/0.85 = 0.47$

$2.50/0.85 = 2.94$

$5.61/0.85 = 6.60$

It is apparent from these diagrams that no shear reinforcement is required.

4.3.3 Design Using Design Aids

Figs. 4.10.5 through 4.10.9 are design aids to assist in determining the shear strength of precast, prestressed members.

Lightweight concrete

When lightweight concrete is used, the shear equations, Eqs. 4.3.1 through 4.3.6, are modified by substituting $\lambda\sqrt{f'_c}$ for $\sqrt{f'_c}$. The coefficient λ is defined as follows:

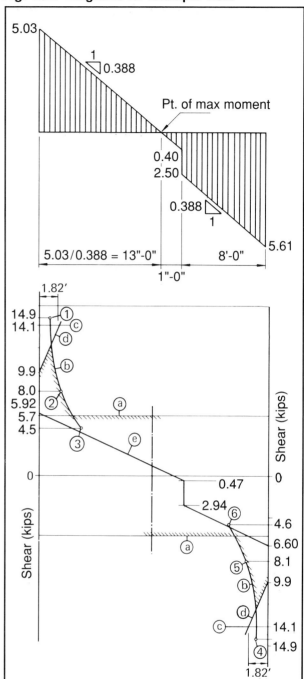

Fig. 4.3.2 Diagrams for Example 4.3.2

$$\lambda = \frac{\frac{f_{ct}}{6.7}}{\sqrt{f'_c}} \leq 1.0$$

In this equation f_{ct} is the splitting tensile strength determined by test (ASTM C496). For normal weight concrete, λ is equal to 1.0. If the value of f_{ct} is not known, $\lambda = 0.85$ for sand-lightweight concrete, and 0.75 for all-lightweight concrete. Figs. 4.10.5 to 4.10.7 provide separate charts for normal weight and lightweight concrete. In these charts, it is assumed that f_{ct} is not known and the material is sand-lightweight, or $\lambda = 0.85$.

Example 4.3.3 Use of Figs. 4.10.5 through 4.10.7— graphical solution of Eq. 4.3.3 (Code Eq. 11-10)

Given:

PCI standard double tee 10LDT24 + 2

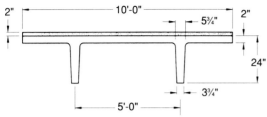

Span = 50 ft

b_w = 9.5 in., d = 21 + 2 = 23 in.

0.8h = 20.8 in.

d (near ends) = 16 in. < 20.8, use 20.8 in.

d (near midspan) = 23 in. > 20.8, use 23 in.

Concrete:

Precast: f'_c = 5000 psi, sand-lightweight

Topping: f'_c = 4000 psi, normal weight

Reinforcement:

Prestressing steel:

270K strand

108-D1 pattern, A_{ps} = 1.53 sq in.

e_e = 7.77 in., e_c = 14.77 in.

Shear reinforcement

f_y = 60 ksi

Loads:

Dead load, w_d = 609 plf

Live load, w_ℓ = 800 plf

Problem:

Find the value of excess shear force, $V_u/\phi - V_c$, along the span using Eq. 4.3.3 for V_c.

Solution (see Fig. 4.3.3):

The parameters needed for use of Figs. 4.10.5 through 4.10.7 are:

Strand drape: 14.77 − 7.77 = 7 in. which is approximately equal to d/3 — thus a shallow drape

$$\ell/d = \frac{50 \times 12}{23} = 26.1$$

V_u/ϕ at support = $[(1.4w_d + 1.7w_\ell)\ell/2]/\phi$

= (1.4 × 609 + 1.7 × 800)(50/2)/0.85(1000)

= 65.1 kips

The graphical solution (Fig. 4.3.3) follows these steps:

(a) Draw a line from V_u/ϕ = 65.1 kips at support to V_u/ϕ = 0 at midspan.

(b) Draw line from $V_c = 3.5\lambda\sqrt{f'_c}\,b_w d$

= 41.6 kips at support to $V_c = 5\lambda\sqrt{f'_c}\,b_w d$

= 59.4 kips at 50 d_b = 25 in. (or 0.042ℓ) from end of member.

(c) Draw a curved line at ℓ/d = 26.1.

(d) Draw a vertical line at h/2 from face of support.

Fig. 4.3.3 Solution of Example 4.3.3

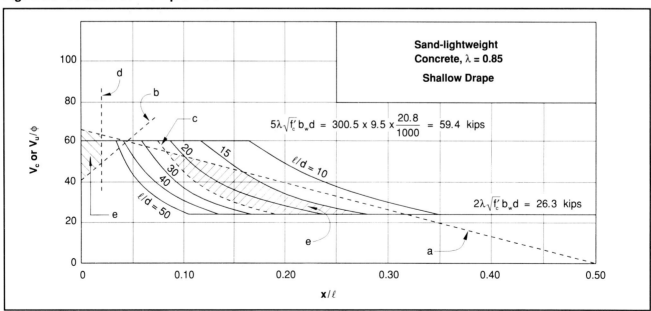

(e) The shaded area is the excess shear, $V_u/\phi - V_c$, for which shear reinforcement is required (see Example 4.3.5 for design of shear reinforcement).

4.3.4 Shear Reinforcement

Shear reinforcement is required in all concrete members, except as noted in Sect. 11.5.5, ACI 318-89. The minimum area required by the ACI Code is determined using Eq. 11-14:

$$A_v = 50\, b_w s/f_y \quad \text{(Eq. 4.3.7)}$$

or, alternatively for prestressed members only, using Eq. 11-15:

$$A_v = \frac{A_{ps}}{80} \frac{f_{pu}}{f_y} \frac{s}{d} \sqrt{\frac{d}{b_w}} \quad \text{(Eq. 4.3.8)}$$

Fig. 4.10.8 is a graphical solution for the minimum shear reinforcement by Eq. 4.3.8.

Example 4.3.4 Minimum shear reinforcement by Eq. 4.3.8 and Fig. 4.10.8

Given:

Double tee of Example 4.3.3

Shear reinforcement: W 4.0 wire each leg

$A_v = 2(0.040) = 0.080$ sq in.

Problem:

Determine the minimum amount of shear reinforcement required by Eq. 4.3.8 (Code Eq. 11-15). Verify the result from Fig. 4.10.8.

Solution:

$b_w d = 9.5(22) = 209$ sq in.

Note: As a simplification, d = 22 in. is used as an average value.

From Eq. 4.3.8:

$$A_v = 0.080 = \frac{1.53}{80}\left(\frac{270}{60}\right)\left(\frac{s}{22}\right)\sqrt{\frac{22}{9.5}}$$

thus, s = 13.4 in.

From Fig. 4.10.8:

For $A_{ps} = 1.53$ sq in., $f_y = 60$ ksi, $f_{pu} = 270$ ksi, and $b_w d = 209$ sq in.,

$A_v = 0.075$ sq in./ft

corresponding s = 12(0.080)/0.075

= 12.8 in. ≈ 13.4 OK

Per Code Sect. 11.5.4:

$s_{max} = \frac{3}{4} h \leq 24$ in.

$\frac{3}{4} h = 0.75(26) = 19.5$ in. > 13.4 OK

Shear reinforcement requirements are defined in ACI 318-89 by Eq. 11-17 which is rewritten as:

$$A_v = \frac{(V_u/\phi - V_c)s}{f_y d} \quad \text{(Eq. 4.3.9)}$$

Fig. 4.10.9 may be used to design shear reinforcement by Eq. 4.3.9 for a given excess shear. Stirrup size, strength or spacing can be varied. Welded wire fabric may also be used for shear reinforcement in accordance with Sects. 11.5.1 and 12.13.2, ACI 318-89.

Example 4.3.5 Use of Fig. 4.10.9—Shear reinforcement

Given:

Double tee of Examples 4.3.3 and 4.3.4

Problem:

Design shear reinforcement for $V_u/\phi - V_c = 10,000$ lb

Solution:

Excess shear per stem

= ½ [$(V_u/\phi - V_c)/d$] = ½(10,000/20.8)

= 240.4 lb/in.

From Fig. 4.10.9:

Use one row per stem of welded wire fabric W4.0, vertical wire spacing = 6 in.

[$(V_u/\phi - V_c)/d$] provided per stem

= 400 lb/in. > 240.4 OK

4.3.5 Horizontal Shear Transfer in Composite Members

Cast-in-place concrete topping is often used on precast members to develop composite structures. The increased stiffness and strength may be required for gravity loads or for developing a diaphragm to transfer lateral loads.

In order for a precast, prestressed member with topping to behave compositely, full transfer of horizontal shear forces must be assured at the interface of the precast member and the cast-in-place topping. This requires that interface surfaces must be clean and free of laitance. In addition, intentional roughening of surfaces and/or horizontal shear ties may also be required depending on the magnitude of shear force to be transferred.

Fig. 4.3.4 Horizontal shear in composite section

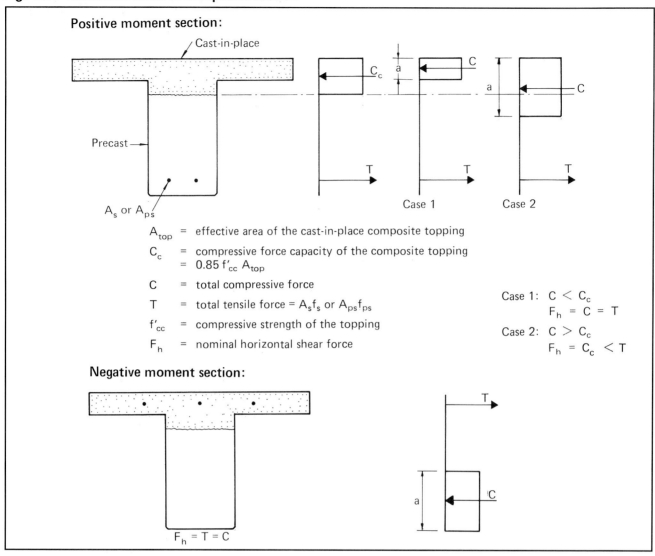

Fig. 4.3.5 Horizontal Shear Length

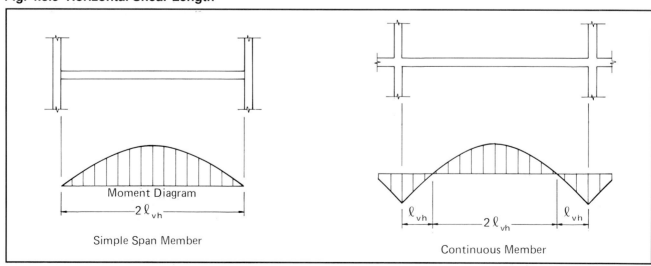

ACI 318-89 includes two alternative methods for design of horizontal shear transfer. The procedure recommended here and described below is given in Code Sect. 17.5.3.

The horizontal shear force, F_h, which must be resisted is the total force in the topping; compression in positive moment regions and tension in negative moment regions as shown in Fig. 4.3.4.

In a composite member which has an interface surface that is intentionally roughened but does not have horizontal shear ties, or where minimum ties are provided in accordance with Code Sect. 17.6 but the surface is not intentionally roughened, F_h should not exceed $\phi 80 b_v \ell_{vh}$, where b_v is the width of the interface surface and ℓ_{vh} is the horizontal shear length as defined in Fig. 4.3.5. (Note: Depending on the method of casting and the method used for finishing, the surface roughness of precast products can vary widely. Furthermore, in some precast products, such as extruded hollow core slabs, it is relatively more difficult to produce surface roughness. Review of various tests available in literature shows a large variability in results. While the limit $F_h = \phi 80 b_v \ell_{vh}$ has proven satisfactory for most precast products, it is recommended that for products where surface roughness is small, such as in extruded hollow core slabs, F_h should not exceed $\phi 40 b_v \ell_{vh}$ unless product-specific tests justify use of a larger value.)

For an interface surface which is both intentionally roughened and includes minimum horizontal shear ties per Code Sect. 17.6, F_h is limited to $\phi 350 b_v \ell_{vh}$. For F_h exceeding this value, the area of horizontal shear ties required in length ℓ_{vh} may be calculated by:

$$A_{cs} = \frac{F_h}{\phi \mu_e f_y} \qquad \text{(Eq. 4.3.10)}$$

where:

A_{cs} = area of horizontal shear ties, sq in.

F_h = horizontal shear force, lb

f_y = yield strength of horizontal shear ties, psi

μ_e = effective shear-friction coefficient as defined in Sect. 6.7

$\quad = \dfrac{1000 \lambda A_{cr} \mu}{V_u}$

ϕ = 0.85

For composite members, $\mu = 1.0\lambda$ and $A_{cr} = b_v \ell_{vh}$, thus:

$$\mu_e = \frac{1000 \lambda^2 b_v \ell_{vh}}{F_h} \leq 2.9 \qquad \text{(Eq. 4.3.11)}$$

(See Table 6.7.1)

The value of F_h is limited to:

$F_h/\phi \text{ (max)} = 0.25 \lambda^2 f'_c b_v \ell_{vh} \leq 1000 \lambda^2 b_v \ell_{vh}$

(Eq. 4.3.12)

where f'_c is the lesser compressive strength of the precast member or the composite topping.

Horizontal shear ties are typically spaced uniformly. While this has proven satisfactory, a nominal adjustment in tie spacing may be made or a nominal number of extra ties provided at member ends to account for the variation of shear forces along member length and thus improving member performance. Sect 17.6.1 of ACI 318-89 also requires that ties, when required, be spaced no more than four times the least dimension of the supported element, nor 24 in., and meet the minimum shear reinforcement requirements of Sect. 11.5.5.3.

$$A_{cs}(\text{min}) = \frac{50 b_v \ell_{vh}}{f_y} \qquad \text{(Eq. 4.3.13)}$$

Anchorage of ties must satisfy Code Sect. 17.6.3. Recent research[4] has shown that this requirement can be satisfied by providing a minimum distance of 1.75, 2.5 and 3.25 in. between the shear transfer interface and the outside ends of standard hooks on No. 3, No. 4 and No. 5 ties respectively.

Example 4.3.6 Horizontal shear design for composite beam

Given:

Inverted tee beam with 2 in. composite topping

(See Example 4.2.6)

Beam length = 20'-0"

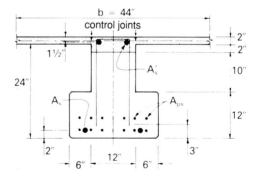

Concrete:

f'_c (precast) = 5000 psi

f'_c (topping) = 3000 psi

Prestressing steel:

12 – ½ in. diameter 270K strands

A_{ps} = 12 × 0.153 = 1.836 sq in.

Tie steel: f_y = 60,000 psi

Problem:

Determine the tie requirements to transfer horizontal shear force.

Solution:

b_v = 12 in.

$\ell_{vh} = \dfrac{20(12)}{2}$ = 120 in.

A_{top} = 2(44) + 2(12) = 112 sq in.

C = $0.85 f'_{cc} A_{top} + A'_s f_y$ = 0.85(3)(112) + 2(60)

  = 285.6 + 120.0 = 405.6 kips

f_{ps} = 260.2 ksi (see Ex.4.2.6)

$A_{ps}f_{ps}$ = 1.836(260.2) = 477.7 kips > 405.6

Therefore, F_h = 405.6 kips

$80\phi b_v \ell_{vh}$ = 80(0.85)(12)(120)/1000

  = 97.9 kips < 405.6

Therefore, ties are required.

λ = 1.0 (normal weight concrete)

$\mu_e = \dfrac{1000\lambda^2 b_v \ell_{vh}}{F_h} = \dfrac{1000(1.0)^2(12)(120)}{405,600}$

  = 3.55 > 2.9 use 2.9

$A_{cs} = \dfrac{F_h}{\phi \mu_e f_y} = \dfrac{405.6}{0.85(2.9)(60)}$

  = 2.74 sq in.

Check minimum requirements:

$A_{cs}(min) = \dfrac{50 b_v \ell_{vh}}{f_y} = \dfrac{50(12)(120)}{60,000}$

  = 1.20 sq in.

Use No. 3 ties, A_{cs} = 2(0.11) = 0.22 sq in.

Maximum tie spacing = 4(4) = 16 in. < 24 in.

For A_{cs} = 2.74 sq in., no. of ties in length
ℓ_{vh} = 2.74/0.22 ≈ 13

Total no. of ties in the beam = 26

Provide 5 No. 3 ties at 6 in. spacing at each end and the remaining 10 No. 3 at approximately 11 in. spacing in the middle portion of the beam.

In this example it is assumed that the full flange width of 44 in. is part of the composite section. Control joints in topping, particularly in parking garages, are often located along the joints between ends of double tees and edges of tee beams for crack control as shown in the example figure. This raises the concern that the extended parts of the topping on each side of the tee beam web (16 in. in this example) may not behave compositely with the rest of the section. However, it can be shown that the usual steel provided in the topping, in most cases, generates sufficient shear friction resistance (see Sect. 6.7) to ensure composite action. The following calculations support this observation.

The topping reinforcement is:

6 x 6 – W2.9 x W2.9,

A_s = 0.058 in.²/ft,

  = 0.58 in.² per ½ span

The portion of the total compression force in the extended part of topping:

$C'_c = \dfrac{16 \times 2}{112}(285.6)$ = 81.6 kips

$\mu_e = \dfrac{(1000)(1.0)^2(2)(120)}{81,600}$ = 2.94 < 3.4

$A'_{cs} = \dfrac{81.6}{0.85(2.94)(60)}$ = 0.54 in.² < 0.58 OK

$(A'_{cs})_{min} = \dfrac{50(2)(120)}{60,000}$ = 0.2 in.² < 0.54 OK

Max. spacing = 4(2) = 8 in.

Thus 6 x 6 – W2.9 x W2.9 is adequate.

4.4 Torsion

ACI 318-89 does not include torsion design provisions for prestressed concrete. The provisions in the Code for non-prestressed concrete will usually be conservative if applied to prestressed concrete. However, methods specifically developed for prestressed concrete are available and have been used for several years and yielded safe and economical designs.

The following two sections illustrate the Code method for non-prestressed concrete members (Sect. 4.4.1) and an updated Zia and McGee[5] method (update by Zia and Hsu[6]) for prestressed concrete members (Sect. 4.4.2)

The critical section for shear and torsion by ACI 318-89 (Sect. 11.1.3) is "d" from the face of the support for non-prestressed members and "h/2" for prestressed members. If a concentrated load occurs within these dimensions, location of the critical section must be re-established to include the load.

4.4.1 Design for Shear and Torsion— Non-Prestressed Members

The torsion design procedures prescribed by ACI 318-89 are illustrated by the following example:

Example 4.4.1 Torsion design—non-prestressed member

Given:

Precast load bearing spandrel beam shown.

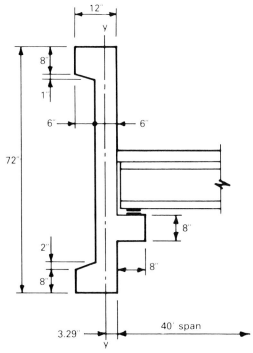

Span of spandrel beam = 30 ft clear
f'_c = 5000 psi, normal weight concrete
Reinforcement f_y = 60,000 psi
d = 69 in.

Loads (kips/ft):

D.L.:

Precast floor	60 psf x (20ft) = 1.2 (1.4)	= 1.68
Topping	25(20) = 0.5 (1.4)	= 0.70
Superimposed	10(20) = 0.2 (1.4)	= 0.28
Window	= 0.05 (1.4)	= 0.07
Spandrel	= 0.63 (1.4)	= 0.88
L.L.:	50 psf x (20 ft) = 1.00 (1.7)	= 1.70
	w_u =	5.31

Problem:

Determine torsion reinforcement requirements.

Solution:

1. Compute torsion moment (T_u) at critical section, located at 5'-9" from face of support:

   $V_u = w_u(15 - 5.75) = 5.31(9.25) = 49.1$ kips

   w_u for torsion = 1.68 + 0.70 + 0.28 + 1.70
   = 4.36 kips/ft

   Eccentricity = 2/3 (8) + 3.29 = 8.62 in.
   $T_u = w_u(e)(9.25) = 4.36(8.62)(9.25)$
   = 348 in.-kips

2. Determine if torsion effects must be considered:

   If $T_u \geq \phi(0.5\sqrt{f'_c}\,\Sigma x^2 y)$ must consider torsion.

   $\Sigma x^2 y = 6^2(72) + 6^2(8)(2) + 8^2(8) = 3680$ in.^{3}

   $\phi(0.5\sqrt{f'_c}\,\Sigma x^2 y) = \dfrac{0.85(0.5)\sqrt{5000}(3680)}{1000}$

   = 110.6 in.-kips

   348 > 110.6, consider torsion

3. Determine the torsion moment strength provided by concrete:

   $T_c = \dfrac{0.8\sqrt{f'_c}\,\Sigma x^2 y}{\sqrt{1 + \left(\dfrac{0.4 V_u}{C_t T_u}\right)^2}}$

   where:

   $C_t = \dfrac{b_w d}{\Sigma x^2 y} = \dfrac{6(69)}{3680} = 0.1125$

   $T_c = \dfrac{0.8\sqrt{5000}(3680)}{1000\sqrt{1 + \left(\dfrac{0.4(49.1)}{0.1125(348)}\right)^2}} = 186$ in.-kips

4. Determine torsion reinforcement requirements:

   $T_u = \phi T_n = \phi(T_c + T_s)$

   $T_s = \dfrac{T_u}{\phi} - T_c = \dfrac{348}{0.85} - 186$

   = 223 in.-kips

   By Sect. 11.6.9.4 (ACI 318-89)
   $T_s \leq 4T_c = 4(186) = 744$ OK

   By Eq. 11-23 of ACI 318-89
   Assume x_1 = 4 in.; y_1 = 70 in.

   $T_s = \dfrac{A_t \alpha_t x_1 y_1 f_y}{s}$

   $A_t = \dfrac{T_s(s)}{\alpha_t x_1 y_1 f_y}$

   $\alpha_t = 0.66 + 0.33(y_1/x_1) \leq 1.5$
   = 0.66 + 0.33(70/4) = 6.44, use α_t = 1.5

   $A_t = \dfrac{223(12)}{1.5(4)(70)(60)} = 0.106$ sq in./ft

   = 0.0088 sq in./in.

   This is the required area of steel in each leg of the closed stirrup for torsion only. The shear steel requirement must be added to A_t. The minimum area of closed stirrups is:

$$A_v + 2A_t = 50b_w s/f_y$$

Arrange reinforcement as follows:

5. Determine the longitudinal bars A_ℓ distributed around the closed stirrups:

$$A_\ell = 2A_t \left(\frac{x_1+y_1}{s}\right) = 2(0.0088)\left(\frac{4+70}{1}\right)$$

$$= 1.30 \text{ sq in.}$$

or

$$A_\ell = \left[\frac{400xs}{f_y}\left(\frac{T_u}{T_u + \frac{V_u}{3C_t}}\right) - 2A_t\left(\frac{x_1+y_1}{s}\right)\right]$$

but $2A_t$ (in this equation) $\geq \dfrac{50b_w s}{f_y}$

$$= \frac{50(6)(1)}{60,000} = 0.005 < 2A_t$$

$$A_\ell = \left[\frac{400(6)(1)}{60,000}\left(\frac{348}{348 + \frac{49.1}{3(0.1125)}}\right)\right.$$

$$\left. - 2(0.0088)\right]\left(\frac{4+70}{1}\right) = 0.78 \text{ sq in.}$$

Use $A_\ell = 1.30$ sq in. distributed around perimeter.

4.4.2 Design for Shear and Torsion—Prestressed Members

The Second Edition of the Handbook included a method which was based on the 1974 Zia and McGee[5] proposal. The Third Edition of this Handbook illustrated a method based on a later development by Collins and Mitchell[7] which utilizes the compression field theory.

Experience of designers with these two methods suggests a strong preference for the Second Edition method. This preference is based on a number of factors, such as its simplicity, similarity with the ACI Code method for the non-prestressed members, and the satisfactory performance of structures for about 15 years. Thus, the method in this edition of the Handbook is based on the Zia and McGee method as later modified by Zia and Hsu[6]. The method is given in terms of forces and moments rather than stresses for consistency with the ACI Code. The step-by-step design procedure is summarized below and Figs. 4.10.10 through 4.10.13 are included to facilitate calculation of various quantities.

Step 1: Determine if torsion can be neglected, i.e., is $T_u \leq T_{u(min)}$:

$$T_{u(min)} = \phi\gamma_t(1.5\beta_t \Sigma x^2 y \lambda \sqrt{f'_c})$$

For $\phi = 0.85$ and $\beta_t = \frac{1}{3}$ (see Note 2 below):

$$T_{u(min)} = 0.425\gamma_t \sqrt{f'_c} \lambda \Sigma x^2 y \quad \text{(Eq. 4.4.1)}$$

where:

T_u = factored torsional moment, in.-lb

ϕ = 0.85

γ_t = a factor dependent on the level of prestress
= $\sqrt{1 + 10f_{pc}/f'_c}$

f_{pc} = average prestress on the section, psi

f'_c = concrete compressive strength, psi

x,y = short side and long side, respectively, of a component rectangle, in.

λ = a conversion factor for lightweight concrete

β_t = $0.35/(0.75 + x/y)$

Note 1: In computing $\Sigma x^2 y$ for a flanged section, the section may be divided into component rectangles such that the quantity $\Sigma x^2 y$ would be the maximum; however, the overhanging flange width used in design shall not be taken greater than three times the flange thickness. Box sections with wall thickness $\geq x/4$ may be treated as solid sections. For wall thickness $\leq x/4$ but $\geq x/10$, the quantity $\Sigma x^2 y$ calculated for a solid section shall be multiplied by $4h/x$.

Note 2: A value of $\beta_t = \frac{1}{3}$ has been suggested in Ref. 6. This value is more conservative (compared with the value determined from the variable parameters above) for deep torsional members, but somewhat less conservative for shallow beams of practical size. Recognizing the variability of various experimental correlations, coupled with the desirability for design simplicity, a value of $\beta_t = \frac{1}{3}$ is reasonable.

Step 2: Check to ensure that the required nominal torsional moment and shear strengths do not exceed the following maximum limits to avoid potential compression failures due to over-reinforcing:

$$T_{n(max)} = \frac{K_t \lambda \sqrt{f'_c} \Sigma x^2 y / 3}{\sqrt{1 + \left(\frac{K_t V_u}{30 C_t T_u}\right)^2}} \quad \text{(Eq. 4.4.2)}$$

$$V_{n(max)} = \frac{10 \lambda \sqrt{f'_c} b_w d}{\sqrt{1 + \left(\frac{30 C_t T_u}{K_t V_u}\right)^2}} \quad \text{(Eq. 4.4.3)}$$

where:

$K_t = \gamma_t (12 - 10 f_{pc}/f'_c)$

$V_u =$ factored shear force, lb

$C_t = \dfrac{b_w d}{\Sigma x^2 y}$

Step 3: If $T_u \leq T_{u\,(max)}$ and $V_u \leq V_{u\,(max)}$, design may be continued. Calculate torsion and shear carried by concrete:

$$T_c = \frac{T'_c}{\sqrt{1 + \left(\frac{T'_c / T_u}{V'_c / V_u}\right)^2}} \quad \text{(Eq. 4.4.4)}$$

$$V_c = \frac{V'_c}{\sqrt{1 + \left(\frac{V'_c / V_u}{T'_c / T_u}\right)^2}} \quad \text{(Eq. 4.4.5)}$$

where:

$T_c, T'_c =$ nominal torsional moment strength of concrete under combined shear and torsion and under pure torsion, respectively.

$V_c, V'_c =$ nominal shear strength of concrete under combined shear and torsion and under pure shear, respectively.

$T'_c = 0.8 \lambda \sqrt{f'_c} \Sigma x^2 y (2.5 \gamma_t - 1.5)$ (Eq. 4.4.6)

$V'_c = V_c$ as calculated in Sect. 4.3.2

Step 4: Determine stirrups for torsional moment in excess of that carried by concrete. Stirrups must be closed and must be spaced not more than 12 in. or $(x_1 + y_1)/4$.

$$A_t = \frac{(T_u/\phi - T_c)s}{\alpha_t x_1 y_1 f_y} \quad \text{(Eq. 4.4.7)}$$

where:

$A_t =$ the required area of one leg of stirrup, sq in.

$x_1 =$ short side of the stirrup, in.

$y_1 =$ long side of the stirrup, in.

$s =$ stirrup spacing $\leq (x_1 + y_1)/4$ or 12 in.

$\alpha_t = [0.66 + 0.33 \, y_1/x_1] \leq 1.5$

$f_y =$ yield strength of stirrup, psi

Note: The required stirrups for torsion are in addition to those required for shear.

If $T_u > T_{u(min)}$ from Eq. 4.4.1, a minimum amount of web reinforcement should be provided to ensure reasonable ductility. The minimum area of closed stirrups which should be provided is:

$$(A_v + 2A_t)_{min} = 50 \frac{b_w s}{f_y} (\gamma_t)^2 \leq 200 \frac{b_w s}{f_y} \quad \text{(Eq. 4.4.8)}$$

where b_w is the web width of the beam.

Step 5: To resist the longitudinal component of the diagonal tension induced by torsion, longitudinal reinforcement approximately equal in volume to that of stirrups for torsion and effectively distributed around the perimeter (spaced $\leq$ 12 in.) should be provided. Prestressing strand may be used in satisfying the longitudinal reinforcement requirement. However, to control diagonal crack width, the stress in the prestressing strand shall be limited to 60,000 psi (see Code Sect. 11.6.7.4.) This longitudinal steel for torsion is in addition to that required for flexure. Thus:

$$A_\ell = \frac{2 A_t (x_1 + y_1)}{s} \quad \text{(Eq. 4.4.9)}$$

or

$$A_\ell = \left[\frac{400 x}{f_y} \left(\frac{T_u}{T_u + \frac{V_u}{3 C_t}}\right) - \frac{2 A_t}{s}\right](x_1 + y_1)$$

(Eq. 4.4.10)

whichever is greater. The value of A_ℓ calculated from Eq. 4.4.10 need not exceed that obtained by substituting $50 (b_w/f_y)(1 + 12 f_{pc}/f'_c) \leq 200 b_w/f_y$ for $2 A_t/s$.

It should be noted that the design of the connections between the torsional member and its supports to provide adequate reactions is critical. Also, the effects of torsion on the overall stability of the frame should be investigated. Guidelines for these design considerations are given in Chapters 5 and 6.

Example 4.4.2 Shear and torsion in a prestressed concrete member

Given:

Precast, prestressed concrete spandrel beam shown in Fig. 4.4.1

D.L. of deck = 89.5 psf

L.L. = 50 psf

Beam properties:

A = 696 sq in.

wt = 725 plf

f'_c = 5000 psi, normal weight

f_y = 60 ksi

Prestressing:

6 – ½ in. dia., 270K strands

A_{ps} = 6 x 0.153 = 0.918 sq in.

d = 69 in.

Problem:

Determine shear and torsion reinforcement for the spandrel beam at critical section.

Solution:

Calculate V_u and T_u

Determine factored loads

D.L. of beam = 1.4 x 0.725 = 1.02 kips/ft

D.L. of deck = 1.4 x 0.0895 x 60/2 x 4

= 15.04 kips per stem

L.L. = 1.7 x 0.050 x 30 x 4

= 10.2 kips per stem

V_u at support

= 1.02 x 14 + 7(15.04 + 10.2) x 1/2

= 102.6 kips

Center of support to h/2 = 0.5 + 6.25/2

= 3.625 ft

Since a concentrated load occurs within h/2, critical section is at face of support.

V_u at face of support

= 102.6 – (1.02 x 0.5)

= 102.1 kips

T_u at support (assume torsion arm = 8 in.)

= (15.04 + 10.2) x 7 x 1/2 x 8

= 707 in.-kips (same at face of support)

(1) Determine if torsion can be neglected:

Max $\Sigma x^2 y = 8^2 \times 63 + 12^2 \times 16 = 6336$

$C_t = \dfrac{b_w d}{\Sigma x^2 y} = \dfrac{8 \times 69}{6336} = 0.087$

Assuming 18% loss:

$f_{pc} = \dfrac{P}{A} = \dfrac{0.918(0.70 \times 270)(0.82)}{696}$

= 0.204 ksi

$\gamma_t = \sqrt{1 + 10 f_{pc}/f'_c} = \sqrt{1 + 10 \times 0.204/5}$

= 1.19

λ = 1.0 (normal weight concrete)

From Eq. 4.4.1:

$T_{u(min)} = 0.425(1.19)\sqrt{5000}(1.0)(6336)/1000$

= 226.6 in.-kips < 707

Torsion must be considered

Alternatively, from Fig. 4.10.10:

For f_{pc} = 204 psi and f'_c = 5000 psi:

$R_{t\ (min)}$ = 36

Therefore, $T_{u\ (min)}$ = 36(1.0)(6336)/1000

= 228.1 in.-kips

vs 226.6

(2) Check max. limits on T_u and V_u:

$K_t = \gamma_t (12 - 10 f_{pc}/f'_c)$

= 1.19(12 – 10 x 0.204/5) = 13.79

From Eq. 4.4.2:

$$T_{n(max)} = \dfrac{13.79(1.0)\sqrt{5000}\,(6336/3)/1000}{\sqrt{1 + \left(\dfrac{13.79 \times 102.1}{30 \times 0.087 \times 707}\right)^2}}$$

= 1637 in.-kips

$\dfrac{T_u}{\phi} = \dfrac{707}{0.85} = 831.8$ in.-kips < 1637 OK

From Eq. 4.4.3

$$V_{n(max)} = \dfrac{10(1.0)\sqrt{5000} \times 8 \times 69 / 1000}{\sqrt{1 + \left(\dfrac{30 \times 0.087 \times 707}{13.79 \times 102.1}\right)^2}}$$

= 236.7 kips

$\dfrac{V_u}{\phi} = \dfrac{102.1}{0.85} = 120.1$ kips < 236.7 OK

Fig. 4.4.1 Structure of Example 4.4.2

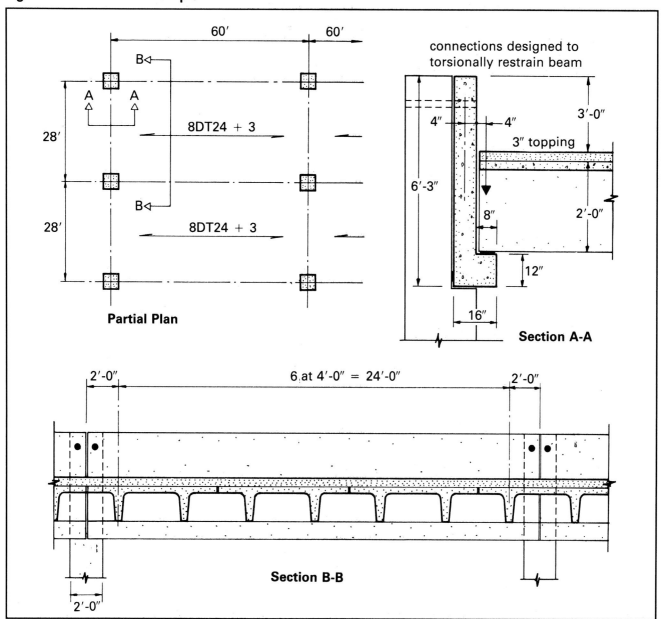

Alternatively, from Fig. 4.10.11 (by interpolation):

For $f_{pc} < 400$ psi, $C_t = 0.087$ and

$V_u/T_u = 102.1/707 = 0.144$

$R_{t\,(max)} = 258$ and $R_{v\,(max)} = 411$

Therefore,

$T_{n\,(max)} = 258(6336)/1000 = 1635$ in.-kips

vs 1637

$V_{n\,(max)} = 411(8 \times 69)/1000 = 226.9$ kips

vs 236.7

(3) Determine torsion and shear strengths under combined loading:

From Eq. 4.4.6:

$T'_c = 0.8(1.0)\sqrt{5000} \times 6336$

$\quad \times (2.5 \times 1.19 - 1.5)/1000$

$\quad = 528.6$ in.-kips

V_{cw} at support face (Eq. 4.3.6):

$V_{cw} = \left[3.5\sqrt{5000} + 0.3\left(\dfrac{6}{50d_b} \times 204\right)\right] \times 8 \times 69 / 1000$

$\quad = 144.7$ kips

Thus, $V'_c = 144.7$ kips

(Note: $6/50d_b$ fraction estimates prestress available at support face, where d_b = strand diameter = ½ in.)

From Eqs. 4.4.4 and 4.4.5:

$$T_c = \frac{528.6}{\sqrt{1+\left(\frac{528.6/707}{144.7/102.1}\right)^2}} = 467.8 \text{ in.-kips}$$

$$V_c = \frac{144.7}{\sqrt{1+\left(\frac{144.7/102.1}{528.6/707}\right)^2}} = 67.6 \text{ kips}$$

Alternatively, from Figs. 4.10.12 and 4.10.13:

For $f_{pc} = 204$ psi, $f'_c = 5000$ psi:

$R'_t = 82$ (Fig. 4.10.13)

Therefore, $T'_c = 82(6336)/1000 = 519$ in.-kips

vs 528.6

For $T'_c/T_u = 519/707 = 0.73$ and

$V'_c/V_u = 144.7/102.1 = 1.42$

$R_t = 0.88$ and $R_v = 0.46$ (Fig. 4.10.12)

Therefore,

$T_c = 0.88(519) = 456$ in.-kips vs 467.8

$V_c = 0.46(144.7) = 66.6$ kips vs 67.6

(Note: V'_c can also be obtained using shear design aids as illustrated in Example 4.3.3.)

(4) Determine stirrup steel:

With 1¼ in. cover

$y_1 = 72.5$ in., $x_1 = 5.5$ in.

$$\alpha_t = \left(0.66 + 0.33\frac{72.5}{5.5}\right) = 5.0 > 1.5, \text{ use } 1.5$$

For torsion (Eq. 4.4.7):

$$\frac{A_t}{s} = \frac{(707/0.85) - 467.8}{1.5(5.5)(72.5)(60)} = 0.0101 \text{ in.}^2/\text{in.}$$

For shear (Eq. 4.3.9):

$$\frac{A_v}{s} = \frac{(102.1/0.85) - 67.6}{69(60)} = 0.0127 \text{ in.}^2/\text{in.}$$

$$\frac{A_v}{s} + \frac{2A_t}{s} = 0.0127 + 2(0.0101) = 0.0329 \text{ in.}^2/\text{in.}$$

$$\left(\frac{A_v}{s} + \frac{2A_t}{s}\right)_{min} = 50\frac{b_w(\gamma_t)^2}{f_y} \leq 200\frac{b_w}{f_y}$$

$$\frac{50(8)(1.19)^2}{60,000} = 0.0094$$

$$200\left(\frac{8}{60,000}\right) = 0.0267 < 0.0329 \text{ OK}$$

For No. 4 stirrup, $A_v = 0.40$ in.^{2}

$$s = \frac{0.40}{0.0329} = 12.2 \text{ in.}$$

max. spacing $= \frac{x_1 + y_1}{4} \leq 12$ in.

$$\frac{5.5 + 72.5}{4} = 19.5 \text{ in.} > 12 \text{ in.}$$

Use 12 in. o.c.

(5) Determine longitudinal steel for torsion:

From Eq. 4.4.9:

$A_\ell = 2(0.0101)(5.5 + 72.5) = 1.576$ in.^{2}

$$\frac{50b_w}{f_y}\left(1 + 12f_{pc}/f'_c\right) = \frac{50 \times 8}{60,000} \times$$

$$(1 + 12 \times 0.204/5) = 0.00993 \text{ in.}^2/\text{in.}$$

$$\frac{2A_t}{s} = 0.0202 > 0.00993$$

use $\frac{2A_t}{s}$ in Eq. 4.4.10:

$$A_\ell = \left[\frac{400(8)}{60,000}\left(\frac{707}{707 + \frac{102.1}{3(0.087)}}\right) - 0.0202\right](78)$$

$= 1.10$ in.$^2 < 1.576$, use 1.576 in.^{2}

A combination of reinforcing bars and prestressing strands may be used to provide the required area of steel. If only reinforcing bars are used, select No. 3 bars.

$n = 1.576/0.11 = 14.3$

Provide 16-No. 3 bars distributed around the perimeter as shown in the figure on page 4-37.

4.5 Loss of Prestress

Loss of prestress is the reduction of tensile stress in prestressing tendons due to shortening of the concrete around the tendons, relaxation of stress within the tendons and external factors which reduce the total initial force before it is applied to the concrete. ACI 318-89 identifies the following sources of loss of prestress:

(a) Anchorage seating loss

(b) Elastic shortening of concrete

(c) Creep of concrete

(d) Shrinkage of concrete

(e) Relaxation of tendon stress

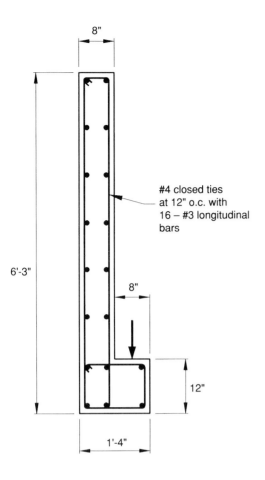

(f) Friction loss due to intended or unintended curvature in post-tensioning tendons.

Accurate determination of losses is more important in some prestressed concrete members than in others. Losses have no effect on the ultimate strength of a flexural member unless the tendons are unbonded or if the final stress after losses is less than 0.50 f_{pu}. Underestimation or overestimation of losses can affect service conditions such as camber, deflection and cracking.

4.5.1 Sources of Stress Loss

Anchorage seating loss and friction

These two sources of loss are mechanical. They represent the difference between the tension applied to the tendon by the jacking unit and the initial tension available for application to the concrete by the tendon. Their magnitude can be determined with reasonable accuracy and, in many cases, they are fully or partially compensated for by overjacking.

Elastic shortening of concrete

The concrete around the tendons shortens as the prestressing force is applied to it. Those tendons which are already bonded to the concrete shorten with it.

Shrinkage of concrete

Loss of stress in the tendon due to shrinkage of the concrete surrounding it is proportional to that part of the shrinkage that takes place after the transfer of prestress force to the concrete.

Creep of concrete and relaxation of tendons

Creep of concrete and relaxation of tendons complicate stress loss calculations. The rate of loss due to each of these factors changes when the stress level changes and the stress level is changing constantly throughout the life of the structure. Therefore, the rates of loss due to creep and relaxation are constantly changing.

4.5.2 Range of Values for Total Loss

Total loss of prestress in typical members will range from about 25,000 to 50,000 psi for normal weight concrete members, and from about 30,000 to 55,000 psi for sand-lightweight members.

The load tables in Chapter 2 have a lower limit on loss of 30,000 psi.

4.5.3 Estimating Prestress Loss

This section is based on the report of a task group sponsored by ACI-ASCE Committee 423, Prestressed Concrete[8]. That report gives simple equations for estimating losses of prestress which would enable the designer to estimate the various types of prestress loss rather than using a lump sum value. It is believed that these equations, intended for practical design applications, provide fairly realistic values for normal design conditions. For unusual design situations and special structures, more detailed analyses may be warranted.

$$T.L. = ES + CR + SH + RE \quad \text{(Eq. 4.5.1)}$$

where:

T.L. = total loss (psi), and other terms are losses due to:

ES = elastic shortening
CR = creep of concrete
SH = shrinkage of concrete
RE = relaxation of tendons

$$ES = K_{es}E_s f_{cir}/E_{ci} \quad \text{(Eq. 4.5.2)}$$

where:

- K_{es} = 1.0 for pretensioned members
- E_s = modulus of elasticity of prestressing tendons (about 28.5×10^6 psi)
- E_{ci} = modulus of elasticity of concrete at time prestress is applied
- f_{cir} = net compressive stress in concrete at center of gravity of tendons immediately after the prestress has been applied to the concrete (see Eq. 4.5.3)

$$f_{cir} = K_{cir}\left(\frac{P_i}{A_g} + \frac{P_i e^2}{I_g}\right) - \frac{M_g e}{I_g} \quad \text{(Eq. 4.5.3)}$$

where:

- K_{cir} = 0.9 for pretensioned members
- P_i = initial prestress force (after anchorage seating loss)
- e = eccentricity of center of gravity of tendons with respect to center of gravity of concrete at the cross section considered
- A_g = area of gross concrete section at the cross section considered
- I_g = moment of inertia of gross concrete section at the cross section considered
- M_g = bending moment due to dead weight of prestressed member and any other permanent loads in place at time of prestressing

$$CR = K_{cr}(E_s/E_c)(f_{cir} - f_{cds}) \quad \text{(Eq. 4.5.4)}$$

where:

- K_{cr} = 2.0 normal weight concrete
  = 1.6 sand-lightweight concrete
- f_{cds} = stress in concrete at center of gravity of tendons due to all superimposed permanent dead loads that are applied to the member after it has been prestressed (see Eq. 4.5.5)
- E_c = modulus of elasticity of concrete at 28 days

$$f_{cds} = M_{sd}(e)/I_g \quad \text{(Eq. 4.5.5)}$$

where:

- M_{sd} = moment due to all superimposed permanent dead loads applied after prestressing

$$SH = (8.2 \times 10^{-6})K_{sh}E_s \times (1 - 0.06 V/S)(100 - R.H.) \quad \text{(Eq. 4.5.6)}$$

where:

- K_{sh} = 1.0 for pretensioned members
- V/S = volume to surface ratio
- R.H. = average ambient relative humidity (see Fig. 3.3.2)

$$RE = [K_{re} - J(SH + CR + ES)]C \quad \text{(Eq. 4.5.7)}$$

where values of K_{re}, J and C are taken from Tables 4.5.1 and 4.5.2.

where:

- f_{pi} = P_i/A_{ps}
- f_{pu} = ultimate strength of prestressing tendons
- A_{ps} = area of prestressing tendons

4.5.4 Critical Locations

Computations for stress losses due to elastic shortening and creep of concrete are based on the compressive stress in the concrete at the center of gravity (cgs) of the tendons.

For bonded tendons, stress losses are computed at that point on the span where flexural tensile stresses are most critical. In members with straight, parabolic or approximately parabolic tendons this is usu-

Table 4.5.1 Values of K_{re} and J

Type of tendon	K_{re}	J
270 Grade stress-relieved strand or wire	20,000	0.15
250 Grade stress-relieved strand or wire	18,500	0.14
240 or 235 Grade stress-relieved wire	17,600	0.13
270 Grade low-relaxation strand	5,000	0.040
250 Grade low-relaxation wire	4,630	0.037
240 or 235 Grade low-relaxation wire	4,400	0.035
145 or 160 Grade stress-relieved bar	6,000	0.05

Table 4.5.2 Values of C

f_{pi}/f_{pu}	Stress-relieved strand or wire	Stress-relieved bar or low-relaxation strand or wire
0.80		1.28
0.79		1.22
0.78		1.16
0.77		1.11
0.76		1.05
0.75	1.45	1.00
0.74	1.36	0.95
0.73	1.27	0.90
0.72	1.18	0.85
0.71	1.09	0.80
0.70	1.00	0.75
0.69	0.94	0.70
0.68	0.89	0.66
0.67	0.83	0.61
0.66	0.78	0.57
0.65	0.73	0.53
0.64	0.68	0.49
0.63	0.63	0.45
0.62	0.58	0.41
0.61	0.53	0.37
0.60	0.49	0.33

ally mid-span. In members with tendons deflected at mid-span only, the critical point is generally near the 0.4 point of the span. Since the tendons are bonded, only the stresses at the critical point need to be considered. Stresses or stress changes at other points along the member do not affect the stresses or stress losses at the critical point.

Example 4.5.1 Loss of prestress

Given:

10LDT 32 + 2 as shown

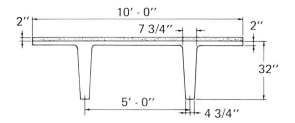

Span = 70 ft

No superimposed dead load except topping

R.H. = 75%

Section properties (untopped):

 A = 615 sq in.

 I = 59,720 in.^{4}

 Z_b = 2717 in.^{3}

 V/S = 615/364 = 1.69 in.

 wt = 491 plf

 wt of topping = 250 plf

Concrete:

Precast: Sand-lightweight

 f'_c = 5000 psi E_c = 3.0 x 10^6 psi

 f'_{ci} = 3500 psi E_{ci} = 2.5 x 10^6 psi

Topping: Normal weight

Prestressing steel:

12 – ½ in. dia. 270K low-relaxation strands

A_{ps} = 12(0.153) = 1.836 sq in.

E_s = 28.5 x 10^6 psi

Depressed at mid-span

 e_e = 12.81 in.

 e_c = 18.73 in.

Problem:

Determine total loss of prestress.

Solution:

For depressed strand, critical section is at 0.4ℓ. Determine moments, eccentricity, and prestress force.

$$M@0.4\ell = \frac{wx}{2}(\ell-x) = \frac{w(0.4\ell)}{2}(\ell-0.4\ell)$$

$$= 0.12\,w\ell^2$$

$$M_g = 0.12(0.491)(70)^2 = 289 \text{ ft-kips}$$

$$M_{sd} = 0.12(0.250)(70)^2 = 147 \text{ ft-kips}$$

e at 0.4ℓ = 12.81 + 0.8(18.73 – 12.81)

 = 17.55 in.

Assume compensation for anchorage seating loss during prestressing.

P_i = 0.75 $A_{ps}f_{pu}$ = 0.75(1.836)(270)

   = 371.8 kips

Determine f_{cir} and f_{cds}:

f_{cir} = $K_{cir}\left(\dfrac{P_i}{A_g} + \dfrac{P_i e^2}{I_g}\right) - \dfrac{M_g e}{I_g}$

   = $0.9\left(\dfrac{371.8}{615} + \dfrac{371.8(17.55)^2}{59,720}\right) - \dfrac{289(12)(17.55)}{59,720}$

   = 1.252 ksi = 1252 psi

f_{cds} = $M_{sd}(e)/I_g$

   = 147(12)(17.55)/59,720

   = 0.518 ksi = 518 psi

ES = $K_{es}E_s f_{cir}/E_{ci}$

   = $(1)(28.5 \times 10^6)(1252)/(2.5 \times 10^6)$

   = 14,272 psi

CR = $K_{cr}(E_s/E_c)(f_{cir} - f_{cds})$

CR = $(1.6)(28.5 \times 10^6/3.0 \times 10^6)(1252 - 518)$

   = 11,157 psi

SH = $(8.2 \times 10^{-6})K_{sh}E_s(1 - 0.06 V/S)(100 - R.H.)$

   = $(8.2 \times 10^{-6})(1)(28.5 \times 10^6)$

   $\times [1 - 0.06(1.69)](100 - 75)$

   = 5249 psi

RE = $[K_{re} - J(SH + CR + ES)]C$

From Table 4.5.1

   K_{re} = 5000

   J = 0.04

f_{pi}/f_{pu} = 0.75

From Table 4.5.2

   C = 1.0

RE = [5000 − 0.04(5249 + 11,157 + 14,272)](1)

   = 3773 psi

T.L. = ES + CR + SH + RE

   = 14,272 + 11,157 + 5249 + 3773

   = 34,451 psi = 34.5 ksi

Final prestress force = 371.8 − 34.5(1.836)

   = 308.5 kips

4.6 Camber and Deflection

Most precast, prestressed concrete flexural members will have a net positive (upward) camber at the time of transfer of prestress, caused by the eccentricity of the prestressing force. This camber may increase or decrease with time, depending on the stress distribution across the member under sustained loads. Camber tolerances are suggested in Chapter 8 of this Handbook.

Limitations on instantaneous deflections and time-dependent cambers and deflections are specified in the ACI Code. Table 9.5(b) of the Code is reprinted for reference (see Table 4.6.1).

The following sections contain suggested methods for computing cambers and deflections. There are many inherent variables that affect camber and deflection, such as concrete mix, storage method, time of release of prestress, time of erection and placement of superimposed loads, relative humidity, etc. *Because of this, calculated long-time values should never be considered any better than estimates.* Nonstructural components attached to members which could be affected by camber variations, such as partitions or folding doors, should be placed with adequate allowance for variation. Calculation of topping quantities should also recognize the imprecision of camber calculations.

It should also be recognized that camber of precast, prestressed members is a result of the placement of the strands needed to resist the design moments and service load stresses. It is not practical to alter the forms of the members to produce a desired camber. Therefore, cambers should not be specified, but their inherent existence should be recognized.

4.6.1 Initial Camber

Initial camber can be calculated using conventional moment-area equations. Fig. 4.10.14 has equations for the camber caused by prestress force for the most common strand patterns used in precast, prestressed members. Design Aids 11.1.3 and 11.1.4 provide deflection equations for typical loading conditions and more general camber equations.

Example 4.6.1 Calculation of initial camber

Given:

8DT24 of Example 4.2.9.

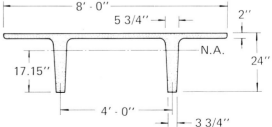

Section Properties

A = 401 in.^{2} Z_b = 1224 in.^{3}

I = 20,985 in.^{4} Z_t = 3063 in.^{3}

Table 4.6.1 Maximum permissible computed deflections

Type of member	Deflection to be considered	Deflection limitation
Flat roofs not supporting or attached to nonstructural elements likely to be damaged by large deflections	Immediate deflection due to live load	$\dfrac{\ell\,*}{180}$
Floors not supporting or attached to nonstructural elements likely to be damaged by large deflections	Immediate deflection due to live load	$\dfrac{\ell}{360}$
Roof or floor construction supporting or attached to nonstructural elements likely to be damaged by large deflections	That part of the total deflection occurring after attachment of nonstructural elements (sum of the long-time deflection due to all sustained loads and the immediate deflection due to any additional live load)‡	$\dfrac{\ell^{\dagger}}{480}$
Roof or floor construction supporting or attached to nonstructural elements not likely to be damaged by large deflections		$\dfrac{\ell^{\S}}{240}$

*Limit not intended to safeguard against ponding. Ponding should be checked by suitable calculations of deflection, including added deflections due to ponded water, and considering long-time effects of all sustained loads, camber, construction tolerances, and reliability of provisions for drainage.

†Limit may be exceeded if adequate measures are taken to prevent damage to supported or attached elements.

‡Long-time deflection shall be determined in accordance with Sections 9.5.2.5 or 9.5.4.2, ACI 318-89, but may be reduced by amount of deflection calculated to occur before attachment of nonstructural elements. This amount shall be determined on the basis of accepted engineering data relating to time-deflection characteristics of members similar to those being considered.

§But not greater than tolerance provided for nonstructural elements. Limit may be exceeded if camber is provided so that total deflection minus camber does not exceed limit.

y_b = 17.15 in.   wt = 418 plf
y_t = 6.85 in.   = 52 psf

Concrete:

f'_c = 5000 psi

Normal weight (150 pcf)

$E_c = 33w^{1.5}\sqrt{f'_c} = 33(150)^{1.5}\sqrt{5000}$
   = 4287 ksi

f'_{ci} = 3500 psi

$E_{ci} = 33(150)^{1.5}\sqrt{3500}$ = 3587 ksi

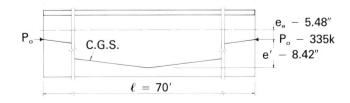

(Note: The values of E_c and E_{ci} could also be read from Design Aid 11.2.2.)

Problem:

Find the initial camber at time of transfer of prestress.

Solution:

The prestress force at transfer and strand eccentricities are calculated in Example 4.2.9 and are shown in the illustration at left.

Calculate the upward component using equations given in Fig. 4.10.14.

$$\Delta\uparrow = \frac{P_o e_e \ell^2}{8 E_{ci} I} + \frac{P_o e' \ell^2}{12 E_{ci} I}$$

$$= \frac{335(5.48)(70 \times 12)^2}{8(3587)(20{,}985)}$$

$$+ \frac{(335)(8.42)(70 \times 12)^2}{12(3587)(20{,}985)}$$

$$= 2.15 + 2.20 = 4.35 \text{ in.}\uparrow$$

Deduct deflection caused by weight of member:

$$\Delta\downarrow = \frac{5w\ell^4}{384E_{ci}I}$$

$$= \frac{5\left(\frac{0.418}{12}\right)(70 \times 12)^4}{384(3587)(20,985)} = 3.00 \text{ in.}\downarrow$$

Net camber at release $= 4.35\uparrow - 3.00\downarrow$

$= 1.35 \text{ in.}\uparrow$

4.6.2 Elastic Deflections

Calculation of instantaneous deflections of both prestressed and non-prestressed members caused by superimposed service loads follows classical methods of mechanics. Design equations for various load conditions are given in Chapter 11 of this Handbook. If the bottom tension in a simple span member does not exceed the modulus of rupture, the deflection is calculated using the uncracked moment of inertia of the section. The modulus of rupture of concrete is defined in Chapter 9 of the Code as:

$$f_r = 7.5\lambda\sqrt{f'_c} \qquad \text{(Eq. 4.6.1)}$$

See Sect. 4.3.3 for definition of λ.

4.6.3 Bilinear Behavior

Sect. 18.4.2 of the Code requires that "bilinear moment-deflection relationships" be used to calculate instantaneous deflections of prestressed concrete members when the bottom tension exceeds $6\sqrt{f'_c}$. This means that the deflection before the member has cracked is calculated using the gross (uncracked) moment of inertia, I_g, and the additional deflection after cracking is calculated using the moment of inertia of the cracked section. This is illustrated graphically in Fig. 4.6.1.

In lieu of a more exact analysis, the empirical relationship:

$$I_{cr} = nA_{ps}d_p^2\left(1 - 1.6\sqrt{n\rho_p}\right) \qquad \text{(Eq. 4.6.2)}$$

may be used to determine the cracked moment of inertia. Fig. 4.10.15 gives coefficients for use in solving this equation.

Example 4.6.2 Deflection calculation using bilinear moment-deflection relationship

Given:

8DT24 of Examples 4.2.9 and 4.6.1.

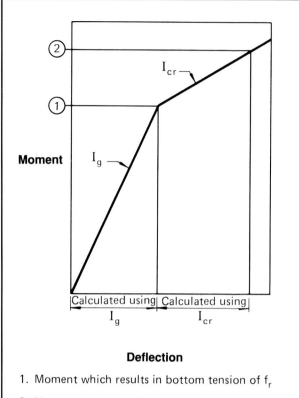

Fig. 4.6.1 Bilinear moment-deflection relationship

1. Moment which results in bottom tension of f_r
2. Moment corresponding to final bottom tension

(Note: Prestress effects not included in illustration.)

Problem:

Determine the total deflection caused by the specified uniform live load.

Solution:

Determine $f_r = 7.5\lambda\sqrt{f'_c} = 530$ psi

From Example 4.2.9 the final tensile stress is 756 psi, which is more than 530 psi, so the bilinear behavior must be considered.

Determine I_{cr} from Fig. 4.10.15:

$A_{ps} = 1.836$ sq in. (See Ex. 4.2.9)

d_p at midspan $= e_c + y_t = 13.90 + 6.85$

$= 20.75$ in.

(Note: It is within the precision of the calculation method and observed behavior to use midspan d_p and to calculate the deflection at midspan, although the maximum tensile stress in this case is assumed at 0.4ℓ.)

$$\rho_p = \frac{A_{ps}}{bd_p} = \frac{1.836}{(96)(20.75)} = 0.00092$$

C = 0.0056

$I_{cr} = Cbd_p^3 = 0.0056(96)(20.75)^3$

     = 4803 in.^{4}

Determine the portion of the live load that would result in a bottom tension of 530 psi.

756 − 530 = 226 psi

The tension caused by live load alone is 1614 psi, therefore, the portion of the live load that would result in a bottom tension of 530 psi is:

$$\frac{1614 - 226}{1614}(0.280) = 0.241 \text{ kips/ft}$$

and:

$$\Delta_g = \frac{5w\ell^4}{384E_cI_g} = \frac{5\left(\frac{0.241}{12}\right)(70 \times 12)^4}{384(4287)(20,985)}$$

     = 1.45 in.

$$\Delta_{cr} = \frac{5\left(\frac{0.039}{12}\right)(70 \times 12)^4}{384(4287)(4803)} = 1.02 \text{ in.}$$

Total deflection = 1.45 + 1.02 = 2.47 in.

4.6.4 Effective Moment of Inertia

The Code allows an alternative to the method of calculation described in the previous section. An effective moment of inertia, I_e, can be determined and the deflection then calculated by substituting I_e for I_g in the deflection calculation.

The equation for effective moment of inertia is:

$$I_e = \left(\frac{M_{cr}}{M_a}\right)^3 I_g + \left[1 - \left(\frac{M_{cr}}{M_a}\right)^3\right] I_{cr} \quad \text{(Eq. 4.6.3)}$$

The difference between the bilinear method and the I_e method is illustrated in Fig. 4.6.2.

The use of I_e with prestressed concrete members is described in a paper by Branson.[9] The value of M_{cr}/M_a for use in determining live load deflections can be expressed as:

$$\frac{M_{cr}}{M_a} = 1 - \left(\frac{f_{t\ell} - f_r}{f_\ell}\right) \quad \text{(Eq. 4.6.4)}$$

where:

$f_{t\ell}$ = the final calculated total stress in the member

f_ℓ = calculated stress due to live load

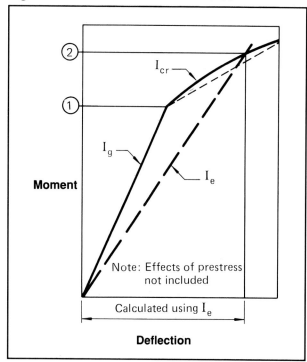

Fig. 4.6.2 Effective moment of inertia

Example 4.6.3 Deflection calculation using effective moment of inertia

Given:

Same section and loading conditions of Examples 4.2.9, 4.6.1 and 4.6.2.

Problem:

Determine the deflection caused by live load using the I_e method.

Solution:

From the table of stresses in Example 4.2.9:

$f_{t\ell}$ = 756 psi (tension)

f_ℓ = 1614 psi (tension)

f_r = $7.5\sqrt{f'_c}$ = 530 psi (tension)

$$\frac{M_{cr}}{M_a} = 1 - \left(\frac{756 - 530}{1614}\right) = 0.860$$

$$\left(\frac{M_{cr}}{M_a}\right)^3 = (0.860)^3 = 0.636$$

$$1 - \left(\frac{M_{cr}}{M_a}\right)^3 = 1 - 0.636 = 0.364$$

I_e = 0.636(20,985) + 0.364(4803)

   = 15,095 in.^{4}

I_e can also be found using Fig. 4.10.16:

$$f_e = 756 - 530 = 226 \text{ psi}$$

$$\frac{f_e}{f_\ell} = \frac{226}{1614} = 0.14$$

$$\frac{I_{cr}}{I_g} = \frac{4803}{20,985} = 0.229$$

Follow arrows on chart:

$$\frac{I_e}{I_g} = 0.715$$

$$I_e = 0.715(20,985) = 15,004 \text{ in.}^4$$

$$\Delta_\ell = \frac{5w\ell^4}{384E_cI_e} = \frac{5\left(\frac{0.280}{12}\right)(70 \times 12)^4}{384(4287)(15,095)}$$

$$= 2.34 \text{ in.}$$

4.6.5 Long-Time Camber/Deflection

ACI 318-89 provides a multiplier, λ, applied to initial deflection for estimating the long-term deflection of non-prestressed reinforced concrete members (Code Sect. 9.5.2.5):

$$\lambda = \frac{\xi}{1 + 50\rho'} \qquad \text{(Eq. 4.6.5)}$$

where ξ is a factor related to length of time, and ρ' is the ratio of compressive reinforcement. No such guide is given for prestressed concrete.

The determination of long-time cambers and deflections in precast, prestressed members is somewhat more complex because of (1) the effect of prestress and the loss of prestress over time, (2) the strength gain of concrete after release of prestress, and because (3) the camber or deflection is important not only at the "initial" and "final" stages, but also at erection, which occurs at some intermediate stage, usually from 30 to 60 days after casting.

It has been customary in the design of precast, prestressed concrete to estimate the camber of a member after a period of time by multiplying the initial calculated camber by some factor, usually based on the experience of the designer. To properly use these "multipliers," the upward and downward components of the initial calculated camber should be separated in order to take into account the effects of loss of prestress, which only affect the upward component.

Table 4.6.2 provides suggested multipliers which can be used as a guide in estimating long-time cambers and deflections for typical members, i.e., those members which are within the span-depth ratios recommended in this Handbook (see Sect. 3.2.2). Derivation of these multipliers is contained in a paper by Martin.[10]

Long-time effects can be substantially reduced by adding non-prestressed reinforcement in prestressed

Table 4.6.2 Suggested multipliers to be used as a guide in estimating long-term cambers and deflections for typical members

	Without Composite Topping	With Composite Topping
At erection:		
(1) Deflection (downward) component—apply to the elastic deflection due to the member weight at release of prestress	1.85	1.85
(2) Camber (upward) component—apply to the elastic camber due to prestress at the time of release of prestress	1.80	1.80
Final:		
(3) Deflection (downward) component—apply to the elastic deflection due to the member weight at release of prestress	2.70	2.40
(4) Camber (upward) component—apply to the elastic camber due to prestress at the time of release of prestress	2.45	2.20
(5) Deflection (downward)—apply to elastic deflection due to superimposed dead load only	3.00	3.00
(6) Deflection (downward)—apply to elastic deflection caused by the composite topping	—	2.30

	(1) Release	Multiplier	(2) Erection	Multiplier	(3) Final
Prestress	4.35 ↑	1.80 x (1)	7.83 ↑	2.45 x (1)	10.66 ↑
w_d	3.00 ↓	1.85 x (1)	5.55 ↓	2.7 x (1)	8.10 ↓
	1.35 ↑		2.28 ↑		2.56 ↑
w_{sd}			0.48 ↓	3.0 x (2)	1.44 ↓
			1.80 ↑		1.12 ↑
w_ℓ					2.34 ↓
					1.22 ↓

concrete members. The reduction effects proposed by Shaikh and Branson[11] can be applied to the approximate multipliers of Table 4.6.2 as follows:

$$C_2 = \frac{C_1 + A_s / A_{ps}}{1 + A_s / A_{ps}} \quad \text{(Eq. 4.6.6)}$$

where:

C_1 = multiplier from Table 4.6.2

C_2 = revised multiplier

A_s = area of non-prestressed reinforcement

A_{ps} = area of prestressed steel

Example 4.6.4 Use of multipliers for determining long-term cambers and deflections

Given:

8DT24 of Examples 4.2.9, 4.6.1, 4.6.2 and 4.6.3.

Non-structural elements are attached, but not likely to be damaged by deflections (light fixtures, etc.).

Problem:

Estimate the camber and deflection and determine if it meets the requirements of Table 9.5(b) of the Code (see Table 4.6.1).

Solution:

Calculate the instantaneous deflections caused by the superimposed dead and live loads.

$$\Delta_d = \frac{5w\ell^4}{384 E_c I} = \frac{5\left(\frac{0.080}{12}\right)(70 \times 12)^4}{384(4287)(20,985)}$$

$= 0.48$ in. ↓

$\Delta_\ell = 2.34$ in. ↓ (see Example 4.6.3)

For convenience, a tabular format is used (above).

The estimated critical cambers and deflections would then be:

At erection of the member
after w_{sd} is applied = 1.80 in.

"Final" long-time camber = 1.12 in.

The deflection limitation of Table 4.6.1 for the above condition is $\ell/240$.

$(70 \times 12)/240 = 3.50$ in.

Total deflection occurring after attachment of non-structural elements:

$\Delta_\ell = (1.80 - 1.12) + 2.34$

$= 3.02$ in. < 3.50 in. OK

4.7 Compression Members

Precast and prestressed concrete columns and load bearing wall panels are usually proportioned on the basis of strength design. Stresses under service conditions, particularly during handling and erection (especially of wall panels) must also be considered. The procedures in this section are based on Chapter 10 of the Code and on the recommendations of the PCI Committee on Prestressed Concrete Columns[12] (referred to in this section as "the Recommended Practice").

4.7.1 Strength Design of Precast Concrete Compression Members

The capacity of a reinforced concrete compression member with eccentric loads is most easily determined by constructing a capacity interaction curve. Points on this curve are calculated using the compatibility of strains and solving the equations of equilibrium as prescribed in Chapter 10 of the Code. Solution of these equations is illustrated in Fig. 4.7.1.

ACI 318-89 waives the minimum vertical reinforcement requirements for compression members if the concrete is prestressed to at least an average of 225 psi after all losses. In addition, the Recommended Practice permits the elimination of column ties, if the nominal capacity is multiplied by 0.85. Interaction curves for typical prestressed square columns and wall panels are provided in Chapter 2.

Construction of an interaction curve usually follows these steps:

Step 1: Determine P_o for $M_n = 0$. (See Fig. 4.7.1(c))

Step 2: Determine M_o for $P_n = 0$. This is normally done by neglecting the reinforcement above the neutral axis and determining the moment capacity by one of the methods described in Sect. 4.2.1.

Step 3: For non-prestressed columns, P_{nb} and M_{nb} at the balance point may be determined (see Fig. 4.7.1(d)). For prestressed columns, the yield point of the prestressed reinforcement is not well defined and the stress-strain relationship is non-linear over a broad range (see Design Aid 11.2.5).

Step 4: For each additional point on the interaction curve, proceed as follows:

a. Select a value of "c" and calculate $a = \beta_1 c$

b. Determine the value of A_{comp} from the geometry of the section (see Fig. 4.7.1 (a)).

c. Determine the strain in the reinforcement assuming that $\varepsilon_c = 0.003$ at the compression face of the column. For prestressed reinforcement, add the strain due to the effective prestress $\varepsilon_{se} = f_{se}/E_{ps}$.

d. Determine the stress in the reinforcement. For non-prestressed reinforcement, $f_s = \varepsilon_s E_s \leq f_y$. For prestressed reinforcement, the stress is determined from a stress-strain relationship (see Design Aid 11.2.5). If the maximum factored moment occurs near the end of a prestressed element, where the strand is not fully developed, an appropriate reduction in the value of f_{ps} should be made as described in Sect. 4.2.3.

e. Calculate P_n and M_n by statics.

f. Calculate ϕP_n and ϕM_n. The Code prescribes that the ϕ-factor for compression elements is 0.7, except that it can vary from 0.7 at a point where $\phi P_n = 0.10 f'_c A_g$ to 0.9 where $\phi P_n = 0$. This ϕ variation is accomplished on the interaction curve by first constructing the curve with $\phi = 0.7$. A straight line is then drawn from the point on that curve where $\phi P_n = 0.10 f'_c A_g$ to a point on the $\phi P_n = 0$ line corresponding to ϕM_o calculated with $\phi = 0.9$.

Step 5: Calculate the maximum factored axial resistance specified by the Code as:

$0.80 \phi P_o$ for tied columns

$0.85 \phi P_o$ for spiral columns.

For cross sections which are not rectangular, it is necessary to determine separate curves for each direction of the applied moment. Further, since most architectural precast column cross sections are not rectangular, the "a" distance only defines the depth of the rectangular concrete stress distribution. Instead of using a/2, as for a rectangular cross-section, it is necessary to calculate the actual centroid of the compression area which is indicated as y'.

As noted in Step 4d, the flexural resistance is reduced for prestressed elements at locations within a distance equal to the strand development length from each end. The flexural resistance of the prestressed reinforcement in this zone can be supplemented by non-prestressed reinforcement that is anchored to end plates, or otherwise developed.

The interaction curves in Chapter 2 are based on a maximum value of $f_{ps} = f_{se}$, which is equivalent to a development length equal to the assumed transfer length. The required area of end reinforcement can be determined by matching interaction curves, or can be approximated by the following equation if the bar locations approximately match the strand locations:

$$A_s = \frac{A_{ps} f_{se}}{f_y} \qquad \text{(Eq. 4.7.1)}$$

where:

A_s = required area of bars

f_{se} = strand stress after losses

f_y = yield strength of bars

The effects of adding end reinforcement to a 24 x 24 in. prestressed concrete column, thus improving moment capacity in the end 2 ft, are shown in Fig. 4.7.2.

Fig. 4.7.1 Equilibrium equations for prestressed and non-prestressed compression members

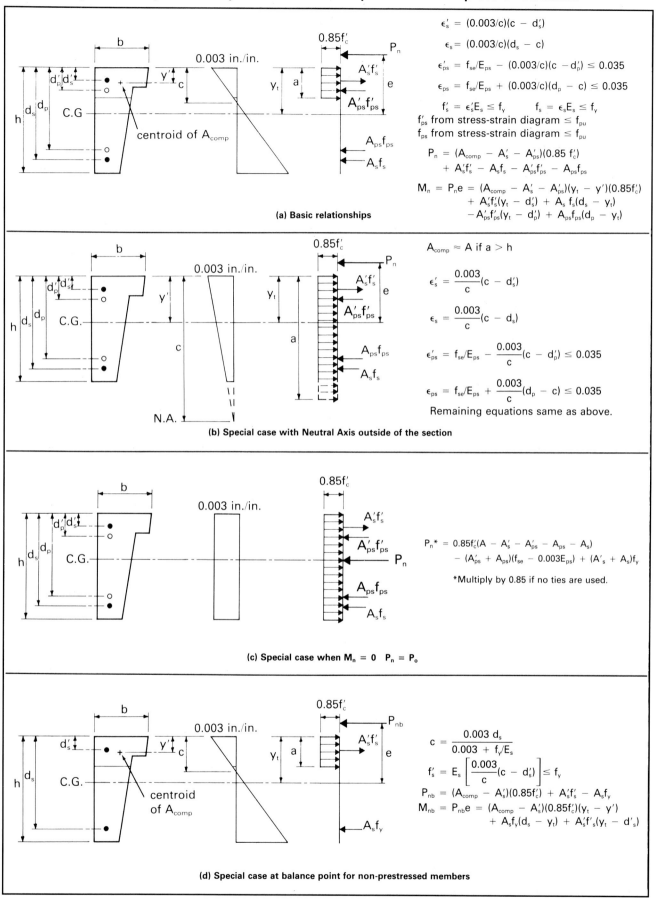

Example 4.7.1 Construction of interaction curve for a precast, reinforced concrete column

Given:

Column cross-section shown

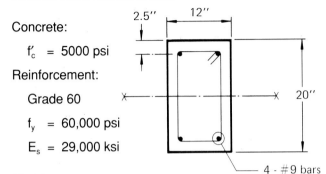

Concrete:

f'_c = 5000 psi

Reinforcement:

Grade 60

f_y = 60,000 psi

E_s = 29,000 ksi

Problem:

Construct interaction curve for bending about x-x axis.

Solution:

Determine following parameters:

β_1 = 0.85 − 0.05 = 0.80

d = 20 − 2.5 = 17.5 in.

d' = 2.5 in.

y_t = 10 in.

$0.85 f'_c$ = 0.85(5) = 4.25 ksi

A_g = 12 × 20 = 240 sq in.

$A_s = A'_s$ = 2.00 sq in.

Step 1 — Determine P_o from Fig. 4.7.1(c):

$P_o = 0.85 f'_c (A_g - A'_s - A_s) + (A'_s + A_s)f_y$

ϕP_o = 0.70[4.25(240 − 4) + (4)(60)]

= 870 kips

Step 2 — Determine P_{nb} and M_{nb} from Fig. 4.7.1(d):

$c = \dfrac{0.003d}{0.003 + f_y / E_s} = \dfrac{0.003(17.5)}{0.003 + 60/29{,}000}$

= 10.36 in.

$f'_s = 29{,}000 \left[\dfrac{0.003}{10.36}(10.36 - 2.5) \right]$

= 66.0 > 60

therefore $f'_s = f_y$ = 60 ksi

A_{comp} = ab = $\beta_1 cb$ = 0.80(10.36)(12)

= 99.5 sq in.

$y' = \dfrac{a}{2} = \dfrac{0.80(10.36)}{2}$ = 4.14 in.

P_{nb} = (99.5 − 2)4.25 + 2(60) − 2(60)

= 414.4 kips

ϕP_{nb} = 0.70(414.4) = 290 kips

M_{nb} = (97.5)(10 − 4.14)(4.25)
+ 2.0(60)(17.5 − 10) + 2.0(60)(10 − 2.5)

ϕM_{nb} = 0.70(2428 + 900 + 900)

= 2960 in.-kips = 247 ft-kips

Step 3 — Determine M_o; use conservative solution neglecting compressive reinforcement:

$a = \dfrac{A_s f_y}{0.85 f'_c b} = \dfrac{2.0(60)}{4.25(12)}$ = 2.35 in.

$M_o = A_s f_y \left(d - \dfrac{a}{2} \right)$ = (2.0)(60)(17.5 − 2.35/2)

= 1959 in.-kips

For ϕ = 0.7, ϕM_o = 1371 in.-kips = 114 ft-kips

(This point is found for curve projection)

For ϕ = 0.9, ϕM_o = 1763 in.-kips = 147 ft-kips

To determine intermediate points on the curve:

Step 4(a) — Set a = 6 in., c = 6/0.80 = 7.5 in.

Step 4(b) — A_{comp} = 6(12) = 72 sq in.

Step 4(d) — Use Fig. 4.7.1(a):

$f'_s = 29{,}000 \left[\dfrac{0.003}{7.5}(7.5 - 2.5) \right]$

= 58.0 ksi < f_y

$f_s = 29{,}000 \left[\dfrac{0.003}{7.5}(17.5 - 7.5) \right]$

= 116 ksi > f_y

Use $f_s = f_y$ = 60 ksi

Steps 4(e) and 4(f) —

P_n = (72 − 2)4.25 + 2.0(58) − 2.0(60)

= 293.5 kips

ϕP_n = 0.7(293.5) = 205 kips

ϕM_n = 0.70[(72 − 2)(10 − 3)4.25
+ 2.0(60)(17.5 − 10) + 2.0(58)(10 − 2.5)]

= 0.70(2082.5 + 900 + 870)

= 2697 in.-kips = 225 ft-kips

(Note: Steps 4a to 4f can be repeated for as many points as desired.)

A plot of these points is shown as Fig. 4.7.3.

Fig. 4.7.2 End reinforcement in a precast, prestressed concrete column

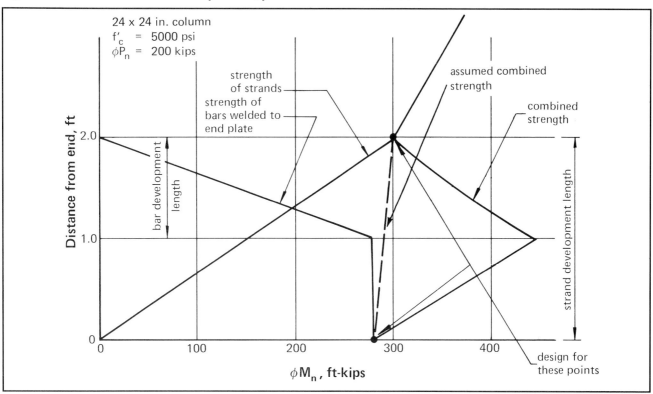

Fig. 4.7.3 Interaction curve for Example 4.7.1

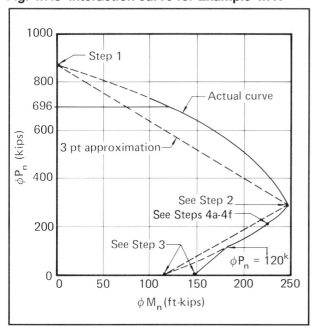

Step 5 — Also calculate:

Maximum design load = $0.80\phi P_o$
   = 0.80(870) = 696 kips

Transition point for $\phi = 0.7$ to $\phi = 0.9$
   = $0.10 f'_c A_g$ = 0.10(5)(240) = 120 kips

Example 4.7.2 Calculation of interaction points for a prestressed concrete compression member

Given:

Hollow-core wall panel shown

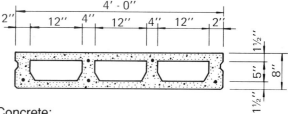

Concrete:
  f'_c = 6000 psi
  A = 204 sq in.

Prestressing steel:
  f_{pu} = 270 ksi
  E_{ps} = 28,500 ksi
  f_{se} = 150 ksi
  5 – ⅜ in. dia., 270K strands
  A_{ps} (bott) = 3(0.085) = 0.255 sq in.
  A'_{ps} (top) = 2(0.085) = 0.170 sq in.

Problem:

Calculate a point on the design interaction curve for a = 2 in.

PCI Design Handbook/Fourth Edition   4–49

Solution:

Step 1: $\beta_1 = 0.85 - 2(.05) = 0.75$

$a = 2$ in., $c = \dfrac{2}{0.75} = 2.67$ in.

Step 2: $A_{comp} = 48(1.5) + 12(2 - 1.5)$

$= 78$ sq in.

$y' = \dfrac{48(1.5)(1.5/2) + 12(0.5)(1.5 + 0.5/2)}{78}$

$= 0.83$ in.

Step 3: $\varepsilon_{se} = \dfrac{f_{se}}{E_{ps}} = \dfrac{150}{28,500} = 0.00526$ in./in.

Step 4: From Fig. 4.7.1(a)

$\varepsilon'_{ps} = 0.00526 - \dfrac{0.003}{2.67}(2.67 - 1.5)$

$= 0.00526 - 0.00131 = 0.00395$ in./in.

From Design Aid 11.2.5, this strain is on the linear part of the curve:

$f'_{ps} = \varepsilon'_{ps} E_s = 0.00395(28,500) = 113$ ksi

$\varepsilon_{ps} = 0.00526 + \dfrac{0.003}{2.67}(6.5 - 2.67)$

$= 0.00526 + 0.00430 = 0.00956$

From Design Aid 11.2.5, $f_{ps} = 252$ ksi

From Fig. 4.7.1(a):

$P_n = (A_{comp})0.85 f'_c - A'_{ps} f'_{ps} - A_{ps} f_{ps}$

$= 78(0.85)(6) - 0.170(113) - 0.255(252)$

$= 397.8 - 19.2 - 64.3 = 314.3$ kips

$\phi P_n = 0.7(314.3) = 220$ kips

$M_n = 397.8(4 - 0.83) - 19.2(4 - 1.5)$

$+ 64.3(6.5 - 4)$

$= 1261.0 - 48.0 + 160.8 = 1373.8$ in.-kips

$= 114.5$ ft-kips

$\phi M_n = 0.7(1373.8) = 961.7$ in.-kips

$= 80.1$ ft-kips

Since no lateral ties are used in this member, the values should be multiplied by 0.85.

$\phi P_n = 0.85(220) = 187$ kips

$\phi M_n = 0.85(80.1) = 68.1$ ft-kips

Note that this is for *fully developed* strand. If the capacity at a point near the end of the transfer zone is desired, then $f_{ps} \leq f_{se} = 150$ ksi.

$\phi P_n = 0.85(0.7)[397.8 - 19.2 - 0.255(150)]$

$= 202.5$ kips

$\phi M_n = 0.85(0.7)[1261.0 - 48.0$

$+ 0.255(150)(6.5 - 4)]$

$= 778.6$ in.-kips $= 64.9$ ft-kips

Note: For compliance with ACI Code Sect. 18.8.3, cracking moment, M_{cr}, must be calculated to verify $\phi M_n \geq 1.2 M_{cr}$ for both faces of the wall panel. The effects of prestressing and its eccentricity should be included in calculating M_{cr} and other aspects of wall panel behavior, such as deflections.

4.7.2 Slenderness Effects

Sects. 10.10 and 10.11 of ACI 318-89 contain provisions for evaluating slenderness effects (buckling) of columns. Use of these provisions is described in Chapter 3 of this Handbook. Additional recommendations are given in the Recommended Practice.

4.7.3 Service Load Stresses

There are no limitations in ACI 318-89 on service load stresses in compression members subject to bending. The Recommended Practice suggests that, for prestressed members, the limitations of Sect. 18.4 of the Code be applied. For non-prestressed members, stresses and crack control are discussed in Sects. 4.2.2.1 and 5.2.4. Handling stresses are nearly always more critical than service load stresses.

4.7.4 Effective Width of Wall Panels

The Recommended Practice specifies that the portion of a wall considered as effective for supporting concentrated loads or for determining the effects of slenderness shall be the least of the following:

a. The center-to-center distance between loads.

b. The length of the loaded portion plus six times the wall thickness on either side (Fig. 4.7.4).

c. The width of the rib (in ribbed wall panels) plus six times the thickness of the wall between ribs on either side of the rib (Fig. 4.7.4).

d. 0.4 times the actual height of the wall.

4.7.5 Varying Section Properties of Compression Members

Architectural wall panels will frequently be of a configuration that varies over the unsupported height of the panel. While there are precise methods of determining the effects of slenderness for such members, the approximate nature of the analysis procedures used do not warrant such precision. Example 4.7.3 illustrates approximate methods for determining section properties used in evaluating slenderness effects.

Example 4.7.3 Approximate section properties of an architectural mullion panel

Given:
The load bearing architectural wall panel shown below.

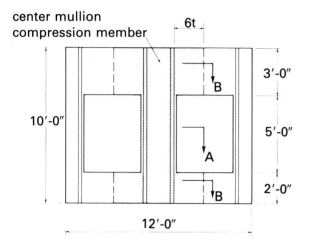

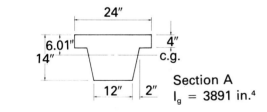

Section A
$I_g = 3891$ in.^{4}

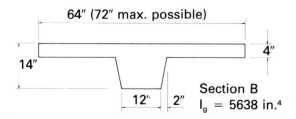

Section B
$I_g = 5638$ in.^{4}

Effective Sections

$f'_c = 5000$ psi; $E_c = 4300$ ksi

Problem:
Determine an approximate moment of inertia for slenderness analysis.

Fig. 4.7.4 Effective width of wall panels

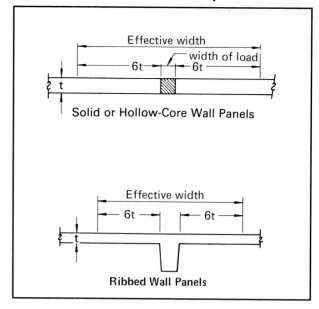

Solution:
One method is to determine a simple span deflection with a uniform load as follows:

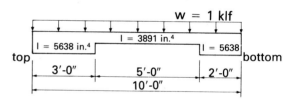

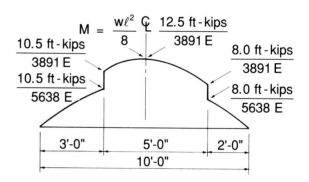

Using the moment area method, the center or mid-height deflection is:

$\Delta_o = 0.013$ in.

$\Delta_o = \dfrac{5wL^4}{384EI_{equiv}}$

$I_{equiv} = \dfrac{5wL^4}{\Delta_o 384E}$

$= \dfrac{5(1/12)(10 \times 12)^4}{0.013(384)(4300)}$

$= 4025$ in.^{4}

A second, more approximate method would be to use a "weighted average" of the moments of inertia of the two sections. In this case:

$$I_{equiv} = \frac{I_1 h_1 + I_2 h_2}{h_1 + h_2}$$

$$= \frac{5638(3+2) + 3891(5)}{10} = 4764 \text{ in.}^4$$

Once an equivalent moment of inertia is determined, slenderness effects are evaluated by one of the methods described in Sect. 3.5.

4.7.6 Piles

4.7.6.1 General

The pile designs considered here are based upon structural capacity alone. The ability of the soil to carry these loads must be established by load tests or evaluated by a geotechnical engineer.

In the following design procedure for pretensioned concrete piles, load capacity is limited by the service load stresses. An overall factor of safety based on the nominal strength ($\phi = 1.0$) of the section is computed, and limits suggested for various loading conditions. Stresses caused by transporting, handling and driving should also be considered. Experience has shown that the frictional and bearing resistance of the soil will control the design more often than the strength and service load stresses on the pile.

The values used in sample calculations are based on a concrete strength of 6,000 psi. In many areas, higher concrete strengths have been effectively used in prestressed pile design. Engineers should check with local prestressed concrete pile producers to determine what concrete strengths are available in their areas as well as sizes and types, i.e., square, octagonal, or round.

The values in the pile load table (Table 2.7.1) may be modified to fit the actual service conditions and to maintain reasonable safety factors consistent with the character of the applied loads and how and where the piles are to be used.

4.7.6.2 Strengths under direct loads

Nominal strength

It is assumed that the concrete stress at failure of a concrete pile will be $0.85f'_c$. The amount of prestress remaining in the tendons must be deducted. Assuming an ultimate concrete strain of 0.003, it can be shown that only about 60% of the effective prestress, f_{pc}, is left in the member when it reaches its nominal strength. Thus, if a 6000 psi concrete pile is prestressed to an effective prestress of 700 psi, the nominal strength can be computed as:

$$P_n = (0.85f'_c - 0.60f_{pc})A_g \quad \text{(Eq. 4.7.2)}$$

$$P_n = (0.85 \times 6000 - 0.60 \times 700)A_g = 4680A_g$$

Where A_g = gross cross sectional area of the concrete in sq in., and P_n = the concentric nominal strength in lb.

Service load

For a concentric load on a short column pile, a factor of safety of between 2.0 and 3.0 is usually adequate.

Based upon a study by the Portland Cement Association, which has been accepted by most current building codes, the formula for service loads on concentrically loaded short column prestressed piles is:

$$N = (0.33f'_c - 0.27f_{pc})A_g \quad \text{(Eq. 4.7.3)}$$

Thus for 6,000 psi concrete and 700 psi prestress

$$N = (0.33 \times 6000 - 0.27 \times 700)A_g = 1790 A_g$$

The overall safety factor will be $4,680/1,790 = 2.61$ which is considered quite adequate.

For a pile with $f'_c = 6,000$ psi and an effective prestress of 1,200 psi, the corresponding overall safety factor will be approximately the same, 2.65. The value of $0.2f'_c$ is considered to be about the desirable upper limit for the prestressing force, and a value of 700 psi is recommended as the desirable lower limit for all piles over about 40 ft in length. The overall safety factors for this range of prestress fall within the acceptable limits as indicated above.

Unsupported length

It is suggested that service loads based on the short column value of Eq. 4.7.3 be limited to a value of h/r not greater than 60.

where:

h = unsupported length of the pile

r = radius of gyration

For piles considered fully fixed at one end and hinged at the other end, it is suggested that h be taken as 0.7 of the length between hinge and assumed point of fixity. For piles fully fixed at both ends, h may be taken as 0.5 of the length between the assumed points of fixity.

4.7.6.3 Moment resisting capacities

Service load stresses

Under the criterion of allowing no tension, if the effective prestress is $f_{pc} = P/A$, the moment capacity is $M = f_{pc}I/c$, where c = distance from extreme fiber to neutral axis. This criterion of "no tension" is much too conservative for prestressed piles under normal conditions. Zero tension will result in an overall safety factor of about 3 or more, which is greater than required for bending and greater than the safety factors used in steel or reinforced concrete design.

The modulus of rupture for 6,000 psi concrete is generally taken to be $7.5\sqrt{f'_c}$ or approximately 600 psi. Allowing tension up to about 50% of the modulus of rupture, the allowable tension for normal bending can be taken as $4\sqrt{f'_c}$ or 300 psi. If the prestress is 700 psi, the total stress available for bending is 1,000 psi and the moment capacity is M = 1000 I/c. This usually gives a safety factor of 2.5 or more.

For earthquake and other transient loads this is conservative and the allowable tension may be increased to 600 psi. This gives a moment capacity of M = 1300 I/c.

Strength design

The nominal moment capacity of a pretensioned pile can be computed by the ultimate strength of the tendons multiplied by a proper lever arm. As an approximation, the total ultimate strength in all the tendons can be used. The lever arm is approximately 0.37 t for solid square piles and 0.32 t for solid circular and octagonal piles. For hollow piles, the lever arm will be a little longer, approximately 0.38 t for square piles and 0.34 t for circular and octagonal piles.

Thus, the approximate nominal moment strength is given by:

$M_n = 0.37\ tA_{ps}f_{pu}$ for solid square piles

$M_n = 0.32\ tA_{ps}f_{pu}$ for solid circular and octagonal piles

$M_n = 0.38\ tA_{ps}f_{pu}$ for hollow square piles

$M_n = 0.34\ tA_{ps}f_{pu}$ for hollow circular and octagonal piles

where A_{ps} = total steel area of all the tendons in sq in., f_{pu} = ultimate strength of the tendons in psi, and t = diameter or thickness of the pile in in.

Allowable service load moment

In general, the allowable moment should be based on an allowable tensile stress based on the modulus of rupture and checked by the nominal moment strength to ensure a factor of safety of 2 for normal loading. For wind, earthquake, or other short-time loads a safety factor of 1.5 is considered adequate. In corrosive conditions the engineer should make an evaluation to determine whether to reduce or eliminate the allowable tension.

4.7.6.4 Combined moment and direct load

Service load stresses

By the elastic theory, the presence of direct load delays the cracking of the concrete piles and thereby increases the moment carrying capacity. For example, if the prestress in the concrete is 700 psi, an external load induces 400 psi, and the allowable tensile stress is 300 psi, the total fiber stress available for bending moment is 1400 psi, and the moment capacity is M = 1400 I/c.

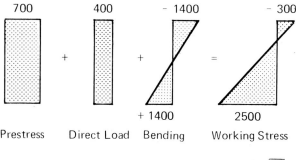

$$\frac{P}{A} + \frac{N}{A} \pm \frac{M}{I/c} = \begin{matrix} \text{tens.} \leq 4\sqrt{f'_c} \\ \text{comp.} \leq 0.45\ f'_c \end{matrix}$$

Piles controlled by compression should also be checked against the interaction formula:

$$\frac{N}{N'} + \frac{M}{M'} \leq 1 \qquad \text{(Eq. 4.7.4)}$$

where N' and M' are the allowable axial load and moment, respectively.

Strength design

The nominal moment capacity of a prestressed pile is reduced by the presence of external direct load, because the area of the compression zone is increased and the available lever arm for the resisting steel is correspondingly reduced.

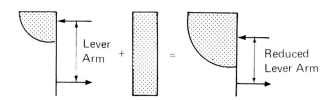

It can be roughly estimated that, for an external load producing 700 psi in the concrete, the nominal moment capacity is given by:

$M_n = 0.29\ t\ A_{ps}\ f_{pu}$ for solid square piles

$M_n = 0.25\ t\ A_{ps}\ f_{pu}$ for solid round and octagonal piles

For hollow piles the lever arm is a little longer, 0.30t and 0.26t respectively.

At the pile head, where combined moment and direct load may be critical, the transfer length in which the prestress increases from zero to full value (about 50 tendon diameters) must be considered.

Allowable service load moment

With the presence of external direct load, the moment capacity of the piles should be first computed by the elastic theory as above, and the nominal strength checked to insure a safety factor of 2. For wind, earthquake, or other short-time loads, a safety factor of 1.5 is considered adequate. In corrosive conditions the engineer should make an evaluation to determine whether to reduce or eliminate the allowable tension.

Example 4.7.4 Bearing pile

For a 12 in. square solid pile, compute the allowable load and moments.

Given:

A = 144 sq in.

I = 1,728 in.^{4}

I/c = 288 in.^{3}

Prestress with 6 – 7/16 in. diameter 270K strands

A_{ps} = 0.69 sq in.

f_{pu} = 270,000 psi or 31,000 lb. per strand

f'_c = 6000 psi

If the strands are stressed initially to 0.7 f_{pu} and total losses are assumed to be 22%, effective prestress = 0.7 x 0.78 x 31,000 = 16,900 lb/strand.

$$f_{pc} = \frac{6 \times 16,900}{144} = 705 \text{ psi}$$

(a) Direct Load

$N = (0.33f'_c - 0.27f_{pc})A$

$N = 1,790 A$

Allowable N = 1,790 x 144/1000 = 258 kips or 129 tons.

Nominal strength is given by:

$P_n = (0.85f'_c - 0.60 f_{pc})A$

= (0.85 x 6000 – 0.60 x 705)144 = 673 kips

Thus, factor of safety = 673/258 = 2.61

(b) Moment Capacity

For an allowable tension of 300 psi:

M = f_c I/c

= (300 + 705)288/1000 = 289 in.-kips

Nominal moment strength:

M_n = 0.37 t A_{ps} f_{pu} = 0.37 x 12 x 0.69 x 270

= 827 in.-kips

Thus, factor of safety = 827/289 = 2.86, which is higher than necessary. Thus, for transient loads, the allowable tension could be increased beyond 300 psi.

(c) Allowable Unsupported Length

This will vary with the load. The maximum h/r for which the short column formula should be used is 60. From pile properties in Table 2.7.1, r = 3.46 in. for a 12-in. square pile.

Thus, for a sustained load of 258 kips, and assuming pile fully fixed at both ends, the allowable unsupported length

$$h = \frac{60r}{0.5} = 415 \text{ in.} = 34.6 \text{ ft}$$

Example 4.7.5 Sheet pile

For a 12 in. x 36 in. sheet pile, compute allowable moment.

Given:

A = 432 sq in.

I = 5184 in.^{4}

I/c = 864 in.^{3} (per pile)

Prestress with 20 – 1/2 in. diameter 270K strands

A_{ps} = 3.06 sq in.

f_{pu} = 270,000 psi or 41,300 lb force per strand

$$f_{pc} = \frac{20 \times 22,560}{432} = 1045 \text{ psi}$$

f'_c = 6000 psi

Effective prestress = 0.7 x 0.78 x 41,300
= 22,560 lb/strand

Moment Capacity

(a) Service loads

For allowable tension of 300 psi:

Allowable M = (300 + 1045) x 864/1000

= 1160 in.-kips

or 1160/3 = 386 in.-kips per foot of wall.

(b) Strength design

Nominal moment strength:

M_n = 0.37 t$A_{ps}f_{pu}$ = 0.37 x 12 x 3.06 x 270

= 3670 in.-kips

Factor of safety = 3670/1160 = 3.1, which is more than sufficient. Thus, the above allowable tension of 300 psi may be increased for transient loads, except in corrosive conditions.

4.8 Special Considerations

This section outlines solutions of special situations which may arise in the design of a precast floor or roof system. Since production methods of products vary, local producers should be consulted. Also, test data may indicate that the conservative guidelines presented here may be exceeded for a specific application.

4.8.1 Load distribution

Frequently, floors and roofs are subjected to line loads, for example from walls, and concentrated loads. The ability of hollow-core systems to transfer or distribute loads laterally through grouted shear keys has been demonstrated in several published reports,[13-15] and many unpublished tests. Based on tests, analysis and experience, the PCI Hollow-Core Slab Producers Committee recommends[16] that line and concentrated loads be resisted by an effective section as described in Fig. 4.8.1. Exception: If the total deck width, perpendicular to the span, is less than the span, modification may be required. Contact local producers for recommendations.

Load distribution in stemmed members may not necessarily follow the same pattern, because of different torsional resistance properties.

Example 4.8.1 Load distribution

Given:

An untopped hollow-core floor with 4 ft wide slabs, and supporting a load bearing wall and concentrated loads as shown in Fig. 4.8.2.

Fig. 4.8.1 Assumed load distribution

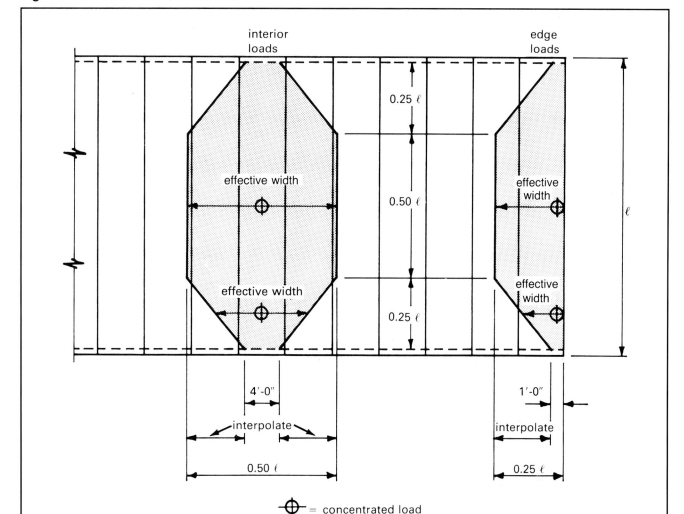

Fig. 4.8.2 Example 4.8.1

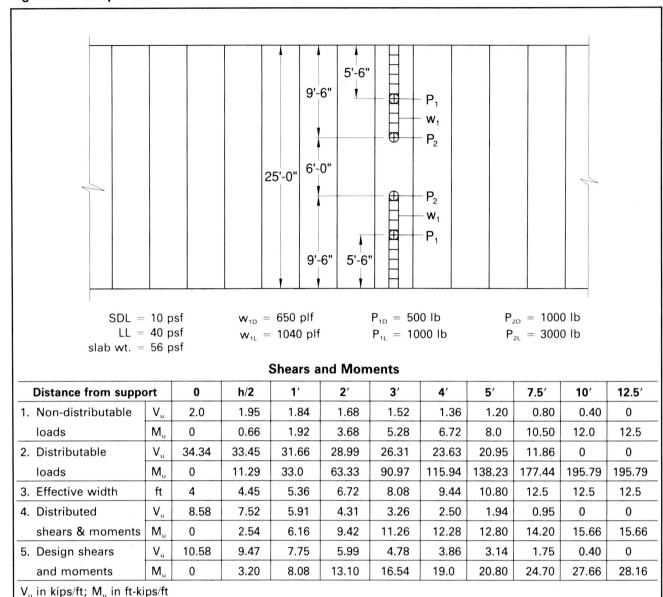

SDL = 10 psf	w_{1D} = 650 plf	P_{1D} = 500 lb	P_{2D} = 1000 lb
LL = 40 psf	w_{1L} = 1040 plf	P_{1L} = 1000 lb	P_{2L} = 3000 lb
slab wt. = 56 psf			

Shears and Moments

Distance from support		0	h/2	1'	2'	3'	4'	5'	7.5'	10'	12.5'
1. Non-distributable	V_u	2.0	1.95	1.84	1.68	1.52	1.36	1.20	0.80	0.40	0
loads	M_u	0	0.66	1.92	3.68	5.28	6.72	8.0	10.50	12.0	12.5
2. Distributable	V_u	34.34	33.45	31.66	28.99	26.31	23.63	20.95	11.86	0	0
loads	M_u	0	11.29	33.0	63.33	90.97	115.94	138.23	177.44	195.79	195.79
3. Effective width	ft	4	4.45	5.36	6.72	8.08	9.44	10.80	12.5	12.5	12.5
4. Distributed	V_u	8.58	7.52	5.91	4.31	3.26	2.50	1.94	0.95	0	0
shears & moments	M_u	0	2.54	6.16	9.42	11.26	12.28	12.80	14.20	15.66	15.66
5. Design shears	V_u	10.58	9.47	7.75	5.99	4.78	3.86	3.14	1.75	0.40	0
and moments	M_u	0	3.20	8.08	13.10	16.54	19.0	20.80	24.70	27.66	28.16

V_u in kips/ft; M_u in ft-kips/ft

Problem:

Determine the design loads for the slab supporting the wall and concentrated loads.

Solution:

(Note: Each step corresponds to a line number in the table in Fig. 4.8.2)

1. Calculate the shears and moments for the non-distributable (uniform) loads:

   w_u = 1.4(56 + 10) + 1.7(40) = 160 psf

2. Calculate the shears and moments for the distributable (concentrated and line) loads:

   w_u = 1.4(650) + 1.7(1040) = 2678 lb/ft
   P_{1u} = 1.4(500) + 1.7(1000) = 2400 lb
   P_{2u} = 1.4(1000) + 1.7(3000) = 6500 lb

3. Calculate effective width along the span:

   At the support, width = 4.0 ft

   At 0.25ℓ (6.25 ft), width = 0.5ℓ = 12.5 ft

   Between x = 0 and x = 6.25 ft:

   width = 4 + (x/6.25)(12.5 − 4) = 4 + 1.36x

4. Divide distributable shears and moments from step 2 by the effective widths from step 3.

5. Add the distributed shears and moments to the non-distributable shears and moments from step 1.

Once the moments and shears are determined, the slabs are designed as described in Sects. 4.2, 4.3, and 4.6.

This method is suitable for computer solution. For manual calculations, the procedure can be simplified by investigating only critical sections. For example, shear may be determined by dividing all distributable loads by 4 ft, and flexure at midspan can be checked by dividing the distributable loads by 0.5ℓ.

4.8.2 Effect of Openings

Openings may be provided in precast decks by: (1) saw cutting after the deck is installed and grouted, (2) forming (blocking out) or sawing in the plant, or (3) using short units with steel headers or other connections. In hollow-core or solid slabs, structural capacity is least affected by orienting the longest dimension of an opening parallel to a span, or by coring small holes to cut the fewest strands. Openings in stemmed members must not cut through the stem, and should be narrower than the distance between stems less the top stem width less 2 in. (see Fig. 9.10.4).

Following are reasonable guidelines regarding design of hollow-core slabs around openings. Some producers may have data to support a different procedure:

1. An opening located near the end of the span and extending into the span less than the lesser of 0.125ℓ or 4 ft may be neglected when designing for flexure in the midspan region.
2. Strand development must be considered on each side of an opening which cuts strand. (See Sect. 4.2.3.)
3. Slabs which are adjacent to openings which are long ($\ell/4$ or more), or occur near midspan, may be considered to have a free edge for flexural design.
4. Slabs which are adjacent to openings closer to the end than $3\ell/8$ may be considered to have a free edge for shear design.

4.8.3 Continuity

Precast deck members are normally used as part of a simple span system. However, when reinforcement is required at supports for structural integrity ties or diaphragm connections, limited continuity can be achieved. The amount of reinforcement is usually too low to develop significant moment capacity, but may be considered for reducing service load deflections. It is recommended that full simple span positive moment capacity be provided for strength design in all deck members due to uncertain moment-curvature conditions at the supports at ultimate loads.

When top steel is required at supports, it may be placed in composite topping (if available), or, in slabs, in the grout keys or concreted into cores.

Advantage may be taken of limited continuity when using rational design procedures for fire resistance (see Sect. 9.3).

4.8.4 Cantilevers

The method by which precast, prestressed members resist cantilever moments depends on (1) the method of production, (2) the length and loading requirements of the cantilever, and (3) the size of the project (amount of repetition).

If a cantilever is not long enough to fully develop top strands, a reduced value of f_{ps} must be used, as discussed in Sect. 4.2.3. As with reinforcing bars, due to settlement of concrete, top strands often do not bond as well as the ACI development equation indicates, especially in dry cast systems. It is often necessary, or at least desirable, to debond the top strands in positive moment regions, and the bottom strands in the cantilever.

In some cases it is preferable to design cantilevers as reinforced concrete members, using deformed reinforcing bars to provide the negative moment resistance. In machine-made products, the steel can be placed in grout keys, composite topping, or concreted into cores.

Top tension under service loads should be limited to $6\sqrt{f'_c}$ so that the section remains uncracked, allowing better prediction of service load deflections.

Consultation with local producers is recommended before choosing a method of reinforcement for cantilevers. Some have developed standard methods that work best with their particular system, and have proven them with tests or experience.

4.9 References

1. "Building Code Requirements for Reinforced Concrete ACI 318-89 and "Commentary," ACI 318R-89," American Concrete Institute, Detroit, MI, 1989.
2. Naaman, A.E., "Ultimate Analysis of Prestressed and Partially Prestressed Sections by Strain Compatibility," *PCI Journal*, V. 22, No. 1, January-February 1977.
3. Gergely, P. and Lutz, L.A., "Maximum Crack Width in Reinforced Concrete Flexural Members," Causes, Mechanism, and Control of Cracking in Concrete, SP-20, American Concrete Institute, Detroit, MI, 1968.

4. Mattock, Alan H., "Anchorage of Stirrups in a Thin Cast-in-Place Topping," *PCI Journal*, V. 32, No. 6, November-December 1987.

5. Zia, Paul and McGee, W.D., "Torsion Design of Prestressed Concrete," *PCI Journal*, V. 19, No. 2, March-April 1974.

6. Zia, Paul and Hsu, T.C., "Design for Torsion and Shear in Prestressed Concrete," Preprint 3424, American Society of Civil Engineers, October 1978.

7. Collins, M.P. and Mitchell, D., "Shear and Torsion Design of Prestressed and Non-Prestressed Concrete Beams," *PCI Journal*, V. 25, No. 5, September-October 1980.

8. Zia, Paul, Preston, H.K., Scott, N.L., and Workman, E.B., "Estimating Prestress Losses," *Concrete International*, V. 1, No. 6, June 1979.

9. Branson, D.E., "The Deformation of Non-composite and Composite Prestressed Concrete Members," Deflections of Concrete Structures, SP-43, American Concrete Institute, Detroit, MI, 1970.

10. Martin, L.D., "A Rational Method for Estimating Camber and Deflection of Precast Prestressed Members," *PCI Journal*, V. 22, No. 1, January-February 1977.

11. Shaikh, A.F. and Branson, D.E., "Non-Tensioned Steel in Prestressed Concrete Beams," *PCI Journal*, V. 15, No. 1, February 1970.

12. PCI Committee on Prestressed Concrete Columns, "Recommended Practice for the Design of Prestressed Concrete Columns and Walls," *PCI Journal*, V. 33, No. 4, July-August 1988.

13. LaGue, David J., "Load Distribution Tests on Precast Prestressed Hollow-Core Slab Construction," *PCI Journal*, V. 16, No. 6, November-December 1971.

14. Johnson, Ted and Ghadiali, Zohair, "Load Distribution Test on Precast Hollow Core Slabs with Openings," *PCI Journal*, V. 17, No. 5, September-October 1972.

15. Pfeifer, Donald W. and Nelson, Theodore A., "Tests to Determine the Lateral Distribution of Vertical Loads in a Long-Span Hollow-Core Floor Assembly," *PCI Journal*, V. 28, No. 6, November-December 1983.

16. "PCI Manual for the Design of Hollow Core Slabs," MNL-126-85, Precast/Prestressed Concrete Institute, Chicago, IL, 1985.

4.10.1 FLEXURE

Fig. 4.10.1 Flexural resistance coefficients for elements with non-prestressed, partially prestressed and prestressed reinforcement

Procedure:

Design:

1. Determine $K_u = \dfrac{M_u/\phi - A'_s f'_y (d - d')}{f'_c b d^2}$

2. Find $\bar{\omega}$ from table

3. For prestressed reinforcement, estimate f_{ps}

4. Select A_{ps}, A_s and A'_s from:
$A_{ps} f_{ps} + A_s f_y - A'_s f'_y = \bar{\omega} b d f'_c$

5. Check assumed value of f_{ps}

Analysis:

1. Determine $\bar{\omega} = \dfrac{A_{ps} f_{ps} + A_s f_y - A'_s f'_y}{b d f'_c}$

2. Find K_u from table

3. Determine $\phi M_n = [K_u f'_c b d^2 + A'_s f'_y (d - d')]$

Basis:

$K_u = \bar{\omega}(1 - 0.59\,\bar{\omega})$

M_u in units of in. lb

f_y, ksi \ f'_c, psi	$\omega_{max} = \bar{\omega}_{max}$ for non-prestressed elements					
	3000	4000	5000	6000	7000	8000
40	0.371	0.371	0.349	0.328	0.306	0.284
50	0.344	0.344	0.324	0.304	0.283	0.263
60	0.321	0.321	0.302	0.283	0.264	0.245

$\bar{\omega}$	Values of K_u									
	0.000	0.001	0.002	0.003	0.004	0.005	0.006	0.007	0.008	0.009
0.00	0.0000	0.0010	0.0020	0.0030	0.0040	0.0050	0.0060	0.0070	0.0080	0.0090
0.01	0.0099	0.0109	0.0119	0.0129	0.0139	0.0149	0.0158	0.0168	0.0178	0.0188
0.02	0.0198	0.0207	0.0217	0.0227	0.0237	0.0246	0.0256	0.0266	0.0275	0.0285
0.03	0.0295	0.0304	0.0314	0.0324	0.0333	0.0343	0.0352	0.0362	0.0371	0.0381
0.04	0.0391	0.0400	0.0410	0.0419	0.0429	0.0438	0.0448	0.0457	0.0466	0.0476
0.05	0.0485	0.0495	0.0504	0.0513	0.0523	0.0532	0.0541	0.0551	0.0560	0.0569
0.06	0.0579	0.0588	0.0597	0.0607	0.0616	0.0625	0.0634	0.0644	0.0653	0.0662
0.07	0.0671	0.0680	0.0689	0.0699	0.0708	0.0717	0.0726	0.0735	0.0744	0.0753
0.08	0.0762	0.0771	0.0780	0.0789	0.0798	0.0807	0.0816	0.0825	0.0834	0.0843
0.09	0.0852	0.0861	0.0870	0.0879	0.0888	0.0897	0.0906	0.0914	0.0923	0.0932
0.10	0.0941	0.0950	0.0959	0.0967	0.0976	0.0985	0.0994	0.1002	0.1011	0.1020
0.11	0.1029	0.1037	0.1046	0.1055	0.1063	0.1072	0.1081	0.1089	0.1098	0.1106
0.12	0.1115	0.1124	0.1132	0.1141	0.1149	0.1158	0.1166	0.1175	0.1183	0.1192
0.13	0.1200	0.1209	0.1217	0.1226	0.1234	0.1242	0.1251	0.1259	0.1268	0.1276
0.14	0.1284	0.1293	0.1301	0.1309	0.1318	0.1326	0.1334	0.1343	0.1351	0.1359
0.15	0.1367	0.1375	0.1384	0.1392	0.1400	0.1408	0.1416	0.1425	0.1433	0.1441
0.16	0.1449	0.1457	0.1465	0.1473	0.1481	0.1489	0.1497	0.1505	0.1513	0.1521
0.17	0.1529	0.1537	0.1545	0.1553	0.1561	0.1569	0.1577	0.1585	0.1593	0.1601
0.18	0.1609	0.1617	0.1625	0.1632	0.1640	0.1648	0.1656	0.1664	0.1671	0.1679
0.19	0.1687	0.1695	0.1703	0.1710	0.1718	0.1726	0.1733	0.1741	0.1749	0.1756
0.20	0.1764	0.1772	0.1779	0.1787	0.1794	0.1802	0.1810	0.1817	0.1825	0.1832
0.21	0.1840	0.1847	0.1855	0.1862	0.1870	0.1877	0.1885	0.1892	0.1900	0.1907
0.22	0.1914	0.1922	0.1929	0.1937	0.1944	0.1951	0.1959	0.1966	0.1973	0.1981
0.23	0.1988	0.1995	0.2002	0.2010	0.2017	0.2024	0.2031	0.2039	0.2046	0.2053
0.24	0.2060	0.2067	0.2074	0.2082	0.2089	0.2096	0.2103	0.2110	0.2117	0.2124
0.25	0.2131	0.2138	0.2145	0.2152	0.2159	0.2166	0.2173	0.2180	0.2187	0.2194
0.26	0.2201	0.2208	0.2215	0.2222	0.2229	0.2236	0.2243	0.2249	0.2256	0.2263
0.27	0.2270	0.2277	0.2283	0.2290	0.2297	0.2304	0.2311	0.2317	0.2324	0.2331
0.28	0.2337	0.2344	0.2351	0.2357	0.2364	0.2371	0.2377	0.2384	0.2391	0.2397
0.29	0.2404	0.2410	0.2417	0.2423	0.2430	0.2437	0.2443	0.2450	0.2456	0.2463
0.30	0.2469									

FLEXURE

Fig. 4.10.2 Coefficients K'_u for determining flexural design strength — bonded prestressing steel

Procedure:

1. Determine $\omega_{pu} = \dfrac{A_{ps}}{bd_p} \dfrac{f_{pu}}{f'_c}$

2. Find K'_u from table

3. Determine $\phi M_n = K'_u \dfrac{bd_p^2}{12{,}000}$ (ft-kips)

Basis:

$$K'_u = \dfrac{\phi f_{ps} f'_c}{f_{pu}}(\omega_{pu})\left[1 - (0.59\omega_{pu})\left(\dfrac{f_{ps}}{f_{pu}}\right) \right]$$

Note: K'_u from this table is approximately equivalent to $\phi K_u f'_c$ from Table 4.10.1.

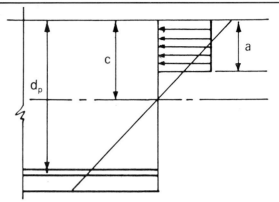

Table values are based on a strain compatibility analysis, using a stress-strain curve for prestressing strand similar to that shown in Design Aid 11.2.5. Asterisk(*) indicates $\omega_p > 0.36\beta_1$ and $\phi M_n = \phi[f'_c bd_p^2 (0.36\beta_1 - 0.08\beta_1^2)]$

Values of K'_u

f'_c	ω_{pu}	.00	.01	.02	.03	.04	.05	.06	.07	.08	.09
3000 psi	0.0	0	27	53	79	105	131	156	180	205	228
	0.1	252	275	298	321	343	364	386	407	427	447
	0.2	467	486	505	524	542	560	577	594	610	626
	0.3	642	657	*670	*670	*670	*670	*670	*670	*670	*670
4000 psi	0.0	0	36	71	106	140	174	207	240	273	305
	0.1	336	367	397	427	457	486	514	542	570	596
	0.2	623	649	674	699	723	746	770	792	814	835
	0.3	856	876	*894	*894	*894	*894	*894	*894	*894	*894
5000 psi	0.0	0	45	89	132	175	217	259	300	341	381
	0.1	420	458	496	534	570	606	642	677	711	744
	0.2	777	809	840	871	901	930	958	986	1013	1039
	0.3	1064	*1066	*1066	*1066	*1066	*1066	*1066	*1066	*1066	*1066
6000 psi	0.0	0	54	107	159	210	261	311	360	409	456
	0.1	503	550	595	640	684	727	769	811	851	891
	0.2	930	968	1005	1041	1077	1111	1144	1177	1208	*1215
	0.3	*1215	*1215	*1215	*1215	*1215	*1215	*1215	*1215	*1215	*1215
7000 psi	0.0	0	63	124	185	245	304	363	420	476	532
	0.1	587	641	693	745	796	846	895	944	991	1037
	0.2	1081	1125	1168	1210	1250	1289	1327	*1341	*1341	*1341
	0.3	*1341	*1341	*1341	*1341	*1341	*1341	*1341	*1341	*1341	*1341
8000 psi	0.0	0	72	142	212	280	348	414	480	544	608
	0.1	670	731	791	850	908	965	1021	1075	1128	1180
	0.2	1231	1280	1328	1374	1419	*1441	*1441	*1441	*1441	*1441
	0.3	*1441	*1441	*1441	*1441	*1441	*1441	*1441	*1441	*1441	*1441

FLEXURE

Fig. 4.10.3 Values of f_{ps} by stress-strain relationship — bonded strand

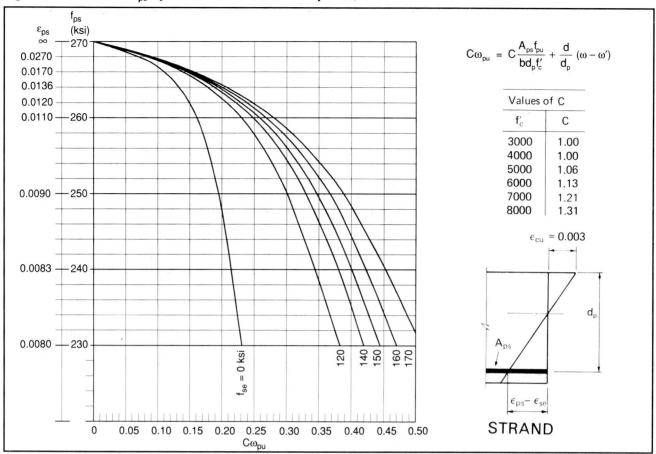

Fig. 4.10.4 Design stress for underdeveloped strand

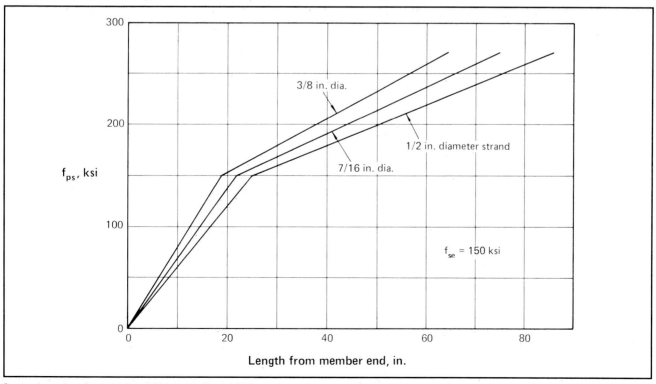

Curves based on Sect. 12.9.1, ACI 318-89. Note ACI Sect. 12.9.3 for strands which are debonded near member ends.

4.10.2 SHEAR

Fig. 4.10.5 Shear design by Eq. 11-10 (ACI 318-89) **Straight strands**

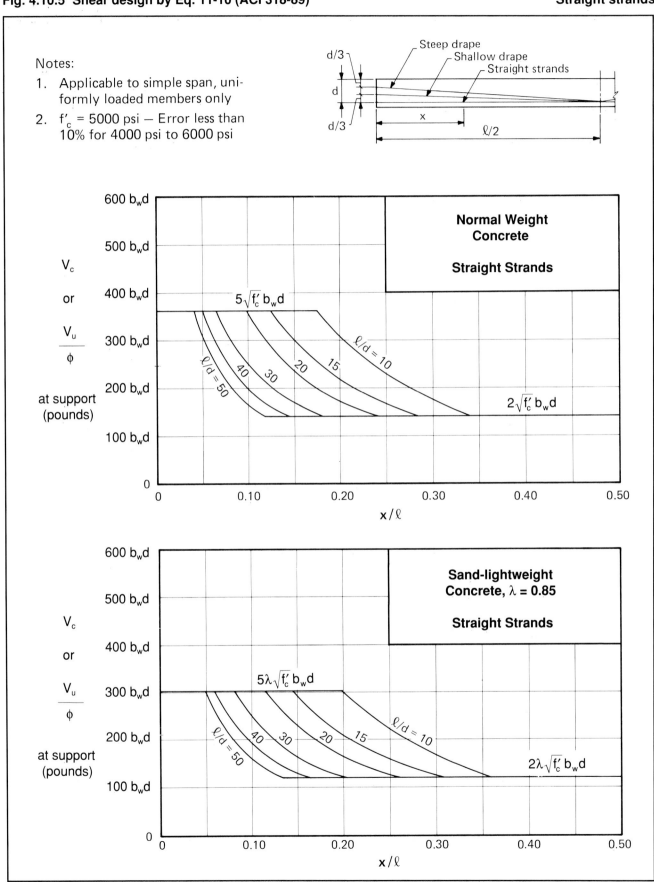

SHEAR

Fig. 4.10.6 Shear design by Eq. 11-10 (ACI 318-89)

Shallow drape

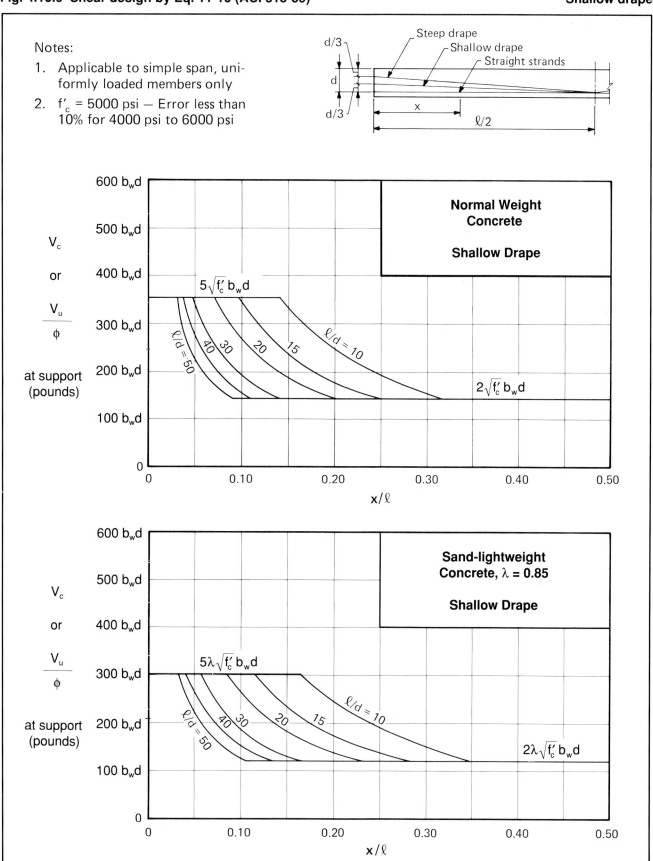

Notes:
1. Applicable to simple span, uniformly loaded members only
2. f'_c = 5000 psi — Error less than 10% for 4000 psi to 6000 psi

SHEAR

Fig. 4.10.7 Shear design by Eq. 11-10 (ACI 318-89) — Steep drape

Notes:
1. Applicable to simple span, uniformly loaded members only
2. f'_c = 5000 psi — Error less than 10% for 4000 psi to 6000 psi

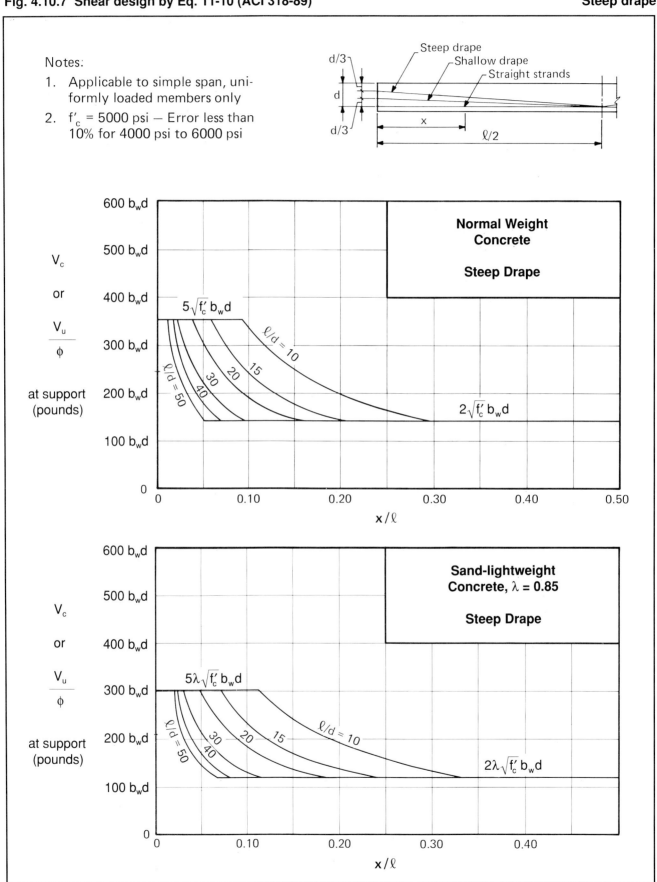

SHEAR

Fig. 4.10.8 Minimum shear reinforcement by Eq. 11-15 (ACI 318-89)

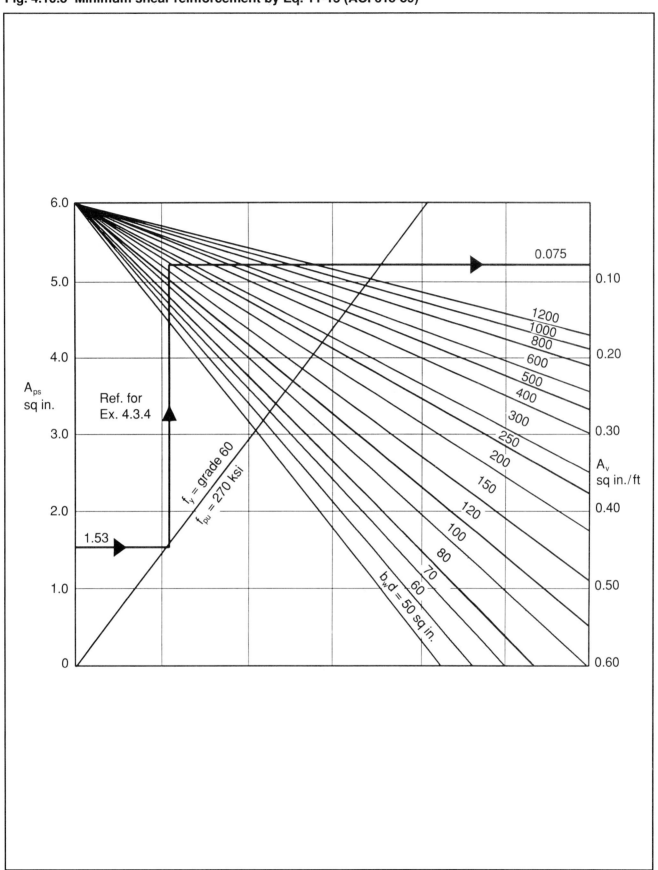

SHEAR

Fig. 4.10.9 Shear reinforcement

$$A_v = \frac{(V_u/\phi - V_c)s}{f_y d} \quad \text{(Eq. 4.3.9)}$$

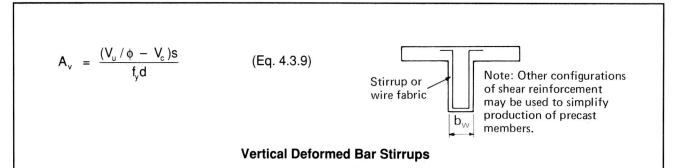

Note: Other configurations of shear reinforcement may be used to simplify production of precast members.

Vertical Deformed Bar Stirrups

Stirrup Spacing (in.)	Maximum values of $(V_u/\phi - V_c)/d$ (lb/in.) $f_y = 60{,}000$ psi						Stirrup Spacing (in.)
	No. 3 $A_v = 0.22$	No. 4 $A_v = 0.40$	No. 5 $A_v = 0.62$	No. 3 $A_v = 0.22$	No. 4 $A_v = 0.40$	No. 5 $A_v = 0.62$	
2.0	6600	12000	18600	1320	2400	3720	10.0
2.5	5280	9600	14880	1200	2182	3382	11.0
3.0	4400	8000	12400	1100	2000	3100	12.0
3.5	3771	6857	10629	1015	1846	2862	13.0
4.0	3300	6000	9300	943	1714	2657	14.0
4.5	2933	5333	8267	880	1600	2480	15.0
5.0	2640	4800	7440	825	1500	2325	16.0
5.5	2400	4364	6764	776	1412	2188	17.0
6.0	2200	4000	6200	733	1333	2067	18.0
7.0	1886	3429	5314	660	1200	1860	20.0
8.0	1650	3000	4650	600	1091	1691	22.0
9.0	1467	2667	4133	550	1000	1550	24.0

Welded Wire Fabric as Shear Reinforcement ($f_y = 60{,}000$ psi)

Spacing of vertical wire (in.)	Maximum values of $(V_u/\phi - V_c)/d$ (lb/in.)								Spacing of vertical wire (in.)
	One row — Vertical wire				Two rows — Vertical wire				
	W7.5 $A_v = 0.075$	W5.5 $A_v = 0.055$	W4 $A_v = 0.040$	W2.9 $A_v = 0.029$	W7.5 $A_v = 0.150$	W5.5 $A_v = 0.110$	W4 $A_v = 0.80$	W2.9 $A_v = 0.058$	
2	2250	1650	1200	870	4500	3300	2400	1740	2
3	1500	1100	800	580	3000	2200	1600	1160	3
4	1125	825	600	435	2250	1650	1200	870	4
6	750	550	400	290	1500	1110	800	580	6

4.10.3 TORSION

Fig. 4.10.10 Determination of $T_{u(min)}$ below which torsion can be neglected — Eq. 4.4.1

$T_{u(min)} = R_{t(min)} \lambda \Sigma x^2 y$ in.-lb

where: $R_{t(min)} = 0.425 \gamma_t \sqrt{f'_c}$ psi (Eq. 4.4.1)

$\gamma_t = \sqrt{1 + 10 f_{pc} / f'_c}$

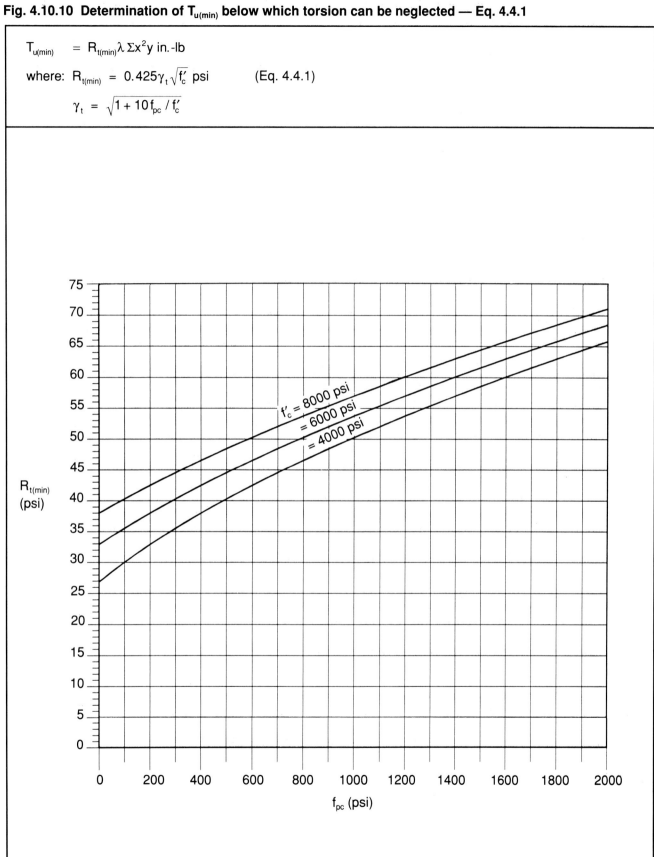

TORSION

Fig. 4.10.11 Determination of upper limits on shear and torsion for design—Eqs. 4.4.2 and 4.4.3

$$T_{n(max)} = R_{t(max)} \lambda \Sigma x^2 y \text{ in.-lb} \quad \text{(Eq. 4.4.2)}$$

where: $R_{t(max)} = \dfrac{K_t \sqrt{f'_c}}{3 \times \sqrt{1 + \left(\dfrac{K_t V_u}{30 C_t T_u}\right)^2}}$ psi

$$V_{n(max)} = R_{v(max)} \lambda b_w d \text{ lb} \quad \text{(Eq. 4.4.3)}$$

where: $R_{v(max)} = \dfrac{10\sqrt{f'_c}}{\sqrt{1 + \left(\dfrac{30 C_t T_u}{K_t V_u}\right)^2}}$ psi

Note: The $R_{t(max)}$ and $R_{v(max)}$ values given in the table correspond to three ranges of f_{pc}. The values given are estimated to be conservative within an approximation of less than ten percent in each range. If desired, exact values may be obtained using Eqs. 4.4.2 and 4.4.3.

		$f_{pc} < 400$ psi				$400 \leq f_{pc} < 1000$ psi				$f_{pc} \geq 1000$ psi			
V_u/T_u	C_t	0.04	0.06	0.08	0.10	0.04	0.06	0.08	0.10	0.04	0.06	0.08	0.10
		\multicolumn{12}{c	}{$R_{t(max)}$, psi}
$f'_c = 4000$ psi	0.1	193	234	256	269	200	248	275	291	210	267	301	322
	0.2	116	160	193	216	118	164	200	228	120	169	210	242
	0.3	81	116	147	172	82	118	150	177	82	120	154	184
	0.4	62	90	116	140	62	91	118	142	62	92	120	146
	0.5	50	73	96	116	50	74	97	118	50	74	98	120
	0.6	42	62	81	99	42	62	82	100	42	62	82	101
	0.7	36	53	70	86	36	53	71	87	36	54	71	88
		\multicolumn{12}{c	}{$R_{v(max)}$, psi}
	0.1	481	390	320	269	500	414	344	291	524	444	376	322
	0.2	582	533	481	433	590	547	501	455	599	564	524	483
	0.3	608	582	550	516	613	590	563	532	617	599	577	551
	0.4	619	602	582	558	621	608	590	570	624	613	599	583
	0.5	623	613	599	582	625	616	604	590	627	620	611	599
	0.6	626	619	608	596	627	621	612	602	628	624	617	609
	0.7	627	622	614	605	629	624	618	610	630	626	621	615
		\multicolumn{12}{c	}{$R_{t(max)}$, psi}
$f'_c = 5000$ psi	0.1	213	257	281	294	221	272	300	317	232	294	331	354
	0.2	130	177	213	239	131	182	221	250	134	188	232	267
	0.3	91	130	163	191	91	131	166	196	92	134	171	207
	0.4	69	101	130	155	69	102	131	158	70	103	134	162
	0.5	56	82	107	130	56	83	108	131	56	83	109	134
	0.6	47	69	91	111	47	69	91	112	47	70	92	113
	0.7	40	60	78	96	40	60	79	97	40	60	79	98
		\multicolumn{12}{c	}{$R_{v(max)}$, psi}
	0.1	532	429	351	294	553	453	375	317	581	490	414	354
	0.2	648	591	532	477	657	607	553	500	668	627	581	535
	0.3	679	648	611	572	683	657	624	589	689	668	642	613
	0.4	691	672	648	621	693	677	657	633	697	684	668	649
	0.5	697	684	667	648	698	688	674	657	700	692	681	668
	0.6	700	691	679	664	701	693	683	671	702	697	689	679
	0.7	702	695	686	675	703	697	689	680	704	699	694	686

TORSION

Fig. 4.10.11 (cont.) Determination of upper limits on shear and torsion for design—Eqs. 4.4.2 and 4.4.3

		$f_{pc} < 400$ psi				$400 \leq f_{pc} < 1000$ psi				$f_{pc} \geq 1000$ psi			
V_u/T_u	C_t	0.04	0.06	0.08	0.10	0.04	0.06	0.08	0.10	0.04	0.06	0.08	0.10
		\multicolumn{12}{c}{$R_{t(max)}$, psi}
$f'_c = 6000$ psi	0.1	231	278	303	317	240	293	332	340	253	318	356	380
	0.2	142	193	231	259	143	198	240	270	146	205	253	289
	0.3	99	142	178	207	100	143	181	213	100	146	187	222
	0.4	76	110	142	169	76	111	143	172	76	112	146	177
	0.5	61	90	117	141	61	90	118	143	61	91	119	146
	0.6	51	76	99	121	51	76	100	122	51	76	100	124
	0.7	44	65	86	105	44	65	86	106	44	66	87	107
		\multicolumn{12}{c}{$R_{v(max)}$, psi}
	0.1	578	464	379	317	599	489	403	340	631	530	446	380
	0.2	708	644	578	518	717	660	599	541	730	683	631	579
	0.3	743	708	666	622	747	717	680	640	754	730	700	667
	0.4	756	735	708	677	759	740	717	689	763	748	730	708
	0.5	763	748	730	708	764	752	736	717	767	757	745	730
	0.6	766	756	742	726	767	759	747	733	769	763	754	743
	0.7	768	761	751	738	769	763	754	743	771	766	759	751
		\multicolumn{12}{c}{$R_{t(max)}$, psi}
$f'_c = 7000$ psi	0.1	248	298	324	339	257	313	343	361	271	339	379	403
	0.2	152	208	248	277	154	213	257	289	157	220	271	309
	0.3	107	152	191	223	107	154	195	229	108	157	201	238
	0.4	82	119	152	182	82	120	154	185	82	121	157	190
	0.5	66	97	126	152	66	97	127	154	66	98	129	157
	0.6	55	82	107	131	55	82	107	132	55	82	108	133
	0.7	47	70	93	114	47	71	93	115	48	71	94	116
		\multicolumn{12}{c}{$R_{v(max)}$, psi}
	0.1	621	496	405	339	641	521	429	361	676	565	473	403
	0.2	762	693	621	555	772	708	641	578	786	734	676	619
	0.3	801	763	717	669	806	772	731	686	813	786	753	715
	0.4	816	792	762	729	819	798	772	741	823	807	786	761
	0.5	823	808	787	762	825	811	793	772	828	817	803	786
	0.6	827	816	801	783	829	819	806	790	831	823	813	800
	0.7	830	821	810	796	831	823	814	801	832	827	819	810
		\multicolumn{12}{c}{$R_{t(max)}$, psi}
$f'_c = 8000$ psi	0.1	264	316	343	356	272	331	362	380	287	358	399	423
	0.2	163	221	264	295	165	226	272	306	168	234	287	327
	0.3	114	163	204	237	115	166	207	243	116	168	214	253
	0.4	87	127	163	194	87	128	165	197	88	129	168	203
	0.5	70	104	134	163	71	104	135	165	71	105	137	168
	0.6	59	87	114	139	59	87	115	140	59	88	116	142
	0.7	51	75	99	121	51	75	99	122	51	76	100	124
		\multicolumn{12}{c}{$R_{v(max)}$, psi}
	0.1	660	527	429	359	681	551	453	380	718	597	499	423
	0.2	814	738	660	589	823	754	681	612	838	718	781	655
	0.3	855	814	764	712	860	823	778	730	868	838	801	760
	0.4	872	846	814	777	875	852	823	789	879	862	838	811
	0.5	880	863	840	814	882	866	846	823	885	873	857	838
	0.6	884	872	855	836	886	875	860	843	888	879	868	854
	0.7	887	878	865	850	888	880	869	856	889	883	875	864

TORSION

Fig. 4.10.12 Shear and torsion strengths of concrete — Eqs. 4.4.4 and 4.4.5

$$T_c = R_t T_c' \quad \text{(Eq. 4.4.4)}$$

where: $R_t = \dfrac{1}{\sqrt{1 + \left(\dfrac{T_c'/T_u}{V_c'/V_u}\right)^2}}$

$$V_c = R_v V_c' \quad \text{(Eq. 4.4.5)}$$

where: $R_v = \dfrac{1}{\sqrt{1 + \left(\dfrac{V_c'/V_u}{T_c'/T_u}\right)^2}}$

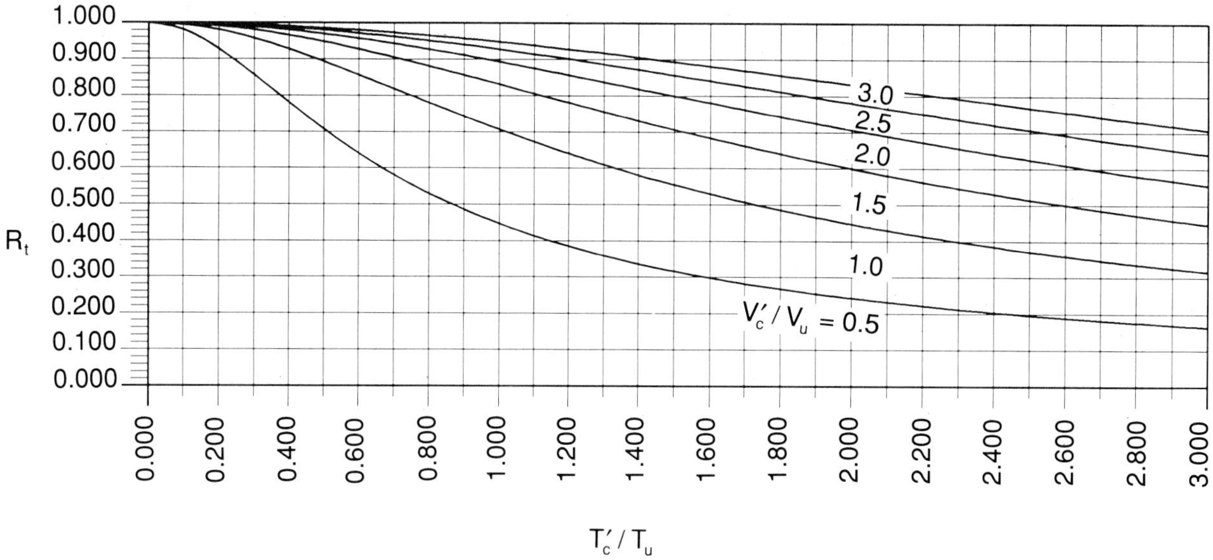

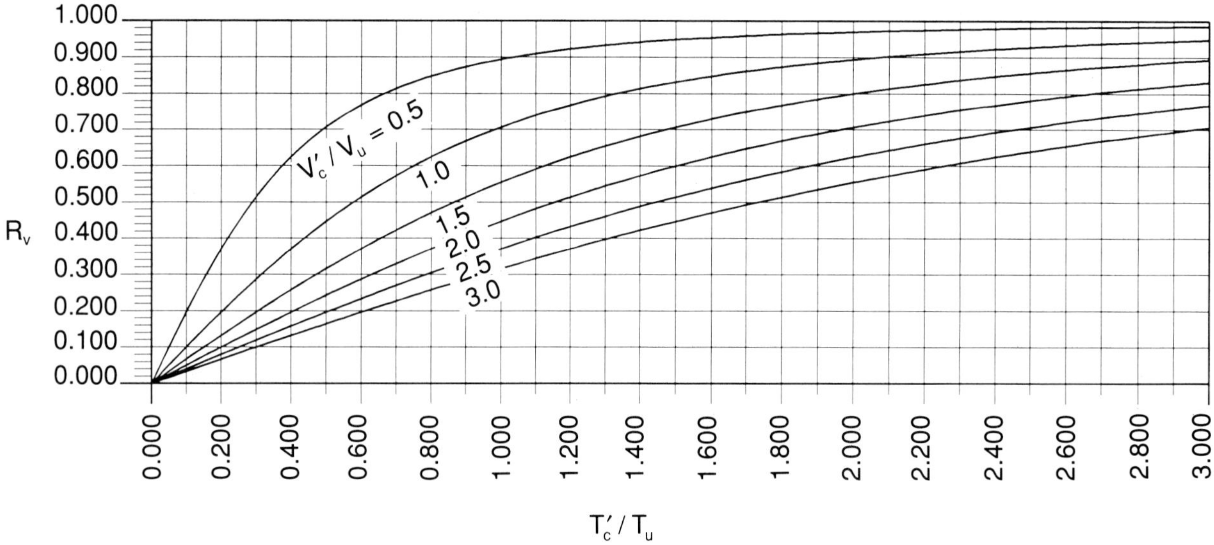

TORSION

Fig. 4.10.13 Determination of T'_c — Eq. 4.4.6

$$T'_c = R'_t \lambda \Sigma x^2 y \text{ in.-lb} \quad \text{(Eq. 4.4.6)}$$

where: $R'_t = 0.8\sqrt{f'_c}(2.5\gamma_t - 1.5)$

$\gamma_t = \sqrt{1 + 10 f_{pc}/f'_c}$

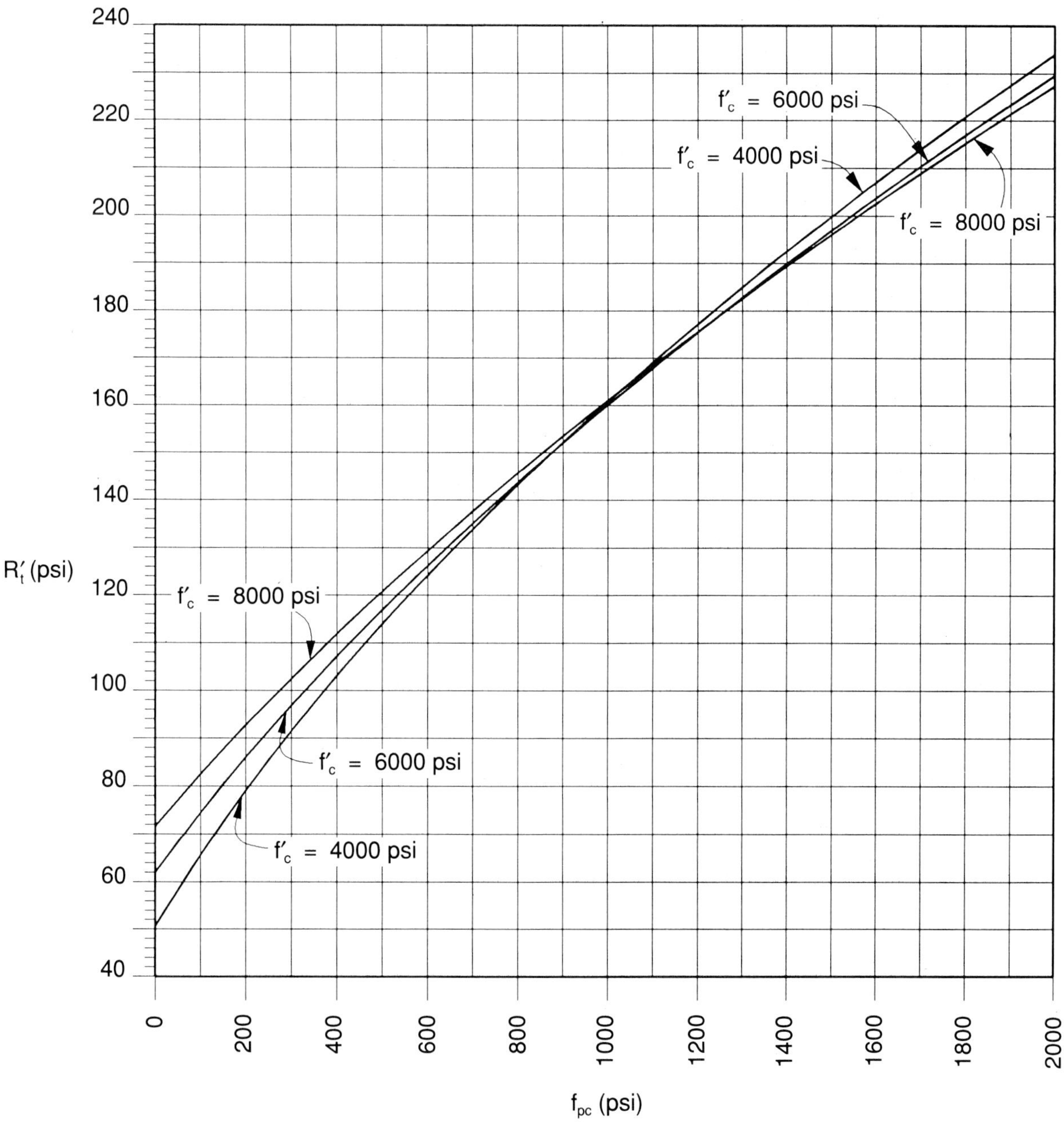

4.10.4 CAMBER AND DEFLECTION

Fig. 4.10.14 Camber equations for typical strand profiles

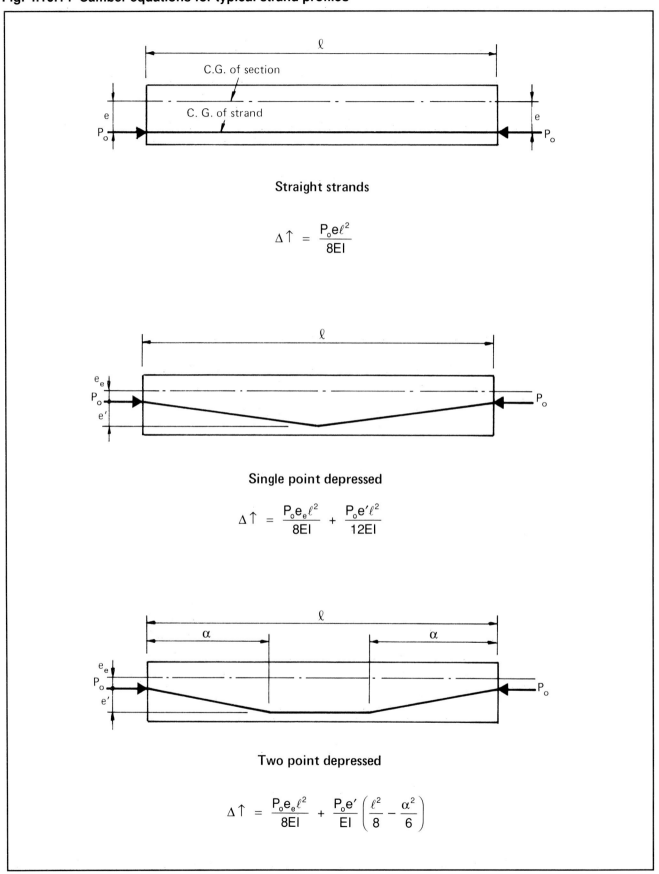

Straight strands

$$\Delta \uparrow = \frac{P_o e \ell^2}{8EI}$$

Single point depressed

$$\Delta \uparrow = \frac{P_o e_e \ell^2}{8EI} + \frac{P_o e' \ell^2}{12EI}$$

Two point depressed

$$\Delta \uparrow = \frac{P_o e_e \ell^2}{8EI} + \frac{P_o e'}{EI}\left(\frac{\ell^2}{8} - \frac{\alpha^2}{6}\right)$$

CAMBER AND DEFLECTION

Fig. 4.10.15 Moment of inertia of transformed section —prestressed members

$$I_{cr} = nA_{ps}d_p^2\left(1 - 1.6\sqrt{n\rho_p}\right)$$

$$= n\rho_p\left(1 - 1.6\sqrt{n\rho_p}\right) \times bd_p^3$$

$$= C \text{ (from table)} \times bd_p^3$$

where: $\rho_p = \dfrac{A_{ps}}{bd_p}$, $n = \dfrac{E_s}{E_c}$

$E_s = 28.5 \times 10^6$ psi

$E_c = 33w_c^{1.5}\sqrt{f'_c}$ (ACI Sect. 8.5.1)

$w_c = 145$ lb/ft³ ...Normal weight concrete

$w_c = 115$ lb/ft³ ...Sand-lightweight concrete

Values of Coefficient, C

	ρ_p	f'_c, psi					
		3000	4000	5000	6000	7000	8000
Normal Weight Concrete	.0005	.0041	.0036	.0032	.0029	.0027	.0026
	.0010	.0077	.0068	.0061	.0056	.0052	.0049
	.0015	.0111	.0098	.0089	.0082	.0076	.0072
	.0020	.0143	.0126	.0115	.0106	.0099	.0093
	.0025	.0173	.0153	.0139	.0129	.0120	.0113
	.0030	.0201	.0179	.0163	.0151	.0141	.0133
	.0035	.0228	.0203	.0185	.0172	.0161	.0152
	.0040	.0254	.0226	.0207	.0192	.0180	.0170
	.0045	.0278	.0248	.0227	.0211	.0198	.0188
	.0050	.0300	.0270	.0247	.0230	.0216	.0205
	.0055	.0322	.0290	.0266	.0248	.0233	.0221
	.0060	.0343	.0309	.0284	.0265	.0250	.0237
	.0065	.0362	.0327	.0302	.0282	.0266	.0253
	.0070	.0381	.0345	.0319	.0298	.0281	.0267
	.0075	.0398	.0362	.0335	.0314	.0296	.0282
	.0080	.0415	.0378	.0350	.0329	.0311	.0296
	.0085	.0430	.0393	.0365	.0343	.0325	.0309
	.0090	.0445	.0408	.0380	.0357	.0338	.0323
	.0095	.0459	.0422	.0393	.0370	.0351	.0335
	.0100	.0472	.0435	.0406	.0383	.0364	.0348
Sand-Lightweight Concrete	.0005	.0056	.0049	.0044	.0040	.0038	.0035
	.0010	.0105	.0092	.0083	.0077	.0071	.0067
	.0015	.0149	.0132	.0120	.0110	.0103	.0097
	.0020	.0190	.0169	.0153	.0142	.0133	.0125
	.0025	.0228	.0203	.0185	.0172	.0161	.0152
	.0030	.0263	.0235	.0215	.0200	.0187	.0177
	.0035	.0296	.0265	.0243	.0226	.0213	.0201
	.0040	.0326	.0294	.0270	.0252	.0237	.0224
	.0045	.0354	.0320	.0295	.0275	.0260	.0246
	.0050	.0381	.0345	.0319	.0298	.0281	.0267
	.0055	.0405	.0368	.0341	.0320	.0302	.0288
	.0060	.0427	.0390	.0362	.0340	.0322	.0307
	.0065	.0448	.0411	.0382	.0360	.0341	.0325
	.0070	.0466	.0430	.0401	.0378	.0359	.0343
	.0075	.0484	.0447	.0419	.0395	.0376	.0359
	.0080	.0499	.0464	.0435	.0412	.0392	.0375
	.0085	.0513	.0479	.0451	.0428	.0408	.0391
	.0090	.0526	.0493	.0465	.0442	.0422	.0405
	.0095	.0537	.0506	.0479	.0456	.0436	.0419
	.0100	.0547	.0518	.0492	.0469	.0449	.0432

CAMBER AND DEFLECTION

Fig. 4.10.16 Effective moment of inertia by Eq. 9-7 (ACI 318-89)

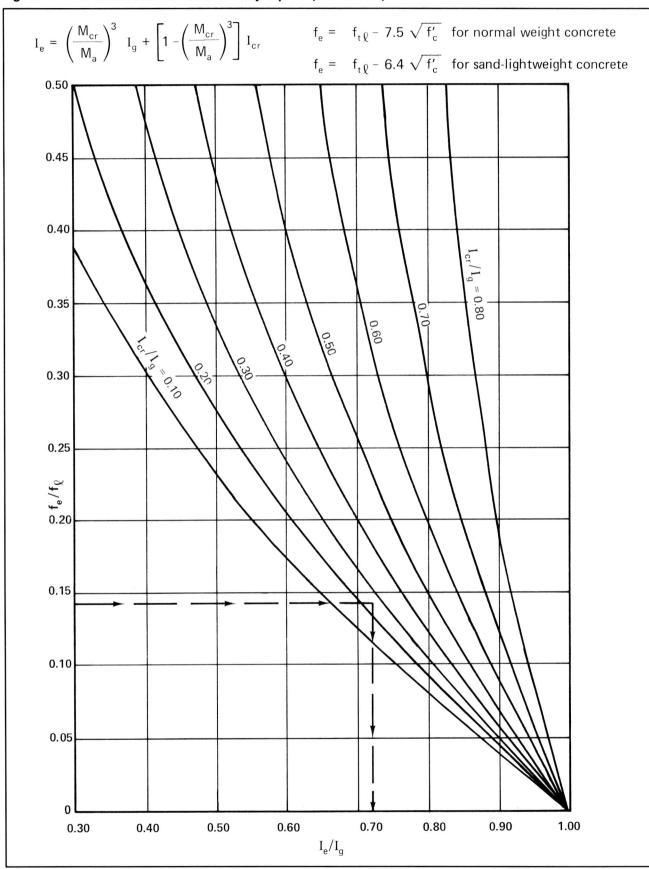

$$I_e = \left(\frac{M_{cr}}{M_a}\right)^3 I_g + \left[1 - \left(\frac{M_{cr}}{M_a}\right)^3\right] I_{cr}$$

$f_e = f_{t\ell} - 7.5\sqrt{f'_c}$ for normal weight concrete

$f_e = f_{t\ell} - 6.4\sqrt{f'_c}$ for sand-lightweight concrete

CHAPTER 5
PRODUCT HANDLING AND ERECTION BRACING

	Page No.
5.1 General	5-2
5.1.1 Notation	5-2
5.1.2 Introduction	5-2
5.2 Product Handling	5-2
5.2.1 Introduction	5-2
5.2.2 Structural Design Criteria	5-3
5.2.3 Form Suction and Impact Factors	5-3
5.2.4 Stress Limitations	5-3
5.2.5 Safety Factors	5-5
5.2.6 Prestressed Wall Panels	5-5
5.2.7 Handling Considerations	5-6
5.2.8 Handling Devices	5-8
5.2.9 Lateral Stability	5-14
5.2.10 Storage	5-16
5.2.11 Transportation	5-17
5.2.12 Erection	5-18
5.2.13 Design Example—Flat Panel	5-18
5.3 Erection Bracing	5-23
5.3.1 Introduction	5-23
5.3.2 Loads	5-24
5.3.3 Factors of Safety	5-24
5.3.4 Bracing Equipment and Materials	5-24
5.3.5 Erection Analysis	5-27
5.4 References	5-35

PRODUCT HANDLING AND ERECTION BRACING

5.1 General

5.1.1 Notation

A	=	area; average effective area around one reinforcing bar
A_s	=	area of reinforcement
A'_s	=	area of compression reinforcement
a	=	dimension defined in section used
b	=	dimension defined in section used
d	=	depth from extreme compression fiber to centroid of tension reinforcement
d_c	=	concrete cover to the center of the reinforcement
e	=	eccentricity of force about center of gravity
e_i	=	initial lateral eccentricity of the center of gravity from the roll axis
E_c	=	modulus of elasticity of concrete
F	=	multiplication factor (Fig. 5.2.7)
f_b, f_t	=	stress in bottom and top fiber, respectively
f'_c	=	concrete compressive strength
f'_{ci}	=	concrete compressive strength at the time considered
f_{ct}	=	splitting tensile strength
f'_r	=	allowable flexural tensile stress computed using the gross concrete section
f_s	=	stress in steel
f_y	=	yield strength of steel
h_1	=	distance from centroid of tensile reinforcement to neutral axis
h_2	=	distance from extreme tension fiber to neutral axis
I	=	moment of inertia (with subscripts)
I_y	=	moment of inertia about y-axis
ℓ	=	span; length of precast unit
ℓ_e	=	embedment length
M_x, M_y, M_z	=	see Figs. 5.2.4, 5.2.5, and 5.2.9
P, P_H, P_V	=	see Figs. 5.2.8 and 5.2.9
R	=	reaction (with subscripts)
T	=	tension on cable
t	=	thickness
W	=	total load
w	=	weight per unit length or area; maximum crack width
w_b	=	wind load on beam
w_c	=	wind load on column
y_b, y_c, y_t	=	see Figs. 5.2.8 and 5.2.9
y_{max}	=	instantaneous maximum displacement
y_t	=	time dependent displacement; height of roll axis above center of gravity of beam
Z_b, Z_t	=	section modulus with respect to bottom and top respectively
z_o	=	distance from roll axis to center of gravity of the deflected arc
β_y	=	midspan beam deflection with dead weight applied laterally (Sect. 5.2.9)
λ	=	correction factor related to unit weight of concrete; deflection amplification factor
ϕ	=	angle of lift line from vertical in transverse direction; strength reduction factor
ρ'	=	reinforcement ratio for non-prestressed compression reinforcement
θ	=	angle of lift lines in longitudinal direction; angle of tilt
θ_{max}	=	maximum permissible tilt angle

5.1.2 Introduction

This Chapter discusses those aspects of precast product manufacture and erection which are of importance to the design engineer, the product engineer, and the engineer responsible for safe erection procedures.

5.2 Product Handling

5.2.1 Introduction

The loads and forces on precast and prestressed members, especially wall panels, during production, transportation or erection will frequently require a separate analysis because concrete strengths are lower and support points and orientation are usually different than the panel in its final position.

Most structural products are manufactured in standard steel molds with fixed dimensions. Standard product dimensions are shown in Chapter 2 and in manufacturer's catalogs. Architectural wall panels are formed in a variety of sizes and shapes, usually designed by the project architect, and are manufactured in molds designed especially for the panel.

The most economical element for a project is usually the largest, considering:

1. Stability and stresses on the element during handling.
2. Transportation size and weight regulations and equipment restrictions.
3. Available crane capacity at both the plant and the project site. Position of the crane must be considered, since capacity is a function of reach.
4. Storage space, truck turning radius, and other site restrictions.

Fig. 5.2.1 Draft on sides of forms

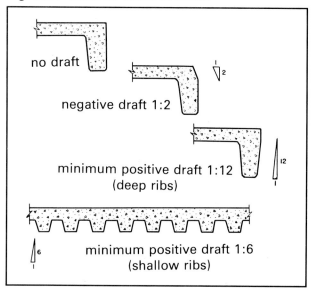

Shape of member must be such that reinforcement and concrete placement can be accomplished easily.

To remove a member from a form without partially dismantling the form, and to avoid trapping air bubbles, the sides must have adequate slope or draft (see Fig. 5.2.1).

5.2.2 Structural Design Criteria

Precast products must be designed for the loadings which occur during each phase of their existence, as shown in Fig. 5.2.2. The items which affect the forces imposed during each phase are listed below:

1. *Stripping.*
   a. Orientation of member—horizontal, vertical or some angle between.
   b. Form suction and impact—see Table 5.2.1.
   c. Number and location of handling devices.
   d. Member weight and weight of any additional items which must be lifted, such as forms which remain with the member during stripping.

2. *Yard Handling and Storage.*
   a. Orientation of the member.
   b. Location of temporary support points.
   c. Location with respect to other stored members.
   d. Orientation with respect to the sun.

3. *Transportation to the Job Site.*
   a. Orientation of the member.
   b. Location of horizontal and vertical supports.
   c. Condition of the transporting vehicle, roads and site.
   d. Dynamic considerations during movement.

4. *Erection.*
   a. Lifting point locations.
   b. Orientation and tripping (rotating).
   c. Location of temporary supports.
   d. Temporary loadings.

5. *In-place.*
   See Chapters 3 and 4.

5.2.3 Form Suction and Impact Factors

To account for the forces on the member caused by form suction and impact, it is common practice to apply a multiplier to the member weight and treat the resulting force as an equivalent static service load. The multipliers cannot be quantitatively derived, so are based on experience. Table 5.2.1 provides typical values.

5.2.4 Stress Limitations

Stress limits for prestressed members during production are specified in Chapter 18 of ACI 318-89, and are discussed in Chapter 4 of this Handbook. However, codes do not restrict stresses on conventionally reinforced members. When exposed to view precast products may be designed for handling to (a) limit stresses so that cracks are not visible, or (b) control cracking in accordance with Sect. 4.2.2.1, considering the lower concrete strengths and actual flexural conditions. When not exposed to view, design is in accordance with the strength design requirements for reinforced concrete in ACI 318-89.

Handling without cracking

Under this criterion, surfaces remain free of discernible cracks by limiting the flexural tension to the modulus of rupture modified by a safety factor. If the modulus of rupture is $7.5\lambda\sqrt{f'_{ci}}$ and a safety factor of 1.5 is used:

$$f'_r = 5\lambda\sqrt{f'_{ci}} \qquad \text{(Eq. 5.2.1)}$$

where:

f'_r = allowable flexural tensile stress computed using the gross concrete section

f'_{ci} = concrete compressive strength at the time considered

For normal weight concrete:

$\lambda = 1.0$

For lightweight concrete, if the splitting tensile strength, f_{ct}, is known:

$$\lambda\sqrt{f'_{ci}} = \frac{f_{ct}}{6.7} \leq \sqrt{f'_{ci}}$$

If f_{ct} is not known:

λ = 0.75 for all-lightweight concrete
  = 0.85 for sand-lightweight concrete

Handling with controlled cracking

The amount and location of reinforcing steel has a negligible effect on performance until a crack develops. As flexural tension increases above the modulus of rupture, hairline cracks will develop and extend a distance into the element. If cracks are narrow and closely spaced, the structural adequacy of the element will remain unimpaired because reinforcement will be fully protected.

For members exposed to view, it is recommended that crack widths be limited to the following:

Exposed to weather: 0.005 in.

Not exposed to weather: 0.010 in.

Design procedures are illustrated in Examples 4.2.7 and 5.2.2.

Whether or not the product is designed to prevent or to control cracking, a minimum reinforcement ratio of 0.001 for wall panels should be provided.[7]

Fig. 5.2.2 Typical handling methods

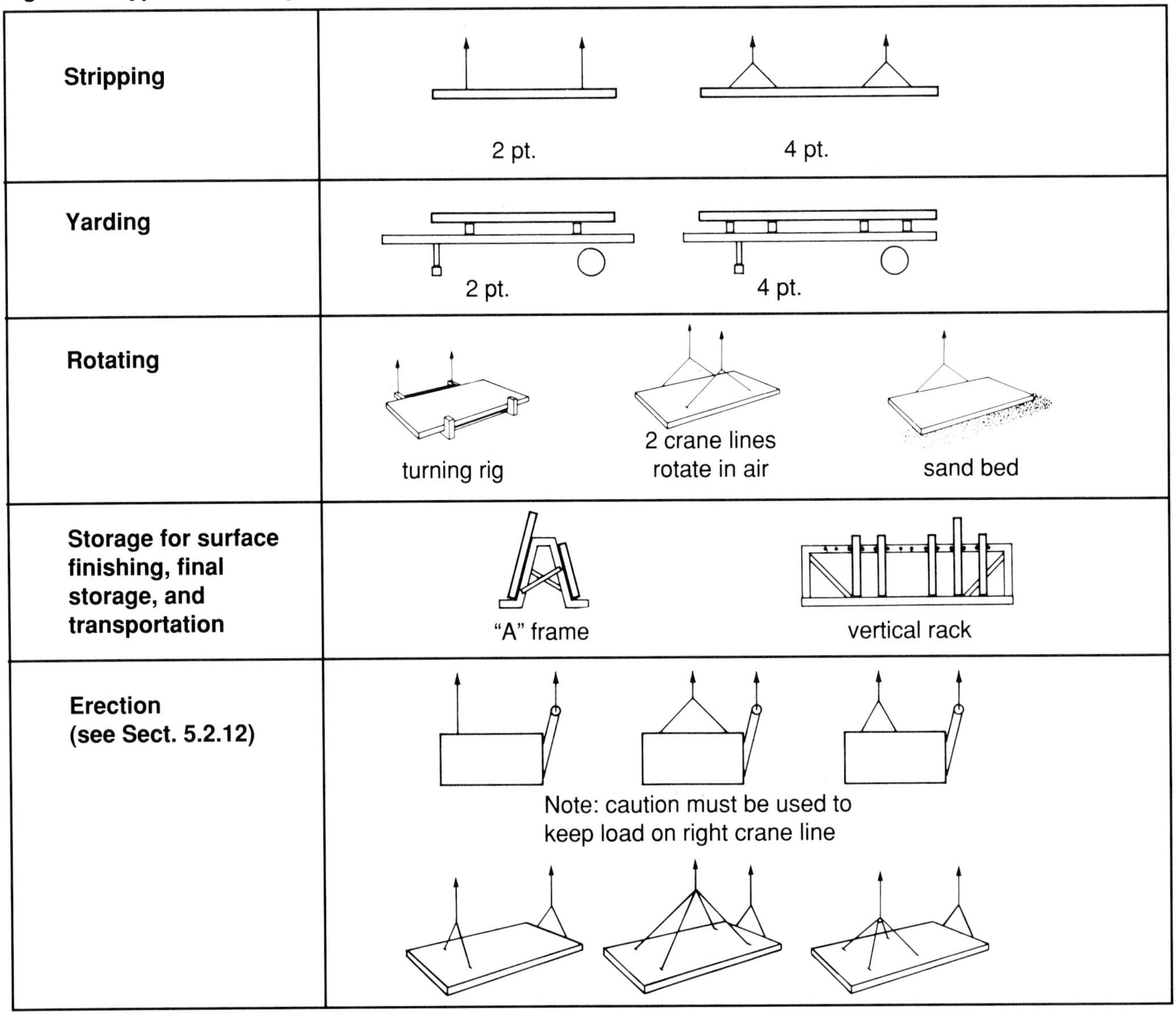

Stripping	2 pt. 4 pt.
Yarding	2 pt. 4 pt.
Rotating	turning rig 2 crane lines rotate in air sand bed
Storage for surface finishing, final storage, and transportation	"A" frame vertical rack
Erection (see Sect. 5.2.12)	Note: caution must be used to keep load on right crane line

Table 5.2.1 Equivalent static load multipliers[1] to account for stripping and dynamic forces

Product Type	Stripping	
	Finish	
	Exposed aggregate with retarder	Smooth mold (form oil only)
Flat, with removable side forms, no false joints or reveals	1.2	1.3
Flat, with false joints and/or reveals	1.3	1.4
Fluted, with proper draft[4]	1.4	1.6
Sculptured	1.5	1.7
Yard handling[2] and erection[3]		
All products	1.2	
Travel[2]		
All products	1.5	

1. These factors are used in flexural design of panels and are not to be applied to required safety factors on lifting devices. At stripping, suction between product and form introduces forces, which are treated here by introducing a multiplier on product weight. It would be more accurate to establish these multipliers based on the actual contact area and a suction factor independent of product weight.
2. Certain unfavorable conditions in road surface, equipment, etc., may require use of higher values.
3. Under certain circumstances may be higher.
4. For example, tees, channels and fluted panels.

5.2.5 Safety Factors

When designing for stripping and handling, the following safety factors are recommended:

1. Use embedded inserts and erection devices with a pullout strength at least equal to 4 times the actual weight lifted.
2. For members designed "without cracking", use a factor of 1.5 applied to the modulus of rupture as discussed in Sect. 5.2.4.

5.2.6 Prestressed Wall Panels

When the handling procedures for wall panels cause the limitations of Sect. 5.2.4 to be exceeded, the panel can be prestressed, using either pretensioning or post-tensioning. Design is based on Chapter 18 of ACI 318-89, as described in Chapter 4, with the further restriction that tensile stresses during handling should not exceed f'_r given by Eq. 5.2.1.

It is recommended that the average stress due to prestressing, after losses, be limited to a range of 125 to 800 psi. The prestressing force should be concentric with the effective cross section in order to minimize camber, although some manufacturers prefer to have a slight inward bow in the in-place position to counteract thermal bow (see Sect. 3.3.2). It should be noted that, since concentrically prestressed members do not camber, the form adhesion may be larger than members which camber.

In order to minimize the possibility of splitting cracks in thin pretensioned members, the strand diameter should not exceed those shown in Table 5.2.2. Additional light transverse reinforcement may be required to control cracking.

Table 5.2.2 Suggested maximum strand diameter

Concrete thickness, in.	Strand diameter, in.
2	3/8
2½	3/8
2½ to 3½	7/16
3½ and thicker	½ and larger

When wall panels are post-tensioned, care must be taken to assure proper transfer of force at the anchorage and protection of anchors and tendons against corrosion. Straight strands or bars may be used, or, to reduce the number of anchors, the method shown in Fig. 5.2.3 may be used. Plastic coated tendons with a low coefficient of angular friction (μ = 0.03 to 0.05) are looped within the panel, and anchors installed at only one end. The tendons remain unbonded.

It should be noted that if an unbonded tendon is cut, the prestress is lost. This can sometimes happen if an unplanned opening is put in at a later date.

5.2.7 Handling Considerations

The number and location of lifting devices are chosen to keep stresses within the allowable limits, which depends on whether the "no cracking" or "controlled cracking" criteria are used. It is desirable to use the same lifting devices for both stripping and erection; however, additional devices may be required to rotate the member to its final position.

Panels that are stripped by rotating about one edge with lifting devices at the opposite edge will develop moments as shown in Fig. 5.2.4. When panels are stripped this way, care should be taken to prevent spalling of the edge along which rotation occurs. A compressible material or sand bed will help protect this edge.

Members that are stripped flat from the mold will develop the moments shown in Fig. 5.2.5.

To determine stresses in flat panels for either rotation or flat stripping, the calculated moment may be assumed to be resisted by the effective widths shown.

In some plants, tilt tables or turning rigs are used to reduce stripping stresses, as shown in Fig. 5.2.6.

When a panel is ribbed or is of a configuration or size such that stripping by rotation or tilting is not practical, vertical pick-up points on the top surface can be used. These lift points should be located to minimize the tensile stresses on the exposed face.

Since the section modulus with respect to the top and bottom faces may not be the same, the designer must select the controlling design limitation:

1. Tensile stresses on both faces to be less than that which would cause cracking.

2. Tensile stress on one face to be less than that which would cause cracking, with controlled cracking permitted on the other face.

3. Controlled cracking permitted on both faces.

If only one of the faces is exposed to view, this face will generally control the design.

Fig. 5.2.3 Example of post-tensioned wall panel

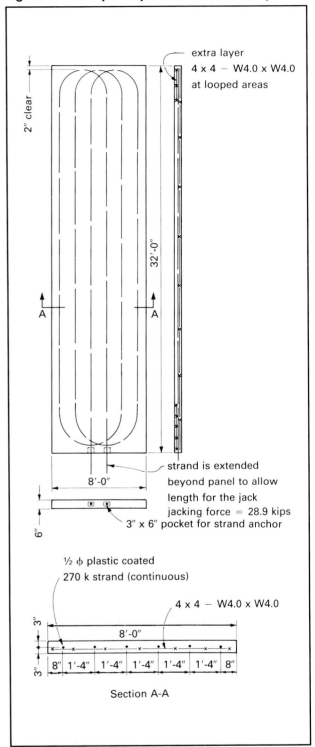

Lift line forces for a two point lift, using inclined lines, is shown in Fig. 5.2.7. When the angle of lift is small, the component of force parallel to the longitudinal axis may generate a significant moment. While this effect can and should be accounted for, it is not recommended that it be allowed to dominate design moments. Rather, consideration should be given to using spreader beams, two cranes or other

Fig. 5.2.4 Moments developed in panels stripped by rotating about one edge

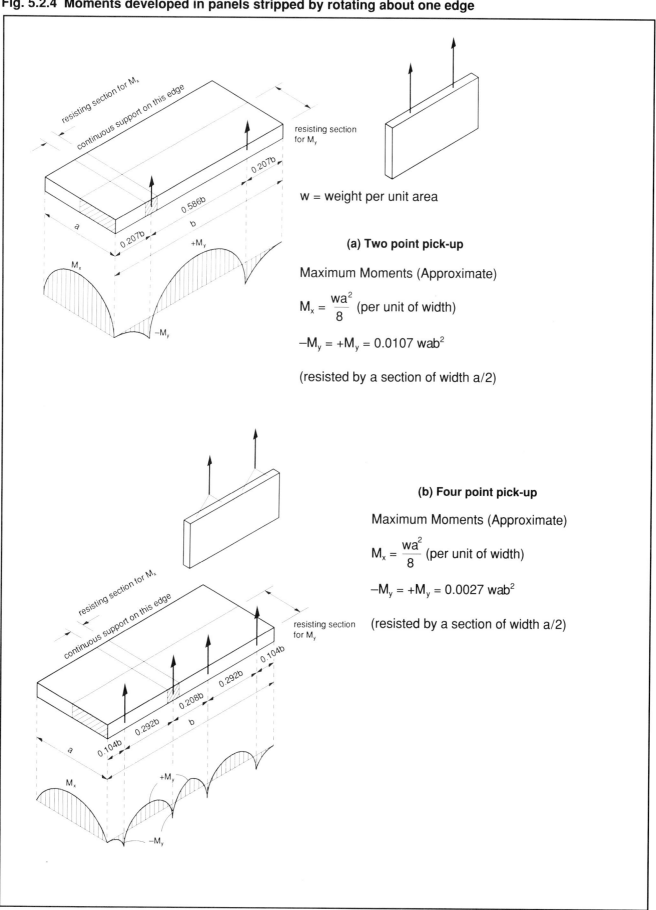

w = weight per unit area

(a) Two point pick-up

Maximum Moments (Approximate)

$$M_x = \frac{wa^2}{8} \text{ (per unit of width)}$$

$$-M_y = +M_y = 0.0107\, wab^2$$

(resisted by a section of width $a/2$)

(b) Four point pick-up

Maximum Moments (Approximate)

$$M_x = \frac{wa^2}{8} \text{ (per unit of width)}$$

$$-M_y = +M_y = 0.0027\, wab^2$$

(resisted by a section of width $a/2$)

Fig. 5.2.5 Moments developed in panels stripped flat

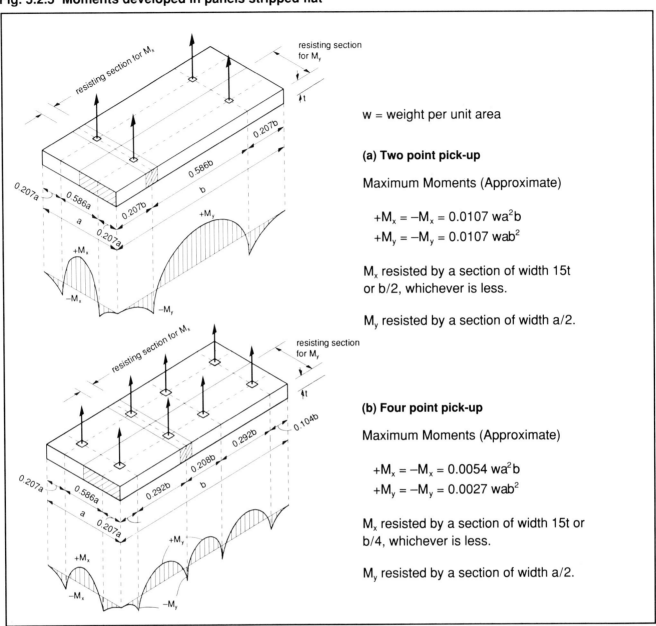

w = weight per unit area

(a) Two point pick-up

Maximum Moments (Approximate)

$+M_x = -M_x = 0.0107\ wa^2b$
$+M_y = -M_y = 0.0107\ wab^2$

M_x resisted by a section of width 15t or b/2, whichever is less.

M_y resisted by a section of width a/2.

(b) Four point pick-up

Maximum Moments (Approximate)

$+M_x = -M_x = 0.0054\ wa^2b$
$+M_y = -M_y = 0.0027\ wab^2$

M_x resisted by a section of width 15t or b/4, whichever is less.

M_y resisted by a section of width a/2.

---

mechanisms to increase the angle of lift. Any such special handling requirements should be clearly shown on the shop drawings.

In addition to longitudinal bending moments, a transverse bending moment may be caused by the orientation of the pick-up points with respect to the transverse dimension (Fig. 5.2.8). For the section shown, a critical moment could occur between the ribs because of the thin cross-section.

The design guidelines listed above apply to elements of constant cross-section. For elements of varying cross-section, the location of lift points is usually determined by trial and error. Rolling blocks can be used on long elements of varying section (Fig. 5.2.10), which makes the forces in the lifting lines equal. The member can then be analyzed as a beam with varying load supported by equal reactions.

The force in inclined lift lines can be determined from Fig. 5.2.7.

5.2.8 Handling Devices

The most common lifting devices are prestressing strand or cable loops projecting from the concrete, threaded inserts, or special proprietary devices.

Since lifting devices are subject to dynamic loads, ductility of the material is part of the design

Fig. 5.2.6 Stripping from a tilt table

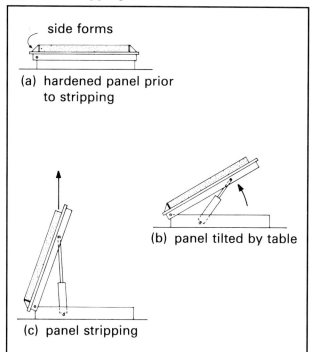

(a) hardened panel prior to stripping

(b) panel tilted by table

(c) panel stripping

Fig. 5.2.7 Determination of force in inclined lift lines

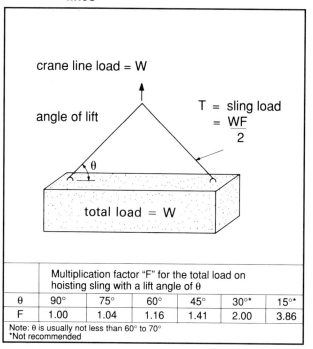

Multiplication factor "F" for the total load on hoisting sling with a lift angle of θ						
θ	90°	75°	60°	45°	30°*	15°*
F	1.00	1.04	1.16	1.41	2.00	3.86

Note: θ is usually not less than 60° to 70°
*Not recommended

Fig. 5.2.8 Pick-up points for equal stresses of a ribbed member

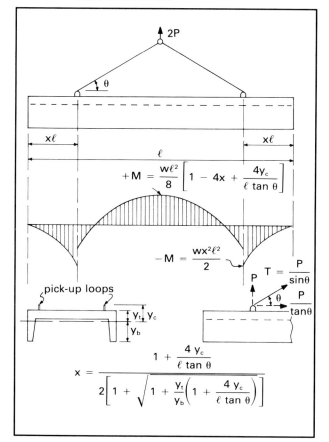

$$+M = \frac{w\ell^2}{8}\left[1 - 4x + \frac{4y_c}{\ell \tan\theta}\right]$$

$$-M = \frac{wx^2\ell^2}{2}$$

$$T = \frac{P}{\sin\theta}$$

$$\frac{P}{\tan\theta}$$

$$x = \frac{1 + \frac{4y_c}{\ell \tan\theta}}{2\left[1 + \sqrt{1 + \frac{y_t}{y_b}\left(1 + \frac{4y_c}{\ell \tan\theta}\right)}\right]}$$

Fig. 5.2.9 Moments caused by eccentric lifting

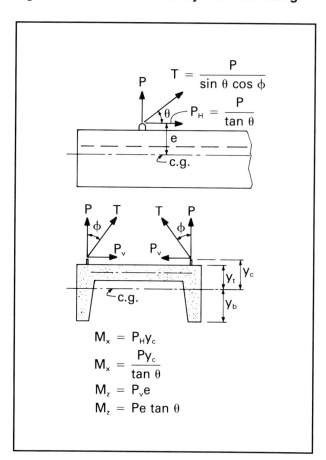

$$T = \frac{P}{\sin\theta \cos\phi}$$

$$P_H = \frac{P}{\tan\theta}$$

$$M_x = P_H y_c$$

$$M_x = \frac{P y_c}{\tan\theta}$$

$$M_z = P_v e$$

$$M_z = Pe \tan\theta$$

Fig 5.2.10 Arrangement for equalizing lifting loads

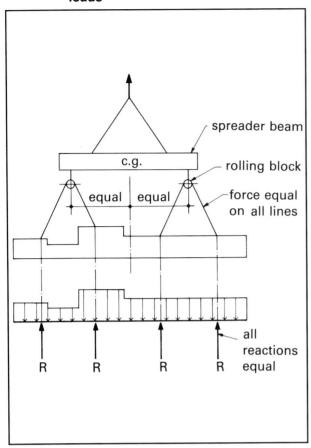

requirement. Deformed reinforcing bars should not be used since the deformations result in stress concentrations from the shackle pin. Also, reinforcing bars may be hard grade or re-rolled rail steel with little ductility and low impact strength at cold temperatures. Smooth bars of a known steel grade may be used if adequate embedment or mechanical anchorage is provided. The diameter must be such that localized failure will not occur by bearing on the shackle pin.

Prestressing strand, both new and used, may be used for lifting loops. The capacity of a lifting loop embedded in concrete is dependent upon the length of embedment, the condition of the strand, the diameter of the loop, and the strength of the concrete. Precast producers' tests and experience offer the best guidelines as to the load limit to apply. A safety factor of 4 is recommended against breakage or slippage. In the absence of tests or experience, it is recommended that the safe load on a single ½-in.dia.-270K strand loop not exceed 10 kips, with a minimum embedment of 16 in. The safe working load of multiple loops may be obtained by multiplying the safe load for one loop by the number of loops. To avoid overstress in one loop when using multiple loops, care should be taken in the fabrication to ensure that all strands are bent the same. Thin-wall conduit over the strands in the region of the bend has been used to reduce this overstress.

Table 5.2.3 Safe working load of 7 x 19 aircraft cable used as lifting loops[1]

Diameter (in.)	Safe Load (kips)[2]
3/8	3.6
7/16	4.4
1/2	5.7

1. 7 strands with 19 wires each.
2. Based on a single strand with a factor of safety of 4 applied to the minimum breaking strength of galvanized cable. The user should consider embedment, loop diameter and other factors discussed for strand loops. Aircraft cables are usually wrapped around reinforcement in the precast product.

Some precast concrete producers also use aircraft cable and standard wire rope for lifting devices (see Table 5.2.3).

Lifting heavy members with threaded inserts should be carefully assessed. When properly designed for both insert and concrete capacities, threaded inserts have many advantages. However, correct usage is sometimes difficult to inspect during handling operations. In order to ensure that an embedded insert acts primarily in tension, a swivel plate as indicated in Fig. 5.2.11 should be used. Sufficient threads must be engaged to develop the strength of the bolt. Some manufacturers use a long bolt with a nut between the bolt head and the swivel plate. After the bolt has "bottomed out", the nut is turned against the swivel plate.

Connection hardware should not be used for lifting or handling any but the lightest units, unless approved by the designer. In order to prevent field error, inserts used for lifting should be of a different type or size than those used for the final connection.

Example 5.2.1 Stripping forces

Given:

A ribbed wall panel as shown. Span ℓ = 40'-0".
f'_{ci} at stripping = 3000 psi (normal weight).

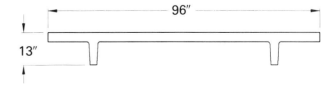

Fig. 5.2.11 Swivel plate

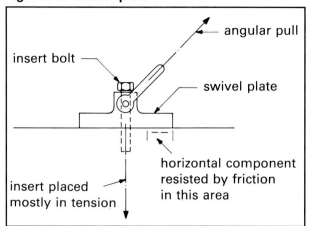

Problem:

Determine stripping forces and corresponding stresses.

Solution:

Half section properties:
y_b = 10.18 in.
y_t = 2.81 in.
Z_b = 156 in.³
Z_t = 565 in.³
wt.= 192 plf
$\dfrac{y_t}{y_b}$ = 0.276

Stripping load:

Assume a load multiplier of 1.4 (Table 5.2.1)
w = 1.4(192) = 269 plf

Case No. 1— Neglecting the moment due to eccentric pick-up points.

Pick-up location for equal tension on each face Fig. 5.2.8, with y_c = 0:

$$x = \dfrac{1}{2\left[1+\sqrt{1+\dfrac{y_t}{y_b}}\right]} = \dfrac{1}{2(1+\sqrt{1.276})}$$

$$= 0.235$$

$$-M = \dfrac{wx^2\ell^2}{2} = \dfrac{0.269}{2}(0.235 \times 40)^2$$

$$= 11.88 \text{ ft-kips}$$

$$f_t = \dfrac{-M}{Z_t} = \dfrac{11,880(12)}{565} = 252 \text{ psi}$$

$$< 5\sqrt{3000} = 274 \text{ psi} \quad \text{OK}$$

Case No. 2— Accounting for moments due to eccentric pick-up points

For θ = 45 deg

Assume: $y_c = y_t + 3"$ = 5.81 in. (Fig. 5.2.8)

$$\dfrac{4y_c}{\ell\tan\theta} = \dfrac{4(5.81)}{12(40)\tan 45} = 0.048$$

$$x = \dfrac{1+\dfrac{4y_c}{\ell\tan\theta}}{2\left[1+\sqrt{1+\dfrac{y_t}{y_b}\left(1+\dfrac{4y_c}{\ell\tan\theta}\right)}\right]}$$

$$= \dfrac{1.048}{2\left[1+\sqrt{1+0.276(1.048)}\right]} = 0.245$$

$$-M = \dfrac{0.269}{2}(0.245 \times 40)^2 = 12.92 \text{ ft-kips}$$

$$f_t = \dfrac{12,920(12)}{565} = 274 \text{ psi} = 5\sqrt{3000} \quad \text{OK}$$

Example 5.2.2 Locating pick-up points

Given:

The window unit shown is to be cast face down and stripped vertically.

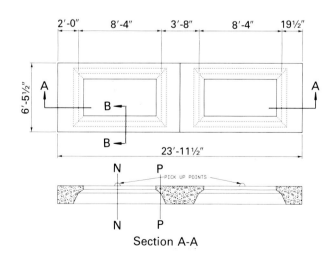

Problem:

Locate pick-up points to minimize tensile stresses in the concrete.

Solution:

Dead load of member—assume 1.6 multiplier (Table 5.2.1):

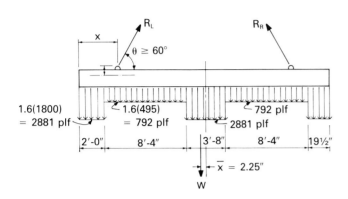

$W = 16.67(792) + 7.292(2881) = 34{,}211$ lb

Lifting loops should be placed symmetrically about the center of gravity of the member.

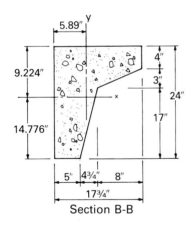

Section B-B

Assume critical cracking stress will occur in the narrow sections of the unit (Section B-B):

Section Properties

$A = 237.6$ in.^{2}

$I = 10{,}969$ in.^{4}

$y_t = 9.224$ in.

$y_b = 14.776$ in.

For equal stresses on each face:

$$\frac{-My_t}{I} = \frac{+My_b}{I}$$

$$-M = \frac{y_b}{y_t}(+M) = \frac{14.776}{9.224}(+M) = 1.60(+M)$$

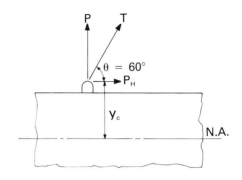

$$P_H = \frac{P}{\tan\theta} = \frac{34{,}211/2}{\tan 60} = 9876 \text{ lb}$$

See Fig. 5.2.9

$y_c = y_t + 3 = 12.22$ in.

$$M_x = \frac{12.22(9876)}{12} = 10{,}060 \text{ ft-lb}$$

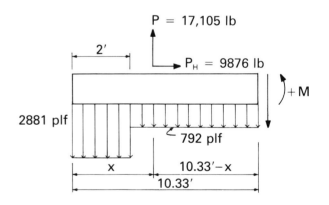

$$+M = 17{,}105(10.33 - x) - 792\frac{(8.33)^2}{2}$$

$$- 2881(2)9.33 + 10{,}060$$

$$= -17{,}105x + 105{,}520$$

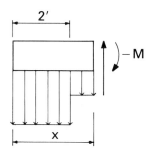

$$-M = 2881(2)(x-1) + 792\frac{(x-2)^2}{2}$$

$$= 396x^2 + 4178x - 4178$$

$$396x^2 + 4178x - 4178$$

$$= 1.60(-17,105x + 105,520)$$

$$x^2 + 91.14x = 499$$

$$x = 5.15 \text{ ft}$$

Use: $x = 5$ ft

$$+M = 105,520 - 17,105(5)$$

$$= 19,996 \text{ ft-lb}$$

$$= 239.9 \text{ in.-kips}$$

$$-M = 396(5)^2 + 4178(5) - 4178$$

$$= 26,614 \text{ ft-lb} = 319.3 \text{ in.-kips}$$

$$f_t = \frac{(-M)y_t}{I} = \frac{319,300(9.224)}{2(10,969)} = 134 \text{ psi}$$

$$f_b = \frac{(+M)y_b}{I} = \frac{239,900(14.776)}{2(10,969)} = 162 \text{ psi}$$

Using an allowable stress of $5\sqrt{f'_{ci}}$, a stripping strength as low as 1050 psi would theoretically be permitted.

To illustrate a "controlled cracking" design allowing a crack width of 0.005 in., assume the moments on each narrow section are:

$$-M = 270 \text{ in.-kips}$$

$$+M = 165 \text{ in.-kips}$$

Referring to Sect. 4.2.2.1:

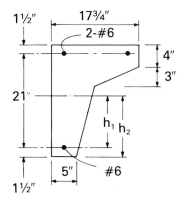

$d = 22.5$ in.

For +M:

$h_1 = 14.78 - 1.5 = 13.28$ in.

$h_2 = 14.78$ in.

$h_2/h_1 = 14.78/13.28 = 1.11$

$A \approx 5(2)(1.5) = 15$ in.^{2}

$d_c = 1.5$ in.

From Eq. 4.2.9 with $w = 0.005$ in.

$$f_s = 21.0 \text{ ksi}$$

From Eq. 4.2.10:

$$A_s = \frac{+M}{0.9f_s d} = \frac{165}{0.9(21.0)(22.5)} = 0.39 \text{ in.}^2$$

$$\rho_{s\ min} = \frac{200}{f_y} = \frac{200}{60,000} = 0.00333$$

$$A_{s\ min} = 0.0033(5)22.5 = 0.375 \text{ in.}^2$$

Use 1 – #6 bar $A_s = 0.44$ in.^{2}

For –M:

$h_2 = 9.22$ in.

$h_1 = 9.22 - 1.5 = 7.72$ in.

$h_2/h_1 = 9.22/7.72 = 1.19$

For 2 bars,

$A = 17.75(2)(1.5)/2 = 26.6$ in.^{2}

$f_s = 16.2$ ksi

$$A_s = \frac{-M}{0.9f_s d} = \frac{270}{0.9(16.2)(22.5)} = 0.82 \text{ in.}^2$$

Use 2 – #6 bars, $A_s = 0.88$ in.^{2}

5.2.9 Lateral Stability

Prestressed members generally are sufficiently stiff as to preclude lateral bucking. However, during handling and transportation, support flexibility may result in lateral roll of the beam, thus producing lateral bending.

The following procedure is developed in detail in Ref. 5. The equilibrium conditions for a hanging beam are shown in Fig. 5.2.12.

When a beam hangs from lifting points, it may roll about an axis through the lifting points. The safety and stability of long beams subject to roll is shown in Ref. 5 to be dependent upon:

1. e_i, initial lateral eccentricity of the gravity center with respect to the roll axis.
2. y_t, height of the roll axis above the gravity center of the beam.
3. z_o, theoretical lateral deflection of the center of gravity of the beam, computed with the full weight applied as a lateral load, measured to the center of gravity of the deflected arc of the beam.
4. θ_{max}, maximum permissible tilt angle of the beam.

For a beam with overall length ℓ and equal overhangs a at each end:

$$z_o = \frac{w}{12EI_y\ell}\left[0.1(\ell_1)^5 - a^2(\ell_1)^3 + 3a^4(\ell_1) + \frac{6}{5}(a^5)\right]$$

(Eq. 5.2.2)

for a beam with no overhangs (a = 0, $\ell_1 = \ell$):

$$z_o = \frac{w(\ell)^4}{120EI_y} \quad \text{(Eq. 5.2.3)}$$

As a reasonable approximation, z_o may be taken as ⅔ of the weak axis maximum deflection of the beam.

For a hanging beam without initial lateral eccentricity, the factor of safety is given by:[5]

$$SF = \frac{y_t}{z_o} \quad \text{(Eq. 5.2.4)}$$

Adopting a minimum safety factor of 2.5 for safe handling and transportation gives $y_t = 2.5z_o$. For beams with initial lateral imperfections, consideration may have to be given separately to the maximum permitted roll angle θ_{max}. This is treated in detail in Ref. 5.

Camber raises the centroid of the member, thus decreasing y_t. This can be approximately accounted for by assuming the centroid of mass is shifted upward by ⅔ of the midspan camber (i.e., $y_t = y_t - ⅔ \times$ camber).

For safe handling of long members, resistance can be improved by several methods. Listed in order of effectiveness and relative ease of accomplishment:

1. Move the lifting points inward (see Fig. 5.2.13). Decreasing the distance between lifting points by just a small amount can significantly increase the safety factor. Stresses must be checked; temporary post-tensioning can be introduced to control stresses.
2. Increase the distance between the center of mass and the lifting point (y_t) by use of a rigid yoke.
3. Provide temporary lateral bracing, in the form of stiffening trusses composed of structural steel shapes.
4. Revise the shape of the member.
5. Increase the stiffness of the member by increasing the concrete strength (and thus E_c).

In the previous edition of the PCI Handbook, a simplified analysis method was used. In that method, the lateral deflection β_y of the beam was calculated assuming the total beam weight acting as a uniformly distributed horizontal load. The distance y_t from the centroid to the top of the beam was set as $> 2\beta_y$, in order to provide a safety factor of 2. A more accurate analysis demonstrates that this method was conservative and that the safety factor was about 3.

The benefit gained by reducing the distance between supports is illustrated in Fig. 5.2.13.

Example 5.2.3 Lifting point locations

Given:

PCI BT-72 bridge beam, 136 ft. long. Determine lifting point locations as governed by lateral bending stability.

ℓ = 136 ft, w = 825.6 lb/ft, I_y = 37,634 in.[4] E_c = 4070 ksi, y_t = 35.4 in. For lifting at ends:

$$z_o = \frac{w\ell^4}{120E_cI_y} = \frac{0.8256 \times 136^4 \times 1728}{120 \times 4070 \times 37,634} = 26.55 \text{ in.}$$

Camber ≈ 4 in.

$$y_t = 35.4 - \frac{2}{3}(4) = 32.73 \text{ in.}$$

$$SF = \frac{y_t}{z_o} = \frac{32.73}{26.55} = 1.23 \quad \text{Too low}$$

For SF = 2.5, z_o = 32.73/2.5 = 13.09

$z_o / (z_o$ when a = 0) = 13.09 / 26.55 = 0.49

Enter Fig. 5.2.13 a = 0.062 (136) = 8.43 ft

Use a = 9 ft. Locate lifting loops 9 ft from each end.

Fig. 5.2.12 Equilibrium of beam in tilted position

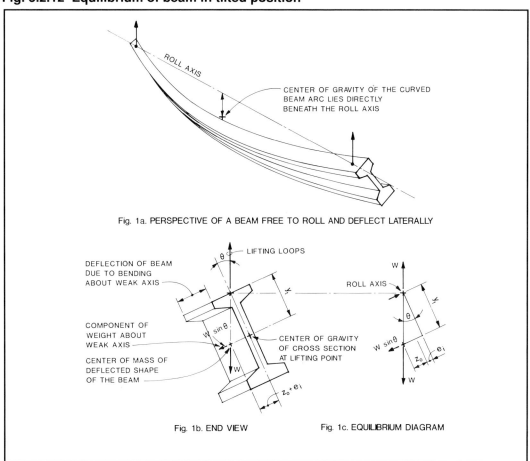

Fig. 5.2.13 Reduction of z_o with overhangs

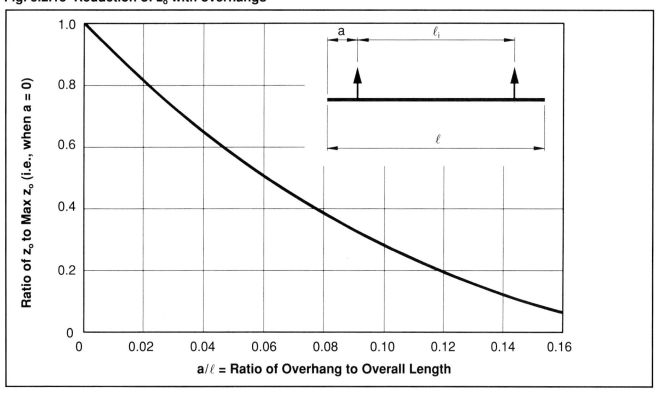

5.2.10 Storage

Wherever possible, an element should be stored on only two points of support located at or near those used for stripping and handling. Thus, the design for stripping and handling will usually control. Where points other than those used for stripping or handling are used for storage, the storage condition must be checked.

If support is provided at more than two points, and the design based on more than two supports, precautions must be taken so that the element does not bridge over one of the supports due to differential support settlement. Particular care must be taken for prestressed elements, with consideration made for the effect of prestressing. Designing for equal stresses on both faces will help to minimize deformations in storage.

Fig. 5.2.14 Panel warpage in storage

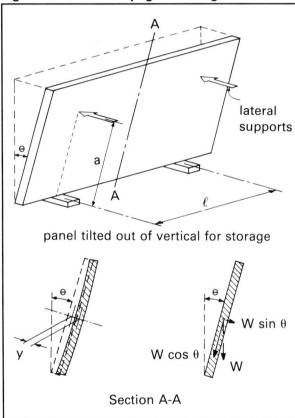

Warpage in storage may be caused by temperature or shrinkage differential between surfaces, creep and storage conditions. Warpage cannot be totally eliminated, although it can be minimized by providing blocking so that the panel remains plane. Where feasible, the member should be oriented in the yard so that the sun does not overheat one side. (See Sect. 3.3.2 for a discussion of thermal bowing.) Storing members so that flexure is resisted about the strong axis will minimize stresses and deformations.

For the support conditions shown in Fig. 5.2.14, warping can occur in both directions. By superposition, the total instantaneous deflection, y_{max}, at the maximum point can be estimated by:

$$y_{max} = \frac{5w \sin\theta}{384E_c}\left[\frac{a^4}{I_c} + \frac{\ell^4}{I_b}\right] \quad \text{(Eq. 5.2.5)}$$

where:

w = panel weight, lb/in.^{2}

E_c = modulus of elasticity of concrete, psi

a = panel support height, in.

ℓ = horizontal distance between supports, in.

I_c, I_b = moment of inertia of uncracked section in the respective directions for 1 in. width of panel, in.^{4}

Fig. 5.2.15 Effect of compression reinforcement on creep

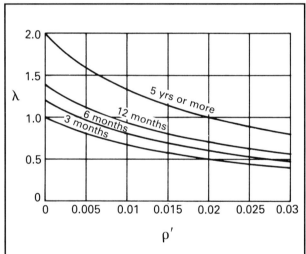

This instantanous deflection should be modified by a factor to account for the time-dependent effects of creep and shrinkage. ACI 318-89 suggests the total deformation, y_t, at any time can be estimated as:

$$y_t = y_{max}(1 + \lambda) \quad \text{(Eq. 5.2.6)}$$

where:

y_t = time dependent displacement

y_{max} = instantaneous displacement

λ = amplification due to creep and shrinkage (Fig. 5.2.15)

ρ = reinforcement ratio for nonprestressed compression reinforcement, A'_s/bt

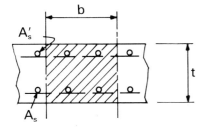

5.2.11 Transportation

The method used for transporting precast concrete products can affect the structural design because of size and weight limitations and the dynamic effects imposed by road conditions.

Except for long prestressed deck members, most products are transported on either flatbed or low-boy trailers. These trailers deform during hauling. Thus, support at more than two points can be achieved only after considerable modification of the trailer, and even then results may be doubtful.

Size and weight limitations vary from one state to another, so a check of local regulations is necessary when large units are moved. Loads are further restricted on some secondary roads during spring thaws.

The common payload for standard trailers without special permits is 20 tons with width and height restricted to 8 ft and length to 40 ft. Low-boy trailers permit the height to be increased to about 10 to 12 ft. However, low-boys cost more to operate and have a shorter bed length. In some states, a total height (roadbed to top of load) of 13 ft 6 in. is allowed without special permit. This height may require special routing to avoid low overpasses and overhead wires.

Maximum width with permit varies among states, and even among cities—from 10 to 14 ft. Some states allow lengths over 70 ft with only a simple permit, while others require, for any load over 55 ft, a special permit, escorts front and rear and travel limited to certain times of the day. In some states, weights of up to 100 tons are allowed with permit, while in other states there are very severe restrictions on loads over 25 tons.

These restrictions add to the cost of precast concrete units, and should be compared with savings realized by combining smaller units into one large unit. When possible, a precast unit, or several units combined, should approximate the usual payload of 20 tons. For example, an 11 ton unit may not be economical, because only one can be shipped per load, while two 10 ton units could be shipped on one load.

Erection is facilitated when members are transported in the same orientation they will have in the structure. For example, single-story wall panels can be transported on A-frames with the panels upright (Fig. 5.2.16). A-frames also provide good lateral support and the desired two points of vertical support. Longer units can be transported on their sides to take advantage of the increased stiffness compared with flat shipment (Fig. 5.2.17). In all cases, the panel support locations should be consistent with the panel design. Panels with large openings frequently require strongbacks, braces or ties to keep stresses within the design values (Fig. 5.2.18).

During transportation, units are usually supported with one or both ends cantilevered. For members not symmetrical with respect to the bending axis, the expressions given in Fig. 5.2.19 can be used for determining the location of supports to give equal tensile stresses.

Fig. 5.2.16 Transportation of single-story panels

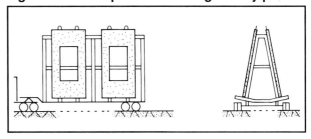

Fig. 5.2.17 Transportation of multi-story panels

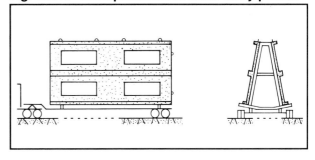

Fig. 5.2.18 Methods of temporary strengthening of panels with significant openings

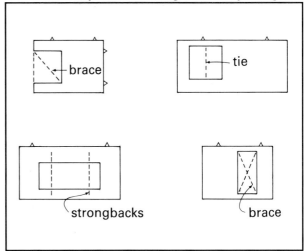

Fig. 5.2.19 Equations for equal tensile stresses top and bottom—unsymmetrical members

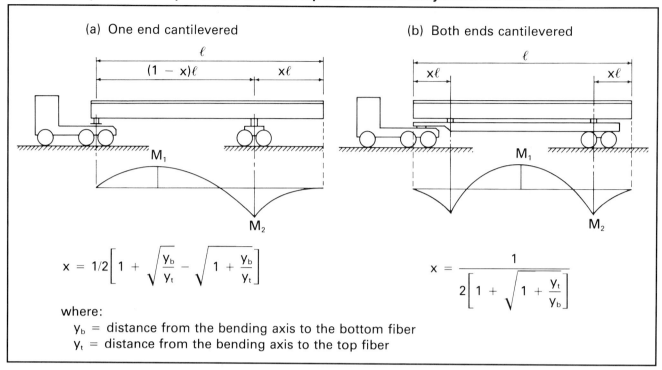

where:
y_b = distance from the bending axis to the bottom fiber
y_t = distance from the bending axis to the top fiber

5.2.12 Erection

Precast concrete members frequently must be re-oriented from the position used to transport to that which it will be in the final construction. The analysis for this "tripping" (rotating) operation is similar to that used during other handling stages. Fig. 5.2.20 shows maximum moments for several commonly used tripping techniques.

When using two crane lines the center of gravity must be between them in order to prevent a sudden shifting of the load while it is being rotated. To ensure that this is avoided, the stability condition shown in Fig. 5.2.21 must be met. The capacities of lifting devices must be checked for the forces imposed during the tripping operation, since the directions vary.

5.2.13 Design Example—Flat Panel

This example illustrates the use of many of the recommendations in this section. It is intended to be illustrative and general only. Each plant will have its own preferred methods of manufacture.

Given:

A flat panel used as a non-load bearing facade on a two-story structure, as shown in Fig. 5.2.22.

Wind Load—20 psf pressure or suction
f'_c = 5000 psi @ 28 days
f'_{ci} = 2000 psi @ stripping

Cracks in rear face permitted without width restriction. Crack width in exposed face limited to 0.005 in.

Solution:

Establish Handling Procedures

Casting: Face down. Use same mix for exposed aggregate surface (retarded) and smooth white side bands. Use gray concrete backup.

Fig. 5.2.21 Stability during erection

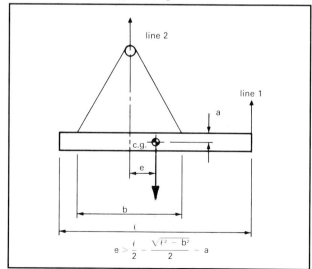

Fig. 5.2.20 Typical tripping (rotating) positions for erection of wall panels

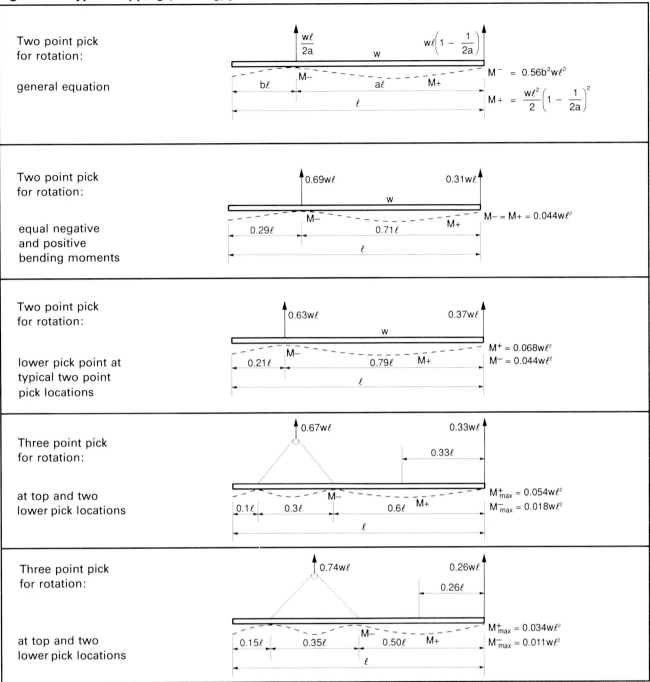

Stripping: Due to edge detail and inside crane headroom, panel cannot be turned on edge directly in mold, therefore strip flat and move to sand bed (or turning equipment) for turning.

Storage: Since panels will be stored for several months and storage yard is subject to settlement, thus negating possible four-point support, and since bowing must be avoided, store on edge.

Determine Handling Multipliers (Table 5.2.1)

Stripping: Exposed flat surface has deep exposure (heavy retarder); side rails removed prior to stripping; drafts on edge detail are good: use 1.2.

Use strand loops in back of panel (plant practice).

Yard handling: Turning: use 1.2; transport to storage: use 1.2.

Shipping: Distance traveled is 150 miles and jobsite roadways are bumpy: use 1.7 (exceeds the value 1.5 recommended in Table 5.2.1 since travel conditions are considered rather severe and cracking is limited).

Fig. 5.2.22 Example of Sect. 5.2.13

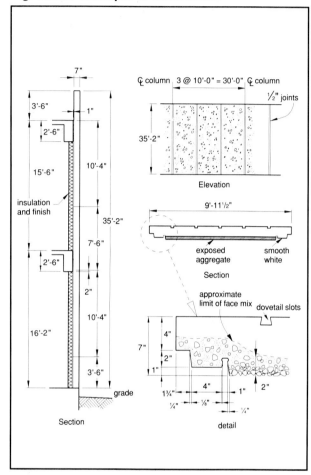

Erection: Use 1.5 (based on engineer's judgment instead of value from Table 5.2.1).

Handling devices: Use 4 (from Sect. 5.2.5).

Section Properties (use 7" thick x 9'-11½" wide)

$A = 836$ in.^{2}

$Z_b = Z_t = 976$ in.^{3}

$I = 3416$ in.^{4}

Unit weight @ 150 pcf = 87 psf, or 870 plf

Total weight = 30.6 kips

Establish Allowable Tensile Stresses

@ stripping, yard handling and storage:

$f'_r = \dfrac{5\sqrt{2000}}{1000} = 0.224$ ksi (from Sect. 5.2.4)

@ shipping and erection:

$f'_r = \dfrac{5\sqrt{5000}}{1000} = 0.354$ ksi

Check Handling Stresses—Stripping

a. Longitudinal bending

Two point pick-up, Fig. 5.2.5(a)

$a = 10.0$ ft $b = 35.2$ ft
$a/2 = 60$ in.

$Z = \dfrac{60(7)^2}{6} = 490$ in.^{3}

$M_y = 0.0107\, wab^2$
$= 0.0107(0.087)(10)(1.2)(35.2)^2(12)$
$= 166$ in.-kips

$f_t = f_b = \dfrac{166}{490} = 0.339 > 0.224$ ksi

Therefore, 2-point stripping no good

Four point pick-up, Fig. 5.2.5(b)

$M_y = 0.0027\, wab^2$
$= 0.0027(0.087)(10)(1.2)(35.2)^2(12)$
$= 41.9$ in.-kips

Additional moment due to lifting angle (Fig. 5.2.9)

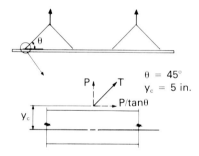

$\theta = 45°$
$y_c = 5$ in.

$M_y = \dfrac{Py_c}{\tan\theta} = \left[\dfrac{0.87(1.2)(35.2)}{4(2)}\right]\dfrac{5}{\tan 45°}$

$M_y = 23$ in.-kips

$M_{total} = 41.9 + 23 = 64.9$ in.-kips

$f_t = f_b = 0.132 < 0.224$ ksi   OK

Use 4-point pick-up for stripping

Note: By inspection of support structure and 4-point pick-up stripping locations, it is determined that shifting lifting loops toward ends slightly will avoid interference between loops and edge beams thus avoiding a delay in erection while loops are burned off. Effect on stresses is minor.

b. Transverse bending—beam strip properties

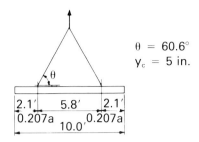

$15t = 105$ in., $b/2 = 211$ in.
$Z = 105(7)^2/6 = 857$ in.^{3}
$M_x = 0.0054(0.087)(1.2)(10)^2(12)(35.2)$
$\quad = 23.8$ in.-kips

$$M_x \text{(lifting angle)} = \left[\frac{1.2(0.087)(10)(35.2)}{(4)(2)}\right]$$
$$\times \frac{5}{\tan 60.6°} = 12.9 \text{ in.-kips}$$

Total $M_x = 23.8 + 12.9 = 36.7$ in.-kips
$f_t = f_b = 36.6/857 = 0.043$ ksi < 0.224 ksi OK

Final stripping loop locations:

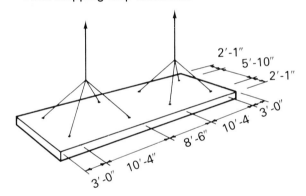

Check Handling Stresses—Turning:

Stresses for turning from edge to edge are excessive. Note also that edge detail may restrict this type of turning, due to excessive shear loading on insert cast into edge.

Therefore turn as follows:

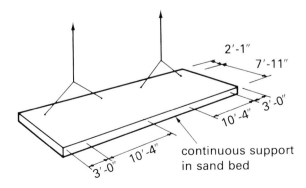

Transverse moments:

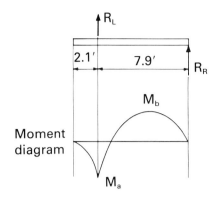

$w = 0.087(1.2)(35.2/4) = 0.92$ k/ft

$$R_L = \frac{0.92(10^2/2)}{7.9} = 5.81 \text{ kips}$$

$R_R = 0.92(10) - 5.81 = 3.39$ kips
$M_a = (0.92)(2.1)^2(12)/2 = 24.3$ in.-kips
M_b maximum at $3.39/0.92 = 3.68$ ft
$M_b = [3.39(3.68) - 0.92(3.68)^2/2]12$
$\quad = 74.9$ in.-kips

Using same resisting section as for stripping:
$\quad f_t = 24.3/857 = 0.028$ ksi < 0.224
$\quad f_b = 74.9/857 = 0.087$ ksi < 0.224

Therefore, concrete should have strength of 2000 psi at stripping.

Longitudinal bending similar to stripping.

Use 4-point turning with one edge in sand bed.

Check Handling Stresses—Shipping:

The following factors were considered in determining shipping method:

Alternatives

(1) Ship flat:
- strength with 2-point support
- permits required in 3 states for each load (since panel weight of 30,600 lb restricts loads to one panel each)

(2) Ship vertical:
- requires low-boy trailer with 35 ft well with maximum height of 3 ft which would restrict total height to 13.5 ft (6 in. support material)

(3) Special frame:
- fabricate special frame so panel could be set at about a 45° angle and be within non-permit restrictions of 13.5 ft high by 8.0 ft wide

Other factors
- Total number of pieces—132
- Erection rate—10 pieces per day
- Drivers not permitted to drop loads on job after working hours (union regulations at job site)
- Permit loads on bridge restricted to traveling between 9:30 A.M. and 2:30 P.M.

To avoid disrupting normal production in plant, loading must be done in P.M. (same cranes used for stripping in A.M., yarding and loading).

These considerations led to the conclusion that 30 trailers would be necessary to properly supply the job. This can be demonstrated as follows with each group of trailers being 10 (A,B,C):

	Mon	Tues	Wed	Thurs	Fri
Load	A	B	C	A	B
Ship		A	B	C	A
Erect			A	B	C
Return Trailer			A	B	C

3 groups of 10 required = 30 trailers

Reconsider alternatives

(2) Vertical—30 low-boys not available

(3) Special frames @ 45°—with 1 panel per trailer

Check panel bending about the weak axis.

Bending in longitudinal direction:

$$w = 0.87(1.7)(\sin 45°) = 1.05 \text{ kips/ft}$$

From Fig. 5.2.5 for full panel width:

$$M_y = 2(0.0107)(1.05)(35.2)^2(12)$$
$$= 334 \text{ in.-kips}$$
$$f = 334/976 = 0.342 \text{ ksi} < 0.354 \text{ OK}$$

Since stress is high and a long panel traveling 150 miles is subject to possible dynamic forces, provide a 3rd frame for lateral support only, to avoid possible harmonic motion.

30(3) = 90 frames required
Cost estimated @ $250 each or $22,500

(1) Flat—cost of permits:
$65 x 132 loads = $8,580
possible saving = $13,920

Provide proper support to achieve this saving. Two-point support is desirable since flexing of trailer normally will lower or raise any additional supports thus causing bridging and unanticipated stresses.

4-Point Support

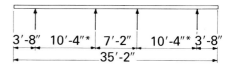

This is too far apart to consider for shipping since the trailer will deflect and cause 2 or 3 point support at times during transportation. Try adjusting support points so that 10'-4" is reduced to 5'-0" which is a practical limit to ensure 4 points will have support at all times during transportation.

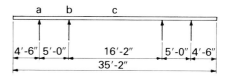

A moment distribution of this condition results in moments as follows:

$M_a = 0.0082w\ell^2 = 0.0082(0.87)(1.7)(35.2)^2(12)$
$= 180.3$ in.-kips
tension in back face

$M_b = 0.014w\ell^2 = 307.8$ in.-kips
tension in back face

$M_c = 0.0125w\ell^2 = 274.8$ in.-kips
tension in front face

Stresses:

$$f_a = \frac{180.3}{976} = 0.185 \text{ ksi} < 0.354 \text{ ksi OK}$$

$$f_b = \frac{307.8}{976} = 0.315 \text{ ksi OK}$$

$$f_c = \frac{274.8}{976} = 0.281 \text{ ksi OK}$$

Ship with supports as shown.

Note: Had stress been excessive, supports could be shifted toward center of panel until cantilever condition is such that it produces a stress of 0.354 ksi.

An alternate solution provides 4-point support for the precast piece with 2 supports on the truck bed as shown:

Check Handling Stresses—Erection:

Try 3-point pick as follows:

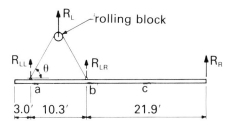

Longitudinal bending: with rolling block, reactions at stripping loops are equal.

$$w = 0.87(1.5) = 1.31 \text{ kips/ft}$$

Distance from R_L to $R_R = 35.2 - 3.0 - \dfrac{10.3}{2}$
$$= 27.0 \text{ ft}$$

$$R_R = 35.2(1.31) - 30.0 = 16.1 \text{ kips}$$

M_c maximum at $16.1/1.31 = 12.3$ ft

$$M_c = [16.1(12.3) - 1.31(12.3)^2/2]12$$
$$= 1187 \text{ in.-kips}$$

$$f = \frac{1187}{976} = 1.2 \text{ ksi} \quad \text{too high}$$

Since this stress is much too high, it is apparent that it cannot be brought within limits by adjusting pick points. Therefore, erect as follows (stresses less critical than stripping):

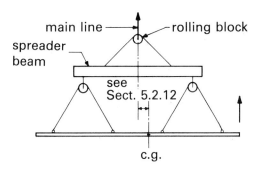

Lift from truck in horizontal position with one line and rotate in air to vertical position with 2nd crane line.

Note: Must use tag line opposing 2nd crane line to avoid rapid movement, or a more suitable solution is to shift lifting points toward bottom of piece to ensure that the main line is below the center of gravity of the piece. This then ensures that a vertical force is required at the top to rotate the piece. See Sect. 5.2.12.

5.3 Erection Bracing

5.3.1 Introduction

This section deals with the temporary bracing which may be necessary to maintain structural stability of a precast structure during construction. When possible, the final connections should be used to provide at least part of the erection bracing, but additional bracing apparatus is frequently required to resist all of the temporary loads. These temporary loads include wind, seismic, eccentric dead loads including construction loads, unbalanced conditions due to erection sequence and incomplete connections. Due to the low probability of design loads occurring during erection, engineering judgment should be used to establish a reasonable load.

Proper planning of the construction process is essential for efficient and safe erection. Sequence of erection must be established early, and the effects accounted for in the bracing analysis and the preparation of shop drawings.

The responsibility for the erection of precast concrete may vary as follows:

1. The precast concrete manufacturer supplies the product erected, either with his own forces, or by an independent erector.

2. The manufacturer is responsible only for supplying the product, F.O.B. plant or jobsite. Erection is done either by the general contractor or by an independent erector under a separate agreement.

3. The products are purchased by an independent erector who has a contract to furnish the complete precast concrete package.

Responsibility for stability during erection must be clearly understood. Design for erection conditions must be in accordance with all local, state and federal regulations. It is desirable that this design be directed or approved by a professional engineer.

Erection drawings define the sequence of erection and the procedure on how to assemble the components into the final structure. The erection drawings must also address the stability of the structure during construction and, where necessary, include temporary connections, shoring, guying and bracing. Additional guidelines are available in Ref. 3. For large and/or complex projects, a pre-job conference prior to the preparation of erection drawings may be warranted, in order to discuss erection methods and to coordinate with other trades.

Handling equipment

The type of jobsite handling equipment selected may influence the erection sequence, and hence affect the temporary bracing requirements. Several types of erection equipment are available, including truck-mounted and crawler mobile cranes, hydraulic cranes, tower cranes, monorail systems, derricks and others. The PCI *Recommended Practice for Erection of Precast Concrete*[3] provides more information on the uses of each.

Surveying and layout

Before products are shipped to the jobsite, a field check of the project should be made to ensure that prior construction is suitable to accept the precast units. This check should include location, line and grade of bearing surfaces, notches, blockouts, anchor bolts, cast-in hardware, and dimensional deviations. Site conditions such as access ramps, overhead electrical lines, truck access, etc., should also be checked. Any discrepancies between actual conditions and those shown on drawings should be corrected before erection is started.

Surveys should be required before, during and after erection:

1. Before, so that the starting point is clearly established and any potential difficulties with the support structure are determined early.
2. During, to maintain alignment.
3. After, to ensure that the products have been erected within tolerances.

5.3.2 Loads

Wind: Wind loads used for erection design should be based on local codes, tempered by engineering judgment. For example, in hurricane regions, the maximum code load is usually not used, since there is sufficient warning of a hurricane so that additional bracing can be installed. Note that it is possible that more surface area is exposed to wind pressure than when the structure is totally enclosed.

Earthquake: In seismic regions, the degree to which earthquake loads are considered for construction design is a decision which should be made by the engineer designing the temporary bracing system, unless the subject is covered by local codes or project specifications. Often seismic loading is neglected unless the project is expected to be shut down in a temporary condition for an extended period of time.

Construction loads: This includes materials stored on floor members such as masonry, drywall or other finishing materials, and construction equipment such as buggies used for concrete placement. A value of 25 psf has been used without creating an excessive burden on the bracing requirements.

Others: Fig. 5.3.1 shows other temporary loading conditions which affect stability and bracing design.

5.3.3 Factors of Safety

Safety factors used for temporary loading conditions are a matter of engineering judgment, and should consider failure mode (brittle or ductile), predictability of loads, quality control of products and construction, opportunity for human error, and economics. The total factor of safety also depends on load factors and capacity reduction factors used in the design of the entire bracing system. These must be consistent with applicable code requirements.

The values shown in Table 5.3.1 are suggested safety factors.

Table 5.3.1 Suggested factors of safety for construction loads

Bracing for wind loads	2
Bracing inserts cast into precast members	3
Reusable hardware	5
Lifting inserts	4

5.3.4 Bracing Equipment and Materials

For most one- and two-story high components that require bracing, steel pipe braces similar to those shown in Fig. 5.3.2 are used. A wide range of bracing types are available from a number of suppliers, who should be consulted for dimensions and capacities. Pipe braces resist both tension and compression. When long braces are used in compression, it may be necessary to provide lateral restraint to the brace to prevent buckling.

Cable guys with turnbuckles are normally used for higher structures. Since wire rope used in cable guys can resist only tension, they are usually used in combination with other cable guys in an opposite direction. Compression struts, which may be the precast concrete components, are needed to complete truss action of the bracing system.

A number of wire rope types are also available.[4] Typically, wire rope is constructed of three basic

Fig. 5.3.1 Temporary loading conditions that affect stability

(a) Columns with eccentric loads from other framing members produce sidesway which means the columns lean out of plumb. Cable or other type of bracing can be used to keep the columns plumb.

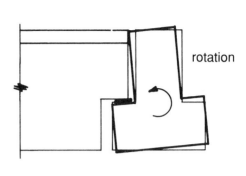

(b) Unbalanced loads due to partially complete erection may result in beam rotation. The erection drawings should address these conditions.

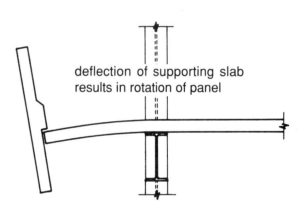

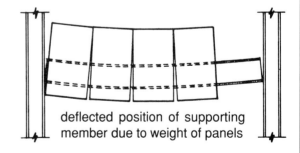

(c) Loading one entire elevation of a structural frame with cladding panels tends to make the building lean out of plumb.

(d) Examples of rotations and deflections of framing members caused by cladding panels. May result in alignment problems and require temporary connections for re-alignment.

components: (1) wires that form the strands, (2) multi-wire strands laid helically around a core and (3) the core (see Fig. 5.3.3).

The wire may be iron, stainless steel, monel, or bronze, but for construction uses it is nearly always high carbon steel. The core, which is the foundation for the wire rope, is made of either fiber or steel. The most commonly used cores are: fiber core (FC), independent wire rope core (IWRC), and wire strand core (WSC).

Rope is classified by the construction type. For example, a 6 x 7 FC consists of 6 strands of 7 wires each, wrapped around a fiber core. A 6 x 19 IWRC has 6—19-wire strands wrapped around an independent wire rope core. Strength of wire rope is dependent on the component materials. Grades include: traction steel (TX), mild plow steel (MPS), plow steel (PS), improved plow steel (IPS), and extra improved plow steel (EIPS).

Table 5.3.2 shows the properites of several commonly used wire rope sizes. It is recommended that the minimum size rope used for bracing be ½ in. diameter.

The elongation or "stretch" of wire ropes must be considered in designing bracing. Elongation comes from two sources: constructional stretch and elastic stretch. Constructional stretch is dependent on the classification and results primarily from a reduction in diameter as load is applied and the strands compact against each other. Approximate ranges of constructional stretch are shown in Table 5.3.2. Wire ropes may be pre-stretched to remove some of the constructional stretch.

Elastic stretch is caused by the deformation of the metal itself when load is applied. As with constructional stretch, a precise value is difficult to establish, but the following equation gives adequate results:

$$\text{Elastic stretch} = \frac{PL}{AE} \qquad \text{(Eq. 5.3.1)}$$

where:

P = change in load
L = length
A = area of wire rope
E = modulus of elasticity

Example 5.3.1 Stretch of wire rope*

Given:

A 70 ft long, ¾ in. diameter, 6 x 7 FC wire rope resisting a tension force of 12 kips.

Problem:

Determine the total stretch.

Solution:

Constructional stretch (use 0.75%) =
0.0075(70)(12) = 6.3 in.

$$\text{Elastic stretch} = \frac{12(70)(12)}{0.288(10,000)} = 3.5 \text{ in.}$$

Total = 6.3 + 3.5 = 9.8 in.

*So that stretch does not result in unacceptable movement of the braced structure, readjustment of the wire rope tension may be necessary.

Fig. 5.3.2 Typical pipe braces

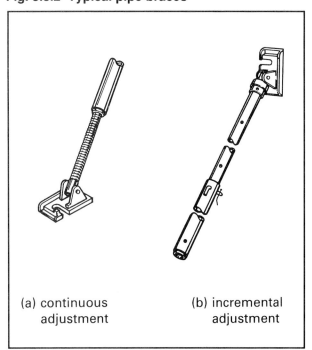

(a) continuous adjustment (b) incremental adjustment

Fig. 5.3.3 Wire rope

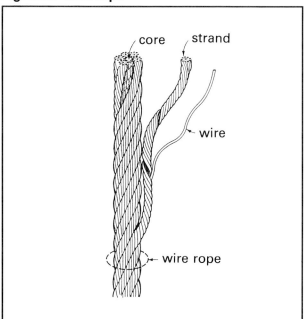

5.3.5 Erection Analysis

The following examples demonstrate a suggested procedure to ensure structural stability and safety during various stages of construction. Actual loads, factors of safety, equipment used, etc., must be evaluated for each project.

Example 5.3.2—Load bearing wall panel structure

Given:

18-story hotel with plan as shown in Fig. 5.3.4.
Floor to floor—8'-8."
Floor system—8 in. hollow-core planks.
Wind load—15 psf. The code requires higher loads at upper levels, but the engineer has judged that for temporary conditions, 15 psf is adequate.
No expansion joint.
Final stability in the east-west direction depends on the stair and elevator walls at the ends, plus the exterior precast panels on the north and south walls (lines A and D). Diaphragm action of the floor distributes lateral loads to these shear walls.

Problem:

Determine sequence of erection and temporary bracing system.

Solution:

After consultation between precaster and erector, it has been decided to erect the floors and load bearing walls through the sixth floor one floor at a time. Then a second crew and lighter crane will be used to erect the precast wall panels on lines A and D. Erection of structural precast floors and walls will never be more than six levels ahead of the shear walls, lines A and D. Design tasks include:

1. Design single panel with two braces at upper level of any erection phase.
2. Determine diaphragm loads and check if elevator walls can resist these for six levels. If not, reduce the number of levels the floor erection can be ahead of the shear wall placement.

Design single panel:

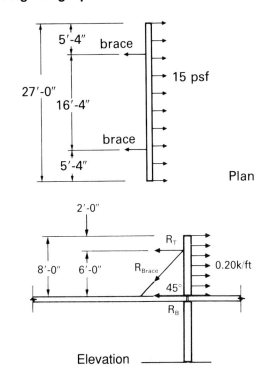

Table 5.3.2 Properties of wire rope

Nominal diameter in.	Fiber Core (FC) E = 10,000 ksi Constructional stretch: 0.5–0.75%		Wire Core (IWRC) E = 13,000 ksi Constructional stretch: 0.25–0.5%	
	Area, in²	Nominal strength (kips)	Area, in²	Nominal strength (kips)
½	0.096	20.6	0.113	22.2
⅝	0.150	31.8	0.176	34.2
¾	0.288	45.4	0.254	48.8
⅞	0.294	61.4	0.345	66.0
1	0.384	79.4	0.451	85.4
1⅛	0.486	99.6	0.571	107.0
1¼	0.600	122.0	0.705	131.2
1⅜	0.726	146.2	0.853	157.2
1½	0.864	172.4	1.015	185.4

Properties based on 6 x 7 classification, 6 x 19 classification approximately 4% higher.
Based on "Improved Plow Steel." "Extra Improved Plow Steel" approximately 15% higher.

Fig. 5.3.4 Schematic plan of typical floor—Example 5.3.2

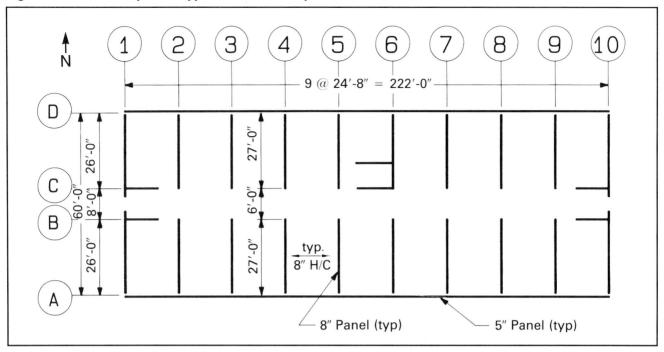

Wind load per brace = 0.015(27)/2 = 0.20 kips/ft

$$R_T = \frac{0.20(8^2/2)}{6} = 1.07 \text{ kips}$$

$R_s = 0.20(8) - 1.07 = 0.53$ kips

$R_{brace} = 1.07\sqrt{2} = 1.51$ kips

These loads and reactions can act in either direction. Typically, critical directions would be suction for the inserts in the panel and floor, and pressure for the brace. Forces on top connection:

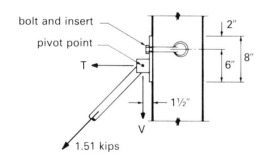

V = 1.07 kips

$$T \approx \frac{1.51}{\sqrt{2}} + \frac{1.07(1.5)}{6-1} = 1.39 \text{ kips}$$

Design of inserts and bolts is discussed in Chapter 6. Design of the connection at the bottom of the brace is similar. Expansion bolts in hollow-core plank must be placed to avoid the cores. Vertical tie connections between wall panels is usually adequate to take the wind shear at the base of the panel.

Check panel section:

Consider the horizontal span:
- w = 0.015(8) = 0.12 kips/ft
- $-M$ = $wa^2/2$ = 0.12(5.33)²(12)/2 = 20.5 in.-kips
- $+M$ = $w\ell^2/8 - (-M)$
  = 0.12(16.33)²(12)/8 − 20.5 = 27.5 in.-kips
- Z = $bd^2/6$ = 96(8)²/6 = 1024 in.³
- f = 27.5/1024 = 0.028 ksi < $5\sqrt{f'_c}$ OK

Determine diaphragm loads:

Shielding of the wind from adjacent structures is not considered in building design, because of the possibility that the shield may eventually be razed. However wind shielding by adjacent panels during erection is more predictable, and may be used with judgment realizing that erection sequences can change.

Various wind directions must also be considered. For this structure, temporary loading from north or south wind is less critical than the final condition, so it can be neglected. Two possible wind directions will be considered:

1. Wind from east or west. Apply full wind to end wall, with the other walls assumed to be 50% shielded (some wind can flow over the tops of the walls), except at the ends. Assume full wind applied to 25% of the length of the interior panels as wind flows along the sides of the building. See Fig. 5.3.5.

Fig. 5.3.5 Distribution of east-west wind load—Example 5.3.2

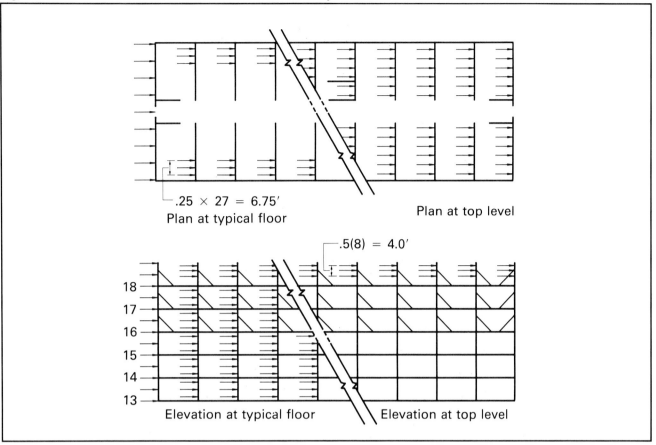

Fig. 5.3.6 Distribution of wind from 45° angle—Example 5.3.2

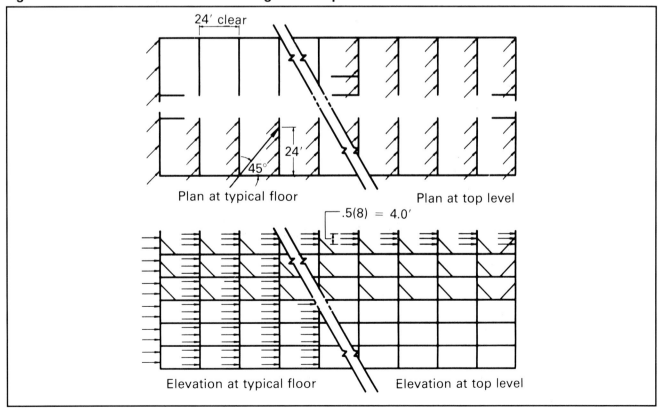

PCI Design Handbook/Fourth Edition

At 18th level:

0.015(8)(60) + 0.015(4)(60)(9) = 39.6 kips

At other levels:

0.015(8)(60) + 0.015(6.75)(2)(8)(9)
= 21.7 kips

2. Wind from any 45° angle. Apply full wind component to end wall with 50% shielding at upper level. Shielding of lower interior panels is determined from the geometry (see Fig. 5.3.6).

At 18th level:

0.015(0.707)(8)(60) + 0.015(0.707)(4)(60)(9)
= 28.0 kips

At other levels:

0.015(0.707)(8)(60) + 0.015(0.707)(24)(8)(9)
= 23.4 kips

The above shielding effects are based on judgment, and will vary among structures and designers. Note that the resultant of loads for condition 1 acts at the center of the structure, while for condition 2 it acts eccentrically.

The design of a diaphragm and shear walls is discussed in Chapter 3, and connections in Chapter 6.

Example 5.3.3—Single-story industrial building

Given:

A single-story building with the plan shown schematically in Fig. 5.3.7.

Totally precast structure—columns, inverted tee beams, double tee roof, load bearing and non-load bearing wall panels.

Wind load = 10 psf

Final stability by exterior panels acting as shear walls.

No expansion joint.

Erection sequence: With a crane in the center bay, start at column line 8 and move toward line 1, erecting all three bays progressively.

Problem:

Determine temporary erection bracing systems.

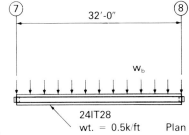

Solution:

To demonstrate the thought process the designer should go through, the following temporary conditions will be considered:

1. Erect columns B8 and B7 with inverted tee beam between them and check as free standing on the base plate.

w_b = 0.01(2.33) = 0.023 kips/ft

P = 0.023(32.5)/2 = 0.37 kips

w_c = 0.01(1) = 0.01 kips/ft

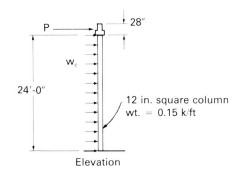

Elevation

Moment at column base = 0.37(24)+0.01(24)²/2
= 11.8 ft-kips

Dead load at column base = 0.5(32.5)/2+0.15(24)
= 11.7 kips

Design base plate and anchor bolts as described in Sect. 6.10.

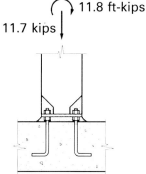

2. Erect wall panels on line A from 7 to 8. There are two cases to consider:

Case a: If the base of the wall has some moment resisting capacity, the base should be designed for:

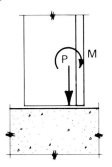

M = 0.01(8)(0.5)(28²/2) = 15.7 ft-kips each stem
P = 0.34(28)/2 = 4.75 kips each stem

Fig. 5.3.7 Schematic plan of structure of Example 5.3.3

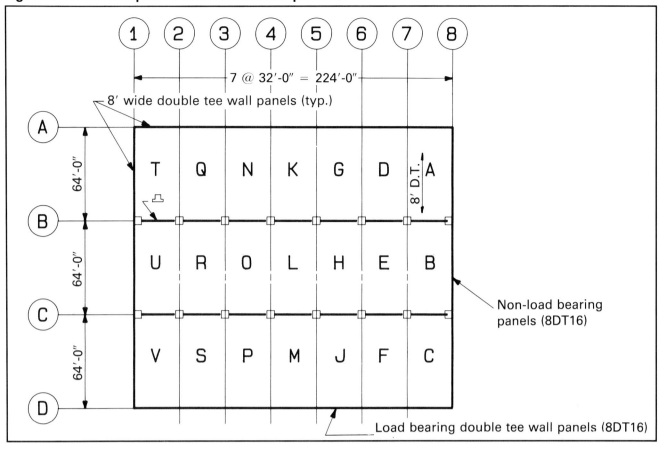

Case b: If the wall panel has no moment resisting capacity, a brace must be provided, as shown in Example 5.3.2. Check the unsupported length of the brace to prevent buckling under compressive loads. Also check the kickout reaction at the base of the panel.

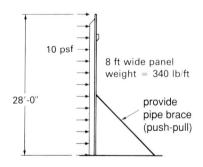

3. Erect double tee roof members in bay A. Assuming the wall panel has moment resisting capacity

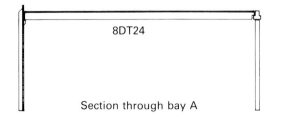

Section through bay A

at the base, two loading conditions must be checked:

Case a: Dead loads only:
Load from roof tee = 0.052(4)(64/2)
= 6.66 kips/stem

$M = 6.66(9/12) = 5.0$ ft-kips
$P = 6.66 + 4.75 = 11.4$ kips

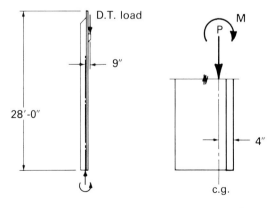

Check beam/column connection. This will be critical when all four roof tees in bay A are in place:

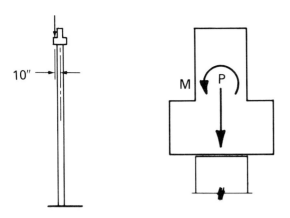

Load from double tees = 6.66 kips/stem
   times 4 stems = 26.6 kips
M = 26.6(10/12) = 22.2 ft-kips
P = 26.6 + 0.5(32/2) = 34.6 kips

Check column base connection:

M = 22.2 ft-kips (from above)
P = 34.6 + 0.15(24) = 38.2 kips

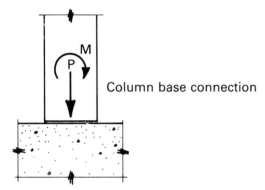

Column base connection

Case b: Dead load plus wind load: In this example, it is apparent that the loading for 1 and 2 above will be more critical than this condition.

The designer should always try to use the permanent connections for erection stability. In this example, the final connections between the wall and roof deck and the beam and roof deck will provide some degree of moment resistant capacity. This will reduce the moments on some of the connections considered above.

4. Erect wall panels on line 8 from A to B. (Note: If the crane reach is limited, these may be erected immediately after the first roof tee is placed.) These will be connected to the roof with permanent diaphragm transfer connections. This provides a rigid "box" which will provide stability to the remainder of the structure.

5. Continue erection working from the rigid box in bay A. It is unlikely that other temporary conditions will be more critical than those encountered in bay A. However, each should be considered and analyzed if appropriate.

Example 5.3.4 Multi-level parking structure

Given:

A typical 8-story parking structure shown schematically in Fig. 5.3.8.

Structural system—pretopped double tees on L-shaped and inverted tee interior beams and load bearing spandrel panels. Multi-level columns are spliced at level 4. No expansion joint.

Loads: Wind—10 psf. Construction loads—5 psf. This is lower than recommended in Sect. 5.3.2 because there are virtually no interior finishing materials, such as masonry or drywall in this construction.

Final stability will be provided by:
   Long direction: The ramped floors acting as a truss.
   Short direction: Shear walls as shown.

Problem:

Outline critical design conditions during erection, and show a detailed erection sequence.

Solution:

Outline of critical erection design conditions:

1. Free standing columns—design columns and base plates in accordance with Chapters 3, 4 and 6.
2. Determine bracing forces for wind loads from either direction.
3. Select wire rope sizes.
4. Determine forces on inserts used for bracing and select anchorages in accordance with Chapter 6.
5. Check column designs with temporary loading and bracing.
6. Check diaphragm design and determine which permanent connections must be made during erection. Determine need for temporary connections.
7. Check inverted tee beams and their connections for loading on one side only.

Sequence of erection: Start at column line 1 and, with crane in north bay, erect vertically and back out of structure at line 9 in the following sequence:

1. Erect columns A1, B1, A2 and B2—(lower tier). Check as free standing columns with 10 psf wind on surface.
2. Erect 3 levels in bay A.
3. Install X-bracing between A1 and B1. Check capacity required with full wind load on exposed surfaces.

Fig. 5.3.8 Parking structure of Example 5.3.4

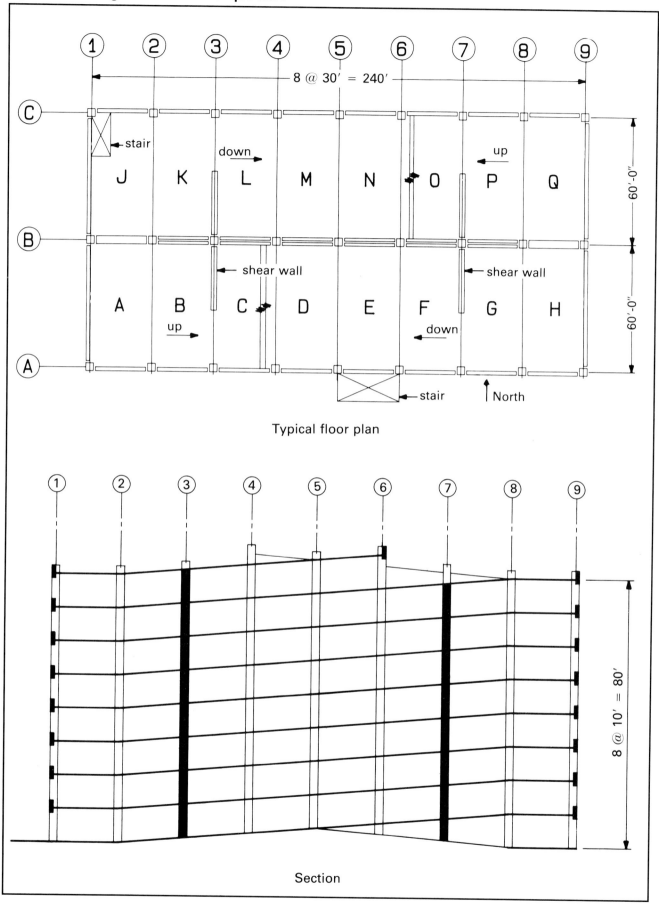

Typical floor plan

Section

4. Erect column A3 and shearwall.
5. Erect 3 levels in bay B.
6. Erect 4th level in bay A.
7. Install X-bracing at line B between B2 and shear wall. Check capacity required with full wind load on exposed surfaces.
8. Splice 2nd tier columns at A1, A2, A3, B1 and B2. Check capacity of welded splice with upper tier as free standing columns with 10 psf wind on surfaces.
9. Erect columns C1 and C2.
10. Erect 4 levels in bay J.
11. Install X-bracing from columns B1 to C1. Check capacity required with full load on exposed surfaces.
12. Erect columns A4 and B4.
13. Erect to 3rd level in bay C.
14. Install X-bracing between B3 and B4.
15. Erect bay A to 6th level.
16. Erect bay J to 6th level.
17. Erect bay A to roof.
18. Erect bay B to 6th level.
19. Erect column C3.
20. Erect bay K to 4th level.
21. Erect bay J to roof.
22. Erect columns A5 and B5.
23. Erect bay D to 3rd level.
24. Install X-bracing from A5 to B5. Check capacity required.
25. Erect bay C to 5th level.
26. Erect column C4.
27. Erect bay L to 4th level.
28. Erect bay K to 6th level.
29. Erect bay B to roof.
30. Continue in same sequence.

Additional notes:
1. After each level is erected weld loose plate from spandrel to plate in double tee. (See Fig. 5.3.9).
2. Install additional X-bracing on line B from 6 to 7 and 7 to 8 after framing is erected.
3. Install X-bracing on line 9 from A to B and B to C when framing is erected.
4. Install cables to upper tier columns as required to keep columns plumb.
5. Check upper tier columns as free standing with full wind load on exposed surfaces.
6. Maintain tightness of cables.

Fig. 5.3.9 Typical load-bearing spandrel—Example 5.3.4

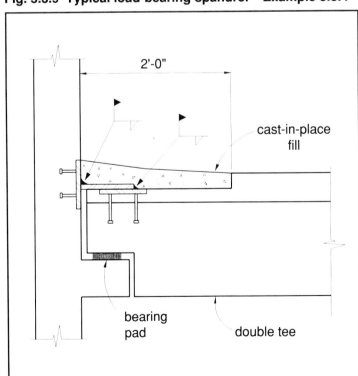

5.4 References

1. Tanner, John, "Architectural Panel Design and Production Using Post-Tensioning," *PCI Journal*, V. 22, No. 3, May-June 1977.

2. Anderson, A.R., "Lateral Stability of Long Prestressed Concrete Beams," *PCI Journal*, V. 16, No. 3, May-June 1971.

3. "Recommended Practice for Erection of Precast Concrete," MNL-127-85, Precast/Prestressed Concrete Institute, Chicago, IL, 1985.

4. *Wire Rope Users Manual*, Committee of Wire Rope Producers, American Iron and Steel Institute, Washington, D.C., 1985.

5. Mast, Robert F. "Lateral Stability of Long Prestressed Concrete Beams," *PCI Journal*, V. 34, No.1, January-February 1989.

6. Salmons, J.R. and McCrate, T.E., "Bond Characteristics of Untensioned Prestressing Strand," *PCI Journal*, V. 22, No. 1, January-February 1977.

7. Ghosh, S.K. and Fintel, M., "Exceptions of Precast, Prestressed Members to Minimum Reinforcement Requirements," R&D 2, Precast/Prestressed Concrete Institute, Chicago, IL, 1986.

CHAPTER 6
DESIGN OF CONNECTIONS

	Page No.
6.1 Notation	6-2
6.2 General	6-4
6.3 Loads and Load Factors	6-4
6.4 Connection Design Criteria	6-4
6.5 Connection Hardware and Load Transfer Devices	6-5
6.5.1 Reinforcing Bars	6-5
6.5.2 Welded Headed Studs	6-7
6.5.3 Deformed Bar Anchors	6-15
6.5.4 Bolts and Threaded Connectors	6-15
6.5.5 Inserts Cast in Concrete	6-15
6.5.6 Structural Steel	6-16
6.5.7 Post-Tensioning Steel	6-19
6.5.8 Bearing Pads	6-19
6.5.9 Connection Angles	6-20
6.6 Friction	6-23
6.7 Shear-Friction	6-23
6.8 Bearing on Plain Concrete	6-24
6.9 Reinforced Concrete Bearing	6-25
6.10 Column Base Plates	6-26
6.11 Concrete Brackets or Corbels	6-28
6.12 Structural Steel Haunches	6-30
6.13 Dapped-End Connections	6-32
6.13.1 Flexure and Axial Tension in the Extended End	6-33
6.13.2 Direct Shear	6-33
6.13.3 Diagonal Tension at Reentrant Corner	6-34
6.13.4 Diagonal Tension in the Extended End	6-34
6.13.5 Anchorage of Reinforcement	6-34
6.13.6 Other Considerations	6-34
6.14 Ledger Beam	6-36
6.14.1 Shear Strength of the Ledge	6-36
6.14.2 Transverse (Cantilever) Bending of the Ledge	6-36
6.14.3 Longitudinal Bending of the Ledge	6-36
6.14.4 Attachment of the Ledge to the Web	6-37
6.14.5 Out-of-Plane Bending Near Beam End	6-37
6.15 Hanger Connections	6-38
6.15.1 Cazaly Hanger	6-38
6.15.2 Loov Hanger	6-44
6.16 Moment Connections	6-45
6.17 Connection of Non-Load Bearing Wall Panels	6-46
6.18 Connection of Load Bearing Wall Panels	6-46
6.18.1 Vertical Joints	6-49
6.18.2 Horizontal Joints	6-51
6.18.3 Typical Details	6-54
6.19 References	6-55
6.20 Design Aids	6-57

DESIGN OF CONNECTIONS

6.1 Notation

a	=	shear span
a	=	depth of equivalent rectangular compression stress block
A_b	=	area of bar or stud
A_{cr}	=	area of crack interface
A_f	=	area of flexural reinforcement in a corbel or dap
A_h	=	area of shear reinforcement parallel to flexural tension reinforcement in a corbel
A_ℓ	=	area of longitudinal reinforcement in beam ledge
A_n	=	area of reinforcement required to resist axial tension
A_o	=	area of concrete failure surface
A_s	=	area of reinforcement
A_s'	=	area of vertical reinforcement near end of steel haunch
A_{sh}	=	area of vertical reinforcement for horizontal or diagonal cracks
A_v	=	diagonal tension reinforcement in dapped end
A_{vf}	=	area of shear-friction reinforcement
A_w	=	area of weld
$A_{w\ell}$	=	area of steel in horizontal direction at beam end for torsional equilibrium
A_{wv}	=	area of steel in vertical direction at beam end for torsional equilibrium
A_1	=	loaded area
A_2	=	maximum area of the portion of the support that is geometrically similar to and concentric with the loaded area
b	=	width of compression stress block
b	=	length of an angle
b	=	center-to-center distance between the outermost studs in back row of a group
b	=	width of welds in weld groups (see Fig. 6.5.12)
b_ℓ	=	ledger beam width at ledge
b_n	=	net length of an angle
b_t	=	width of bearing area under concentrated loads on ledges
b_w	=	net width of hollow-core slab on bearing wall
b_1, b_2	=	width (see specific application)
C	=	compressive force
C_w, C_t, C_c	=	adjustment factors for stud groups
C	=	symbol for element Carbon
C_r	=	reduction coefficient (see Eq. 6.8.1)
C_{es}	=	reduction coefficient for edge distance (see Eq. 6.5.5)
Cr	=	symbol for element Chromium
Cu	=	symbol for element Copper
d	=	depth to centroid of reinforcement
d	=	depth of welds in weld groups (see Fig. 6.5.12)
d_b	=	bar or stud diameter
d_c	=	distance from free edge of concrete to the centerline of nearest stud measured perpendicular to direction of load
d_e	=	distance from center of load to beam end
d_e	=	distance from edge of member to centerline of single stud or of back row of stud group in direction of load
d_h	=	head diameter of stud
d_ℓ	=	depth of centroid of A_s reinforcement in ledger beam ledges
d_w	=	depth of $A_{w\ell}$ and A_{wv} reinforcement from outside face of ledger beam
D	=	durometer (shore A hardness)
e	=	eccentricity of load
e_i	=	center of bolt to horizontal reaction
e_v	=	eccentricity of vertical load
e_x, e_y	=	eccentricity of load in x,y directions
f	=	unit stress
f_{bu}	=	factored bearing stress
f_{ct}	=	splitting tensile strength of concrete
f_c'	=	compressive strength of concrete
f_r	=	resultant stress on weld
f_{ue}	=	see Sect. 6.18.2
f_w	=	design strength of weld (see Table 6.20.1)
f_x	=	combined shear and torsion stress in horizontal direction
f_y	=	combined shear and torsion stress in vertical direction

f_y	=	yield strength of reinforcement
F_s	=	factored friction force
ΣF	=	greatest sum of factored anchor bolt forces on one side of the column
F_y	=	yield strength of structural steel
g	=	gage of angle
g	=	gap between beam end and face of support element
h	=	total depth
h	=	depth of a dapped end member above the dap
h_ℓ	=	depth of a ledger beam ledge
H	=	total depth of a dapped end member
I_p	=	polar moment of inertia
I_{xx}, I_{yy}	=	moment of inertia of weld segment with respect to its own axes
j_u	=	lever arm coefficient, used in $j_u d$
k	=	distance from back face of angle to web fillet toe
ℓ	=	length of joint
ℓ_b	=	bearing length
ℓ_d	=	development length
ℓ_e	=	embedment length
ℓ_ℓ	=	angle leg length
ℓ_p	=	bearing length of exterior cantilever in hanger connections
ℓ_p	=	projection of corbel or dapped end
ℓ_w	=	length of weld
m	=	modification factor for hanger steel calculation
Mo	=	symbol for element Molybdenum
M_t	=	torsional moment
Mn	=	symbol for element Manganese
M_u	=	factored moment
n	=	number of studs in a group
n_s	=	number of studs in back row of a group
N	=	unfactored horizontal or axial force
Ni	=	symbol for element Nickel
N_u	=	factored horizontal or axial force
P	=	applied load
P_c	=	nominal tensile strength of concrete element
P_s	=	nominal tensile strength of steel element
P_u	=	applied factored load
P_u	=	factored tension load
P_x, P_y	=	applied force in x, y direction
R_e	=	reduction factor for load eccentricity
s	=	distance from free edge to center of bearing
s	=	spacing of concentrated loads
S	=	shape factor
S	=	section modulus of weld groups (see Fig. 6.5.12)
t	=	thickness
t_g	=	grout thickness
t_w	=	effective throat thickness of weld
T	=	tensile force
T_c	=	nominal torsional strength of concrete element
T_u	=	factored torsional moment
V	=	symbol for element Vanadium
V	=	unfactored vertical or shear force
V_c	=	nominal shear strength of concrete element
V_d	=	unfactored vertical dead load
V_n	=	nominal bearing or shear strength of an element
V_r	=	nominal strength provided by reinforcement
V_u	=	factored shear force
V_{ud}	=	factored dead load force normal to the friction face
w	=	dimension (see specific application)
w	=	uniform load
x, y	=	horizontal and vertical distance respectively, from c.g. of weld group to point under consideration
x, y	=	overall dimensions (width and length) of a stud group
x_c	=	distance from centerline of bolt to face of column
x_o	=	base plate projection
x_t	=	distance from centerline of bolt to centerline of column reinforcement
x_1, y_1	=	dimensions of flat bottom part of the truncated pyramid failure—stud groups
Z	=	section modulus
Z_s	=	plastic section modulus of structural steel section
Δ	=	horizontal deformation of bearing pad

α = with subscript: modification factor for development length

γ_t = ratio of T_c to T_u (see Table 6.14.1)

θ = angle of assumed crack plane

λ = coefficient for use with lightweight concrete (see Sect. 5.2.4)

μ = shear-friction coefficient

μ_e = effective shear-friction coefficient

μ_s = static coefficient of friction

ϕ = strength reduction factor

6.2 General

The design of connections is one of the most important considerations in the structural design of a precast concrete structure. There may be several successful solutions to each connection problem, and the design methods and examples included in this chapter are not the only acceptable ones. Information is included on the design of common precast concrete connections. It is intended for use by those with an understanding of engineering mechanics and structural design, and in no case should it replace good engineering judgment.

The purpose of a connection is to transfer load and/or provide stability. Within any one connection, there may be several load transfers; each one must be considered by the designer. In the sections that follow, different methods of transferring load will be examined separately, then it will be shown how some of these are combined in typical connection situations. More complete information on connections can be found in Refs. 1 and 2 listed in Sect. 6.19.

6.3 Loads and Load Factors

With noted exceptions, such as bearing pads, the design methods in this chapter are based on strength design relationships, incorporating the load factors and strength reduction factors (ϕ-factors) specified in ACI 318-89. Local codes may have more stringent requirements and must be checked.

In addition to gravity, wind and seismic loads, forces resulting from restraint of volume changes as well as those required for compatibility of deformations must be considered. Determination of these forces is covered in Chapter 3. For flexural members, it is recommended that the connections be designed for a minimum horizontal tensile force, acting parallel to the span, of 0.2 times the dead load transferred at the bearing unless a smaller value can be justified by using properly designed bearing pads (see Sect. 6.5.8).

It is undesirable for the connection to be the weak link in a precast concrete structure because connection failures are typically brittle as they are not preceded by large deformations. To ensure that the overall safety of the connection is adequate, the use of an additional load factor in the range 1.0 to 1.33 has historically been used by the industry. The need and the magnitude of this additional load factor for a particular connection must depend on the Engineer's judgment and consideration of:

1. *Mode of Failure:* A larger load factor may be appropriate for a brittle failure. Such failures are typically precipitated by failure of concrete due to insufficient anchorage of connection reinforcement and inserts, such as short studs.

2. *Consequences of Failure:* If failure of a connection is likely to produce catastrophic results, the connection should have a larger overall factor of safety.

3. *Sensitivity of Connection to Tolerances:* Production and erection tolerances as well as movements due to volume changes and applied loads produce changes in load transfer positions on the connection. Certain connections, for example corbels and dapped ends, are more sensitive to load transfer positions than other connections, such as base plates. The magnitude of the additional load factor should be consistent with this sensitivity.

4. *Accumulation of Live Loads:* Certain connections in multi-story structures, such as column base plates, are unlikely to be subjected to the full accumulated live load from all stories. Thus, unless a live load reduction is considered, additional load factor may not be necessary.

5. *Requirements of Local Codes:* Additional load factor may not be necessary if the intent of the applicable local code required load factors is the same as that described above, namely that the connection is not the weak link.

6.4 Connection Design Criteria

Precast concrete connections must meet a variety of design and performance criteria, and not all connections are required to meet the same criteria. These criteria include:

1. *Strength:* A connection must have the strength to transfer the forces to which it will be subjected during its lifetime, including those caused by volume change restraint and those required to maintain stability.

2. *Ductility:* This is the ability to undergo relatively large inelastic deformations without failure. In connections, ductility is achieved by designing and detailing so that steel devices yield prior to

weld failure or concrete failure. Concrete typically fails in a brittle manner unless it is confined.

3. *Volume change accommodation:* Restraint of creep, shrinkage and temperature change strains can cause large stresses in precast concrete members and their supports. These stresses must be considered in the design. It is usually far better if the connection allows some movement to take place, thus relieving the stresses.

4. *Durability:* When exposed to weather, or used in a corrosive environment, steel elements should be adequately covered by concrete, or be painted, epoxy coated or galvanized. Stainless steel is sometimes used.

5. *Fire resistance:* Connections which could jeopardize the structure's stability if weakened by fire should be protected to the same degree as that required for the members that they connect.

6. *Constructability:* The following items should be reviewed when designing connections:

    a. Standardize products or connections
    b. Avoid reinforcement and hardware congestion
    c. Check material and size availability
    d. Avoid penetration of forms, where possible
    e. Reduce post-stripping work
    f. Be aware of material sizes and limitations
    g. Consider clearances and tolerances
    h. Avoid non-standard production and erection tolerances
    i. Use standard hardware items and as few sizes as possible
    j. Use repetitious details
    k. Plan for the shortest possible hoist hook-up time
    l. Provide for field adjustment
    m. Provide accessibility
    n. Use connections that are not susceptible to damage in handling

6.5 Connection Hardware and Load Transfer Devices

A wide variety of hardware including reinforcing bars, studs, coil inserts, structural steel shapes, bolts, threaded rods and other materials are used in connections. These devices provide load transfer to concrete via anchorage in the concrete by bond or by a shear cone resistance mechanism. It is preferable to have a steel material failure, typically defined by yielding, govern the connection strength because such failures are more predictable and ductile. Load transfer should be as direct as possible to reduce the complexity and increase the efficiency of the connection.

6.5.1 Reinforcing Bars

Reinforcing bars are usually anchored by bonding to the concrete. Very often, there is insufficient length available to anchor the bars by bond alone, and supplemental mechanical anchorage is required. This can be accomplished by hooks or welded cross-bars as shown in Fig. 6.11.1. Load transfer between bars may be achieved by welding, lap splices or mechanical couplers. Required development lengths and standard hook dimensions are given in Chapter 11.

Reinforcing bars may be anchored by embedment in flexible metallic interlocking conduit using grout as shown in Fig. 6.5.1. The conduit must have sufficient concrete aound it as shown in Fig. 6.5.1 for adequate confinement. This scheme can be used to transfer tension or compression forces and is convenient for certain connections, such as column to footing and column to column connections.

For No. 8 and smaller uncoated reinforcing bars, where the bar is forced into the grout-filled flexible conduit, the embedment length is given by:

$$\ell_e = 0.04 A_b f_y / \sqrt{f'_c} \geq 12 \text{ in.} \qquad \text{(Eq. 6.5.1)}$$

where:

ℓ_e = embedement length, in.

A_b = area of bar, sq in.

f_y = steel yield strength, psi

f'_c = concrete strength, psi

Fig. 6.5.1 Anchorage in grout-filled conduit

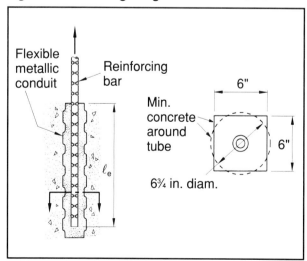

Fig. 6.5.2 Typical reinforcing bar welds

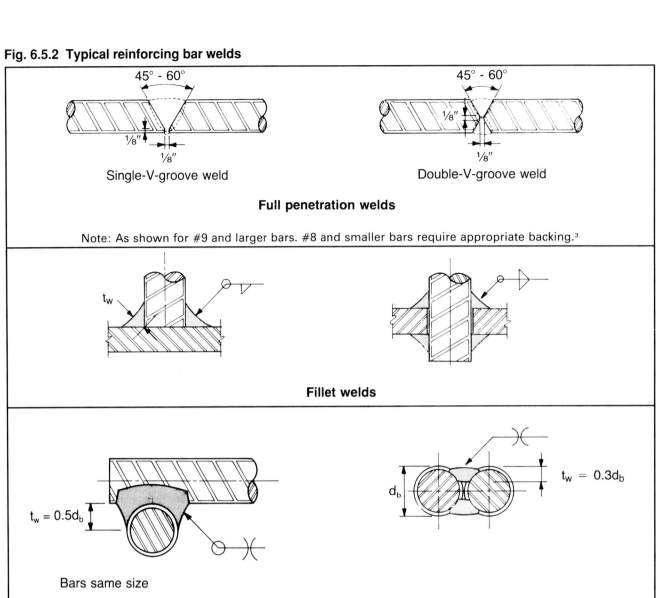
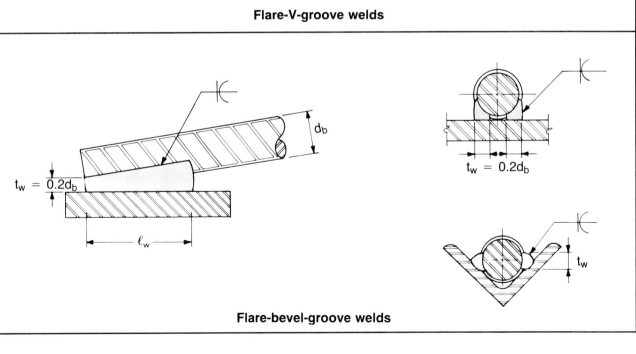

Reinforcing bar welding

Welding is covered by AWS D1.4-79, "Structural Welding Code—Reinforcing Steel,"[3] and by AWS D1.1-79, "Structural Welding Code—Steel,"[4] by the American Welding Society. Weldability is defined by AWS as a function of the chemical composition of steel as shown in the mill report by the following formula:

$$C.E. = \%C + \frac{\%Mn}{6} + \frac{\%Cu}{40} + \frac{\%Ni}{20} + \frac{\%Cr}{10} + \frac{\%Mo}{50} - \frac{\%V}{10} \quad \text{(Eq. 6.5.2)}$$

where C.E. = carbon equivalent

The last three elements usually appear only as trace elements, so they are often not included in the mill report. For reinforcing bars that are to be welded, the carbon equivalent should be requested with the order from the mill.

AWS D1.4-79 indicates that most reinforcing bars can be welded. However, stringent preheat and other quality control measures are required for bars with high carbon equivalents. Except for welding shops with proven quality control procedures that meet AWS D1.4-79, it is recommended that carbon equivalents be less than 0.45% for No. 7 and larger bars, and 0.55% for No. 6 and smaller bars.

Most reinforcing bars which meet ASTM A615, Grade 60, will not meet the above chemistry specifications. A615, Grade 40 bars may or may not meet the above specifications. Bars which meet ASTM A706 are specially formulated to be weldable.

Fig. 6.5.2 shows the most common welds used with reinforcing bars. Full penetration groove welds can be considered to have the same nominal strength as the bar when matching weld metal is used. The design strength of the other weld types can be calculated using the values from Tables 6.20.1 and 6.20.2. The total design strength of the weld is $f_w \ell_w t_w$.

where:

$f_w = \phi(0.6 F_{exx})$, see Table 6.20.1 for values

$\phi = 0.75$

F_{exx} = classification strength of weld metal

ℓ_w = length of weld

t_w = effective throat thickness of weld
(Fig. 6.5.2)

Tables 6.20.3 through 6.20.5 show welding required to develop the full strength of reinforcing bars.

The welded cross-bar detail shown in Fig. 6.5.2 is not included in AWS D1.4. However, it has been used in numerous structures and verified by tests[5,6] and, when the diameter of the cross bar is at least the same size as the main bar, the full strength of the main bar has been shown to be developed.

Reinforcing bars should not be welded within 2 bar diameters of a bend to avoid potential crystalization.

AWS D1.4 requires that tack welds be made using the same preheat and quality control requirements as permanent welds, and prohibits them unless authorized by the engineer.

Reinforcing bar couplers

Proprietary bar coupling devices are available as an alternative to lap splices or welding. Manufacturers of these devices furnish design information and test data. Refs. 7. and 8 contain more detailed information.

6.5.2 Welded Headed Studs

Welded headed studs are designed to resist direct tension, shear or a combination of the two. The design equations given below are applicable to studs which are welded to steel plates or other structural members, and embedded in unconfined concrete. Confinement of the concrete, either from applied compressive loads or from reinforcement, is known to increase the capacity, however due to limited research, acceptable design equations which include confinement are not available.

Where feasible, headed stud connections should be designed and detailed such that the connection failure is precipitated by failure (typically defined as yielding) of the stud material rather than failure of the surrounding concrete. The in-place strength should be taken as the smaller of the values based on concrete and steel.

Fig. 6.5.3 Shear cone development for welded headed studs

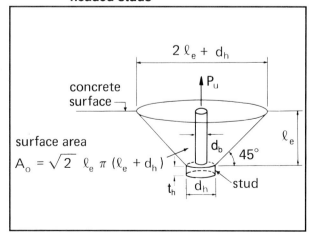

6.5.2.1 Tension

The design tensile strength governed by concrete failure is[9]:

$$\phi P_c = \phi A_o \left(2.8\lambda\sqrt{f'_c}\right) \quad \text{(Eq. 6.5.3)}$$

where:

$\phi = 0.85$

A_o = area of the assumed failure surface which, for a single stud not located near a free edge, is taken to be that of a 45° truncated cone as shown in Fig. 6.5.3.

Using the 45° cone area and $\phi = 0.85$, Eq 6.5.3 may be written as:

$$\phi P_c = 10.7\ell_e(\ell_e + d_h)\lambda\sqrt{f'_c} \quad \text{(Eq. 6.5.4)}$$

Note: The stud length is often used in place of the actual embedment length, ℓ_e, which is equal to the stud length minus the thickness of the head. This simplification is generally acceptable except in short studs. In short studs (length ≤ 4 in.), the use of actual embedment length is recommended. It should also be noted that short stud capacities are also sensitive to fabrication tolerances. Thus, use of a larger overall factor of safety may be appropriate for short studs— see Sect. 6.3.

For a stud located closer to a free edge than the embedment length, ℓ_e, the design tensile strength given by Eq. 6.5.4, should be reduced by multiplying it by C_{es}:

$$C_{es} = \frac{d_e}{\ell_e} \leq 1.0 \quad \text{(Eq. 6.5.5)}$$

where d_e is the distance measured from the stud axis to the free edge. If a stud is located in the corner of a concrete member, Eq. 6.5.5 should be applied twice, once for each edge distance. Table 6.20.6 lists values based on Eqs. 6.5.4 and 6.5.5.

For a group of studs, the concrete failure surface may be along a truncated pyramid rather than separate shear cones, as shown in Fig. 6.5.4.

For this case, the design tensile strength is:

$$\phi P_c = \phi\lambda\sqrt{f'_c}(2.8A_{slope} + 4A_{flat}) \quad \text{(Eq. 6.5.6)}$$

where:

A_{slope} = sum of the areas of the sloping sides

A_{flat} = area of the flat bottom of the truncated pyramid

For stud groups in thin members, the failure surface may penetrate the thickness of the member as shown in Fig. 6.5.5. This type of failure is likely when the thickness of the member is less than a certain minimum thickness, h_{min}, given in Fig. 6.5.6 and listed in Table 6.20.7B. The pull-out strength corresponding to $h < h_{min}$ is then based on area of the sloping sides only.

Nominal pull-out strengths for both conditions (i.e., $h \geq h_{min}$ and $h < h_{min}$) for different edge vicinity cases are given in Fig. 6.5.6.

Fig. 6.5.4 Truncated pyramid failure

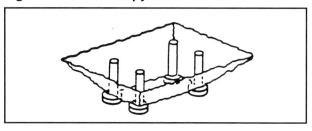

The design tensile strength per stud as governed by steel failure* is:

$$\phi P_s = \phi A_b f_y = \phi A_b(0.9 f_s) = 54,000 A_b \quad \text{(Eq. 6.5.7)}$$

where:

$\phi = 1.0$

$f_s = 60,000$ psi

Table 6.20.6 lists the design strength values from the above equation.

Example 6.5.1 Tension strength of stud groups

Given:

A base plate with four headed studs embedded in a corner of a foundation slab.

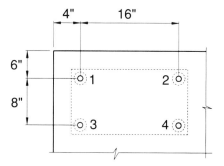

4 – ¾ in. diam. headed studs

embedment, ℓ_e = 8 in.

slab thickness, h = 10 in.

f'_c = 4000 psi (normal weight)

Problem:

Determine the tension strength of the stud group.

Solution:

1. Check for edge effect:

---
*The minimum tensile strength of steel for Type B studs, f_s is typically 60,000 psi. The yield strength in tension may be taken as $0.9 f_s$ and the yield strength in shear may be taken as $0.75 f_s$.

Fig. 6.5.5 Pull-out surface areas for stud groups in thin sections

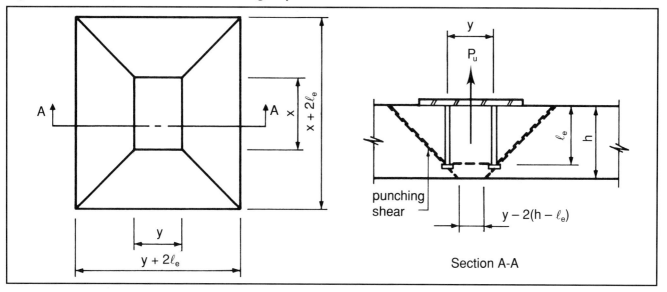

For the given problem (see Fig. 6.5.6)
$x = 16$ in., $y = 8$ in., $d_{e1} = 4$ in.,
$d_{e2} > \ell_e$, $d_{e3} = 6$ in., $d_{e4} > \ell_e$

Thus the effects of vicinity to two edges apply (Case 4 in Fig. 6.5.6).

2. Check for member thickness (see Fig. 6.5.6 or Table 6.20.7B):

   $h_{min} = (z + 2\ell_e)/2$

   where z is lesser of x and y

   $h_{min} = [8 + 2(8)]/2 = 12$ in. (Note: Same result can be read from Table 6.20.7B)

   Since h (= 10 in.) < h_{min}, failure surface is likely to penetrate through the slab.

3. Calculate tension strength based on concrete for the studs as a group:

   The applicable equation is given in Fig. 6.5.6 (Case 4, h < h_{min}):

   $\phi P_c = \phi 4\lambda\sqrt{f'_c}[(x + \ell_e + d_{e1})(y + \ell_e + d_{e3}) - A_R]$

   where: $A_R = (x + 2\ell_e - 2h)(y + 2\ell_e - 2h)$

   substituting given values:

   $x + \ell_e + d_{e1} = 16 + 8 + 4 = 28$ in.

   $y + \ell_e + d_{e3} = 8 + 8 + 6 = 22$ in.

   $A_R = (16 + 2(8) - 2(10))(8 + 2(8) - 2(10))$
   $= 12(4) = 48$ in.^{2}

   $\phi P_c = 0.85(4)(1.0)\sqrt{4000}\,[(28)(22) - 48]/1000$

   $= 132.5 - 10.3 = 122.2$ kips

   Alternatively, from Tables 6.20.7A (Case 4) and 6.20.7C:

   For $\ell_e = 8$ in., $x = 16$ in., $x_1 = 20$ in., $y = 8$ in.
   $y_1 = 14$ in., $h - \ell_e = 2$ in.

   $\phi P_{c1} = 148\sqrt{\dfrac{4000}{5000}} = 132.4$

   $\phi P_{c2} = 12\sqrt{\dfrac{4000}{5000}} = 10.7$

   Therefore, $\phi P_c = \phi P_{c1} - \phi P_{c2} = 132.4 - 10.7$
   $= 121.7$ kips

   Note: Pull-out strength based on individual studs should also be checked.

4. Check capacity based on steel failure:

   From Table 6.20.6, for ¾ in. diam. stud:

   $\phi P_s = 23.9$ kips/stud

   Total $\phi P_s = 4(23.9) = 95.6$ kips < 122.2

   Thus the tension strength of the group = 95.6 kips

6.5.2.2 Shear

The design shear strength governed by concrete failure is based on the concepts and results given in Refs. 9 and 10. Some modifications were made to improve correlation with test data obtained under a PCI Research Fellowship, the results of which are reported in Ref. 11.

The design shear strength limited by concrete, ϕV_c, is calculated as follows:

For $d_e \geq 15d_b$, where d_e is the edge distance (see Fig. 6.5.7) and d_b is the stud diameter:

$$\phi V_c = \left(\phi 800 A_b \lambda \sqrt{f'_c}\right)n \qquad \text{(Eq. 6.5.8)}$$

where:
$\phi = 0.85$
A_b = area of stud
n = number of studs in the group

Fig. 6.5.6 Design tensile strength of stud groups

Case 1: Not Near a Free Edge [1]

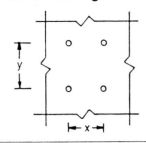 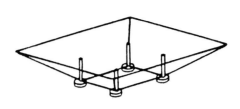

$h \geq h_{min}$ [2]	$\phi P_c = \phi \, 4 \lambda \sqrt{f'_c} \, (x + 2l_e)(y + 2l_e)$
$h < h_{min}$	$\phi P_c = \phi 4 \lambda \sqrt{f'_c} \, [(x + 2l_e)(y + 2l_e) - A_R]$ [3]

Case 2: Free Edge on One Side

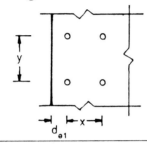 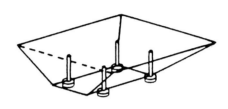

$h \geq h_{min}$	$\phi P_c = \phi \, 4 \lambda \sqrt{f'_c} \, (x + l_e + d_{e1})(y + 2l_e)$
$h < h_{min}$	$\phi P_c = \phi \, 4 \lambda \sqrt{f'_c} \, [(x + l_e + d_{e1})(y + 2l_e) - A_R]$

Case 3: Free Edges on 2 Opposite Sides

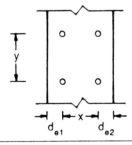 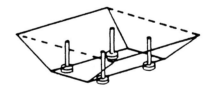

$h \geq h_{min}$	$\phi P_c = \phi \, 4 \lambda \sqrt{f'_c} \, (x + d_{e1} + d_{e2})(y + 2l_e)$
$h < h_{min}$	$\phi P_c = \phi \, 4 \lambda \sqrt{f'_c} \, [(x + d_{e1} + d_{e2})(y + 2l_e) - A_R]$

1. Near a free edge implies $d_e < \ell_e$.
2. $h_{min} = (z + 2\ell_e)/2$, where z is lesser of x and y (see Table 6.20.7B).
3. $A_R = (x + 2\ell_e - 2h)(y + 2\ell_e - 2h)$

Fig. 6.5.6 Design tensile strength of stud groups (continued)

Case 4: Free Edges on 2 Adjacent Sides

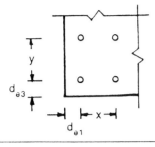
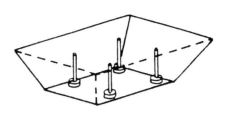

$h \geq h_{min}$	$\phi P_c = \phi\, 4\, \lambda \sqrt{f'_c}\, (x + l_e + d_{e1})(y + l_e + d_{e3})$
$h < h_{min}$	$\phi P_c = \phi\, 4\, \lambda \sqrt{f'_c}\, [(x + l_e + d_{e1})(y + l_e + d_{e3}) - A_R]$

Case 5: Free Edges on 3 Sides

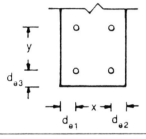
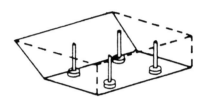

$h \geq h_{min}$	$\phi P_c = \phi\, 4\, \lambda \sqrt{f'_c}\, (x + d_{e1} + d_{e2})(y + l_e + d_{e3})$
$h < h_{min}$	$\phi P_c = \phi\, 4\, \lambda \sqrt{f'_c}\, [(x + d_{e1} + d_{e2})(y + l_e + d_{e3}) - A_R]$

Case 6: Free Edges on 4 Sides

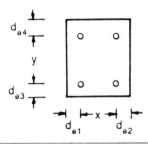
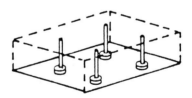

$h \geq h_{min}$	$\phi P_c = \phi\, 4\, \lambda \sqrt{f'_c}\, (x + d_{e1} + d_{e2})(y + d_{e3} + d_{e4})$
$h < h_{min}$	$\phi P_c = \phi\, 4\, \lambda \sqrt{f'_c}\, [(x + d_{e1} + d_{e2})(y + d_{e3} + d_{e4}) - A_R]$

or, in terms of stud diameter d_b:

$$\phi V_c = \left(\phi 628 d_b^2 \lambda \sqrt{f'_c}\right) n \quad \text{(Eq. 6.5.8a)}$$

For $d_e < 15d_b$, ϕV_c given by Eq. 6.5.8 or 6.5.8a shall not exceed the value given by Eq. 6.5.9:

$$\phi V_c = \phi V'_c C_w C_t C_c \quad \text{(Eq. 6.5.9)}$$

where:

$\phi V'_c$ = design shear strength of a single stud in the back row (Eq. 6.5.10).

C_w, C_t, C_c = adjustment factors for group width, member thickness and vicinity to member corner effects, respectively (see Fig. 6.5.7). These quantities are given by Eqs. 6.5.11 through 6.5.13.

$$\phi V'_c = \phi 12.5 d_e^{1.5} \lambda \sqrt{f'_c} \quad \text{(Eq. 6.5.10)}$$

where:

ϕ = 0.85

d_e = distance from free edge of concrete to back row of studs in direction of load

$$C_w = \left(1 + \frac{b}{(3.5 d_e)}\right) \leq n_s \quad \text{(Eq. 6.5.11)}$$

where:

b = center-to-center distance between the outer-most studs in the back row of the group

n_s = number of studs in the back row

$$C_t = \frac{h}{(1.3 d_e)} \leq 1.0 \quad \text{(Eq. 6.5.12)}$$

where:

h = thickness of the concrete member

$$C_c = \left[0.4 + 0.7 \frac{d_c}{d_e}\right] \leq 1.0 \quad \text{(Eq. 6.5.13)}$$

where:

d_c = the distance, measured perpendicular to the direction of the load, from free edge of concrete to the centerline of the nearest stud.

The design shear strength as governed by steel is:

$$\phi V_s = \phi 0.75 f_s A_b n = 45,000 A_b n \quad \text{(Eq. 6.5.14)}$$

where:

ϕ = 1.0

or, in terms of stud diameter, d_b:

$$\phi V_s = (35,344 d_b^2) n \quad \text{(Eq. 6.5.14a)}$$

Fig. 6.5.7 Shear on stud group

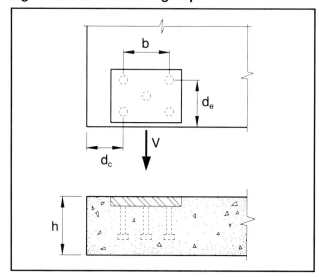

Table 6.20.8 gives values for ϕV_c (Eq. 6.5.8a), $\phi V'_c$ (Eq. 6.5.10) and ϕV_s (Eq. 6.5.14a).

Example 6.5.2 Shear strength of stud groups

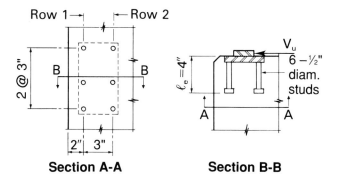

Section A-A　　　　Section B-B

Given:

A stud group in a column subject to the shear force shown.

f'_c = 5000 psi (normal weight)

Problem:

Find the design shear strength.

Solution:

The design parameters are:

d_b = 0.5 in., b = 6 in., d_e = 5 in.

n = 6, n_s = 3,

h > 1.3d_e,

No vicinity to corner.

From Eq. 6.5.8a:

ϕV_c = 0.85(628)(0.5)²(1.0)($\sqrt{5000}$)(6)/1000

= 56.6 kips

Alternatively, from Table 6.20.8:

ϕV_c per stud = 9.45

Therefore, for 6 studs:

ϕV_c = 9.45(6) = 56.7 kips

From Eq. 6.5.10:

$\phi V'_c = 0.85(12.5)(5)^{1.5}\sqrt{5000}/1000 = 8.4$ kips

(Note: Same value is obtained from Table 6.20.8)

From Eq. 6.5.11: (width effect)

$C_w = 1 + 6/(3.5(5)) = 1.34 < 3$ OK

From Eq. 6.5.12: (thickness effect)

$C_t = 1.0$

From Eq. 6.5.13: (corner effect)

$C_c = 1.0$

From Eq. 6.5.9, the design shear strength based on concrete is:

ϕV_c = (8.4)(1.34)(1.0)(1.0)
= 11.3 kips

From Eq. 6.5.14a, the design shear strength based on steel is:

$\phi V_s = 35{,}344(0.5)^2(6)/1000$
= 53.0 kips > 11.3

(Note: A value of 8.8 kips per stud is given in Table 6.20.8. For 6 studs, ϕV_s = 52.8 kips)

Thus, design shear strength is governed by concrete failure.

ϕV_c = 11.3 kips

6.5.2.3 Combined shear and tension

The design strength of studs under combined tension and shear should satisfy the following interaction equations:

Concrete: $\dfrac{1}{\phi}\left[\left(\dfrac{P_u}{P_c}\right)^2 + \left(\dfrac{V_u}{V_c}\right)^2\right] \le 1.0$ (Eq. 6.5.15)

where ϕ = 0.85

Steel: $\dfrac{1}{\phi}\left[\left(\dfrac{P_u}{P_s}\right)^2 + \left(\dfrac{V_u}{V_s}\right)^2\right] \le 1.0$ (Eq. 6.5.16)

where ϕ = 1.0

P_u and V_u are the factored tension and shear loads.

6.5.2.4 Plate thickness

Thickness of plates to which studs are attached should be at least ⅔ of the diameter of the stud.

Example 6.5.3 Design of welded headed studs for combined loads

Given:

A plate with headed studs for attachment of a steel bracket to a column as shown in the figure.

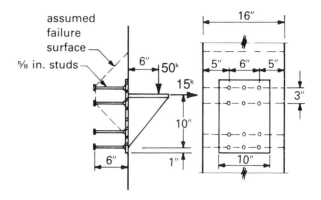

f'_c = 5000 psi (normal weight), λ = 1.0
12 – ⅝ in. diam. studs; A_b/stud = 0.307 sq in.
stud length = 6 in.
d_h = 1.25 in.
f_s = 60,000 psi
V_u = 50 kips, N_u = 15 kips
(all load factors included)
column size 16 in. x 16 in.

Problem:

Determine if the studs are adequate for the connection and design the bracket plate.*

Solution:

Check studs

1. Strength based on concrete:

    a. Tension (group of top six studs)

    (1) Strength based on individual cones

    (Eqs. 6.5.4 and 6.5.5):

    $\phi P_c = 10.7 \ell_e(\ell_e + d_h)\lambda\sqrt{f'_c}(d_e/\ell_e)/1000$
    $= 10.7(6)(6 + 1.25)(\sqrt{5000})(5/6)/1000$
    $= 27.4$ k/stud

    (Note: The value obtained from Table 6.20.6 is 25.9 kips. The small difference is due to the use of actual embedment length i.e. stud length minus the stud head thickness in Table 6.20.6.)

    For six studs:

    $\phi P_c = 6(27.4) = 164.4$ kips

*Bracket plate may be designed using the plastic strength design method. See "Steel Structures — Design and Behavior", 3rd Edition, 1990, Charles G. Salmon and John E. Johnson, Harper and Row, New York.

(2) Strength based on truncated pyramid failure:

$h_{min} = (3 + 2(6))/2 = 7.5$ in. < 16

Use Case 3 in Fig. 6.5.6 corresponding to $h > h_{min}$.

$\phi P_c = \phi 4\lambda\sqrt{f'_c}(x + d_{e1} + d_{e2})(y + 2\ell_e)$
$= 0.85(4)(1.0)(\sqrt{5000})(6 + 5 + 5) \times (3 + 2(6))/1000$
$= 57.7$ kips

or, $P_c = 57.7/0.85 = 67.9$ kips

Note: Using Table 6.20.7A—Case 3, for $\ell_e = 6$ in. and $x_1 = 16$ in. ϕP_c values of 54 kips and 62 kips are given for $y_1 = 2$ in. and $y_1 = 4$ in. respectively. An average value of 58 kips for $y_1 = 3$ in. is thus obtained.

(3) Required tension strength, P_u for group of top six studs:

With reference to the free-body diagram shown, the required tension strength is (see reference included in footnote on previous page):

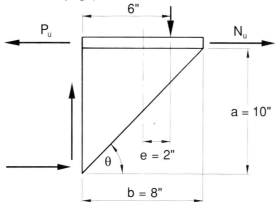

$P_u = V_u \cot\theta + N_u$
$= 50(0.8) + 15$
$= 55$ kips < 57.7 OK

b. Shear (all 12 studs)
Since d_e is large $(\geq 15d_b)$, Eq. 6.5.8 will apply:
$\phi V_c = 0.85(800)(0.307)(\sqrt{5000})(12)/1000$
$= 177.1$ kips
(Note: Table 6.20.8 yields $14.72(12) = 176.6$ kips)
$V_c = 177.1/0.85 = 208.4$ kips

c. Combined tension and shear
Using Eq. 6.5.15

$$\frac{1}{\phi}\left[\left(\frac{P_u}{P_c}\right)^2 + \left(\frac{V_u}{V_c}\right)^2\right] \leq 1.0$$

$$\frac{1}{0.85}\left[\left(\frac{55.0}{67.9}\right)^2 + \left(\frac{50.0}{208.4}\right)^2\right] = 0.84 < 1.0 \text{ OK}$$

2. Strength based on steel:
a. Tension (group of top six studs)
Using Eq. 6.5.7:
$\phi P_s = P_s$ (since $\phi = 1.0$)
$= 54,000(0.307)/1000$
$= 16.6$ k/stud
For six studs, $P_s = 6(16.6) = 99.6$ kips

b. Shear (all 12 studs)
Using Eq. 6.5.14:
$\phi V_s = V_s = 45,000(0.307)(12)/1000$
$= 165.8$ kips

c. Combined tension and shear
Using Eq. 6.5.16:

$$\frac{1}{\phi}\left[\left(\frac{P_u}{P_s}\right)^2 + \left(\frac{V_u}{V_s}\right)^2\right] \leq 1.0$$

$$\frac{1}{1.0}\left[\left(\frac{55.0}{99.6}\right)^2 + \left(\frac{50.0}{165.8}\right)^2\right] = 0.40 < 1.0 \text{ OK}$$

Thus, studs are adequate for the connection.

Bracket plate design

Note: The plastic steel design method used here assures yielding of the plate prior to buckling of the plate. This method may also be used to size gusset plates of connection angles (Sect. 6.5.9).

The bracket plate thickness, "t" is taken as the greater of (see free-body diagram at left for nomenclature):

$$t = \frac{V_u}{\phi\left[F_y \sin^2\theta\left(\sqrt{4e^2 + b^2} - 2e\right)\right]}$$

and, $t = b\dfrac{\sqrt{F_y}}{\phi(125)}$, for $0.5 \leq \dfrac{b}{a} \leq 1.0$

$t = b\dfrac{\sqrt{F_y}}{\phi\left[125\left(\dfrac{b}{a}\right)\right]}$, for $1.0 \leq \left(\dfrac{b}{a}\right) \leq 2.0$

where:
$\phi = 0.9$
F_y = yield strength of steel, ksi

substituting the given values:

$$t = \frac{50}{(0.9)\left[(36)\left(\dfrac{10}{12.8}\right)^2\left(\sqrt{4(2)^2 + (8)^2} - 2(2)\right)\right]}$$

$= 0.51$ in.

$\dfrac{b}{a} = 0.8$, $t = \dfrac{8\sqrt{36}}{0.9(125)}$

$t = 0.43$ in. < 0.51

use $\frac{9}{16}$ in. thick plate.

6.5.3 Deformed Bar Anchors

Deformed bar anchors are generally automatically welded to steel plates, similar to headed studs. They are anchored to the concrete by bond, and the development length can be taken the same as for Grade 60 reinforcing bars per ACI 318-89, Sect. 12.2 (see Design Aid 11.2.8).

6.5.4 Bolts and Threaded Connectors

In most connections, bolts are shipped loose and threaded into inserts. Embedded inserts are designed for concrete strength similar to that for studs. Occasionally a precast concrete member will be cast with a threaded connector projecting from the face. This is usually undesirable because of possible damage during handling.

High strength bolts are used infrequently in precast concrete connections because it is questionable as to whether the bolt pretension can be maintained when tightened against concrete. When used, AISC recommendations[12,13] should be followed.

Table 6.20.9 gives allowable working and design strengths for bolts in most commonly used threaded fasteners.

High strength threaded rods

Rods with threads and specially designed nuts and couplers are available with properties similar to Grade 60 reinforcing bars and post-tensioning bars. Design information is given in Chapter 11.

6.5.5 Inserts Cast in Concrete

Loop inserts of the type shown in Fig. 6.5.8 can be investigated in a manner similar to that for welded studs, using Eq. 6.5.3 for tensile strength, and Eqs. 6.5.8 and 6.5.9 for the shear strength. The strength as controlled by steel may be calculated based on wire strengths shown in Table 6.20.11 or the strength of the bolt or threaded rod shown in Tables 6.20.9 and 6.20.10, or taken from manufacturer's catalogs.

Manufacturer's data may include recommendations on factors of safety. The factor of safety used should reflect the actual insert application and whether it is applied to yield or to ultimate strength.

An evaluation of published test results[5] leads to certain characteristics common to most of the available inserts:

1. Controlling strength conditions of various types of inserts are similar.
2. Pullout strength decreases with decreasing unit weight of concrete.
3. For inserts located in zones of potential flexural cracking, the pullout strength should be reduced. Depending on the degree of cracking, a reduction in the range of 20 to 50 percent should be considered.

Fig. 6.5.8 Shear cone development for loop inserts

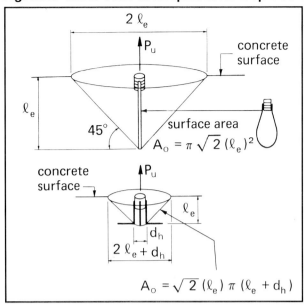

4. Sustained vibrations decrease strength by up to 30 percent.

Expansion inserts

All expansion inserts are proprietary, and design values should be taken from manufacturers' catalogs, although tensile strength can be estimated from Eq. 6.5.4 where ℓ_e = minimum recommended length. However, it should be noted that because of slip of the anchor in the hole, deeper embedment does not proportionally increase the capacity. Edge distances for expansion inserts are more critical than for cast-in inserts. Expanding the insert in the direction of the edge should be avoided (Fig. 6.5.9).

The performance of expansion inserts when subjected to stress reversal, vibrations or earthquake loading is not sufficiently known. Their use for these load conditions should be carefully considered by the designer.

Anchorage strength depends entirely on the lateral force (wedge action) on the concrete. Therefore, it is advisable to limit their use to connections with more shear than tension applied to them.

Fig. 6.5.9 Expansion inserts

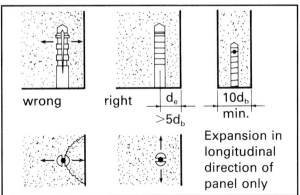

PCI Design Handbook/Fourth Edition   6–15

6.5.6 Structural Steel

Structural steel plates, angles, wide flange beams, channels, tubes, etc., are often used in connections (see Sect. 6.12). When designed using factored loads, the AISC LRFD Manual[13] is the appropriate design reference. For typical steel sections used in connections, plastic section properties and yield strengths with appropriate strength reduction factors are used. A dilemma exists where AISC load factors are different from those used by ACI. While it may be appropriate to carry two sets of load factors through a design, it would also be very cumbersome. It is thus suggested that ACI load factors be used for connection design and the AISC provisions be used to determine the usable strength to be provided by the steel.

In basic terms, flexural strength of a steel section can be determined as:

$\phi M_n = \phi F_y Z_s$

$\phi V_n = \phi (0.6 F_y) wh$

where:

$\phi = 0.9$

w, h = steel section dimensions

The plastic section modulus can be calulated using Table 6.20.12 or may be obtained from the AISC LRFD Manual[13]. For thin plates bent about their major axis, buckling provisions may govern the design.

Welding of structural steel

Nearly all structural steel used in precast concrete connections is ASTM A-36 steel. Thus, it is readily weldable with standard equipment.[4]

For corrosion resistance, exposed components of connections are sometimes hot-dip galvanized. Welding of hot-dip galvanized steel requires special care[2,6], such as thorough removal of galvanizing material. Stainless steel is an alternative for exposed connection parts where corrosion protection is important. Stainless steel plates are weldable to other stainless steel or to low carbon steel. The general procedure for welding low carbon steels should be followed, taking into account the stainless steel characteristics that differ, such as higher thermal expansion and lower thermal conductivity. Ref. 6 contains information on weldability of stainless steel.

The design strength of welds for use with factored loads can be determined from the provisions of the AISC LRFD Manual. The most commonly used welds in connections are full penetration welds or fillet welds. Full penetration strength is typically taken as the strength of the base metal. A "matching" weld metal per AWS D1.1 must be used. Full penetration welds loaded in shear may also be governed by the weld metal where the design weld stress is to be limited to $\phi (0.6 F_{exx})$ where $\phi = 0.8$. For fillet welds, as most commonly loaded in connections, the design strength may be limited by either the weld strength or the base metal strength. The weld design stress is limited to $\phi(0.6 F_{exx})$ where $\phi = 0.75$. The base metal design stress is limited to $\phi(0.6 F_u)$ where $\phi = 0.75$ and F_u is the tensile strength of the base metal.

Partial penetration groove welds have various design strength limits. The most typical application of this weld is in lap splices of reinforcing bars as discussed in section 6.5.1 When loaded in shear parallel to the weld axis, the limiting design stress is $\phi (0.6 F_{exx})$ where $\phi = 0.75$. Table 6.20.1 provides design strengths for fillet welds and partial penetration groove welds loaded in shear. Table 6.20.2 gives strengths of 45° fillet welds.

Minimization of cracking in concrete around welded connections

When welding is performed on components that are embedded in concrete, thermal expansion and distortion of the steel may destroy bond between the steel and concrete or induce cracking or spalling in the surrounding concrete.

The extent of cracking and distortion is dependent on the amount of heat generated during welding and the stiffness of the steel member. Heat may be reduced by: (1) use of low-heat welding rods of small size; (2) use of intermittent rather than continuous welds; or (3) using smaller welds in multiple passes.

Distortion can be minimized by using thicker steel sections. Providing a space around the metal on the surface with sealing foam or weather stripping may also reduce potential for damage.

Weld groups

Weld groups are more efficient than linear welds in resisting bending moments and torsion created by eccentric loads. Examples of this type of loading are shown in Fig. 6.5.10. Design procedure is illustrated by Fig. 6.5.11, where:

f_x = combined shear and torsion stress in horizontal direction

$= \dfrac{P_x}{A_w} + \dfrac{M_t y}{I_p}$

f_y = combined shear and torsion stress in vertical direction

$= \dfrac{P_y}{A_w} + \dfrac{M_t x}{I_p}$

f_r = resultant stress on the weld

$= \sqrt{(f_x)^2 + (f_y)^2}$

P_x = applied force in x direction

P_y = applied force in y direction

y = vertical distance from c.g. of weld group to point under investigation

x = horizontal distance from c.g. of weld group to point under investigation

I_p = polar moment of inertia
= $I_x + I_y = \Sigma I_{xx} + \Sigma A_w \bar{y}^2 + \Sigma I_{yy} + \Sigma A_w \bar{x}^2$

M_t = torsional moment = $P_x e_y + P_y e_x$

t_w = effective throat thickness of weld

A_w = area of weld = weld length x t_w

I_{xx}, I_{yy} = moment of inertia of weld segment with respect to its own axes

Fig. 6.5.10 Typical eccentric loadings and weld groups

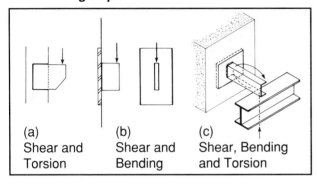

(a) Shear and Torsion (b) Shear and Bending (c) Shear, Bending and Torsion

Fig. 6.5.11 Eccentric bracket connection

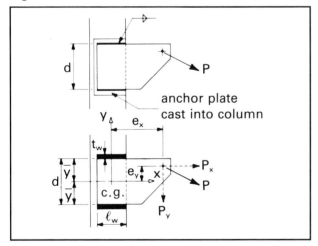

For computing nominal stresses the locations of the lines of weld are defined by edges along which the fillets are placed, rather than to the center of the effective throat. This makes little difference, since the throat dimension is usually small. By treating the welds as lines with t_w = 1, the physical properties of weld groups are simplified. The most commonly occurring weld groups are listed in Fig. 6.5.12.

The equations given above and in Fig. 6.5.12 are for elastic section properties, which is inconsistent, but conservative, when used with factored loads and design strength of weld material. When it is desirable to take maximum advantage of weld groups, AISC—LRFD Manual[13] should be used.

Example 6.5.4 Design of a weld group

Given:

Corner angle connection as shown

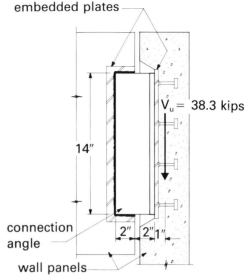

Angle size = 4 x 4 x ½ x 1'-2"

F_y = 36 ksi, E70 electrodes

Applied factored load = 38.3 kips

Problem:

Determine required weld size.

Solution:

Apply additional factor of 1.3 recognizing sensitivity of the connection.

$V_u = P_y = 1.3(38.3) = 49.8$ kips

Find c.g. of weld group:

$$\bar{x} = \frac{2(2) + 14(0)}{14 + 2(2)} = 0.22 \text{ in.}$$

by symmetry, $\bar{y}$ = 7 in.

$A_w = [2(2) + 14]t_w = 18 t_w$

$I_p = 2[t_w(2)^3/12 + 2t_w(0.78)^2 + 2(t_w)^3/12$
$\quad + 2(t_w)(72)] + t_w(14)^3/12 + 14(t_w)^3/12$
$\quad + 14(t_w)(0.22)^2$

$\quad = 429.1 t_w + 1.50(t_w)^3$

Since second term is small, it may be neglected. Alternatively, using Fig. 6.5.12, case 5:

b = 2 in., d = 14 in.

$\bar{x} = b^2/(2b+d) = 4/(4+14) = 0.22$ in.

$$I_p = \frac{8b^3 + 6bd^2 + d^3}{12} - \frac{b^4}{2b+d}$$

$$= \frac{8(2)^3 + 6(2)(14)^2 + 14^3}{12} - \frac{2^4}{2(2)+14}$$

$$= 429.1 \text{ in.}^4$$

Fig. 6.5.12 Properties of weld groups treated as lines ($t_w = 1$)

Section h = width; d = depth	Section modulus* $I_x/\bar{y}$	Polar moment of inertia, I_p, about center of gravity
1. 	$S = \dfrac{d^2}{6}$	$I_p = \dfrac{d^3}{12}$
2. 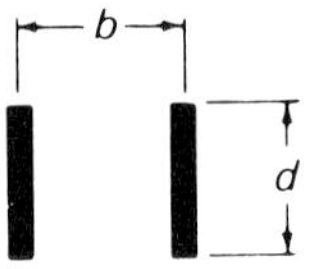	$S = \dfrac{d^2}{3}$	$I_p = \dfrac{d(3b^2 + d^2)}{6}$
3. 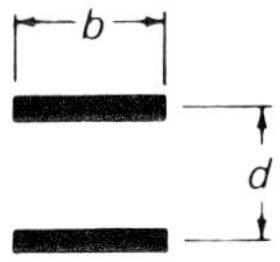	$S = bd$	$I_p = \dfrac{b(3d^2 + b^2)}{6}$
4. 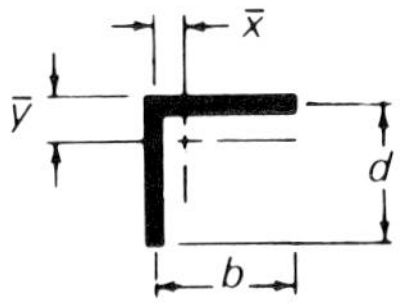 $\bar{y} = \dfrac{d^2}{2(b+d)}$; $\bar{x} = \dfrac{b^2}{2(b+d)}$	$S = \dfrac{4bd + d^2}{6}$	$I_p = \dfrac{(b+d)^4 - 6b^2d^2}{12(b+d)}$
5. 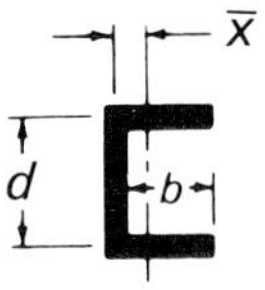 $\bar{x} = \dfrac{b^2}{2b+d}$	$S = bd + \dfrac{d^2}{6}$	$I_p = \dfrac{8b^3 + 6bd^2 + d^3}{12} - \dfrac{b^4}{2b+d}$
6. 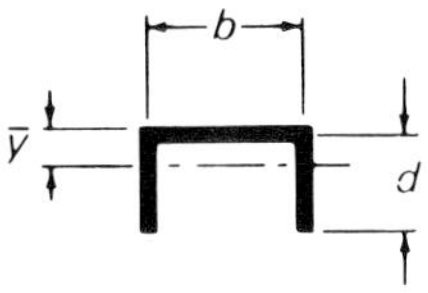 $\bar{y} = \dfrac{d^2}{b+2d}$	$S = \dfrac{2bd + d^2}{3}$	$I_p = \dfrac{b^3 + 6b^2d + 8d^3}{12} - \dfrac{d^4}{2d+b}$
7. 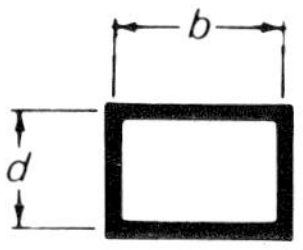	$S = bd + \dfrac{d^2}{3}$	$I_p = \dfrac{(b+d)^3}{6}$
8. 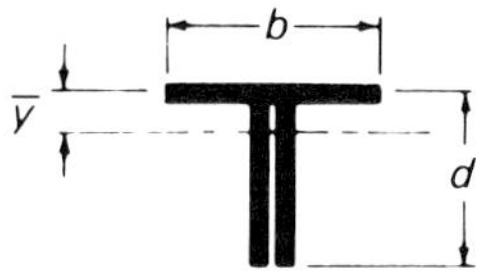 $\bar{y} = \dfrac{d^2}{b+2d}$	$S = \dfrac{2bd + d^2}{3}$	$I_p = \dfrac{b^3 + 8d^3}{12} - \dfrac{d^4}{b+2d}$
9. 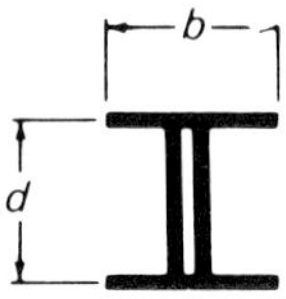	$S = bd + \dfrac{d^2}{3}$	$I_p = \dfrac{b^3 + 3b^2d + d^3}{6}$
10. 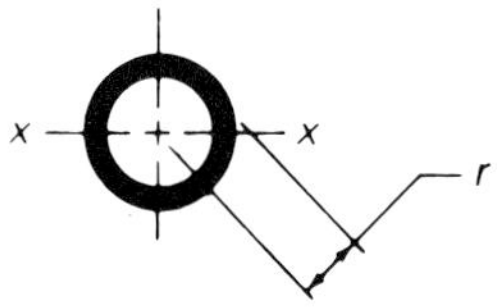	$S = \pi r^2$	$I_p = 2\pi r^3$

*Relates to top fiber. To obtain bottom fiber section modulus, multiply given values by $(d - \bar{y})/\bar{y}$.

$e_x = 1 + 4 - 0.22 = 4.78$ in.

$M_t = P_y e_x = 49.8(4.78) = 238$ in.-kips

$x = 2 - 0.22 = 1.78$

$$f_y = \frac{49.8}{18 t_w} + \frac{238(1.78)}{429.1 t_w}$$

$$= \frac{2.77}{t_w} + \frac{0.99}{t_w} = \frac{3.76}{t_w}$$

$$f_x = 0 + \frac{238(7)}{429.1 t_w} = \frac{3.88}{t_w}$$

$$f_r = \sqrt{\left(\frac{3.88}{t_w}\right)^2 + \left(\frac{3.76}{t_w}\right)^2}$$

$$= \frac{5.40}{t_w}$$

or, $t_w = \dfrac{5.40}{f_r}$

From Table 6.20.1, design strength of E70 weld
   = 31.5 ksi

$$t_w = \frac{5.40}{31.5} = 0.171 \text{ in.}$$

for a 45° fillet weld,

$$\text{leg size} = \frac{0.171}{0.707}$$

$$= 0.242 \text{ in.}$$

Use ¼ in. fillet weld (E70 electrode)

6.5.7 Post-Tensioning Steel

Post-tensioning is often used in connections, particularly those which are subjected to high tensile forces, such as the connections in moment-resisting frames. The post-tensioning is done using either 7-wire strand (ASTM A416) or bars (ASTM A722). The tensile strength of strand is either 250,000 or 270,000 psi, while for bars, it ranges from 150,000 to 160,000 psi.

In accounting for the effect of prestressing on connections due to post-tensioning, a reliable measurement of the prestressing force is necessary. Typically, 15 to 20 ft. length of strand or bar is desirable for this purpose, although shorter lengths can be used where anchor set is well defined and considered in prestress loss calculations. Some designers discount the effect of prestressing when short tendons are used and consider only the ductility and the tensile strength aspects of post-tensioning steel.

6.5.8 Bearing Pads

Bearing pads are used to distribute concentrated loads and reactions over the bearing area and to allow limited horizontal and rotational movements to provide stress relief. Their use has proven beneficial and often may be necessary for satisfactory performance of precast concrete structures.

Several materials are commonly used for bearing pads:

1. AASHTO-grade chloroprene pads are made with 100 percent chloroprene (neoprene) as the only elastomer and conform to the requirements of the AASHTO Standard Specifications for Highway Bridges (1989), Sect. 25. Inert fillers are used with the chloroprene and the resulting pad is black in color and of a smooth uniform texture. While allowable compressive stresses are somewhat lower than other pad types, these pads allow the greatest freedom in movement at the bearing. Note: chloroprene pads which do not meet the AASHTO Specifications are not recommended for use in precast concrete structures.

2. Pads reinforced with randomly oriented fibers have been used successfully for many years. These pads are usually black, and the short reinforcing fibers are clearly visible. Vertical load capacity is increased by the reinforcement, but capability of rotations and horizontal movement is somewhat less than chloroprene pads. Some random oriented fiber pads possess different properties in different directions in the plane of the pad. Therefore, unless proper planning and care is used in their installation, it may be prudent to specify those pads that have been tested to exhibit similar properties in different directions. There are no national standard specifications for this material. Manufacturers have developed appropriate design and performance documentation.

3. Cotton duck fabric reinforced pads are generally used where a higher compressive strength is desired. These pads are often yellow-orange in color and are reinforced with closely spaced, horizontal layers of fabric, bonded in the elastomer. The horizontal reinforcement layers are easily observed at the edge of the pad. Section 2.10.3(L) of the AASHTO Standard Specifications for Bridges and Military Specification MIL-C-882C discuss this material.

4. Chloroprene pads laminated with alternate layers of bonded steel or fiberglass are often used in bridges, but seldom in building construction. The above mentioned AASHTO Specifications cover these pads.

5. A multimonomer plastic bearing strip is manufactured expressly for bearing purposes. It is a commonly used material for the bearing sup-

port of hollow-core slabs, and is highly suitable for this application. It is also often used for bearing of architectural precast concrete cladding panels.

6. Tempered hardboard strips are also used with hollow-core slabs. These pads should be used with caution under moist conditions. In addition to progressive deterioration of the pad, staining of the precast concrete unit may occur.

7. TFE (trade name Teflon) coated materials are often used in bearing areas when large horizontal movements are anticipated, for example at "slip" joints or expansion joints. The TFE is normally reinforced by bonding to an appropriate backing material, such as steel. Fig. 6.5.13 shows a typical bearing detail using TFE, and Fig. 6.5.14 shows the range of friction coefficients that may be used for design. Typical allowable stress is about 1000 psi for virgin TFE and up to 2000 psi for filled material with reinforcing agents such as glass fibers.

Design recommendations

Bearing pads provide stress relief due to a combination of slippage and pad deformation. In elastomeric bearing pads (1 through 4 above), research[14] has shown that slippage is the more significant factor. This research has also shown that the ratio of shear to compressive stress on the pad reduces significantly under slow cyclic movements, such as those produced by temperature variations. Following are recommendations which, along with Figs. 6.5.15 and 6.5.16, can be used to select bearing pads.

1. Use unfactored service loads for design.

2. At the suggested maximum uniform compressive stress, instantaneous vertical strains of 10 to 20% can be expected. This number may double if the bearing surfaces are not parallel. In addition, the time-dependent creep will typically increase the instantaneous strains by 25 to 100%, depending on the magnitude of sustained dead load.

3. For stability of the pad, the length and width of unreinforced pads should be at least five times the thickness.

4. A minimum thickness of ¼ in. for joists and double tee stems, and ⅜ in. for beams is recommended.

5. Fig. 6.5.16 may be used to estimate shear resistance of chloroprene, random fiber reinforced and cotton duck pads.

6. The portion of pad outside of the covered bearing surface as well as the portion which is not

Fig. 6.5.13 Typical TFE bearing pad detail

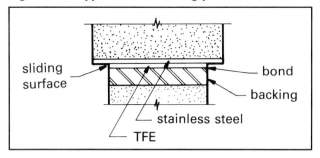

Fig. 6.5.14 TFE friction coefficients

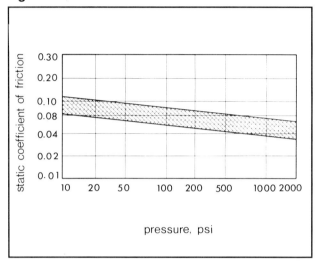

under load because of rotation of member should be ignored in calculating shape factors, pad stresses, stability and movements.

7. Shape factors, S, for unreinforced pads should be greater than 2 when used under tee stems, and greater than 3 under beams.

8. The sustained dead load compressive stress on unreinforced chloroprene pads should be limited to the range of 300 to 500 psi.

9. The volume change strains shown in Sect. 3.3.1 may be reduced by one half when calculating horizontal movement, Δ, because of compensating creep and slip in the bearing pad.

6.5.9 Connection Angles

Angles used to support precast members can be designed by principles of mechanics as shown in Fig. 6.5.17. In addition to the applied vertical and horizontal loads, the design should include all loads induced by restraint of relative movement between the precast member and the supporting member.

The minimum thickness of non-gusseted angles loaded in shear as shown in Fig. 6.5.18 can be determined by:

Fig. 6.5.15 Single layer bearing pads free to slip

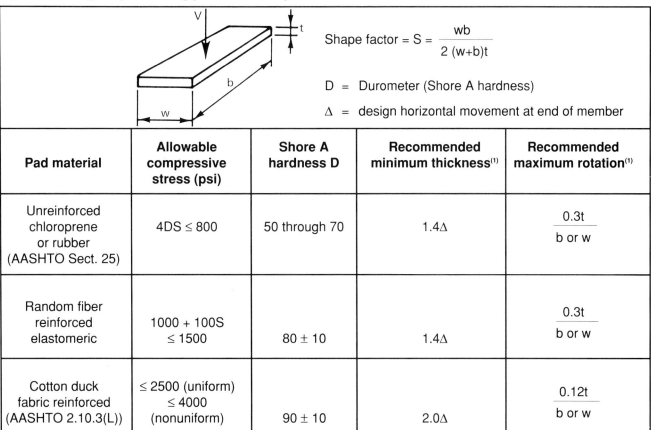

Shape factor = $S = \dfrac{wb}{2(w+b)t}$

D = Durometer (Shore A hardness)

Δ = design horizontal movement at end of member

Pad material	Allowable compressive stress (psi)	Shore A hardness D	Recommended minimum thickness[1]	Recommended maximum rotation[1]
Unreinforced chloroprene or rubber (AASHTO Sect. 25)	$4DS \leq 800$	50 through 70	1.4Δ	$\dfrac{0.3t}{b \text{ or } w}$
Random fiber reinforced elastomeric	$1000 + 100S$ ≤ 1500	80 ± 10	1.4Δ	$\dfrac{0.3t}{b \text{ or } w}$
Cotton duck fabric reinforced (AASHTO 2.10.3(L))	≤ 2500 (uniform) ≤ 4000 (nonuniform)	90 ± 10	2.0Δ	$\dfrac{0.12t}{b \text{ or } w}$

(1) The values in the table are based on sliding criteria. If sliding is not critical or testing indicates more advantageous conditions, thinner pads may be used. The minimum thickness and maximum rotation values for the cotton duck pad account for the effects of creep.

Fig. 6.5.16 Shear resistance of bearing pads

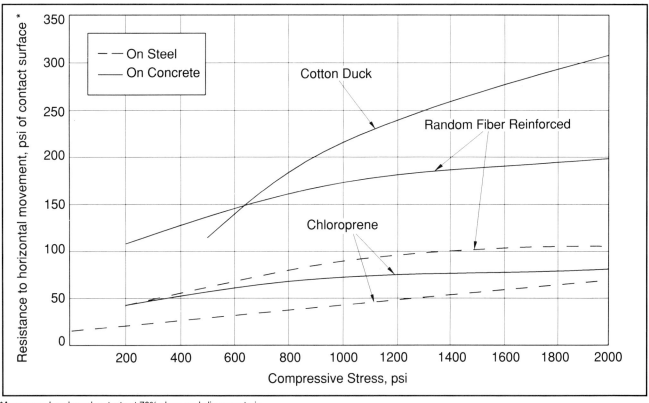

*Average values based on tests at 70% shear and slippage strain.

Fig. 6.5.17 Design parameters for connection angles

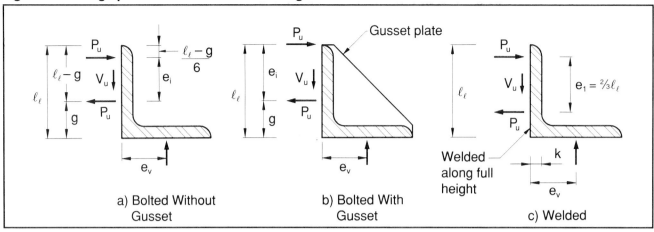

$$t = \sqrt{\frac{4V_u e_v}{\phi F_y b_n}} \quad \text{(Eq. 6.5.17)}$$

where:

$\phi = 0.90$

b_n = length of the angle − bolt hole diam. − 1/16 in.

design e_v = specified e_v + ½ in.

Note: For welded angles—Fig. 6.5.17 (c), design e_v in Eq. 6.5.17 may be taken equal to actual e_v minus k, where k is the distance from the back face of angle to web toe of fillet.

The tension on the bolt can be calculated by:

$$P_u = V_u \frac{e_v}{e_i} \quad \text{(Eq. 6.5.18)}$$

For angles loaded axially, Fig. 6.5.19, the minimum thickness of non-gusseted angles can be calculated by:

$$t = \sqrt{\frac{4N_u g}{\phi F_y b_n}} \quad \text{(Eq. 6.5.19)}$$

and the tension on the bolt can be calculated by:

$$P_u = N_u\left(1 + \frac{g}{e_i}\right) \quad \text{(Eq. 6.5.20)}$$

where

$\phi = 0.90$

g = gage of the angle (see Fig. 6.5.17)

The minimum edge distance for bolt holes is shown in Table 6.5.1.

Tables 6.20.13 and 6.20.14 may be used for the design of connection angles.

Fig. 6.5.18 Vertical load on angle

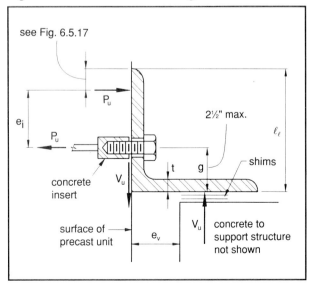

Fig. 6.5.19 Horizontal load on angle

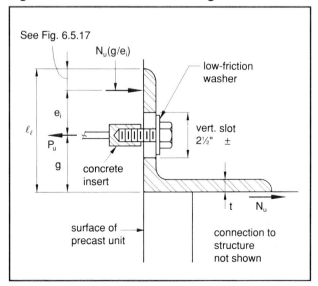

For connections made by welding instead of bolting, the welds should be designed in accordance with Sect. 6.5.6.

Table 6.5.1 Minimum edge distance for punched, reamed or drilled holes, measured from center of hole, in.***

Bolt diameter (in.)	At sheared edges	At rolled edges of plates, shapes or bars or gas cut edges **
½	⅞	¾
⅝	1⅛	⅞
¾	1¼	1
⅞	1½ *	1⅛
1	1¾ *	1¼
1¼	2¼	1⅝

*These may be 1¼ in. at the ends of beam connection angles.
**All edge distances in this column may be reduced ⅛ in. when the hole is at a point where stress does not exceed 25% of the maximum allowable stress in the element.
***When oversized or slotted holes are used, edge distances should be increased to provide the same distance from edge of hole to free edge as given by the tabulated edge distances.

6.6 Friction

The static coefficients of friction shown in Table 6.6.1 are conservative values for use in determining the upper limit of volume change forces in members wherein the connections would allow insufficient movement for stress relief. Thus, the maximum force resulting from the frictional restraint of axial movements can be determined by:

$$F_s = \mu_s V_{ud} \quad \text{(Eq. 6.6.1)}$$

where:

F_s = factored friction force

μ_s = static coefficient of friction as given in Table 6.6.1

V_{ud} = factored dead load force normal to the friction face

Friction may be depended upon to resist temporary construction loads as approved by the Engineer. In such cases, the coefficients of friction shown in Table 6.6.1 should be divided by 5.

Friction may also be depended upon to contribute to the resistance to design loads in specific situations where analysis and/or testing justifies its use. One such situation is the horizontal joints of precast wall panels which is illustrated in Ref. 2.

Table 6.6.1 Upper limits of static coefficients of friction of dry materials*

Material	μ_s
Elastomeric to steel or concrete	Fig. 6.5.16
Concrete to concrete	0.8
Concrete to steel	0.4
Steel to steel (not rusted)	0.25
TFE to stainless steel	Fig. 6.5.14
Hardboard to concrete	0.5
Multimonomer plastic (non-skid) to concrete	1.2
Multimonomer plastic (smooth) to concrete	0.4

*Reduce by 25% for wet conditions.

6.7 Shear-Friction

Shear-friction is an extremely useful tool in connection design, and certain other applications in precast and prestressed concrete structures.

Use of the shear-friction theory is recognized by Sect. 11.7 of ACI 318-89, which states that shear friction is "to be applied where it is appropriate to consider shear transfer across a given plane, such as an existing or potential crack, an interface between dissimilar materials, or an interface between two concretes cast at different times."

A basic assumption used in applying the shear-friction concept is that concrete within the direct shear area of the connection will crack in the most undesirable manner. Ductility is achieved by placing reinforcement across this anticipated crack so that the tension developed by the reinforcing bars will provide a force normal to the crack. This normal force in combination with "friction" at the crack interface provides the shear resistance. The shear-friction analogy can be adapted to designs for reinforced concrete bearing, corbels, daps, composite sections, and other connections.

An "effective shear-friction coefficient," μ_e, may be used[15] when the concept is applied to precast concrete connections. The shear-friction reinforcement nominally perpendicular to the assumed crack plane can be determined by:

$$A_{vf} = \frac{V_u}{\phi f_y \mu_e} \quad \text{(Eq. 6.7.1)}$$

where:

ϕ = 0.85

A_{vf} = area of reinforcement nominally perpendicular to the assumed crack plane, sq in.

f_y = yield strength of A_{vf}, psi (equal to or less than 60,000 psi)

Table 6.7.1 Shear-friction coefficients

Crack interface condition	Recommended μ	Maximum μ_e	Maximum $V_n = V_u/\phi$, lb
1. Concrete to concrete, cast monolithically	1.4λ	3.4	$0.30\lambda^2 f'_c A_{cr} \leq 1000\lambda^2 A_{cr}$
2. Concrete to hardened concrete, with roughened surface	1.0λ	2.9	$0.25\lambda^2 f'_c A_{cr} \leq 1000\lambda^2 A_{cr}$
3. Concrete to concrete	0.6λ	2.2	$0.20\lambda^2 f'_c A_{cr} \leq 800\lambda^2 A_{cr}$
4. Concrete to steel	0.7λ	2.4	$0.20\lambda^2 f'_c A_{cr} \leq 800\lambda^2 A_{cr}$

V_u = applied factored shear force, parallel to the assumed crack plane, lb (limited by the values given in Table 6.7.1)

$$\mu_e = \frac{1000 \lambda A_{cr} \mu}{V_u} \leq \text{values in Table 6.7.1} \quad \text{(Eq. 6.7.2)}$$

λ = 1.0 for normal weight concrete
  = $(f_{ct}/6.7)/\sqrt{f'_c}$ for sand-lightweight and all-lightweight concrete. If f_{ct} is unknown:
     0.85 for sand-lightweight concrete
     0.75 for all-lightweight concrete

f_{ct} = splitting tensile strength of concrete, psi

μ = shear-friction coefficient (values in Table 6.7.1)

A_{cr} = area of the crack interface, sq in. (varies depending on the type of connection)

When axial tension is present, additional reinforcement area should be provided:

$$A_n = \frac{N_u}{\phi f_y} \quad \text{(Eq. 6.7.3)}$$

where:

A_n = area of reinforcement required to resist axial tension, sq in.

N_u = applied factored horizontal tensile force nominally perpendicular to the assumed crack plane, lb

ϕ = 0.85

All reinforcement, either side of the assumed crack plane, should be properly anchored to develop the design forces. The anchorage may be achieved by extending the reinforcement for the required development length (with or without hooks), or by welding to reinforcing bars, angles or plates.

6.8 Bearing on Plain Concrete

Plain concrete bearing may be used in situations where the bearing is uniform and the bearing stresses are low as is typical, for example, in hollow-core and solid slabs. In other situations, and in thin stemmed members where the bearing area is smaller than 20 sq in., a minimum reinforcement equal to $N_u/\phi f_y$ (but not less than one No. 3 bar) is recommended.

The design bearing strength of plain concrete may be calculated as:

$$\phi V_n = \phi C_r (0.85 f'_c A_1) \sqrt{\frac{A_2}{A_1}} \leq 1.2 f'_c A_1 \quad \text{(Eq. 6.8.1)}$$

where:

ϕV_n = design bearing strength

ϕ = 0.70

Fig. 6.8.1 Bearing on plain concrete

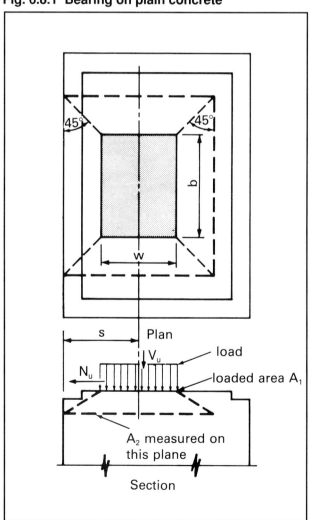

$$C_r = \left(\frac{sw}{200}\right)^{\frac{N_u}{V_u}}$$

= 1.0 when reinforcement is provided in direction of N_u, in accordance with Sect. 6.9, or when N_u is zero. sw should not be taken greater than 9.0 sq in.

A_1 = loaded area

A_2 = maximum area of the portion of the supporting surface that is geometrically similiar to and concentric with the loaded area

6.9 Reinforced Concrete Bearing

If the applied load, V_u, exceeds the design bearing strength, ϕV_n, as calculated by Eq. 6.8.1, reinforcement is required in the bearing area. This reinforcement can be designed by shear-friction as discussed in Sect. 6.7. Referring to Fig. 6.9.1, the reinforcement $A_{vf} + A_n$ nominally parallel to the direction of the axial load, N_u, is determined by Eqs. 6.7.1 through 6.7.3 with A_{cr} equal to b x h (i.e., $\theta \approx 0°$, see Ref. 2). This reinforcement must be appropriately developed by hooks or by welding to anchor bar, bearing plate or angle.

Vertical reinforcement across potential horizontal cracks can be calculated by:

$$A_{sh} = \frac{(A_{vf} + A_n)f_y}{\mu_e f_{ys}} \qquad \text{(Eq. 6.9.1)}$$

where:

$$\mu_e = \frac{1000\lambda A_{cr}\mu}{(A_{vf} + A_n)f_y}$$

f_{ys} = yield strength of A_{sh}, psi

A_{cr} = ℓ_d b, sq in.

b = average member width, in.

ℓ_d = development length of $A_{vf} + A_n$ bars, in.

Stirrups or mesh used for diagonal tension reinforcement can be considered to act as A_{sh} reinforcement.

When members are subjected to bearing stresses in excess of the limits indicated in Sect. 10.15 of ACI 318-89, confinement reinforcement in all directions may be necessary.

Example 6.9.1 Reinforced bearing for a rectangular beam

Given:

PCI standard rectangular beam 16RB28

V_u = 94 kips (includes all load factors)

Fig. 6.9.1 Reinforced concrete bearing

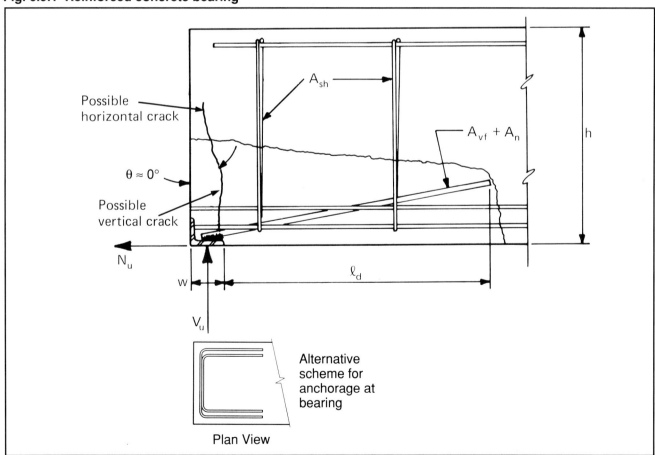

N_u = 20 kips (includes all load factors)
bearing pad = 4 in. x 14 in.
f_y = 60,000 psi (all reinforcement)
f'_c = 5000 psi (normal weight)

Problem:

Determine reinforcement requirements for the end of the member.

Solution:

A_{cr} = b(h) = 16(28) = 448 sq in.

Check max V_n from Table 6.7.1

$1000\lambda^2 A_{cr}$ = 1000(1.0)²(448)/1000 = 448 kips

max V_u = 0.85(448) = 381 kips > 94 OK

By Eq. 6.7.2:

$$\mu_e = \frac{1000\lambda A_{cr}\mu}{V_u} = \frac{1000(1)(448)(1.4)(1)}{94,000}$$

= 6.67 > 3.4, use 3.4

By Eq. 6.7.1:

$$A_{vf} = \frac{V_u}{\phi f_y \mu_e} = \frac{94,000}{0.85(60,000)(3.4)}$$

= 0.54 sq in.

By Eq. 6.7.3:

$$A_n = \frac{N_u}{\phi f_y} = \frac{20,000}{0.85(60,000)}$$

= 0.39 sq in.

$A_{vf} + A_n$ = 0.54 + 0.39 = 0.93 sq. in.

Use 3 – #5 = 0.93 sq in.

Determine ℓ_d from Design Aid 11.2.8:

For α_A = 1.0 = α_B, $(\ell_d)_1$ = 15.9 in.

For α_C = 1.0 = α_E and α_{mt} = 1.0,

ℓ_d = 15.9(1)(1)(1)(1) = 15.9 in.

$A_{cr} = \ell_d b$ = (15.9)(16) = 254 sq in.

$$\mu_e = \frac{1000(1)(254)(1.4)(1)}{0.93(60,000)} = 6.35 > 3.4$$

Use μ_e = 3.4

$$A_{sh} = \frac{(A_{vf} + A_n)f_y}{\mu_e f_{ys}} = \frac{0.93(60,000)}{3.4(60,000)}$$

= 0.27 sq in.

Use 2 – #3 stirrups = 0.44 sq in.

6.10 Column Base Plates

Column bases must be designed for both erection loads and loads which occur in service, the former often being more critical. Two commonly used base plate details are shown in Fig. 6.10.1 although many other details are also used.

If in the analysis for erection loads or temporary construction loads, before grout is placed under the plate, all the anchor bolts are in compression, the base plate thickness required to satisfy bending is determined from:

$$t = \sqrt{\frac{(\Sigma F)4x_c}{\phi b F_y}} \quad \text{(Eq. 6.10.1)}$$

where:

ϕ = 0.90

x_c, b = dimensions as shown in Fig. 6.10.1, in.

F_y = yield strength of the base plate, psi

ΣF = greatest sum of anchor bolt factored forces on one side of the column, lb

If the analysis indicates the anchor bolts on one or both sides of the column are in tension, the base plate thickness is determined by:

$$t = \sqrt{\frac{(\Sigma F)4x_t}{\phi b F_y}} \quad \text{(Eq. 6.10.2)}$$

where:

ϕ = 0.9

x_t = dimension as shown in Fig. 6.10.1, in.

Under loads which occur at service, the base plate thickness may be controlled by bearing on the concrete or grout. In this case, the base plate thickness is determined by:

$$t = x_o \sqrt{\frac{2f_{bu}}{\phi F_y}} \quad \text{(Eq. 6.10.3)}$$

where:

ϕ = 0.9

x_o = dimension as shown in Fig. 6.10.1, in.

f_{bu} = bearing stress on concrete or grout under factored loads $\leq \phi(0.85)f'_c$, where ϕ = 0.7

Table 6.20.15 may be used for base plate design. Nominal base plate shearing stresses should not exceed $0.9(0.6 F_y)$.

The anchor bolt diameter is determined by the tension or compression on the stress area of the threaded portion of the bolt. Anchor bolts may be ASTM A307 bolts or, more frequently, threaded rods of ASTM A36 steel. The requirements for structural integrity (Sect. 3.10) must also be satisfied.

Fig. 6.10.1 Column base connections

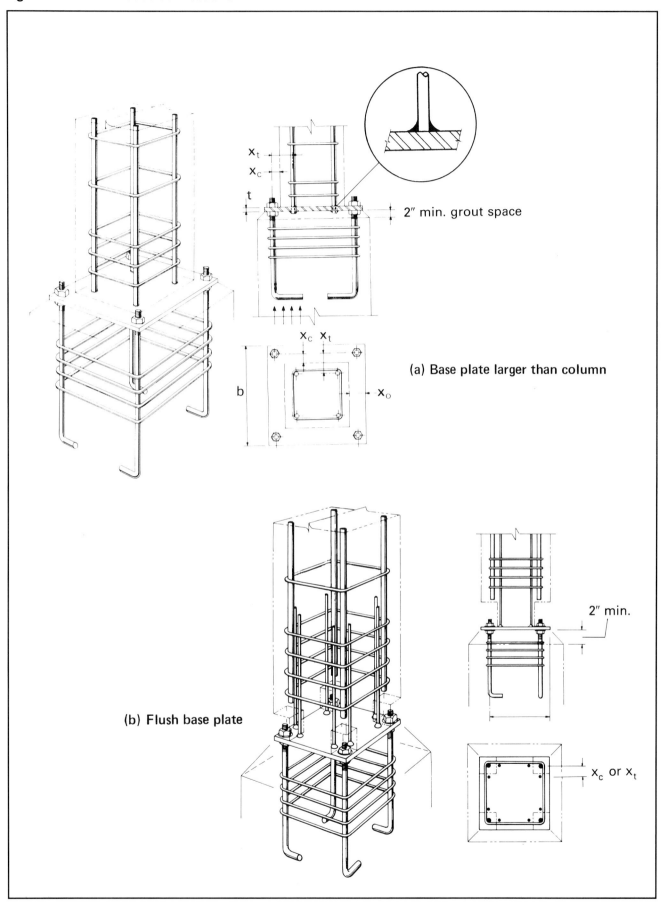

In most cases, both base plate and anchor bolt stresses can be significantly reduced by using properly placed shims during erection.

When the bolts are near a free edge, as in a pier or wall, the buckling of the bolts before grouting may be a consideration. Confinement reinforcement, as shown in Fig. 6.10.1, should be provided in such cases. A minimum of 4 No. 3 ties at about 3 in. centers is recommended for confinement.

The in-place tension strength of the bolt should be taken as the lesser of the strengths calculated based on concrete failure and steel failure (typically yield.) The calculation of concrete based strength will depend upon the type of anchor bolt used. For anchor bolts with head/washer of adequate stiffness, similar to headed studs, the strength can be determined by assuming a shear cone pull-out failure described in Sect. 6.5.2. Otherwise, and for hooked anchor bolts, the strength should be determined by adding the bond resistance of the bolt shank and the bearing resistance of the bolt head or the hook. If necessary, the bearing area of the bolt head can be increased by welding a washer or steel plate. Nominal bond stress on smooth anchor bolts should not exceed 250 psi. The confined bearing stress on the hook or bolt head should not exceed $\phi(0.85 f'_c)\sqrt{A_2/A_1}$ as per ACI 318-89, Sect. 10.15. Typically, $\sqrt{A_2/A_1}$ will be larger than 2.0 in such cases, thus the limiting value for design bearing strength of $0.7(0.85)(f'_c) \times 2.0 = 1.2 f'_c$ may be used.

The bottom of the bolt should be a minimum of 4 in. above the bottom of the footing, and also above the footing reinforcement.

6.11 Concrete Brackets or Corbels

ACI 318-89 prescribes the design method for corbels, based on Refs. 16 and 17. The equations in this section follow those recommendations, and are subject to the following limitations (see Figs. 6.11.1 and 6.11.2):

1. $a/d \leq 1$
2. $N_u \leq V_u$
3. $\phi = 0.85$ for all calculations
4. Anchorage at the front face must be provided by welding or other positive means.
5. Concentrated loads on continuous corbels may be distributed as for beam ledges, Sect. 6.14.

The area of primary tension reinforcement, A_s, is the greater of $A_f + A_n$ as calculated below, or $\tfrac{2}{3}A_{vf} + A_n$ as calculated in Sect. 6.7;

$$A_f = \frac{V_u a + N_u(h-d)}{\phi f_y d} \quad \text{(Eq. 6.11.1)}$$

Fig. 6.11.1 Design of concrete corbels

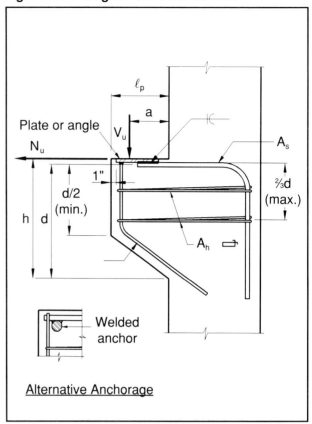

$$A_n = \frac{N_u}{\phi f_y} \quad \text{(Eq. 6.11.2)}$$

For convenience, the equations can be rewritten so that A_s shall be the greater of Eq. 6.11.3 and 6.11.4:

$$A_s = \frac{1}{\phi f_y}\left[V_u\left(\frac{a}{d}\right) + N_u\left(\frac{h}{d}\right)\right] \quad \text{(Eq. 6.11.3)}$$

$$A_s = \frac{1}{\phi f_y}\left[\frac{2V_u}{3\mu_e} + N_u\right] \quad \text{(Eq. 6.11.4)}$$

$$A_s \text{ (min.)} = 0.04(f'_c/f_y)bd \quad \text{(Eq. 6.11.5)}$$

$$A_h \geq 0.5(A_s - A_n) \quad \text{(Eq. 6.11.6)}$$

A_h should be distributed within the upper $\tfrac{2}{3}$ d.

The shear strength of a corbel is limited by the maximum values given in Table 6.7.1.

Example 6.11.1 Reinforced concrete corbel

Given:

A concrete corbel similar to that shown in Fig. 6.11.1

V_u = 80 kips (includes all load factors)

N_u = 15 kips (includes all load factors)

f_y = Grade 60 (weldable)
f'_c = 5000 psi (normal weight)
Bearing pad—12 in. x 6 in.
b = 14 in.
ℓ_p = 8 in.

Problem:

Find corbel depth and reinforcement.

Solution:

Try h = 14 in., d = 13 in.

a = ¾ ℓ_p = 6 in.

From Table 6.7.1:

max V_n = 1000 λ^2 A_{cr}
    = 1000(1)²(14 x 14)/1000
    = 196 kips

max V_u = 0.85(196) = 166.6 kips > 80.0 OK

By Eq. 6.11.3:

$$A_s = \frac{1}{\phi f_y}\left[V_u\left(\frac{a}{d}\right) + N_u\left(\frac{h}{d}\right)\right]$$

$$= \frac{1}{0.85(60)}\left[80\left(\frac{6}{13}\right) + 15\left(\frac{14}{13}\right)\right]$$

= 1.04 sq in.

By Eq. 6.7.2:

$$\mu_e = \frac{1000\lambda bh\mu}{V_u} = \frac{1000(1)(14)(14)(1.4)(1)}{80,000}$$

= 3.43 > 3.4, Use 3.4

By Eq. 6.11.4:

$$A_s = \frac{1}{\phi f_y}\left[\frac{2V_u}{3\mu_e} + N_u\right]$$

$$= \frac{1}{0.85(60)}\left[\frac{2(80)}{3(3.4)} + 15\right]$$

= 0.60 sq in. < 1.04

By Eq. 6.11.5:

min. A_s = 0.04bd(f'_c/f_y) = 0.04(14)(13)(5/60)
    = 0.61 sq in. < 1.04

Provide 2 – #7 bars = 1.20 sq in.

The A_s reinforcement could also be estimated from Table 6.20.16:

For b = 14 in., ℓ_p = 8 in., and h = 14 in., the corbel would have a strength of about 65 kips with 2 – No. 6 bars and 89 kips with 2 – No. 7 bars. Use 2 – No. 7 bars for V_u = 80 kips.

Fig. 6.11.2 Corbel force diagrams and typical reinforcement — wall panels

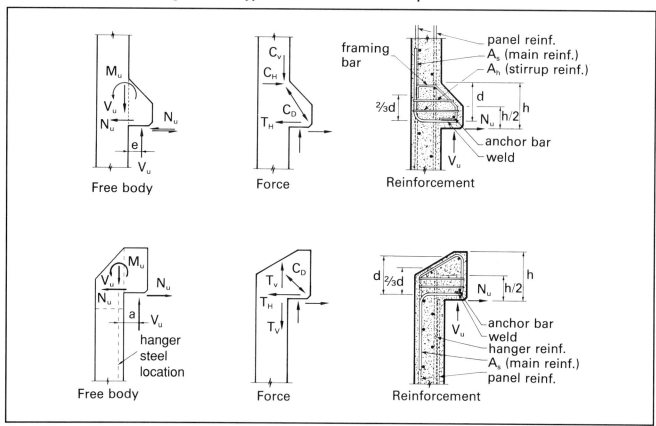

By Eq. 6.11.6:

$$A_h = 0.5(A_s - A_n) = 0.5\left[1.04 - \frac{15}{(0.85)(60)}\right]$$

$$= 0.37 \text{ sq in.}$$

Provide 2 – #3 closed ties = 0.44 sq in.

6.12 Structural Steel Haunches

Structural steel shapes, such as wide flange beams, double channels, tubes or vertical plates, often serve as haunches or brackets as illustrated in Fig. 6.12.1. The concrete-based capacity of these members can be calculated by statics, using the assumptions shown in Figs. 6.12.2 and 6.12.3.[18]

The nominal strength of the section is:

$$V_c = \frac{0.85 f'_c b \ell_e}{1 + 3.6 e/\ell_e} \quad \text{(Eq. 6.12.1)}$$

where:
- V_c = nominal strength of the section controlled by concrete, lb
- e = $a + \ell_e/2$, in.
- a = shear span, in.
- ℓ_e = embedment length, in.
- b = effective width of the compression block, in.

For the additional contribution of reinforcement welded to the embedded shape, and properly developed in the concrete, and with $A'_s = A_s$:

$$V_r = \frac{2A_s f_y}{1 + \frac{6e/\ell_e}{4.8s/\ell_e - 1}} \quad \text{(Eq. 6.12.2)}$$

where:
- f_y = yield strength of reinforcement

then $V_n = (V_c + V_r)$; and $V_u \leq \phi V_n$, where $\phi = 0.85$.

Other notation for Eqs. 6.12.1 and 6.12.2 are shown in Fig. 6.12.3.

The following assumptions and limitations are recommended:

1. In a column with closely spaced ties (spacing ≤ 3 in.) above and below the haunch, the effective width, b, can be assumed as the width of the confined region, i.e., outside to outside of ties, or 2.5 times the width of the steel section, whichever is less.
2. Thin-walled members, such as the tube shown in Fig. 6.12.3, may require filling with concrete to prevent local buckling.
3. When the supplemental reinforcement, A_s and A'_s, is anchored both above and below the members, as in Fig. 6.12.3, it can be counted twice.
4. The critical section for bending of the steel member is located a distance $V_u/(0.85 f'_c b)$ inward from the face of the column.

If the steel section projects from both sides, as in Fig. 6.12.2(a), the eccentricity factor, e/ℓ_e in Eq. 6.12.1 should be calculated from the total unbalanced live load. Conservatively, e/ℓ_e may be taken equal to 0.5.

The design strength of the steel section can be determined by:

Flexural design strength:

$$\phi V_n = \frac{\phi Z_s F_y}{a + V_u/(0.85 f'_c b)} \quad \text{(Eq. 6.12.3)}$$

Shear design strength:

$$\phi V_n = \phi(0.6 F_y h t) \quad \text{(Eq. 6.12.4)}$$

where:
- Z_s = plastic section modulus of the steel section (see Table 6.20.12)
- F_y = yield strength of the steel
- h, t = depth and thickness of steel web, respectively
- ϕ = 0.90

Horizontal forces, N_u, are resisted by bond on the perimeter of the section. If the bond stress resulting from factored loads exceeds 250 psi, headed studs or reinforcing bars should be welded to the embedded steel section.

Example 6.12.1 Design of structural steel haunch

Given:

The structural steel haunch shown.

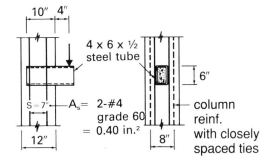

Fig. 6.12.1 Structural steel haunches

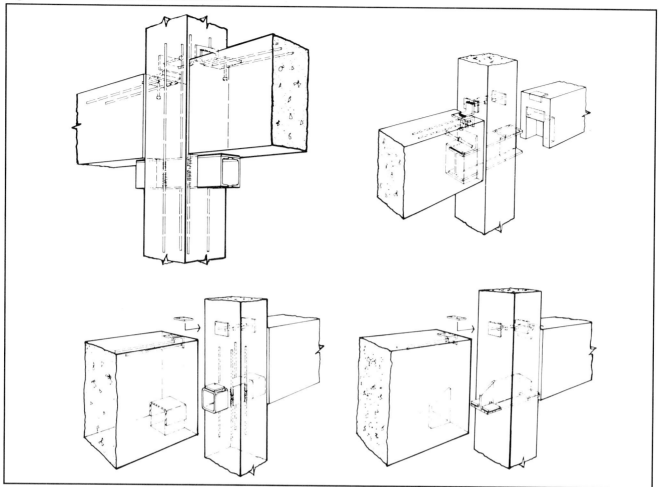

Fig. 6.12.2 Stress-strain relationships

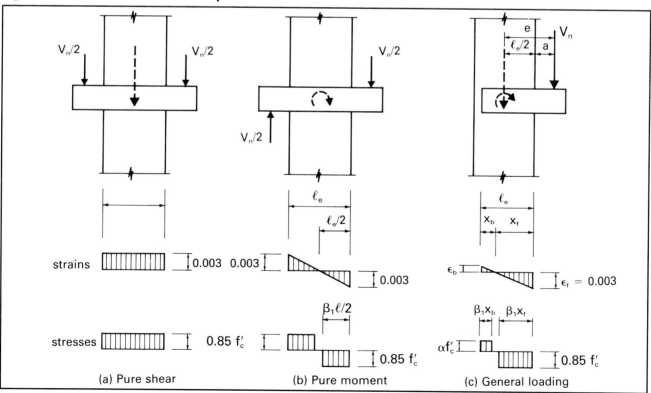

(a) Pure shear (b) Pure moment (c) General loading

Fig. 6.12.3 Assumptions and notation— steel haunch design

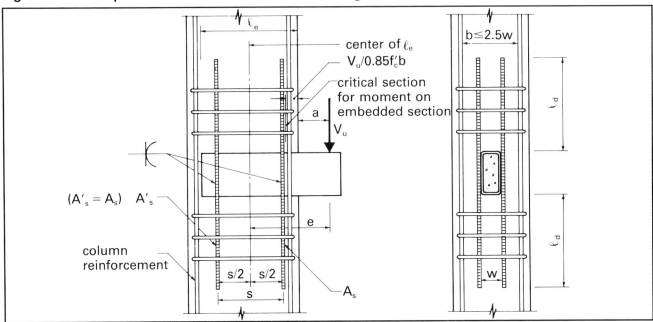

f'_c = 5000 psi
f_y (reinforcement) = 60,000 psi (weldable)
F_y (structural steel) = 46,000 psi

Problem:
Find the design strength.

Solution:
Effective width, b = confined width (8 in.)
or b = 2.5w = 2.5(4) = 10 in.
Use b = 8 in.   e = 4 + 10/2 = 9 in.

$$V_c = \frac{0.85f'_c b\ell_e}{1 + 3.6e/\ell_e} = \frac{0.85(5)(8)(10)}{1 + 3.6(9)/(10)}$$

$$= 80.2 \text{ kips}$$

Since the A_s bars are anchored above and below, they can be counted twice.

A_s = 2 – #4 = 2(2)(0.2) = 0.80 sq in.

$$V_r = \frac{2A_s f_y}{1 + \frac{6e/\ell_e}{4.8s/\ell_e - 1}} = \frac{2(0.80)(60)}{1 + \frac{6(9)/10}{4.8(7)/(10) - 1}}$$

$$= 29.2 \text{ kips}$$

ϕV_n = 0.85(80.2 + 29.2) = 93.0 kips

Alternative solution using Tables 6.20.17 and 6.20.18:
For b = 8 in., a = 4 in., ℓ_e = 10 in.
Read ϕV_c = 68 kips
For A_s = 2 – #4, anchored above and below
Read V_r = 28 kips

$\phi V_n = \phi V_c + \phi V_r$ = 68 + 0.85(28) = 92.0 kips

Steel section shear capacity:
$\phi V_n = \phi(0.6 F_y ht)$ = 0.9(0.6)(46)(6)(2)(0.5)
= 149 kips

Steel section flexure capacity:
From AISC—LRFD Manual[13]:
Z_s = 15.4
Assume V_u = 85 kips,

$$\phi V_n = \frac{\phi Z_s F_y}{a + V_u / 0.85 f'_c b} = \frac{0.9(15.4)(46)}{4 + 2.50}$$

= 98.0 kips

Steel flexure controls.

6.13 Dapped-End Connections

Design of connections which are recessed or dapped into the end of the member greater than 0.2 times the height (H in Fig. 6.13.1), requires the investigation of several potential failure modes. These are numbered and shown in Fig. 6.13.1 and listed below along with the reinforcement required for each consideration. It should be noted that the design equations given in this section are based primarily on Refs. 19 and 20 and are appropriate for cases where shear span-to-depth ratio (a/d in Fig. 6.13.1) is not more than 1.0.*

---

*For a/d > 1.0, until PCI design guidelines are available, information given in the Proceedings of the U.S.-Japan Joint Seminar on Precast Concrete, "Behavior and Design of Dapped End Members" by A.H. Mattock, 1986 may be used.

Fig. 6.13.1 Potential failure modes and required reinforcement in dapped-end connections

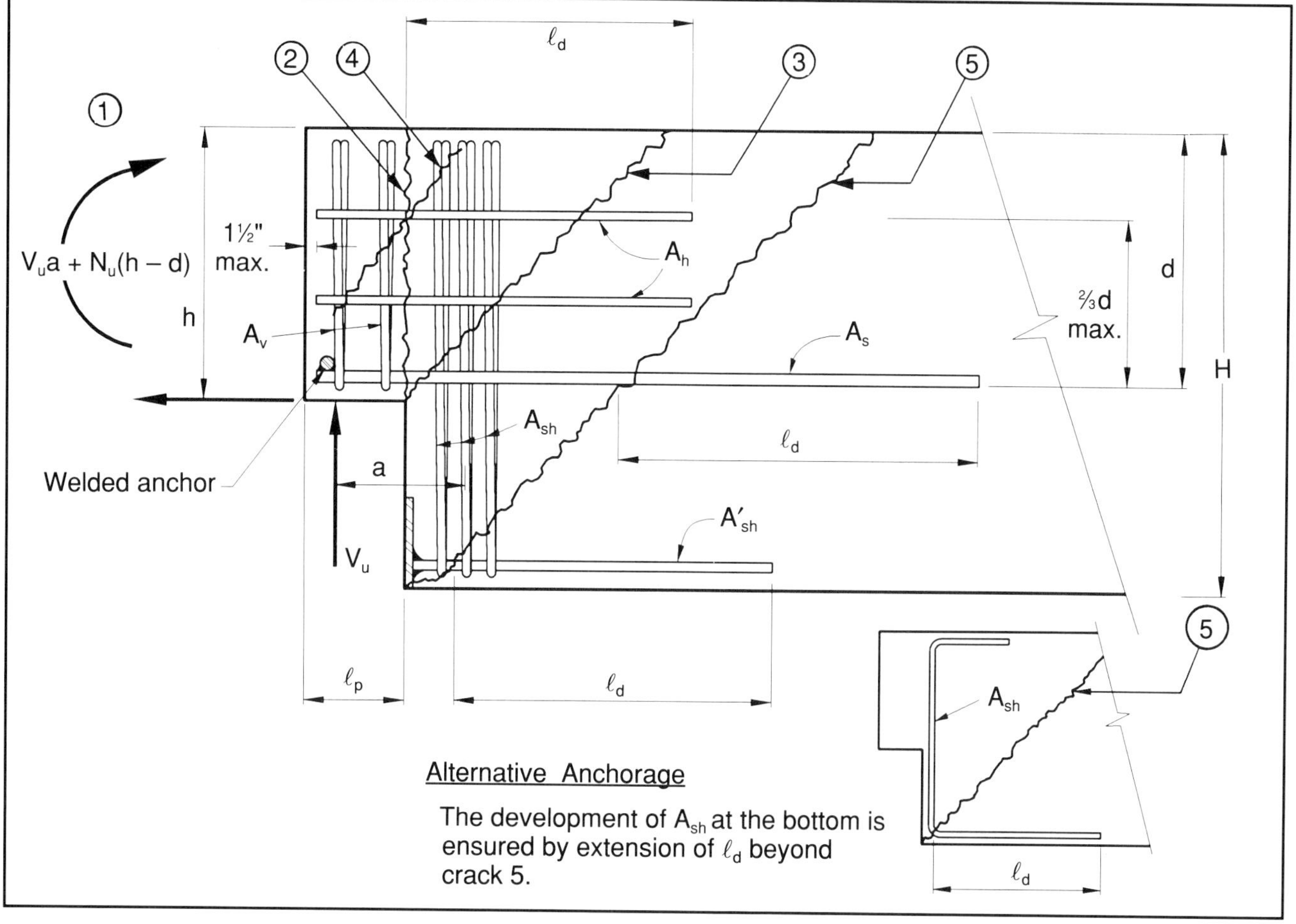

1. Flexure (cantilever bending) and axial tension in the extended end. Provide flexural reinforcement, A_f, plus axial tension reinforcement, A_n.
2. Direct shear at the junction of the dap and the main body of the member. Provide shear-friction reinforcement composed of A_{vf} and A_h, plus axial tension reinforcement, A_n.
3. Diagonal tension emanating from the reentrant corner. Provide shear reinforcement, A_{sh}.
4. Diagonal tension in the extended end. Provide shear reinforcement composed of A_h and A_v.
5. Diagonal tension in the undapped portion. This is resisted by providing a full development length for A_s beyond the potential crack.[19]

Each of these potential failure modes should be investigated separately. The reinforcement requirements are not cumulative, that is, A_s is the greater of that required by 1 or 2, not the sum. A_h is the greater of that required by 2 or 4, not the sum.

6.13.1 Flexure and Axial Tension in the Extended End

The horizontal reinforcement is determined in a manner similar to that for column corbels, Sect. 6.11.

Thus:
$$A_s = A_f + A_n = \frac{1}{\phi f_y}\left[V_u\left(\frac{a}{d}\right) + N_u\left(\frac{h}{d}\right)\right] \quad \text{(Eq. 6.13.1)}$$

where:
- ϕ = 0.85*
- a = shear span, in., measured from load to center of A_{sh}
- h = depth of the member above the dap, in.
- d = distance from top to center of the reinforcement, A_s, in.
- f_y = yield strength of the flexural reinforcement, psi

6.13.2 Direct Shear

The potential vertical crack shown in Fig. 6.13.1 is resisted by a combination of A_s and A_h. This reinforcement can be calculated by Eqs. 6.13.2 through 6.13.4.

*To be exact, Eq. 6.13.1 should have $j_u d$ in the denominator. The use of $\phi = 0.85$ instead of 0.90 (flexure) compensates for this approximation. Also, for uniformity $\phi = 0.85$ is used throughout Sect. 6.13 even though $\phi = 0.90$ would be more appropriate and may be used where flexure and direct tension are addressed.

$$A_s = \frac{2V_u}{3\phi f_y \mu_e} + A_n \quad \text{(Eq. 6.13.2)}$$

$$A_n = \frac{N_u}{\phi f_y} \quad \text{(Eq. 6.13.3)}$$

$$A_h = 0.5(A_s - A_n) \quad \text{(Eq. 6.13.4)}$$

where:

$\phi = 0.85$

f_y = yield strength of A_s, A_n, A_h, psi

$\mu_e = \dfrac{1000\lambda bh\mu}{V_u} \leq$ values in Table 6.7.1

The shear strength of the extended end is limited by the maximum values given in Table 6.7.1.

6.13.3 Diagonal Tension at Reentrant Corner

The reinforcement required to resist diagonal tension cracking starting from the reentrant corner, shown as 3 in Fig. 6.13.1, can be calculated from:

$$A_{sh} = \frac{V_u}{\phi f_y} \quad \text{(Eq. 6.13.5)}$$

where:

$\phi = 0.85$

V_u = applied factored load

A_{sh} = vertical or diagonal bars across potential diagonal tension crack, sq in.

f_y = yield strength of A_{sh}

6.13.4 Diagonal Tension in the Extended End

Additional reinforcement for crack 4 in Fig. 6.13.1 is required in the extended end, such that:

$$\phi V_n = \phi\left(A_v f_y + A_h f_y + 2\lambda bd\sqrt{f_c'}\right) \quad \text{(Eq. 6.13.6)}$$

At least one half of the reinforcement required in this area should be placed vertically. Thus:

$$\text{min. } A_v = \frac{1}{2f_y}\left(\frac{V_u}{\phi} - 2\lambda bd\sqrt{f_c'}\right) \quad \text{(Eq. 6.13.7)}$$

6.13.5 Anchorage of Reinforcement

With reference to Fig. 6.13.1:

1. Horizontal bars A_s should be extended a minimum of ℓ_d past crack 5, and anchored at the end of the beam by welding to cross bars, plates or angles.

2. Horizontal bars A_h should be extended a minimum of ℓ_d past crack 2 and anchored at the end of the beam by hooks or other suitable means.

3. To ensure development of hanger reinforcement, A_{sh}, it may be bent and continued parallel to the beam bottom, or separate horizontal reinforcement, $A'_{sh} \geq A_{sh}$ must be provided. The extension of A_{sh} or A'_{sh} reinforcement at beam bottom must be at least ℓ_d beyond crack 5. The A'_{sh} reinforcement may be anchored on the dap side by welding it to a plate (as shown in Fig. 6.13.1), angle or cross bar. The beam flexure reinforcement may also be used to ensure development of A_{sh} reinforcement provided that the flexure reinforcement is adequately anchored on the dap side.

4. Vertical reinforcement A_v should be properly anchored by hooks as required by ACI 318-89.

5. Welded wire fabric in place of bars may be used for reinforcement. It should be anchored in accordance with ACI 318-89.

6.13.6 Other Considerations

1. The depth of the extended end should not be less than one-half the depth of the beam, unless the beam is significantly deeper than necessary for other than structural reasons.

2. The hanger reinforcement, A_{sh}, should be placed as close as practical to the reentrant corner. This reinforcement requirement is not additive to other shear reinforcement requirements.

3. If the flexural stress in the full depth section immediately beyond the dap, using factored loads and gross section properties, exceeds $6\sqrt{f_c'}$, longitudinal reinforcement should be placed in the beam to develop the required flexural strength.

4. The Ref. 20 study, found that, due to formation of the critical diagonal tension crack (crack 5 in Fig. 6.13.1), it was not possible to develop a full depth beam shear strength greater than the diagonal tension cracking shear in the vicinity of the dap. It is therefore suggested that, for a length of the beam equal to the overall depth, H, of the beam, the nominal shear strength of concrete, V_c, be taken as the lesser of V_{ci} and V_{cw} calculated at H/2 from the end of the full depth web.

Example 6.13.1 Reinforcement for dapped-end beam

Given:

The 16RB28 beam with a dapped end as shown in Fig. 6.13.2.

V_u = 100 kips (includes all load factors)

N_u = 15 kips (includes all load factors)

Fig. 6.13.2 Dapped-end beam of Example 6.13.1

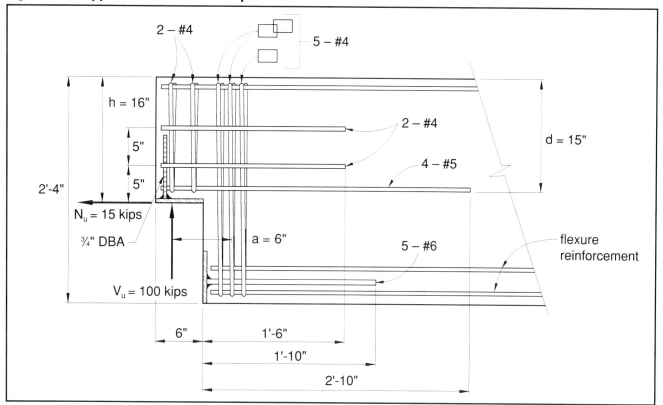

f'_c = 5000 psi (normal weight)
f_y for all reinforcement = 60 ksi

Problem:

Determine the required reinforcements A_s, A_h, A_{sh}, and A_v shown in Fig. 6.13.1

Solution:

Assume: Shear span, a = 6 in.
d = 15 in.

Check shear strength, Table 6.7.1:

$\phi V_n = \phi(1000\lambda^2 bd)$
 = 0.85(1000)(1)²(16)(15)/1000
 = 204 kips > 100 OK

1. Flexure in extended end:
   By Eq. 6.13.1:
   $$A_s = \frac{1}{\phi f_y}\left[V_u\left(\frac{a}{d}\right) + N_u\left(\frac{h}{d}\right)\right]$$
   $$= \frac{1}{0.85 \times 60}\left[100\left(\frac{6}{15}\right) + 15\left(\frac{16}{15}\right)\right]$$
   = 1.10 sq in.

2. Direct shear:
   $$\mu_e = \frac{1000\lambda bh\mu}{V_u} = \frac{1000(1)(16)(16)(1.4)(1)}{100,000}$$
   = 3.58 > 3.4 Use 3.4

   By Eqs. 6.13.2 and 6.13.3:
   $$A_s = \frac{2V_u}{3\phi f_y \mu_e} + \frac{N_u}{\phi f_y}$$
   $$= \frac{2(100)}{3(0.85)(60)(3.4)} + \frac{15}{0.85(60)}$$
   = 0.38 + 0.29 = 0.67 sq in. < 1.10
   Therefore A_s = 1.10 sq in.
   Use 4 – #5, A_s = 1.24 sq in.
   By Eq. 6.13.4:
   A_h = 0.5($A_s - A_n$) = 0.5(1.10 – 0.29)
    = 0.41 sq in.
   Use 2 – #3 U-bars. A = 0.44 sq in.

3. Diagonal tension at reentrant corner:
   By Eq. 6.13.5:
   $$A_{sh} = \frac{V_u}{\phi f_y} = \frac{100}{0.85(60)} = 1.96 \text{ sq in.}$$
   Use 5 – #4 closed ties = 2.00 sq in.
   For A'_{sh} (min. area = A_{sh})
   Use 5 – #6 = 2.2 sq in. > 1.96 OK

4. Diagonal tension in the extended end:
   Concrete capacity = $2\lambda\sqrt{f'_c}\, bd$
    = $2(1)\sqrt{5000}(16)(15)/1000$ = 33.9 kips

By Eq. 6.13.7:
$$A_v = \frac{1}{2f_y}\left[\frac{V_u}{\phi} - 2\lambda bd\sqrt{f'_c}\right]$$
$$= \frac{1}{2(60)}\left(\frac{100}{0.85} - 33.9\right) = 0.70 \text{ sq in.}$$

Try 2 – #4 stirrups = 0.80 sq in.

Check Eq. 6.13.6:
$$\phi V_n = \phi\left(A_v f_y + A_h f_y + 2\lambda\sqrt{f'_c}\,bd\right)$$
$$= 0.85[0.80(60) + 0.44(60) + 33.9]$$
$$= 92.1 \text{ kips} < 100$$

Change A_h to 2 – #4

$\phi V_n = 110.4$ kips > 100 OK

Check anchorage requirements:

A_s bars:

From Table 11.2.8:

$f_y = 60{,}000$ psi, $f'_c = 5000$ psi, #5 bars

For $\alpha_A = 1.0 = \alpha_B = \alpha_D = \alpha_E$; $\alpha_C = 1.3$ and $\alpha_{mt} = 1.0$

$\ell_d = 15.9(1.3) = 20.7$ in.

Extension past dap = $H - d + \ell_d$
$$= 28 - 15 + 20.7$$
$$= 33.7 \text{ in. (say, 2 ft-10 in.)}$$

A_h bars:

From Table 11.2.8, for #4 bars:

$\ell_d = 12.8(1.3) = 16.6$ in. (say, 1 ft-6 in.)

A'_{sh} bars:

From Table 11.2.8, for #6 bars:

$\ell_d = 19.1$ in.

bar length = $H - D + \ell_d$
$$= 28 - 26 + 19.1$$
$$= 21.1 \text{ in. (say 1 ft-10 in.)}$$

6.14 Ledger Beam

The flexure, shear and torsion design of ledger beams are covered in Chapter 4 and the bearing design in Sects. 6.8 and 6.9. This section covers additional design items related to the beam end, and the ledge and its attachment to the web. These items are discussed in Refs. 21 through 24 and are covered with reference to Fig. 6.14.1.

6.14.1 Shear Strength of the Ledge

The design shear strength of continuous beam ledges supporting concentrated loads, can be determined by the lesser of Eq. 6.14.1 and 6.14.2:

for $s > b_t + h_\ell$

$$\phi V_n = 3\phi\lambda\sqrt{f'_c}\,h_\ell\left[2(b_\ell - b) + b_t + h_\ell\right] \quad \text{(Eq. 6.14.1)}$$

$$\phi V_n = \phi\lambda\sqrt{f'_c}\,h_\ell\left[2(b_\ell - b) + b_t + h_\ell + 2d_e\right] \quad \text{(Eq. 6.14.2)}$$

for $s < b_t + h_\ell$, and equal concentrated loads, use the lesser of Eq. 6.14.1a, 6.14.2a or 6.14.3

$$\phi V_n = 1.5\phi\lambda\sqrt{f'_c}\,h_\ell\left[2(b_\ell - b) + b_t + h_\ell + s\right] \quad \text{(Eq. 6.14.1a)}$$

$$\phi V_n = \phi\lambda\sqrt{f'_c}\,h_\ell\left[(b_\ell - b) + \frac{(b_t + h_\ell)}{2} + d_e + s\right] \quad \text{(Eq. 6.14.2a)}$$

where:

- $\phi = 0.85$
- h_ℓ = depth of the beam ledge, in.
- b = beam web width
- b_ℓ = width of web and one ledge, in.
- b_t = width of bearing area, in.
- s = spacing of concentrated loads, in.
- d_e = distance from center of load to the end of the beam, in.

If the ledge supports a continuous load or closely spaced concentrated loads, the design shear strength is:

$$\phi V_n = 24\phi h_\ell \lambda \sqrt{f'_c} \quad \text{(Eq. 6.14.3)}$$

where ϕV_n is the design shear strength in pounds per foot.

If the applied factored load exceeds the strength as determined by Eqs. 6.14.1, 6.14.2 or 6.14.3, the ledge should be designed for shear transfer and diagonal tension in accordance with Sects. 6.13.2 through 6.13.4.

6.14.2 Transverse (Cantilever) Bending of the Ledge

Transverse (cantilever) bending of the ledge requires flexural reinforcement, A_s, which is computed by Eq. 6.13.1. Such reinforcement may be uniformly spaced over a width of $6h_\ell$ on either side of the bearing, but not to exceed half the distance to the next load. Bar spacing should not exceed the ledge depth, h_ℓ, or 18 in.

6.14.3 Longitudinal Bending of the Ledge

Longitudinal reinforcement, calculated by Eq. 6.14.4, should be placed in both the top and bottom of the ledge portion of the beam:

$$A_\ell = 200(b_\ell - b)d_\ell/f_y \quad \text{(Eq. 6.14.4)}$$

Where d_ℓ is the design depth of A_ℓ reinforcement.

Fig. 6.14.1 Design of beam ledges

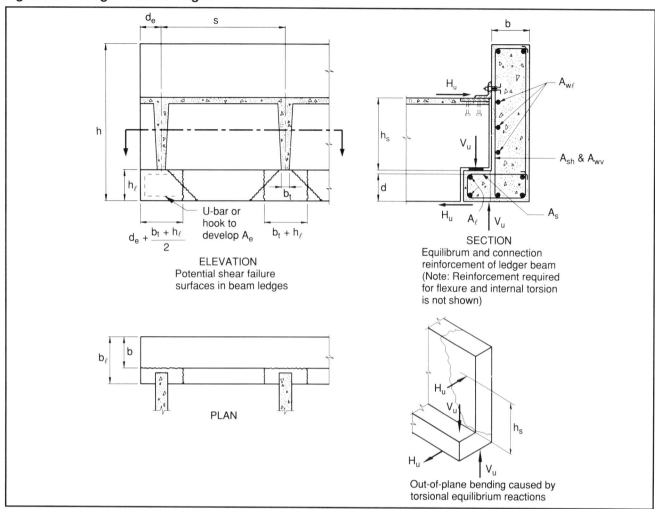

6.14.4 Attachment of the Ledge to the Web

Hanger steel, A_{sh}, computed by Eq. 6.14.5 is required for attachment of the ledge to the web. Distribution and spacing of A_{sh} reinforcement should follow the same guidelines as for A_s reinforcement in Sect. 6.14.2. A_{sh} is not additive to shear and torsion reinforcement designed in accordance with Sects. 4.3 and 4.4.

$$A_{sh} = \frac{V_u}{\phi f_y}(m) \qquad \text{(Eq. 6.14.5)}$$

where:
- V_u = applied factored load
- ϕ = 0.85
- f_y = yield strength of A_{sh} reinforcement
- m = a modification factor which can be derived from Eqs. 6 through 9 of Ref. 24 and is dependent on beam section geometry (see Table 6.14.1) and a factor γ_t which accounts for proportioning of applied torsion between the ledge and the web. If closed stirrups are provided in the ledge, γ_t may be taken as 1.0.

In other cases, γ_t is calculated as the ratio of torsional moment strength provided by concrete, T_c (see Sect. 4.4) and factored torsional moment, T_u at critical section. Alternatively, γ_t = 0.0 may be used in such cases for a conservative solution. Table 6.14.1 gives values of the modification factor, m, for commonly used ledger beam sizes and several values of γ_t.

6.14.5 Out-of-Plane Bending Near Beam End

In Ref. 24 study, it was found that when the reaction is not co-linear with applied loads, the resulting out-of-plane bending may require additional vertical and horizontal reinforcement. These are computed by Eq. 6.14.6 and provided on the inside face of the beam. This reinforcement is not additive to the reinforcement for internal torsion. The $A_{w\ell}$ and A_{wv} bars should be evenly distributed over a height and width equal to h_s (see Fig. 6.14.1)

$$A_{wv} = A_{w\ell} = \frac{T_u}{2\phi f_y d_w} \qquad \text{(Eq. 6.14.6)}$$

where:

T_u = factored torsional moment at critical section

ϕ = 0.85 (Note: The use of ϕ = 0.85 instead of 0.90 (flexure) compensates for the use of d in place of the actual, somewhat smaller, lever arm.)

f_y = yield strength

d_w = depth of A_{wv} and $A_{w\ell}$ reinforcement from outside face of beam

Example 6.14.1 Ledger beam end and ledge design

Given:

8 ft wide double tees resting on a standard L-beam similar to that shown in Fig. 6.14.1. Layout of tees is irregular so that a stem can be placed at any point on the ledge.

V_u = 18 kips per stem
N_u = 3 kips per stem
h_ℓ = 12 in.
d = 11 in.
b_t = 3 in.
b_ℓ = 14 in.
h = 36 in.
b = 8 in.
s = 48 in.
f'_c = 5000 psi (normal weight)
f_y = 60 ksi

Problem:

Investigate strength of the ledge and its attachment to the web. Determine required reinforcement.

Solution:

Min. $d_e = b_t/2 = 1.5$ in.

Since $s > b_t + h_\ell$ and $d_e < 2(b_\ell - b) + b_t + h_\ell$

use Eq. 6.14.2:

$\phi V_n = \phi h_\ell \lambda \sqrt{f'_c}\,(2(b_\ell - b) + b_t + h_\ell + 2d_e)$

$= 0.85(12)(1)\sqrt{5000}\,[2(6) + 3 + 12 + 2(1.5)]$

$= 21{,}600$ lb $= 21.6$ kips > 18 OK

Determine reinforcement for flexure and axial tension:

Shear span, $a = \tfrac{3}{4}(b_\ell - b) + 1\tfrac{1}{2} = 6$ in.

By Eq. 6.13.1:

$A_s = \dfrac{1}{\phi f_y}\left[V_u\left(\dfrac{a}{d}\right) + N_u\left(\dfrac{h_\ell}{d}\right)\right]$

$= \dfrac{1}{0.85(60)}[18(6/11) + 3(12/11)]$

$= 0.28$ sq in.

$6h = 6$ ft $> s/2$

Therefore distribute reinforcement over $s/2$ each side of the load.

$(s/2)(2) = 4$ ft

Maximum bar spacing $= h_\ell = 12$ in.

Use #3 bars @ 12 in. = 0.44 sq in. in each 4 ft

Place 2 additional bars at the beam end to provide equivalent reinforcement for stem placed near the end.

By Eq. 6.14.5:

$A_{sh} = \dfrac{V_u}{\phi f_y}(m) = \dfrac{18}{0.85(60)}(1.35) = 0.48$ sq in.

Note: m = 1.33 is obtained from Table 6.14.1 corresponding to: $b_t/b = 14/8 = 1.75$, $h_\ell/h = 12/36 = 0.33$, and $\lambda_t = 1.0$ (closed stirrups in the ledge).

$A_{sh} = \dfrac{0.48}{4} = 0.12$ sq in./ft

Maximum bar spacing $= h_\ell = 12$ in.

A_{sh} = #3 @ 10 in. = 0.132 sq in./ft

By Eq. 6.14.4:

$A_\ell = \dfrac{200(b_\ell - b)d_\ell}{f_y} = \dfrac{200(14 - 8)(11)}{60{,}000} = 0.22$ sq in.

Use 2 – #3 top and bottom = 0.22 sq in.

By Eq. 6.14.6

$A_{wv} = A_{w\ell} = \dfrac{T_u}{2\phi f_y d_w}$

Assume: T_u = 820 k-in.
$d_w = b - 1.25 = 6.75$ in.
$h_s = 18$ in.

$A_{wv} = A_{w\ell} = \dfrac{820}{2(0.85)((60)(6.75)}$

$= 1.19$ in.^{2}

This amount should be compared with the reinforcement for internal shear and torsion design (Sects. 4.3 and 4.4) and only the excess provided over a width and height of 18 in.

6.15 Hanger Connections

Hangers are similar to dapped ends, except that the extended or bearing end is steel instead of concrete. They are used when it is desired to keep the structural depth very shallow. Examples are shown in Fig. 6.15.1.

6.15.1 Cazaly Hanger[25]

The Cazaly hanger has three basic components (Fig. 6.15.2a). Design assumptions are as follows (Fig. 6.15.2b):

Table 6.14.1 Modification factor, m, for design of hanger steel

$$A_{sh} = \frac{V_u}{\phi f_y}(m) \quad \text{(Eq. 6.14.5)}$$

where:

$$m = \frac{\left[(d+a) - \left(3 - 2\frac{h_\ell}{h}\right)\left(\frac{h_\ell}{h}\right)^2\left(\frac{b_\ell}{2}\right) - e\gamma_t\frac{(x^2y)_\ell}{\Sigma x^2 y}\right]}{d}$$

x, y = shorter and longer sides respectively, of the component rectangles forming the ledge and the web parts of the beam.

$\gamma_t = T_c/T_u$ (Sect. 4.4)

= 1.0 when closed ties are used in the ledge

The table values are based on the following assumptions:

(1) $d = b - 1.25$

(2) V_u is applied at a distance equal to ⅔ of the ledge projection from the inside face of web.

(3) $\dfrac{(x^2y)_\ell}{\Sigma x^2 y} = \dfrac{(b_p - b)^2\dfrac{h_\ell}{h}}{(b_\ell - b)^2\dfrac{h_\ell}{h} + b^2}$

This is a conservative assumption for commonly used ledger beam sizes.

(Note: A lower limit of m = 0.6 is suggested to account for the limited variability in beam sizes tested. Values of m smaller than 0.6 are flagged with * in the table.)

Modification factor, m

b_ℓ/b	h_ℓ/h	γ_t b = 6 in. 0.00	0.25	0.50	0.75	1.00	b = 8 in. 0.00	0.25	0.50	0.75	1.00	b = 10 in. 0.00	0.25	0.50	0.75	1.00
1.25	0.10	1.45	1.45	1.45	1.45	1.45	1.36	1.36	1.36	1.36	1.36	1.31	1.31	1.31	1.31	1.31
	0.15	1.43	1.42	1.42	1.42	1.42	1.34	1.34	1.34	1.33	1.33	1.29	1.29	1.29	1.29	1.28
	0.20	1.39	1.39	1.39	1.38	1.38	1.31	1.30	1.30	1.30	1.30	1.26	1.26	1.26	1.25	1.25
	0.25	1.35	1.35	1.34	1.34	1.34	1.27	1.26	1.26	1.26	1.26	1.22	1.22	1.22	1.21	1.21
	0.30	1.30	1.30	1.30	1.29	1.29	1.22	1.22	1.22	1.21	1.21	1.18	1.18	1.17	1.17	1.17
	0.35	1.25	1.25	1.24	1.24	1.23	1.18	1.17	1.17	1.16	1.16	1.13	1.13	1.12	1.12	1.12
	0.40	1.20	1.19	1.19	1.18	1.18	1.12	1.12	1.11	1.11	1.10	1.05	1.08	1.07	1.07	1.06
	0.45	1.14	1.13	1.13	1.12	1.12	1.07	1.06	1.06	1.05	1.05	1.03	1.03	1.02	1.01	1.01
	0.50	1.08	1.07	1.07	1.06	1.05	1.01	1.01	1.00	1.00	0.99	0.98	0.97	0.97	0.96	0.95
	0.55	1.02	1.01	1.01	1.00	0.99	0.96	0.95	0.94	0.94	0.93	0.92	0.92	0.91	0.90	0.90
	0.60	0.96	0.96	0.95	0.94	0.93	0.90	0.90	0.89	0.88	0.87	0.87	0.86	0.86	0.85	0.84
	0.65	0.91	0.90	0.89	0.88	0.87	0.85	0.84	0.84	0.83	0.82	0.82	0.81	0.81	0.80	0.79
	0.70	0.86	0.85	0.84	0.83	0.82	0.80	0.79	0.79	0.78	0.77	0.77	0.77	0.76	0.75	0.74
	0.75	0.81	0.80	0.79	0.78	0.77	0.76	0.75	0.74	0.73	0.72	0.73	0.72	0.71	0.71	0.70
1.50	0.10	1.66	1.65	1.65	1.64	1.63	1.56	1.55	1.55	1.54	1.53	1.50	1.50	1.49	1.48	1.48
	0.15	1.63	1.62	1.61	1.60	1.59	1.53	1.52	1.51	1.50	1.49	1.47	1.46	1.46	1.45	1.44
	0.20	1.59	1.58	1.56	1.55	1.54	1.49	1.48	1.47	1.45	1.44	1.44	1.43	1.41	1.40	1.39
	0.25	1.54	1.52	1.51	1.49	1.48	1.44	1.43	1.41	1.40	1.38	1.39	1.38	1.36	1.35	1.34
	0.30	1.48	1.46	1.44	1.43	1.41	1.39	1.37	1.36	1.34	1.32	1.34	1.32	1.31	1.29	1.27
	0.35	1.42	1.40	1.38	1.36	1.33	1.33	1.31	1.29	1.27	1.25	1.28	1.26	1.25	1.23	1.21
	0.40	1.35	1.33	1.30	1.28	1.26	1.27	1.25	1.22	1.20	1.18	1.22	1.20	1.18	1.16	1.14
	0.45	1.28	1.26	1.23	1.20	1.18	1.20	1.18	1.15	1.13	1.10	1.16	1.14	1.11	1.09	1.06
	0.50	1.21	1.18	1.15	1.12	1.10	1.14	1.11	1.08	1.06	1.03	1.10	1.07	1.04	1.02	0.99
	0.55	1.14	1.11	1.08	1.05	1.01	1.07	1.04	1.01	0.98	0.85	1.03	1.00	0.98	0.95	0.92
	0.60	1.07	1.04	1.00	0.97	0.93	1.01	0.97	0.94	0.91	0.88	0.97	0.94	0.91	0.88	0.85
	0.65	1.01	0.97	0.93	0.89	0.86	0.94	0.91	0.87	0.84	0.80	0.91	0.88	0.84	0.81	0.78
	0.70	0.94	0.90	0.86	0.83	0.79	0.89	0.85	0.81	0.77	0.74	0.85	0.82	0.78	0.75	0.71
	0.75	0.89	0.85	0.80	0.76	0.72	0.83	0.79	0.75	0.71	0.68	0.80	0.76	0.73	0.69	0.65

Table 6.14.1 Modification factor, m, for design of hanger steel (continued)

b_ℓ/b	h_ℓ/h	γ_t b = 6 in. 0.00	0.25	0.50	0.75	1.00	b = 8 in. 0.00	0.25	0.50	0.75	1.00	b = 10 in. 0.00	0.25	0.50	0.75	1.00
1.75	0.10	1.87	1.85	1.83	1.82	1.80	1.75	1.74	1.72	1.70	1.69	1.69	1.67	1.66	1.64	1.63
	0.15	1.83	1.81	1.78	1.76	1.73	1.72	1.69	1.67	1.65	1.63	1.66	1.63	1.61	1.59	1.57
	0.20	1.78	1.75	1.72	1.69	1.65	1.67	1.64	1.61	1.58	1.55	1.61	1.58	1.56	1.53	1.50
	0.25	1.73	1.69	1.65	1.61	1.57	1.62	1.58	1.55	1.51	1.47	1.56	1.53	1.49	1.45	1.42
	0.30	1.66	1.61	1.57	1.52	1.48	1.56	1.51	1.47	1.43	1.38	1.50	1.46	1.42	1.38	1.34
	0.35	1.59	1.53	1.48	1.43	1.38	1.49	1.44	1.39	1.34	1.29	1.44	1.39	1.34	1.29	1.25
	0.40	1.51	1.45	1.39	1.33	1.28	1.42	1.36	1.31	1.25	1.20	1.37	1.31	1.26	1.21	1.15
	0.45	1.43	1.36	1.30	1.24	1.17	1.34	1.28	1.22	1.16	1.10	1.29	1.23	1.18	1.12	1.06
	0.50	1.35	1.28	1.21	1.14	1.07	1.26	1.20	1.13	1.07	1.00	1.22	1.15	1.09	1.03	0.96
	0.55	1.26	1.19	1.11	1.04	0.96	1.18	1.11	1.04	0.97	0.90	1.14	1.07	1.01	0.94	0.87
	0.60	1.18	1.10	1.02	0.94	0.86	1.11	1.03	0.96	0.88	0.81	1.07	1.00	0.92	0.85	0.78
	0.65	1.10	1.02	0.93	0.85	0.76	1.04	0.96	0.88	0.80	0.72	1.00	0.92	0.85	0.77	0.69
	0.70	1.03	0.94	0.85	0.76	0.67	0.97	0.88	0.80	0.72	0.63	0.93	0.85	0.77	0.69	0.61
	0.75	0.97	0.87	0.78	0.68	0.59*	0.91	0.82	0.73	0.64	0.55*	0.87	0.79	0.70	0.62	0.53*
2.00	0.10	2.07	2.04	2.01	1.97	1.94	1.95	1.91	1.88	1.85	1.82	1.88	1.85	1.82	1.79	1.75
	0.15	2.03	1.98	1.94	1.89	1.84	1.91	1.86	1.82	1.77	1.73	1.84	1.80	1.75	1.71	1.66
	0.20	1.98	1.92	1.85	1.79	1.73	1.86	1.80	1.74	1.59	1.62	1.79	1.73	1.68	1.62	1.57
	0.25	1.91	1.84	1.76	1.69	1.62	1.79	1.72	1.66	1.59	1.52	1.73	1.66	1.60	1.53	1.46
	0.30	1.84	1.75	1.67	1.58	1.49	1.72	1.64	1.56	1.48	1.40	1.66	1.58	1.51	1.43	1.35
	0.35	1.75	1.66	1.56	1.47	1.37	1.65	1.56	1.47	1.38	1.28	1.59	1.50	1.41	1.33	1.24
	0.40	1.66	1.56	1.45	1.35	1.24	1.56	1.46	1.36	1.26	1.16	1.51	1.40	1.31	1.22	1.12
	0.45	1.57	1.46	1.34	1.23	1.11	1.48	1.37	1.26	1.15	1.04	1.42	1.32	1.21	1.11	1.01
	0.50	1.48	1.35	1.23	1.11	0.98	1.39	1.27	1.15	1.04	0.92	1.34	1.23	1.11	1.00	0.89
	0.55	1.38	1.25	1.12	0.99	0.86	1.30	1.17	1.05	0.93	0.80	1.25	1.13	1.01	0.89	0.78
	0.60	1.29	1.15	1.01	0.87	0.74	1.21	1.08	0.95	0.82	0.69	1.17	1.04	0.92	0.79	0.67
	0.65	1.20	1.06	0.91	0.76	0.62	1.13	0.99	0.85	0.72	0.58*	1.09	0.96	0.82	0.69	0.56*
	0.70	1.12	0.97	0.81	0.66	0.51*	1.05	0.91	0.76	0.62	0.48*	1.01	0.87	0.74	0.60	0.46*
	0.75	1.04	0.88	0.73	0.57*	0.41*	0.98	0.83	0.68	0.53*	0.38*	0.94	0.80	0.66	0.51*	0.37*
2.25	0.10	2.28	2.22	2.17	2.11	2.05	2.14	2.09	2.03	1.98	1.93	2.06	2.01	1.96	1.91	1.86
	0.15	2.23	2.15	2.07	1.99	1.91	2.10	2.02	1.95	1.87	1.80	2.02	1.95	1.88	1.80	1.73
	0.20	2.17	2.07	1.97	1.87	1.77	2.04	1.94	1.85	1.76	1.66	1.97	1.88	1.78	1.69	1.60
	0.25	2.10	1.98	1.86	1.74	1.62	1.97	1.86	1.75	1.63	1.52	1.90	1.79	1.68	1.58	1.47
	0.30	2.01	1.88	1.74	1.61	1.47	1.89	1.76	1.64	1.52	1.38	1.82	1.70	1.58	1.46	1.33
	0.35	1.92	1.77	1.62	1.47	1.32	1.80	1.66	1.52	1.38	1.24	1.74	1.60	1.47	1.33	1.20
	0.40	1.82	1.66	1.50	1.33	1.17	1.71	1.56	1.40	1.25	1.10	1.65	1.50	1.35	1.21	1.06
	0.45	1.72	1.54	1.37	1.19	1.02	1.61	1.45	1.28	1.12	0.95	1.55	1.40	1.24	1.08	0.92
	0.50	1.61	1.42	1.24	1.05	0.87	1.51	1.34	1.16	0.99	0.81	1.46	1.29	1.12	0.95	0.79
	0.55	1.50	1.31	1.11	0.92	0.72	1.41	1.23	1.04	0.86	0.68	1.36	1.18	1.01	0.83	0.65
	0.60	1.40	1.20	0.99	0.79	0.58*	1.31	1.12	0.93	0.74	0.55*	1.27	1.08	0.90	0.71	0.53*
	0.65	1.30	1.09	0.87	0.66	0.45*	1.22	1.02	0.82	0.62	0.42*	1.18	0.98	0.79	0.60	0.40*
	0.70	1.21	0.99	0.76	0.54*	0.32*	1.13	0.92	0.72	0.51*	0.30*	1.09	0.89	0.69	0.49*	0.29*
	0.75	1.12	0.89	0.67	0.44*	0.21*	1.05	0.84	0.62	0.41*	0.20*	1.02	0.81	0.60	0.40*	0.19*

b_ℓ/b	h_ℓ/h	γ_t b = 12 in. 0.00	0.25	0.50	0.75	1.00	b = 14 in. 0.00	0.25	0.50	0.75	1.00	b = 16 in. 0.00	0.25	0.50	0.75	1.00
1.25	0.10	1.28	1.28	1.28	1.28	1.28	1.26	1.26	1.26	1.26	1.26	1.25	1.25	1.25	1.24	1.24
	0.15	1.26	1.26	1.26	1.26	1.25	1.24	1.24	1.24	1.24	1.23	1.23	1.22	1.22	1.22	1.22
	0.20	1.23	1.23	1.23	1.22	1.22	1.21	1.21	1.21	1.20	1.20	1.20	1.19	1.19	1.19	1.19
	0.25	1.19	1.19	1.19	1.19	1.18	1.17	1.17	1.17	1.17	1.16	1.16	1.16	1.15	1.15	1.15
	0.30	1.15	1.15	1.15	1.14	1.14	1.13	1.13	1.13	1.12	1.12	1.12	1.12	1.11	1.11	1.11
	0.35	1.11	1.10	1.10	1.09	1.09	1.09	1.08	1.08	1.08	1.07	1.08	1.07	1.07	1.06	1.06
	0.40	1.06	1.05	1.05	1.04	1.04	1.04	1.04	1.03	1.03	1.02	1.03	1.02	1.02	1.01	1.01
	0.45	1.01	1.00	1.00	0.99	0.99	0.99	0.99	0.98	0.98	0.97	0.98	0.97	0.97	0.96	0.96
	0.50	0.85	0.95	0.94	0.94	0.93	0.94	0.93	0.93	0.92	0.92	0.93	0.92	0.92	0.91	0.91
	0.55	0.90	0.90	0.89	0.88	0.88	0.89	0.88	0.88	0.87	0.86	0.88	0.87	0.86	0.86	0.85
	0.60	0.85	0.84	0.84	0.83	0.82	0.84	0.93	0.82	0.82	0.81	0.83	0.82	0.81	0.81	0.80
	0.65	0.80	0.79	0.79	0.78	0.77	0.79	0.78	0.77	0.77	0.76	0.78	0.77	0.77	0.76	0.75
	0.70	0.76	0.75	0.74	0.73	0.72	0.74	0.74	0.73	0.72	0.71	0.73	0.73	0.72	0.71	0.70
	0.75	0.71	0.71	0.70	0.69	0.68	0.70	0.69	0.69	0.68	0.67	0.69	0.69	0.68	0.67	0.66

Table 6.14.1 Modification factor, m, for design of hanger steel (continued)

b_ℓ/b	h_ℓ/h	γ_t b = 12 in. 0.00	0.25	0.50	0.75	1.00	b = 14 in. 0.00	0.25	0.50	0.75	1.00	b = 16 in. 0.00	0.25	0.50	0.75	1.00
1.50	0.10	1.47	1.46	1.46	1.45	1.44	1.44	1.44	1.43	1.43	1.42	1.43	1.42	1.41	1.41	1.40
	0.15	1.44	1.43	1.42	1.41	1.41	1.42	1.41	1.40	1.39	1.38	1.40	1.39	1.38	1.37	1.37
	0.20	1.40	1.39	1.38	1.37	1.36	1.38	1.37	1.36	1.35	1.34	1.36	1.35	1.34	1.33	1.32
	0.25	1.36	1.35	1.33	1.32	1.30	1.34	1.32	1.31	1.30	1.28	1.32	1.31	1.29	1.28	1.27
	0.30	1.31	1.29	1.28	1.26	1.24	1.29	1.27	1.26	1.24	1.22	1.27	1.26	1.24	1.22	1.21
	0.35	1.25	1.24	1.22	1.20	1.18	1.23	1.22	1.20	1.18	1.16	1.22	1.20	1.18	1.16	1.15
	0.40	1.20	1.17	1.15	1.13	1.11	1.18	1.16	1.13	1.11	1.09	1.16	1.14	1.12	1.10	1.08
	0.45	1.13	1.11	1.09	1.06	1.04	1.12	1.09	1.07	1.05	1.02	1.10	1.08	1.06	1.03	1.01
	0.50	1.07	1.05	1.02	0.99	0.97	1.05	1.03	1.00	0.98	0.95	1.04	1.02	0.99	0.97	0.94
	0.55	1.01	0.98	0.95	0.92	0.90	0.99	0.96	0.94	0.91	0.88	0.98	1.95	0.93	0.90	0.87
	0.60	0.95	0.92	0.89	0.86	0.83	0.93	0.90	0.87	0.84	0.81	0.92	0.89	0.86	0.93	0.80
	0.65	0.89	0.86	0.82	0.79	0.76	0.87	0.84	0.81	0.78	0.75	0.86	0.83	0.80	0.77	0.74
	0.70	0.83	0.80	0.76	0.73	0.69	0.82	0.79	0.75	0.72	0.68	0.81	0.78	0.74	0.71	0.67
	0.75	0.78	0.75	0.71	0.67	0.64	0.77	0.73	0.70	0.66	0.63	0.76	0.73	0.69	0.65	0.62
1.75	0.10	1.65	1.63	1.62	1.61	1.59	1.62	1.61	1.59	1.58	1.56	1.60	1.59	1.57	1.56	1.55
	0.15	1.62	1.60	1.57	1.55	1.53	1.59	1.57	1.55	1.53	1.51	1.57	1.55	1.53	1.51	1.49
	0.20	1.58	1.55	1.52	1.49	1.46	1.55	1.52	1.49	1.47	1.44	1.53	0.50	1.48	1.45	1.42
	0.25	1.52	1.49	1.46	1.43	1.39	1.50	1.47	1.43	1.40	1.36	1.48	1.45	1.41	1.38	1.35
	0.30	1.47	1.43	1.39	1.34	0.30	1.44	1.40	1.36	1.32	1.28	1.52	1.39	1.35	1.31	1.27
	0.35	1.40	1.36	1.31	1.26	1.22	1.38	1.33	1.29	1.24	1.20	1.36	1.32	1.27	1.23	1.18
	0.40	1.33	1.28	1.23	1.18	1.13	1.31	1.26	1.21	1.16	1.11	1.30	1.23	1.20	1.15	1.10
	0.45	1.26	1.21	1.15	1.09	1.04	1.24	1.19	1.13	1.07	1.02	1.23	1.17	1.12	1.06	1.01
	0.50	1.19	1.13	1.07	1.00	0.94	1.17	1.11	1.05	0.99	0.93	1.16	1.10	1.04	0.98	0.92
	0.55	1.12	1.05	0.98	0.92	0.85	1.10	1.03	0.97	0.90	0.84	1.08	1.02	0.96	0.89	0.83
	0.60	1.04	0.97	0.90	0.83	0.76	1.03	0.96	0.89	0.82	0.75	1.01	0.95	0.88	0.81	0.74
	0.65	0.98	0.90	0.83	0.75	0.68	0.96	0.89	0.81	0.74	0.66	0.95	0.88	0.80	0.73	0.66
	0.70	0.91	0.83	0.75	0.67	0.59*	0.90	0.82	0.74	0.66	0.58*	0.89	0.81	0.73	0.65	0.58*
	0.75	0.85	0.77	0.69	0.60	0.52*	0.84	0.76	0.68	0.59*	0.51*	0.83	0.75	0.67	0.59*	0.51*
2.00	0.10	1.83	1.80	1.77	1.74	1.71	1.80	1.77	1.74	1.72	1.69	1.78	1.75	1.72	1.69	1.67
	0.15	1.80	1.75	1.71	1.67	1.63	1.77	1.73	1.68	1.64	1.60	1.75	1.70	1.66	1.62	1.58
	0.20	1.75	1.69	1.64	1.58	1.53	1.72	1.67	1.61	1.56	1.50	1.70	1.65	1.59	1.54	1.49
	0.25	1.69	1.62	1.56	1.49	1.43	1.66	1.60	1.53	1.47	1.40	1.64	1.58	1.51	1.45	1.39
	0.30	1.62	1.55	1.47	1.40	1.32	1.60	1.52	1.45	1.37	1.30	1.58	1.50	1.43	1.36	1.28
	0.35	1.55	1.46	1.38	1.30	1.21	1.52	1.44	1.36	1.27	1.19	1.51	1.42	1.34	1.26	1.28
	0.40	1.47	1.38	1.28	1.19	1.10	1.45	1.36	1.26	1.17	1.08	1.43	1.34	1.25	1.16	1.07
	0.45	1.39	1.29	1.19	1.08	0.98	1.37	1.27	1.17	1.07	0.97	1.35	1.25	1.15	1.05	0.96
	0.50	1.31	1.20	1.09	0.98	0.87	1.28	1.18	1.07	0.96	0.86	1.27	1.16	1.06	0.95	0.85
	0.55	1.22	1.11	0.99	0.87	0.76	1.20	1.09	0.97	0.86	0.75	1.19	1.08	0.96	0.85	0.74
	0.60	1.14	1.02	0.90	0.77	0.65	1.12	1.00	0.88	0.76	0.64	1.11	0.99	0.87	0.75	0.63
	0.65	1.06	0.93	0.80	0.68	0.55*	1.05	0.82	0.79	0.66	0.54*	1.03	0.91	0.78	0.66	0.63
	0.70	0.99	0.85	0.72	0.58*	0.45*	0.97	0.84	0.71	0.58*	0.44*	0.96	0.83	0.70	0.57*	0.44*
	0.75	0.92	0.78	0.64	0.50*	0.36*	0.91	0.77	0.63	0.49*	0.36*	0.90	0.76	0.62	0.49*	0.35*
2.25	0.10	1.98	1.93	1.88	1.83	1.78	1.96	1.91	1.86	1.81	1.76	1.94	1.89	1.84	1.79	1.75
	0.15	1.94	1.87	1.80	1.73	1.66	1.92	1.85	1.78	1.71	1.64	1.90	1.83	1.76	1.70	1.63
	0.20	1.89	1.80	1.71	1.63	1.54	1.87	1.78	1.69	1.61	1.52	1.85	1.76	1.68	1.59	1.51
	0.25	1.82	1.72	1.62	1.51	1.41	1.80	1.70	1.60	1.50	0.39	1.79	1.68	1.58	1.48	1.38
	0.30	1.75	1.63	1.52	1.40	1.28	1.73	1.61	1.50	1.38	1.27	1.71	1.60	1.48	1.37	1.25
	0.35	1.67	1.56	1.41	1.28	1.15	1.65	1.52	1.39	1.26	1.14	1.63	1.51	1.38	1.25	1.12
	0.40	1.58	1.44	1.30	1.16	1.02	1.56	1.42	1.28	1.14	1.00	1.55	1.41	1.27	1.13	1.00
	0.45	1.79	1.34	1.19	1.04	0.88	1.47	1.32	1.17	1.02	0.87	1.46	1.31	1.16	1.01	0.87
	0.50	1.40	1.24	1.08	0.92	0.75	1.38	1.22	1.06	0.90	0.75	1.37	1.21	1.05	0.90	0.74
	0.55	1.31	1.14	0.97	0.80	0.63	1.29	1.12	0.96	0.79	0.62	1.28	1.11	0.95	0.78	0.61
	0.60	1.22	1.04	0.86	0.68	0.50*	1.20	1.03	0.85	0.67	0.50*	1.19	1.02	0.84	0.67	0.49*
	0.65	1.13	0.94	0.76	0.57*	0.39*	1.12	0.93	0.75	0.57*	0.38*	1.11	0.92	0.74	0.56*	0.38*
	0.70	1.05	0.86	0.66	0.47*	0.28*	1.04	0.85	0.66	0.47*	0.28*	1.03	0.84	0.65	0.46*	0.27*
	0.75	0.98	0.78	0.58*	0.38*	0.18*	0.96	0.77	0.57*	0.38*	0.18*	0.95	0.76	0.57*	0.37*	0.18*

1. The cantilevered bar is usually proportioned so that the interior reaction from the concrete is 0.33 V_u. The hanger strap should then be proportioned to yield under a tension of 1.33 V_u.

$$A_s = \frac{1.33V_u}{\phi F_y} \quad \text{(Eq. 6.15.1)}$$

where:
F_y = yield strength of the strap material
ϕ = 0.90

2. V_u may be assumed to be applied $\ell_p/2$ from the face of the seat. The remaining part of the moment arm is the width of the joint, g, which is measured from the edge of the strap to the end of the supporting member. Since moment is sensitive to this dimension, it is important that this dimension is kept as small as feasible and that the value used in analysis is not exceeded in the field. Most hangers in practice have exterior cantilever lengths, (ℓ_p + g), of 2½ to 3½ in.

3. The moment in the cantilevered bar is then given by:

$$M_u = V_u(\ell_p/2 + g + 0.375s) \quad \text{(Eq. 6.15.2)}$$

where:
ℓ_p = bearing length of the exterior cantilever

Other notation is shown in Fig. 6.15.2b

Note: The bar should be proportioned to carry this moment in combination with shear and tensile forces. Alternatively, if the bar is proportioned to take this moment at the yield stress, but using elastic section properties, (i.e., $M_u = \phi F_y bd^2/6$) the shear and tensile forces can usually be neglected.

4. The bearing pressure creating the interior reaction may be calculated as in Sect. 6.8. Conservatively, if the width of the member in which the hanger is cast = b_1, then:

$$f_{bu} = 0.85\phi f'_c \sqrt{b_1/b} \leq 1.2 f'_c \quad \text{(Eq. 6.15.3)}$$

where ϕ = 0.7

The bearing length, ℓ_b, is then given by:

$$\ell_b = \frac{V_u/3}{bf_{bu}} \quad \text{(Eq. 6.15.4)}$$

5. To maintain the conditions of equilibrium assumed, the interior cantilever must have a length:

$(1.5\ell_p + 3.0g + s + 0.5\ell_b)$ in.

6. The minimum total length of bar is then:

$(2.5\ell_p + 4.0g + 2.0s + 0.5\ell_b)$ in. (Eq. 6.15.5)

7. Longitudinal dowels, A_n, are welded to the cantilevered bar to transmit the axial force, N_u:

$$A_n = \frac{N_u}{\phi f_y} \quad \text{(Eq. 6.15.6)}$$

where:
f_y = yield strength of the dowel
ϕ = 0.90

8. The lower dowel area, A_{vf}, can be proportioned using effective shear-friction described in Sect. 6.7.

$$A_{vf} = \frac{1.33V_u}{\phi f_y \mu_e} \quad \text{(Eq. 6.15.7)}$$

where:
ϕ = 0.85
f_y = yield strength of lower dowels, psi

$$\mu_e = \frac{1000\lambda^2 b_1 h\mu}{V_u} \leq \text{values in Table 6.7.1} \quad \text{(Eq. 6.15.8)}$$

The nominal shear strength ($1.33V_u/\phi$) is limited by the values in Table 6.7.1

Example 6.15.1 Design of a Cazaly hanger

Given:
Hanger similar to that shown in Fig. 6.15.1b
f'_c = 5000 psi (both member and support)
f_y (reinforcing bars) = 60 ksi
F_y (structural steel) = 36 ksi
V_u = 24 kips (includes all load factors)
N_u = 4 kips (includes all load factors)
b_1 = 6 in.
g = 1 in.
ℓ_p = 2 in.

Problem:
Size the hanger components.

Solution:
By Eq. 6.15.1:

$$A_s(\text{strap}) = \frac{1.33V_u}{\phi F_y}$$

$$= \frac{1.33(24)}{0.9(36)} = 0.99 \text{ in.}^2$$

Use ¼ x 2 in. strap; A_s = 0.25(2)(2) = 1.00 in.²
Use 3/16 in. fillet weld; Table 6.20.2, E70 electrode:

$$\ell_w = \frac{1.33(24)}{2(4.18)} = 3.10 \text{ in.}$$

Weld 2 in. across top, ¾ in. down sides = 3.50 in.

Fig. 6.15.1 Hanger Connections

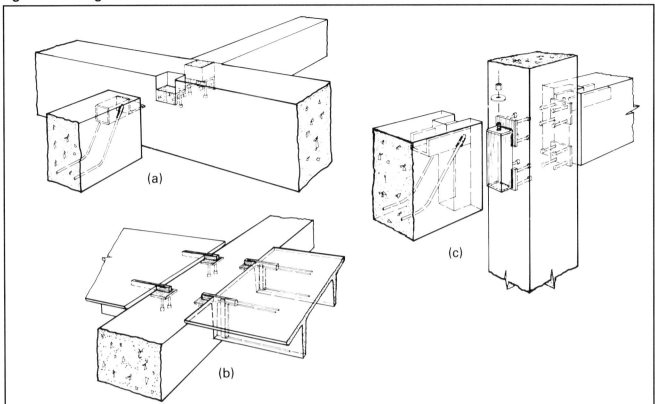

By Eq. 6.15.2:
$$M_u = V_u(0.5\ell_p + g + 0.375s) = 24(2.75)$$
$$= 66 \text{ in.-kips}$$

$$Z_{req'd} = \frac{M_u}{\phi F_y} = \frac{66}{0.9(36)} = 2.04 \text{ in.}^3 = \frac{bd^2}{6}$$

Try 2 in. wide bar:

$$d = \sqrt{\frac{6(2.04)}{2}} = 2.47 \text{ in. min.}$$

Use 2 x 2 ½ in. bar

By Eqs. 6.15.3 and 6.15.4:
$$f_{bu} = 0.85\phi f'_c \sqrt{b_1/b} = 0.85(0.7)(5)\sqrt{6/2}$$
$$= 5.15 \text{ ksi}$$

$$\ell_b = \frac{V_u/3}{f_{bu}(b)} = \frac{24/3}{5.15(2)} = 0.78 \text{ in.}$$

Min. interior cantilever
$$= 1.5\ell_p + 3.0g + s + 0.5\ell_b$$
$$= 3.0 + 3.0 + 2.0 + 0.78/2 = 8.39 \text{ in.}$$

Min. total length (Eq. 6.15.5)
$$= 2.5\ell_p + 4.0g + 2.0s + 0.5\ell_b = 13.39 \text{ in.}$$

Use bar 2 x 2 ½ x 14 in.

By Eq. 6.15.6:
$$A_n = \frac{N_u}{\phi f_y} = \frac{4}{0.9(60)} = 0.07 \text{ sq in.}$$

Use 1 – #3 dowel

Fig. 6.15.2 Cazaly hanger

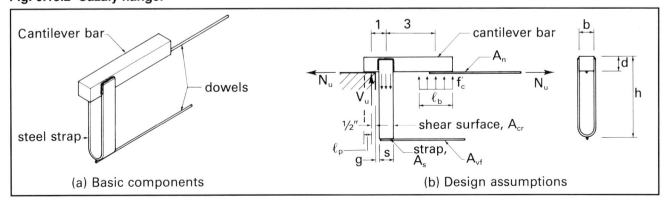

(a) Basic components   (b) Design assumptions

Try h = 16 in.; by Eqs. 6.15.7 and 6.15.8:

$$\mu_e = \frac{1000\lambda b_1 h\mu}{V_u} = \frac{1000(1.0)(6)(16)(1.4)(1.0)}{24,000}$$

$$= 5.6 > 3.4$$

use $\mu_e = 3.4$

$$A_{vf} = \frac{1.33V_u}{\phi f_y \mu_e} = \frac{1.33(24)}{0.85(60)(3.4)} = 0.18 \text{ sq in.}^2$$

Use 1 – #6 dowel

Also check welding requirements.

6.15.2 Loov Hanger[26]

The hanger illustrated in Fig. 6.15.3 is designed using the following equations:

$$A_{sh} = \frac{V_u}{\phi f_y \cos\alpha} \quad \text{(Eq. 6.15.9)}$$

where:

$\phi = 0.85$

f_y = yield strength of A_{sh}

$$A_n = \frac{N_u}{\phi f_y}\left(1 + \frac{h-d}{d-a/2}\right) \quad \text{(Eq. 6.15.10)}$$

where:

$\phi = 0.90$

f_y = yield strength of A_n

The steel bar is proportioned so that the bearing strength of the concrete is not exceeded, and to provide sufficient weld length to develop the diagonal bars. Bearing strength is discussed in Sect. 6.8. However, if the bar is at the top of the member as in Fig. 6.15.3, there is no "geometrically similar" area larger than the edge of the bar, and:

$$f_{bu} = 0.85\phi f'_c = 0.6 f'_c \quad \text{(Eq. 6.15.11)}$$

where:

$\phi = 0.7$

The connection should be detailed so that the reaction, the center of compression and the center of the diagonal bars meet at a common point, as shown in Fig. 6.15.3b. The compressive force, C_u, is assumed to act at a distance a/2 from the top of the

Fig. 6.15.3 Loov hanger

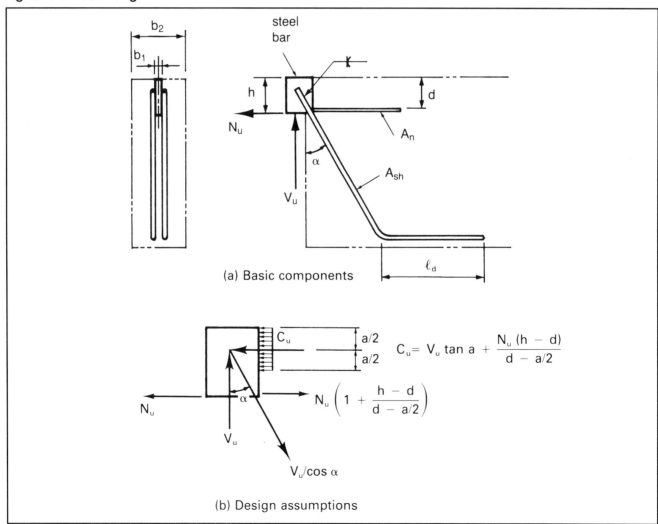

(a) Basic components

$C_u = V_u \tan\alpha + \dfrac{N_u(h-d)}{d-a/2}$

$N_u\left(1 + \dfrac{h-d}{d-a/2}\right)$

(b) Design assumptions

bearing plate. Thus:

$$a = \frac{C_u}{b_1 f_{bu}} \quad \text{(Eq. 6.15.12)}$$

where:

$$C_u = V_u \tan\alpha + \frac{N_u(h-d)}{d-a/2} \quad \text{(Eq. 6.15.13)}$$

For most designs, the horizontal reinforcement, A_n, is placed very close to the bottom of the steel bar. Thus, the term (h – d) can be assumed equal to zero, simplifying Eqs. 6.15.10 and 6.15.13.

Tests have indicated a weakness in shear in the vicinity of the hangers, so it is recommended that stirrups in the beam end be designed to carry the total shear.

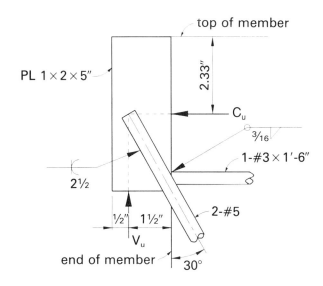

Example 6.15.2 Design of a Loov hanger

Given:

Hanger similar to that shown in Fig. 6.15.3.
Design for same data as Example 6.15.1
$\alpha = 30°$

Problem:

Size the hanger components.

Solution:

$$A_{sh} = \frac{V_u}{\phi f_y \cos\alpha} = \frac{24}{0.85(60)\cos 30°} = 0.54 \text{ sq in.}$$

Use 2 – #5 bars

Min. weld length, #5 bar, E70 electrode (Table 6.20.3) = 2½ in.

Detail A_n so it is near the bottom of the steel bar
h – d ≈ 0

$$A_n = \frac{N_u}{\phi f_y} = \frac{4}{0.9(60)} = 0.07 \text{ sq in.}$$

Use 1 – #3 dowel

By Eq. 6.15.11:

$$f_{bu} = 0.85\phi f'_c = 0.85(0.7)(5) = 2.98 \text{ ksi}$$
$$C_u = V_u \tan\alpha = 24 \tan 30° = 13.9 \text{ kips}$$

Assume b_1 = 1 in.

$$a = \frac{C_u}{b_1 f_{bu}} = \frac{13.9}{1(2.98)} = 4.65 \text{ in.}$$

a/2 = 2.33

Provide end bearing plate as shown at top right:

6.16 Moment Connections

When lateral stability of precast, prestressed concrete buildings is achieved by frame action or by a combination of shear wall and frame action, the connections to develop frame action must be designed for appropriate moment and shear transfer capabilities.

The tension force for the moment resistance within a connection can be provided by various types of inserts, such as headed studs and deformed bar anchors. These inserts must be properly anchored to preclude failure of the concrete and thus ensuring a ductile mode of failure. Post-tensioning can also be used to develop moment resistance at joints between interconnected members. Where a high degree of moment resistance and ductility are required, composite construction is frequently used to achieve connections that are similar to monolithic concrete joints in their behavior.

Achieving "full rigid" connections can be costly. In most cases, it may not be desirable to build-in a high degree of fixity, since the restraint of volume changes could result in large forces in the connections and the members. It is therefore preferable that the design of moment-resisting connections be based on the concept of "partial fixity", wherein the desired moment resistance is achieved with some deformation/rotation at the connection. The deformation should be controlled to provide for the desired ductility.

A few examples of different types of moment-resisting connections are shown in Fig. 6.16.1. Ref. 2 provides additional examples and also discusses relative degrees of fixity.

Moment-curvature analysis of precast and prestressed concrete members is readily done based on established analytical methods. PCI funded research,[27] as well as other research in progress, are expected to lead to an adequate knowledge base on moment-resisting connections and enable formulation of improved analytical procedures.

6.17 Connection of Non-Load Bearing Wall Panels

The design of connections of non-load bearing architectural wall panels follows the same principles as structural connections, except the loads are generally lighter. Usually, the connections are detailed to minimize the volume change forces and thus, are primarily designed for the self-weight of the panel and for the lateral loads. Attention to details and proper protection of any exposed hardware are most important to ensure satisfactory performance of the connections for the service life of the structure.

A discussion of connection design fundamentals and production and erection considerations is given in Refs. 2 and 28 and a summary is included in Chapter 7 (Sect. 7.2). A few examples of typical details illustrating the three primary categories of connections, namely bearing, tie-back and alignment connections, are shown in Figs. 6.17.1 through 6.17.3.

6.18 Connection of Load Bearing Wall Panels

Connections for load bearing wall panels are an essential part of the structural support system and the stability of the structure may depend upon them. In addition to the weight of the panels, the connections must resist and transfer dead, live, wind and earthquake loads, and effects of volume changes.

Erected load-bearing walls may have both horizontal and/or vertical joints across which forces must

Fig. 6.16.1 Moment Connections

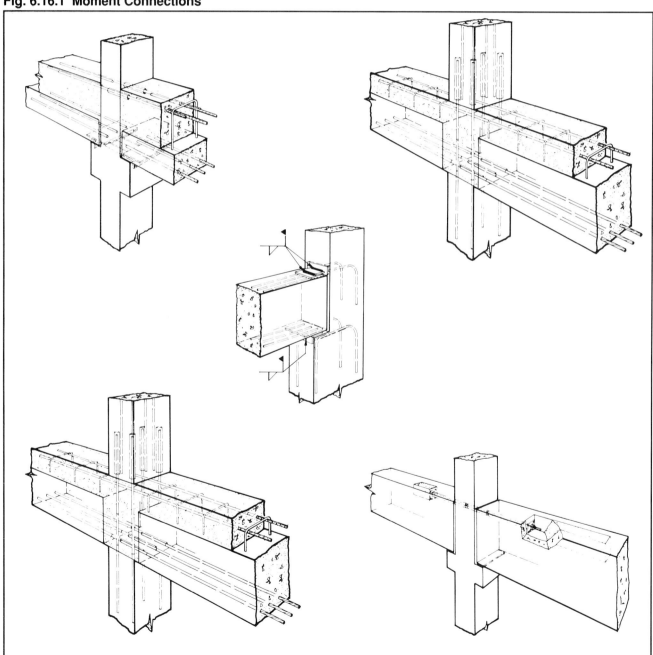

Fig. 6.17.1 Bearing Connections

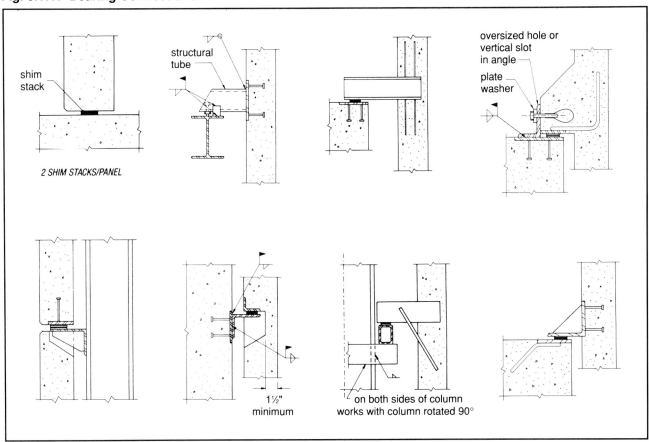

Fig. 6.17.2 Tie-back connections

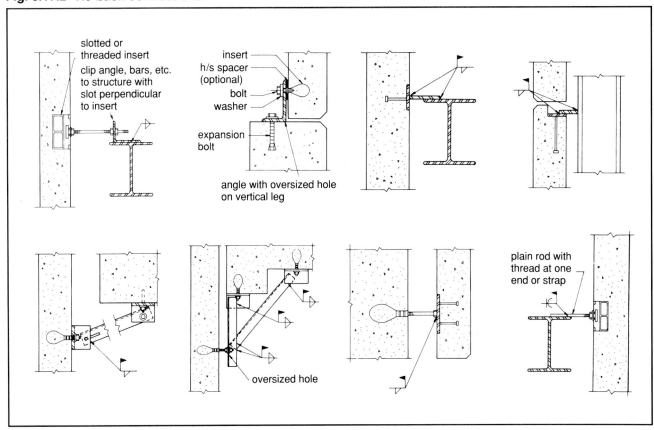

PCI Design Handbook/Fourth Edition   6-47

Fig. 6.17.3 Alignment connections

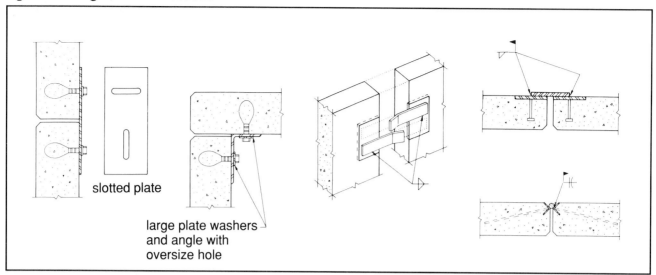

be transferred. Fig. 6.18.1 indicates, for separate cases, the principal applied forces and the resulting joint force systems. In buildings, superposition of forces and various combinations of panel and joint assemblies must be considered.

Distribution of lateral forces to shear walls depends largely on adequate connections of floors to walls. In addition to the transfer of vertical shear forces due to lateral loads, vertical joints may also be subject to shear forces induced by differential loads on adjacent panels. Joint and connection details of exterior bearing walls are especially critical since the floor elements are usually connected at this elevation and a waterproofing detail must be incorporated.

Fig. 6.18.1 Exterior forces and joint force systems

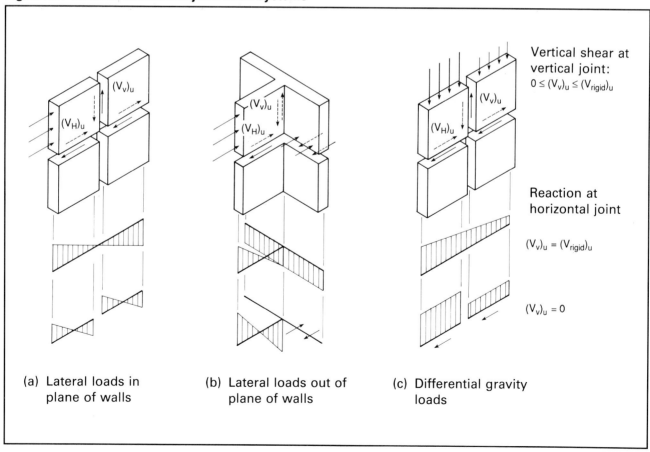

(a) Lateral loads in plane of walls

(b) Lateral loads out of plane of walls

(c) Differential gravity loads

6-48 PCI Design Handbook/Fourth Edition

Fig. 6.18.2 Wall to wall hinge connections at floor levels

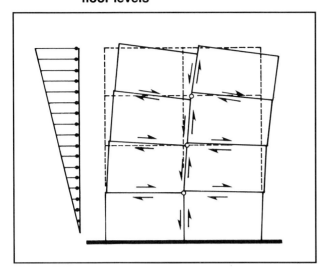

Fig. 6.18.3 Grooved joint connections

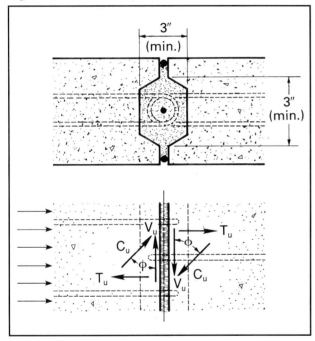

Fig. 6.18.4 Mechanical connections

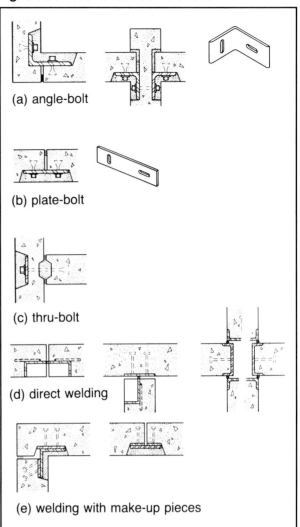

(a) angle-bolt

(b) plate-bolt

(c) thru-bolt

(d) direct welding

(e) welding with make-up pieces

Grooved joint connection

Grooved joints are continuous and usually filled with grout. The minimum groove dimension should be 1½ in. deep and 3 in. wide (Fig. 6.18.3). The joint strength can be evaluated by shear-friction even if shrinkage, creep, and temperature movements have caused a crack at the wall-grout interface.

Mechanical connection

Mechanical connections consist of anchorage devices cast into the wall panels and steel sections (plates, angles, bars, etc.) crossing the joint. The strength is usually controlled by the capacity of the cast-in anchorage (Fig. 6.18.4); connection of the steel section to the anchorage device can be made by bolting, welding, or grouting.

Tie beam connections at floor levels may act with the mechanical connections. The relative participation in resisting applied forces depends on their force-deformation characteristics. The ultimate capacity is the sum of the strength of the tie beams and the mechanical connections.

6.18.1 Vertical Joints

Vertical joints may be designed so that the wall panels form one structural unit or act as independent wall units.

Hinge connection

A hinge connection transfers compression and tension forces but not moments. This is usually done at floor levels through floor diaphragms and tie beams. The joint between floor levels usually is "open" so the panels resist lateral loads independently according to their relative rigidity (Fig. 6.18.2). Sound and waterproofing details may also have to be considered.

Once the connection forces have been established, evaluation of connection strength is made according to strength of materials and the principles developed in other sections of this Handbook.

Keyed joint connection

Keyed joints can either be reinforced or non-reinforced (Fig. 6.18.5). Test results indicate similar deformation behavior, but also show that reinforced joints are stronger. Reinforcement is required in high seismic zones.

As shown in Fig. 6.18.5, the resistance can be limited by:

(a) cracking of grout concrete parallel to joint,

(b) diagonal cracks across joints,

(c) crushing of key edges or joint concrete at key edges, or

(d) slippage along contact area.

Fig. 6.18.5 Forces in keyed joint

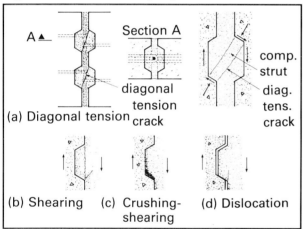

For (a) the shear-friction concept applies (Sect. 6.7). For (b), (c) and (d) the strength of the connection is usually a function of the compressive strength of the grout, the bond strength of the grout to the precast concrete, and the profile of the keys. As shown in Fig. 6.18.6, the vertical shear force can be resolved into tension and compression components with ϕ as the apparent friction coefficient and α the angle of the key.

Depending on the number of keys per floor, the unit forces per key resulting from the vertical shear force V_u are:

$$J_u = V_u \sin\alpha$$
$$C_u = V_u \cos\alpha$$

The joint force J_u is resisted by the shear-friction force R_u developed in the plane of J_u, with:

$$R_u = C_u \tan\phi$$

Assuming a conservative value of $\tan\phi = 0.60$, a sliding along J_u will not occur if:

$$R_u > J_u$$

which is the case for $\alpha \leq 30°$.

For $\alpha > 30°$ and $R_u < J_u$, a tension force T_u develops which must be absorbed by horizontal reinforcing of the joint. According to Fig. 6.18.6:

$$\Delta T_u = \frac{\Delta J_u}{\cos\alpha} = \frac{J_u - R_u}{\cos\alpha} = \frac{V_u(\sin\alpha - \cos\alpha \tan\phi)}{\cos\alpha}$$
$$= V_u(\tan\alpha - \tan\phi) \quad \text{(Eq. 6.18.1)}$$

The sum of the unit tension forces can be absorbed at each floor level by horizontal ties or by uniformly distributed horizontal reinforcing bars protruding from the wall.

Fig. 6.18.6 Keyed joint connection

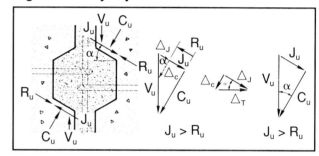

Fig. 6.18.7 Typical interior horizontal connection

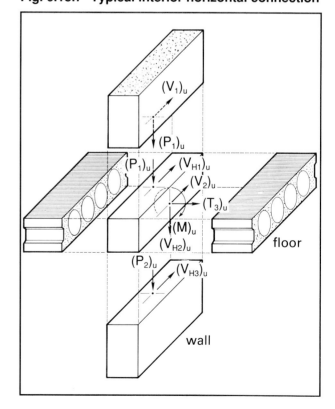

Fig. 6.18.8 Typical joints in a bearing wall building

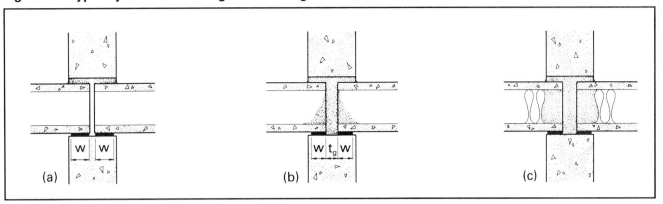

6.18.2 Horizontal Joints

Horizontal joints in load bearing wall construction occur at floor levels and at the transition to foundation or transfer beams. The principal forces to be transferred are vertical and horizontal loads from panels above and from the diaphragm action of floor slabs. The resulting forces are listed below and shown in Fig. 6.18.7:

(a) normal to joint—compression or tension, $(P_1)_u$;

(b) horizontal to joint—horizontal shear, $(V_{H1})_u$;

(c) vertical to joint at face—vertical shear, $(V_{H2})_u$; and

(d) perpendicular to joint—compression or tension from floor to diaphragm, $(T_3)_u$.

Because of the limited frame action that can be developed perpendicular to a wall, moment stresses in the joint are normally only of minor importance.

Axial load transfer through horizontal joints[29]

Fig. 6.18.8 shows three joint details used in multi-story load bearing buildings with hollow-core slabs used for the floors. For the condition of Fig. 6.18.8 (a), the joint strength is:

$$\phi P_n = \phi 0.85 A_e f'_c R_e \quad \text{(Eq. 6.18.2)}$$

where:

P_n = nominal strength of the joint

A_e = effective slab bearing area = $2wb_w$

w = bearing length

b_w = net web width of slab

f'_c = design concrete strength of slab

R_e = reduction factor for eccentricity of load

= $1 - (2e/h)$

e = eccentricity of load (occurs when floor spans or loads on either side of wall are unequal, or at end walls)

h = slab thickness

ϕ = 0.7

When the space between slab ends is grouted, load is shared by the slab ends and grout columns according to their stiffnesses. The splitting strength of the wall may also limit the joint capacity. The effect of grout flowing solidly into the slab ends is to confine the grout column. If the space between slab ends is less than about 1½ in., an unconfined grout column will add little strength.

Table 6.18.1 Equivalent bearing strength, f_{ue}(ksi)

	Grout strength, psi		
	3000	4000	5000
Slab cores not filled	4.5	5.9	5.9
Slab cores filled	5.8	6.5	7.1

Valid for f'_c = 5000 psi or higher; slabs supported on multi-monomer plastic bearing strips. For other conditions see Ref. 29.

The strength of the connection can be determined by:

$$\phi P_n = \phi t_g \ell f_{ue} R_e \quad \text{(Eq. 6.18.3)}$$

where:

t_g = grout thickness

ℓ = length of joint (parallel to wall) being considered

f_{ue} = equivalent bearing strength from Table 6.18.1. Accounts for distribution of load between grout column and slab ends

R_e = $1 - 2e/h$

ϕ = 0.7

Example 6.18.1 Design of grouted horizontal joint

Given:

An 18-story bearing wall building has 8 in. precast concrete walls and 8 in. hollow-core floor and roof units. Floor slabs span 28 ft and bear on multi-monomer plastic bearing strips.

f'_c (precast concrete) = 5000 psi

Fig. 6.18.9 Connections through horizontal joints

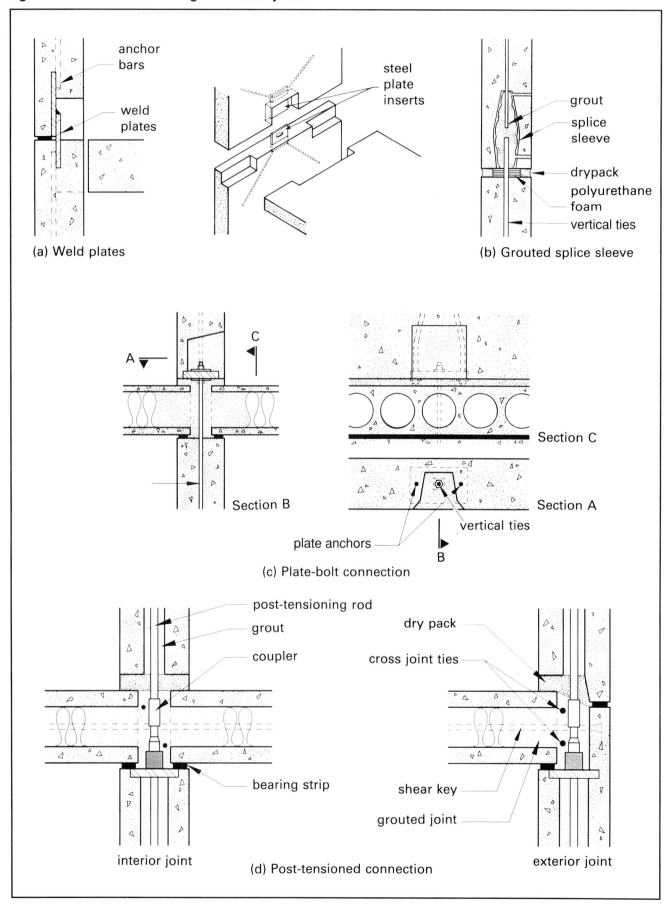

Fig. 6.18.10 Floor to bearing wall connections

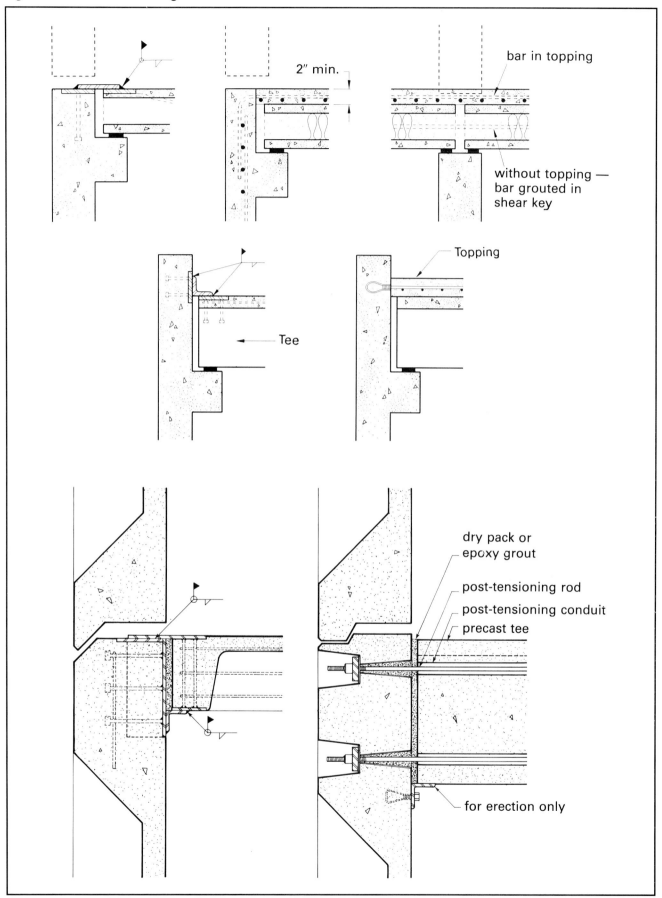

Loads:
- Roof: DL = 15 psf; LL = 30 psf
- Floors: DL = 10 psf; LL = 40 psf
- Hollow-core = 60 psf
- Walls = 800 plf/story
- No LL reduction for example

Problem:

Find grouting requirements for interior joint.

Solution:

Loads:

Roof: w_u = 28[1.4(60 + 15) +1.7(30)]/1000
= 4.37 klf

Floors: w_u = 28[1.4(60 + 10) + 1.7 (40)]/1000
= 4.65 klf

Walls: w_u = 1.4(800)/1000
= 1.12 klf/story

Accumulate loads above floor noted:

Floor	w_u	Σw_u
18	4.37 + 1.12	5.49
17	4.65 + 1.12	11.26
16	5.77	17.03
15	5.77	22.80
14	5.77	28.57
13	5.77	34.34
12	5.77	40.11
11	5.77	45.88
10	5.77	51.65
9	5.77	57.42
8	5.77	63.19
7	5.77	68.96
6	5.77	74.73
5	5.77	80.50
4	5.77	86.27
3	5.77	92.04
2	5.77	97.81

Evaluate capacity of ungrouted joint:

Assume the ratio of web width to total width of the slabs = 0.3 and bearing length, w = 3 in.

$\phi P_n = \phi 0.85 A_e f'_c R_e$

$= 0.7(0.85)(3 + 3)(0.3 \times 12)(5)\left(1 - \frac{2(0)}{8}\right)$

= 64.26 kips/ft

Adequate floors 8 through roof.

Evaluate capacity of grouted joint:

Try f'_c of grout = 3000 psi

Use t_g = 2 in. R_e = 1.0 as above

From Table 6.18.1:

Slab cores not filled, f_{ue} = 4.5 ksi

From Eq. 6.18.3:

ϕP_n = 0.7(2)(12)(4.5)(1) = 75.6 kips

Adequate for floors 6 and 7.

Slab cores filled, f_{ue} = 5.8 ksi

ϕP_n = 0.7(2)(12)(5.8)(1)
= 97.4 kips ≈ 97.81 say OK

Typical examples of connections through horizontal joints are shown in Fig. 6.18.9. Since forces are concentrated at a few points, they must be redistributed into the panels above and below. Connections should have more ductility and strength than the vertical tie fastened to it.

To satisfy structural integrity requirements, minimum tensile ties should be provided at the joints to resist the forces given in Chapter 3, Sect. 3.10.

6.18.3 Typical Details

A few typical floor to wall and wall to foundation details are shown in Figs. 6.18.10 through 6.18.12. For additional details, see Ref. 2.

Fig. 6.18.11 Floor to shear wall connections

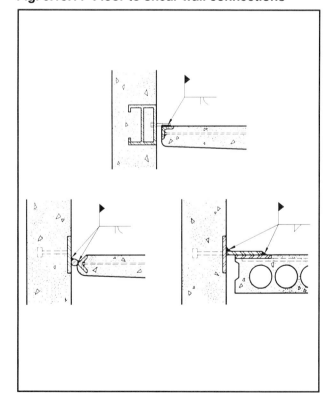

Fig. 6.18.12 Wall to foundation connections

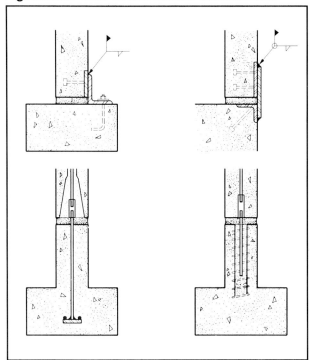

6.19 References

1. Martin, L.D. and Korkosz, W.J., "Connections for Precast Prestressed Concrete Buildings—Including Earthquake Resistance," Technical Report No. 2, Precast/Prestressed Concrete Institute, Chicago, IL, 1982.

2. "Design and Typical Details of Connections for Precast and Prestressed Concrete," Second Edition, MNL-123-88, Precast/Prestressed Concrete Institute, Chicago, IL, 1988.

3. "Structural Welding Code—Reinforcing Steel," AWS D1.4-79, American Welding Society, Miami, FL, 1979.

4. "Structural Welding Code—Steel," AWS D1.1-79, American Welding Society, Miami, FL, 1979.

5. Reichard, T.W., Carpenter, E.F. and Leyendecker, E.V., "Design Loads for Inserts Embedded in Concrete," Building Science Series 42. N.B.S (U.S.) Code:BSSNVB, Superintendent of Documents, U.S. Government Printing Office, Washington, D.C., May 1972.

6. "The Procedure Handbook of Arc Welding", Thirteenth Edition, The Lincoln Electric Company, Cleveland, Ohio, June 1988.

7. "Reinforcement: Anchorages, Lap Splices and Connections," Concrete Reinforcing Steel Institute, Schaumburg, IL, 1990.

8. ACI Committee 439, "Mechanical Connections of Reinforcing Bars," *Concrete International,* V.5, No.1, January 1983.

9. Shaikh, A.F. and Yi, W., "In-Place Strength of Welded Headed Studs," *PCI Journal,* Vol. 30, No. 2, March-April 1985.

10. Ehligehausen, R. and Fuchs, W., "Load-bearing Behavior of Anchor Fastenings Under Shear, Combined Tension and Shear or Flexural Loading," *Betonwerk + Fertigteil-Technik,* Heft 2, 1988.

11. Wong, T.L., Donahey, R.C. and Lloyd, J.P., "Stud Groups Loaded in Shear Near a Free Edge," *PCI Journal* (scheduled for publication in 1992).

12. "Manual of Steel Construction Allowable Stress Design," Ninth Edition, 1989, American Institute of Steel Construction, Chicago, IL.

13. "Manual of Steel Construction Load and Resistance Factor Design," First Edition, 1986, American Institute of Steel Construction, Chicago, IL.

14. Iverson, J.K. and Pfeifer, D.W., "Criteria for Design of Bearing Pads," *Technical Report,* TR-4-85, Precast/Prestressed Concrete Institute, Chicago, IL, 1985.

15. Shaikh, A.F., "Proposed Revisions to Shear-Friction Provisions," *PCI Journal,* Vol. 23, No. 2, March-April 1978.

16. Kriz, L.B. and Raths, C.H., "Connections in Precast Concrete Structures—Strength of Corbels," *PCI Journal,* V.10, No.1, February 1965.

17. Mattock, A.H., "Design Proposals for Reinforced Concrete Corbels," *PCI Journal,* Vol. 21, No. 3, May-June 1976.

18. Marcakis, K. and Mitchell, D., "Precast Concrete Connections with Embedded Steel Members," *PCI Journal,* V. 25, No. 4, July-August 1980.

19. Mattock, A.H. and Chan, T.C., "Design and Behavior of Dapped-End Beams," *PCI Journal,* V. 24, No. 6, November-December 1979.

20. Mattock, A.H. and Theryo, T.S., "Strength of Members with Dapped Ends," Research Project No. 6, Precast/Prestressed Concrete Institute, Chicago, IL, 1986; summary paper in *PCI Journal,* Vol. 31, No. 5, September-October 1986.

21. Mirza, S.A. and Furlong, R.W., "Serviceability Behavior and Failure Mechanisms of Concrete Inverted T-Beam Bridge Bentcaps," *Journal of the American Concrete Institute,* V.80, No.4, July-August 1983.

22. Mirza S.A. and Furlong, R.W., "Strength Criteria for Concrete Inverted T-Girders", *ASCE Journal of Structural Engineering,* V. 109, No. 8, August 1983.

23. Raths, Charles H., "Spandrel Beam Behavior and Design," *PCI Journal,* V.29, No.2, March-April 1984.

24. Klein, G.J., "Design of Spandrel Beams," Research Project No. 5, Precast/Prestressed Concrete Institute, Chicago, IL, 1986; summary paper in *PCI Journal,* Vol. 31, No. 5, September-October 1986.

25. Cazaly, L. and Huggins, M.W., "Canadian Prestressed Concrete Institute Handbook," Canadian Prestressed Concrete Institute, Ottawa, Ontario, Canada, 1964.

26. Loov, Robert, "A Precast Beam Connection Designed for Shear and Axial Load," *PCI Journal,* V.13, No.3, June 1968.

27. Stanton, J.F., Anderson, R.G., Dolan, C.W. and McCleary, D.E., "Moment Resistant Connections and Simple Connections," Research Project No 1/4, Precast/Prestressed Concrete Institute, Chicago, Illinois, 1986.

28. "Architectural Precast Concrete," Second Edition, MNL 122-89, Precast/Prestressed Concrete Institute, Chicago, IL, 1989.

29. Johal L.S. and Hanson, N.W., "Design for Vertical Load on Horizontal Connections in Large Panel Structures," *PCI Journal,* V.27, No.1, January-February, 1982.

CONNECTIONS

Table 6.20.1 Allowable and design stress for fillet and partial penetration welds[1]

Electrode	Allowable[2] working stress (ksi)	Design[3] strength (ksi)
E60	18	27
E70	21	31.5
E80	24	36
E90	27	40.5
E100	30	45

(1) For partial penetration welds loaded in shear parallel to the axis of the weld.
(2) Based on AISC Allowable Stress Design Manual of Steel Construction.[12]
(3) Based on AISC Load and Resistance Factor Design Manual of Steel Construction.[13] Includes $\phi = 0.75$

Table 6.20.2 Strength of fillet welds for building construction

Fillet[1] weld size	E60 Electrode		E70 Electrode	
	Allowable stress design (k/in.)	Design[2] strength (k/in.)	Allowable stress design (k/in.)	Design[2] strength (k/in.)
1/8	1.59	2.39	1.86	2.78
3/16	2.39	3.58	2.78	4.18
1/4	3.18	4.77	3.71	5.57
5/16	3.98	5.96	4.64	6.69
3/8	4.77	7.16	5.57	8.35
7/16	5.57	8.35	6.50	9.74
1/2	6.36	9.54	7.42	11.14
9/16	7.16	10.74	8.35	12.53
5/8	7.95	11.93	9.28	13.92

(1) Assumes 45° fillet.
(2) Based on AISC Load and Resistance Factor Design Manual of Steel Construction.[13] Includes $\phi = 0.75$

CONNECTIONS

Table 6.20.3 Minimum length of weld to develop full strength of bar. Weld parallel to bar length*

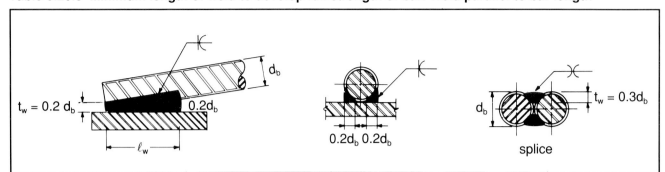

Electrode	Bar size	Plate thickness in.					Min. splice length,
		¼	⁵⁄₁₆	⅜	⁷⁄₁₆	½	
		Minimum length of weld, in.*					
E70	3	1½	1½	1½	1½	1½	1
	4	2	2	2	2	2	1½
	5	2½	2½	2½	2½	2½	1¾
	6	3	3	3	3	3	2
	7	3¼	3¼	3¼	3¼	3¼	2¼
	8	3¾	3¾	3¾	3¾	3¾	2½
	9	4¾	4¼	4¼	4¼	4¼	3
	10	6	4¾	4¾	4¾	4¾	3¼
	11	7¼	5¾	5¼	5¼	5¼	3½
E80	3	1¼	1¼	1¼	1¼	1¼	1
	4	1¾	1¾	1¾	1¾	1¾	1¼
	5	2¼	2¼	2¾	2½	2½	1½
	6	2½	2½	2½	2½	2½	1¾
	7	3	3	3	3	3	2
	8	3½	3½	3½	3½	3½	2¼
	9	4¾	3¾	3¾	3¾	3¾	2½
	10	6	4¾	4¼	4¼	4¼	3
	11	7½	5¾	5¾	4¾	4¾	3¾
E90	3	1¼	1¼	1¼	1¼	1¼	1
	4	1½	1½	1½	1½	1½	1
	5	2	2	2	2	2	1¼
	6	2¼	2¼	2¼	2¼	2¼	1½
	7	3	2¾	2¾	2¾	2¾	1¾
	8	3¾	3	5	5	5	2
	9	4¾	3¾	3½	3½	3½	2¼
	10	6	4¾	4	3¾	3¾	2½
	11	7¼	5¾	5	4¼	4¼	2¾

*Lengths below heavy line are governed by plate shear. Lengths above heavy line are governed by weld strength.
Basis: Bar f_y = 60 ksi Plate F_y = 36 ksi Shear on net section limited to 0.75(0.6)(58) ksi.

CONNECTIONS

Table 6.20.4 Size of fillet weld required to develop full strength of bar

Bar perpendicular to plate, welded one side

Plate F_y = 36 ksi

$$\ell_w = \pi\left(d_b + \frac{\text{weld size}}{2}\right)$$

Plate area = $\pi(d_b + 2 \times \text{weld size})$

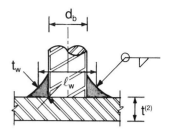

	Grade 40 Bar					
	E70 Electrode		E80 Electrode		E90 Electrode	
Bar Size	Nominal weld size[1] (in.)	Min. plate thickness[2] (in.)	Nominal weld size[1] (in.)	Min. plate thickness[2] (in.)	Nominal weld size[1] (in.)	Min. plate thickness[2] (in.)
3	3/16	1/4	3/16	1/4	3/16	1/4
4	1/4	1/4	3/16	1/4	3/16	1/4
5	1/4	1/4	1/4	1/4	1/4	1/4
6	5/16	1/4	5/16	1/4	1/4	1/4
7	3/8	1/4	5/16	1/4	5/16	1/4
8	7/16	1/4	3/8	1/4	5/16	1/4
9	7/16	1/4	7/16	1/4	3/8	1/4
10	1/2	5/16	7/16	5/16	7/16	5/16
11	9/16	5/16	1/2	3/8	7/16	3/8
	Grade 60 Bar					
3	1/4	1/4	3/16	1/4	3/16	1/4
4	5/16	1/4	1/4	1/4	1/4	1/4
5	3/8	1/4	5/16	1/4	5/16	1/4
6	7/16	1/4	3/8	1/4	3/8	1/4
7	1/2	1/4	7/16	5/16	7/16	5/16
8	9/16	5/16	1/2	5/16	7/16	5/16
9	5/8	5/16	9/16	3/8	1/2	3/8
10	11/16	3/8	5/8	3/8	9/16	7/16
11	3/4	7/16	11/16	7/16	5/8	7/16

(1) A minimum of 3/16 in. weld size is suggested.

(2) Theoretical thickness for shear stress on base metal = 0.75(0.6)(58) ksi. A more practical thickness might be taken as 2/3 d_b as used with headed studs.
    A minimum of 1/4 in. plate thickness is suggested.

CONNECTIONS

Table 6.20.5 Size of fillet weld required to develop full strength of bar

Bar perpendicular to plate,
 welded both sides

Plate F_y = 36 ksi

see Table 6.20.4 for other information

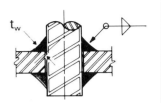

Bar Size	Grade 40 Bar					
	E70 Electrode		E80 Electrode		E90 Electrode	
	Nominal weld size[1] (in.)	Min. plate thickness[2] (in.)	Nominal weld size[1] (in.)	Min. plate thickness[2] (in.)	Nominal weld size[1] (in.)	Min. plate thickness[2] (in.)
3	3/16	1/4	3/16	1/4	3/16	1/4
4	3/16	1/4	3/16	1/4	3/16	1/4
5	3/16	1/4	3/16	1/4	3/16	1/4
6	3/16	1/4	3/16	1/4	3/16	1/4
7	3/16	1/4	3/16	1/4	3/16	1/4
8	1/4	5/16	3/16	5/16	3/16	5/16
9	1/4	5/16	1/4	5/16	3/16	3/8
10	5/16	3/8	1/4	3/8	1/4	3/8
11	5/16	7/16	5/16	3/8	1/4	7/16
	Grade 60 Bar					
3	3/16	1/4	3/16	1/4	3/16	1/4
4	3/16	1/4	3/16	1/4	3/16	1/4
5	3/16	1/4	3/16	1/4	3/16	1/4
6	1/4	5/16	1/4	5/16	3/16	5/16
7	5/16	5/16	1/4	3/8	1/4	3/8
8	5/16	3/8	5/16	3/8	1/4	7/16
9	3/8	7/16	5/16	7/16	5/16	7/16
10	3/8	1/2	3/8	1/2	5/16	1/2
11	7/16	1/2	3/8	9/16	3/8	9/16

(1) A minimum of 3/16 in. weld size is suggested.
(2) Theoretical thickness for shear stress on base metal = 0.75(0.6)(58) ksi. A more practical thickness might be taken as 2/3d_b as used with headed studs.
 A minimum of 1/4 in. plate thickness is suggested.

CONNECTIONS

Table 6.20.6 Dimensions and design tensile strength of welded headed studs[1] (see Fig. 6.5.3)

Dimensions of headed studs							
Stud diameter, d_b, in.		1/4	3/8	1/2	5/8	3/4	7/8
Head diameter, d_h, in.		1/2	3/4	1	1 1/4	1 1/4	1 3/8
Head thickness, t_h, in.		3/16	9/32	5/16	5/16	3/8	3/8

Design tensile strength limited by concrete strength ϕP_c (kips)—Eqs. 6.5.4 and 6.5.5

Edge dist. d_e[3] (in.)	Stud length[4] (in.)	Normal weight concrete[2] ($\lambda = 1.0$) Diameter, d_b, in.						Sand-lightweight concrete[2] ($\lambda = 0.85$) Diameter, d_b, in.					
		1/4	3/8	1/2	5/8	3/4[6]	7/8[6]	1/4	3/8	1/2	5/8	3/4[6]	7/8[6]
2	3.0	5.0	5.2	5.5	5.9	5.8	6.0	4.2	4.4	4.7	5.0	4.9	5.1
	4.0	6.4	6.7	7.0	7.4	7.3	7.5	5.5	5.7	6.0	6.3	6.2	6.4
	5.0	7.9	8.2	8.5	8.9	8.8	9.0	6.8	7.0	7.2	7.5	7.5	7.6
	6.0	9.4	9.7	10.0	10.4	10.3	10.5	8.0	8.2	8.5	8.8	8.7	8.9
	7.0	10.9	11.2	11.5	11.9	11.8	12.0	9.3	9.5	9.8	10.1	10.0	10.2
	8.0	12.4	12.7	13.0	13.4	13.3	13.5	10.6	10.8	11.0	11.4	11.3	11.4
3	3.0	7.0	7.1	7.4	7.9	7.6	7.9	5.9	6.0	6.3	6.7	6.5	6.7
	4.0	9.7	10.0	10.5	11.1	10.9	11.2	8.2	8.5	8.9	9.4	9.3	9.5
	5.0	11.9	12.3	12.8	13.3	13.2	13.5	10.1	10.4	10.8	11.3	11.2	11.4
	6.0	14.2	14.5	15.0	15.6	15.4	15.7	12.0	12.3	12.8	13.2	13.1	13.3
	7.0	16.4	16.8	17.2	17.8	17.7	17.9	13.9	14.2	14.7	15.1	15.0	15.3
	8.0	18.6	19.0	19.5	20.0	19.9	20.2	15.8	16.1	16.6	17.0	16.9	17.2
4	3.0	7.0	7.1	7.4	7.9	7.6	7.9	5.9	6.0	6.3	6.7	6.5	6.7
	4.0	12.3	12.4	12.9	13.6	13.2	13.6	10.5	10.6	11.0	11.6	11.2	11.5
	5.0	15.9	16.4	17.0	17.8	17.6	17.9	13.5	13.9	14.5	15.1	14.9	15.3
	6.0	18.9	19.3	20.0	20.8	20.6	20.9	16.0	16.4	17.0	17.6	17.5	17.8
	7.0	21.9	22.3	23.0	23.7	23.8	23.9	18.6	19.0	19.5	20.2	20.0	20.3
	8.0	24.9	25.3	26.0	26.7	26.5	26.9	21.1	21.5	22.1	22.7	22.6	22.9
5	3.0	7.0	7.1	7.4	7.9	7.6	7.9	5.9	6.0	6.3	6.7	6.5	6.7
	4.0	12.3	12.4	12.9	13.6	13.2	13.6	10.5	10.6	11.0	11.6	11.2	11.5
	5.0	19.1	19.3	19.9	20.8	20.3	20.7	16.3	16.4	16.9	17.7	17.3	17.6
	6.0	23.6	24.2	25.0	25.9	25.7	26.2	20.1	20.6	21.3	22.0	21.8	22.2
	7.0	27.3	27.9	28.7	29.7	29.4	29.9	23.2	23.7	24.4	25.2	25.0	25.4
	8.0	31.1	31.7	32.5	33.4	33.2	33.6	26.4	26.9	27.6	26.4	28.2	28.6
6	3.0	7.0	7.1	7.4	7.9	7.6	7.9	5.9	6.0	6.3	6.7	6.5	6.7
	4.0	12.3	12.4	12.9	13.6	13.2	13.6	10.5	10.6	11.0	11.6	11.2	11.5
	5.0	19.1	19.3	19.9	20.8	20.3	20.7	16.3	16.4	16.9	17.7	17.3	17.6
	6.0	27.4	27.7	28.4	29.5	28.9	29.4	23.3	23.5	24.2	25.1	24.6	25.0
	7.0	32.8	33.5	34.5	35.6	35.3	35.9	27.9	28.5	29.3	30.3	30.0	30.5
	8.0	37.3	38.0	39.0	40.1	39.8	40.4	31.7	32.3	33.1	34.1	33.8	34.3
7	3.0	7.0	7.1	7.4	7.9	7.6	7.9	5.9	6.0	6.3	6.7	6.5	6.7
	4.0	12.3	12.4	12.9	13.6	13.2	13.6	10.5	10.6	11.0	11.6	11.2	11.5
	5.0	19.1	19.3	19.9	20.8	20.3	20.7	16.3	16.4	16.9	17.7	17.3	17.6
	6.0	27.4	27.7	28.4	29.5	28.9	29.4	23.3	23.5	24.2	25.1	24.6	25.0
	7.0	37.3	37.5	38.4	39.7	39.0	39.6	31.7	31.9	32.7	33.7	33.2	33.7
	8.0	43.5	44.3	45.5	46.8	46.5	47.1	37.0	37.7	38.6	39.8	39.5	40.0
8	3.0	7.0	7.1	7.4	7.9	7.6	7.9	5.9	6.0	6.3	6.7	6.5	6.7
	4.0	12.3	12.4	12.9	13.6	13.2	13.6	10.5	10.6	11.0	11.6	11.2	11.5
	5.0	19.1	19.3	19.9	20.8	20.3	20.7	16.3	16.4	16.9	17.7	17.3	17.6
	6.0	27.4	27.7	28.4	29.5	28.9	29.4	23.3	23.5	24.2	25.1	24.6	25.0
	7.0	37.3	37.5	38.4	39.7	39.0	39.6	31.7	31.9	32.7	33.7	33.2	33.7
	8.0	48.6	48.9	49.9	51.4	50.6	51.3	41.3	41.5	42.4	43.7	43.0	43.6

Design tensile strength ϕP_s of studs, limited by steel strength[5] (kips)—Eq. 6.5.7	Diameter	1/4	3/8	1/2	5/8	3/4	7/8
	ϕP_s	2.7	6.0	10.6	16.6	23.9	32.5

(1) For stud groups, also check Table 6.20.7. The table values for ϕP_c which are larger than ϕP_s values are for use in combined tension and shear equations.
(2) f'_c = 5000 psi, for other strengths multiply by $\sqrt{f'_c/5000}$
(3) Edge effect considers vicinity to only one edge.
(4) The embedment length is equal to the stud length minus the thickness of the stud head (see Fig. 6.5.3).
(5) Some of the values for 3/4 in. and 7/8 in. studs are smaller than those for the 5/8 in. stud. This apparent contradiction is due to the interaction of the stud head diameter with other design dimensions. The values given in the table are correct.
(6) See Table 6.20.9 for bolts.

CONNECTIONS

Table 6.20.7 Design tensile strength of welded headed stud groups limited by concrete

The design tensile strength (Fig. 6.5.6, Cases 1 through 6) may be written as:

$$\phi P_c = \phi P_{c1} - \phi P_{c2}$$

where:

ϕP_c = net design tensile strength

ϕP_{c1} = design tensile strength for $h \geq h_{min}$ (Table 6.20.7A). Note: $\phi P_{c2} = 0$ for $h \geq h_{min}$.

ϕP_{c2} = strength reduction when $h < h_{min}$ (Table 6.20.7C)

h = member thickness

h_{min} = minimum member thickness to ensure truncated pyramid failure shown in Fig. 6.5.4 (Table 6.20.7B)

The following pages include:

Table 6.20.7A for values of ϕP_{c1}

Table 6.20.7B for values of h_{min}

Table 6.20.7C for values of ϕP_{c2}

CONNECTIONS

Table 6.20.7A Design tensile strength for $h \geq h_{min}$, ϕP_{c1}—Case 1

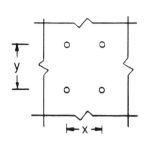

x and y are the overall dimensions (width and length) of the stud group.

Case 1: Not near a free edge

$\phi P_{c1} = \phi 4\lambda \sqrt{f'_c}(x_1 + 2\ell_e)(y_1 + 2\ell_e)$

$\phi = 0.85$

where: x_1 and y_1 are the dimensions of the flat bottom of the part of the truncated pyramid.

For Case 1: $x_1 = x$, $y_1 = y$

Note: Table values are based on $\lambda = 1.0$ and $f'_c = 5000$ psi; for different material properties, multiply table values by $\lambda\sqrt{f'_c/5000}$

ℓ_e, in.	y_1, in. \ x_1, in.	2	4	6	8	10	12	14	16	18	20	22	24	26	28	30
3	0	12	14	17	20	23	26	29	32	35	38	40	43	46	49	52
	2	15	19	23	27	31	35	38	42	46	50	54	58	62	65	69
	4	19	24	29	34	38	43	48	53	58	63	67	72	77	82	87
	6	23	29	35	40	46	52	58	63	69	75	81	87	92	98	104
	8	27	34	40	47	54	61	67	74	81	88	94	101	108	114	121
	10	31	38	46	54	62	69	77	85	92	100	108	115	123	131	138
	12	35	43	52	61	69	78	87	95	104	113	121	130	138	147	156
	14	38	48	58	67	77	87	96	106	115	125	135	144	154	163	173
	16	42	53	63	74	85	95	106	116	127	138	148	159	169	180	190
4	0	19	23	27	31	35	38	42	46	50	54	58	62	65	69	73
	2	24	29	34	38	43	48	53	58	63	67	72	77	82	87	91
	4	29	35	40	46	52	58	63	69	75	81	87	92	98	104	110
	6	37	40	47	54	61	67	74	81	88	94	101	108	114	121	128
	8	38	46	54	62	69	77	85	92	100	108	115	123	131	138	146
	10	43	52	61	69	78	87	95	104	113	121	130	138	147	156	164
	12	48	58	67	77	87	96	106	115	125	135	144	154	163	173	183
	14	53	63	74	85	95	106	116	127	138	148	159	169	180	190	201
	16	58	69	81	92	104	115	127	138	150	162	173	185	196	208	219
6	0	40	46	52	58	63	69	75	81	87	92	98	104	110	115	121
	2	47	54	61	67	74	81	88	94	101	108	114	121	128	135	141
	4	54	62	69	77	85	92	100	108	115	123	131	138	146	154	162
	6	61	69	78	87	95	104	113	121	130	138	147	156	164	173	182
	8	67	77	87	96	106	115	125	135	144	154	163	173	183	192	202
	10	74	85	95	106	116	127	138	148	159	169	180	190	201	212	222
	12	81	92	104	115	127	138	150	162	173	185	196	208	219	231	242
	14	88	100	113	125	138	150	163	175	188	200	213	225	238	250	263
	16	94	108	121	135	148	162	175	188	202	215	229	242	256	269	283
8	0	69	77	85	92	100	108	115	123	131	138	146	154	162	169	177
	2	78	87	95	104	113	121	130	138	147	156	164	173	182	190	199
	4	87	96	106	115	125	135	144	154	163	173	183	192	202	212	221
	6	95	106	116	127	138	148	159	169	180	190	201	212	222	233	243
	8	104	115	127	138	150	162	173	185	196	208	219	231	242	254	265
	10	113	125	138	150	163	175	188	200	213	225	238	250	263	275	288
	12	121	135	148	162	175	188	202	215	229	242	256	269	283	296	310
	14	130	144	159	173	188	202	216	231	245	260	274	288	303	317	332
	16	138	154	169	185	200	215	231	246	262	277	292	308	323	339	354
10	0	106	115	125	135	144	154	163	173	183	192	202	212	221	231	240
	2	116	127	138	148	159	169	180	190	201	212	222	233	243	254	264
	4	127	138	150	162	173	185	196	208	219	231	242	254	265	277	288
	6	138	150	163	175	188	200	213	225	238	250	263	275	288	300	313
	8	148	162	175	188	202	215	229	242	256	269	283	296	310	323	337
	10	159	173	188	202	216	231	245	260	274	288	303	317	332	346	361
	12	169	185	200	215	231	246	262	277	292	308	323	339	354	369	385
	14	180	196	213	229	245	262	278	294	311	327	343	360	376	392	409
	16	190	208	225	242	260	277	294	312	329	346	364	381	398	415	433

CONNECTIONS

Table 6.20.7A (continued) Design tensile strength for $h \geq h_{min}$, ϕP_{c1}—Case 2

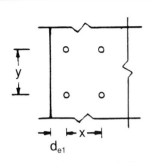

x and y are the overall dimensions (width and length) of the stud group.

Case 2: Free edge on one side

$\phi P_{c1} = \phi 4\lambda\sqrt{f'_c}\,(x_1 + \ell_e)(y_1 + 2\ell_e)$

$\phi = 0.85$

where: x_1 and y_1 are the dimensions of the flat bottom of the part of the truncated pyramid.

For Case 2: $x_1 = x + d_{e1}$, $y_1 = y$

Note: Table values are based on
$\lambda = 1.0$ and $f'_c = 5000$ psi;
for different material properties, multiply table values by $\lambda\sqrt{f'_c}/5000$

ℓ_e, in.	x_1, in. / y_1, in.	Design tensile strength, ϕP_{c1} (kips)													
		2	4	6	8	10	12	14	16	18	20	22	24	26	28	30
3	0	7	10	13	16	19	22	25	27	30	33	36	39	42	45	48
	2	10	13	17	21	25	29	33	37	40	44	48	52	56	60	63
	4	12	17	22	26	31	36	41	46	50	55	60	65	70	75	79
	6	14	20	26	32	38	43	49	55	61	66	72	78	84	89	95
	8	17	24	30	37	44	50	57	64	71	77	84	91	98	104	111
	10	19	27	35	42	50	58	65	73	81	88	96	104	112	119	127
	12	22	30	39	48	56	65	74	82	91	100	108	117	125	134	143
	14	24	34	43	53	63	72	82	91	101	111	120	130	139	149	159
	16	26	37	48	58	69	79	90	100	111	122	132	143	153	164	175
4	0	12	15	19	23	27	31	35	38	42	46	50	54	58	62	65
	2	14	19	24	29	34	38	43	48	53	58	63	67	72	77	82
	4	17	23	29	35	40	46	52	58	63	69	75	81	87	92	98
	6	20	27	34	40	47	54	61	67	74	81	88	94	101	108	114
	8	23	31	38	46	54	62	69	77	85	92	100	108	115	123	131
	10	26	35	43	52	61	69	78	87	95	104	113	121	130	138	147
	12	29	38	48	58	67	77	87	96	106	115	125	135	144	154	163
	14	32	42	53	63	74	85	95	106	116	127	138	148	159	169	180
	16	35	46	58	69	81	92	104	115	127	138	150	162	173	185	196
6	0	23	29	35	40	46	52	58	63	69	75	81	87	92	98	104
	2	27	34	40	47	54	61	67	74	81	88	94	101	108	114	121
	4	31	38	46	54	62	69	77	85	92	100	108	115	123	131	138
	6	35	43	52	61	69	78	87	95	104	113	121	130	138	147	156
	8	38	48	58	67	77	87	96	106	115	125	135	144	154	163	173
	10	42	53	63	74	85	95	106	116	127	138	148	159	169	180	190
	12	46	58	59	81	92	104	115	127	138	150	162	173	185	196	208
	14	50	63	75	88	100	113	125	138	150	163	175	188	200	213	225
	16	54	67	81	94	108	121	135	148	162	175	188	202	215	229	242
8	0	38	46	54	62	69	77	85	92	100	108	115	123	131	138	146
	2	43	53	61	69	78	87	95	104	113	121	130	138	147	156	164
	4	48	58	67	77	87	96	106	115	125	135	144	154	163	173	183
	6	53	63	74	85	95	106	116	127	138	148	159	169	180	190	201
	8	58	69	81	92	104	115	127	138	150	162	173	185	196	208	219
	10	63	75	88	100	113	125	138	150	163	175	188	200	213	225	238
	12	67	81	94	108	121	135	148	162	175	188	202	215	229	242	256
	14	72	87	101	115	130	144	159	173	188	202	216	231	245	260	274
	16	77	92	108	123	138	154	169	185	200	215	231	246	262	277	292
10	0	58	67	77	87	96	106	115	125	135	144	154	163	173	183	192
	2	63	74	85	95	106	116	127	138	148	159	169	180	190	201	212
	4	69	81	92	104	115	127	138	150	162	173	185	196	208	219	231
	6	75	88	100	113	125	138	150	163	175	188	200	213	225	238	250
	8	81	94	108	121	135	148	162	175	188	202	215	229	242	256	269
	10	87	101	115	130	144	159	173	188	202	216	231	245	260	274	288
	12	92	108	123	138	154	169	185	200	215	231	246	262	277	292	308
	14	98	114	131	147	163	180	196	213	229	245	262	278	294	311	327
	16	104	121	138	156	173	190	208	225	242	260	277	294	312	329	346

CONNECTIONS

Table 6.20.7A (continued) Design tensile strength for $h \geq h_{min}$, ϕP_{c1}—Case 3

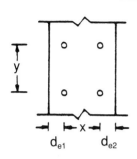

x and y are the overall dimensions (width and length) of the stud group.

Case 3: Free edges on two opposite sides

$\phi P_{c1} = \phi 4\lambda\sqrt{f'_c}(x_1)(y_1 + 2\ell_e)$

$\phi = 0.85$

where: x_1 and y_1 are the dimensions of the flat bottom of the part of the truncated pyramid.

For Case 3: $x_1 = x + d_{e1} + d_{e2}$, $y_1 = y$

Note: Table values are based on $\lambda = 1.0$ and $f'_c = 5000$ psi;

for different material properties, multiply table values by $\lambda\sqrt{f'_c/5000}$

ℓ_e, in.	x_1, y_1, in.	\multicolumn{15}{c}{Design tensile strength, ϕP_{c1} (kips)}														
		2	4	6	8	10	12	14	16	18	20	22	24	26	28	30
3	0	3	6	9	12	14	17	20	23	26	29	32	35	38	40	43
	2	4	8	12	15	19	23	27	31	35	38	42	46	50	54	58
	4	5	10	14	19	24	29	34	38	43	48	53	58	63	67	72
	6	6	12	17	23	29	35	40	46	52	58	63	69	75	81	87
	8	7	13	20	27	34	40	47	54	61	67	74	81	88	94	101
	10	8	15	23	31	38	46	54	62	69	77	85	92	100	108	115
	12	9	17	26	35	43	52	61	69	78	87	95	104	113	121	130
	14	10	19	29	38	48	58	67	77	87	96	106	115	125	135	144
	16	11	21	32	42	53	63	74	85	95	106	116	127	138	148	159
4	0	4	8	12	15	19	23	27	31	35	38	42	46	50	54	58
	2	5	10	14	19	24	29	34	38	43	48	53	58	63	67	72
	4	6	12	17	23	29	35	40	46	52	58	63	69	75	81	87
	6	7	13	20	27	34	40	47	54	61	67	74	81	88	94	101
	8	8	15	23	31	38	46	54	62	69	77	85	92	100	108	115
	10	9	17	26	35	43	52	61	69	78	87	95	104	113	121	130
	12	10	19	29	38	48	58	67	77	87	96	106	115	125	135	144
	14	11	21	32	42	53	63	74	85	95	106	116	127	138	148	159
	16	12	23	35	46	58	69	81	92	104	115	127	138	150	162	173
6	0	6	12	17	23	29	35	40	46	52	58	63	69	75	81	87
	2	7	13	20	27	34	40	47	54	61	67	74	81	88	94	101
	4	8	15	23	31	38	46	54	62	69	77	85	92	100	108	115
	6	9	17	26	35	43	52	61	69	78	87	95	104	113	121	130
	8	10	19	29	38	48	58	67	77	87	96	106	115	125	135	144
	10	11	21	32	42	53	63	74	85	95	106	116	127	138	148	159
	12	12	23	35	46	58	69	81	92	104	115	127	138	150	162	173
	14	13	25	38	50	63	75	88	100	113	125	138	150	163	175	188
	16	13	27	40	54	67	81	94	108	121	135	148	162	175	188	202
8	0	8	15	23	31	38	46	54	62	69	77	85	92	100	108	115
	2	9	17	26	35	43	52	61	69	78	87	95	104	113	121	130
	4	10	19	29	38	48	58	67	77	87	96	106	115	125	135	144
	6	11	21	32	42	53	63	74	85	95	106	116	127	138	148	159
	8	12	23	35	46	58	69	81	92	104	115	127	138	150	162	173
	10	13	25	38	50	63	75	88	100	113	125	138	150	163	175	188
	12	13	27	40	54	67	81	94	108	121	135	148	162	175	188	202
	14	14	29	43	58	72	87	101	115	130	144	159	173	188	202	216
	16	15	31	46	62	77	92	108	123	138	154	169	185	200	215	231
10	0	10	19	29	38	48	58	67	77	87	96	106	115	125	135	144
	2	11	21	32	42	53	63	74	85	95	106	116	127	138	148	159
	4	12	23	35	46	58	69	81	92	104	115	127	138	150	162	173
	6	13	25	38	50	63	75	88	100	113	125	138	150	163	175	188
	8	13	27	40	54	67	81	94	108	121	135	148	162	175	188	202
	10	14	29	43	58	72	87	101	115	130	144	159	173	188	202	216
	12	15	31	46	62	77	92	108	123	138	154	169	185	200	215	231
	14	16	33	49	65	82	98	114	131	147	163	180	196	213	229	245
	16	17	35	52	69	87	104	121	138	156	173	190	208	225	242	260

CONNECTIONS

Table 6.20.7A (continued) Design tensile strength for $h \geq h_{min}$, ϕP_{c1}—Case 4

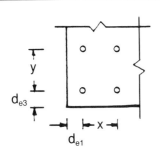

x and y are the overall dimensions (width and length) of the stud group.

Case 4: Free edges on two adjacent sides

$\phi P_{c1} = \phi 4\lambda\sqrt{f'_c}(x_1 + \ell_e)(y_1 + \ell_e)$

$\phi = 0.85$

where: x_1 and y_1 are the dimensions of the flat bottom of the part of the truncated pyramid.

For Case 4: $x_1 = x + d_{e1}$, $y_1 = y + d_{e3}$

Note: Table values are based on $\lambda = 1.0$ and $f'_c = 5000$ psi; for different material properties, multiply table values by $\lambda\sqrt{f'_c}/5000$

ℓ_e, in.	x_1, y_1, in.	\multicolumn{15}{c}{Design tensile strength, ϕP_{c1} (kips)}														
		2	4	6	8	10	12	14	16	18	20	22	24	26	28	30
3	0	4	5	6	8	9	11	12	14	15	17	18	19	21	21	24
	2	6	8	11	13	16	18	20	23	25	28	30	32	35	35	40
	4	8	12	15	19	22	25	29	32	35	39	42	45	49	49	56
	6	11	15	19	24	28	32	37	41	45	50	54	58	63	63	71
	8	13	19	24	29	34	40	45	50	56	61	66	71	77	77	87
	10	16	22	28	34	41	47	53	59	66	72	78	84	91	91	103
	12	18	25	32	40	47	54	61	69	76	83	90	97	105	105	119
	14	20	29	37	45	53	61	69	78	86	94	102	110	119	119	135
	16	23	32	41	50	59	69	78	87	96	105	114	123	132	132	151
4	0	6	8	10	12	13	15	17	19	21	23	25	27	29	31	33
	2	9	12	14	17	20	23	26	29	32	35	38	40	43	46	49
	4	12	15	19	23	27	31	35	38	42	46	50	54	58	62	65
	6	14	19	24	29	34	38	43	48	63	58	63	67	72	77	82
	8	17	23	29	35	40	46	52	58	63	69	75	81	87	92	98
	10	20	27	34	40	47	54	61	67	74	81	88	94	101	108	114
	12	23	31	38	46	54	52	69	77	85	92	100	108	115	123	131
	14	26	35	43	52	61	59	78	87	95	104	113	121	130	138	147
	16	29	38	48	58	67	77	87	96	106	115	125	135	144	154	163
6	0	12	14	17	20	23	26	29	32	35	38	40	43	46	49	52
	2	15	19	23	27	31	35	38	42	46	50	54	58	62	65	69
	4	19	24	29	34	38	43	48	53	58	63	57	72	77	82	87
	6	23	29	35	40	46	52	58	63	69	75	81	87	92	98	104
	8	27	34	40	47	54	61	67	74	81	88	94	101	108	114	121
	10	31	38	46	54	62	69	77	85	92	100	108	115	123	131	138
	12	35	43	52	61	69	78	87	95	104	113	121	130	138	147	156
	14	38	48	58	67	77	87	96	106	115	125	135	144	154	163	173
	16	42	53	63	74	85	95	106	116	127	138	148	159	169	180	190
8	0	19	23	27	31	35	38	42	46	50	54	58	62	65	69	73
	2	24	29	34	38	43	48	53	58	63	67	72	77	82	87	91
	4	29	35	40	46	52	58	63	69	75	81	87	92	98	104	110
	6	34	40	47	54	61	67	74	81	88	94	101	108	114	121	128
	8	38	46	54	62	69	77	85	82	100	108	115	123	131	138	146
	10	43	52	61	69	78	87	95	104	113	121	131	130	138	147	156	164
	12	48	58	67	77	87	96	106	115	125	138	144	154	163	173	183
	14	53	63	74	85	95	106	116	127	138	148	159	169	180	190	201
	16	58	69	81	92	104	115	127	138	150	162	173	185	196	208	219
10	0	29	34	38	43	48	53	58	63	67	72	77	82	87	91	96
	2	35	40	46	52	58	63	69	75	81	87	92	98	104	110	115
	4	40	47	54	61	67	74	82	88	94	101	108	114	121	128	135
	6	46	54	62	69	77	85	92	100	108	115	123	131	138	146	154
	8	52	61	69	78	87	95	104	113	121	130	138	147	156	164	173
	10	58	67	77	87	96	106	115	127	135	144	154	163	173	183	193
	12	63	74	85	95	106	116	127	138	148	159	169	180	190	201	212
	14	69	81	92	104	115	127	138	150	162	173	185	196	208	219	231
	16	75	88	100	113	125	138	150	163	175	188	200	213	225	238	250

CONNECTIONS

Table 6.20.7A (continued) Design tensile strength for $h \geq h_{min}$, ϕP_{c1} — Case 5

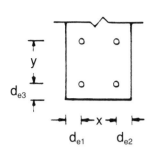

x and y are the overall dimensions (width and length) of the stud group.

Case 5: Free edges on three sides

$\phi P_{c1} = \phi 4\lambda\sqrt{f'_c}\,(x_1)(y_1 + \ell_e)$

$\phi = 0.85$

where: x_1 and y_1 are the dimensions of the flat bottom of the part of the truncated pyramid.

For Case 5: $x_1 = x + d_{e1} + d_{e2}$, $y_1 = y + d_{e3}$

Note: Table values are based on $\lambda = 1.0$ and $f'_c = 5000$ psi; for different material properties, multiply table values by $\lambda\sqrt{f'_c}/5000$

ℓ_e, in.	x_1, y_1, in.	\multicolumn{15}{c}{Design tensile strength, ϕP_{c1} (kips)}
		2	4	6	8	10	12	14	16	18	20	22	24	26	28	30
3	0	1	3	4	6	7	9	10	12	13	14	16	17	19	20	22
	2	2	5	7	10	12	14	17	19	22	24	26	29	31	34	36
	4	3	7	10	13	17	20	24	27	30	34	37	40	44	47	50
	6	4	9	13	17	22	26	30	35	39	43	48	52	56	61	65
	8	5	11	16	21	26	32	37	42	48	53	58	63	69	74	79
	10	6	13	19	25	31	38	44	50	56	63	69	75	81	88	94
	12	7	14	22	29	36	43	50	58	65	72	79	87	94	101	108
	14	8	16	25	33	41	49	57	65	74	82	90	98	106	114	123
	16	9	18	27	37	46	55	64	73	82	91	100	110	119	128	137
4	0	2	4	6	8	10	12	13	15	17	19	21	23	25	27	29
	2	3	6	9	12	14	17	20	23	26	29	32	35	38	40	43
	4	4	8	12	15	19	23	27	31	35	38	42	46	50	54	58
	6	5	10	14	19	24	29	34	38	43	48	53	58	63	67	72
	8	6	12	17	23	29	35	40	46	52	58	63	69	75	81	87
	10	7	13	20	27	34	40	47	54	61	67	74	81	88	94	101
	12	8	15	23	31	38	46	54	62	69	77	85	92	100	108	115
	14	9	17	26	35	43	52	61	69	78	87	95	104	113	121	130
	16	10	19	29	38	48	58	67	77	87	96	106	115	125	135	144
6	0	3	6	9	12	14	17	20	23	26	29	32	35	38	40	43
	2	4	8	12	15	19	23	27	31	35	38	42	46	50	54	58
	4	5	10	14	19	24	29	34	38	43	48	53	58	63	67	72
	6	6	12	17	23	29	35	40	46	52	58	63	69	75	81	87
	8	7	13	20	27	34	40	47	54	61	67	74	81	88	94	101
	10	8	15	23	31	38	46	54	62	69	77	85	92	100	108	115
	12	9	17	26	35	43	52	61	69	78	87	95	104	113	121	130
	14	10	19	29	38	48	58	67	77	87	96	106	115	125	135	144
	16	11	21	32	42	53	63	74	85	95	106	116	127	138	148	159
8	0	4	8	12	15	19	23	27	31	35	38	42	46	50	54	58
	2	5	10	14	19	24	29	34	38	43	48	53	58	63	67	72
	4	6	12	17	23	29	35	40	45	52	58	63	69	75	81	87
	6	7	13	20	27	34	40	47	54	61	67	74	81	88	94	101
	8	8	15	23	31	38	46	54	62	69	77	85	92	100	108	115
	10	9	17	26	35	43	52	61	69	78	87	95	104	113	121	130
	12	10	19	29	38	48	58	67	77	87	96	106	115	125	135	144
	14	11	21	32	42	53	63	74	85	95	106	116	127	138	148	159
	16	12	23	35	46	58	69	81	92	104	115	127	138	150	162	173
10	0	5	10	14	19	24	29	34	38	43	48	53	58	63	67	72
	2	6	12	17	23	29	35	40	46	52	58	63	69	75	81	87
	4	7	13	20	27	34	40	47	54	61	67	74	81	88	94	101
	6	8	15	23	31	38	46	54	62	69	77	85	92	100	108	115
	8	9	17	26	35	43	52	61	69	78	87	95	104	113	121	130
	10	10	19	29	38	48	58	67	77	87	96	106	115	125	135	144
	12	11	21	32	42	53	63	74	85	95	106	116	127	138	148	159
	14	12	23	35	46	58	69	81	92	104	115	127	138	150	162	173
	16	13	25	38	50	63	75	88	100	113	125	138	150	163	175	188

PCI Design Handbook/Fourth Edition

CONNECTIONS

Table 6.20.7A (continued) Design tensile strength for h ≥ h_{min}, ϕP_{c1}—Case 6

Case 6: Free edges on four adjacent sides

$\phi P_{c1} = \phi 4\lambda\sqrt{f'_c}\,(x_1)(y_1)$

$\phi = 0.85$

where: x_1 and y_1 are the dimensions of the flat bottom of the part of the truncated pyramid.

For Case 6: $x_1 = x + d_{e1} + d_{e2}$, $y_1 = y + d_{e3} + d_{e4}$

Note: Table values are based on
$\lambda = 1.0$ and $f'_c = 5000$ psi;
for different material properties, multiply table values by $\lambda\sqrt{f'_c/5000}$

x and y are the overall dimensions (width and length) of the stud group.

ℓ_e, in.	y_1, in. \ x_1, in.	2	4	6	8	10	12	14	16	18	20	22	24	26	28	30
									Design tensile strength, ϕP_{c1} (kips)							
3	0	0	0	0	0	0	0	0	0	0	0	0	0	0	0	0
	2	1	2	3	4	5	6	7	8	9	10	11	12	13	13	14
	4	2	4	6	8	10	12	13	15	17	19	21	23	25	27	29
	6	3	6	9	12	14	17	20	23	26	29	32	35	38	40	43
	8	4	8	12	15	19	23	27	31	35	38	43	46	50	54	58
	10	5	10	14	19	24	29	34	38	43	48	53	58	63	67	72
	12	6	12	17	23	29	35	40	46	52	58	63	69	75	81	87
	14	7	13	20	27	34	40	47	54	61	67	74	81	88	94	101
	16	8	15	23	31	38	46	54	62	69	77	52	92	100	108	115
4	0	0	0	0	0	0	0	0	0	0	0	0	0	0	0	0
	2	1	2	3	4	5	6	7	8	9	10	11	12	13	13	14
	4	2	4	6	8	10	12	13	15	17	19	21	23	25	27	29
	6	3	6	9	12	14	17	20	23	26	29	32	35	38	40	43
	8	4	8	12	15	19	23	27	31	35	38	42	46	50	54	58
	10	5	10	14	19	24	29	34	38	43	48	53	58	63	67	72
	12	6	12	17	23	29	35	40	46	52	58	63	69	75	81	87
	14	7	13	20	27	34	40	47	54	61	67	74	81	88	94	101
	16	8	15	23	31	38	46	54	62	69	77	85	92	100	108	115
6	0	0	0	0	0	0	0	0	0	0	0	0	0	0	0	0
	2	1	2	3	4	5	6	7	8	9	10	11	12	13	13	14
	4	2	4	6	8	10	12	13	15	17	19	21	23	25	27	29
	6	3	6	9	12	14	17	20	23	26	29	32	35	38	40	43
	8	4	8	12	15	19	23	27	31	35	38	42	46	50	54	58
	10	5	10	14	19	24	29	34	38	43	48	53	58	63	67	72
	12	6	12	17	23	29	35	40	46	52	58	63	69	75	81	87
	14	7	13	20	27	34	40	47	54	61	67	74	81	88	94	101
	16	8	15	23	31	38	46	54	62	69	77	85	92	100	108	115
8	0	0	0	0	0	0	0	0	0	0	0	0	0	0	0	0
	2	1	2	3	4	5	6	7	8	9	10	11	12	13	13	14
	4	2	4	6	8	10	12	13	15	17	19	21	23	25	27	29
	6	3	6	9	12	14	17	20	23	26	29	32	35	38	40	43
	8	4	8	12	15	19	23	27	31	35	38	42	46	50	54	58
	10	5	10	14	19	24	29	34	38	43	48	53	58	63	67	72
	12	6	12	17	23	29	35	40	46	52	58	63	69	75	81	87
	14	7	13	20	27	34	40	47	54	61	67	74	81	88	94	101
	16	8	15	23	31	38	46	54	62	69	77	85	92	100	108	115
10	0	0	0	0	0	0	0	0	0	0	0	0	0	0	0	0
	2	1	2	3	4	5	6	7	8	9	10	11	12	13	13	14
	4	2	4	6	8	10	12	13	15	17	19	21	23	25	27	29
	6	3	6	9	12	14	17	20	23	26	29	32	35	38	40	43
	8	4	8	12	15	19	23	27	31	35	38	42	46	50	54	58
	10	5	10	14	19	24	29	34	38	43	48	53	58	63	67	72
	12	6	12	17	23	29	35	40	46	52	58	63	69	75	81	87
	14	7	13	20	37	34	40	47	54	61	67	74	81	88	94	101
	16	8	15	23	31	38	46	54	62	69	77	85	92	100	108	115

CONNECTIONS

Table 6.20.7B Minimum member thickness for truncated pyramid failure—stud groups

Minimum thickness (see Fig. 6.5.6)

$h_{min} = (z + 2\ell_e)/2$

where:

z = lesser of x and y

ℓ_e = embedment length

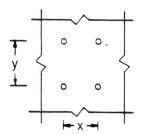

ℓ_e, in.	x, in. / y, in.	Minimum Thickness, h_{min} (in.)								
		2	4	6	8	10	12	14	16	18
3	0	3	3	3	3	3	3	3	3	3
	2	4	4	4	4	4	4	4	4	4
	4	4	5	5	5	5	5	5	5	5
	6	4	5	6	6	6	6	6	6	6
	8	4	5	6	7	7	7	7	7	7
	10	4	5	6	7	8	8	8	8	8
	12	4	5	6	7	8	9	9	9	9
4	0	4	4	4	4	4	4	4	4	4
	2	5	5	5	5	5	5	5	5	5
	4	5	6	6	6	6	6	6	6	6
	6	5	6	7	7	7	7	7	7	7
	8	5	6	7	8	8	8	8	8	8
	10	5	6	7	8	9	9	9	9	9
	12	5	6	7	8	9	10	10	10	10
6	0	6	6	6	6	6	6	6	6	6
	2	7	7	7	7	7	7	7	7	7
	4	7	8	8	8	8	8	8	8	8
	6	7	8	9	9	9	9	9	9	9
	8	7	8	9	10	10	10	10	10	10
	10	7	8	9	10	11	11	11	11	11
	12	7	8	9	10	11	12	12	12	12
8	0	8	8	8	8	8	8	8	8	8
	2	9	9	9	9	9	9	9	9	9
	4	9	10	10	10	10	10	10	10	10
	6	9	10	11	11	11	11	11	11	11
	8	9	10	11	12	12	12	12	12	12
	10	9	10	11	12	13	13	13	13	13
	12	9	10	11	12	13	14	14	14	14
10	0	10	10	10	10	10	10	10	10	10
	2	11	11	11	11	11	11	11	11	11
	4	11	12	12	12	12	12	12	12	12
	6	11	12	13	13	13	13	13	13	13
	8	11	12	13	14	14	14	14	14	14
	10	11	12	13	14	15	15	15	15	15
	12	11	12	13	14	15	16	16	16	16
12	0	12	12	12	12	12	12	12	12	12
	2	13	13	13	13	13	13	13	13	13
	4	13	14	14	14	14	14	14	14	14
	6	13	14	15	15	15	15	15	15	15
	8	13	14	15	16	16	16	16	16	16
	10	13	14	15	16	17	17	17	17	17
	12	13	14	15	16	17	18	18	18	18

CONNECTIONS

Table 6.20.7C Reduction in design tensile strength of stud groups for $h < h_{min}$

$\phi P_{c2} = \phi 4\lambda\sqrt{f'_c}\,[x - 2(h - \ell_e)][y - 2(h - \ell_e)]$

where $\phi = 0.85$:

 x and y = overall dimensions (width and length) of stud groups
 h = member thickness
 ℓ_e = embedment length

Note: For $h \geq h_{min}$, $\phi P_{c2} = 0$. See Table 6.20.7B for h_{min} values.

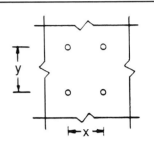

Reduction in strength, ϕP_{c2} (kips)

$h - \ell_e$, in.	y, in. \ x, in.	2	4	6	8	10	12	14	16	18	20	22	24	26	28	30
1	0	0	0	0	0	0	0	0	0	0	0	0	0	0	0	0
	2	0	0	0	0	0	0	0	0	0	0	0	0	0	0	0
	4	0	1	2	3	4	5	6	7	8	9	10	11	12	13	13
	6	0	2	4	6	8	10	12	13	15	17	19	21	23	25	27
	8	0	3	6	9	12	14	17	20	23	26	29	32	35	38	40
	10	0	4	8	12	15	19	23	27	31	35	38	42	46	50	54
	12	0	5	10	14	19	24	29	34	38	43	48	53	58	63	67
	14	0	6	12	17	23	29	35	40	46	52	58	63	69	75	81
	16	0	7	13	20	27	34	40	47	54	61	67	74	81	88	94
2	0	2	0	0	0	0	0	0	0	0	0	0	0	0	0	0
	2	1	0	0	0	0	0	0	0	0	0	0	0	0	0	0
	4	0	0	0	0	0	0	0	0	0	0	0	0	0	0	0
	6	0	0	1	2	3	4	5	6	7	8	9	10	11	12	13
	8	0	0	2	4	6	8	10	12	13	15	17	19	21	23	25
	10	0	0	3	6	9	12	14	17	20	23	26	29	32	35	38
	12	0	0	4	8	12	15	19	23	27	31	35	38	42	46	50
	14	0	0	5	10	14	19	24	29	34	38	43	48	53	58	63
	16	0	0	6	12	17	23	29	35	40	46	52	58	63	69	75
3	0	6	3	0	0	0	0	0	0	0	0	0	0	0	0	0
	2	4	2	0	0	0	0	0	0	0	0	0	0	0	0	0
	4	2	1	0	0	0	0	0	0	0	0	0	0	0	0	0
	6	0	0	0	0	0	0	0	0	0	0	0	0	0	0	0
	8	0	0	0	1	2	3	4	5	6	7	8	9	10	11	12
	10	0	0	0	2	4	6	8	10	12	13	15	17	19	21	23
	12	0	0	0	3	6	9	12	14	17	20	23	26	29	32	35
	14	0	0	0	4	8	12	15	19	23	27	31	35	38	42	46
	16	0	0	0	5	10	14	19	24	29	34	38	43	48	53	58
4	0	12	8	4	0	0	0	0	0	0	0	0	0	0	0	0
	2	9	6	3	0	0	0	0	0	0	0	0	0	0	0	0
	4	6	4	2	0	0	0	0	0	0	0	0	0	0	0	0
	6	3	2	1	0	0	0	0	0	0	0	0	0	0	0	0
	8	0	0	0	0	0	0	0	0	0	0	0	0	0	0	0
	10	0	0	0	0	1	2	3	4	5	6	7	8	9	10	11
	12	0	0	0	0	2	4	6	8	10	12	13	15	17	19	21
	14	0	0	0	0	3	6	9	12	14	17	20	23	26	29	32
	16	0	0	0	0	4	8	12	15	19	23	27	31	35	38	42
5	0	19	14	10	5	0	0	0	0	0	0	0	0	0	0	0
	2	15	12	8	4	0	0	0	0	0	0	0	0	0	0	0
	4	12	9	6	3	0	0	0	0	0	0	0	0	0	0	0
	6	8	6	4	2	0	0	0	0	0	0	0	0	0	0	0
	8	4	3	2	1	0	0	0	0	0	0	0	0	0	0	0
	10	0	0	0	0	0	0	0	0	0	0	0	0	0	0	0
	12	0	0	0	0	0	1	2	3	4	5	6	7	8	9	10
	14	0	0	0	0	0	2	4	6	8	10	12	13	15	17	19
	16	0	0	0	0	0	3	6	9	12	14	17	20	23	26	29

CONNECTIONS

Table 6.20.8 Shear strength of welded headed studs

I—Design shear strength limited by concrete:

Use smaller of the values from Eqs. 6.5.8a and 6.5.9

$\phi V_c = (\phi 628 d_b^2 \lambda \sqrt{f'_c}) n$ (Eq. 6.5.8a)

Table A gives values for n = 1, ϕ = 0.85

$\phi V_c = \phi V'_c C_w C_t C_c$ (Eq. 6.5.9)

where:

$\phi V'_c = \phi 12.5 d_e^{1.5} \lambda \sqrt{f'_c}$

$C_w = \left(1 + \dfrac{b}{3.5 d_e}\right) \leq n_s$

$C_t = \dfrac{h}{1.3 d_e} \leq 1.0$

$C_c = \left[0.4 + 0.7\left(\dfrac{d_c}{d_e}\right)\right] \leq 1.0$

Table B gives values for ϕ = 0.85

where: n_s = number of studs in back row; see figure for other notation

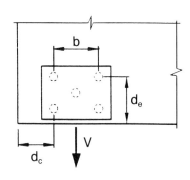
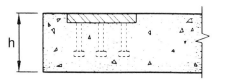

II—Design shear strength limited by steel:

$\phi V_s = (\phi 35,344 d_b^2) n$ (Eq. 6.5.14a)

Table C gives value for n = 1, ϕ = 1.0

Table A—ϕV_c, kips										
f'_c, psi	4000		5000		6000		7000		8000	
d_b, in. \ λ	1.0	0.85	1.0	0.85	1.0	0.85	1.0	0.85	1.0	0.85
¼	2.15	1.83	2.40	2.04	2.63	2.24	2.85	2.42	3.04	2.58
⅜	4.74	4.03	5.30	4.51	5.81	4.94	6.28	5.33	6.71	5.70
½	8.43	7.16	9.45	8.03	10.32	8.78	11.15	9.48	11.82	10.13
⅝	13.19	11.21	14.72	11.79	16.15	13.72	17.44	14.83	18.65	15.85
¾	19.00	16.14	21.23	18.04	23.26	19.77	25.12	21.35	26.85	22.82
⅞	25.85	21.97	28.90	24.56	31.66	26.91	34.19	26.09	36.55	31.07
Table B—$\phi V'_c$, kips										
d_e, in. \ λ	1.0	0.85	1.0	0.85	1.0	0.85	1.0	0.85	1.0	0.85
2	1.90	1.62	2.12	1.81	1.77	1.51	2.51	2.14	2.69	2.29
3	3.49	2.97	3.90	3.31	4.26	3.63	4.62	3.82	4.94	4.20
4	6.38	4.57	6.00	5.11	6.58	5.59	7.11	6.04	7.60	6.46
5	7.51	6.38	8.39	7.14	9.19	7.82	9.94	8.45	10.62	9.03
6	9.88	8.40	11.04	9.39	12.09	10.29	13.08	11.11	13.97	11.87
7	12.45	10.98	13.80	11.82	15.24	12.95	16.46	13.99	17.60	14.96
8	15.20	12.82	16.99	14.44	18.61	15.81	16.44	17.08	21.50	18.27
9	18.14	15.44	20.28	17.24	22.21	18.88	23.99	20.40	25.65	21.80
10	21.25	18.06	23.75	20.18	26.01	22.11	18.10	23.88	30.04	25.53
11	24.52	20.84	27.41	23.30	30.03	25.52	32.43	27.57	34.67	29.47
12	27.94	23.74	31.22	26.53	34.20	29.07	36.94	31.40	39.49	33.67

Table C—ϕV_s, kips						
Diameter, in.	¼	⅜	½	⅝	¾	⅞
ϕV_s	2.2	5.0	8.8	13.8	19.9	27.1

CONNECTIONS

Table 6.20.9 Allowable loads on bolts[1]

Bolt. diam. (in.)	A, in.² Nominal	A-36				A-307			
		Tension (kips)		Shear[2] kips		Tension (kips)		Shear[2] (kips)	
		Design	Service	Design	Service	Design	Service	Design	Service
½	0.196	6.4	3.7	3.3	2.0	6.6	3.9	3.2	2.0
⅝	0.307	10.0	5.9	5.2	3.0	10.4	6.1	5.0	3.1
¾	0.442	14.4	8.4	7.5	4.4	14.9	8.8	7.2	4.4
⅞	0.601	19.6	11.5	10.2	6.0	20.3	12.0	9.7	6.0
1	0.785	25.6	15.0	13.3	7.8	26.5	15.7	12.7	7.9
1¼	1.227	40	23.4	20.8	12.1	41.4	24.5	19.9	12.3

(1) AISC Allowable Stress Design[12] or AISC Load and Resistance Factor Design[13]. See these manuals for shear-tension interaction.
(2) For threads included in shear plane.

Table 6.20.10 Capacity of coil bolts and threaded coil rods

Bolt diameter (in.)	Minimum coil penetration (in.)	Tensile strength, P_o (lb)	Shear strength, V_o (lb)
½	1½	13,500	8,100
¾	2	18,470	11,080
1	2½	37,870	22,720
1¼	2½	54,960	32,980
1½	3	83,340	50,000

Table 6.20.11 Capacity of wire used in concrete inserts

Leg wire diameter (in.)	Wire grade	Yield strength (lb)
0.218	C1008	2000
0.223	C1038	3900
0.225	C1038	3700
0.240	C1008	2900
0.260	C1008	3550
0.281	C1035	6000
0.306	C1035	6900
0.340	C1035	7500
0.375	C1008	7450
0.440	C1005	12,000

CONNECTIONS

Table 6.20.12 Plastic section moduli and shape factors

Section	Plastic modulus, Z_s, in.^{3}	Shape factor
Rectangle (b × h)	$\dfrac{bh^2}{4}$	1.5
I-section	x-x axis: $bt(h-t) + \dfrac{w}{4}(h-2t)^2$	1.12 (approx)
	y-y axis: $\dfrac{b^2 t}{2} + \dfrac{(h-2t)w^2}{4}$	1.55 (approx)
Channel	$bt(h-t) + \dfrac{(h-2t)^2}{4}$	1.12 (approx)
Circle	$\dfrac{h^3}{6}$	1.70
Hollow circle	$\dfrac{h^3}{6}\left[1-\left(1-\dfrac{2t}{h}\right)^3\right]$ th^2 for $t \ll h$	$\dfrac{16}{3\pi}\left[\dfrac{1-\left(1-\dfrac{2t}{h}\right)^3}{1-\left(1-\dfrac{2t}{h}\right)^4}\right]$ 1.27 for $t \ll h$
Hollow rectangle (see footnote)	$\dfrac{bh^2}{4}\left[1-\left(1-\dfrac{2w}{b}\right)\left(1-\dfrac{2t}{h}\right)^2\right]$	1.12 (approx) for thin walls
Diamond	$\dfrac{bh^2}{12}$	2

For TS shapes, refer to AISC—LFRD Manual[13] for Z_s values

CONNECTIONS

Table 6.20.13 Shear strength of connection angles

$t = \sqrt{\dfrac{4V_u e_v}{\phi F_y b_n}}$, in.

$\phi = 0.90$

b_n = net length of angle, in.

F_y = yield strength of angle steel = 36,000 psi

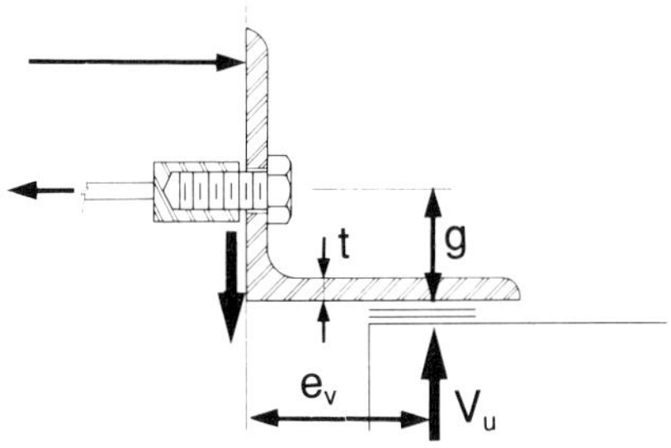

Angle thickness t	$V_u = \phi V_n$, lb. per inch of length				
	$e_v = \tfrac{3}{4}"$	$e_v = 1"$	$e_v = 1\tfrac{1}{2}"$	$e_v = 2"$	$e_v = 2\tfrac{1}{2}"$
5/16"	1055	791	527	396	316
3/8"	1519	1139	759	570	456
7/16"	2067	1550	1034	775	620
1/2"	2700	2025	1350	1013	810
9/16"	3417	2563	1709	1281	1025
5/8"	4219	3164	2109	1582	1266

Table 6.20.14 Axial strength of connection angles

$t = \sqrt{\dfrac{4N_u g}{\phi F_y b_n}}$

$\phi = 0.90$

b_n = net length of angle, in.

F_y = yield strength of angle steel = 36,000 psi

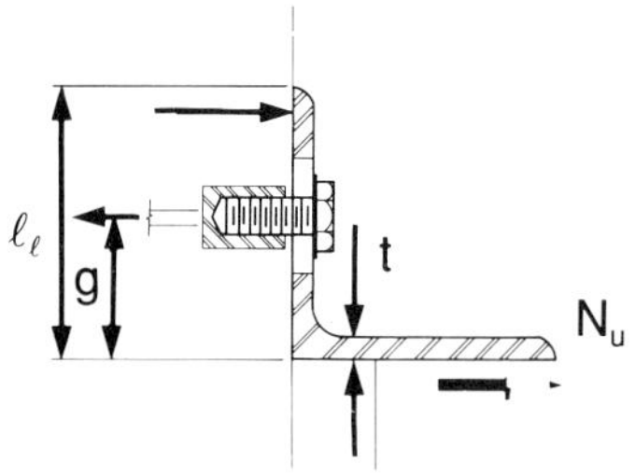

Angle thickness t	$N_u = \phi N_n$, lb. per inch of length			
	$\ell_\ell = 5"$, g = 3"	$\ell_\ell = 6"$, g = 4"	$\ell_\ell = 7"$, g = 5"	$\ell_\ell = 8"$, g = 6"
5/16"	264	198		
3/8"	380	285	228	
7/16"	517	388	310	258
1/2"	675	506	405	338
9/16"	854	641	513	427
5/8"	1055	791	633	527

CONNECTIONS

Table 6.20.15 Column base plate thickness requirements

Thickness required for concrete bearing

f_{bu} (psi)	$x_o = 3"$	$x_o = 4"$	$x_o = 5"$
500	5/8	3/4	1
1000	3/4	1	1 3/8
1500	1	1 3/8	1 5/8
2000	1 1/8	1 1/2	1 7/8
2500	1 1/4	1 5/8	2
3000	1 3/8	1 7/8	2 1/4
3500	1 1/2	2	2 1/2
4000	1 5/8	2 1/8	2 5/8

Thickness required for bolt loading

Tension on external anchor bolts

b (in.)	No. & diameter of A-36 or A-307 anchor bolts per side							
	2 – 3/4" $x_t = 3.75"$	2 – 3/4" $x_t = 4.25"$	2 – 1" $x_t = 3.75"$	2 – 1" $x_t = 4.25"$	2 – 1 1/4" $x_t = 3.75"$	2 – 1 1/4" $x_t = 4.25"$	2 – 1 1/2" $x_t = 3.75"$	2 – 1 1/2" $x_t = 4.25"$
12	1	1 1/8	1 3/8	1 1/2	1 3/4	1 7/8	2 1/8	2 1/4
14	1	1	1 3/8	1 3/8	1 5/8	1 3/4	2	2 1/8
16	7/8	1	1 1/4	1 3/8	1 1/2	1 5/8	1 7/8	2
18	7/8	1	1 1/8	1 1/4	1 1/2	1 1/2	1 3/4	1 7/8
20	7/8	7/8	1 1/8	1 1/8	1 3/8	1 1/2	1 5/8	1 3/4
22	3/4	7/8	1	1 1/8	1 3/8	1 3/8	1 5/8	1 5/8
24	3/4	3/4	1	1 1/8	1 1/4	1 3/8	1 1/2	1 5/8
26	3/4	3/4	1	1	1 1/4	1 1/4	1 1/2	1 1/2
28	3/4	3/4	1	1	1 1/8	1 1/4	1 3/8	1 1/2

Compression on anchor bolts or tension on internal anchor bolts

b (in.)	No. & diameter of A-36 or A-307 anchor bolts per side							
	2 – 3/4" $x_c = 1.5"$	2 – 3/4" $x_c = 2.0"$	2 – 1" $x_c = 1.5"$	2 – 1" $x_c = 2.0"$	2 – 1 1/4" $x_c = 1.5"$	2 – 1 1/4" $x_c = 2.0"$	2 – 1 1/2" $x_c = 1.5"$	2 – 1 1/2" $x_c = 2.0"$
12	3/4	3/4	7/8	1	1 1/4	1 3/8	1 3/8	1 5/8
14	3/4	3/4	7/8	1	1	1 1/4	1 1/4	1 1/2
16	3/4	3/4	3/4	7/8	1	1 1/8	1 1/4	1 3/8
18	3/4	3/4	3/4	7/8	1	1 1/8	1 1/8	1 1/4
20	3/4	3/4	3/4	7/8	7/8	1	1 1/8	1 1/4
22	3/4	3/4	3/4	3/4	7/8	1	1	1 1/8
24	3/4	3/4	3/4	3/4	7/8	1	1	1 1/8
26	3/4	3/4	3/4	3/4	3/4	7/8	1	1 1/8
28	3/4	3/4	3/4	3/4	3/4	7/8	7/8	1

External anchor bolts   Internal anchor bolts

CONNECTIONS

Table 6.20.16 Design strength of concrete brackets, corbels or haunches

Design strength by Eqs. 6.11.3 or 6.11.4 for following criteria:

- f_y = 60,000 psi
- N_u = 0.2 V_u
- b = width of bearing, in.
- d = h − 1.25 (h in inches)
- ϕ = 0.85
- a = ¾ ℓ_p
- λ = 1.0 (normal wieght concrete)
- $\phi V_n \leq \phi 1000 \lambda^2 bd$

(The design strength value listed in the table immediately above a blank entree corresponds to this limit. The blank entree(s) in a given column will have this same limiting value.)

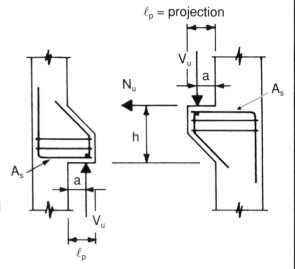

ℓ_p = projection

			\multicolumn{22}{c}{Values of ϕV_n (kips)}
	ℓ_p (in.)				4							6								8				
b	A_s	h	4	6	8	10	12	14	16	18	6	8	10	12	14	16	18	20	8	10	12	14	16	18	20	22
6	2 – #4		14	23	29	35	40	44	47	48	17	22	27	31	35	38	41	0	18	22	26	29	32	35	0	0
	2 – #5			24	34	45	55	59	62	65	24	34	42	49	55	60	65	67	28	34	40	45	50	55	59	62
	2 – #6							65	75	81			45	55	65	75	81	84	34	45	55	65	72	79	84	87
	2 – #7									85							85	96					75	85	96	106
8	2 – #4		14	23	29	35	40	44	0	0	17	22	27	31	35	0	0	0	18	22	26	29	0	0	0	0
	2 – #5		19	32	46	55	62	65	69	72	26	35	42	49	55	60	65	69	28	34	40	45	50	55	59	62
	2 – #6				60	73	82	86	90		32	46	60	70	79	86	90	94	40	49	58	65	72	79	84	90
	2 – #7						87	100	109					73	87	100	109	114		60	73	87	98	107	114	118
	2 – #8								114								114	128					100	114	128	139
	2 – #9																									141
10	2 – #4		14	23	29	35	0	0	0	0	17	22	27	0	0	0	0	0	18	22	0	0	0	0	0	0
	2 – #5		23	35	46	55	62	69	74	77	26	35	42	49	55	60	65	0	28	34	40	45	50	55	0	0
	2 – #6			40	57	74	84	89	94	98	38	50	61	70	79	86	93	99	40	49	58	65	72	79	84	90
	2 – #7					91	108	114	119		40	57	74	91	107	114	119	124	54	67	78	89	98	107	115	122
	2 – #8							125	140						108	125	140	146	57	74	91	108	125	140	146	152
	2 – #9								142								142	159						142	159	175
12	2 – #4		14	23	29	0	0	0	0	0	17	22	0	0	0	0	0	0	18	0	0	11	11	72	0	0
	2 – #5		23	35	46	55	62	69	0	0	26	35	42	49	55	0	0	0	28	34	40	0	6	98	0	17
	2 – #6		28	48	66	79	90	96	100	105	38	50	61	70	79	86	93	99	40	49	58	0	13	12	79	1
	2 – #7				69	89	109	116	122	128	48	68	83	96	107	117	127	133	54	67	78	45	0	8	10	0
	2 – #8						110	130	144	151		69	89	10	130	144	151	157	69	88	10	65	0	15	7	0
	2 – #9								150	171						150	171	181		89	3	89	0	0	14	84
	3 – #4		22	34	44	53	60	66	0	0	25	33	40	47	52	0	0	0	27	33	38	44	0	0	0	0
	3 – #5		28	48	69	82	93	98	103	107	39	52	63	73	82	90	97	104	42	51	60	68	75	82	88	94
	3 – #6				89	110	123	130	136		48	69	89	105	118	129	136	141	60	74	87	98	108	118	127	135
	3 – #7						130	150	164					110	130	150	164	171	69	89	110	130	148	161	171	177
	3 – #8								171								171	191					150	171	191	209
	3 – #9																									212

CONNECTIONS

Table 6.20.16 Design strength of concrete brackets, corbels or haunches (continued)

			Values of ϕV_n (kips)			
	ℓ_p (in.)		4	6	8	
b	A_s \ h	4 6 8 10 12 14 16 18	6 8 10 12 14 16 18 20	8 10 12 14 16 18 20 22

b	A_s	h: 4 6 8 10 12 14 16 18	h: 6 8 10 12 14 16 18 20	h: 8 10 12 14 16 18 20 22
14	2 – #4	14 23 29 0 0 0 0 0	17 22 0 0 0 0 0 0	18 0 0 0 0 0 0 0
	2 – #5	23 35 46 55 62 0 0 0	26 35 42 49 0 0 0 0	28 34 40 0 0 0 0 0
	2 – #6	33 51 66 79 90 99 106 110	38 50 61 70 79 86 93 0	40 49 58 65 72 79 0 0
	2 – #7	57 80 104 116 123 129 135	51 68 83 96 107 117 127 135	54 67 78 89 98 107 115 122
	2 – #8	128 145 153 160	57 80 104 125 140 153 160 166	71 88 103 116 128 140 150 160
	2 – #9	152 176 185	128 152 176 185 192	80 104 128 147 163 177 190 199
	3 – #4	22 34 44 53 60 0 0 0	25 33 40 47 0 0 0 0	27 33 38 0 0 0 0 0
	3 – #5	33 53 69 82 93 103 109 113	39 52 63 73 82 90 97 0	42 51 60 68 75 82 0 0
	3 – #6	57 80 104 123 131 137 143	56 75 91 105 118 129 140 149	60 74 87 98 108 118 127 135
	3 – #7	128 152 166 174	57 80 104 128 152 166 174 181	80 101 118 133 148 161 173 184
	3 – #8	176 199	176 199 213	104 128 152 176 199 213 222
	3 – #9		223	223 247
16	2 – #6	33 51 66 79 90 99 107 0	38 50 61 70 79 86 0 0	40 49 58 65 72 0 0 0
	2 – #7	37 65 90 107 122 129 136 141	51 68 83 96 107 117 127 135	54 67 78 89 98 107 115 122
	2 – #8	92 119 144 153 161 168	65 89 108 125 140 153 166 174	71 88 103 116 128 140 150 160
	2 – #9	146 173 186 194	92 119 146 173 186 194 202	90 111 130 147 163 177 190 202
	3 – #6	130 137 144 151	75 91 105 118 129 140 149	60 74 87 98 108 118 127 135
	3 – #7	146 166 175 183	92 119 143 161 175 183 190	82 101 118 133 148 161 173 184
	3 – #8	173 201 216	146 173 201 216 225	92 119 146 173 193 210 225 233
	3 – #9	228	228 255	201 228 255 269
	4 – #6	164 173 181	140 157 173 181 188	80 99 115 131 144 157 169 180
	4 – #7	201 219	146 173 201 219 228	92 119 146 173 197 214 228 236
	4 – #8	228	228 255	201 228 255 278
	4 – #9			282
18	2 – #6	33 51 66 79 90 99 0 0	38 50 61 70 79 0 0 0	40 49 58 65 0 0 0 0
	2 – #7	42 69 90 107 122 135 141 147	51 68 83 96 107 117 127 135	54 67 78 89 98 107 115 0
	2 – #8	73 103 134 151 116 168 175	67 89 108 125 140 153 166 177	71 88 103 116 128 140 150 160
	2 – #9	164 185 194 203	73 103 134 158 177 194 203 211	90 111 130 147 163 177 190 202
	3 – #6	99 118 135 151 157	56 75 91 105 118 129 140 149	60 74 87 98 108 118 127 135
	3 – #7	103 134 164 174 183 191	73 102 124 143 161 176 190 199	82 101 118 133 148 161 173 184
	3 – #8	195 216 226	103 134 164 195 216 226 235	103 131 154 174 193 210 225 240
	3 – #9	226 256	226 256 272	134 164 195 226 256 272 282
	4 – #6	162 172 181 189	100 121 140 157 173 186 196	80 99 115 131 144 157 169 180
	4 – #7	164 195 219 229	103 134 164 195 219 229 238	103 134 157 178 197 214 230 245
	4 – #8	226 256	226 256 281	164 195 226 256 281 291
	4 – #9		287	287 317
20	2 – #6	33 51 66 79 90 0 0 0	38 50 61 70 0 0 0 0	40 49 58 0 0 0 0 0
	2 – #7	44 69 90 107 122 135 146 153	51 68 83 96 107 117 127 0	54 67 78 89 98 107 0 0
	2 – #8	47 81 115 140 157 166 174 182	67 89 108 125 140 153 166 177	71 88 103 116 128 140 150 160
	2 – #9	149 181 192 202 211	81 112 136 158 177 194 210 220	90 111 130 147 163 177 190 202
	3 – #6	76 99 118 135 149 156 163	56 75 91 105 118 129 140 0	60 74 87 98 108 118 0 0
	3 – #7	81 115 149 171 181 190 199	77 102 124 143 161 176 190 203	82 101 118 133 148 161 173 184
	3 – #8	183 213 225 235	81 115 149 183 210 225 235 245	107 131 154 174 193 210 225 240
	3 – #9	217 251 272	217 251 272 283	115 149 183 217 244 265 283 294
	4 – #6	169 179 188 196	75 100 121 140 157 173 186 199	80 99 115 131 144 157 169 180
	4 – #7	183 216 228 238	81 115 149 183 214 228 238 248	109 134 157 178 197 214 230 245
	4 – #8	217 251 281	217 251 281 292	115 149 183 217 251 280 292 303
	4 – #9	285	285 319	285 319 330

PCI Design Handbook/Fourth Edition

CONNECTIONS

Table 6.20.16 Design strength of concrete brackets, corbels or haunches (continued)

		Values of ϕV_n (kips)																						
ℓ_p (in.)		**10**								**12**							**14**							
b	h → A_s ↓	10	12	14	16	18	20	22	24	12	14	16	18	20	22	24	26	14	16	18	20	22	24	26	28
6	2 – #4	18	22	25	28	30	0	0	0	19	22	24	27	0	0	0	0	19	22	24	0	0	0	0	0
	2 – #5	29	34	39	43	47	51	55	58	30	34	38	42	45	48	52	55	30	34	37	40	44	47	49	52
	2 – #6	42	49	56	62	68	73	79	83	42	49	54	60	65	70	74	79	43	49	54	58	63	67	71	75
	2 – #7	45	55	65	75	85	96	106	109	55	65	74	82	88	95	101	107	59	66	73	79	85	91	97	102
	2 – #8								116			75	85	96	106	116	126	65	75	85	96	106	116	126	133
	2 – #9																								136
8	2 – #4	18	22	25	0	0	0	0	0	19	22	0	0	0	0	0	0	19	0	0	0	0	0	0	0
	2 – #5	29	34	39	43	47	51	55	0	30	34	38	42	45	48	0	0	30	34	37	40	44	0	0	0
	2 – #6	42	49	56	62	68	73	79	83	42	49	54	60	65	70	74	79	43	49	54	58	63	67	71	75
	2 – #7	56	67	76	85	93	100	107	113	58	66	74	82	88	95	101	107	59	66	73	79	85	91	97	102
	2 – #8	60	73	87	100	114	128	139	144	73	87	97	106	116	124	132	140	77	86	95	104	112	119	126	133
	2 – #9							141	155			100	114	128	141	155	168	87	100	114	128	141	151	160	168
10	2 – #4	18	0	0	0	0	0	0	0	0	0	0	0	0	0	0	0	0	0	0	0	0	0	0	0
	2 – #5	29	34	39	43	47	0	0	0	30	34	38	42	0	0	0	0	30	34	37	0	0	0	0	0
	2 – #6	42	49	56	62	68	73	79	83	42	49	54	60	65	70	74	79	43	49	54	58	63	67	71	0
	2 – #7	56	67	76	85	93	100	107	113	58	66	74	82	88	95	101	107	59	66	73	79	85	91	97	102
	2 – #8	74	87	99	110	121	131	140	148	76	87	97	106	116	124	132	140	77	86	95	104	112	119	126	133
	2 – #9		91	108	125	142	159	175	181	91	108	123	135	146	157	167	177	97	109	120	131	141	151	160	168
12	2 – #4	0	0	0	0	0	0	0	0	0	0	0	0	0	0	0	0	0	0	0	0	0	0	0	0
	2 – #5	29	34	39	0	0	0	0	0	30	34	0	0	0	0	0	0	30	0	0	0	0	0	0	0
	2 – #6	42	49	56	62	68	73	79	0	42	49	54	60	65	70	0	0	43	49	54	58	63	0	0	0
	2 – #7	56	67	76	85	93	100	107	113	58	66	74	82	88	95	101	107	59	66	73	79	85	91	97	102
	2 – #8	74	87	99	110	121	131	140	148	76	87	97	106	116	124	132	140	77	86	95	104	112	119	126	133
	2 – #9	89	110	126	140	153	165	177	188	96	110	123	135	146	157	167	177	97	109	120	131	141	151	160	168
	3 – #4	28	33	37	0	0	0	0	0	28	32	0	0	0	0	0	0	29	0	0	0	0	0	0	0
	3 – #5	43	51	58	65	71	77	82	0	44	51	57	62	68	73	0	0	45	51	56	61	65	0	0	0
	3 – #6	62	73	84	93	102	110	118	125	64	73	82	90	97	105	111	118	65	73	80	87	94	101	107	112
	3 – #7	85	100	114	127	139	150	160	170	87	99	111	122	133	142	152	160	88	99	109	119	128	137	145	153
	3 – #8	89	110	130	150	171	191	209	216	110	130	145	160	173	186	198	209	115	129	143	155	167	179	189	200
	3 – #9							212	232			150	171	191	212	232	252	130	150	171	191	212	226	240	253
14	2 – #4	0	0	0	0	0	0	0	0	0	0	0	0	0	0	0	0	0	0	0	0	0	0	0	0
	2 – #5	29	34	0	0	0	0	0	0	30	0	0	0	0	0	0	0	0	0	0	0	0	0	0	0
	2 – #6	42	49	56	62	68	0	0	0	42	49	54	60	0	0	0	0	43	49	54	0	0	0	0	0
	2 – #7	56	67	76	85	93	100	107	113	58	66	74	82	88	95	101	0	59	66	73	79	85	91	0	0
	2 – #8	74	87	99	110	121	131	140	148	76	87	97	106	116	124	132	140	77	86	95	104	112	119	126	133
	2 – #9	93	110	126	140	153	165	177	188	96	110	123	135	146	157	167	177	97	109	120	131	141	151	160	168
	3 – #4	28	33	0	0	0	0	0	0	28	0	0	0	0	0	0	0	0	0	0	0	0	0	0	0
	3 – #5	43	51	58	65	71	0	0	0	44	51	57	62	0	0	0	0	45	51	56	0	0	0	0	0
	3 – #6	62	73	84	93	102	110	118	125	64	73	82	90	97	105	111	118	65	73	80	87	94	101	107	112
	3 – #7	85	100	114	127	139	150	160	170	87	99	111	122	133	142	152	160	88	99	109	119	128	137	145	153
	3 – #8	104	128	149	166	181	196	210	222	113	130	145	160	173	186	198	209	115	129	143	155	167	179	189	200
	3 – #9			152	176	199	223	247	265	128	152	176	199	219	236	253	265	146	164	181	197	212	226	240	253
16	2 – #6	0	0	0	0	0	0	0	0	0	0	0	0	0	0	0	0	0	0	0	0	0	0	0	0
	2 – #7	29	34	39	0	0	0	0	0	30	34	0	0	0	0	0	0	30	0	0	0	0	0	0	0
	2 – #8	42	49	56	62	68	73	79	0	42	49	54	60	65	70	0	0	43	49	54	58	63	0	0	0
	2 – #9	56	67	76	85	93	100	107	113	58	66	74	82	88	95	101	107	59	66	73	79	85	91	97	102
	3 – #6	74	87	99	110	121	131	140	148	76	87	97	106	116	124	132	140	77	86	95	104	112	119	126	133
	3 – #7	89	110	126	140	153	165	177	188	96	110	123	135	146	157	167	177	97	109	120	131	141	151	160	168
	3 – #8	28	33	37	0	0	0	0	0	28	32	0	0	0	0	0	0	29	0	0	0	0	0	0	0
	3 – #9	43	51	58	65	71	77	82	0	44	51	57	62	68	73	0	0	45	51	56	61	65	0	0	0
	4 – #6	62	73	84	93	102	110	118	125	64	73	82	90	97	105	111	118	65	73	80	87	94	101	107	112
	4 – #7	85	100	114	127	139	150	160	170	87	99	111	122	133	142	152	160	88	99	109	119	128	137	145	153
	4 – #8	89	110	130	150	171	191	209	216	110	130	145	160	173	186	198	209	115	129	143	155	167	179	189	200
	4 – #9							212	232			150	171	191	212	232	252	130	150	171	191	212	226	240	253

CONNECTIONS

Table 6.20.16 Design strength of concrete brackets, corbels or haunches (continued)

			Values of ϕV_n (kips)																				
	ℓ_p (in.)		10							12							14						
b	A_s \ h		10	12	14	16	18	20	22	24	12	14	16	18	20	22	24	26	14	16	18	20	22	24	26	28

b	A_s	10	12	14	16	18	20	22	24	12	14	16	18	20	22	24	26	14	16	18	20	22	24	26	28
18	2 – #4	42	49	56	0	0	0	0	0	42	49	0	0	0	0	0	0	43	0	0	0	0	0	0	0
	2 – #5	56	67	76	85	93	100	0	0	58	66	74	82	88	0	0	0	59	66	73	79	0	0	0	0
	2 – #6	74	87	99	110	121	131	140	148	76	87	97	106	116	124	132	140	77	86	95	104	102	119	126	0
	2 – #7	93	110	126	140	153	165	177	188	96	110	123	135	146	157	167	177	97	109	120	131	141	151	160	168
	2 – #8	62	73	84	93	102	110	118	0	64	73	82	90	97	105	0	0	65	73	80	87	94	0	0	0
	2 – #9	85	100	114	127	139	150	160	170	87	99	111	122	133	142	152	160	88	99	109	119	128	137	145	153
	3 – #4	111	130	149	166	181	196	210	222	113	130	145	160	173	186	198	209	115	129	143	155	167	179	189	200
	3 – #5	134	164	188	210	229	248	265	281	143	164	184	202	219	236	251	265	146	164	181	197	212	226	240	253
	3 – #6	83	98	112	124	136	147	157	167	85	97	109	120	130	140	149	157	86	97	107	117	126	134	142	150
	3 – #7	113	133	152	169	185	200	214	227	116	133	148	163	177	190	202	214	118	132	146	159	171	182	193	204
	3 – #8	134	164	195	221	242	261	279	296	151	173	194	213	231	248	264	279	154	173	190	207	223	238	253	266
	3 – #9				226	256	287	317	348	164	195	226	256	287	314	334	353	194	218	241	262	282	302	320	337
20	2 – #4	42	49	0	0	0	0	0	0	42	0	0	0	0	0	0	0	0	0	0	0	0	0	0	0
	2 – #5	56	67	76	85	93	0	0	0	58	66	74	82	0	0	0	0	59	66	73	0	0	0	0	0
	2 – #6	74	87	99	110	121	131	140	0	76	87	97	106	116	124	0	0	77	86	95	104	112	0	0	0
	2 – #7	93	110	126	140	153	165	177	188	96	110	123	135	143	157	167	177	97	109	120	131	141	151	160	168
	2 – #8	62	73	84	93	102	0	0	0	64	73	82	90	0	0	0	0	65	73	80	0	0	0	0	0
	2 – #9	85	100	114	127	139	150	160	170	87	99	111	122	133	142	152	160	88	99	109	119	128	137	145	0
	3 – #4	111	130	149	166	181	196	210	222	113	130	145	160	173	186	198	209	115	129	143	155	167	179	189	200
	3 – #5	140	165	188	210	229	248	265	281	143	164	184	202	219	236	251	265	146	164	181	197	212	226	240	253
	3 – #6	83	98	112	124	136	147	157	167	85	97	109	120	130	140	149	157	86	97	107	117	126	134	142	0
	3 – #7	113	133	152	169	185	200	214	227	116	133	148	163	177	190	202	214	118	132	146	159	171	182	193	204
	3 – #8	148	174	198	221	242	261	279	296	151	173	194	213	231	248	264	279	154	173	190	207	223	238	253	266
	3 – #9	149	183	217	251	285	319	350	362	183	217	245	270	292	314	334	353	194	218	241	262	282	302	320	337

| | ℓ_p (in.) | | 4 | | | | | | | 6 | | | | | | | 8 | | | | | | |

b	A_s	4	6	8	10	12	14	16	18	6	8	10	12	14	16	18	20	8	10	12	14	16	18	20	22
22	2 – #6	33	51	66	79	90	0	0	0	38	50	61	70	0	0	0	0	40	49	58	0	0	0	0	0
	2 – #7	44	69	90	107	122	135	146	0	51	68	83	96	107	117	0	0	54	67	78	89	98	0	0	0
	2 – #8	51	89	118	140	159	172	180	188	67	89	108	125	140	153	166	177	71	88	103	116	128	140	150	0
	2 – #9			126	164	188	199	210	219	84	112	136	158	177	194	210	224	90	111	130	147	163	177	190	202
	3 – #6	49	76	99	118	135	149	161	168	56	75	91	105	118	129	140	0	60	74	87	98	108	118	0	0
	3 – #7	51	89	126	161	177	188	197	206	77	102	124	143	161	176	190	203	82	101	118	133	148	161	173	184
	3 – #8			164	201	222	233	244		89	126	162	187	210	230	244	253	107	131	154	174	193	210	225	240
	3 – #9					238	269	282				164	201	238	269	282	294	126	164	195	220	244	265	285	303
	4 – #6				158	175	185	195	203	75	100	121	140	157	173	186	199	80	99	115	131	144	157	169	180
	4 – #7				164	201	224	236	247	89	126	164	191	214	235	247	257	109	134	157	178	197	214	230	245
	4 – #8					238	276	291				201	238	276	291	303		126	164	201	232	257	280	300	315
	4 – #9							313							313	350				238	276	313	350	364	
24	2 – #6	33	51	66	79	0	0	0	0	38	50	61	0	0	0	0	0	40	49	0	0	0	0	0	0
	2 – #7	44	69	90	107	122	135	0	0	51	68	83	96	107	0	0	0	54	67	78	89	0	0	0	0
	2 – #8	56	91	118	140	159	176	186	194	67	89	108	125	140	153	166	0	71	88	103	116	128	140	0	0
	2 – #9		97	138	177	194	206	216	226	84	112	136	158	177	194	210	224	90	111	130	147	163	177	190	202
	3 – #6	49	76	99	118	135	149	161	0	56	75	91	105	118	129	0	0	60	74	87	98	108	0	0	0
	3 – #7	56	97	135	161	183	194	203	212	77	102	124	143	161	176	190	203	82	101	118	133	148	161	173	184
	3 – #8			138	179	216	229	241	252	97	133	162	187	210	230	248	262	107	131	154	174	193	210	225	240
	3 – #9				219	260	279	292			138	179	219	260	279	292	304	135	166	195	220	244	265	285	303
	4 – #6			132	158	179	191	201	209	75	100	121	140	157	173	186	199	80	99	115	131	144	157	169	180
	4 – #7			138	179	219	232	244	255	97	136	165	191	214	235	254	265	109	134	157	178	197	214	230	245
	4 – #8					260	288	301			138	179	219	260	288	301	314	138	175	205	232	257	280	300	320
	4 – #9					301	342							301	342	362			179	219	260	301	342	362	376

PCI Design Handbook/Fourth Edition

CONNECTIONS

Table 6.20.16 Design strength of concrete brackets, corbels or haunches (continued)

			\multicolumn{22}{c}{Values of ϕV_n (kips)}

b	A_s \\ h		\multicolumn{8}{c}{ℓ_p (in.) = 10}	\multicolumn{7}{c}{12}	\multicolumn{7}{c}{14}
			10	12	14	16	18	20	22	24	12	14	16	18	20	22	24	26	14	16	18	20	22	24	26	28
22	2–#6		42	49	0	0	0	0	0	0	42	0	0	0	0	0	0	0	0	0	0	0	0	0	0	0
	2–#7		56	67	76	85	0	0	0	0	58	66	74	0	0	0	0	0	59	66	0	0	0	0	0	0
	2–#8		74	87	99	110	121	131	0	0	76	87	97	106	116	0	0	0	77	86	95	104	0	0	0	0
	2–#9		93	110	126	140	153	165	177	188	96	110	123	135	146	157	167	177	97	109	120	131	141	151	160	0
	3–#6		62	73	84	93	102	0	0	0	64	73	82	90	0	0	0	0	65	73	80	0	0	0	0	0
	3–#7		85	100	114	127	139	150	160	170	87	99	111	122	133	142	152	0	88	99	109	119	128	137	0	0
	3–#8		111	130	149	166	181	196	210	222	113	130	145	160	173	186	198	209	115	129	143	155	167	179	189	200
	3–#9		140	165	188	210	229	248	265	281	143	164	184	202	219	236	251	265	146	164	181	197	212	226	240	253
	4–#6		83	98	11	124	136	147	157	167	85	97	109	120	130	140	149	0	86	97	107	117	126	134	0	0
	4–#7		113	133	152	169	185	200	214	227	116	133	148	163	177	190	202	214	118	132	146	159	171	182	193	204
	4–#8		148	174	198	221	242	261	279	296	151	173	194	213	231	248	264	279	154	173	190	207	223	238	253	266
	4–#9		164	201	238	276	306	331	354	375	191	219	245	270	292	314	334	353	194	218	241	262	282	302	320	337
24	2–#6		42	0	0	0	0	0	0	0	0	0	0	0	0	0	0	0	0	0	0	0	0	0	0	0
	2–#7		56	67	76	0	0	0	0	0	58	66	0	0	0	0	0	0	59	0	0	0	0	0	0	0
	2–#8		74	87	99	110	121	0	0	0	76	87	97	106	0	0	0	0	77	86	95	0	0	0	0	0
	2–#9		93	110	126	140	153	165	177	188	96	110	123	135	146	157	157	0	97	109	120	131	141	151	0	0
	3–#6		62	73	84	93	0	0	0	0	64	73	82	0	0	0	0	0	65	73	0	0	0	0	0	0
	3–#7		85	100	114	127	139	150	160	0	87	99	111	122	133	142	0	0	89	99	109	119	128	0	0	0
	3–#8		111	130	149	166	181	196	210	222	113	130	145	160	173	186	198	209	115	129	143	155	167	179	189	200
	3–#9		140	165	188	210	229	248	265	281	143	164	184	202	219	236	251	265	146	164	181	197	212	226	240	253
	4–#6		83	98	112	124	136	147	157	0	85	97	109	120	130	140	0	0	86	97	107	117	126	0	0	0
	4–#7		113	133	152	169	185	200	214	227	116	133	148	163	177	190	202	214	118	132	146	159	171	182	193	204
	4–#8		148	174	198	221	242	261	279	296	151	173	194	213	231	248	264	279	154	173	190	207	223	238	253	266
	4–#9		179	219	251	280	306	331	354	375	191	219	245	270	292	314	334	353	194	218	241	262	282	302	320	337

CONNECTIONS

Table 6.20.17 Design of structural steel haunches – concrete

Values are for design strength of concrete by Eq. 6.12.1 for following criteria:

f'_c = 5000 psi, for other strengths multiply values by $f'_c/5000$

Adequacy of structural steel section should be checked

Additional design strength, ϕV_r, can be obtained with reinforcing bars – see Table 6.20.18.

$V_u \le \phi(V_c + V_r)$

$\phi = 0.85$

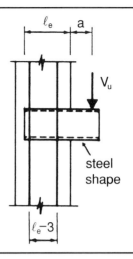

Shear span a (in.)	Embedment ℓ_e (in.)	Values of ϕV_c (kips) — Effective width of section (in.)											
		6	7	8	9	10	11	12	13	14	15	16	17
2	6	33	38	43	49	54	60	65	70	76	81	87	92
	8	47	55	62	70	78	86	94	102	109	117	125	133
	10	62	72	82	92	103	113	123	133	144	154	164	174
	12	76	89	102	115	128	140	153	166	179	191	204	217
	14	92	107	122	137	153	168	183	198	214	229	244	259
	16	107	124	142	160	178	196	213	231	249	267	284	302
	18	122	142	163	183	203	224	244	264	284	305	325	345
	20	137	160	183	206	229	252	274	297	320	343	366	389
	22	152	178	203	229	254	280	305	330	356	381	407	432
4	6	25	29	33	38	42	46	50	54	58	63	67	71
	8	38	44	50	57	63	69	75	82	88	94	101	107
	10	51	60	68	77	85	94	102	111	119	128	136	145
	12	65	76	87	98	108	119	130	141	152	163	173	184
	14	79	92	106	119	132	145	159	182	185	198	211	225
	16	94	109	125	141	156	172	187	203	219	234	250	266
	18	108	126	145	163	181	199	217	235	253	271	289	307
	20	123	144	164	185	205	226	246	267	287	308	328	349
	22	138	161	184	207	230	253	276	299	322	345	368	391
6	6	20	24	27	30	34	37	41	44	47	51	54	58
	8	32	37	42	47	53	58	63	68	74	79	84	89
	10	44	51	58	66	73	80	87	95	102	109	117	124
	12	57	66	75	85	94	104	113	123	132	141	151	160
	14	70	82	93	105	116	128	140	151	163	175	186	198
	16	84	97	111	125	139	153	167	181	195	209	223	237
	18	98	114	130	146	163	179	195	211	228	244	260	276
	20	112	130	149	168	186	205	223	242	261	279	298	317
	22	136	147	168	189	210	231	252	273	294	315	336	357
8	6	17	20	23	26	29	31	34	37	40	43	46	48
	8	27	32	36	41	45	50	54	59	63	68	72	77
	10	38	45	51	57	64	70	76	83	89	95	102	108
	12	50	58	67	75	83	92	100	108	117	125	133	142
	14	62	73	83	94	104	115	125	135	146	156	167	177
	16	75	88	101	113	126	138	151	163	176	188	201	214
	18	89	103	118	133	148	163	177	192	207	222	236	251
	20	102	119	136	153	170	187	204	222	239	256	273	290
	22	116	135	155	174	193	213	232	251	271	290	309	329

CONNECTIONS

Table 6.20.18 Design of structural steel haunches – reinforcement

Values are for additional design strength of concrete obtained from reinforcement by Eq. 6.12.2 for following criteria:

A_s = 2 bars welded to steel shape
$A'_s = A_s$

Reinforcement anchored in only one direction.

When reinforcement, A_s and A'_s, is anchored both above and below steel shape, it can be counted twice (values may be doubled).

For design strength of concrete, ϕV_c – see Table 6.20.17.

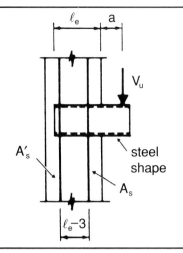

Shear span a (in.)	Embedment ℓ_e (in.)	Values of V_r (kips)											
		Reinforcing bar size f_y = 40 ksi						Reinforcing bar size f_y = 60 ksi					
		#4	#5	#6	#7	#8	#9	#4	#5	#6	#7	#8	#9
2	6	7	11	15	21	27	35	10	16	23	32	41	52
	8	10	15	22	30	39	49	14	23	33	44	58	73
	10	11	18	25	35	45	57	17	26	38	52	68	86
	12	12	19	28	38	50	63	19	29	42	57	74	94
	14	13	21	30	40	53	66	20	31	44	60	79	100
	16	14	21	31	42	55	69	21	32	46	63	82	104
	18	14	22	32	43	57	72	21	33	48	65	85	107
	20	14	23	33	44	58	73	22	34	49	67	87	110
	22	15	23	33	45	59	75	22	35	50	68	89	112
4	6	5	8	12	16	21	27	8	12	18	24	31	40
	8	8	12	18	24	31	40	12	18	27	36	47	60
	10	10	15	21	29	38	48	14	22	32	44	57	73
	12	11	17	24	33	43	54	16	25	36	49	64	82
	14	12	18	26	36	47	59	17	27	39	53	70	88
	16	12	19	28	38	79	62	18	29	42	57	74	93
	18	13	20	29	39	51	65	19	30	43	59	77	98
	20	13	21	30	41	53	67	20	31	45	61	80	101
	22	14	21	31	42	55	69	20	32	46	63	82	104
6	6	4	7	10	13	17	21	6	10	14	19	25	32
	8	7	10	15	20	26	33	10	16	22	30	40	50
	10	8	13	19	25	33	42	12	19	28	38	50	63
	12	9	15	21	29	38	48	14	22	32	44	57	72
	14	10	16	23	32	42	53	16	24	35	48	63	79
	16	11	17	25	34	45	57	17	26	38	51	67	85
	18	12	18	27	36	47	60	18	28	40	54	71	89
	20	12	19	28	38	49	62	18	29	41	56	74	93
	22	13	20	29	39	51	64	19	30	43	58	76	96
8	6	4	6	8	11	14	18	5	8	12	16	21	27
	8	6	9	13	17	23	29	9	13	19	26	34	43
	10	7	11	16	22	29	37	11	17	25	34	44	55
	12	9	13	19	26	37	43	13	20	29	39	51	65
	14	9	15	21	29	38	48	14	22	32	43	57	72
	16	10	16	23	31	41	52	15	24	35	47	61	78
	18	11	17	24	33	43	55	16	25	37	50	65	83
	20	11	18	26	35	46	58	17	27	39	52	68	87
	22	12	19	27	36	47	60	18	28	40	55	70	90

CHAPTER 7
SELECTED TOPICS FOR ARCHITECTURAL PRECAST CONCRETE

		Page No.
7.1	Introduction	7-2
	7.1.1 Structural Design and Analysis	7-2
	7.1.2 Load Transfer—General Methods	7-3
	7.1.3 Application	7-3
	7.1.4 Design Analysis	7-3
	7.1.5 Design Objectives	7-4
7.2	Connections of Non-Load Bearing Panels	7-4
	7.2.1 Design Fundamentals	7-4
	7.2.2 Production Considerations	7-5
	7.2.3 Erection Considerations	7-5
7.3	Precast Concrete Used as Forms	7-5
	7.3.1 General	7-5
	7.3.2 Design Considerations	7-5
	7.3.3 Construction Considerations	7-8
7.4	Column Covers and Mullions	7-11
7.5	Veneered Panels	7-11
	7.5.1 General	7-11
	7.5.2 Clay Products	7-12
	7.5.2.1 Clay product properties	7-13
	7.5.2.2 Clay product selection	7-13
	7.5.2.3 Design considerations	7-13
	7.5.3 Natural Stone	7-15
	7.5.3.1 Properties	7-15
	7.5.3.2 Sizes	7-16
	7.5.3.3 Anchorage of stone facing	7-17
	7.5.3.4 Veneer jointing	7-18
	7.5.3.5 Samples	7-18
7.6	Design Example—Window Wall Panel	7-20
7.7	References	7-24

SELECTED TOPICS FOR ARCHITECTURAL PRECAST CONCRETE

7.1 Introduction

Architectural precast concrete products are generally of special shape and, through finish, shape, color or texture, contribute to the architectural form and finished appearance of the structure.

Architectural precast concrete is available in complex shapes which serve not only as non-load bearing walls, but may also combine an attractive appearance with the ability to serve as load bearing members. The successful and economical use of architectural precast concrete depends on an understanding of the production and erection limitations, as well as the structural behavior of the element.

This chapter provides a brief discussion of selected topics of relevance to the structural engineer. A complete treatment of all aspects of this subject is provided in Ref. 1.

7.1.1 Structural Design and Analysis

Architectural precast concrete construction can be considered in three parts:

1. The precast elements.
2. The support of the precast element, i.e., the beam, wall, column, foundation or any other part of the structure which provides vertical and horizontal support for the element.
3. The connection that joins the precast element to its support system.

Other parts of this Handbook give detailed information and procedures for analyzing and designing architectural and structural precast concrete elements and structures. The following sections are also applicable to architectural precast concrete:

Chapter or Section	Title
1	Materials
2.6	Load Bearing Wall Panels
3.3	Volume Changes
3.4	Component Analysis
3.7	Shear Wall Buildings
3.11	Earthquake Analysis
4.2	Flexure
4.3	Shear
4.4	Torsion
4.7	Compression Members
5	Product Handling and Erection Bracing
6.4	Connection Design Criteria
6.5	Connection Hardware and Load Transfer Devices
6.8	Bearing on Plain Concrete
6.11	Concrete Brackets or Corbels
6.12	Structural Steel Haunches
6.18	Connection of Load Bearing Wall Panels
8	Tolerances for Precast and Prestressed Concrete
9	Thermal, Acoustical, Fire and Other Considerations
10.2	Guide Specifications for Architectural Precast Concrete
10.3	Code of Standard Practice for Precast Concrete
10.4	Recommendations on Responsibility for Design and Construction of Precast Concrete Structures

Following is a brief summary of structural engineering principles discussed in detail in the above listed sections:

The potential for element volume change requires consideration. Volume changes may be a result of:

1. Elastic and inelastic (creep) deformations.
2. Shrinkage.
3. Expansion and contraction resulting from temperature and moisture change.

Movement in the support system must also be considered. Movement can result from:

1. Elastic and inelastic deformation from gravity loads.
2. Horizontal displacement resulting from wind and earthquake.
3. Foundation movement.
4. Temperature change.
5. Shrinkage of concrete.

In most cases, movements may be estimated by analysis, and provisions made to accommodate these movements.

The structural design of architectural precast concrete requires that, after all loads are determined, the following be considered:

1. Forces and strains caused by handling, transportation and erection.
2. Acceptable crack width and location.
3. Strain gradients and restraint forces from thermal and moisture differentials through the panel.
4. Localized wind forces and the response of a precast element to these transient loads.
5. Forces transferred to the element and connections due to distortion of the structural frame.

6. Differential deflections between the precast element and the supporting structure.
7. Accommodation of tolerances allowed in the supporting structure.
8. Experience with various types of connections.

The designer must recognize that loads and behavior cannot be established precisely, particularly for members exposed to the environment. This imprecision will generally not affect the safety of the member, provided that reasonable values have been established and the above factors considered.

7.1.2 Load Transfer—General Methods

The forces which must be considered in the design of architectural precast concrete components are:

1. Those caused by the precast member itself, e.g., self weight and earthquake forces.
2. Those caused by loads, such as wind, snow, floor live loads, soil or fluid pressure, or construction loads, that are externally applied to the element or transferred to the element by the behavior of the supporting structure.
3. Those that are a result of restraint of volume change or support system movement. They are generally concentrated at the connections.

All non-load bearing elements should be designed to accommodate movement freely and, whenever possible, with no redundant supports, except where necessary to restrain bowing. Relatively simple analyses provide the forces required for connection design. The calculations required for movement accommodation are more complex. The designer can use the simplified methods discussed in Chapter 3, or computer analysis programs.

When redundant supports are necessary or when movement is to be resisted, the load-deformation characteristics of the element, connections and support system should be taken into account. Design of connections that restrain support system elements may need to consider the load-deformation characteristics of the connections.

7.1.3 Application

The most common applications of architectural precast concrete in building construction are those in which the precast elements function as walls. Concrete is a nearly ideal wall material, since it has excellent sound transmission characteristics, is fire resistant, and is durable. In addition, architectural precast concrete provides almost unlimited potential for economically achieving aesthetic design objectives. Architectural precast elements are capable of functioning as a wall without backing. Large elements provide the complete wall, and often are made with integral window openings.

Precast walls which are of sufficient mass to satisfy manufacturing, sound transmission, and fire resistance requirements are also capable of supporting significant loads. Thus, using precast concrete walls, which separate or enclose space, to carry loads is an economical alternative to the use of separate structural systems and wall materials.

In this Handbook, a distinction is made between load bearing and non-load bearing elements:

1. A non-load bearing (cladding) element is one which resists and transfers negligible load from other elements of the structure. It is generally used only to enclose space, but must be designed to resist wind, seismic forces generated from self weight, and forces required to transfer its weight to the structure which supports it.
2. A load bearing element resists and transfers loads applied from other elements. Therefore, a load bearing member cannot be removed without affecting the strength or stability of the building.

The use of architectural precast concrete walls as shear walls to resist and transfer horizontal loads is another efficient application. In this case, the connections between panels and other structural elements are the primary design consideration. Precast panels can also be attached to existing frames to improve lateral load resistance, such as in upgrading structures in seismic zones.

Architectural precast concrete elements also often function as beams to transfer gravity loads to supports. The beams may also participate in the transfer of lateral loads. For example, the beams may serve as struts between lateral load resisting elements, such as columns or shear walls, transferring the load as an axial force. In other cases, precast beams interact with columns to form a moment frame as a lateral load resisting system.

In composite construction, precast concrete elements may be used as forms for cast-in-place concrete (Sect. 7.3). This is especially suitable for combining architectural and structural functions in load bearing facades, or for improving ductility in locations of high seismic risk.

Suggested connection details for cladding elements are shown in Chapter 6.

7.1.4 Design Analysis

In some cases, the structural design of a single precast element can be completed with very little consideration of other materials and elements in the structure. The weight of the element and the superimposed loads are simply transferred to supports, and the element can be considered independently of the structure. Occasionally, however, it is necessary to consider the characteristics of other materials and elements within the structure. For example, neglecting the relative movement between two support points may lead to inaccurate estimates of connection forces with the attendant possibility of distress.

In other cases, architectural precast elements may interact with other parts of the structure in the transfer of loads. The design of the interacting system and the individual precast elements must be based on analysis of the whole system.

The designer of a precast building can choose to transfer loads through architectural precast concrete elements or intentionally avoid significant load transfers through the precast elements. In the preliminary design phase, the structural engineer should therefore recognize that he is able to choose the structural characteristics rather than simply analyze a predetermined set of criteria.

7.1.5 Design Objectives

Structural integrity of the completed structure is the primary objective. Deflections must be limited to acceptable levels, and distress that could result in instability of an individual element or of the complete structure must be prevented. The inherent stiffness of architectural precast concrete panels will significantly stiffen a structure, thereby reducing deflections and improving stability.

Economy is an important design objective. The total cost of a completed structure is the determining factor. What seems to be a relatively expensive precast element may result in the most economical building because of the reduction or elimination of other costs. In some cases, precast elements are not economical due to improper application or the necessity for complex on-site construction procedures. The designer should attempt to optimize the structure by using precast panels to serve several functions, and take advantage of the economy gained by standardization.

Standardization reduces costs because fewer molds are required. Also, productivity in all phases of manufacture and erection is improved through repetition of familiar tasks; there is also less chance of error.

A very strict discipline is required of the designer to avoid a large number of non-standard units. Preliminary planning and budgeting should recognize the probability that the number of different units will increase as the design progresses. If non-standard units are unavoidable, costs can be minimized if they can be cast from a master mold with simple modifications for the special pieces, rather than requiring special molds.[1]

The aesthetic design objectives for the structure should be a matter of concern to the structural engineer. A precast element or system may achieve all other design objectives but fall short of aesthetic objectives through treatment of structural features only.

7.2 Connections of Non-Load Bearing Panels

7.2.1 Design Fundamentals

To assure the satisfactory performance of architectural precast concrete elements, it is important that connections be selected and arranged to transfer applied loads without producing unintended restraint. The design of connections follows the procedures given in Chapter 6.

The designer should provide simple load paths through the connections and ductility within the connections. This will reduce the sensitivity of the connection and the necessity to precisely calculate loads and forces from, for example, volume changes and frame distortions. The number of load transfer points should be kept to a practical minimum. It is recommended that no more than two connections per panel be used to transfer gravity loads. More than two results in a statically indeterminate system; since the stiffness of the panel and its support may be significantly different, and the elevation of supports different than assumed in the design, the reactions on an indeterminate system may not be accurately known. Furthermore, the fewer the connections the easier it is to provide for the various movements required to accommodate volume changes, drift, etc. Load transfer should be as direct as possible. Fig. 7.2.1 shows examples of several load transfer mechanisms.

Connections and assemblies should develop sufficient ductility to preclude brittle failures. For example, inserts in concrete should be attached to and/or hooked around reinforcing steel, provided with confinement reinforcement or otherwise be terminated to effectively transfer forces to the concrete and/or reinforcement. It is desirable that the pullout strength of concrete and strength of welds be greater than the tension or bending strength of the connection steel.

The design of the panel connection and the supporting frame are interdependent. For example, if a supporting spandrel beam lacks sufficient rotational stiffness, the precast connections should be designed to transmit the vertical loads to the centerline of the beam, or the supporting members should be braced to avoid torsional rotation.

Connection hardware must be permanently protected from the elements, or corrosion resistant materials used. For fire-rated structures, exposed connections should be protected as required to assure the proper rating for the entire assembly.

Adequate tolerances and clearances, as discussed in Chapter 8, are required.

In zones of seismic activity, present codes treat precast concrete walls subjected to lateral loads as non-ductile, with an appropriately high multiplier to establish design base shear. Continuity and ductility may be achieved by casting in place spandrel beams and columns using the wall panels as forms. The ductility of walls partially depends on the location of reinforcement. Ductile behavior is significantly improved if the reinforcement is located at the ends of the walls. Thus, a usually inactive curtain wall can become a major lateral load (seismic) resisting element.

7.2.2 Production Considerations

Connections should allow economical production of the precast elements. The hardware should not interfere with concrete placement or cause finishing problems, nor require penetration through molds. The size of reinforcing bars should be limited, because large bars require anchorage lengths and hook sizes that may be impractical. It is often better to use welded cross bars or other types of mechanical anchorage than to rely on bond. Connection details with reinforcing bars crossing each other require careful dimensional checking to ensure sufficient cover.

Plates, angles or other steel shapes embedded in precast concrete must be securely attached to the forms to prevent their becoming misaligned or skewed.

Inserts must be placed accurately because their capacity depends on the depth of embedment, spacing and distance from free edges. They should be kept free of dirt and protected so that concrete does not enter during casting.

When possible, connections should be dimensioned to the nearest ½ in. The minimum shim space between elements of a connection should not be less than ½ in., with 1 in. preferred. It should be remembered that the real dimensions of reinforcing bars are approximately ⅛ in. greater than the nominal, because of the deformations. Prestressed panels require special form considerations to permit product movement at release.

7.2.3 Erection Considerations

Ease and speed of erection are heavily dependent on simplicity and ruggedness of assembly details. Field patching and finishing should be kept to a minimum. If possible, the details should allow erection to proceed in nearly all kinds of weather. The panels should be capable of being erected without temporary shoring. Ideally, it should be possible to complete the connection by working downhand from the top of the erected member or from a stable deck.

The maximum feasible adjustability should be provided in all directions, preferably at least 1 in. For example, the supporting beam may rotate or deflect under the weight of the precast panels, making it necessary to adjust the connections during erection. Significant support rotation and deflection are more common in structural steel than in concrete frames. The designer of the support system should verify that eccentrically loaded support beams are of sufficient stiffness to prevent excessive rotation.

Connections should be detailed so that hoisting equipment can be quickly released. It may be necessary to provide temporary connections that are released after final adjustments are made. Loading conditions during erection may be more critical than other phases, due to eccentricities, construction loads or impact.

When cast-in-place concrete, grout or drypack is required to complete a connection, the detail should provide for self-forming if possible. When not practical, the connection should allow for easy placement and removal of formwork. Temporary shoring may be required.

7.3 Precast Concrete Used as Forms

7.3.1 General

This section provides design and detailing information for structures in which architectural precast concrete is used as the formwork for cast-in-place structural concrete.[3]

In most cases, the architectural precast concrete is considered solely as a form, serving only decorative purposes after the cast-in-place concrete has achieved design strength. This is accomplished by providing open or compressible joints between abutting precast panels, and neglecting or eliminating bond at the interface of the precast and cast-in-place concrete. When the architectural precast concrete unit is non-composite with the cast-in-place concrete, the reinforcing steel extending from the precast into the cast-in-place concrete need only be of sufficient strength to support the formwork unit.

In other cases, it may be desirable to detail the structure so that the precast and cast-in-place concrete act compositely, thus combining the strength of both. It is then necessary to provide shear transfer as for other composite assemblies. Sect. 4.3.5 describes design procedures.

7.3.2 Design Considerations

Following are recommendations for design of precast panels used as forms for cast-in-place concrete:

Fig. 7.2.1 Connections illustrating number of force transfers

(a) Horizontal Load Transfer in X and Y Direction
- Concrete to Insert
- Insert to Bolt
- Bolt to Washer (X directional)
- Washer to Angle (X directional)
- Bolt to Angle (Y directional)
- Angle to Weld
- Weld to Angle
- Angle to Studs
- Studs to Concrete
(9 Force Transfers)

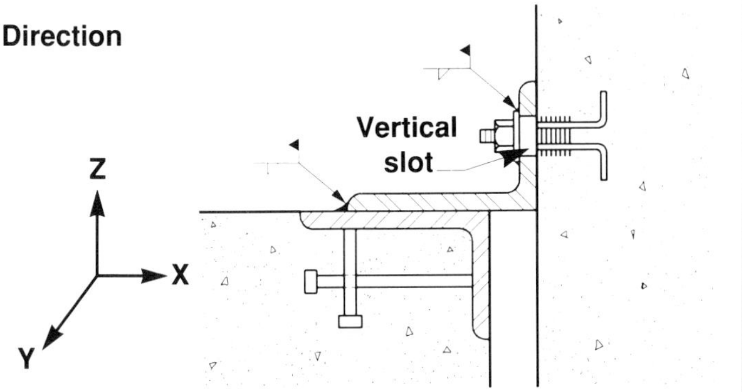

(b) Horizontal Load Transfer in X Direction and Vertical Load Transfer
- Concrete to Studs
- Studs to Plate
- Plate to Weld
- Weld to Angle
- Angle to Shim
- Shim to Concrete
- Angle to Bolt
- Bolt to Insert
- Insert to Concrete
(9 Force Transfers)

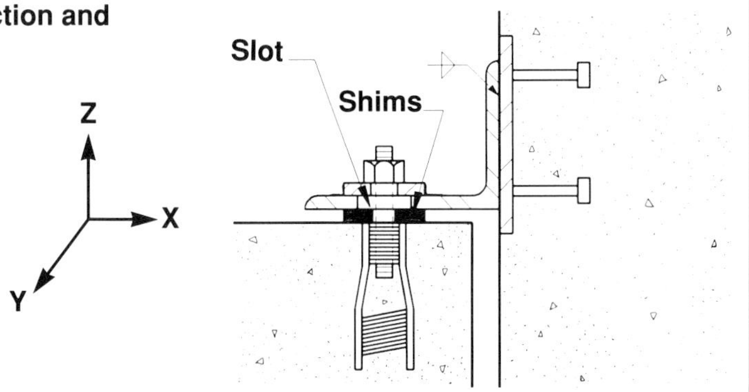

(c) Horizontal Load Transfer in X or Y Direction
- Concrete to Bolt
- Bolt to Angle
- Angle to Weld
- Weld to Support
(4 Force Transfers)

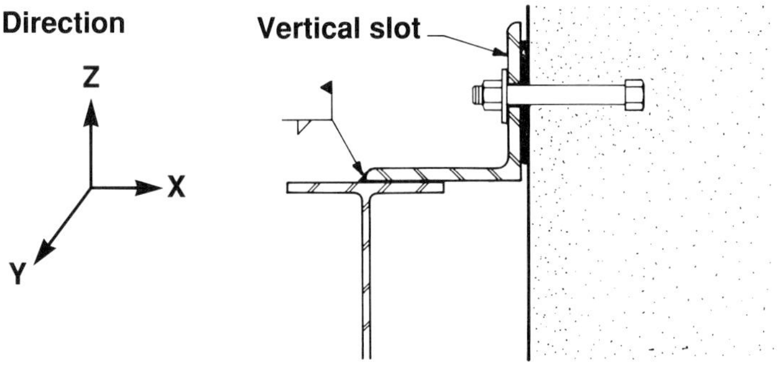

(d) Vertical Load Transfer
- Concrete to Reinforcement
- Reinforcement to Welds
- Welds to Knife Edge Plate
- Plate to Shims
- Shims to Bearing Plate
- Bearing Plate to Concrete
(6 Force Transfers)

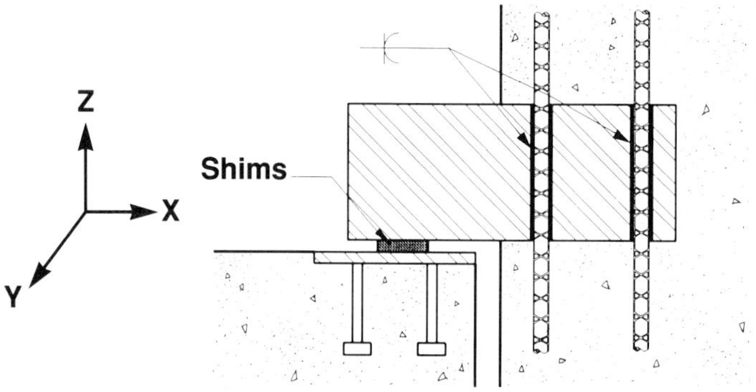

(e) Horizontal Load Transfer in X and Y Direction and Vertical Load Transfer
- Concrete to Grout
- Grout to Support
(2 Force Transfers)

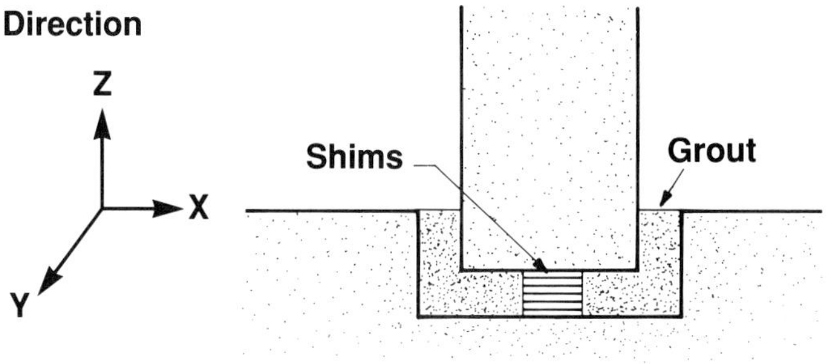

Lateral pressure of fresh concrete. With the following limitations, the formulas below may be used to determine the forces to be resisted by forms, ties, and bracing:[4]

1. Material is normal weight concrete, with a unit weight of 145 to 155 pcf.
2. Vibration is by internal means only; form vibrators, if used, will require a modification of the formulas.
3. The depth vibrated, or revibrated, does not exceed 4 ft below the concrete surface.
4. Retarding admixtures, if used, will require a modification of the formulas.
5. For heights greater than 18 ft, an interval of 2 hr should elapse after each 18 ft lift prior to continuation.
6. Slump is not greater than 4 in. at point of placement.

Columns, placement rate up to 25 ft/hr:

$p = 150 + 9000R/T$, maximum 3000 psf or 150h, whichever is least (Eq. 7.3.1)

Walls, placement rate less than 7 ft/hr:

$p = 150 + 9000R/T$, maximum 2000 psf or 150h, whichever is least (Eq. 7.3.2a)

Walls, placement rate 7 to 10 ft/hr:

$p = 150 + 43,400/T + 2800R/T$, maximum 2000 psf or 150h, whichever is least (Eq. 7.3.2b)

Walls, placement rate greater than 10 ft/hr:

$p = 150h$ (Eq. 7.3.2c)

where:

p = lateral pressure, psf
R = placement rate, ft/hr
T = temperature of concrete in forms, °F
h = height of fresh concrete above point considered, ft

Lateral forces on precast concrete forms. In addition to pressure induced by fresh concrete, the applied lateral forces on forms and bracing should be in accordance with the local building code, but not less than:

1. Wind—15 psf applied to all exposed surfaces, unless local codes specifically permit less.
2. Columns—2% of the total dead load supported by the form, applied as a horizontal load to the top of the form.
3. Walls—100 lb per lin ft of wall, applied to the top of the wall form.

Vertical loads on precast concrete forms. The vertical load supported by precast concrete forms should consist of all superimposed dead loads, including an allowance for storage of construction materials, and a live load of not less than 25 psf.

Formwork accessories. In the design of formwork accessories such as form ties and form anchors, a minimum safety factor of 2 based on the ultimate strength of the accessory is recommended except that yield point must not be exceeded.

Design assumptions

1. Forms are generally supported in a manner which will permit them to act as continuous beams. It is generally sufficiently accurate to assume that the form acts as a fixed beam between adjacent lateral supports. The engineer must judge whether to include deformation of the form supports in the analysis.
2. Partially open concrete shapes are often used to form columns. These U-shapes will develop internal stresses which are a function of the stiffness of the sides. Design assumptions are shown in Fig. 7.3.1.
3. The internal stresses, produced by the temperature rise of the form as the cast-in-place concrete is placed, are usually neglected. Moments developed due to uniform temperature rise for U-shaped forms are shown in Fig. 7.3.1.
4. The cross-sectional area of column form panels is usually neglected when determining either the immediate (elastic) or long-term (creep) column shortening. The thickness of panels is usually included when investigating the effect of various temperature changes across the column.

Deflection. Form deflection should generally be limited to 1/360 of the unsupported height or length. Deflection of beam forms, and warping of wall forms, may result from differential shrinkage of precast and cast-in-place concrete, as well as the dead load or lateral pressure of the cast-in-place concrete.

Cambering of architectural precast forms to compensate for deflections is expensive and should be avoided.

Crack control. In order to minimize form cracking due to the pressures induced by fresh concrete, the design procedures for crack control in Chapters 4 and 5 should be followed. Where the member is long enough to develop bonded strand, pretensioning may be used in the precast form units.

Composite design. Interaction between precast forms and cast-in-place concrete can be achieved by providing for the transfer of shear forces at the interface. Effective composite behavior will significantly increase the strength of members and reduce deflections. Design methods and shear transfer requirements are discussed in Chapter 4.

Fig. 7.3.1 Forces due to internal pressure

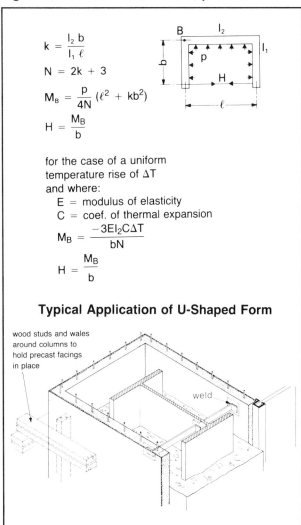

$$k = \frac{l_2 b}{l_1 \ell}$$

$$N = 2k + 3$$

$$M_B = \frac{p}{4N}(\ell^2 + kb^2)$$

$$H = \frac{M_B}{b}$$

for the case of a uniform temperature rise of ΔT and where:

E = modulus of elasticity
C = coef. of thermal expansion

$$M_B = \frac{-3EI_2 C\Delta T}{bN}$$

$$H = \frac{M_B}{b}$$

Typical Application of U-Shaped Form

When the precast and cast-in-place concrete are designed to act compositely, the form joints must be located away from points of high moment.

In compression members, axial loads will tend to be distributed initially to the two components in accordance with their individual axial stiffness. Over time, there will be some redistribution of load from the more highly stressed components to the other due to creep.

7.3.3 Construction Considerations

Realistic assumptions with regard to construction techniques are required. It must be determined (or specified) how the precast panels will be supported during concreting in order to design them.

Panel forms of concrete should be erected and temporarily braced to proper grade and alignment in such a way that the tolerances specified for the finished structure can be met. Temporary bracing for the panels generally consists of adjustable pipe bracing from panel to floor slab. Supports, braces, and form ties must be stiff enough so that their elastic deformation will not significantly affect the assumed load distribution. Form ties may be attached as shown in Fig. 7.3.2, or welded to plates cast in the panels. Column forms may use column clamps or be wrapped with steel bands to aid in resisting hydrostatic pressure.

Attachments between the precast form and other elements, such as steel columns, must be detailed to provide the necessary field adjustments. Contact area between precast concrete and external braces, clamps or bands should be protected from staining and chipping.

Joints. Architectural precast concrete surfaces require tight joints so that concrete does not leak and mar decorative facings. Methods used to close joints include buttering on the inside with mortar (units may also be bedded on mortar when designed to accept load transfer through the joint), and gasketing with low density, closed cell neoprene rubber or other durable, permanent, resilient materials.

Where form panels are non-composite, the joint material should prevent load transfer. Where form panels are intended to act compositely with the cast-in-place concrete, the joint material must be mortar or other non-staining material of sufficient strength

Fig. 7.3.2 Typical form ties

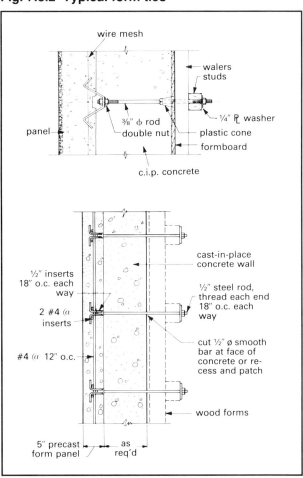

to transfer the intended loads. For joints exposed to the environment, mortar is usually raked back from the face, and the joint caulked.

Some precaution and special details may be required to take care of differential shrinkage or creep between cast-in-place concrete and precast concrete used as forms for columns or load bearing walls.[3,17] Stress relief may be simply handled in the design of the joints in the precast forms at the top and bottom of vertical structural components carrying axial loads.

Horizontal construction joints in the cast-in-place concrete are generally 3 in. below the top edge of the panels used as permanent forms rather than in line with horizontal form joints. This reduces the possibility of water leakage through the construction joints.

Example 7.3.1 Precast panel used as a wall form

Given:

The wall panel section shown below.
Panel concrete strength: f'_c = 5000 psi
C.i.p. concrete placement rate: 4 ft per hr
Maximum height of c.i.p. concrete: 4 ft
C.i.p. concrete temperature: 90°F

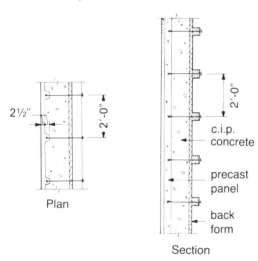

Plan / Section

Problem:

Determine panel reinforcement requirements for pressure of wet concrete.

Solution:

Assume a 5 in. panel thickness is necessary to provide anchorage for the form ties. Either a 5 in. flat panel could be used or, as in this case, a ribbed section with a 5 in. rib depth. The rib spacing is set by the strength of the ties. Since the placement rate is less than 7 ft per hr:

p = 150 + 9000R/T = 150 + 9000(4)/90
= 550 psf

or a maximum of:

p = 150(4) = 600 psf. Use 550 psf.

Check flexure:

Neglecting effect of ribs, assume the 2½ in. panel acts as a fixed beam of 24 in. span:

$M = p\ell^2 / 12 = 550(2)^2 / 12 = 183$ ft-lb

$f = \dfrac{6M}{bt^2} = \dfrac{6(183)(12)}{12(2.5)^2} = 176$ psi

Maximum allowable tension to prevent cracking

= $5\sqrt{5000}$

= 353 psi > 176 OK

Provide reinforcement to develop an ultimate moment capacity > 1.4M

M_u = 1.4(183) = 256 ft-lb

Try 6 x 6 – W4.0 x W4.0 in center of panel.

A_s = 0.08 in.²/ft

$a = \dfrac{A_s f_y}{0.85 f'_c b} = \dfrac{0.08(60)}{0.85(5)(12)} = 0.09$ in.

d = 2.5/2 = 1.12 in.

$\phi M_n = \phi A_s f_y (d - a/2)$

= 0.90(0.08)(60,000)(1.12 − 0.09/2)/12

= 385 ft-lb > 256 OK

Check shear:

V_u = 1.4(550)2/2 = 770 lb

$\phi V_n = \phi 2\sqrt{f'_c}\, b_w d = 0.85(2)\sqrt{5000}(12)(1.12)$

= 1616 lb > 770 OK

Also check panel for stripping and handling—see Chapter 5.

Example 7.3.2 Column cover acting as a form

Given:

The column shown on the next page.
Precast concrete strength: f'_c = 5000 psi
C.i.p. concrete placement rate: 8 ft per hr
Maximum height of c.i.p. concrete: 6 ft
C.i.p. concrete temperature: 90°F

Problem:

Determine panel reinforcement and tie requirements.

Solution:

Design for pressure of wet concrete:

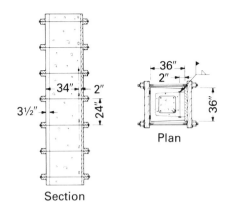

$p = 150 + 9000 R/T = 150 + 9000(8)/90$
$\quad = 950 \text{ psf}$

or a maximum of:

$p = 150(6) = 900 \text{ psf} < 950$. Use 900 psf

Check flexure:

$b = 36 - 2 - 3.50/2$
$\quad = 32.25 \text{ in.} = 2.69 \text{ ft}$
$\ell = 36 - 3.50$
$\quad = 32.50 \text{ in.} = 2.71 \text{ ft}$
$I_2 = I_1; \ k = I_2/I_1(b/\ell) = 0.99$
$N = 2k + 3 = 4.98$
$M_B = \dfrac{p}{4N}(\ell^2 + kb^2)$
$\quad = \dfrac{900}{4(4.98)}[(2.71)^2 + (0.99)(2.69)^2]$
$\quad = 653 \text{ ft-lb}$
$f = \dfrac{6M}{bt^2} = \dfrac{6(653)(12)}{12(3.5)^2}$
$\quad = 321 \text{ psi} < 5\sqrt{5000} = 353 \text{ psi} \text{ OK}$

Provide reinforcement to develop design strength > 1.4M:

$M_u = 1.4(653) = 914 \text{ ft-lb}$

Try W4 wire at 4 in. o.c.

$A_s = 0.12 \text{ in.}^2/\text{ft}$
$a = \dfrac{A_s f_y}{0.85 f'_c b} = \dfrac{0.12(60)}{0.85(5)(12)} = 0.14 \text{ in.}$
$d = 3.50/2 = 1.75 \text{ in.}$
$\phi M_n = \phi A_s f_y (d - a/2)$
$\quad = 0.90(0.12)(60,000)(1.75 - 0.14/2)/12$
$\quad = 907 \text{ ft-lb} \approx 914 \text{ ft-lb} \text{ say OK}$

Tie force:

$H = M_B/b = 653/2.69 = 243 \text{ lb/ft}$

If ties are spaced 2 ft o.c.,

tie force $= 2(243) = 486 \text{ lb}$

Use 3/16 x 1 in. A-36 steel strap. Weld to embedded plates.

Check shear:

$V_u = 1.4(900)(2.71/2) = 1707 \text{ lb}$
$\phi V_n = \phi 2\sqrt{f'_c}\, b_w d = 0.85(2)\sqrt{5000}(12)(1.75)$
$\quad = 2524 \text{ lb} > 1707 \text{ OK}$

Also check form for stripping and handling.

Example 7.3.3 Precast fascia as form for beam

Given:

Precast concrete fascia forming a beam supporting 24-ft span hollow-core floor as shown below. Beam is supported on 2 ft square columns at 20 ft on center.

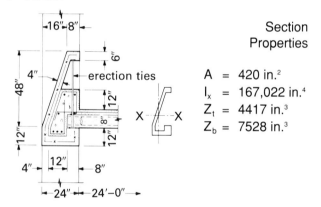

Section Properties
$A = 420 \text{ in.}^2$
$I_x = 167,022 \text{ in.}^4$
$Z_t = 4417 \text{ in.}^3$
$Z_b = 7528 \text{ in.}^3$

Problem:

Check stresses of fascia panel during erection.

Solution:

Loads:

Hollow-core slabs:
$\quad 0.055(24/2 + 8/12) = 0.70 \text{ k/ft}$
3 in. lightweight topping: $0.03(24/2) = 0.36$
C.i.p. lightweight concrete beam: $= 0.30$
Precast concrete fascia:
$\quad$ area = 420 sq in. $= 0.44$
$\quad\quad$ Total dead load $= 1.80 \text{ k/ft}$
Construction live load: $0.025(24/2) = 0.30 \text{ k/ft}$
Clear span $= 20 - 2 = 18 \text{ ft}$
$M_{d\ell} = 1.80(18)^2/8 = 72.9 \text{ ft-kip}$
$M_{\ell\ell} = 0.30(18)^2/8 = \underline{12.2}$
$\quad\quad\quad\quad$ Total $= 85.1 \text{ ft-kip}$

Stresses:

$f_t = 85,100(12)/4417 = 231$ psi compression

$f_b = 85,100(12)/7528$

$\quad = 136$ psi $< 5\sqrt{5000} = 353$ psi OK

$V_u = [1.4(1.80) + 1.7(0.3)](18/2) = 27.3$ kips

$b_w = 4$ in., $d = 57$ in.

$\phi V_n = \phi 2\sqrt{f'_c}\, b_w d$

$\quad = 0.85(2)\sqrt{5000}(4)(57)/1000$

$\quad = 27.4$ kips > 27.3 kips OK

Also check for stripping and handling.

7.4 Column Covers and Mullions

The use of precast concrete panels as covers over steel or cast-in-place concrete columns and beams, and as mullions, is a common method of achieving architectural expression, special shapes, or fire rating.[1]

Column covers are usually supported by the structural column or floor beam, and are themselves designed to transfer no vertical load other than their own weight. The vertical load of each length of column cover section is usually supported at one elevation, and tied back top and bottom for lateral load transfer and stability. In order to minimize erection costs and horizontal joints, it is desirable to make the cover or mullion as long as possible, subject to limitations imposed by weight and handling.

With adequate shear connectors between the cover and the column, the cover can add stiffness to the column. In tall buildings, this combined stiffness may significantly reduce the calculated drift.

Mullions are vertical elements serving to separate glass areas. They generally resist only wind loads applied from the adjacent glass, and must be stiff enough to maintain deflections within the limitations imposed by the window manufacturer. Since mullions are often thin, they are sometimes prestressed to prevent cracking.

Column covers and mullions are usually major focal points in a structure, and aesthetic success requires that careful thought be given to all facets of design and erection. Following are some items which should be considered:

1. Since column covers and mullions are often isolated elements forming a long vertical line, any variation from a vertical plane is readily observable. This variation is usually the result of the tolerances allowed in the structural frames. To some degree these variations can be handled by precast connections with adjustability. The designer should plan a clearance of at least 1½ in. between the panel and structure. For steel columns, the designer should consider the clearances around splice plates and projecting bolts.

2. Provide support at only one elevation for vertical loads, and at additional locations for lateral loads and stability. When access is available, consider providing an intermediate connection for lateral support and restraint of bowing.

3. Column covers and mullions which project from the facade will be subjected to shearing wind loads. Connection design must account for these forces.

4. Members which are exposed to the environment will be subjected to temperature and humidity change. Horizontal joints between abutting precast column covers and mullions should be wide enough to permit length changes and rotation from temperature gradients. The behavior of thin flexible members will be improved by prestressing.

5. Due to vertical loads and the effects of creep and shrinkage in cast-in-place concrete columns, structural columns will tend to shorten. The width of the horizontal joint between abutting precast covers should be sufficient to permit this shortening to occur freely.

6. The designer must envision the erection process. Column cover and mullion connections are often difficult to reach and, once made, difficult to adjust. This problem of access is compounded when all four sides of a column are covered for a height exceeding the length of a precast cover. Sometimes this problem can be solved by welding the lower piece to the column and anchoring the upper piece to the lower with dowels set in front, or by a mechanical device that does not require access.

7. Use of insulation on the interior face of the column cover reduces heat loss at these locations. Such insulation will also minimize temperature differentials between exterior columns and the interior of the structure.

Typical connection details are illustrated in Chapter 6, Fig. 6.17.1-Fig. 6.17.3.

7.5 Veneered Panels

7.5.1 General

Precast concrete panels faced with brick, tile, terra cotta or natural stone combine the rich beauty of traditional materials with the strength, versatility, and economy of precast concrete. Some applications are shown in Fig. 7.5.1.

Structural design of veneered precast concrete units is the same as for other precast concrete wall panels, except that consideration must be given to the veneer material and its attachment to the concrete. The physical properties of the facing material must be compared with the properties of the concrete back-up. These properties include:

1. Tensile, compressive and shear strength.
2. Modulus of elasticity.
3. Coefficient of thermal expansion.
4. Volume change.

Some natural stone veneers exhibit different properties depending on the orientation of the applied force with respect to the natural planes of the material. If feasible, all blocks should be slabbed in the same direction in relation to the bedding plane.

Cover depth of reinforcing steel should be a minimum of ½ in. at the veneer surface, maintained by non-corrosive spacers. Galvanized or epoxy coated reinforcement is recommended at cover depths less than ¾ in. Precautions should be taken to avoid any materials that could corrode and cause discoloration of the concrete or veneer.

Because of the difference in material properties, veneered panels are somewhat more susceptible to bowing than all-concrete panels. This is a consideration in the reinforcement design. If thickness is sufficient, two layers of reinforcement may be used; this can help reduce bowing. In some cases, reinforcing trusses are used to add stiffness; in others, concrete ribs are formed on the back of the panel.

Minimum thickness of back up concrete of flat panels to control bowing is usually 5 to 6 in., but 4 in. can be used where the panel is small, or when it has adequate rigidity provided by panel shape.

Generally, stone veneers should be connected to the backup concrete with flexible mechanical anchors; bond of veneer surface to concrete should be avoided. The details of anchorage will depend on:

1. Shrinkage of concrete during curing.
2. Stresses imposed during handling and erection.
3. Thermal response caused by different coefficients of expansion and by thermal gradients (see Chapter 3).
4. Moisture expansion of veneer.
5. Service loads.

For those veneers rigidly attached or bonded to the backup, the differential shrinkage of the concrete and veneer will cause outward bowing in a simple span panel. The flat surfaces of some veneers, such as cut stone, reveal bowing more prominently than other finishes. Therefore, bowing may be a critical consideration even though the rigidity of the cut stone helps to resist bowing. Mid-point tie-back connections will reduce bowing.

Cracking in the veneer may occur if the bonding or anchoring details force the veneer pieces to follow the bowing. This is particularly critical where the face materials are large (cut stone) and the differential movement between concrete and veneer is significant. A good mix design, quality control, and prolonged curing will help reduce shrinkage.

Even with concrete shrinkage kept to the lowest possible level, there may still be some interaction with the facing material either through bond or the mechanical anchors of the facing units. This interaction will be minimized if a bond-breaker is used between the facing material and the concrete. Connections of natural stone to the concrete should be made with mechanical anchors which can accommodate some relative in-plane movement, a necessity if bond-breakers are used. One exception is the limestone industry which uses rigid, rather than flexible connectors. See Ref. 1 for recommendations on bond breakers and anchorages.

7.5.2 Clay Products

Clay products which are bonded directly to concrete include brick, tile, and architectural terra cotta

Fig. 7.5.1 Applications of veneer faced precast concrete

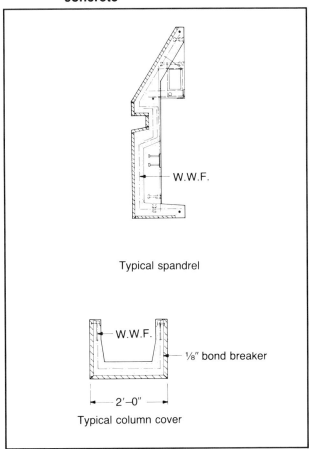

(ceramic veneer). The clay product facing may cover the entire exposed panel surface or only part of the face, serving as an accent band.

7.5.2.1 Clay product properties. Physical properties of brick vary considerably depending on the source and grade of brick. Table 7.5.1 shows the range of physical properties of clay products. Since clay products are subject to local variation, the designer should seek property values from suppliers that are being considered.

As the temperature or length of burning period is increased, clays burn to darker colors, and compressive strength and modulus of elasticity are increased. In general, the modulus of elasticity of brick increases with compressive strength to a compressive value of approximately 5000 psi, after which, there is little change. The thermal expansions of individual clay units is not the same as the thermal expansion of clay product-faced precast concrete panels due to mortar joints.

Table 7.5.1 Range of physical properties of clay products

Type of unit	Compressive strength, psi	Modulus of elasticity, psi (10^6)	Tensile strength, psi*	Coefficient of thermal expansion in./in./deg.F
Brick	3,000-15,000	1.4-5.0	*see note	4.0×10^{-6}
Quarry Tile[5]	10,000-30,000	7.0	*see note	$2.2\text{-}4.1 \times 10^{-6}$
Glazed Wall Tile[5]	8,000-22,000	1.4-5.0	*see note	$4.0\text{-}4.7 \times 10^{-6}$
Terra Cotta	8,000-11,000	2.8-6.1	*see note	4.0×10^{-6}

*Usually approximated at 10% of the compressive strength.

7.5.2.2 Clay product selection. Clay product manufacturers or distributors should be consulted early in the design stage to determine available colors, textures, shapes, sizes, and size deviations as well as manufacturing capability for special shapes, sizes and tolerances. In addition to standard facing brick shapes and sizes (conforming to ASTM C216), thin brick veneer units ⅜ to ¾ in. thick are available in various sizes, colors and textures. Thin brick units should conform to Type TBX of ASTM C1088.

Many bricks are not manufactured accurately enough to permit their use in a preformed grid used to position bricks for a precast concrete panel. Tolerances in an individual brick of ± 3/32 in. or more cause problems for the precast concrete producer. Brick may be available from some suppliers with closer tolerances (+0, -⅛ in.) necessary for precasting, or close tolerances may be obtained by saw cutting each brick, which increases costs.

Thin brick should preferably be a minimum of ¾ in. thick (many precasters have used ½ in. thick brick) to ensure proper location and secure fit in the template during casting operations, and to minimize the misalignment or tilting of individual units.

Glazed and unglazed ceramic tile units should conform to American National Standards Institute (ANSI) A1371, which includes American Society for Testing and Materials (ASTM) test procedures and provides a standardized system for evaluating a tile's key characteristics. Tiles are typically ½ in. thick with a 1% tolerance on the length and width measurements. Within one shipment, the maximum tolerance is ± 1/16 in. When several sizes or sources of tile are used to produce a pattern on a panel, the tiles must be manufactured on a modular sizing system in order to have grout joints of the same width.

Architectural terra cotta is a custom product and, within limitations, is produced in sizes for specific jobs. Two thicknesses and sizes of units are usually manufactured: 1¼ in. thick units, including dovetails spaced 5 in. on centers, size may be 20 x 30 in.; 2¼ in. thick units including dovetails spaced 7 in. on centers, size may be 32 x 48 in. Other sizes used are 4 or 6 ft x 2 ft. Tolerances on length and width are a maximum of ± 1/16 in. with a warpage tolerance on the exposed face (variation from a plane surface) of not more than 0.005 in. per in. of length.

7.5.2.3 Design considerations. The nature of the surface of clay products is important for bond to the backup concrete. Textures which give good bond include: scored finish, in which the surface is grooved as it comes from the die; combed finish where the surface is altered by parallel scratches; and roughened finish produced by wire cutting or wire brushing to remove the smooth surface or die skin from the extrusion process.

The back side of clay product units should preferably have a keyback or dovetail configuration in order to develop adequate bond to concrete. Grooved or rib back units will also develop adequate bond.

With thin clay products used to face precast concrete, metal ties are not required to attach them to the concrete. If whole brick, die skin, or heavily sanded brick must be used, a mechanical anchor with corrosion-resistant metal ties may be required, see Fig. 7.5.2.

Where ties are required, there should be a minimum of one metal tie for each 4½ sq ft of wall area. Ties in alternate courses should be staggered. The maximum vertical distance between ties should not exceed 24 in. and the maximum horizontal distance should not exceed 36 in. See Fig. 7.5.3.

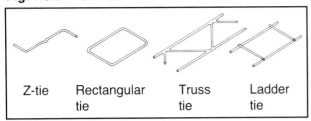

Fig. 7.5.2 Wall ties

Z-tie, Rectangular tie, Truss tie, Ladder tie

Additional ties should be provided at all openings, spaced not more than 3 ft apart around the perimeter and within 12 in. of the opening. Ties should be of corrosion-resistant metal such as stainless steel or should be coated with a corrosion-resistant protective coating.

Corrosion resistance is usually provided by coating the metal with copper or zinc. To ensure adequate resistance to corrosion, coatings should conform to the following ASTM specifications:

1. Zinc coated ties—A153, class B1, B2, or B3.
2. Zinc coated wire—A116, class 2 or class 3.
3. Copper coated wire—B227, grade 30HS.

When ties are used, the clay product joints are grouted and the ties are placed into the horizontal joint as the wet grout is placed. The required concrete reinforcing steel is anchored in place before the grout has received initial set. The concrete is then placed and cured.

Latex additives in the concrete or latex bonding materials provide high bond and high strength, but have limitations. They are water sensitive, losing as much as 50% of their strength when wet (although they regain that strength when dry). The lowered strength of the concrete is usually sufficient to sustain low shear stress like the dead weight of the clay product, but when differential movements cause additional stress, problems can occur.

Generally, clay products cast integrally with the concrete have bond strengths exceeding that obtained when laying units in the conventional manner in the field (clay product to mortar). It is necessary, in either case, to use care to avoid entrapped air or excess water-caused voids which could reduce the area of contact between the units and the concrete, thereby reducing bond.

Bond between the facing and the concrete varies depending on the absorption of the clay product. Low absorption will result in poor bond, as will high absorption due to the rapid loss of the mixing water preventing proper hydration of the cement and the development of good bond strength. Bricks with a water absorption by boiling (ASTM C216) of about 6% to 9% provide good bonding potential. Bricks with an initial rate of absorption (suction) less than 30 g per min. per 30 sq in., when tested in accordance with ASTM C67, are not required to be wetted. However, bricks with high suction or with an initial rate of absorption in excess of 30 g per min. per 30 sq in. should be wetted prior to placement of the concrete, to reduce the amount of mix water absorbed, and thereby improve bond. Terra cotta units should be soaked in water for at least one hour prior to placement in order to reduce suction. They should be damp at the time of concrete placement.

Clay bricks, when removed from the kiln after firing, will begin to permanently increase in size as a result of absorption of atmospheric moisture. The design coefficient moisture expansion of clay bricks as recommended by the Brick Institute of America is 0.0005 in. per in. but is specified as 0.0003 by Ref. 6. A value of 0.0005 is typical for ceramic quarry tile. The environmental factors affecting moisture expansion are:

1. Time of exposure. Expansion increases linearly with the logarithm of time. It is estimated that approximately 25% of the total potential moisture expansion of bricks will have occurred within two weeks after the bricks have been fired and that approximately 60% of the total potential moisture expansion will have occurred approximately one year after the bricks have been fired.

2. Time of placement. How much the brick will expand subsequent to placement in the panel depends upon how much expansion has already occurred and what proportion this represents of the total potential for expansion.

3. Temperature. The rate of expansion increases with increased temperature when moisture is present.

4. Humidity. The rate of expansion increases with an increase in relative humidity. Bricks exposed to a relative humidity of 70% have a moisture expansion two to four times as large as those exposed to RH of 50% over a 4-month interval. The 70% RH bricks also exhibit almost all of their expansion within the first twelve months of exposure, while the 50% RH bricks generally exhibit a gradual continuous moisture expansion.

In addition to continuous permanent growth due to moisture absorption, seasonal reversible expansion and contraction of clay bricks will occur due to changes in the ambient air temperature. It is not uncommon for the exterior surface to reach temperatures of 165°F with dark colored brick, 145°F for medium color, or 120°F for light colored brick on a hot summer day when directly exposed to solar radiation. Surface temperatures as low as −30°F can be reached on a cold night.

The expansion of clay products can be absorbed

by dimensional changes of the clay product and grout (mortar) or concrete due to:

1. Drying shrinkage of the grout.
2. Elastic deformation of the grout under stress.
3. Creep of the grout under stress.
4. Elastic deformation of the clay product under stress.

In general, strains imposed slowly and evenly will not cause problems. Consider the first 6 months to a year after panel production (see Fig. 7.5.4). Tile expansion is small (rate of strain application is slow) but mortar shrinkage is nearly complete. The mortar or concrete creeps under load to relieve the tensile stress generated in the tile by the mortar or concrete shrinkage since the tile are relatively rigid. After this time period, the tile has years to accommodate the additional moisture expansion.

Failures occur when strain rates exceed creep relief rates. This can occur when:

1. Total shrinkage is higher than normal because overly rich concrete or mortar was used.
2. Sudden rise in temperature or drop in humidity causes shrinkage to proceed faster than creep relieves the stresses that are generated.
3. Bond between clay product and concrete was never adequately achieved.
4. Sudden temperature drop imposes a sudden differential strain because the clay product and mortar (or concrete) have different thermal coefficients of expansion.

The difference in creep characteristics between concrete and clay product, along with the differences in their respective modulus elasticity, do not pose a problem to the production of small (less than 30 ft) panels when good quality clay products are used. Some producers have used larger panels after first conducting static load tests simulating differential creep.

Clay product faced precast panels may be designed as concrete members, neglecting for design purposes, the structural action of the face veneer. The thickness of the precast panel is reduced by the thickness of the veneer and design assumptions usually exclude consideration of differential shrinkage or differential thermal expansion. However, if the panel is to be prestressed, the effect of composite behavior and the resulting prestress eccentricity should be considered in design.

7.5.3 Natural Stone [8,9,10,11]

Natural stone facings are used in various sizes, shapes and colors to provide an infinite number of patterns and color possibilities.

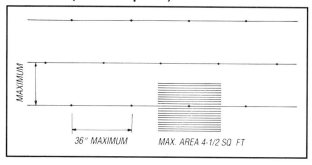

Fig. 7.5.3 Spacing and staggering of metal ties (where required)

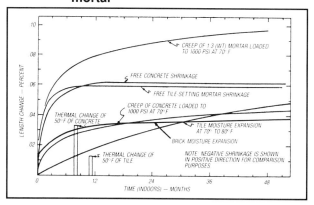

Fig. 7.5.4 Relative temperature and moisture movements of concrete, tile, brick and mortar[7]

7.5.3.1 Properties. The strength of natural stone depends on several factors: the size, rift and cleavage of crystals, the degree of cohesion, the interlocking geometry of crystals, and the nature of cementing materials present. The properties of the stone will vary with the locality from which it is quarried. Sedimentary and metamorphic rocks such as limestone and marble will exhibit different strengths when measured parallel and perpendicular to their original bedding planes. Igneous rocks such as granite may exhibit relatively uniform strength characteristics on the various planes. In addition, the surface finish, freezing and thawing, and large temperature fluctuations will affect the strength and in turn influence the anchorage system.

Information on the durability of the specified stone should be obtained from the supplier or from observations of existing installations of that particular stone. This information should include such factors as tendency to warp, reaction to weathering forces, resistance to chemical pollutants, resistance to chemical reaction from adjacent materials, and reduction in strength from the effects of weathering.

Testing for physical property should be done on stone with the same finish and thickness as will be used on the structure. An adequate number of test samples, usually 20, should be selected and statistical methods should be used to evaluate the physical properties and obtain design values. Also, modulus of rupture tests (ASTM C99) are required to demonstrate compliance of the stone to minimum properties specified by ASTM for the particular stone type [limestone (ASTM C568); marble (ASTM C503); granite (ASTM C615)].

The process used to obtain a thermal or flame finish on granite veneers reduces the effective thickness by about 1/8 in. and the physical strength to a measurable degree.[12] Bushhammered and other similar surface finishes also reduce the effective thickness. For 1 1/4 in. thick veneers, a reduction in thickness of 1/8 in. increases flexural stress by about 20% and increases the elastic deflection under wind loads by about 37%. Laboratory tests on 1 1/4 in. thick specimens of unaged, thermally finished granite, revealed that the effects of the thermal finish reduced the bending strength of the specimens by as much as 25 to 30 percent.[10] The loss of strength depends mainly on the physical properties of the stone forming minerals, on the coherence of the stone, and on the presence of micro and macro fractures in the stone.

Thermal finishing of granite surfaces causes microfracturing, particularly of quartz and feldspars. These microcracks permit absorption of water to a depth of about 1/4 in. in the distress surface region of the stone which can result in degradation by cyclic freezing and a further reduction in bending strength.

All natural stone loses strength as a result of exposure to the environment (thermal cycling, i.e., heating to 150° F and cooling to −10° F, and wet/dry cycling). The modulus of rupture of building stone can also be affected by freezing and thawing of the stone. Flexural tests (ASTM C880) should be conducted on the selected stone, at the thickness and surface finish to be used, in both the new condition and the condition after 50 cycles of laboratory freeze/thaw testing to determine the reduction in strength, if any. [Suggested freeze-thaw test procedure include (1) dry cycling between 170° F and −10° F, and (2) freezing in water at −10° F and thawing in water at room temperature.] Also, stones with high absorption should also be tested in a saturated condition as their flexural, shear and tensile properties may be significantly lower when wet. A wet to dry ratio of the modulus of rupture of 70% gives an approximate indication of good durability. Accelerated cyclic temperature tests should also be conducted on the stone-concrete assembly to determine the effect of strength loss on the shear and tensile strengths of the anchors.

For most types of stone, temperature induced movements are theoretically reversible. However, certain stones, particularly uniform-textured, fine-grained, relatively pure marble, when subjected to a large number of thermal cycles, develop an irreversible expansion in the material amounting to as much as 20% of the total original thermal expansion. This residual growth is caused by slipping of individual calcite crystals with respect to each other.[13,14] Such growth, if not considered in the stone size, design of the anchors, or the stone veneer joints may result in curling or bowing of thin marble. For relatively thick marble veneers, the expansion effects are restrained or accommodated by the unaffected portion of the veneer. Tests should be performed to establish the minimum thickness required to obtain satisfactory serviceability.

Volume changes due to moisture changes in most stones are relatively small and not a critical item in design, except that bowing of the stone can occur. Moisture permeability of stone veneers is generally not a problem (see Table 7.5.2). However, as stone veneers become thinner, water may penetrate in greater amounts and at faster rates than normally expected, and damp appearing areas of moisture on the exterior surface of thin stone veneers may occur. These damp areas result when the rate of evaporation of water from the stone surface is slower than the rate at which the water moves to the surface.

7.5.3.2 Sizes. Stone veneers used for precast facing are usually thinner than those used for conventionally set stone with the maximum size generally determined by strength of stone.

Table 7.5.2 Permeability of commercial building stones,[13] cu in./sq ft/hr/1/2 in. thickness

Stone type	Pressure, psi		
	1.2	50	100
Granite	0.06-0.08	0.11	0.28
Limestone	0.36-2.24	4.2-44.8	0.9-109
Marble	0.06-0.35	1.3-16.8	0.9-28.0
Sandstone	4.2-174.0	51.2	221
Slate	0.006-0.008	0.08-0.11	0.11

Thicknesses of marble veneer used on precast concrete have been 7/8, 1, 1 1/4, 1 1/2 and 2 in. Thicknesses of 7/8 in. or less are not desirable as it is probable that the anchor will be reflected on the surface and the development of adequate anchor capacity is questionable. Lengths of marble pieces are typically 3 to 5 ft and widths are 2 to 5 ft with a maximum area of 25 sq ft.[11]

Travertine has been used in thicknesses of 3/4, 1, 1 1/4 and 1 1/2 inches with the surface voids filled front and back on the thinner pieces. Thicknesses of 3/4 and 1 in. have resulted in excessive breakage and are not recommended. Lengths varied between 2 to

5 ft. and widths between 1 to 4 ft. with a maximum area of 16 sq ft.[11]

Granite veneer thicknesses used have been 3/4, 7/8, 1, 1 1/4, 1 5/8, 2, and 2 1/2 in. Thicknesses greater than 1 1/4 in. are recommended, unless strength of stone determined from testing indicated otherwise. Lengths vary from 3 to 7 ft. and widths between 1 to 5 ft. with a maximum area of 30 sq ft.[11]

Limestone thicknesses used for veneer on precast concrete have typically been 1 1/4, 1 1/2, 1 3/4 and 2 in., although stones as thick as 5 in. have been used. The performance of veneers of thicknesses less than 1 3/4 in. is questionable and their use is not recommended because of potential permeability and strength problems. Length has varied from 4 to 5 ft, and the width between 2 to 4 ft. with a maximum area of 15 sq ft.[11]

The length and width of veneer materials should be sized to a tolerance of +0, −1/8 in. since a plus tolerance can present problems on precast panels. This tolerance becomes important when trying to line up the false joints on one panel with the false joints on the panel above or below, particularly when there are a large number of pieces of stone on a panel. Tolerance allowance for out-of-square is ±1/16 in. difference in length of the two diagonal measurements. Flatness tolerances for finished surfaces depend on the type of finish. For example, the granite industry tolerances vary from 3/64 in. for a polished surface to 3/16 in. for flame (thermal) finish when measured with a 4 ft. straightedge.[15] Thickness variations are less important since concrete will provide a uniform back face, except at corner butt joints. In such cases, the finished edges should be within ±1/16 in. of specified thickness. However, large thickness variations may lead to the stone being encased with concrete and thus being unable to move.

7.5.3.3 Anchorage of stone facing. It is recommended that there be no bonding between stone veneer and concrete backup in order to minimize bowing, cracking and staining of the veneer. Mechanical anchors should be used to secure the veneer.

The following methods have been used to break the bond between the veneer and concrete: (1) a liquid bondbreaker, of sufficient thickness to provide a low shear modulus, applied to the veneer back surface prior to placing the concrete; (2) a 6 mil polyethylene sheet; (3) a 1/8 in polyethylene foam pad; and (4) a one component polyurethane coating. The use of a compressible bondbreaker is preferred in order to have movement capability with uneven stone surfaces, either on individual pieces or between stone pieces on a panel.

Stone veneer is supplied with holes predrilled in the back surface for the attachment of mechanical anchors. Preformed anchors fabricated from stainless steel, Type 304, are usually used. The number and location of anchors should be determined by shear and tension tests conducted on the anchors embedded in a stone/precast concrete test sample and the anticipated wind and shear loads applied to the panel. Anchor size and spacing in veneers of questionable strengths or with natural planes of weakness may require special analysis.

Four anchors are usually used per stone piece with a minimum of 2 recommended. The number of anchors has varied from 1 per 1 1/2 sq ft of stone to 1 per 6 sq ft with 1 per 2 to 3 sq ft being the most common.[1] Anchors should be 6 to 9 in. from an edge with not over 30 in. between anchors. The shear capacity of the spring clip (hairpin) anchors perpendicular to the anchor legs is greater than when they are parallel and depends on the strength of the stone. A typical marble veneer anchor detail with a toe-in spring clip (hairpin) is shown in Fig 7.5.5(a) and a typical granite veneer anchor detail is shown in Fig. 7.5.5(b). The toe-out anchor in granite may have as much as 50% more tensile capacity than a toe-in anchor depending on the stone strength.

Depth of anchor holes should be approximately one-half the thickness of the veneer (minimum depth of 3/4 in.), and are often drilled at an angle of 30° to 45° to the plane of the stone. Holes which are approximately 50% oversize have been used to allow for differential movement between the stone and the concrete. However, in most cases, holes 1/16 to 1/8 in. larger than the anchor are common as excessive looseness in hole reduces holding power. Anchor holes should be within ±3/16 in. of the specified hole spacing, particularly for the spring clip anchors.

Stainless steel dowels, smooth or threaded, are installed to a depth of 2/3 the stone thickness with a maximum depth of 2 in. at angles of 45° to 75° to the plane of the stone. Dowel size varies from 3/16 to 5/8 in. for most stones, except that it varies from 1/4 to 5/8 in. for soft limestone and sandstone and depends on thickness and strength of stone. The dowel hole is usually 1/16 to 1/8 in. larger in diameter than the anchor, see Figs. 7.5.5(c) and 7.5.5(d).

Limestone has traditionally been bonded and anchored to the concrete, because it has the lowest coefficient of expansion. Limestone has also traditionally been used in thicknesses of 3 to 5 in. but it is now being used as thin as 1 1/4 in. When limestone is 2 in. or thinner, it is prudent to use a bondbreaker, along with mechanical anchors. Dowels and spring clip anchors have been used to anchor limestone. Typical dowel details for limestone veneers are shown in Figs. 7.5.5 (c) and 7.5.5(d). The dowels in Fig. 7.5.5(c) should be inserted at angles alternately up and down to secure stone facing to backup concrete.

Some flexibility should be introduced with all anchors of stone veneer to precast concrete panels, e.g., by keeping the diameter of the anchors to a minimum, to allow for the inevitable relative movements which occur with temperature variations and concrete shrinkage. Unaccommodated relative movements can result in excessive stresses and eventual failure at an anchor location.

Some designers use epoxy to fill the spring clip anchor or dowel holes in order to eliminate intrusion of water into the holes and the possible dark, damp appearance of moisture on the exposed stone surface. The epoxy increases the shear capacity and rigidity of the anchor. The rigidity may be partially overcome by using ½ in. compressible rubber or elastomeric grommets or sleeves on the anchor at the back surface of the stone. Differential thermal expansion of the stone and epoxy may cause cracking of the stone veneer; this may be overcome by keeping the oversizing of the hole to a minimum, thereby reducing epoxy volume. It may be preferable to fill the anchor hole with an elastic, fast-curing silicone which has been proven to be non-staining to light colored stones, or a low modulus polyurethane sealant. The overall effect of either epoxy or sealant materials on the behavior of the entire veneer should be evaluated prior to their use. At best, the long-term service of epoxy is questionable, therefore, any increase in shear value should not be used in calculating long-term anchor capacity.

Design of anchorage and size of the stone should be based on specific test values for the actual stone to be installed. Anchor test procedures have not been standardized. Test samples for anchor tests should be a typical panel section of about 1 sq ft and approximate, as closely as possible, actual panel anchoring conditions. A bondbreaker should be placed between stone and concrete during sample manufacture to eliminate any bond between veneer and concrete surface. Each test sample should contain one anchor connecting stone to concrete backup and a minimum of 5 tests are needed to determine tensile (pull-out) and shear strength of each anchor. Depending on the size of the project it may be desirable to perform shear and tensile tests of the anchors at intervals during the fabrication period.

The outward wind suction forces on walls parallel to the airstream and on walls at the leeward side of the building are normally the governing design consideration.

Safety factors are recommended by the stone trade associations and the suppliers of different kinds of building stones. Because of the expected variation in the physical properties of natural stones and the effects of weathering, recommended safety factors are larger than those used for manufactured building materials, such as steel and concrete. The minimum recommended safety factor, based on the average of the test results, is 3 for granite, 4 for anchorage components in granite,[15] 8 for limestone veneers,[16] and 5 for marble veneers[14]. If the range of test values exceeds the average by more than ± 20% then the safety factor should be applied to the lower bound value.

7.5.3.4 Veneer jointing. Joints between veneer pieces on a precast element should be a minimum of ¼ in. The veneer pieces may be spaced with a non-staining, compressible spacing material, or a chemically neutral, resilient, non-removable gasket which will not stain the veneer nor adversely affect the sealant to be applied later. The gaskets are of a size and configuration that will provide a pocket to receive the sealant and also prevent any backing concrete from entering the joint between the veneer units. Shore A hardness of the gasket should be less than 20.

In many cases, stone veneer is used as an accent or feature strip on precast concrete panels. The major difference is the stone attachment, i.e., a ½ in. space is left between the edge of the stone and the precast concrete to allow for differential movements of the materials. This space is then caulked as if it were a conventional joint.

Caulking should be of a type that will not stain the veneer material. In some projects, caulking may be installed more economically and satisfactorily at the same time as the caulking between precast elements.

7.5.3.5 Samples. There is now a good background of experience in the production and erection of stone veneer-faced precast concrete panels. However, it is recommended that, for new and major applications, full scale mockup units be manufactured to check out the feasibility of the production and erection process. Tests on sample panels should be made to confirm the suitability of the stone and anchors and the effects of bowing on the panel's performance. Tests on the behavior of the unit for anticipated temperature changes may be required. Mockups should be built to test wall, window and joint performance under the most severe wind and rain conditions. Acceptance criteria for the stone, as well as the anchorage, should be established in the project specifications.

Fig. 7.5.5 Typical anchor details

(a) Typical anchor for marble veneer

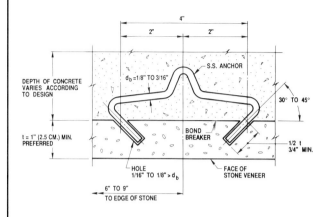

(b) Typical anchor for granite veneer

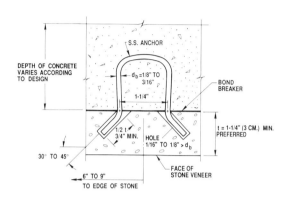

(c) Typical anchor for limestone veneer

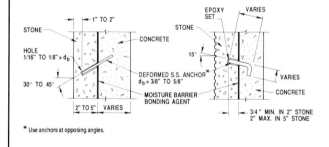

(d) Typical cross anchor dowels for stone veneer

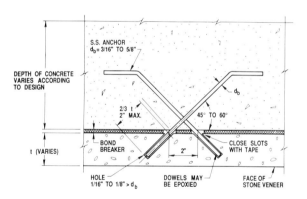

7.6 Design Example—Window Wall Panel

This example illustrates the design of a 2-story high non-load bearing window wall with deep reveals. The project is a 4-story building with a structural steel frame and cast-in-place concrete floor slabs. Precast concrete window wall panels are supported at their lowest points, and tied back to the structure at the floor and roof levels. They transfer only horizontal loads to the steel frame. The vertical loads are transferred directly to the foundation (see Fig. 7.6.1).

Given:

Wind load = 15 psf, both during erection and after

f'_c = 5000 psi at 28 days

f'_{ci} = 2000 psi at stripping

Fig. 7.6.1 Window wall panel example

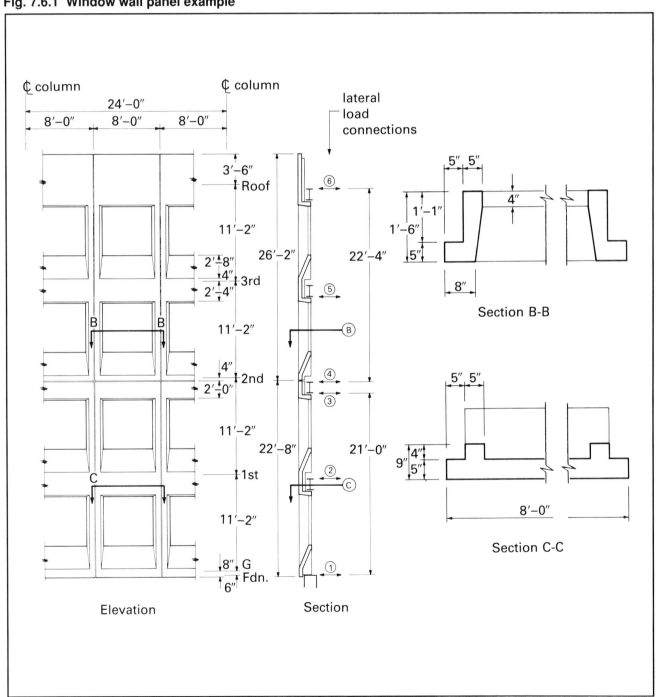

Problem:

Exposed face to be designed "crack free"
Cracks in rear face permitted without width restriction.

Solution:

Step 1

Design panel for handling—see Chapter 5.
Each piece will be cast, stripped, stored, and shipped in the flat, face-down position.

Determine handling multipliers (Table 5.2.1)
Stripping: surface is not retarded, use 1.7
Yard handling: 1.2
Shipping: 1.5
Erection: 1.5

Determine allowable tensile stress in exposed face (Sect. 5.2.4)

At stripping, yard handling and storage:

$f'_r = 5\sqrt{2000} = 224$ psi

At shipping and erection:

$f'_r = 5\sqrt{5000} = 353$ psi

Check handling stresses

Since all handling except erection is to be in the flat position, using the same lift points, it is apparent that stripping is the critical design condition. After several trials and adjustment of lift point locations, it is determined that the tensile stresses at the critical section (Section C-C, Fig. 7.6.1) are too sensitive to slight relocations. Therefore, select locations so that there is compression in the face at this section. The loading and moment diagrams are shown in Fig. 7.6.2.

Section properties at Section B-B:

$A = 202$ in.^{2} $I = 6044$ in.^{4}
$y_b = 8.00$ in. $Z_b = 755$ in.^{3}
$y_t = 10.00$ in. $Z_t = 604$ in.^{3}

Section properties at Section C-C:

$A = 520$ in.^{2} $I = 1801$ in.^{4}
$y_b = 2.85$ in. $Z_b = 632$ in.^{3}
$y_t = 6.15$ in. $Z_t = 293$ in.^{3}

Stress Summary:

Location	Moments	(in.-kips)	Tensile
	(1)	(2)	Stress (psi)
5	−252	−428	709 high
4	−176	−299	1,020 high
3	−72	−122	416 high
2	+23	+39	52 > 0
1	−30	−51	174 < 224

(1) No multiplier
(2) 1.7 multiplier

Design panel reinforcement for stripping

Stresses at points 3, 4, and 5 exceed $5\sqrt{f'_{ci}}$. Therefore, more than minimum reinforcement is required. Since these points are on the rear face of the panel, where crack control is not required, select reinforcement using the "z" factor, Sect. 4.2.2.1, as well as to satisfy the strength requirements, Sect. 4.2.1. From Table 4.2.1, the maximum allowable z = 175.

Reinforcement required at point 5, Fig. 7.6.2:

M_u per window jamb = 428/2 = 214 in.-kips

$d = 18 − 2 = 16$ in.; $b = 8$ in.

Try 1 – #5, grade 60, $A_s = 0.31$ in.

$a = \dfrac{A_s f_y}{0.85 b f'_{ci}} = \dfrac{0.31(60)}{0.85(8)(2)}$

$= 1.37$ in.

$\phi M_n = \phi A_s f_y (d - a/2) = 0.9(0.31)(60)(16 - 0.68)$

$= 256$ in.-kips > 214 OK

$z = f_s \sqrt[3]{d_c A}$

$f_s = 0.6 f_y = 36$ ksi; $d_c = 2$ in.

$A = (2)(2)(5) = 20$ in.^{2}

$z = 36 \sqrt[3]{2(20)} = 123$ kips/in. < 175 OK

Fig. 7.6.2 Loads and moments on panel at stripping

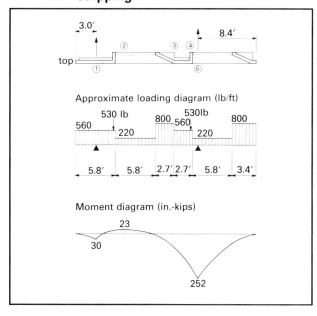

PCI Design Handbook/Fourth Edition

Reinforcement required at point 4, Fig. 7.6.2 (use same for point 3):

M_u = 299 in.-kips

d = 9 − 1.5 = 7.5 in.; b = 96 in.

Try 2 − #6 (one each rib), grade 60, A_s = 0.88 in.^{2}

$$a = \frac{0.88(60)}{0.85(96)(2)} = 0.32 \text{ in.}$$

ϕM_n = 0.9(0.88)(60)(7.5 − 0.16)

= 349 in.-kips > 299   OK

d_c = 1.5 in.; A = (2)(1.5)(5) = 15 in.^{2}

z = 36 $\sqrt[3]{1.5(15)}$ = 102 kips/in. < 175   OK

Note: Although the section at points 3 and 4 is adequate without a strongback, it would be prudent to provide temporary stiffening arranged so as not to interfere with the structural beams.

Reinforcement required at points 1 and 2:

Since the tensile stress is below $5\sqrt{2000}$, only minimum reinforcement is required.

At point 1:
minimum A_s = 0.001 bt = 0.001(96)(5)
= 0.48 in.^{2}

At point 2:
minimum A_s = 0.001 bt = 0.001 (area)
= 0.001(202) = 0.202 in.^{2}

Transverse Bending

By inspection, the section is very stiff in the transverse direction. Therefore, use minimum reinforcement as shown in the summary, Fig. 7.6.3.

Step 2

Design for condition during erection:

Overall construction sequence:
1. Erect structural steel frame.
2. Cast floor slabs.
3. Set precast concrete wall panels.

Precast concrete erection sequence:
1. Set bottom panel with lateral connections at points 1 and 3 only (Fig. 7.6.1).
2. Make lateral connection at point 2.
3. Set upper panel with lateral connection at points 4 and 6 only.
4. Make lateral connection at point 5.

Wind load on panel:

w = 15(8) = 120 lb/ft = 0.12 kips/ft

Determine loads on connections (see Fig. 7.6.4)

Before connection at point 2 is made:

a. Due to eccentric panel weight (Detail B):

Weight = 14 kips; e = 5 in.

$$R_1 = (-R_3) = \frac{14(5)}{21(12)} = 0.28 \text{ kips}$$

b. Due to wind:

Fig. 7.6.3 Reinforcement summary

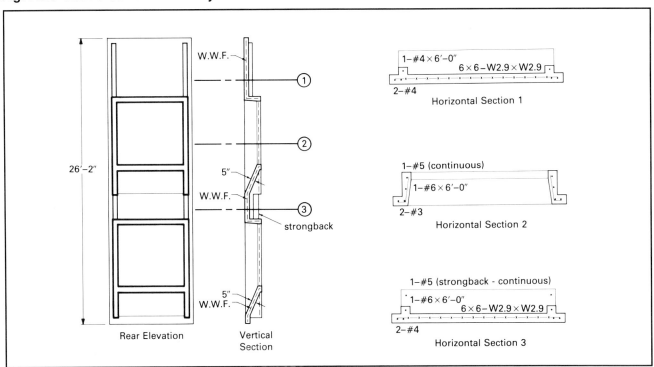

$$R_1 = R_3 = \frac{0.015(8)(22.67)}{2} = 1.36 \text{ kips}$$

c. Load from upper panel on lower panel:
With connection 2 made before setting the upper panel, the lateral loads at 1, 2 and 3 are indeterminate. To simplify, conservatively assume connection 2 has not been made.

Weight of upper panel = 13.3 kips

e (detail A) = 13 − 5 = 8 in.

$$R_1 = (-R_3) = \frac{13.3(8)}{21(12)} = 0.42 \text{ kips}$$

d. Design loads (ACI 318-89, Sect. 9.2):
Two connections per panel

$U = 1.4D + 1.7L$
$= 1.4(0.28 + 0.42)/2 = 0.49 \text{ kips}$

$U = 0.75(1.4D + 1.7L + 1.7W)$
$= 0.75[1.4(0.70) + 1.7(1.36)]/2 = 1.23 \text{ kips}$

e. Loads on upper panel connections:
An analysis similar to a through d on the upper panels shows that the maximum design load on the connection is 0.92 kips. For simplicity, design all connections for a lateral load of 1.23 kips. An additional load factor of 1.3 (see Sect. 6.3) is considered appropriate for this connection.

$1.3(1.23) = 1.60 \text{ kips}$

Check panel bending
Before connection at point 2 is made:

$M = w\ell^2/8 = 0.12(21.0)^2/8$
$= 6.62 \text{ ft-kips} = 79.4 \text{ in.-kips}$

$M_u = 0.75(1.7W) = 0.75(1.7)(79.4) = 101 \text{ in.-kips}$

This is less critical than condition at stripping.

Connection design (Fig. 7.6.4)
Assume the panel weight is transferred through shims at the second floor and foundation (Details A and B). The shims must be large enough to transfer load in plain concrete bearing (Sect. 6.8).

Weight of upper panel = 13.3 kips
Weight of lower panel = 14.0
                        27.3 kips

Use 1.3 connection load factor.

V_u (each connection) = $27.3(1.4)(1.3)/2$
= 24.8 kips

For more efficient erection, use same size shims throughout.

Foundation concrete f'_c = 3000 psi.

Rearrange Eq. 6.8.1. Conservatively assume $A_2 = A_1$, and $C_r = 1$, since other connections will be designed to take lateral loads.

$$A_1 = \frac{V_u}{\phi(0.85)f'_c} = \frac{24.8}{0.7(0.85)(3)}$$
$= 13.9 \text{ in.}^2$

Use standard size plastic or steel shims with a minimum area of 13.9 sq. in.

Fig. 7.6.4 Connection design—example problem

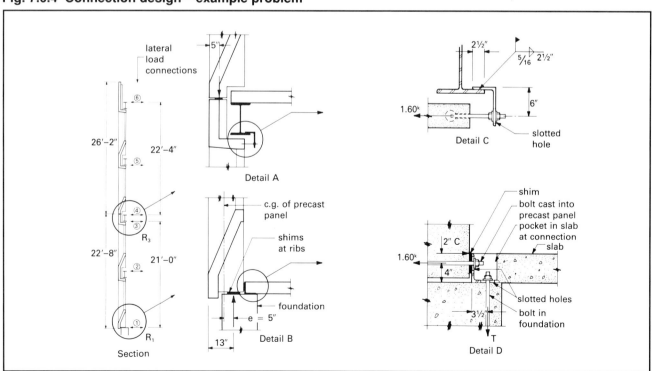

Connection at point 1:

Angle and bolt design: Choose a 6 in. long angle and minimum ⅝ in. diameter bolts. Note: Smaller bolts are likely to be damaged during handling.

From Eq. 6.5.19, with g = 4 in.; b = 6 in.;
$$F_y = 36 \text{ ksi}$$
$$t = \sqrt{\frac{4N_u g}{\phi F_y b}} = \sqrt{\frac{4(1.60)(4)}{0.9(36)(6)}} = 0.36 \text{ in.}$$

Use 6 x 6 x ⅜ angle 6 in. long.

From Table 6.20.19, the minimum ⅝ in. bolt is adequate. The bolt in the foundation resists both tension and shear:

Taking moments about the upper toe of the angle:

$3.5T_u = 2(1.60)$; $T_u = 0.91$ kips

The interaction equation, Eq. 6.5.16, is applicable:

From Table 6.20.9, $P_s = 10.4$ kips; $V_s = 5.0$ kips

$$\left(\frac{P_u}{P_s}\right)^2 + \left(\frac{V_u}{V_s}\right)^2 = \left(\frac{0.91}{10.4}\right)^2 + \left(\frac{1.60}{5.0}\right)^2$$
$$= 0.008 + 0.102 = 0.11 < 1.0 \quad \text{OK}$$

(Note: The connection at Point 4 would be similar to Point 1. See Chapter 6 for examples of these "clip angle" type connections.)

Connections 2, 3 and 5:

Determine angle thickness:

Try 5 in. long angle.

From Eq. 6.5.19 with g = 6 in.; b = 5 in.

$$t = \sqrt{\frac{4N_u g}{\phi F_y b}} = \sqrt{\frac{4(1.60)(6)}{0.9(36)(5)}} = 0.49 \text{ in.}$$

Use angle 8 x 4 x ½ x 5 in. long.

Design weld to beam flange:

Refer to Sect. 6.5.6.

Try 5/16 in. weld by 2½ in. long each side of angle

$t_w = 0.707(0.31) = 0.22$ in.

Conservatively use elastic section properties.

From Fig. 6.5.12, Case 2:

$S = d^2 t_w/3 = (2.5)^2 \, 0.22/3 = 0.46$ in.^{3}

From Table 6.20.1:

Use E70 electrodes, allowable strength = 31.5 ksi

$$f = \frac{P}{A} + \frac{M}{S} = \frac{1.60}{0.22(2.5)(2)} + \frac{1.60(6)}{0.46}$$
$$= 1.45 + 20.87 = 22.32 < 31.5 \quad \text{OK}$$

7.7 References

1. "Architectural Precast Concrete," Second Edition, MNL-122-89, Precast/Prestressed Concrete Institute, Chicago, IL, 1989.

2. Sheppard, D.A. and Phillips, W.R., "Plant-Cast Precast and Prestressed Concrete—A Design Guide," Third Edition, McGraw-Hill Publishing Co., New York, NY, 1989.

3. ACI Committee 347, "Precast Concrete Units Used as Forms for Cast-in-Place Concrete," *Journal of the American Concrete Institute*, V. 66, No. 10, October 1969.

4. "Formwork for Concrete," Fifth Edition, Special Publication SP-4, American Concrete Institute, Detroit, MI, 1989.

5. Fitzgerald, J.V. and Kastenbein, E.L., "Tests for Engineering Properties of Ceramic Tile," ASTM Bulletin No. 231, July 1958, American Society for Testing and Materials, Philadelphia, PA.

6. "Building Code Requirements for Masonry Structures," ACI 530-88/ASCE 5-88, and "Specifications for Masonry Structures," ACI 530.1-88/ASCE 6-88, American Concrete Institute, Detroit, MI/American Society of Civil Engineers, New York, NY, 1989.

7. Bernett, Frank E., "Effects of Moisture Expansion in Installed Quarry Tile," *Ceramic Bulletin*, The American Ceramic Society, V. 5, No. 12, 1976.

8. McDaniel, W. Bryant, "Marble-Faced Precast Panels," *PCI Journal*, V. 12, No. 4, August 1967.

9. "Marble-Faced Precast Panels," National Association of Marble Producers, 1966.

10. Chin, I.R., Stecich, J.P. and Erlin, B., "Design of Thin Stone Veneers on Buildings," *Building Stone Magazine*, May/June 1986.

11. "Stone Veneer-Faced Precast Concrete Panels," PR-22, Precast/Prestressed Concrete Institute, Chicago, IL, 1988.

12. "Italian Marble Technical Guide," V. 1, 1982 Edition, Italian Institute for Foreign Trade, New York, NY.

13. "Building Construction Handbook," Third Edition, Frederick S. Merritt, Editor, McGraw-Hill Book Co., New York, NY, 1975.

14. "Dimensional Stone: Design Manual III," Marble Institute of America, Farmington, MI, 1985.

15. "Specifications for Architectural Granite," 1986 Edition, National Building Granite Quarries Association, West Chelmsford, MA.

16. "Indiana Limestone Handbook," 1984-1985, Indiana Limestone Institute of America, Bedford, IN.

17. Kulka, Felix, Lin, T.Y. and Yang, Y.C., "Prestressed Concrete Building Construction Using Precast Wall Panels," *PCI Journal*, V. 20, No. 1, January-February 1975.

18. "Clay Product-Faced Precast Concrete Panels," *PCI Journal* (scheduled for publication in 1992).

CHAPTER 8
TOLERANCES FOR PRECAST AND PRESTRESSED CONCRETE

		Page No.

8.1 General 8-2
    8.1.1   Definitions 8-2
    8.1.2   Purpose 8-2
    8.1.3   Responsibility 8-2
    8.1.4   Tolerance Acceptability Range 8-3
    8.1.5   Relationships between Different Tolerances 8-3

8.2 Product Tolerances 8-3
    8.2.1   General 8-3
    8.2.2   Overall Dimensions 8-3
    8.2.3   Sweep or Horizontal Alignment 8-3
    8.2.4   Position of Strands 8-4
    8.2.5   Camber and Differential Camber 8-5
    8.2.6   Weld Plates 8-5
    8.2.7   Haunches of Columns and Wall Panels 8-5
    8.2.8   Warping and Bowing 8-5
    8.2.9   Smoothness 8-7
    8.2.10 Architectural Panels vs. Structural Walls 8-7

8.3 Erection Tolerances 8-7
    8.3.1   General 8-7
    8.3.2   Recommended Erection Tolerances 8-7
    8.3.3   Mixed Building Systems 8-7
    8.3.4   Connections and Bearing 8-13

8.4 Clearances 8-13
    8.4.1   General 8-13
    8.4.2   Joint Clearance 8-13
    8.4.3   Procedure for Determining Clearance 8-13
    8.4.4   Clearance Examples 8-14

8.5 Interfacing Tolerances 8-17
    8.5.1   General 8-17
    8.5.2   Interface Design Approach 8-17
    8.5.3   Characteristics of the Interface 8-18

8.6 References 8-19

TOLERANCES FOR PRECAST AND PRESTRESSED CONCRETE

8.1 General

8.1.1 Definitions

Tolerance—
- The permitted variation from a basic dimension or quantity, as in the length or width of a member
- The range of variation permitted in maintaining a basic dimension, as in an alignment tolerance
- A permitted variation from location or alignment

Product Tolerances—Those variations in dimensions relating to individual precast concrete members.

Erection Tolerances—Those variations in dimensions required for acceptable matching of precast members after they are erected.

Interfacing Tolerances—Those variations in dimensions associated with other materials in contact with or in close proximity to precast concrete.

Variation—The difference between the actual and the basic dimension. Variations may be either negative (less) or positive (greater).

Basic Dimension—Those shown on contract drawings or in specifications. Basic dimensions apply to size and location. May also be called "nominal" dimensions.

Working Dimension—The planned dimension of a member which considers both its basic dimension and joints or clearances. For example, a member with a basic width of 8'-0" may have a working width of 7'-11". Product tolerances are applied to working dimensions.

Actual Dimension—The measured dimension of the member after casting. This may differ from the working dimension due to construction- and material-induced variation.

Alignment Face—The face of a precast element which is to be set in alignment with the face of adjacent elements or features.

Primary Control Surface—A surface on a precast element which is dimensionally controlled during erection. Clearance is generally allowed to vary so the primary control surface can be set within tolerance.

Secondary Control Surfaces—A surface on a precast element, the location of which is dependent on the locational tolerance of the primary control surface plus the product tolerances.

Feature Tolerance—The locational or dimensional tolerance of a feature, such as a corbel or a blockout, with respect to the overall member dimensions.

8.1.2 Purpose

Tolerances are normally established by economical and practical production, erection and interfacing considerations. Once established, they should be shown in the project documents, and used in design and detailing of components and connections. Architectural and structural concepts should be developed with the practical limitations of dimensional control in mind, as the tolerances will affect the dimensions of the completed structure.

Tolerances are required for the following reasons:

Structural. To ensure that structural design accounts for factors sensitive to variations in dimension. Examples include eccentric loadings, bearing areas, and locations of reinforcement and embedded items.

Feasibility. To ensure acceptable performance of joints and interfacing materials in the finished structure.

Visual. To ensure that the variations will be controllable and result in an acceptable looking structure.

Economic. To ensure ease and speed of production and erection.

Legal. To avoid encroaching on property lines and to establish a standard against which the work can be compared.

Contractual. To establish an acceptability range and also to establish responsibility for developing and maintaining specified tolerances.

8.1.3 Responsibility

While the responsibility for specifying and maintaining tolerances of the various elements may vary among projects, it is important that these responsibilities be clearly assigned. The conceptual design of a precast project is the place to begin consideration of dimensional control. The established tolerances or required performance should fall within generally accepted limits and should not be made more rigid than necessary.

Once the tolerances have been specified, and connections which consider those tolerances have been designed, the production of the elements must be organized to assure tolerance compliance.

An organized quality control program which emphasizes dimensional control is necessary. Likewise, an erection quality assurance effort which includes a clear definition of responsibilities will aid in assuring that the products are assembled in accordance with the specified erection tolerances.

Responsibility should include dimension verification and adjustment, when necessary, of both precast components and any interfacing structure.

8.1.4 Tolerance Acceptability Range

Tolerances must be used as guidelines for acceptability and not limits for rejection. If specified tolerances are met, the member should be accepted. If not, the member may be accepted if it meets any of the following criteria:

1. Exceeding the tolerance does not affect the structural integrity or architectural performance of the member.
2. The member can be brought within tolerance by structurally and architecturally satisfactory means.
3. The total erected assembly can be modified to meet all structural and architectural requirements.

8.1.5 Relationships between Different Tolerances

A precast member is erected so that its primary control surface is in conformance with the established erection and interfacing tolerances. The secondary control surfaces are generally not directly positioned during erection, but are controlled by the product tolerances. Thus, if the primary control surfaces are within erection and interfacing tolerances, and the secondary surfaces are within product tolerances, the member is erected within tolerance. The result is that the tolerance limit for the secondary surface may be the sum of the product and erection tolerances.

Since tolerances for some features of a precast member may be additive, it must be clear to the erector which are the primary control surfaces. If both primary and secondary control surfaces must be controlled, provisions for adjustment should be included. The accumulated tolerance limits may have to be accommodated in the interface clearance. Surface and feature control requirements should be clearly outlined in the plans and specifications.

On occasion, the structure may not perform properly if the tolerances are allowed to accumulate. Which tolerance takes precedence is a question of economics. The costs associated with each of the three tolerances must be evaluated, recognizing unusual situations. This may include difficult erection requirements, connections which are tolerance sensitive, or production requirements which are set by the available equipment. Any special tolerance requirements should be clearly noted in the contract documents.

8.2 Product Tolerances

8.2.1 General

Product tolerances are listed in the full report of the PCI Committee on Tolerances[1] as well as in the *Manual for Quality Control for Plants and Production of Precast and Prestressed Concrete Products*[2] and *Manual for Quality Control for Plants and Production of Architectural Precast Concrete Products.*[3] The products included are listed in Table 8.2.1. Discussion of the more critical tolerances are given in the following sections. The values shown have become the consensus standards of the precast concrete industry. More rigid tolerances may significantly increase costs, so should not be specified unless absolutely necessary.

8.2.2 Overall Dimensions

Tolerances for the overall dimensions of most common products are given in Table 8.2.1. Architectural precast concrete panels have plan dimension tolerances that vary with panel size from ± ⅛ in. for a dimension under 10 ft to ± ¼ in. for a dimension of 20 to 40 ft. Overall thickness of wall panels are listed in Table 8.2.1 under depth tolerances.

Top and bottom slabs (flanges) of box beams and hollow-core slabs are dependent on position of cores. Flange thickness tolerances are not given for hollow-core slabs. Instead, measured flange areas cannot be less than 85% of nominal calculated area.

The committee report emphasizes that the recommended tolerances are only guidelines. Different values may be applicable in some cases, and each project should be considered individually.

8.2.3 Sweep or Horizontal Alignment

Horizontal misalignment, or sweep, usually occurs as a result of form and member width tolerances. It can also result from prestressing with lateral eccentricity, which should be considered in the design. Joints should be dimensioned to accommodate such variations.

Sweep tolerances generally vary with length of unit, for example, ± ⅛ in. per 10 ft. The upper limit of sweep varies from ± ⅜ in. for wall panels and hollow-core slabs to ± ¾ in. for joists usually used in composite construction.

Table 8.2.1 Typical tolerances for precast, prestressed concrete products*

Product Tolerances	Products
Length**—	
± ½ in.	6,7,8,9,13
± ¾ in.	3,5
± 1 in.	1,2,4,11,12
Width**—	
± ¼ in.	1,2,3,5,6,7,8,9,12
+ ⅜ in., − ¼ in.	4
± ⅜ in.	11,13
Depth—	
+ ¼ in., − ⅛ in.	10
± ¼ in.	1,2,3,5,6,7,8,9,12
+ ½ in., − ¼ in.	4
± ⅜ in.	11
± ½ in.	13
Flange Thickness—	
+ ¼ in., − ⅛ in.	1,2,8,10,12
± ¼ in.	3,4,13
Web Thickness—	
± ⅛ in.	1,8,10,12
± ¼ in.	2,3
+ ⅜ in., − ¼ in.	4
± ⅜ in.	5
Position of Tendons—	
± ¼ in.	1,2,3,4,5,6,8,9,11,12
± ⅛ in.	10
Camber, variation from design—	
± ¼ in. per 10 ft, ± ¾ in. max.	1,2,12
± ⅛ in. per ft, ± 1 in. max.	4
± ¾ in. max.	3
± ½ in. max.	5
Camber, differential—	
¼ in. per 10 ft, ¾ in. max.	1,2,5
Bearing Plates, position—	
± ½ in.	1,2,3,12
± ⅝ in.	4
Bearing Plates, tipping and flushness—	
± ⅛ in.	1,2,3,4,12

*For more details such as graphic descriptions of features to which tolerances apply and tolerances for sleeves, blockouts, inserts, plates, end squareness, surface smoothness, etc., see committee report.[1]

**See Sect. 8.2.2 for dimensional tolerances for architectural wall panels.

Key:
1 = double tee
2 = single tee
3 = building beam (rect. and ledger)
4 = I-beam
5 = box beam
6 = column
7 = hollow-core slab
8 = ribbed wall panel
9 = insulated wall panel
10 = architectural wall panel
11 = pile
12 = joist
13 = step unit

Table 8.2.2 Erection tolerances for interface design

Item	Recommended Tolerances (in.)
Variation in plan location (any column or beam, any location)	± ½
Variation in plan parallel to specified building lines	1/40 per ft, any beam less than 20 ft or adjacent columns less than 20 ft apart; ½, adjacent columns 20 ft or more apart
Difference in relative position of adjacent columns from specified relative position (at any deck level)	½
Variation from plumb	¼, any 10 ft of height; 1, maximum for the entire height
Variation in elevation of bearing surfaces from specified elevation (any column or beam, any location)	± ½
Variation of top of spandrel from specified elevation (any location)	± ½
Variation in elevation of bearing surfaces from lines parallel to specified grade lines	1/40 per ft, any beam less than 20 ft or adjacent columns less than 20 ft apart; ½, maximum any beam 20 ft or more in length or adjacent columns 20 ft or more apart
Variation from specified bearing length on support	± ¾
Variation from specified bearing width on support	± ½
Jog in alignment of matching edges	½, maximum

Table 8.2.3 Recommended minimum clearances

Item	Recommended Minimum Clearance (in.)
Precast to precast	½ (1 preferred)
Precast to cast-in-place	1 (2 preferred)
Precast to steel	1 (2 preferred)
Precast column covers	1½ (3 preferred for tall buildings)

8.2.4 Position of Strands

It is a common practice to use ⅝ in. diameter holes in end dividers (bulkheads or headers) for ⅜ to ½ in. diameter strands, since it is costly to switch end dividers for different strand diameters. Thus, better accuracy is achieved when using larger diameter strands.

Generally, individual strands must be positioned within ± ¼ in. of design position and bundled strands within ± ½ in. Hollow-core slabs have greater individual strand tolerances as long as the center of gravity of the strand group is within ± ¼ in. and a minimum cover of ¾ in. is maintained.

8.2.5 Camber and Differential Camber

Design camber is generally based on camber at release of prestress; thus, camber measurements on products should be made as soon after stripping as possible. Differential camber refers to the final in-place condition of adjacent products.

It is important that cambers are measured at the same time of day, preferably in the early hours before the sun has begun to warm the members. Cambers for all units used in the same assembly should be checked at the same age.

If a significant variation in camber from calculated values is observed, the cause should be determined and the effect of the variation on the performance of the member evaluated. If differential cambers exceed recommended tolerances, additional effort is often required to erect the members in a manner which is satisfactory for the intended use.

The final installed differential between two adjacent cambered members erected in the field may be the combined result of member differential cambers, variations in support elevations, and any adjustments made to the members during erection.

For most flexural members, maximum camber variation from design camber is ± ¾ in., and maximum differential camber between adjacent units of the same design is ¾ in. This may be increased for joists used in composite construction. Recommendations for camber and differential camber of hollow-core slabs are not definitive, because of production variations between hollow-core systems.

8.2.6 Weld Plates

In general, it is easier to hold plates to closer tolerances at the bottom of the member (or against the side form) than with plates cast on top of the member. Bottom and side plates can be fastened to the form and hence are less susceptible to movement caused by vibration. This applies to position of weld plates as well as tipping and flushness.

The tolerance on weld plates is less restrictive than for bearing plates. The position tolerance is ± 1 in. for all products except for hollow-core slabs where the tolerance is ± 2 in.; tipping and flushness tolerance is ± ¼ in.

8.2.7 Haunches of Columns and Wall Panels

The importance of haunch location tolerances depends on the connection at the base of the member. Since base connections usually allow some flexibility, it is more important to control dimensions from haunch to haunch in multilevel columns or walls than from haunch to end of member.

The haunch to haunch tolerance is ± ⅛ in. Bearing surface squareness tolerance is ± ⅛ in. per 18 in. with a maximum of ± ¼ in., except for architectural precast concrete panels, with a tolerance of ± ⅛ in., and columns, with a maximum of ± ⅛ in. in short direction and ± ⅜ in. in long directions.

8.2.8 Warping and Bowing

Warping and bowing tolerances affect panel edge matchup during erection, and the appearance of the erected members. They are especially critical with

Fig. 8.2.1 Definition of panel warping

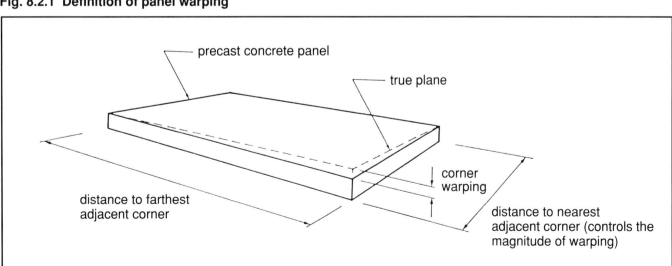

architectural panels.

Warping is a variation from plane in which the corners of the panel do not fall within the same plane. Warping tolerances are given in terms of corner variations, as shown in Fig. 8.2.1. The allowable variation from the nearest adjacent corner is 1/16 in. per ft.

Bowing differs from warping in that two opposite edges of a panel may fall in the same plane, but the portion between is out of plane (Fig. 8.2.2). Bowing tolerance is L/360, where L is the length of bow. Maximum tolerance on differential bowing between panels of the same design is 1/2 in.

The effects of differential temperature and moisture absorption between the inside and outside of a panel, and prestress eccentricity, should be considered in design of the panel and its connections to minimize bowing and warping. Pre-erection storage conditions may also affect warping and bowing (see Sects. 3.3.2 and 5.2.10).

Thin panels are more likely to bow, and the tolerances should be more liberal. Table 8.2.4 gives thicknesses, related to panel dimensions, below which the warping and bowing tolerances given above may not apply. (Note: Table 8.2.4 is not intended to limit panel thickness.) For example, a panel that is 16 x 8 ft, and less than 6 in. thick, may require greater warping and bowing tolerance than indicated above.

Fig. 8.2.2 Definition of panel bowing

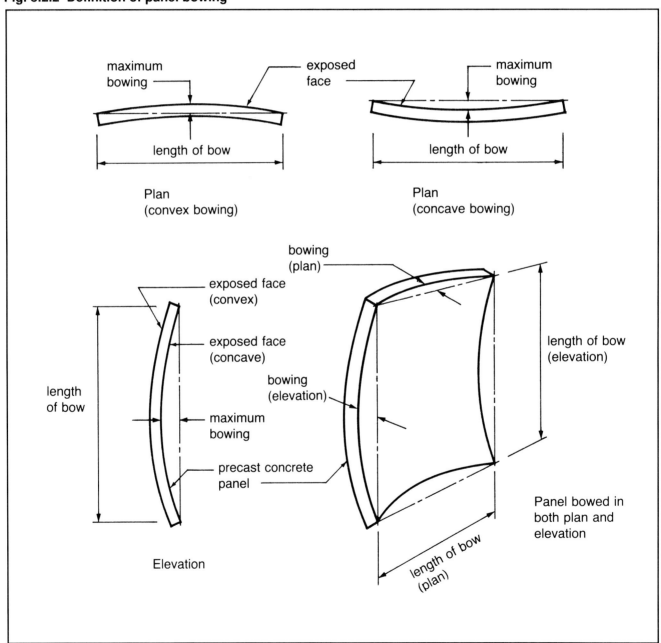

Table 8.2.4 Panel thickness values to maintain bowing and warping within suggested normal tolerances*, in.**

Panel dimensions, ft	8	10	12	16	20	24	28	32
4	3	4	4	5	5	6	6	7
6	3	4	4	5	6	6	6	7
8	4	5	5	6	6	7	7	8
10	5	5	6	6	7	7	8	8

*This table should not be used for panel thickness selection.
**For ribbed panels, the overall thickness of ribs may be used for comparison with this table if the ribs are continuous from one end of the panel to the other.

Similarly, panels made from concrete with over ¾ in. aggregate, panels using two significantly different concrete mixes, and veneered and insulated panels may require special consideration. In all cases, the local precaster should be consulted regarding overall economic and construction feasibility.

8.2.9 Smoothness

Local smoothness describes the condition where small areas of the surface may be out of plane, as shown in Fig. 8.2.3. The tolerance for this type of variation is ¼ in. per 10 ft for all products. The tolerance is usually checked with a 10 ft straight edge or the equivalent, as explained in Fig. 8.2.3.

8.2.10 Architectural Panels vs. Structural Walls

When discussing tolerances, "architectural panel" refers to a class of tolerances specified, and not necessarily to the use of the member in the final structure. Architectural panels usually require more rigid tolerances than structural members for aesthetic reasons.

Double tees, hollow-core slabs and solid slabs are often used for exterior facades, but are not classed as "architectural panels". Since the above listed products are manufactured the same whether they are architectural or structural products, the manufacturing accuracy for the product when used architecturally should not be expected to meet "architectural panel" tolerances. If more rigid tolerances are required, they should be clearly indicated in the contract documents, and subsequent higher costs anticipated.

8.3 Erection Tolerances

8.3.1 General

Erection tolerance values are those to which the primary control surfaces of the member should be set. The final location of other features and surfaces

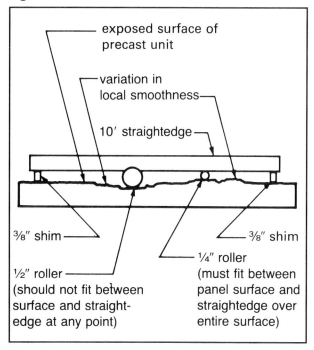

Fig. 8.2.3 Local smoothness variation

will be the result of the combination of the erection tolerances and the product tolerances given in Sect. 8.2.

Because erection is equipment and site dependent, there may be good reason to vary some of the recommended tolerances to account for unique project conditions. Combining liberal product tolerances with restrictive erection tolerances may place an unrealistic burden on the erector. Thus, the designer should review proposed tolerances with manufacturers and erectors prior to deciding on the final project tolerances.

To minimize erection problems, the dimensions of the in-place structure should be checked prior to starting precast erection. After erection, and before other trades interface with the precast concrete members, it should be verified that the precast elements are erected within tolerances.

8.3.2 Recommended Erection Tolerances

Figs. 8.3.1 to 8.3.5 show erection tolerances for:

- Precast element to precast element
- Precast element to cast-in-place concrete
- Precast element to masonry
- Precast element to structural steel construction

8.3.3 Mixed Building Systems

Mixed building systems combine precast and prestressed concrete with other materials, usually cast-in-place concrete, masonry or steel. Each in-

Fig. 8.3.1 Erection tolerances—beams and spandrels

a	=	Plan location from building grid datum	± 1 in.
a_1	=	Plan location from centerline of steel*	± 1 in.
b	=	Bearing elevation** from nominal elevation at support	
		Maximum low	½ in.
		Maximum high	¼ in.
c	=	Maximum plumb variation over height of element	
		Per 12 in. height	⅛ in.
		Maximum	½ in.
d	=	Maximum jog in alignment of matching edges	
		Architectural exposed edges	¼ in.
		Visually non-critical edges	½ in.
e	=	Joint width	
		Architectural exposed joints	± ¼ in.
		Hidden joints	± ¾ in.
		Exposed structural joint *not* visually critical	± ½ in.
f	=	Bearing length*** (span direction)	± ¾ in.
g	=	Bearing width***	± ½ in.

*For precast elements erected on a steel frame, this tolerance takes precedence over tolerance dimension "a".

**Or member top elevation where member is part of a frame without bearings.

***This is a setting tolerance and should not be confused with structural performance requirements set by the architect/engineer. The nominal bearing dimensions and the allowable variations in bearing length and width should be specified by the engineer and shown on the contract drawings.

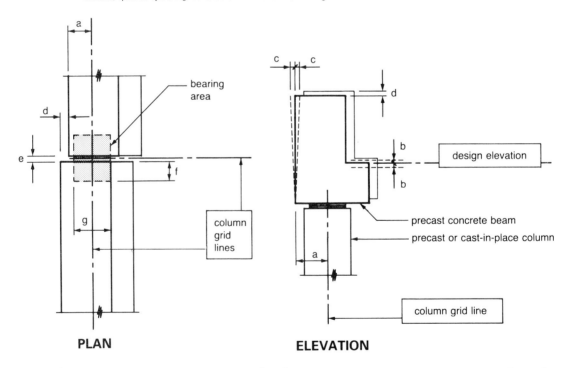

PLAN   **ELEVATION**

Precast element to precast concrete, cast-in-place concrete, masonry, or structural steel

Fig. 8.3.2 Erection tolerances—floor and roof members

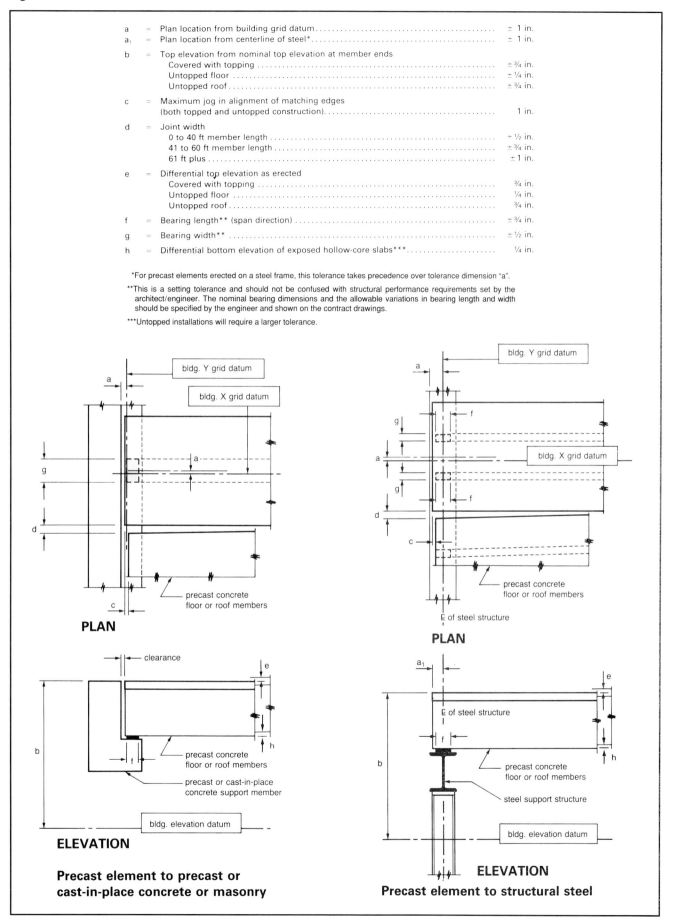

Fig. 8.3.3 Erection tolerances—columns

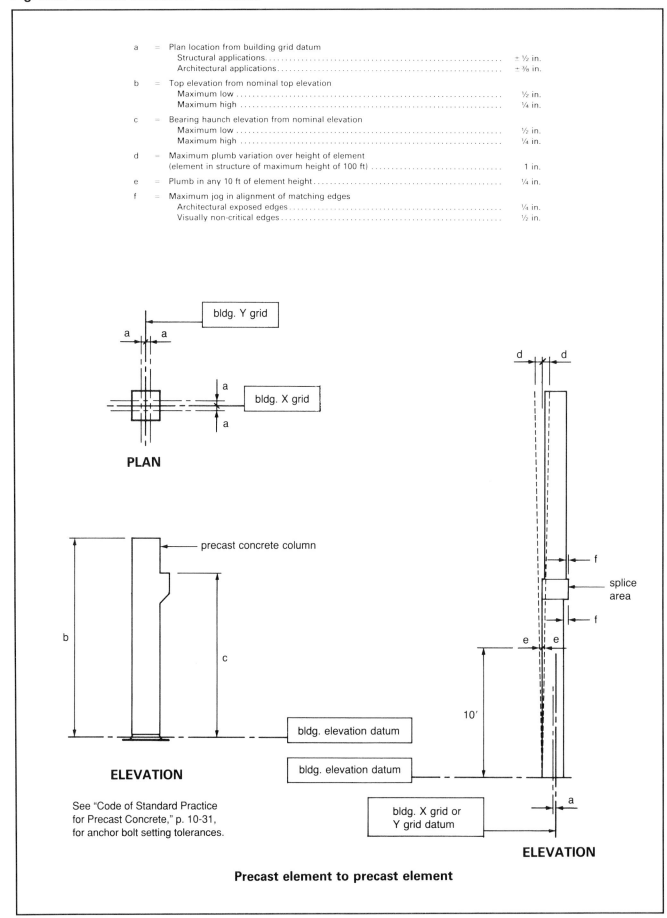

Fig. 8.3.4 Erection tolerances—structural wall panels

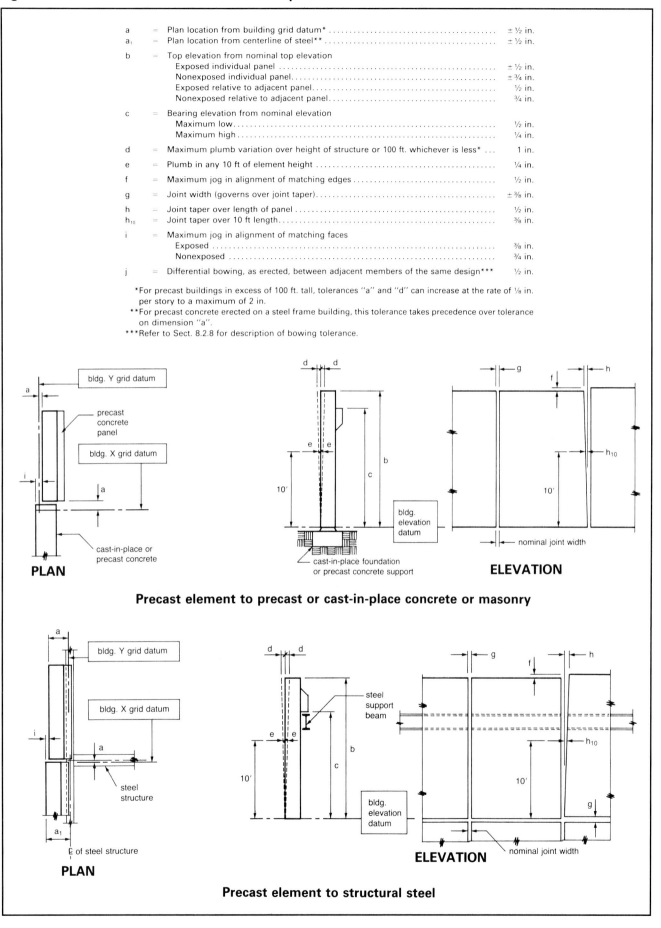

Fig. 8.3.5 Erection tolerances—architectural wall panels

a	= Plan location from building grid datum*	± ½ in.
a_1	= Plan location from centerline of steel**	± ½ in.
b	= Top elevation from nominal top elevation	
	Exposed individual panel	± ¼ in.
	Nonexposed individual panel	± ½ in.
	Exposed relative to adjacent panel	¼ in.
	Nonexposed relative to adjacent panel	½ in.
c	= Support elevation from nominal elevation	
	Maximum low	½ in.
	Maximum high	¼ in.
d	= Maximum plumb variation over height of structure or 100 ft. whichever is less*	1 in.
e	= Plumb in any 10 ft of element height	¼ in.
f	= Maximum jog in alignment of matching edges	¼ in.
g	= Joint width (governs over joint taper)	± ¼ in.
h	= Joint taper maximum	⅜ in.
h_{10}	= Joint taper over 10 ft. length	¼ in.
i	= Maximum jog in alignment of matching faces	¼ in.
j	= Differential bowing or camber as erected between adjacent members of the same design***	¼ in.

*For precast buildings in excess of 100 ft. tall, tolerances "a" and "d" can increase at the rate of ⅛ in. per story to a maximum of 2 in.

**For precast concrete erected on a steel frame building, this tolerance takes precedence over tolerance on dimension "a".

***Refer to Sect. 8.2.8 for description of bowing tolerance.

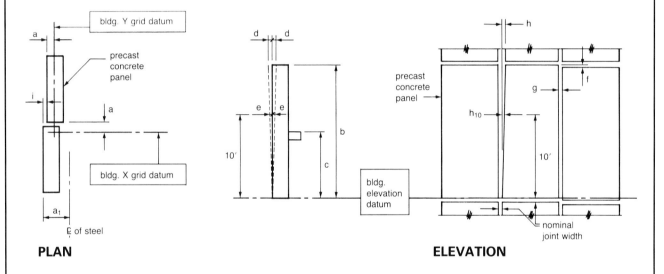

Precast element to precast or cast-in-place concrete, masonry, or structural steel

dustry has its own recommended erection tolerances which apply when its products are used exclusively. The compatibility of those tolerances with the precast tolerances should be checked and adjusted when necessary.

Example 8.4.2 shows one problem that can occur when erection tolerances are chosen for each system without considering the project as a whole.

8.3.4 Connections and Bearing

The details of connections must be considered when specifying erection tolerances. Space must be provided to make the connection under the most adverse combination of tolerances.

Bearing length is measured in the direction of the span, and bearing width is measured perpendicular to the span. Bearing length is often not the same as the length of the end of a member over the support, as shown in Fig. 8.3.6. When they differ, it should be noted on erection drawings.

The Engineer may wish to specify a minimum bearing for various precast products. For further information, see Ref. 4.

8.4 Clearances

8.4.1 General

Clearance is the space between adjacent members and provides a buffer area where erection and production tolerance variations can be absorbed. The following items should be addressed when determining the appropriate clearance to provide in the design:

- Product tolerance
- Type of member
- Size of member
- Location of member
- Member movement
- Function of member
- Erection tolerance
- Fireproofing of steel
- Thickness of plates, bolt heads, and other projecting elements.

Of these factors, product tolerances and member movement are the most significant. As shown in the examples, it may not always be practical to account for all possible factors in the clearance provided.

8.4.2 Joint Clearance

Joints between architectural panels must accommodate variations in the panel dimensions and the erection tolerances for the panels. They must also

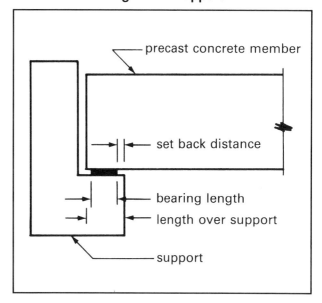

Fig. 8.3.6 Relationship between bearing length and length over support

provide a good visual line and sufficient width to allow for effective sealing. Generally, the larger the panel, the wider the joints should be. For most situations, architectural panel joints should be designed to be not less than ¾ in. wide. Tolerances in overall building width and length are normally accommodated in panel joints.

8.4.3 Procedure for Determining Clearance

Following is a systematic approach to selecting an appropriate clearance:

Step 1:
Determine the maximum size of the members involved (basic or nominal dimension plus additive tolerances). This should include not only the precast and prestressed members, but also other materials.

Step 2:
Add the minimum space required for member movement.

Step 3:
Check if this clearance allows the member to be erected within the erection and interfacing tolerances, such as plumbness, face alignment, etc. Adjust the clearance as required to meet all the needs.

Step 4:
Check if the member can physically be erected with this clearance. Consider the size and location of members in the structure and how connections will be made. Adjust the clearance as required.

Fig. 8.4.1 Clearance Example 8.4.1

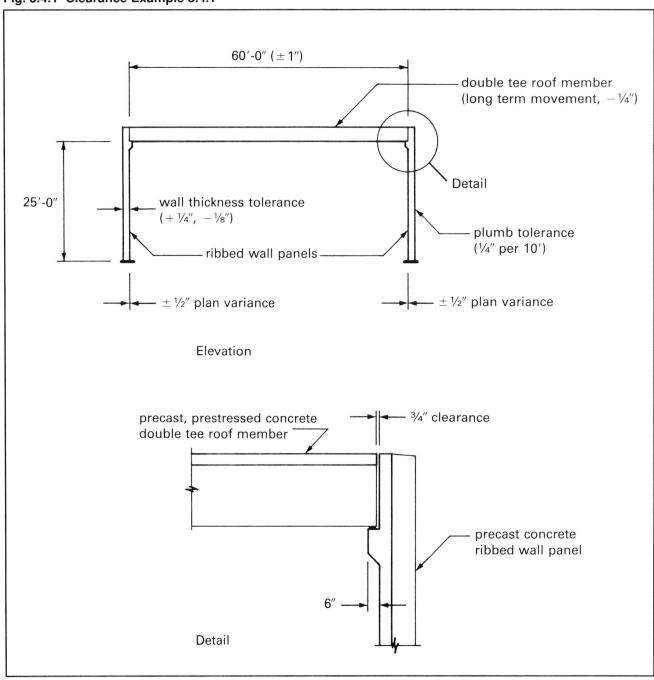

Step 5:
Review the clearance to see whether increasing its dimensions will allow easier, more economical erection without adversely affecting aesthetics. Adjust the clearance as required.

Step 6:
Review structural considerations such as types of connections involved, sizes required, bearing area requirements, and other structural issues.

Step 7:
Check design to assure adequacy in the event that minimum member size should occur. Adjust clearance as required for minimum bearing and other structural considerations.

Step 8:
Select final clearance.

8.4.4 Clearance Examples

The following examples are given to show the thought process, and may not be the only correct solutions for the situations described.

Example 8.4.1 Clearance determination—single story industrial building

Given:

A double tee roof member 60 ft long, ± 1 in. length tolerance, bearing on ribbed wall members 25 ft high, maximum plan variance ± ½ in., variation from plumb ¼ in. per 10 ft, haunch depth 6 in. beyond face of panel, long term roof movement –¼ in. Refer to Fig. 8.4.1.

Problem:

Find the minimum acceptable clearance.

Solution:

Step 1: Determine Maximum Member Sizes (refer to product tolerances)

Maximum tee length	+ 1 in.
Wall thickness	+ ¼ in.
Initial clearance chosen	¾ in. per end

Step 2: Member Movement

Long term shrinkage and creep will increase the clearance so this movement can be neglected in the initial clearance determination, although it must be considered structurally.

Required clearance adjustment as a result of member movement	0
Clearance chosen	¾ in.

Step 3: Other Erection Tolerances

If the wall panel is set inward toward the building interior ½ in. and erected plumb, the clearance should be increased by ½ in. If the panel is erected out of plumb outward ½ in., no clearance adjustment is needed.

Clearance adjustment required for erection tolerances	0
Clearance chosen	¾ in.

Step 4: Erection Considerations

If all members are fabricated perfectly, then the joint clearance is ¾ in. at either end (1½ in. total). This is ample space for erection. If all members are at maximum size variance, maximum inward plan variance, and maximum inward variance from plumb, then the total clearance is zero. This is undesirable as it would require some rework during erection. A judgment should be made as to the likelihood of maximum product tolerances all occurring in one location. If the likelihood is low, the ¾ in. clearance needs no adjustment, but, if the likelihood is high, the engineer might increase the clearance to 1 in. In this instance, the likelihood has been judged low; therefore, no adjustment has been made.

Clearance chosen	¾ in.

Step 5: Economy

In single-story construction, increasing the clearance beyond ¾ in. is not likely to speed up erection as long as product tolerances remain within allowables. No adjustment is required for economic considerations.

Step 6: Review Structural Considerations

Allowing a setback from the edge of the corbel, assumed in this instance to have been set by the engineer at 1 in., plus the clearance, the bearing is 4 in. and there should be space to allow member movement. The engineer judges this to be acceptable from a structural and architectural point of view and no adjustment is required for structural considerations.

Step 7: Check for Minimum Member Sizes (refer to product tolerances)

Tee length	–1 in. (½ in. each end)
Wall thickness	–⅛ in.
Bearing haunch	No change
Clearance chosen	¾ in.
Minimum bearing without setback	4⅝ in.

OK in this instance.

Wall plumbness would also be considered in an actual application.

Step 8: Final Solution

Minimum clearance used	¾ in. per end

Satisfies all conditions considered.

(Note: For simplicity in this example, end rotation, flange skew, and global skew tolerances have not been considered. In an actual situation, these issues should also be considered.)

Example 8.4.2 Clearance determination—high rise frame structure

Given:

A 36-story steel frame structure, precast concrete cladding, steel tolerances per AISC, member movement negligible. In this example, precast tolerance for variation in plan is ± ¼ in. Refer to Fig. 8.4.2.

Problem:

Determine whether or not the panels can be erected plumb and determine the minimum acceptable clearance at the 36th story.

Fig. 8.4.2 Clearance Example 8.4.2

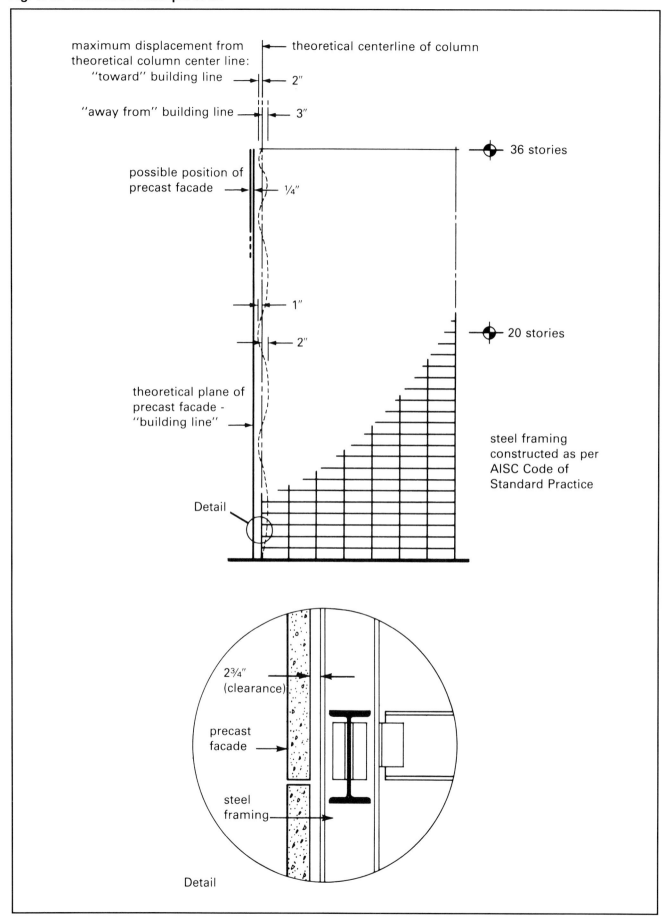

Solution:

Step 1: Product Tolerances
(refer to product tolerances)

Precast cladding thickness	+ 1/4 in., – 1/8 in.
Steel width	+ 1/4 in., – 3/16 in.
Steel sweep (varies)	1/4 in. assumption
Initial clearance chosen	3/4 in.

Step 2: Member Movement

For simplification, assume this can be neglected in this example.

Step 3: Other Erection Tolerances

Steel variance in plan, maximum	2 in.
Initial clearance	3/4 in.
Clearance chosen	2 3/4 in.

Step 4: Erection Considerations

No adjustment required for erection considerations.

Step 5: Economy

Clearance chosen	2 3/4 in.

Increasing clearance will not increase economy. No adjustment for economic considerations.

Step 6: Structural Considerations

Clearance chosen	2 3/4 in.

Expensive connection but possible. No adjustment.

Step 7: Check Minimum Member Sizes at 36th Story (refer to product tolerances)

Initial clearance	2 3/4 in.
Precast thickness	– 1/8 in.
Steel width	– 3/16 in.
Steel sweep	– 1/4 in.
Steel variance in plan minimum	– 3 in.
Clearance calculated	6 5/16 in.

Step 8: Final Solution

A clearance of over 6 in. would require an extremely expensive connection for the precast panel, and would produce high torsional stresses in the steel supporting beams. The 6 in. clearance is not practical, although the 2 3/4 in. minimum initial clearance is still needed. Either the precast panels should be allowed to follow the steel frame or the tolerances for the exterior columns need to be made more stringent, such as the AISC requirements for elevator columns. The most economical will likely be for the panels to follow the steel frame.

Minimum clearance used	2 3/4 in.
Allow panels to follow the steel frame.	

8.5 Interfacing Tolerances

8.5.1 General

In interfacing with other materials, the tolerances may be very system-dependent. For example, different brands of windows may have different tolerance requirements. If substitutions are made after the initial design is complete, the interface details must be reviewed for the new system.

Following is a partial checklist for consideration in determining interfacing requirements:

- Structural requirements
- Volume change
- Weathering and corrosion protection
- Waterproofing
- Drainage
- Architectural requirements
- Dimensional considerations
- Vibration considerations
- Fire-rating requirements
- Acoustical considerations
- Economics
- Manufacturing/erection considerations

8.5.2 Interface Design Approach

The following approach is one method of organizing the task of designing the interface between two systems:

Step 1:

Define the interface between the two systems, show shape and location, and determine contractual responsibility. For example, the precast panel furnished by the precaster, the window furnished and installed by the window manufacturer, and the sealant between the window and the precast concrete furnished and installed by the precast manufacturer.

Step 2:

Define the functional requirements of each interfacing system. For example, the building drain line must have a flow line which allows adequate drainage. This will place limits on where the line must penetrate precast units. Note whether this creates problems such as conflict with prestressing strands.

Step 3:

List the tolerances of each interfacing system. For example, determine from manufacturer's specifications what are the external tolerances on the door jamb. Determine from the precast concrete product tolerances what is the tolerance on a large panel door blockout. For the door installation, determine what is the floor surface tolerance in the area of the door.

Step 4:

Select the operational clearances required. For example, determine the magnitude of operational clearances which are needed to align the door to function properly. Then choose dimensions which include necessary clearances.

Step 5:

Check compatibility of the interface tolerances. Starting with the least precise system, check the tolerance requirements and compare against the minimum and maximum dimensions of the interfacing system. If interferences result, alter the nominal dimension of the appropriate system. For example, it is usually more economical to make a larger opening than to specify a non-standard window size.

Step 6:

Establish assembly and installation procedures for the interfacing systems to assure compatibility. Show the preferred adjustments to accommodate the tolerances of the systems. Specify such things as minimum bearing areas, minimum and maximum joint gaps, and other dimensions which will vary as a result of interface tolerances. Consider economic trade-offs such as in-plant work vs. field work, and minor fit-up rework vs. tighter tolerances.

Step 7:

Check the final project specifications as they relate to interfacing. Be aware of subsystem substitutions which might be made during the final bidding and procurement.

8.5.3 Characteristics of the Interface

The following list of questions will help to define the nature of the interface:

1. What specifically is to be interfaced?
2. How does the interface function?
3. Is there provision for adjustment upon installation?
4. How much adjustment can occur without rework?
5. What are the consequences of an interface tolerance mismatch?
    —Rework requirements (labor and material)
    —Rejection limits
6. What are the high material cost elements of the interface?
7. What are the high labor cost elements of the interface?
8. What are the normal tolerances associated with the system to interfaced?
9. Are the system interface tolerances simple planar tolerances or are they more complex and three dimensional?
10. Do all of the different products of this type have the same interface tolerance requirements?
11. Do you as the designer of the precast system have control over all the aspects of the interfaces involved? If not, what actions do you need to take to accommodate this fact?

Listed below are common characteristics of most systems of the type listed:

1. Windows and Doors
    —No gravity load transfer through window element
    —Compatible with air and moisture sealant system
    —Open/close characteristics (swing or slide)
    —Compatibility with door locking mechanisms

2. Mechanical Equipment
    —Duct clearances for complex prefabricated ductwork
    —Large diameter prefabricated pipe clearance requirements
    —Deflections from forces associated with large diameter piping and valves
    —Expansion/contraction allowances for hot/cold piping
    —Vibration isolation/transfer considerations
    —Acoustical shielding considerations
    —Hazardous gas/fluid containment requirements

3. Electrical Equipment
    —Multiple mating conduit runs
    —Prefabricated cable trays
    —Embedded conduits and outlet boxes
    —Corrosion related to DC power
    —Special insert placement requirements for isolation
    —Location requirements for embedded grounding cables
    —Shielding clearance requirements for special "clean" electrical lines

4. Elevators and Escalators
    —Elevator guide location requirements
    —Electrical conduit location requirements
    —Elevator door mechanism clearances
    —Special insert placement requirements

5. Architectural Cladding
    —Joint tolerance for sealant system

- —Flashing and reglet fit-up. (Lining up reglets from panel to panel is very difficult and often costly. Surface-mounted flashing should be considered.)
- —Expansion and contraction provisions for dissimilar materials
- —Effects of rotation, deflection, and differential thermal gradients

6. Structural Steel and Miscellaneous Steel
   - —Details to prevent rust staining of concrete
   - —Details to minimize potential for corrosion at field connections between steel and precast concrete
   - —Coordination of structural steel expansion/contraction provisions with those of the precast system
   - —Special provisions for weld plates or other attachment features for steel structures

7. Masonry
   - —Coordination of masonry expansion/contraction provisions with those of the precast system
   - —Detailing to ensure desired contact bearing between masonry and precast units
   - —Detailing to ensure desired transfer of load between masonry shear wall and precast frame

8. Roofing
   - —Roof camber, both upon erection and long term, as it relates to roof drain placement
   - —Fit-up of prefabricated flashing
   - —Dimensional effects of added material during reroofing
   - —Coordination of structural control joint locations with roofing system expansion/contraction provisions
   - —Location of embedded HVAC unit supports

9. Waterproofing
   - —Location and dimensions of flashing reglets
   - —Location and shape of window gasket grooves
   - —Coordination of waterproofing system requirements with structural system expansion/provisions
   - —Special details around special penetrations

10. Interior Finishes—Floors, Walls, and Ceilings
    - —Joints between precast members for direct carpet overlay
    - —Visual appearance of joints for exposed ceilings
    - —Fit-up details to assure good appearance of interior corners
    - —Appearance of cast-in-place to precast concrete interfaces

11. Interior Walls and Partitions
    - —Clearance for prefabricated cabinetry
    - —Interfacing of mating embedded conduit runs

8.6 References

1. PCI Committee on Tolerances, "Tolerances for Precast and Prestressed Concrete," *PCI Journal*, V. 30, No. 1, January-February, 1985.

2. "Manual for Quality Control for Plants and Production of Precast and Prestressed Concrete Products," Third Edition, MNL-116-85, Precast/Prestressed Concrete Institute, Chicago, IL, 1985.

3. "Manual for Quality Control for Plants and Production of Architectural Precast Concrete Products," MNL-117-77, Precast/Prestressed Concrete Institute, Chicago, IL, 1977.

4. "Proposed Design Requirements for Precast Concrete," *PCI Journal*, V. 31, No. 6, November-December, 1986.

CHAPTER 9
THERMAL, ACOUSTICAL, FIRE AND OTHER CONSIDERATIONS

	Page No.
9.1 Thermal Properties of Precast Concrete	9-3
9.1.1 Notation	9-3
9.1.2 Glossary	9-3
9.1.3 General	9-4
9.1.4 Thermal Properties of Materials, Surfaces, and Air Spaces	9-4
9.1.5 Computation of Thermal Transmittance Values	9-5
9.1.6 Thermal Storage Effects	9-13
9.1.7 Condensation Control	9-15
9.1.8 Thermal Bridges	9-19
9.1.9 References	9-19
9.2 Acoustical Properties of Precast Concrete	9-20
9.2.1 Glossary	9-20
9.2.2 General	9-20
9.2.3 Approaching the Design Process	9-20
9.2.4 Dealing with Sound Levels	9-21
9.2.5 Sound Transmission Loss	9-21
9.2.6 Impact Noise Reduction	9-21
9.2.7 Absorption of Sound	9-23
9.2.8 Acceptable Noise Criteria	9-25
9.2.9 Establishment of Noise Insulation Objectives	9-28
9.2.10 Composite Wall Considerations	9-29
9.2.11 Leaks and Flanking	9-30
9.2.12 References	9-31
9.3 Fire Resistance	9-33
9.3.1 Notation	9-33
9.3.2 Glossary	9-33
9.3.3 Introduction	9-33
9.3.4 Standard Fire Tests	9-34
9.3.5 Fire Tests of Prestressed Concrete Assemblies	9-35
9.3.6 Designing for Heat Transmission	9-36
9.3.7 Designing for Structural Integrity	9-42
9.3.8 Protection of Connections	9-50
9.3.9 Precast Concrete Column Covers	9-50
9.3.10 Code and Economic Considerations	9-52
9.3.11 References	9-52

Page No.

9.4 Sandwich Panels 9-53
    9.4.1   General 9-53
    9.4.2   Structural Design 9-53
    9.4.3   Connections 9-54
    9.4.4   Insulation 9-56
    9.4.5   Thermal Bowing 9-56
    9.4.6   Typical Details 9-57

9.5 Quality Control 9-58
    9.5.1   Introduction 9-58
    9.5.2   Plant Certification 9-58
    9.5.3   PCI Quality Control Manuals 9-58
    9.5.4   References 9-58

9.6 Concrete Coatings and Joint Sealants 9-59
    9.6.1   Coatings for Horizontal Deck Surfaces 9-59
    9.6.2   Clear Surface Sealers for Architectural Precast Panels 9-59
    9.6.3   Joint Sealants 9-59
    9.6.4   References 9-60

9.7 Vibration in Concrete Structures 9-61
    9.7.1   Human Response to Building Vibrations 9-61
    9.7.2   Vibration Isolation for Mechanical Equipment 9-62
    9.7.3   References 9-63

9.8 Cracking, Repair and Maintenance 9-64
    9.8.1   Cracking 9-64
    9.8.2   Repair and Maintenance 9-64
    9.8.3   References 9-64

9.9 Precast Segmental Construction 9-65
    9.9.1   General 9-65
    9.9.2   Joints 9-65
    9.9.3   Design 9-66
    9.9.4   Post-Tensioning 9-66
    9.9.5   References 9-66

9.10 Coordination with Mechanical, Electrical and Other Sub-Systems 9-68
    9.10.1   Introduction 9-68
    9.10.2   Lighting and Power Distribution 9-68
    9.10.3   Electrified Floors 9-68
    9.10.4   Ductwork 9-68
    9.10.5   Other Sub-Systems 9-70
    9.10.6   Systems Building 9-71

9.1 THERMAL PROPERTIES OF PRECAST CONCRETE

9.1.1 Notation

a	=	thermal conductance of an air space
Btu	=	British thermal unit
C	=	thermal conductance for the specified thickness, Btu/(hr)(ft²)(deg. F)
d	=	clear cover to ends of metal ties in sandwich panels
D	=	heating degree day (65°F base)
f	=	film or surface conductance
k	=	thermal conductivity, (Btu-in.)/(hr)(ft²)(deg. F)
ℓ	=	thickness of a material, in.
m	=	width or diameter of metal ties in sandwich panels
M	=	permeance, perms
R	=	thermal resistance, (hr)(ft²)(deg. F)/Btu
R_a	=	thermal resistance of air space
R_{fi}, R_{fo}	=	thermal resistances of inside and outside surfaces
R_t	=	total thermal resistance
$R_{1,2,...,n}$	=	thermal resistance of material layer 1, 2,...,n
$R_{materials}$	=	summation of thermal resistance of opaque material layers
R.H.	=	relative humidity, %
t_i	=	indoor temperature, deg F
t_o	=	outdoor temperature, deg F
t_s	=	dew-point temperature, room air, at design maximum relative humidity, deg. F
U	=	heat transmittance value [all heat and average transmittance values are Btu/(hr)(ft²)(deg F)]
$U_{a,b,...,n}$	=	heat transmittance values of parallel conductance paths in sandwich panels
U_o	=	average heat transmittance value
W	=	width of Zone A in sandwich panels, Sect. 9.1.8
μ	=	permeability, permeance of a unit thickness, given material, perm-in.

9.1.2 Glossary

British thermal unit (Btu)—Approximately the amount of heat to raise one pound of water from 59°F to 60°F.

Degree day (D)—A unit, based on temperature difference and time, used in estimating fuel consumption and specifying nominal heating load of a building in winter. For any one day, when the mean temperature is less than 65°F, there are as many degree days as there are degrees F difference in temperature between the mean temperature for the day and 65°F.

Dew-point temperature (t_s)—The temperature at which condensation of water vapor begins for a given humidity and pressure as the vapor temperature is reduced. The temperature corresponding to saturation (100% R.H.) for a given absolute humidity at constant pressure.

Film or surface conductance (f)—The time rate of heat exchange by radiation, conduction, and convection of a unit area of a surface with its surroundings. Its value is usually expressed in Btu per (hr)(sq ft of surface area)(deg F temperature difference). Subscripts "i" and "o" are usually used to denote inside and outside surface conductances, respectively.

Heat capacity (H_c)—The amount of heat necessary to raise the temperature of a given mass one degree. Numerically, the mass multiplied by the specific heat.

Heat transmittance (U)—Overall coefficient of heat transmission or thermal transmittance (air-to-air); the time rate of heat flow usually expressed in Btu per (hr) (sq ft of surface area)(deg F temperature difference between air on the inside and air on the outside of a wall, floor, roof, or ceiling). The term is applied to the usual combinations of materials and also single materials such as window glass, and includes the surface conductance on both sides. This term is frequently called the U-value.

Perm—A unit of permeance. A perm is 1 grain per (sq ft of area)(hr)(in. of mercury vapor pressure difference).

Permeability, water vapor (μ)—The property of a substance which permits the passage of water vapor. It is equal to the permeance of 1 in. of a substance. Permeability is measured in perm-inches. The permeability of a material varies with barometric pressure, temperature and relative humidity conditions.

Permeance (M)—The water vapor permeance of any sheet or assembly is the ratio of the water vapor flow per unit area per hour to the vapor pressure difference between the two surfaces. Permeance is measured in perms. Two commonly used test methods are the Wet Cup and Dry Cup Tests. Specimens are sealed over the tops of cups containing either water or desiccant, placed in a controlled atmosphere, usually at 50% R.H., and weight changes measured.

Relative humidity (R.H.)—The ratio of water vapor present in air to the water vapor present in saturated air at the same temperature and pressure.

Thermal conductance (C)—The time rate of heat flow expressed in Btu per (hr) (sq ft of area) (deg F average temperature difference between two surfaces). The term is applied to specific materials as used, either homogeneous or heterogeneous for the thickness of construction stated, not per in. of thickness.

Thermal conductance of an air space (a)—The time rate of heat flow through a unit area of an air space per unit temperature difference between the boundary surfaces. Its value is usually expressed in Btu per (hr)(sq ft of area)(deg F).

Thermal conductivity (k)—The time rate of heat flow by conduction only through a unit thickness of a homogeneous material under steady-state conditions per unit area per unit temperature gradient in the direction perpendicular to the isothermal surface. Its unit is (Btu-in.) per (hr)(sq ft of area) (deg F).

Thermal mass—Characteristic of materials with mass heat capacity and surface area capable of affecting building loads by storing and releasing heat as the interior and/or exterior temperature and radiant conditions fluctuate.

Thermal resistance (R)—The reciprocal of a heat transmission coefficient, as expressed by U, C, f, or a. Its unit is (deg F)(hr)(sq ft of area) per Btu. For example, a wall with a U-value of 0.25 would have a resistance value of R = 1/U = 1/0.25 = 4.0.

9.1.3 General

Thermal codes and standards specify the heat transmission requirements for buildings in many different ways. Prescriptive standards specify U or R values for each building component, whereas with performance standards, two buildings are equivalent if they use the same amount of energy, regardless of the U or R values of the components. This allows the designer to choose conservation strategies that provide the required performance at the least cost.

In ASHRAE Standard 90.1-1989[2], these two approaches are used to determine the maximum allowable U_{ow} or U_{or} for a wall or roof assembly:

1. Prescriptive Criteria (Sect. 8.5)—Table of Alternate Component Packages provides a limited number of complying combinations of building variables for a set of climate variable ranges. For most climate locations and building assemblies, the Prescriptive Criteria may be slightly more stringent than the System Performance Criteria.

2. System Performance Criteria (Sect 8.6)—mathematical models and computer programs based on these models provide a system approach (criteria and compliance values) to comply with the envelope requirements of the Standard. It provides more flexibility than the prescriptive approach, but it requires more analysis.

Precast and prestressed concrete construction, with its thermal inertia and thermal storage properties, has an advantage over lightweight materials. Procedures to account for the benefits of heavier materials are presented in ASHRAE Standard 90.1-1989.

The trend is toward more insulation with little regard given to its total impact or energy used. Mass effects, glass area, air infiltration, ventilation, building orientation, exterior color, shading or reflections from adjacent structures, surrounding surfaces or vegetation, building aspect ratio, number of stories, wind direction and speed, all have an effect on insulation requirements.

This section is condensed from a more complete treatment given in Ref. 4. Except where noted, the information and design criteria are taken or derived from the ASHRAE Handbook[1], and from the ASHRAE Standard 90.1-1989.[2] All design criteria are not given in this section, and the criteria may change as the ASHRAE Standard and Handbook are revised. Local codes and latest references must be used for specified values and procedures.

9.1.4 Thermal Properties of Materials, Surfaces, and Air Spaces

The thermal properties of materials and air spaces are based on steady state tests, which measure the heat that passes from the warm side to the cool side of the test specimen. The tests determine the conductivity, k, or, for non-homogeneous sections, compound sections and air spaces, the conductance, C, for the total thickness. The values of k and C do not include surface conductances, f_i and f_o.

The overall thermal resistance of wall, floor, and roof sections is the sum of the resistances, R (reciprocal of k, C, f_i, and f_o). The R-values of construction materials are not influenced by the direction of heat flow, but the R-values of surfaces and air spaces differ depending on whether they are vertical, sloping, or horizontal. Also, the R-values of surfaces are affected by the velocity of air at the surfaces and by their reflective properties.

Tables 9.1.1 and 9.1.2 give the thermal resistances of surfaces and 3½ in. air spaces. Table 9.1.3 gives the thermal properties of most commonly used building materials. Only U-values are given for glass because the surface resistances and air space between panes account for nearly all of the U-value. Table 9.1.4 gives the thermal properties of various weight concretes and some standard precast, prestressed concrete products in the "normally dry" condition. Normally dry is the condition of concrete containing an equilibrium amount of free water after extended exposure to warm air at 35 to 50% R.H.

Thermal conductances and resistances of other building materials are usually reported for oven dry conditions. Normally dry concrete in combination with insulation generally provides about the same R-value as equally insulated oven dry concrete, but because of the moisture content, has the ability to store a greater amount of heat than oven dry concrete. However, higher moisture content in concrete causes higher thermal conductance.

9.1.5 Computation of Thermal Transmittance Values

The heat transmittance (U-values) of a building wall, floor or roof is computed by adding together the R-values of the materials in the section, the surfaces (R_{fi} and R_{fo}) and air spaces (R_a) within the section. The reciprocal of the R's is the U-value:

$$U = \frac{1}{R_{fi} + R_{materials} + R_a + R_{fo}} \quad \text{(Eq. 9.1.1)}$$

where $R_{materials}$ is the sum of all opaque materials in the wall. A number of typical wall and roof U-values are given in Tables 9.1.5, 9.1.6, and 9.1.7.

The following examples show use of Tables 9.1.1 through 9.1.4 in calculating R- and U-values for wall and roof assemblies:

Example 9.1.1—Wall

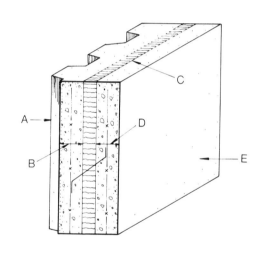

	R Winter	R Summer	Table
A. Surface, outside	0.17	0.25	9.1.1
B. Concrete, 2 in. (110 pcf)	0.38	0.38	9.1.4
C. Polystyrene, 1½ in.	6.00	6.00	9.1.3
D. Concrete, 2½ in. (110 pcf)	0.48	0.48	9.1.4
E. Surface, inside	0.68	0.68	9.1.1
Total R =	7.71	7.79	
U = 1/R =	0.13	0.13	

Example 9.1.2—Roof

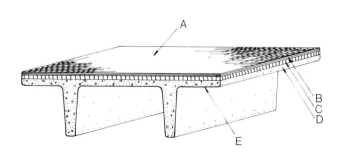

	R Winter	R Summer	Table
A. Surface, outside	0.17	0.25	9.1.1
B. Roofing, built-up	0.33	0.33	9.1.3
C. Polystyrene, 2 in.	8.00	8.00	9.1.3
D. Concrete, 2 in. (145 pcf)	0.15	0.15	9.1.4
E. Surface, inside	0.61	0.92	9.1.1
Total R =	9.26	9.65	
U = 1/R =	0.11	0.10	

Table 9.1.1 Thermal resistances, R_f, of surfaces

Position of surface	Direction of heat flow	Still air, R_{fi}			Moving air, R_{fo}	
		Non reflective surface	Reflective surface		Non reflective surface	
			Aluminum painted paper	Bright aluminum foil	15 mph winter design	7½ mph summer design
Vertical	Horizontal	0.68	1.35	1.70	0.17	0.25
Horizontal	Up	0.61	1.10	1.32	0.17	0.25
	Down	0.92	2.70	4.55	0.17	0.25

Table 9.1.2 Thermal resistances, R_a, of air spaces[1]

Position of air space	Direction of heat flow	Air space		Non reflective surfaces	Reflective surfaces		
		Mean temp. °F	Temp. diff. °F		One side[2]	One side[3]	Both sides[3]
Vertical	**Winter** Horizontal (walls)	50 50	10 30	1.01 0.91	2.32 1.89	3.40 2.55	3.69 2.67
	Summer Horizontal (walls)	90	10	0.85	2.15	3.40	3.69
Horizontal	**Winter** Up (roofs)	50 50	10 30	0.93 0.84	1.95 1.58	2.66 2.01	2.80 2.09
	Down (floors)	50	30	1.22	3.86	8.17	9.60
	Summer Down (roofs)	90	10	1.00	3.41	8.19	10.07

1. For 3½ in. air space thickness. The values with the exception of those for reflective surfaces, heat flow down, will differ about 10% for air space thickness of ¾ in. to 16 in. Refer to Table 2, Chapter 22 of the ASHRAE Handbook for values of other thicknesses, reflective surfaces, heat flow down.
2. Aluminum painted paper
3. Bright aluminum foil

Table 9.1.3 Thermal properties of various building materials[1,6]

Material	Unit weight, pcf	Resistance, R		Trans-mittance, U	Specific heat, Btu/(lb)(°F)
		Per inch of thickness, 1/k	For thickness shown, 1/C		
Insulation, rigid					
Cellular glass	8.5	2.86			0.18
Glass fiber, organic bonded	4-9	4.00			0.23
Mineral fiber, resin binder	15	3.45			0.17
Mineral fiberboard, wet felted, roof insulation	16-17	2.94			–
Cement fiber slabs (shredded wood with magnesia oxysulfide binder)	22	1.75			0.31
Expanded polystyrene extruded smooth skin surface	1.8-3.5	5.00			0.29
Expanded polystyrene molded bead	1.0	3.85-4.17			–
Cellular polyurethane	1.5	6.25-5.56			0.38
Miscellaneous					
Acoustical tile (mineral fiberboard, wet felted)	18	2.86			0.19
Carpet, fibrous pad			2.08		0.34
Carpet, rubber pad			1.23		0.33
Floor tile, asphalt, rubber, vinyl			0.05		0.30
Gypsum board	50	0.88[2]			0.26
Particle board	50	1.06			0.31
Plaster					
cement, sand agg.	116	0.20			0.20
gypsum, L.W. agg.	45	0.63[2]			–
gypsum, sand agg.	105	0.18			0.20
Roofing, ⅜ in. built-up	70		0.33		0.35
Wood, hard	38-47	0.94-0.80			0.39
Wood, soft	24-41	1.00-0.89			0.33
Plywood	34	1.25			0.29
Glass doors and windows[3]					
Single, winter				1.11	
Single, summer				1.04	
Double, winter[4]				0.50	
Double, summer[4]				0.61	
Doors, metal[5]					
Insulated				0.40	

1. See Table 9.1.4 for all concretes, including insulating concrete for roof fill.
2. Average value.
3. Does not include correction for sash resistance. Refer to Chapter 27 of the ASHRAE Handbook for sash correction.
4. ¼ in. air space; coating on either glass surface facing air space.
5. Urethane foam core without thermal break.
6. See manufacturers' data for specific values.

Table 9.1.4 Thermal properties of concrete[1]

Description	Concrete weight, pcf	Thickness, in.	Resistance, R		Specific heat,[3] Btu/(lb)(°F)
			Per inch of thickness, 1/k	For thickness shown, 1/C	
Concretes including normal weight, lightweight and lightweight insulation concretes	140 120 100 80 60 40 30 20		0.10-0.05 0.18-0.09 0.27-0.17 0.40-0.29 0.63-0.56 1.08-0.90 1.33-1.10 1.59-1.20		0.19
Normal weight tees[2] and solid slabs	145	2 3 4 5 6 8		0.15 0.23 0.30 0.38 0.45 0.60	0.19
Normal weight hollow-core slabs	145	6 8 10 12		1.07 1.34 1.73 1.91	0.19
Structural lightweight tees[2] and solid slabs	110	2 3 4 5 6 8		0.38 0.57 0.76 0.95 1.14 1.52	0.19
Structural lightweight hollow-core slabs	110	8 12		2.00 2.59	0.19

1. Based on normally dry concrete (see Chapter 4 of Ref. 3).
2. Thickness for tees is thickness of slab portion including topping, if used. The effect of the stems generally is not significant, therefore, their thickness and surface area may be disregarded.
3. The specific heat shown is the mean value from test data compiled in Ref. 4.

Table 9.1.5 Wall U-values: prestressed tees, hollow-core slabs, solid and sandwich panels; winter and summer conditions[1]

Concrete weight, pcf	Type of wall panel	Thickness, t, and resistance, R, of concrete		Winter $R_{fo} = 0.17$, $R_{fi} = 0.68$					Summer $R_{fo} = 0.25$, $R_{fi} = 0.68$				
				Insulation resistance, R									
		t	R	None	4	6	8	10	None	4	6	8	10
145	Solid walls, tees,[2] and sandwich panels	2	0.15	1.00	.20	.14	.11	.09	.93	.20	.14	.11	.09
		3	0.23	.93	.20	.14	.11	.09	.86	.19	.14	.11	.09
		4	0.30	.87	.19	.14	.11	.09	.81	.19	.14	.11	.09
		5	0.38	.81	.19	.19	.11	.09	.76	.19	.14	.11	.09
		6	0.45	.77	.19	.14	.11	.09	.72	.19	.14	.11	.09
		8	0.60	.69	.18	.13	.11	.09	.65	.18	.13	.10	.09
145	Hollow core slabs[3]	6(o)	1.07	.52	.17	.13	.10	.08	.50	.17	.13	.10	.08
		(f)	1.86	.37	.15	.11	.09	.08	.36	.15	.11	.09	.08
		8(o)	1.34	.46	.16	.12	.10	.08	.44	.16	.12	.10	.08
		(f)	3.14	.25	.13	.10	.08	.07	.25	.12	.10	.08	.07
		10(o)	1.73	.39	.15	.12	.09	.08	.38	.15	.12	.09	.08
		(f)	4.05	.20	.11	.09	.08	.07	.20	.11	.09	.08	.07
		12(o)	1.91	.36	.15	.11	.09	.08	.35	.15	.11	.09	.08
		(f)	5.01	.17	.10	.08	.07	.06	.17	.10	.08	.07	.06
110	Solid walls, tees,[2] and sandwich panels	2	0.38	.81	.19	.14	.11	.09	.76	.19	.14	.11	.09
		3	0.57	.70	.18	.13	.11	.09	.67	.18	.13	.11	.09
		4	0.76	.62	.18	.13	.10	.09	.59	.18	.13	.10	.09
		5	0.95	.56	.17	.13	.10	.09	.53	.17	.13	.10	.08
		6	1.14	.50	.17	.13	.10	.08	.48	.16	.12	.10	.08
		8	1.52	.42	.16	.12	.10	.08	.41	.16	.12	.10	.08
110	Hollow core slabs[3]	8(o)	2.00	.35	.15	.11	.09	.08	.34	.14	.11	.09	.08
		(f)	4.41	.19	.11	.09	.08	.07	.19	.11	.09	.07	.07
		12(o)	2.59	.29	.13	.11	.09	.07	.28	.13	.11	.09	.07
		(f)	6.85	.13	.09	.07	.06	.06	.13	.08	.07	.06	.06

1. When insulations having other R-values are used, U-values can be interpolated with adequate accuracy, or U can be calculated as shown in Sect. 9.1.5. When a finish, air space or any other material layer is added, the new U-value is:

$$\frac{1}{\frac{1}{U \text{ from table}} + R \text{ of added finish, air space or material}}$$

2. Thickness for tees is thickness of slab portion. For sandwich panels, t is the sum of the thicknesses of the wythes.
3. For hollow panels (o) and (f) after thickness designates cores open or cores filled with insulation.

Table 9.1.6 Roof U-values: concrete units with built-up roofing, winter conditions, heat flow upward[1]

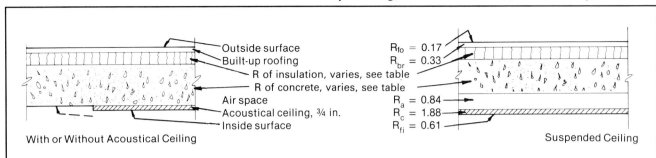

$R_{fo} = 0.17$
$R_{br} = 0.33$
$R_a = 0.84$
$R_c = 1.88$
$R_{fi} = 0.61$

With or Without Acoustical Ceiling — Suspended Ceiling

Concrete weight, pcf	Prestressed concrete member	Thickness, t and resistance, R of concrete		Without ceiling				With ceiling							
								Applied direct				Suspended			
				Top insulation resistance, R											
		t	R	None	4	10	16	None	4	10	16	None	4	10	16
145	Solid slabs and tees[2]	2	0.15	.79	.19	.09	.06	.29	.13	.07	.05	.24	.12	.07	.05
		3	0.23	.75	.19	.09	.06	.29	.13	.07	.05	.23	.12	.07	.05
		4	0.30	.71	.18	.09	.06	.28	.13	.07	.05	.23	.12	.07	.05
		5	0.38	.67	.18	.09	.06	.27	.13	.07	.05	.22	.12	.07	.05
		6	0.45	.64	.18	.09	.06	.27	.13	.07	.05	.22	.12	.07	.05
		8	0.60	.58	.18	.09	.06	.26	.13	.07	.05	.21	.11	.07	.05
145	Hollow core slabs[3]	6(o)	1.07	.46	.16	.08	.06	.23	.12	.07	.05	.19	.11	.07	.05
		(f)	1.86	.34	.14	.08	.05	.20	.11	.07	.05	.17	.10	.06	.05
		8(o)	1.34	.41	.16	.08	.05	.22	.12	.07	.05	.18	.11	.06	.05
		(f)	3.14	.24	.12	.07	.05	.16	.10	.06	.04	.14	.09	.06	.04
		10(o)	1.73	.35	.15	.08	.05	.20	.11	.07	.05	.17	.10	.06	.05
		(f)	4.05	.19	.11	.07	.05	.14	.09	.06	.04	.12	.08	.06	.04
		12(o)	1.91	.33	.14	.08	.05	.19	.11	.07	.05	.17	.10	.06	.05
		(f)	5.01	.16	.10	.06	.05	.12	.08	.05	.04	.11	.08	.05	.04
110	Solid slabs and tees[2]	2	0.38	.67	.18	.09	.06	.27	.13	.07	.05	.22	.12	.07	.05
		3	0.57	.60	.18	.09	.06	.26	.13	.07	.05	.21	.12	.07	.05
		4	0.76	.53	.17	.08	.06	.25	.12	.07	.05	.21	.11	.07	.05
		5	0.95	.49	.17	.08	.06	.24	.12	.07	.05	.20	.11	.07	.05
		6	1.14	.44	.16	.08	.05	.23	.12	.07	.05	.19	.11	.07	.05
		8	1.52	.38	.15	.08	.05	.21	.11	.07	.05	.18	.10	.06	.05
110	Hollow core slabs[3]	8(o)	2.00	.32	.14	.08	.05	.19	.11	.07	.05	.16	.10	.06	.05
		(f)	4.41	.18	.11	.06	.05	.13	.09	.06	.04	.12	.08	.05	.04
		12(o)	2.59	.27	.13	.07	.05	.17	.10	.06	.05	.15	.09	.06	.04
		(f)	6.85	.13	.08	.06	.04	.10	.07	.05	.04	.09	.07	.05	.04

1. When insulations having other R-values are used, U-values can be interpolated with adequate accuracy, or U can be calculated as shown in Sect. 9.1.5. When a finish, air space or any material layer is added, the new U-value is:

$$\frac{1}{\frac{1}{U \text{ from table}} + R \text{ of added finish, air space or material}}$$

2. Thickness for tees is thickness of slab portion.
3. For hollow panels (o) and (f) after thickness designates cores open or cores filled with insulation.

Table 9.1.7 Roof U-values: concrete units with built-up roofing, summer conditions, heat flow downward[1]

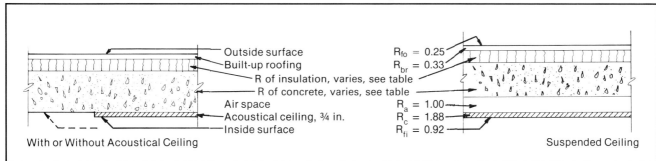

Concrete weight, pcf	Prestressed concrete member	Thickness, t and resistance, R of concrete		Without ceiling				With ceiling							
								Applied direct				Suspended			
				Top insulation resistance, R											
		t	R	None	4	10	16	None	4	10	16	None	4	10	16
145	Solid slabs and tees[2]	2	0.15	.61	.18	.09	.06	.26	.13	.07	.05	.21	.11	.07	.05
		3	0.23	.58	.17	.09	.06	.26	.13	.07	.05	.20	.11	.07	.05
		4	0.30	.56	.17	.08	.06	.25	.13	.07	.05	.20	.11	.07	.05
		5	0.38	.53	.17	.08	.06	.25	.12	.07	.05	.20	.11	.07	.05
		6	0.45	.51	.17	.08	.06	.24	.12	.07	.05	.20	.11	.07	.05
		8	0.60	.48	.16	.08	.06	.24	.12	.07	.05	.19	.11	.07	.05
145	Hollow core slabs[3]	6(o)	1.07	.39	.15	.08	.05	.21	.11	.07	.05	.17	.10	.06	.05
		(f)	1.86	.30	.14	.07	.05	.18	.11	.06	.05	.15	.10	.06	.04
		8(o)	1.34	.35	.15	.08	.05	.20	.11	.07	.05	.17	.10	.06	.05
		(f)	3.14	.22	.12	.07	.05	.15	.09	.06	.04	.13	.08	.06	.04
		10(o)	1.73	.31	.14	.08	.05	.19	.11	.07	.05	.16	.10	.06	.04
		(f)	4.05	.18	.10	.06	.05	.13	.09	.06	.04	.11	.08	.06	.04
		12(o)	1.91	.29	.13	.07	.05	.18	.10	.06	.05	.15	.09	.06	.04
		(f)	5.01	.15	.10	.06	.04	.12	.08	.05	.04	.10	.07	.05	.04
110	Solid slabs and tees[2]	2	0.38	.53	.17	.08	.06	.25	.12	.07	.05	.20	.11	.07	.05
		3	0.57	.48	.16	.08	.06	.24	.12	.07	.05	.19	.11	.07	.05
		4	0.76	.44	.16	.08	.05	.23	.12	.07	.05	.18	.11	.06	.05
		5	0.95	.41	.16	.08	.05	.22	.12	.07	.05	.18	.10	.06	.05
		6	1.14	.38	.15	.08	.05	.21	.11	.07	.05	.17	.10	.06	.05
		8	1.52	.33	.14	.08	.05	.19	.11	.07	.05	.16	.10	.06	.05
110	Hollow core slabs[3]	8(o)	2.00	.29	.13	.07	.05	.18	.10	.06	.05	.15	.09	.06	.04
		(f)	4.41	.17	.10	.06	.05	.12	.08	.06	.04	.11	.08	.05	.04
		12(o)	2.59	.24	.12	.07	.05	.16	.10	.06	.04	.14	.09	.06	.04
		(f)	6.85	.12	.08	.05	.04	.10	.07	.05	.04	.09	.06	.05	.04

1. When insulations having other R-values are used, U-values can be interpolated with adequate accuracy, or U can be calculated as shown in Sect. 9.1.5. When a finish, air space or any material layer is added, the new U-value is:

$$\frac{1}{\frac{1}{U \text{ from table}} + R \text{ of added finish, air space or material}}$$

2. Thickness for tees is thickness of slab portion.
3. For hollow panels (o) and (f) after thickness designates cores open or cores filled with insulation.

Fig. 9.1.1 Heating and cooling load comparisons

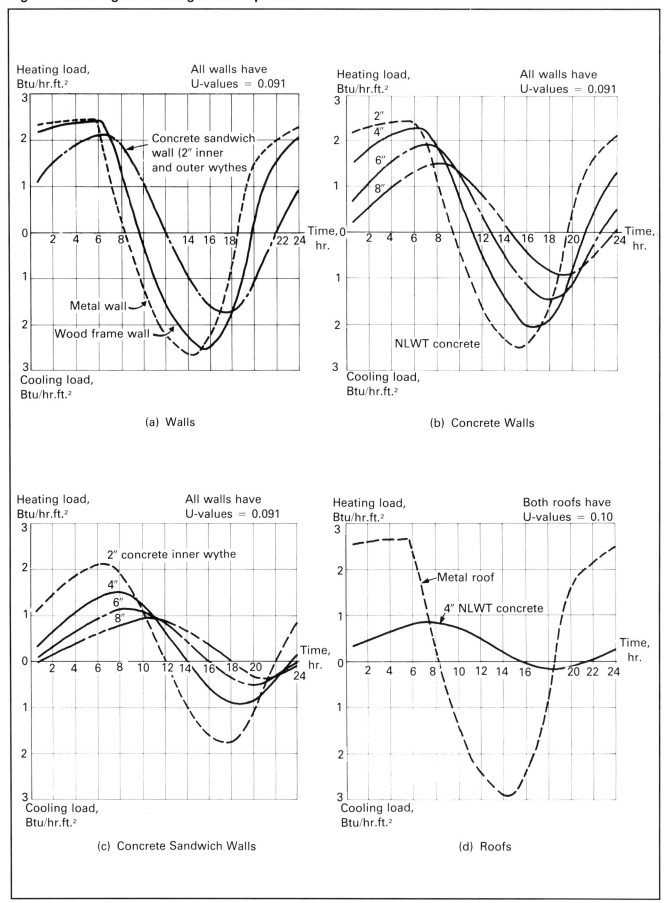

Table 9.1.8 Design considerations for building with high available free heat[1]

Climate Classification		Relative Importance of Design Considerations[2]						
		Thermal Mass	Increase Insulation	External Fins[3]	Surface Color Light	Surface Color Dark	Daylighting	Reduce Infiltration
Winter								
Long Heating Season (6000 degree days or more)	With sun[4] and wind[5]	1	2	2		2	1	3
	With sun without wind	1	2			2	1	3
	Without sun and wind		2			1		3
	Without sun with wind	1	2	2		1		3
Moderate Heating Season (3000-6000 degree days)	With sun and wind	2	2	1	1		2	2
	With sun without wind	2	2		1		2	2
	Without sun and wind	1	2					2
	Without sun with wind	1	2	1				2
Short Heating Season (3000 degree days or less)	With sun and wind	3	1				2	1
	With sun without wind	3	1				2	1
	Without sun and wind	2	1					1
	Without sun with wind	2	1					1
Summer								
Long Cooling Season (1500 hr @ 80°F)	Dry or humid	3		3	3		3	3
Moderate Cooling Season (600-1500 hr @ 80°F)	Dry or humid	3		2	2		2	3
Short Cooling Season (Less than 600 hr @ 80°F)	Dry or humid	2		1	1		1	2

1. Includes office buildings, factories, and commercial buildings.
2. Higher numbers indicate greater importance.
3. Provide shading and protection from direct wind.
4. With sun: Sunshine during at least 60% of daylight time.
5. With wind: Average wind velocity over 9 mph.

9.1.6 Thermal Storage Effects

In years past, the U-factor was considered the most significant indication of heat gain, principally because laboratory tests have shown that thermal transmission is directly proportional to the U-factor during steady-state heat flow. However, the steady-state condition is rarely realized in actual practice.

External conditions (temperature, position of the sun, presence of shadows, etc.) vary throughout a 24 hr. day, and heat gain is not instantaneous through most solid materials, resulting in the phenomenon of time lag (thermal inertia). As temperatures rise on one side of a wall, heat begins to flow towards the cooler side. Before heat transfer can be achieved, the wall must undergo a temperature increase. The Btu's of thermal energy necessary to achieve this increase are directly proportional to the specific heat and density of the wall.

Due to its density, concrete has the capacity to absorb and store large quantities of heat. This thermal mass allows concrete to react very slowly to changes in outside temperature. This characteristic of thermal mass reduces peak heating and cooling loads and delays the time at which these peak loads occur by several hours. This delay improves the performance of heating and cooling equipment, since the peak cooling loads are delayed until the evening hours, when the outside temperature has dropped.

Energy use differences between light and heavy materials are illustrated in the hour-by-hour computer analyses shown in Fig. 9.1.1.

Fig. 9.1.1(a) compares the heat flow through three walls having the same U-value, but made of different materials. The concrete wall consisted of a layer of insulation sandwiched between inner and outer wythes of 2 in. concrete and weighed 48.3 psf. The metal wall, weighing 3.3 psf, had insulation sandwiched between an exterior metal panel and ½ in. drywall. The wood frame wall weighed 7.0 psf and had wood siding on the outside, insulation between 2 x 4 studs, and ½ in. drywall on the inside. The walls were exposed to simulated outside temperatures that represented a typical spring day in a moderate climate. The massive concrete wall had lower peak loads by about 13% for heating and 30% for cooling than the less massive walls.

Table 9.1.9 Design considerations for building with low available free heat[1]

Climate Classification		Relative Importance of Design Considerations[2]					
		Thermal Mass	Increase Insulation	External Fins[3]	Surface Color Light	Surface Color Dark	Reduce Infiltration
Winter							
Long Heating Season (6000 degree days or more)	With sun[4] and wind[5]		3	2		3	3
	With sun without wind		3			3	3
	Without sun and wind		3			2	3
	Without sun with wind		3	2		2	3
Moderate Heating Season (3000-6000 degree days)	With sun and wind	1	2	1		2	3
	With sun without wind	1	2			2	3
	Without sun and wind		2			1	3
	Without sun with wind	1	2	1		1	3
Short Heating Season (3000 degree days or less)	With sun and wind	2	1			1	2
	With sun without wind	2	1			1	2
	Without sun and wind	1	1				2
	Without sun with wind	1	1				2
Summer							
Long Cooling Season (1500 hr @ 80°F)	Dry[6] or humid[7]	3		2	2		3
Moderate Cooling Season (600-1500 hr @ 80°F)	Dry	2		1	1		2
	Humid	2		1	1		3
Short Cooling Season (Less than 600 hr @ 80°F)	Dry or humid	1					1

1. Includes low-rise residential buildings and some warehouses.
2. Higher numbers indicate greater importance.
3. Provide shading and protection from direct wind.
4. With sun: Sunshine during at least 60% of daylight time.
5. With wind: Average wind velocity over 9 mph.
6. Dry: Daily average relative humidity less than 60% during summer.
7. Humid: Daily average relative humidity greater than 60% during summer.

Normal weight concrete walls of various thicknesses that were exposed to the same simulated outside temperatures, are compared in Fig. 9.1.1(b). The walls had a layer of insulation sandwiched between concrete on the outside and ½ in. drywall on the inside; U-values were the same. The figure shows that the more massive the wall the lower the peak loads and the more the peaks were delayed.

Fig. 9.1.1(c) compares concrete sandwich panels having an outer wythe of 2 in., various thicknesses of insulation, and various thicknesses of inner wythes. All walls had U-values of 0.091 and were exposed to the same simulated outside temperatures. The figure shows that by increasing the thickness of the inner concrete wythes, peak loads were reduced and delayed.

A metal roof is compared to a concrete roof in Fig. 9.1.1(d). Both roof systems had built-up roofing on rigid board insulation on the outside and acoustical tile on the inside. The concrete roof weighed 48.3 psf and the metal roof 1.5 psf. The roofs had identical U-values of 0.10 and were exposed to the same simulated outside temperatures. The figure shows that the concrete roof had lower peak loads by 68% for heating and by 94% for cooling, and peaks were delayed by about 1.8 hours for heating and about 4 hours for cooling.

Another factor affecting the behavior of thermal mass is the availability of so called "free heat". This includes heat generated inside the building by lights, equipment, appliances, and people. It also includes heat from the sun entering through windows. Generally, during the heating season, benefits of thermal mass increase with the availability of "free heat" as shown in Tables 9.1.8 and 9.1.9. Thus, office buildings which have high internal heat gains from lights, people, and large glass areas represent an ideal application for thermal mass designs. This is especially true if the glass has been located to take maximum advantage of the sun. Building codes and standards now provide for the benefits of thermal mass. In increasing numbers, they are beginning to acknowledge the effect of the greater heat storage capacity in buildings having high thermal mass.

The rates of many utilities are structured so that lower peaks and delayed peaks can result in significant cost savings.

Other studies have shown that concrete buildings have lower average heating and cooling loads than lightweight buildings for a given insulation level. Thus, life-cycle costs will be lower, or less insulation can be used for equivalent performance.

9.1.7 Condensation Control

Moisture which condenses on the interior of a building is unsightly and can cause damage to the building or its contents. Even more undesirable is the condensation of moisture within a building wall or ceiling assembly where it is not readily noticed until damage has occurred. All air in buildings contains water vapor, with warm air carrying more moisture than cold air. In many buildings moisture is added to the air by industrial processes, cooking, laundering, or humidifiers. If the inside surface temperature of a wall, floor or ceiling is too cold, the air contacting this surface will be cooled below its dew-point temperature and leave its excess water on that surface. Condensation occurs on the surface with the lowest temperature.

Condensation on interior room surfaces can be controlled both by suitable construction and by precautions such as: (1) reducing the interior dew point temperature; (2) raising the temperatures of interior surfaces that are below the dew point, generally by use of insulation; and (3) using vapor retarders.

The interior air dew point temperature can be lowered by removing moisture from the air, either through ventilation or dehumidification. Adequate surface temperatures can be maintained during the winter by incorporating sufficient thermal insulation, using double glazing, circulating warm air over the surfaces, or directly heating the surfaces, and by paying proper attention during design to the prevention of thermal bridging.

Infiltration and exfiltration are air leakage into and out of a building through cracks or joints between infill components and structural elements, interstices around windows and doors, through floors and walls and openings for building services. They are often a major source of energy loss in buildings.

Moisture can move into or across a wall assembly by means of vapor diffusion and air movement. If air, especially exfiltrating, hot, humid air, can leak into the enclosure, then this will be the major source of moisture. Air migration occurs from air pressure differentials independent of moisture pressure differentials.

Atmospheric air pressure differences between the inside and outside of a building envelope exist because of the action of wind, the density difference between outside cold heavy air and inside warm light air creating a "stack effect" and the operation of equipment such as fans. The pressure differences will tend to equalize, and the air will flow through holes or cracks in the building envelope carrying with it the water vapor it contains.

A thorough analysis of air leakage is very complex, involving many parameters, including wall construction, building height and orientation.

Many condensation-related problems in building enclosures are caused by exfiltration and subsequent condensation within the enclosure assembly. Condensation due to air movement is usually much greater than that due to vapor diffusion for most buildings. However, when air leakage is controlled or avoided, the contribution from vapor diffusion can still be significant. In a well designed wall, attention must therefore be paid to the control of air flow and vapor diffusion.

An air barrier and vapor retarder are both needed, and in many instances a single material can be used to provide both of these as well as other functions. The principal function of the air barrier is to stop outside air from entering the building through the walls, windows or roof, and inside air from exfiltrating through the building envelope to the outside. This applies whether the air is humid or dry, since air leakage can result in problems other than the deposition of moisture in cavities. Exfiltrating air carries away heating and cooling energy, while incoming air may bring in pollution as well as reduce the effectiveness of a rain screen wall system.

9.1.7.1 Air barriers

Materials and the method of assembly chosen to build an air barrier must meet several requirements if they are to perform the air leakage control function successfully.

1. There must be continuity throughout the building envelope. The low air permeability materials of the wall must be continuous with the low air barrier materials of the roof (e.g., the roofing membrane) and must be connected to the air barrier material of the window frame, etc.

2. Each membrane or assembly of materials intended to support a differential air pressure load must be designed and constructed to carry that load, inward or outward, or it must receive the necessary support from other elements of the wall. If the air barrier system is made of flexible materials, then it must be supported on both sides by materials capable of resisting the peak air pressure loads; or it must be made of self-supporting materials, such as board products ade-

quately fastened to the structure. If an air pressure difference can not move air, it will act to displace the materials that prevent the air from flowing.

3. The air barrier system must be virtually air-impermeable. A value for maximum allowable air permeability has not yet been determined. However, materials such as polyethylene, gypsum board, precast concrete panels, metal sheeting or glass qualify as low air-impermeable materials when joints are properly sealed, whereas concrete block, acoustic insulation, open cell polystyrene insulation or fiberboard would not qualify. The metal and glass curtain wall industry in the U.S. has adopted a value of 0.06 CFM/ft^2 at 1.57 lb/ft^2 as the maximum allowable air leakage rate for these types of wall construction.

4. The air barrier assembly must be durable in the same sense that the building is durable, and be made of materials that are known to have a long service life or be positioned so that if may be serviced from time to time.

In climates where the heating season dominates, it is strongly recommended that the visible interior surface of a building envelope be installed and treated as the primary air barrier and vapor retarder. Where floors and cross walls are of solid concrete, it is necessary to seal only the joints, as floors and walls themselves do not constitute air paths. Where hollow partitions, such as steel stud or hollow masonry units are used, the interior finish of the envelope should be first made continuous. Where this is impractical, polyethylene film should be installed across these junctions and later sealed to the interior finish material. An interior finish of gypsum wallboard or plaster painted with two coats of enamel paint will provide a satisfactory air barrier/vapor retarder in many instances if the floor/wall and ceiling/ wall joints are tightly fitted and sealed with caulking.

A recent development is an air barrier and vapor retarder system consisting of panel joints sealed from the inside with a foam backer rod and sealant, plus a thermal fusible membrane (TFM) seal around the panel, covering the gap between the structure and the panel. Surfaces should be clean, as dry as possible, smooth, and free of foreign matter which may impede adhesion. The bond between concrete and membrane may be improved by priming the concrete before fusing the membrane to it.

While it is preferable that the air barrier system be placed on the warm side of an insulated assembly, where thermal stresses will be at a minimum, it is not an essential requirement. (This does not necessarily mean on the inside surface of the wall.) The position of the air barrier in a wall is more a matter of suitable construction practice and the type of materials to be used. However, if this barrier is positioned on the outside of the insulation, consideration must be given to its water vapor permeability in case it should also act as a barrier to vapor which is on its way out from inside the wall assembly. This situation may be prevented by choosing an air barrier material that is ten to twenty times more permeable to water vapor diffusion than the vapor barrier material.

In the case of construction assemblies which do not lend themselves to the sealing of interior surfaces, or where it is desirable to limit condensation to very small amounts, such as sandwich wall panel construction (which may have no air leaks through the panels themselves or any air space which can ensure venting and drying out in summer), the use of a separate vapor retarder must be considered. In such cases, the insulation material itself, if it is of a rigid closed-cell type, can be installed on a complete bed of adhesive applied to the interior of the inner wythe of the wall with joints fully sealed with adhesive, to provide a complete barrier to both air and vapor movement.

While the discussion above has been concerned with the flat areas of walls, the joints between them may well present the most important design and construction problems. There are many kinds of joints and the following are considered the most critical: the roof/wall connection, the wall/ foundation, soffit connections, corner details, and connections between different types of exterior wall systems, such as brick and precast concrete, or curtain wall and precast concrete.

9.1.7.2 Vapor retarders

Water vapor diffusion, another way in which indoor water can move through a building envelope to condense in the colder zones, occurs when water vapor molecules diffuse through solid interior materials at a rate dependent on the permeability of the materials, the vapor pressure and temperature differentials. Generally, the colder the outside temperature the greater the pressure of the water vapor in the warm inside air to reach the cooler, drier outside air.

The principal function of a vapor retarder made of low permeability materials is to stop or, more accurately, to retard the passage of moisture as it diffuses through the assembly of materials in a wall. Vapor diffusion control is simple to achieve and is primarily a function of the water vapor diffusion resistance of the chosen materials and their position within the building envelope assembly. The vapor retarder should be clearly identified by the designer and be clearly identifiable by the general contractor.

In temperate climates, vapor retarders should be applied on or near the warm side (inner surface) of assemblies. Vapor retarders may be structural, or in the form of thin sheets, or as coatings. Vapor retarders may also be positioned part way into the insulation but, to avoid condensation, they should be no further than the point at which the dew-point temperature is reached.

In climates with high humidities and high temperatures, especially where air-conditioning is virually continuous, the ingress of moisture may be minimized by a vapor retarder system in the building envelope near the outer surface. For air-conditioned buildings in hot and humid climates without extended cold periods, it may be more economical to use only adequate air infiltration retarding systems rather than vapor retarders since the interior temperature is very rarely below the dew point of the outside air.

Where warm and cold sides may reverse, with resulting reversal of vapor flow, careful analysis of the condition is recommended rather than to ignore the problem and omit any vapor retarders. The designer should refer to the ASHRAE Handbook of Fundamentals[1] or ASTM C755, Selection of Vapor Barriers for Thermal Insulation[5]. In general, a vapor retarder should not be placed at both the inside and outside of wall assemblies.

High thermal conductance paths reaching inward from or near the colder surfaces may cause condensation within the construction. High conductance paths may occur at junctions of floors and walls, walls and ceilings, and walls and roofs; around wall or roof openings; at perimeters of slabs on the ground; and at connections.

Fittings installed in outer walls, such as electrical boxes without holes and conduits, should be completely sealed against moisture and air passage, and they should be installed on the warm side of unbroken vapor retarders or air barriers that are completely sealed.

9.1.7.3 Prevention of condensation on wall surfaces

The U-value of a wall must be such that the surface temperature will not fall below the dew-point temperature of the room air in order to prevent condensation on the interior surface of a wall.

Fig. 9.1.2 gives U-values for any combination of outside temperatures and inside relative humidities above which condensation will occur on the interior surfaces. For example, if a building were located in an area with an outdoor design temperature of 0°F and it was desired to maintain a relative humidity within the building of 25%, the wall must be designed so that all components have a U-value less than 0.78, otherwise there will be a problem with condensation. In many designs the desire to conserve energy will dictate the use of lower U-values than those required to avoid the condensation problem.

The degree of wall heat transmission resistance that must be provided to avoid condensation may be determined from the following relationship:

$$R_t = R_{fi}\left(\frac{t_i - t_o}{t_i - t_s}\right) \quad \text{(Eq. 9.1.2)}$$

Dew-point temperatures to the nearest deg F for various values of t_i and relative humidity are shown in Table 9.1.10.

Example 9.1.3

Determine R_t when the room temperature and relative humidity to be maintained are 70°F and 40%, and t_o during the heating season is −10°F.

From Table 9.1.10 the dew-point temperature t_s is 45°F, and from Table 9.1.1, $R_{fi} = 0.68$.

$$R_t = 0.68\frac{[70 - (-10)]}{[70 - 45]} = 2.18$$

$$U = 1/R_t = 0.46$$

9.1.7.4 Prevention of condensation within wall construction

Water vapor in air behaves as a gas and will diffuse through building materials at rates which depend on vapor permeabilities of materials and vapor pressure differentials. The colder the outside temperature the greater the pressure of the water vapor in the warm inside air to reach the cooler, drier outside air. Also, leakage of moisture laden air into an assembly through small cracks may be a greater problem than vapor diffusion. The passage of water vapor through material is in itself generally not harmful. Water vapor passage becomes harmful when the vapor flow path encounters a temperature below the dew point. Condensation then results within the material.

Building materials have water vapor permeances from very low to very high (see Table 9.1.11). When properly used, low permeance materials keep moisture from entering a wall or roof assembly. Materials with higher permeance allow construction moisture and moisture which enters inadvertently or by design to escape.

When a material such as plaster or gypsum board has a permeance which is too high for the intended use, one or two coats of paint is frequently sufficient to lower the permeance to an acceptable level, or a vapor barrier can be used directly behind such products. Polyethylene sheet, aluminum foil and roofing materials are commonly used. Proprietary vapor barriers, usually combinations of foil and polyethylene or asphalt, are frequently used in freezer and cold storage construction.

Concrete is a relatively good vapor retarder, provided it remains crack free. Permeance is a function of the water-cement ratio of the concrete. A low water-cement ratio, such as that used in most precast concrete members, results in concrete with low permeance.

Where climatic conditions require insulation, a vapor retarder is generally necessary in order to prevent condensation. A closed cell insulation, if properly applied, will serve as its own vapor retarder. For other insulation materials a vapor retarder should be applied to the warm side of the insulation.

For a more complete treatment of the subject of condensation within wall or roof assemblies, see Refs. 1 and 4.

Table 9.1.10 Dew-point temperatures,[1] t_s, °F

Dry bulb or room temperature	Relative humidity (%)									
	10	20	30	40	50	60	70	80	90	100
40	−7	6	14	19	24	28	31	34	37	40
45	−3	9	18	23	28	32	36	39	42	45
50	−1	13	21	27	32	37	41	44	47	50
55	5	17	26	32	37	41	45	49	52	55
60	7	21	30	36	42	46	50	54	57	60
65	11	24	33	40	46	51	55	59	62	65
70	14	27	38	45	51	56	60	63	67	70
75	17	32	42	49	55	60	64	69	72	75
80	21	36	46	54	60	65	69	73	77	80
85	23	40	50	58	64	70	74	78	82	85
90	27	44	55	63	69	74	79	83	85	90

1. Temperatures are based on barometric pressure of 29.92 in. Hg.

Table 9.1.11 Typical permeance (M) and permeability (μ) values[1]

Material	M perms	μ perm-in.
Concrete (1:2:4 mix)[2]	—	3.2
Wood (sugar pine)	—	0.4-5.4
Expanded polystyrene (extruded)	—	1.2
Paint—2 coats		
Asphalt paint on plywood	0.4	
Enamels on smooth plaster	0.5-1.5	
Various primers plus 1 coat flat oil paint on plaster	1.6-3.0	
Expanded polystyrene (bead)	—	2.0-5.8
Plaster on gypsum lath (with studs)	20.00	
Gypsum wallboard, 0.375 in.	50.00	
Polyethylene, 2 mil	0.16	
Polyethylene, 10 mil	0.03	
Aluminum foil, 0.35 mil	0.05	
Aluminum foil, 1 mil	0.00	
Built-up roofing (hot mopped)	0.00	
Duplex sheet, asphalt laminated aluminum foil one side	0.002	

1. ASHRAE Handbook, Chapter 22, Table 7.
2. Permeability for concrete vary depending on the concrete's water-cement ratio and other factors.

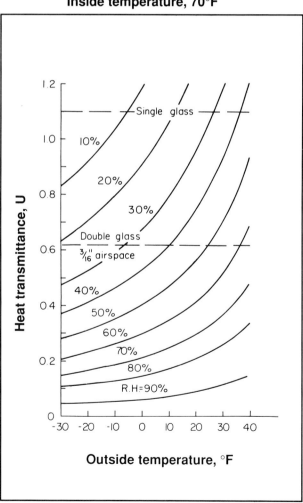

Fig. 9.1.2 Relative humidity at which visible condensation occurs on inside surfaces. Inside temperature, 70°F

9.1.8 Thermal Bridges

Metal ties through walls or solid concrete paths through sandwich panels as described in Sect. 9.4 may cause localized cold spots. The most significant effect of these cold spots is condensation which may cause annoying or damaging wet streaks.

The effect of metal tie thermal bridges on the heat transmittance can be calculated with reasonable accuracy by the zone method described in Chapter 22 of Ref. 1. With the zone method, the panel is divided into Zone A, which contains the thermal bridge, and Zone B, where thermal bridges do not occur, as shown in Fig. 9.1.3. The width of Zone A is calculated as W = m + 2d, where m is the width or diameter of the metal or other conductive bridge material, and d is the distance from the panel surface to the metal. After the width (W) and area of Zone A are calculated, the heat transmissions of the zonal sections are determined and converted to area resistances, which are then added to obtain the total resistance (R_t) of that portion of the panel. The resistance of Zone A is combined with that of Zone B to obtain the overall resistance and the gross transmission value U_0, where U_0 is the overall weighted average heat transmission coefficient of the panel.

The net effect of metal ties is to increase the U-value by 10 or 15%, depending on type, size and spacing. For example, a wall as shown in Fig. 9.1.3 would have a U-value of 0.13 if the effect of the ties is neglected. If the effect of ¼ in. diameter ties at 16 in. on center is included, U = 0.16; at 24 in. spacing, U = 0.15.

For sandwich panels with solid concrete paths, parallel heat flow paths of different conductances result. If there is no lateral heat flow between paths, each path is considered to extend from inside to outside, and transmittance of each path can be calculated using:

$$R = R_1 + R_2 + R_3 + R_4 + ... + R_n \quad \text{(Eq. 9.1.3)}$$

$$R_t = R_{fi} + R + R_{fo} \quad \text{(Eq. 9.1.4)}$$

The average weighted transmittance is then:

$$U_0 = a(U_a) + b(U_b) + ... + n(U_n) \quad \text{(Eq. 9.1.5)}$$

where a, b,...,n are respective fractions of a typical basic area composed of several different paths with transmittances U_a, U_b, ..., U_n.

If heat flows laterally in any continuous layer, creating transverse isothermal planes, total average resistance $R_{t(av)}$ is the sum of the resistances of the layers between these planes. This is a series combination of layers, of which one or more provides parallel paths. The calculated transmittance, assuming parallel heat flow only, is usually considerably lower than that calculated by assuming a combined series-parallel heat flow. The actual transmittance is a value between the two calculated values. In the absence of test values for the combination, an intermediate value should be used; examination of the construction usually indicates whether the value used should be closer to the higher or lower calculated value. Generally, if the construction contains any layer in which lateral conduction is significantly higher than the transmittance through the wall, a value closer to the combined series-parallel calculation should be used. However, if there is no layer of high lateral conductance, a value closer to the calculation for parallel heat flow only should be used.

Fig. 9.1.3 Example of thermal bridges

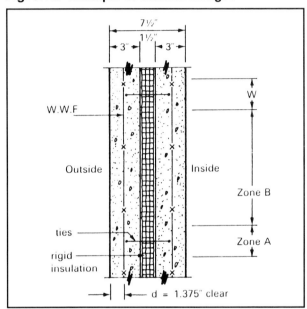

9.1.9 References

1. ASHRAE Handbook 1989, "Fundamentals", American Society of Heating, Refrigerating, and Air-Conditioning Engineers, New York, NY, 1989.

2. ASHRAE Standard 90.1-1989, "Energy Efficient Design of New Buildings Except Low-Rise Residential Buildings," American Society of Heating, Refrigerating, and Air-Conditioning Engineers, New York, NY, 1989.

3. "Simplified Thermal Design of Building Envelopes," *Bulletin EB089*, Portland Cement Association, Skokie, IL, 1981.

4. Balik, J.S., and Barney, G.B., "Thermal Design of Precast Concrete Buildings," *PCI Journal*, V. 29, No. 6, November-December, 1984.

5. "Recommended Practice for Selection of Vapor Barriers for Thermal Insulation," ASTM C755, American Society for Testing and Materials, Philadelphia, PA.

9.2 ACOUSTICAL PROPERTIES OF PRECAST CONCRETE

9.2.1 Glossary

Airborne Sound—sound that reaches the point of interest by propagation through air.

Background Level—the ambient sound pressure level existing in a space.

Decibel (dB)—a logarithmic unit of measure of sound pressure or sound power. Zero on the decibel scale corresponds to a standardized reference pressure (20 µPa) or sound power (10-12 watt).

Flanking Transmission—transmission of sound by indirect paths other than through the primary barrier.

Frequency (Hz)—the number of complete vibration cycles per second.

Impact Insulation Class (IIC)—a single figure rating of the overall impact sound insulation merits of floor-ceiling assemblies in terms of a reference contour (ASTM E989).

Impact Noise—the sound produced by one object striking another.

Noise—unwanted sound.

Noise Criteria (NC)—a series of curves, used as design goals to specify satisfactory background sound levels as they relate to particular use functions.

Noise Reduction (NR)—the difference in decibels between the space-time average sound pressure levels produced in two enclosed spaces by one or more sound sources in one of them.

Noise Reduction Coefficient (NRC)—the arithmetic average of the sound absorption coefficients at 250, 500, 1000 and 2000 Hz expressed to the nearest multiple of 0.05 (ASTM C423).

RC Curves—a revision of the NC curves based on empirical studies of background sounds.

Reverberation—the persistence of sound in an enclosed or partially enclosed space after the source of sound has stopped.

Sabin—the unit of measure of sound absorption (ASTM C423).

Sound Absorption Coefficient (α)—the fraction of randomly incident sound energy absorbed or otherwise not reflected off a surface (ASTM C423).

Sound Pressure Level (SPL)—ten times the common logarithm of the ratio of the square of the sound pressure to the square of the standard reference pressure of 20 µPa. Commonly measured with a sound level meter and microphone, this quantity is expressed in decibels.

Sound Transmission Class (STC)—the single number rating system used to give a preliminary estimate of the sound insulation properties of a partition system. This rating is derived from measured values of transmission loss (ASTM E413).

Sound Transmission Loss (TL)—ten times the common logarithm of the ratio, expressed in decibels, of the airborne sound power incident on the partition that is transmitted by the partition and radiated on the other side (ASTM E90).

Structureborne Sound—sound that reaches the point of interest over at least part of its path by vibration of a solid structure.

9.2.2 General

The basic purpose of architectural acoustics is to provide a satisfactory environment in which desired sounds are clearly heard by the intended listeners and unwanted sounds (noise) are isolated or absorbed.

Under most conditions, the architect/engineer can determine the acoustical needs of the space and then design the building to satisfy those needs. Good acoustical design utilizes both absorptive and reflective surfaces, sound barriers and vibration isolators. Some surfaces must reflect sound so that the loudness will be adequate in all areas where listeners are located. Other surfaces absorb sound to avoid echoes, sound distortion and long reverberation times. Sound is isolated from rooms where it is not wanted by selected wall and floor-ceiling constructions. Vibration generated by mechanical equipment must be isolated from the structural frame of the building.

Most acoustical situations can be described in terms of: (1) sound source, (2) sound transmission path, and (3) sound receiver. Sometimes the sound source strength and path can be controlled and the receiver made more attentive by removing distraction or made more tolerant of disturbance. Acoustical design must include consideration of these three elements.

9.2.3 Approaching the Design Process

Criteria must be established before the acoustical design of a building can begin. Basically a satisfactory acoustical environment is one in which the character and magnitude of all sounds are compatible with the intended space function.

Although a reasonable objective, it is not always easy to express these intentions in quantitative terms.

In addition to the amplitude of sound, the properties such as spectral characteristics, continuity, reverberation and intelligibility must be specified.

People are highly adaptable to the sensations of heat, light, odor, sound, etc., with sensitivities varying widely. The human ear can detect a sound intensity of rustling leaves, 10 dB, and can tolerate, if even briefly, the powerful exhaust of a jet engine at 120 dB, 10^{12} times the intensity of the rustling leaves sound.

9.2.4 Dealing with Sound Levels

The problems of sound insulation are usually considerably more complicated than those of sound absorption. The former involves reductions of sound level, which are of greater orders of magnitude than can be achieved by absorption. These large reductions of sound level from space to space can be achieved only by continuous, impervious barriers. If the problem also involves structureborne sound, it may be necessary to introduce resilient layers or discontinuities into the barrier.

Sound absorbing materials and sound insulating materials are used for different purposes. There is not much sound absorption from an 8-in. concrete wall; similarly, high sound insulation is not available from a porous lightweight material that may be applied to room surfaces. It is important to recognize that the basic mechanisms of sound absorption and sound insulation are quite different.

9.2.5 Sound Transmission Loss

Sound transmission loss measurements are made at 16 frequencies at one-third octave intervals covering the range from 125 to 4000 Hz. The testing procedure is ASTM Specification E90, *Laboratory Measurement of Airborne Sound Transmission Loss of Building Partitions*. To simplify specification of desired performance characteristics the single number Sound Transmission Class (STC) was developed.

Airborne sound reaching a wall, floor or ceiling produces vibration in the wall and is radiated with reduced intensity on the other side. Airborne sound transmission loss of walls and floor-ceiling assemblies is a function of its weight, stiffness and vibration damping characteristics.

Weight is concrete's greatest asset when it is used as a sound insulator. For sections of similar design, but different weights, the STC increases approximately 6 units for each doubling of weight as shown in Fig. 9.2.1.

Precast concrete walls, floors and roofs usually do not need additional treatments in order to provide adequate sound insulation. If desired, greater sound insulation can be obtained by using a resiliently attached layer(s) of gypsum board or other building material. The increased transmission loss occurs because the energy flow path is now increased to include a dissipative air column and additional mass.

The acoustical test results of both airborne sound transmission loss and impact insulation of 4, 6, and 8 in. flat panels, a 14 in. double tee, and 6 and 8 in. hollow-core slabs are shown in Figs. 9.2.2, 9.2.3 and 9.2.4.

Table 9.2.1 presents the ratings for various precast concrete walls and floor-ceiling assemblies. The effects of various assembly treatments on sound transmission can also be predicted from results of previous tests as shown in Table 9.2.2. The improvements are additive, but in some cases the total effect may be slightly less than the sum.

9.2.6 Impact Noise Reduction

Footsteps, dragged chairs, dropped objects, slammed doors, and plumbing generate impact noise. Even when airborne sounds are adequately controlled there can be severe impact noise problems.

The test method used to evaluate systems for impact sound insulation is described in ASTM Specification E492, *Laboratory Measurement of Impact Sound Transmission Using the Tapping Machine*. As with the airborne standard, measurements are made at 16 one-third octave intervals but in the range from 100 to 3150 Hz. For performance specification purposes the single number Impact Insulation Class (IIC) is used.

Fig. 9.2.1 Sound transmission class as a function of weight of floor or wall

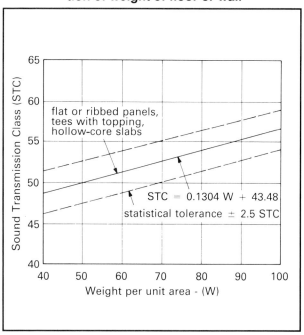

Fig. 9.2.2 Acoustical test data of solid flat concrete panels—normal weight concrete

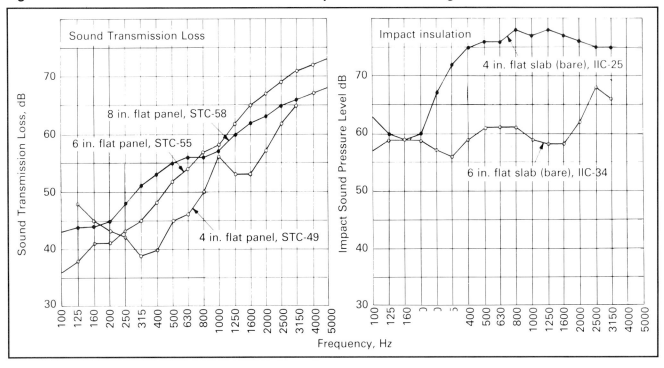

Fig. 9.2.3 Acoustical test data of 14 in. precast double tees with 2 in. topping—normal weight concrete

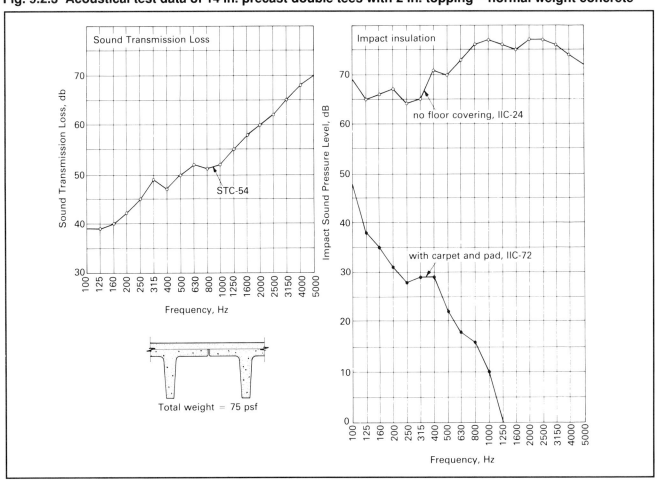

9–22 PCI Design Handbook/Fourth Edition

Fig. 9.2.4 Acoustical test data of hollow-core panels—normal weight concrete

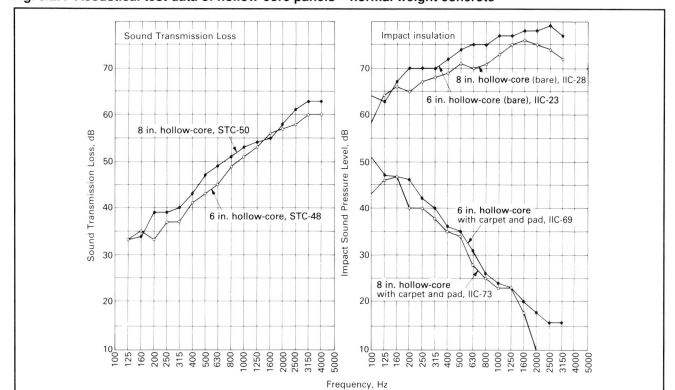

In general, thickness or unit weight of concrete does not greatly affect the transmission of impact sounds as shown in the following table:

Thickness, in.	Unit weight of concrete, pcf	IIC
5	79	23
	114	24
	144	24
10	79	28
	114	30
	144	31

Structural concrete floors in combination with resilient materials effectively control impact sound. One simple solution consists of good carpeting on resilient padding. Table 9.2.2 shows that a carpet and pad over a bare concrete slab will increase the impact noise reduction as much as 56 points. The overall efficiency varies according to the characteristics of the carpeting and padding such as resilience, thickness and weight. So called resilient flooring materials, such as linoleum, rubber, asphalt, vinyl, etc., are not entirely satisfactory directly on concrete, nor are parquet or strip wood floors when applied directly.

Impact sound also may be controlled by providing a discontinuity in the structure such as would be obtained by adding a resilient-mounted plaster or drywall suspended ceiling or a floating floor consisting of a second layer of concrete cast over resilient pads, insulation boards or mastic. The thickness of floating slabs is usually controlled by structural requirements, however, 8 in. is adequate in most instances.

9.2.7 Absorption of Sound

A sound wave always loses part of its energy as it is reflected by a surface. This loss of energy is termed sound absorption. It appears as a decrease in sound pressure of the reflected wave. The sound absorption coefficient is the fraction of energy incident to but not reflected per unit of surface area. Sound absorption can be specified at individual frequencies or as an average of absorption coefficients (NRC).

A dense non-porous concrete surface typically absorbs 1 to 2% of incident sound and has an NRC of 0.015. There are specially fabricated units with porous concrete surfaces which provide greater absorption. In the case where additional sound absorption of precast concrete is desired, a coating of acoustical material can be spray applied, acoustical tile can be applied with adhesive, or an acoustical ceiling can be

Table 9.2.1 Airborne sound transmission and impact insulation class ratings from tests of precast concrete assemblies

Assembly No.	Description	STC	IIC
	Wall Systems		
1	4 in. flat panel, 54 psf	49	–
2	6 in. flat panel, 75 psf	55	–
3	Assembly 2 with "Z" furring channels, 1 in. insulation and ½ in. gypsum board, 75.5 psf	62	–
4	Assembly 2 with wood furring, 1½ in. insulation and ½ in. gypsum board, 73 psf	63	–
5	Assembly 2 with ½ in. space, 1⅝ in. metal stud row, 1½ in. insulation and ½ in. gypsum board	63*	–
6	8 in. flat panel, 95 psf	58	–
7	14 in. prestressed tees with 4 in. flange, 75 psf	54	–
	Floor-Ceiling Systems		
8	8 in. hollow-core prestressed units, 57 psf	50	28
9	Assembly 8 with carpet and pad, 58 psf	50	73
10	8 in. hollow-core prestressed units with ½ in. wood block flooring adhered directly, 58 psf	51	47
11	Assembly 10 except ½ in. wood block flooring adhered to ½ in. sound-deadening board underlayment adhered to concrete, 60 psf	52	55
12	Assembly 11 with acoustical ceiling, 62 psf	59	61
13	Assembly 8 with quarry tile, 1¼ in. reinforced mortar bed with 0.4 in. nylon and carbon black spinerette matting, 76 psf	60	54
14	Assembly 13 with suspended ⅝ in. gypsum board ceiling with 3½ in. insulation, 78.8 psf	61	62
15	14 in. prestressed tees with 2 in. concrete topping, 75 psf	54	24
16	Assembly 15 with carpet and pad, 76 psf	54	72
17	Assembly 15 with resiliently suspended acoustical ceiling with 1½ in. mineral fiber blanket above, 77 psf	59	51
18	Assembly 17 with carpet and pad, 78 psf	59	82
19	4 in. flat slabs, 54 psf	49	25
20	5 in. flat slabs, 60 psf	52*	24
21	5 in. flat slab concrete with carpet and pad, 61 psf	52*	68
22	6 in. flat slabs, 75 psf	55	34
23	8 in. flat slabs, 95 psf	58	34*
24	10 in. flat slabs, 120 psf	59*	31
25	10 in. flat slab concrete with carpet and pad, 121 psf	59*	74

*Estimated values

suspended. Most of the spray applied fire retardant materials used to increase the fire resistance of precast concrete and other floor-ceiling systems can also be used to absorb sound. The NRC of the sprayed fiber types range from 0.25 to 0.75. Most cementitious types have an NRC from 0.25 to 0.50.

If an acoustical ceiling were added to Assembly 15 of Table 9.2.1 (as in Assembly No. 17), the sound entry through a floor or roof would be reduced 5 dB. In addition, the acoustical ceiling would absorb a portion of the sound after entry and provide a few more decibels of quieting. Use of the following expression can be made to determine the intraroom noise or loudness reduction due to the absorption of sound.

Table 9.2.2 Typical improvements for wall, floor, and ceiling treatments used with precast concrete elements

Treatment	Increase in Ratings	
	Airborne (STC)	Impact (IIC)
Wall furring, ¾ in. insulation and ½ in. gypsum board attached to concrete wall	3	0
Separate metal stud system, 1½ in. insulation in stud cavity and ½ in. gypsum board attached to concrete wall	5 to 10	0
2 in. concrete topping (24 psf)	3	0
Carpet and pad	0	43 to 56
Vinyl tile	0	3
½ in. wood block adhered to concrete	0	20
½ in. wood block and resilient fiber underlayment adhered to concrete	4	26
Floating concrete floor on fiberboard	7	15
Wood floor, sleepers on concrete	5	15
Wood floor on fiberboard	10	20
Acoustical ceiling resiliently mounted	5	27
— if added to floor with carpet	5	10
Plaster or gypsum board ceiling resiliently mounted	10	8
— with insulation in space above ceiling	13	13
Plaster direct to concrete	0	0

$$NR = 10 \log \frac{A_o + A_a}{A_o} \quad \text{(Eq. 9.2.1)}$$

where:
NR = sound pressure level reduction, dB
A_o = original absorption, Sabins
A_a = added absorption, Sabins

Values for A_o and A_a are the products of the absorption coefficients of the various room materials and their surface areas.

A plot of this equation is shown in Fig. 9.2.5. For an absorption ratio of 5, the decibel reduction is 7 dB. Note that the decibel reduction is the same, regardless of the original sound pressure level and depends only on the absorption ratio. This is due to the fact that the decibel scale is itself a scale of ratios, rather than difference in sound energy.

While a decibel difference is an engineering quantity which can be physically measured, it is also important to know how the ear judges the change in sound energy due to sound conditioning. Apart from the subjective annoyance factors associated with excessive sound reflection, the ear can make accurate judgments of the relative loudness between sounds. An approximate relation between percentage loudness reduction of reflected sound and absorption ratio is plotted in Fig. 9.2.6.

The percentage loudness reduction does not depend on the original loudness, but only on the absorption ratio. (The curve is drawn for loudness within the normal range of hearing and does not apply to extremely faint sounds.) Referring again to the absorption ratio of 5, the loudness reduction from Fig. 9.2.6 is approximately 40 percent.

Example 9.2.1 Using Table 9.2.2

The performance of a 2 in. concrete topping, carpet and pad added to 8 in. hollow-core prestressed floor units is calculated as follows:

Materials	STC	IIC
Bare slab	50	28
2 in. concrete topping	3	0
Carpet and pad	0	50
Totals	53	78

9.2.8 Acceptable Noise Criteria

As a rule, a certain amount of continuous sound can be tolerated before it becomes noise. An "acceptable" level neither disturbs room occupants nor interferes with the communication of wanted sound.

The most widely accepted and used noise criteria today are expressed as the Noise Criteria (NC) curves Fig. 9.2.7. The figures in Table 9.2.3 represent general acoustical goals. They can also be compared with anticipated noise levels in specific rooms

Fig. 9.2.5 Relation of decibel reduction of reflected sound to absorption ratio

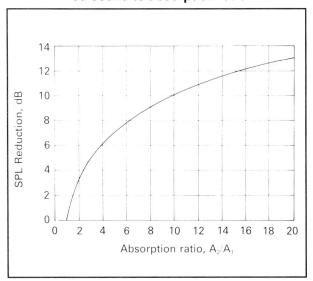

Fig. 9.2.6 Relation of percent loudness reduction of reflected sound to absorption ratio

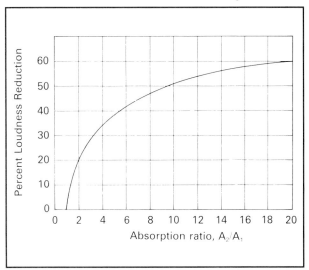

Table 9.2.3 Recommended category classification and suggested noise criteria range for steady background noise as heard in various indoor functional activity areas*

Type of space	NC or RC curve
1. Private residences	25 to 30
2. Apartments	30 to 35
3. Hotels/motels	
a. Individual rooms or suites	30 to 35
b. Meeting/banquet rooms	30 to 35
c. Halls, corridors, lobbies	35 to 40
d. Service/support area	40 to 45
4. Offices	
a. Executive	25 to 30
b. Conference rooms	25 to 30
c. Private	30 to 35
d. Open-plan areas	35 to 40
e. Computer/business machine areas	40 to 45
f. Public circulation	40 to 45
5. Hospitals and clinics	
a. Private rooms	25 to 30
b. Wards	30 to 35
c. Operating rooms	25 to 30
d. Laboratories	30 to 35
e. Corridors	30 to 35
f. Public areas	35 to 40
6. Churches	25 t 30**
7. Schools	
a. Lecture and classrooms	25 to 30
b. Open-plan classrooms	30 to 35**
8. Libraries	30 to 35
9. Concert Halls	**
10. Legitimate theaters	**
11 Recording studios	**
12. Movie theaters	30 to 35

*Design goals can be increased by 5 dB when dictated by budget constraints or when noise intrusion from other sources represents a limiting condition.

**An acoustical expert should be consulted for guidance on these critical spaces.

to assist in evaluating noise reduction problems.

The main criticism of NC curves is that they are too permissive when the control of low or high frequency noise is of concern. For this reason, Room Criteria (RC) Curves were developed (Fig. 9.2.8).[1,2] RC curves are the result of extensive studies based on the human response to both sound pressure level and frequency and take into account the requirements for speech intelligibility.

A low background level obviously is necessary where listening and speech intelligibility is important. Conversely, higher ambient levels can persist in large business offices or factories where speech communication is limited to short distances. Often it is

Fig. 9.2.7 NC (Noise Criteria) curves

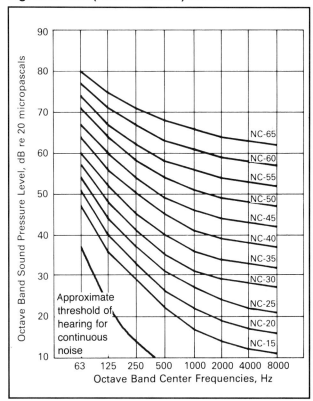

Fig. 9.2.8 RC (Room Criteria) curves

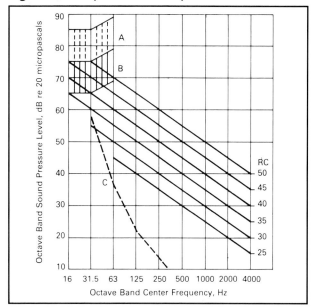

Region A: High probability that noise-induced vibration levels in lightweight wall/ceiling constructions will be clearly feelable; anticipate audible rattles in light fixtures, doors, windows, etc.

Region B: Noise-induced vibration levels in lightweight wall/ceiling constructions may be moderately feelable; slight possibility of rattles in light fixtures, doors, windows, etc.

Region C: Below threshold of hearing for continuous noise.

just as important to be interested in the minimum as in the maximum permissible levels of Table 9.2.3. In an office or residence, it is desirable to have a certain ambient sound level to ensure adequate acoustical privacy between spaces, thus minimizing the transmission loss requirements of unwanted sound (noise).

These undesirable sounds may be from an exterior source such as automobiles or aircraft, or they may be generated as speech in an adjacent classroom or music in an apartment. They may be direct impact-induced sound such as footfalls on the floor above, rain impact on a lightweight roof construction or vibrating mechanical equipment.

Thus, the designer must always be ready to accept the task of analyzing the many potential sources of intruding sound as related to their frequency characteristics and the rate at which they occur. The level of toleration that is to be expected by those who will occupy the space must also be

Table 9.2.4 Noise Criteria (NC) curve values (Fig. 9.2.7)

Noise criteria curves	Octave band center frequency, Hz							
	63	125	250	500	1000	2000	4000	8000
NC-15*	47	36	29	22	17	14	12	11
NC-20*	51	40	33	26	22	19	17	16
NC-25*	54	44	37	31	27	24	22	21
NC-30	57	48	41	35	31	29	28	27
NC-35	60	52	45	40	36	34	33	32
NC-40	64	56	50	45	41	39	38	37
NC-45	67	60	54	49	46	44	43	42
NC-50	71	64	58	54	51	49	48	47
NC-55	74	67	62	58	56	54	53	52
NC-60	77	71	67	63	61	59	58	57
NC-65	80	75	71	68	66	64	63	62

*The applications requiring background levels less than NC-25 are special purpose spaces in which an acoustical consultant should set the criteria.

established. Figs. 9.2.9 and 9.2.10 are the spectral characteristics of common noise sources.

With these criteria, the problem of sound isolation now must be solved, namely the reduction process between the high unwanted noise source and the desired ambient level. For this solution, two related yet mutually exclusive processes must be incorporated, i.e., sound transmission loss and sound absorption.

Fig. 9.2.9 Sound pressure levels—exterior noise sources

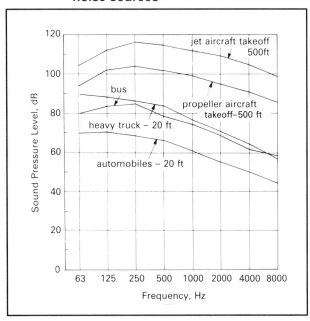

Fig. 9.2.10 Sound pressure levels—interior noise sources

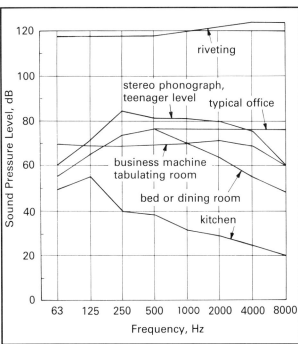

9.2.9 Establishment of Noise Insulation Objectives

Often acoustical control is specified as to the minimum insulation values of the dividing partition system. Municipal building codes, lending institutions and the Department of Housing and Urban Development (HUD) list both airborne STC and impact IIC values for different living environments. For example, the HUD recommendations[3] were:

Location	STC	IIC
Between living units	45	45
Between living units and public space	50	50

Other community ordinances are more specific, listing the sound insulation criteria with relation to particular ambient environments.[4]

	Grade I Suburban	Grade II Residential Urban and Suburban	Grade III Urban
Ambient level	NC or RC 20-25	NC or RC 25-30	NC or RC 35+
Walls	STC 55	STC 52	STC 48
Floor-ceiling assemblies	STC 55 IIC 55	STC 52 IIC 55	STC 48 IIIC 48

Once the objectives are established, the designer then should refer to available data, e.g., Fig. 9.2.1 or Tables 9.2.1 and 9.2.2, and select the system which best meets these requirements. In this respect, concrete systems have superior properties and can with minimal effort comply with these criteria. When the insulation value has not been specified, selection of the necessary barrier can be determined analytically by (1) identifying exterior and/or interior noise sources, and (2) by establishing acceptable interior noise criteria.

Example 9.2.2 Sound insulation criteria

Assume a precast, prestressed concrete office building is to be erected adjacent to a major highway. Private and semiprivate offices will run along the perimeter of the structure. The first step is to determine the degree of insulation of the exterior wall system.

	Sound pressure level—(dB)							
Frequency (Hz)	63	125	250	500	1000	2000	4000	8000
Bus traffic source noise (Fig. 9.2.9)	80	83	85	78	74	68	62	58
Private office noise criteria- NC 35 (Table 9.2.4)	60	52	45	40	36	34	33	32
Required insulation	20	31	40	38	38	34	29	26

	Sound pressure level—(dB)					
Frequency (Hz)	125	250	500	1000	2000	4000
Required insulation	31	40	38	38	34	29
6 in. precast solid concrete wall (Fig. 9.2.2)	38	43	52	59	67	72
Deficiencies	–	–	–	–	–	–

The 500 Hz requirement of 38 dB reduction (above) can be used as the first approximation of the wall STC category. However, if windows are planned for the wall, a system of about 50-55 STC should be selected (see Sect. 9.2.10). Individual transmission loss performance values of this system are then compared to the calculated need.

The selected wall (above) should meet or exceed the insulation needs at all frequencies. However, to achieve the most efficient design conditions, certain limited deficiencies can be tolerated. Experience has shown that the maximum deficiencies are 3 dB at two frequencies or 5 dB at one frequency.

9.2.10 Composite Wall Considerations

Door and windows are often the weak link in an otherwise effective sound barrier. Minimal effects on sound transmission loss will be achieved in most cases by a proper selection of glass (Table 9.2.5).[5] Mounting of the glass in its frame should be done with care to eliminate noise leaks and to reduce the glass plate vibrations.

Sound transmission loss of a door depends upon its material and construction, and the sealing between the door and the frame.[6] There is a mass law dependence of STC on weight (psf) for both wood and steel doors. The approximate relationships are:

For steel doors:
$$STC = 15 + 27 \log W \quad \text{(Eq. 9.2.2)}$$
For wood doors:
$$STC = 12 + 32 \log W \quad \text{(Eq. 9.2.3)}$$
where:
W = weight of the door, psf

These relationships are purely empirical and a large deviation can be expected for any given door.

For best results, the distances between adjacent door and/or window openings should be maximized, staggered when possible and held to a minimum area. Minimizing openings retains the acoustical prop-erties of precast concrete. The design characteristics of a door or window system must be analyzed prior to specification. Such qualities as frame design, door construction and glazing thickness are vital performance criteria. Installation procedures must be exact and care given to the frame of each opening. Gaskets, weatherstripping and raised threshold serve as both excellent thermal and acoustical seals and are recommended.

Fig. 9.2.11 can be used to calculate the effective acoustic isolation of a wall system which contains a composite of elements, each with known individual transmission loss data.

Example 9.2.3 Composite wall insulation criteria

To complete the office building wall acoustical design from Sect. 9.2.9, assume the following:

1. The glazing area represents 10% of the exterior wall area.
2. The windows will be double glazed with a 38 STC acoustical insulation rating.

The problem now becomes the task of determining the combined effect of the concrete-glass combination and a re-determination of criteria compliance. The calculations are shown in the bottom table, page 9-30.

The maximum deficiency is 3 dB and occurs at only one frequency point. The 6 in. precast concrete

Table 9.2.5 Acoustical properties of glass

Type and overall thickness	Inside light	Construction space	Outside light	STC
⅛" Plate or float	–	–	⅛"	23
¼" Plate or float	–	–	¼"	28
½" Plate or float	–	–	½"	31
1" Insulated glass	¼"	½" Air space	¼"	31
¼" Laminated	⅛"	0.030" Vinyl	⅛"	34
1½" Insulated glass	¼"	1" Air space	¼"	35
¾" Plate or float	–	–	¾"	36
1" Insulated glass	¼"	½" Air space	¼" Laminated	38
1" Plate or float	–	–	1"	37
2¾" Insulated glass	¼"	2" Air space	½"	39
4¾" Insulated glass	¼"	4" Air space	½"	40
6½" Insulated glass	¼"	6" Air space	¼" Laminated	42

Transmission loss (Db)

Frequency (Hz)	125	160	200	250	315	400	500	630	800	1000	1250	1600	2000	2500	3150	4000
¼ inch plate glass—28 STC	24	22	24	24	21	23	21	23	26	27	33	36	37	39	40	40
1 inch insulating glass with ½ inch air space—31 STC	25	25	22	20	24	27	27	30	32	33	35	34	29	31	33	36
1 inch insulating glass laminated with ½ inch air space—38 STC	30	29	26	28	31	34	35	37	37	38	38	40	41	40	41	44

wall with double glazed windows will provide the required acoustical insulation.

Floor-ceiling assembly acoustical insulation requirements are determined in the same manner as walls by using Figs. 9.2.2, 9.9.3 or 9.2.4 along with 9.2.10.

9.2.11 Leaks and Flanking

The performance of a building section with an otherwise adequate STC can be seriously reduced by a relatively small hole or any other path which allows sound to bypass the acoustical barrier. All noise which reaches a space by paths other than through the primary barrier is called flanking. Common flanking paths are openings around doors or windows, at electrical outlets, telephone and television connections, and pipe and duct penetrations. Suspended ceilings in rooms where walls do not extend from the ceiling to the roof or floor above allow sound to travel to adjacent rooms.

Anticipation and prevention of leaks begins at the design stage. Flanking paths (gaps) at the perime-

Sound pressure level—(dB)						
Frequency (Hz)	125	250	500	1000	2000	4000
6 in. precast solid concrete wall (Fig. 9.2.2)	38	43	52	59	67	72
Double glazed windows (Table 9.2.5)	30	28	35	38	41	44
Correction (Fig. 9.2.11)	–2	–6	–7	–11	–15	–19
Combined transmission loss	36	37	45	48	52	53
Insulation requirements	31	40	38	38	34	29
Deficiencies	–	3	–	–	–	–

Fig. 9.2.11 Chart for calculating the effective transmission loss of a composite barrier

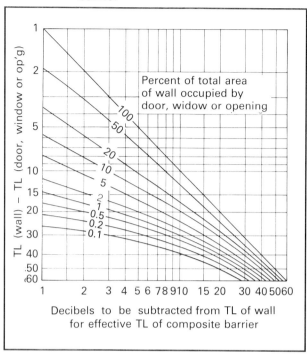

*For purposes of approximation STC values can be used in place of TL values.

Fig. 9.2.12 Effect of safing insulation seals

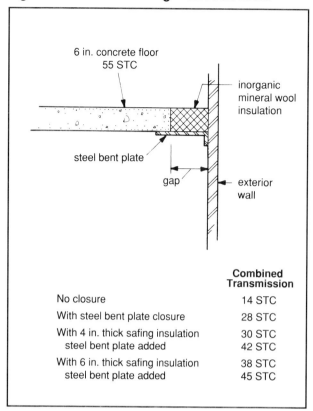

ters of interior precast walls and floors are generally sealed during construction with grout or drypack. In addition, all openings around penetrations through walls or floors should be as small as possible and must be sealed airtight. The higher the STC of the barrier, the greater the effect of an unsealed opening (see Fig. 9.2.11).

Perimeter leakage more commonly occurs at the intersection between an exterior curtain wall and floor slab. It is of vital importance to seal this gap in order to retain the acoustical integrity of the system as well as provide the required fire stop between floors. One way to achieve this seal is to place a 4 pcf density mineral wool blanket between the floor slab and the exterior wall. Fig. 9.2.12 demonstrates the acoustical isolation effects of this treatment.

In exterior walls, the proper application of sealant and backup materials in the joints between units will not allow sound to flank the wall.

If the acoustical design is balanced, the maximum amount of acoustic energy reaching a space via flanking should not equal the energy transmitted through the primary barriers.

Although not easily quantified, an inverse relationship exists between the performance of an element as a primary barrier and its propensity to transmit flanking sound. In other words, the probability of existing flanking paths in a concrete structure is much less than in one of a steel or wood frame.

In addition to using basic structural materials, flanking paths can be minimized by:
1. Interrupting the continuous flow of energy with dissimilar materials, e.g., expansion or control joints or dead airspace.
2. Increasing the resistance to energy flow with floating floor systems, full height and/or double partitions and suspended ceilings.

9.2.12 References

1. "ASHRAE Systems Handbook for 1984," American Society of Heating, Refrigerating and Air-Conditioning Engineers, New York, NY, 1984.

2. Blazier, W. E., "Revised Noise Criteria for Design and Rating of HVAC Systems", paper presented at ASHRAE Semiannual Meeting, Chicago, IL, Jan. 26, 1981.

3. Berendt, R. D., Winzer G. E., Burroughs, C. B., "A Guide to Airborne, Impact and Structureborne Noise Control in Multifamily Dwellings", prepared for Federal Housing Administration, U. S. Government Printing Office, Washington, DC, 1975.

4. Sabine, H. J., Lacher, M. B., Flynn, D. R. and Quindry, T. L., "Acoustical and Thermal Performance of Exterior Residential Walls, Doors and Windows", National Bureau of Standards, U.S. Government Printing Office, Washington, DC, 1975.

5. IITRI, "Compendium of Materials for Noise Control", U. S. Department of Health, Education and Welfare, U. S. Government Printing Office, Washington, DC, 1980.
6. Beranek, L. L., "Noise Reduction," McGraw-Hill Book Co., New York, NY, 1960.

Additional Bibliography

Harris, C. M., "Handbook of Noise Control," McGraw-Hill Book Co., New York, NY, 1967.

Harris, C. M. and Crede, C. E., "Shock and Vibration Handbook", 2nd Edition, McGraw-Hill Book Co., New York, NY, 1976.

Litvin, A. and Belliston, H. W., "Sound Transmission Loss Through Concrete and Concrete Masonry Walls," *Journal of the American Concrete Institute*, V. 75, No. 12, December 1978.

9.3 FIRE RESISTANCE

9.3.1 Notation

Note: Subscript θ indicates the property as affected by elevated temperatures.

a	=	depth of equivalent rectangular compression stress block
A_{ps}	=	area of uncoated prestressing steel
A_s	=	area of non-prestressed reinforcement
A_s^-	=	area of reinforcement in negative moment region
b	=	width of member
d	=	distance from centroid of prestressing steel to the extreme compression fiber
f_c'	=	compressive strength of concrete
f_{ps}	=	stress in the uncoated prestressing steel at nominal strength
f_{pu}	=	ultimate tensile strength of uncoated prestressing steel
h	=	total depth of a member
ℓ	=	span length
M_n	=	nominal moment strength
$M_{n\theta}^+$, $M_{n\theta}^-$	=	positive and negative nominal moment strength at elevated temperatures, respectively
R	=	fire endurance of an element or a composite assembly
R_1, R_2, R_n	=	fire endurance of individual courses
s	=	spacing of ribs in ribbed panels
t	=	minimum thickness of ribbed panels
t_e	=	equivalent thickness of ribbed panels
u	=	distance from prestressing steel to the fire exposed surface
w	=	uniform total load
w_d	=	uniform dead load
w_ℓ	=	uniform live load
x, x_0, x_1, x_2	=	horizontal distances as shown in Figs. 9.3.13, 9.3.14, and 9.3.15
y_s	=	distance from centroid of prestressing steel to the bottom surface
θ_s	=	temperature of steel
ø	=	strength reduction factor

9.3.2 Glossary

Carbonate aggregate concrete—concrete made with aggregates consisting mainly of calcium or magnesium carbonate, e.g., limestone or dolomite.

Fire endurance—a measure of the elapsed time during which a material or assembly continues to exhibit fire resistance under specified conditions of test and performance. As applied to elements of buildings it shall be measured by the methods and to the criteria defined in ASTM E119. (Defined in ASTM E176).

Fire resistance—the property of a material or assembly to withstand fire or to give protection from it. As applied to elements of buildings, it is characterized by the abilty to confine a fire or to continue to perform a given structural function, or both. (Defined in ASTM E176).

Fire resistance rating (sometimes called **fire rating, fire resistance classification,** or **hourly rating**)—a legal term defined in building codes, usually based on fire endurances. Fire resistance ratings are assigned by building codes for various types of construction and occupancies and are usually given in hourly increments.

Lightweight aggregate concrete—concrete made with aggregate of expanded clay, shale, slag, or slate or sintered fly ash, and weighing about 85 to 115 pcf.

Sand-lightweight concrete—concrete made with a combination of expanded clay, shale, slag, or slate or sintered fly ash and natural sand. Its unit weight is generally between 105 and 120 pcf.

Siliceous aggregate concrete—concrete made with normal weight aggregates consisting mainly of silica or compounds other than calcium or magnesium carbonate.

9.3.3 Introduction

Precast and prestressed concrete members can be provided with any degree of fire resistance that may be required by building codes, insurance companies, and other authorities. The fire resistance of building assemblies is determined from standard fire tests defined by the American Society for Testing and Materials.

To ensure that fire resistance requirements are satisfied, the engineer can use tabulated information provided by various authoritative bodies, such as Underwriters Laboratories, Inc., the American Insurance Association and model building codes. This information is based on the results of standard fire tests of assemblies that may include ceilings and other building components. The 1983 edition of the UL *Fire Resistance Directory*[9] alone provides information on more than 120 assemblies incorporating precast, prestressed concrete members.

In the absence of tabulated data, the fire resistance of precast and prestressed concrete members and assemblies can be determined in most cases by calculation. These calculations are based on engineering principles and take into account the conditions of a standard fire test. This is known as the Rational Design Method of determining fire resistance. It is based on extensive research sponsored in part by the Precast/Prestressed Concrete Institute (PCI) and conducted by the Portland Cement Association (PCA) and other laboratories.

After a discussion on fire tests in Sects. 9.3.4 and 9.3.5, calculations using the Rational Design Method are described in the sections that follow. Brief explanations of the underlying principles are also given. For additional examples, design charts and a complete explanation of the method, refer to the PCI manual, MNL-124-89, *Design for Fire Resistance of Precast Prestressed Concrete*,[4] and the references listed in that manual, as well as the CRSI manual *Reinforced Concrete Fire Resistance*.[10]

Each of the model codes recognizes the use of PCI MNL-124 as an alternative to fire tests of specific precast, prestressed concrete units or assemblies in which the units are incorporated, or to general tabulated fire resistance requirements.

9.3.4 Standard Fire Tests

The fire resistance of building components is measured in standard fire tests defined by ASTM E119, *Standard Test Methods for Fire Tests of Building Construction and Materials*. During these tests the building assembly, such as a portion of floors, walls, roofs or columns, is subjected to increasing temperatures that vary with time as shown in Fig. 9.3.1. This time-temperature relation is used as a standard to represent the combustion of about 10 lb of wood (with a heat potential of 8,000 Btu per lb) per sq ft of exposed area per hour of test. Actually, the fuel consumption to maintain the standard time-temperature relation during a fire test depends on the design of the furnace and on the test specimen. When fire tested, assemblies with exposed concrete members require considerably more fuel than other assemblies due to the favorable heat capacity. This fact is not recognized when evaluating fire resistance.

In addition to defining a standard time-temperature relationship, standard fire tests involve regulations concerning the size of the assemblies, the amount of applied load, the region of the assembly to be exposed to fire, and the end point criteria on which fire resistance (duration) is based.

ASTM E119 specifies the minimum sizes of specimens to be exposed in fire tests.* For floors and roofs, at least 180 sq ft must be exposed to fire from beneath, and neither dimension can be less than 12 ft. For tests of walls, either load bearing or non-load bearing, the minimum specified area is 100 sq ft with neither dimension less than 9 ft. The minimum length for columns is specified to be 9 ft, while for beams it is 12 ft.

Fig. 9.3.1 Standard time-temperature curve

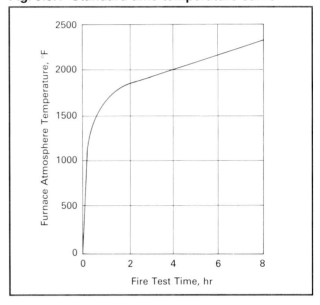

During fire tests of floors, roofs, beams, load bearing walls, and columns, the maximum permissible superimposed load, as required or permitted by nationally recognized standards, is applied. A load other than the maximum load may be applied, but the test results then apply only to the restricted load condition.

Floor and roof specimens are exposed to fire from beneath, beams from the bottom and sides, walls from one side, and columns from all sides.

ASTM E119 distinguishes between "restrained" and "unrestrained" assemblies and defines them as follows:

> "Floor and roof assemblies and individual beams in buildings shall be considered restrained when the surrounding or supporting structure is capable of resisting substantial thermal expansion throughout the range of anticipated elevated temperatures. Constructions not complying with this definition are assumed to be free to rotate and expand and shall therefore be considered as unrestrained."

---
*Much valuable data have been developed from tests on specimens smaller than the ASTM minimum sizes.

ASTM E119 includes a guide for classifying types of construction as restrained or unrestrained. The guide indicates that cast-in-place and most precast concrete constructions are considered to be restrained.

Fire endurance, end point criteria, and fire rating

The *fire resistance* of an assembly is measured by its fire endurance, defined as the period of time elapsed before a prescribed condition of failure or end point is reached during a standard fire test. A *fire rating* or *classification* is a legal term for a fire endurance required by a building code authority.

End point criteria defined by ASTM E119 include:

1. Load bearing specimens must sustain the applied loading. Collapse is an obvious end point (structural end point).
2. Holes, cracks, or fissures through which flames or gases hot enough to ignite cotton waste must not form (flame passage end point).
3. The temperature increase of the unexposed surface of floors, roofs, or walls must not exceed an average of 250°F or a maximum of 325°F at any one point (heat transmission end point).

Unrestrained assembly classifications can be derived from tests of restrained floor, roof, or beam specimens provided that the average temperature of the tension steel at any section must not exceed 800°F for cold-drawn prestressing steel or 1100°F for reinforcing bars.

Additional end point criteria for restrained specimens are:

1. Beams more than 4 ft on centers: the above steel temperatures must not be exceeded for classifications of 1 hr or less; for classifications longer than 1 hr, the above temperatures must not be exceeded for the first half of the classification period or 1 hr, whichever is longer.
2. Beams 4 ft or less on centers or slabs are not subjected to steel temperature limitations.

Walls and partitions must meet the same structural, flame passage, and heat transmission end points described above. In addition, they must withstand a hose stream test (simulating, in a specified manner, a fire fighter's hose stream).

9.3.5 Fire Tests of Prestressed Concrete Assemblies

General

The first fire test of a prestressed concrete assembly in America was conducted in 1953 at the National Institute of Science and Technology (NIST), formerly the National Bureau of Standards. Since that time, more than 150 prestressed concrete assemblies have been subjected to standard fire tests in America. Although many of the tests were conducted for the purpose of deriving specific fire ratings, most of the tests were performed in conjunction with broad research studies whose objectives have been to understand the behavior of prestressed concrete subjected to fire. The knowledge gained from these tests has resulted in the development of (1) lists of fire resistive prestressed concrete building components, and (2) procedures for determining the fire endurance of prestressed concrete members by calculation.

Many different types of prestressed concrete elements have been fire tested. These elements include joists, double tees, monowing tees, single tees, solid slabs, hollow-core slabs, rectangular beams, ledger beams, and I-shaped beams. In addition, roofs with thermal insulation and load bearing wall panels have also been tested. Nearly all of these elements have been exposed directly to fire, but a few tests have been conducted on specimens that received additional protection from the fire by spray-applied coatings, ceilings, etc.

9.3.5.1 Fire tests of flexural elements

Tests have shown that the structural fire endurance of a flexural precast, prestressed concrete element depends on several factors, the most important of which is the method of support, i.e., restrained or unrestrained. Other factors include size and shape of the element, thickness of cover (or more precisely, the distance between the centers of the prestressing tendons and the nearest fire-exposed surface), aggregate type, and load intensity. The fire endurance as determined by the criteria for temperature rise of the unexposed surface (heat transmission) depends primarily on the concrete thickness and aggregate type.

Reports of a number of tests sponsored by the Precast/Prestressed Concrete Institute have been issued by Underwriters Laboratories, Inc. Most of the reports have been reprinted by the PCI, and the results of the tests are the basis for UL's listings and specifications for non-proprietary products such as double tee and single tee floors and roofs, wet-cast hollow-core and solid slabs, and prestressed concrete beams.

The Portland Cement Association conducted many fire tests of prestressed concrete assemblies. PCA's unique furnaces have made it possible to study in depth the effects of support conditions. Four series of tests dealt with simply supported slabs and beams; two series dealt with continuous slabs and beams; and one major series dealt with the effects of restrained thermal expansion on the behavior during fire of prestressed concrete floors and roofs. PCA has also conducted a number of miscellaneous fire tests of prestressed and reinforced concrete assemblies. Reports of these tests have been published and are available from the PCA.

In addition to the tests sponsored by PCI and PCA, a number of fire tests of proprietary products, such as hollow-core slabs, have been sponsored by their manufacturers. Most of these tests have been performed by Underwriters Laboratories, Inc., but some have been conducted by Ohio State University, the Fire Prevention Research Institute and NIST. Reports of proprietary tests are generally available from tests sponsors. The UL *Fire Resistance Directory*[9] includes many proprietary assemblies.

9.3.5.2 Fire tests of walls and columns

Not all the tests conducted by Underwriters Laboratories, Inc., result in listings in UL's publications; some tests are conducted for research purposes. One such test was conducted on a double tee wall assembly. Fire was applied to the flat surface of the flange. The flange was only 1½ in. thick. A load of about 10 kips per ft was applied at the top of the wall. The wall withstood a 2 hr fire and a subsequent hose stream test followed by a double load test without distress. Because the flange was only 1½ in. thick, the heat transmission requirement was exceeded for most of the test. By providing adequate flange thickness or insulation, the heat transmission requirement would have been met in addition to the structural requirement.

Fire tests of reinforced concrete columns have been conducted by the PCA and the National Research Council of Canada. While no tests have been conducted for prestressed concrete columns, results from these tests are considered to be equally applicable to prestressed concrete columns.

9.3.6 Designing for Heat Transmission

As noted in Sect. 9.3.4, ASTM E119 imposes heat transmission criteria for floor, roof, and wall assemblies. Thus, floors, roofs, or walls requiring a fire-resistance rating must satisfy the heat transmission requirements as well as the various structural criteria. The heat transmission fire endurance of a concrete assembly is essentially the same whether the assembly is tested as a floor (oriented horizontally) or as a wall (tested vertically). Because of this, and unless otherwise noted, the information, which follows is applicable to floors, roofs, or walls.

9.3.6.1 Single course slabs or wall panels

For concrete slabs or wall panels, the temperature rise of the unexposed surface depends mainly on the thickness and aggregate type of the concrete. Other less important factors include unit weight, moisture condition, air content, and maximum aggregate size. Within the usual ranges, water-cement ratio, strength, and age have only insignificant effects.

Fig. 9.3.2 shows the fire endurance (heat transmission) of concrete slabs as influenced by aggregate types and thickness. For a hollow-core slab, this thickness may be obtained by dividing the net cross sectional area by its width. The curves represent air-entrained concrete made with air-dry aggregates having a nominal maximum size of ¾ in. and fire tested when the concrete was at the standard moisture condition (75% R.H. at mid-depth). On the graph, concrete aggregates are designated as lightweight, sand-lightweight, carbonate, or siliceous. Lightweight aggregates include expanded clay, shale, and slate which produce concretes having unit weights of about 95 to 105 pcf without sand replacement. Lightweight concretes, in which sand is used as part or all of the fine aggregate and weigh no more than about 120 pcf, are designated as sand-lightweight. Carbonate aggregates include limestone and dolomite, i.e., those consisting mainly of calcium and/or magnesium carbonate. Siliceous aggregates include quartzite, granite, basalt, and most hard rocks other than limestone and dolomite.

Fig. 9.3.2 Fire endurance (heat transmission) of concrete slabs or panels.*

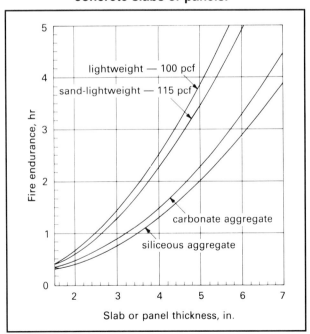

*Interpolation for different concrete unit weights is reasonably accurate.

Fig. 9.3.3 Temperatures on vertical centerline of stemmed units at 1 hr of exposure

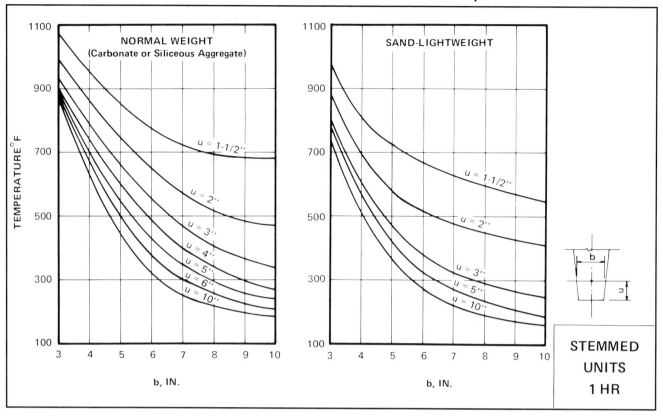

Fig. 9.3.4 Temperatures on vertical centerline of stemmed units at 2 hr of exposure

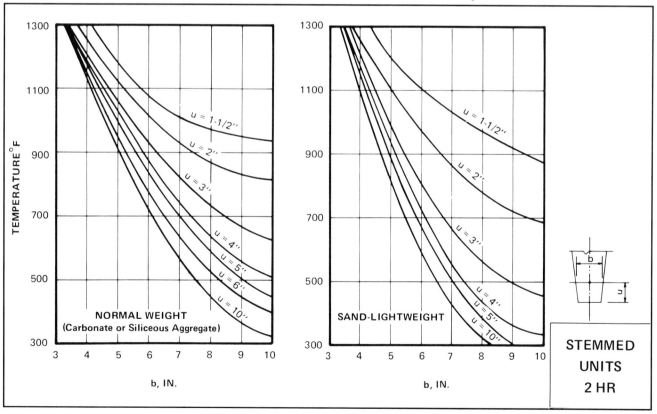

Fig. 9.3.5 Temperatures on vertical centerline of stemmed units at 3 hr of exposure

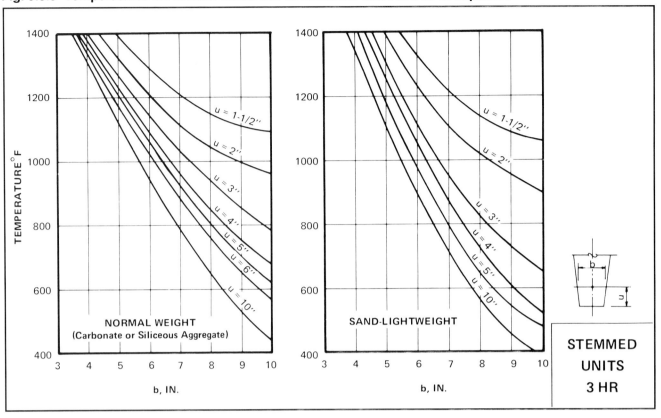

Fig. 9.3.6 Temperatures on vertical centerline of stemmed units at 4 hr of exposure

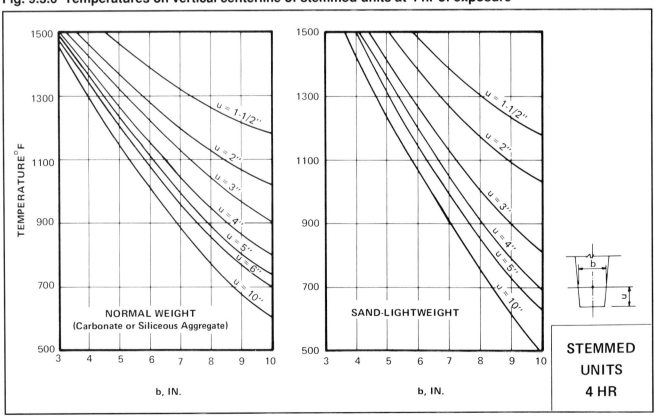

9.3.6.2 Floors, roofs or walls faced with gypsum wall-board

Table 9.3.1 shows the fire endurance of concrete slabs with ⅝ in. gypsum wallboard (Type X) for two cases: (1) a 6-in. air space between the wallboard and slab, and (2) no space between the wallboard and slab. Materials and techniques of attaching the wallboard should be similar to those used in the UL test[1] on which the data are based.

Table 9.3.1 Thickness of concrete slabs or wall panels faced with ⅝-in. Type X gypsum wallboard to provide fire endurances of 2 and 3 hr

	Thickness (in.) of concrete panel for fire endurance of			
	With no air space		With 6-in. air space	
Aggregate	2 hr	3 hr	2 hr	3hr
Sand-lightweight	2.0	3.0	1.2	2.4
Carbonate	2.3	3.7	1.3	2.7
Siliceous	2.5	3.9	1.3	2.8

9.3.6.3 Ribbed panels

Heat transmission through a ribbed panel is influenced by the thinnest portion of the panel and by the panel's "equivalent thickness." Here, equivalent thickness is defined as the net cross-sectional area of the panel divided by the width of the cross-section. In calculating the net cross-sectional area of the panel, portions of ribs that project beyond twice the minimum thickness should be neglected, as shown in Fig. 9.3.7(a).

The heat transmission fire endurance can be governed by either the thinnest section, or by the average thickness, or by a combination of the two. The following rule-of-thumb expressions appear to give a reasonable guide as to when the average thickness governs:

If $t \leq s/4$, fire endurance R is governed by t and is equal to R_t.

If $t \geq s/2$, fire endurance R is governed by t_e and is equal to R_{te}.

If $s/2 > t > s/4$:
$$R = R_t + (4t/s - 1)(R_{te} - R_t)$$

where:

t = minimum thickness
t_e = equivalent thickness of panel
s = rib spacing

and where R is the fire endurance of a concrete panel and subscripts t and t_e relate the corresponding R values to concrete slab thicknesses t and t_e, respectively.

These expressions apply to ribbed and corrugated panels, but for panels with widely spaced grooves or rustications they give excessively low results. Consequently, engineering judgment must be used when applying the above expressions.

Example 9.3.1 Fire endurance of a ribbed panel

Given:

The section of a wall panel shown.

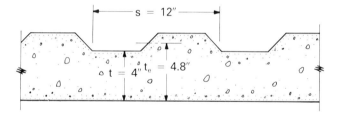

Problem:

Estimate the fire endurance if the minimum thickness is 4 in. and the equivalent thickness is 4.8 in. Assume that the panel is made of sand-lightweight concrete.

Solution:

See Fig. 9.3.2.

t = 4 in., s/2 = 6 in., s/4 = 3 in.

Therefore, s/2 > 4 > s/4

R = $R_t + (4 t/s - 1)(R_{te} - R_t)$
R_t = fire endurance of 4 in. sand-lightweight panel = 135 min
R_{te} = fire endurance of 4.8 in. sand-lightweight panel = 193 min
R = 135 + [4(4)/12 − 1][193 − 135]
  = 154 min

Fig. 9.3.7 Cross sections of ribbed wall panels

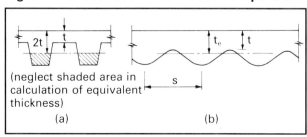

9.3.6.4 Multi-course assemblies

Floors and roofs often consist of concrete base slabs with overlays or undercoatings of other types of concrete or insulating materials. In addition, roofs generally have built-up roofing.

If the fire endurance of the individual courses are known, the fire endurance of the composite assembly can be estimated from the formula:

$$R = (R_1^{0.59} + R_2^{0.59} + \ldots + R_n^{0.59})^{1.7} \quad \text{(Eq. 9.3.1)}$$

where:

R = fire endurance of the composite assembly in minutes

R_1, R_2, R_n = fire endurances of the individual courses in minutes

The following example illustrates the use of this equation.

Example 9.3.2 Fire endurance of an assembly

Problem:

Determine the fire endurance of a slab consisting of a 2 in. base slab of siliceous aggregate concrete with a 2½ in. topping of sand-lightweight concrete (115 pcf).

Solution:

From Fig. 9.3.2, the fire endurances of a 2 in. thick slab of siliceous aggregate concrete and 2½ in. of sand-lightweight aggregate concrete are 25 min and 54 min, respectively.

$$R = [(25)^{0.59} + (54)^{0.59}]^{1.7}$$

$$R = (6.68 + 10.52)^{1.7} = 126 \text{ min} = 2 \text{ hr } 6 \text{ min}$$

Table 9.3.2 Values of R of various insulating materials for use in Eq. 9.3.1

Roof Insulation Material[1]	Thickness (in.)	R (min)
Cellular plastic	≥ 1	5
Glass fiber board	¾	11
Glass fiber board	1½	35
Foam glass	2	55
Mineral board	1	19
Mineral board	2	62
Mineral board	3	123

1. Some of these materials may be used only where combustible construction is permitted.

Table 9.3.2 gives values which can be used in this equation for certain insulating materials. For heat transmission, three-ply built-up roofing contributes 10 min to the fire endurance.

Eq. 9.3.1 has certain shortcomings in that it does not account for the location of the individual courses relative to the fired surface. Also, it is not possible to directly obtain the fire endurances of many insulating materials. Nevertheless, in a series of tests, the formula estimated the fire endurances within about 10% for most assemblies.

A report on two-course floors and roofs[2] gives results of many fire tests. The report also shows graphically the fire endurances of assemblies consisting of various thicknesses of two materials. Tables 9.3.3 through 9.3.5, which are based on test results, can be used to estimate the required thicknesses of two-course materials for various fire endurances.

9.3.6.5 Sandwich panels

Some wall panels are made by sandwiching an insulating material between two face slabs of concrete. See Sect. 9.4.

Several building codes require that where non-combustible construction is specified, combustible elements in walls shall be limited to thermal and sound insulation having a flame spread classification of not more than 75 when the insulation is sandwiched between two layers of non-combustible material such as concrete.

When insulation is not installed in this manner, it is required to have a flame spread of not more than 25. Data on flame spread classification are available from insulation manufacturers.

A fire test was conducted of one such panel that consisted of a 2 in. base slab of carbonate aggregate concrete, a 1 in. layer of cellular polystyrene insulation, and a 2 in. face slab of carbonate aggregate concrete. The resulting fire endurance was 2 hr 00 min. From Eq. 9.3.1, the contribution of the 1 in. layer of polystyrene was calculated to be 5 min.

It is likely that the comparable R value for a 1 in. layer of cellular polyurethane would be somewhat greater than that for a 1 in. layer of cellular polystyrene, but test values are not available. Until more definitive data are obtained, it is suggested that 5 min be used as the value for R for any layer of cellular plastic 1 in. or greater.

It should be noted that the cellular plastics melt and are consumed at about 400 to 600°F. Thus, additional thickness or changes in composition probably have only a minor effect on the fire endurance of sandwich panels. The danger of toxic fumes caused by the burning cellular plastics is practically eliminated when the plastics are completely encased within concrete sandwich panels.[8]

Table 9.3.3 Thickness of spray-applied insulation on fire-exposed surface of concrete[1] slabs or panels to resist transfer of heat through the assemblies

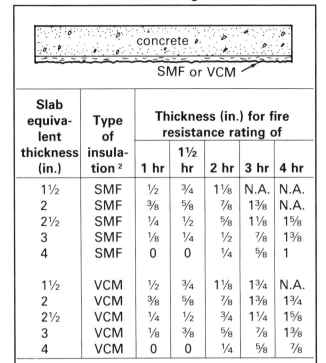

Slab equivalent thickness (in.)	Type of insulation[2]	Thickness (in.) for fire resistance rating of				
		1 hr	1½ hr	2 hr	3 hr	4 hr
1½	SMF	½	¾	1⅛	N.A.	N.A.
2	SMF	⅜	⅝	⅞	1⅜	N.A.
2½	SMF	¼	½	⅝	1⅛	1⅝
3	SMF	⅛	¼	½	⅞	1⅜
4	SMF	0	0	¼	⅝	1
1½	VCM	½	¾	1⅛	1¾	N.A.
2	VCM	⅜	⅝	⅞	1⅜	1¾
2½	VCM	¼	½	¾	1¼	1⅝
3	VCM	⅛	⅜	⅝	⅞	1⅜
4	VCM	0	0	¼	⅝	⅞

1. Values shown are for siliceous aggregate concrete, and are conservative for other concretes.
2. SMF = Sprayed mineral fiber consists of refined mineral fibers with inorganic binders and water added during the spraying operation. The density of the oven-dry material should be at least 13 pcf.
   VCM = Vermiculite cementitious material consists of expanded vermiculite with inorganic binders and water. The density of the oven-dry material should be at least 14 pcf.
   N.A. = Not applicable.

Table 9.3.5 Thickness of roof assemblies consisting of concrete[1] slabs with insulation and built-up roofing

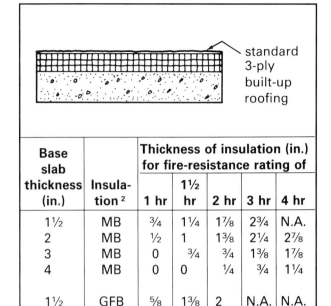

Base slab thickness (in.)	Insulation[2]	Thickness of insulation (in.) for fire-resistance rating of				
		1 hr	1½ hr	2 hr	3 hr	4 hr
1½	MB	¾	1¼	1⅞	2¾	N.A.
2	MB	½	1	1⅜	2¼	2⅞
3	MB	0	¾	¾	1⅜	1⅞
4	MB	0	0	¼	¾	1¼
1½	GFB	⅝	1⅜	2	N.A.	N.A.
2	GFB	¼	⅞	1½	2⅞	N.A.
3	GFB	0	⅜	¾	1½	2⅛
4	GFB	0	0	¼	¾	1¼

1. Values shown are for siliceous aggregate concrete, and are conservative for other concretes.
2. MB = Mineral board insulation composed of spherical cellular beads of expanded aggregate and fibers formed into rigid flat rectangular units with an integral waterproofing treatment.
   GFB = Glass fiber board fibrous glass roof insulation consisting of inorganic glass fibers formed into rigid boards using a binder. The board has a top surface faced with glass fiber reinforced with asphalt and kraft.
   N.A. = Not applicable.

Table 9.3.4 Thickness of two-course roof assemblies consisting of concrete[1] slabs with insulating concrete overlays[2]

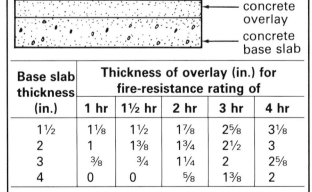

Base slab thickness (in.)	Thickness of overlay (in.) for fire-resistance rating of				
	1 hr	1½ hr	2 hr	3 hr	4 hr
1½	1⅛	1½	1⅞	2⅝	3⅛
2	1	1⅜	1¾	2½	3
3	⅜	¾	1¼	2	2⅝
4	0	0	⅝	1⅜	2

1. Values shown are for siliceous aggregate concrete and are conservative for other concretes.
2. Insulating concrete having a dry density less than 35 pcf.

Table 9.3.6 lists fire endurances of sandwich panels with either cellular plastic, glass fiber board, or insulating concrete used as the insulating material. The fire resistance values were obtained by use of Eq. 9.3.1.

9.3.6.6 Treatment of joints between wall panels

Joints between wall panels should be detailed so that passage of flame or hot gases is prevented, and transmission of heat does not exceed the limits specified in ASTM E119. Concrete wall panels expand when heated, so the joints tend to close during fire exposure. Non-combustible materials that are flexible, such as ceramic fiber blankets, provide thermal, flame, and smoke barriers, and, when used in conjunction with caulking materials, they can provide the necessary weather-tightness while permitting normal volume change movements. Joints that do not move can be filled with mortar.

Table 9.3.6 Fire endurance of precast concrete sandwich walls [calculated, based on Eq. 9.3.1]

Outside and inside wythes	Insulation	Fire endurance hr: min
1½ in. Sil	1 in. CP	1:23
1½ in. Carb	1 in. CP	1:23
1½ in. SLW	1 in. CP	1:45
2 in. Sil	1 in. CP	1:50
2 in. Carb	1 in. CP	2:00
2 in. SLW	1 in. CP	2:32
3 in. Sil	1 in. CP	3:07
1½ in.. Sil	¾ in. GFB	1:39
2 in. Sil	¾ in. GFB	2:07
2 in. SLW	¾ in. GFB	2:52
1½ in. Sil	1½ in. GFB	2:35
2 in. Sil	1½ in. GFB	3:08
2 in. SLW	1½ in. GFB	4:00
1½ in. Sil	1 in. IC	2:12
1½ in. SLW	1 in. IC	2:39
2 in. Carb	1 in. IC	2:56
2 in. SLW	1 in. IC	3:33
1½ in. Sil	1½ in. IC	2:54
1½ in. SLW	1½ in. IC	3:24
2 in. Sil	2 in. IC	4:25
1½ in. SLW	2 in. IC	4:19

Note:
Carb = carbonate aggregate concrete
Sil = siliceous aggregate concrete
SLW = sand-lightweight concrete (115 pcf maximum)
CP = cellular plastic (polystyrene or polyurethane)
IC = lightweight insulating concrete (35 pcf maximum)
GFB = glass fiber board

Joints between wall panels are similar to openings. Most building codes do not require openings to be protected against fire if the openings constitute only a small percentage of the wall area and if the spatial separation is greater than some minimum distance. In those cases, protection of joints would not be required.

In other cases, openings must be protected, but most codes permit a lesser degree of protection. For example, the Uniform Building Code requires that, when openings are permitted and must be protected, the "openings shall be protected by a fire assembly having a ¾-hour fire-protection rating." Where no openings are permitted, the fire resistance required for the wall should be provided at the joints.

Table 9.3.7 is based on results of fire tests of panels with butt joints.[3] The tabulated values apply to one-stage butt joints and are conservative for two-stage and ship-lap joints.

Joints between adjacent precast floor or roof elements may be ignored in calculating the slab thickness provided that a concrete topping at least 1½ in. thick is used. Where no concrete topping is used, joints should be grouted to a depth of at least one-third the slab thickness at the joint, or the joints made fire-resistive in a manner acceptable to the authority having jurisdiction.

9.3.7 Designing for Structural Integrity

It was noted above that many fire tests and related research studies have been directed toward an understanding of the structural behavior of prestressed concrete subjected to fire. The information gained from that work has led to the development of calculation procedures which can be used in lieu of fire tests. The purpose of this section is to present an

Table 9.3.7 Protection of joints between wall panels utilizing ceramic fiber felt

Panel equivalent thickness[2] (in.)	Thickness of ceramic fiber felt (in.) required for fire resistance ratings and joint widths[1] shown							
	Joint width = ⅜ in.				Joint width = 1 in.			
	1 hr	2 hr	3 hr	4 hr	1 hr	2 hr	3 hr	4 hr
4	¼	N.A.	N.A.	N.A.	¾	N.A.	N.A.	N.A.
5	0	¾	N.A.	N.A.	½	2⅛	N.A.	N.A.
6	0	0	1⅛	N.A.	¼	1¼	3½	N.A.
7	0	0	0	1	¼	⅞	2	3¾

N.A. = Not applicable

1. Interpolation may be used for joint width between ⅜ in. and 1 in. The tabulated values apply to one-stage butt joints and are conservative for two-stage and ship-lap joints.
2. Panel equivalent thicknesses are for carbonate concrete. For siliceous aggregate concrete change "4, 5, 6, and 7" to "4.3, 5.3, 6.5, and 7.5." For sand-lightweight concrete change "4, 5, 6, and 7" to "3.3, 4.1, 4.9, and 5.7."

introduction to these calculation procedures. Because the method of support is the most important factor affecting structural behavior of flexural elements during a fire, the discussion that follows deals with three conditions of support: simply supported members, continuous slabs and beams, and members in which restraint to thermal expansion occurs.

9.3.7.1 Simply supported members

Assume that a simply supported prestressed concrete slab is exposed to fire from below, that the ends of the slab are free to rotate, and that expansion can occur without restriction. Also, assume that the reinforcement consists of straight uncoated strands located near the bottom of the slab. With the underside of the slab exposed to fire, the bottom will expand more than the top causing the slab to deflect downward; also, the strength of the steel and concrete near the bottom will decrease as the temperature rises. When the strength of the steel diminishes to that required to support the slab, flexural collapse will occur. In essence, the applied moment remains practically constant during the fire exposure, but the resisting moment capacity is reduced as the steel weakens.

Fig. 9.3.8 illustrates the behavior of a simply supported slab exposed to fire from beneath, as described above. Because strands are parallel to the axis of the slab, the design moment strength is constant throughout the length:

$$\phi M_n = \phi A_{ps} f_{ps} (d - a/2) \qquad \text{(Eq. 9.3.2)}$$

f_{ps} can be determined from Fig. 4.10.3 or Eq. 18-3 of the ACI Building Code, ACI 318-89.[11]

If the slab is uniformly loaded, the moment diagram will be parabolic with a maximum value at midspan of:

$$M = \frac{w\ell^2}{8} \qquad \text{(Eq. 9.3.3)}$$

where:

    w = dead plus live load per unit of length, k/in.
    ℓ = span length, in.

As the material strengths diminish with elevated temperatures, the retained nominal strength becomes:

$$M_{n\theta} = A_{ps} f_{ps\theta} (d - a_\theta/2) \qquad \text{(Eq. 9.3.4)}$$

in which θ signifies the effects of high temperatures. Note that A_{ps} and d are not affected, but f_{ps} is reduced. Similarly, a is reduced, but the concrete strength at the top of the slab, f'_c, is generally not reduced significantly because of its lower temperature.

Flexural failure can be assumed to occur when $M_{n\theta}$ is reduced to M. Strength reduction factor, ϕ, is not applied because a safety factor is included in the required ratings.[4] From this expression, it can be seen that the fire endurance depends on the applied loading and on the strength-temperature characteristics of the steel.

In turn, the duration of the fire before the "critical" steel temperature is reached depends on the protection afforded to the reinforcement.

Fig. 9.3.8 Moment diagrams for simply supported beam or slab

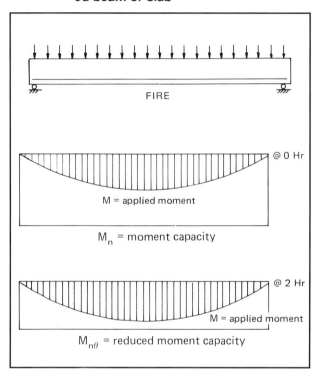

To solve problems involving the above equations, it is necessary to utilize data on the strength-temperature relationships for steel and concrete, and information on temperature distributions within concrete members during fire exposures. Fig. 9.3.9 shows strengths of certain steels at elevated temperatures, and Fig. 9.3.10 shows similar data for various types of concrete.

Data on temperature distribution in concrete slabs during fire tests are shown in Figs. 9.3.11 and 9.3.12. These figures can also be used for beams wider than about 10 in. An "effective u," $\bar{u}$, is used, which is the average of the distances between the centers of the individual strands or bars and the nearest fire-exposed surface. The values for corner strands or bars are reduced one-half to account for the exposure from two sides (see Example 9.3.4). The procedure does not apply to bundled bars or strands.

Example 9.3.3 Calculation of fire endurance of a hollow-core slab

Given:

An 8 in. deep hollow-core slab with a simply supported unrestrained span of 25 ft.

h = 8 in.
u = 1.75 in.
Eight ½ in. 270 ksi strands
A_{ps} = 8(0.153) = 1.224 sq in.
b = 48 in.
d = 8 − 1.75 = 6.25 in.
w_d = 60 psf
Carbonate aggregate concrete
f'_c = 5000 psi
ℓ = 25 ft

Problem:

Determine the maximum safe superimposed load that can be supported for a fire endurance of 3 hr.

Solution:

(a) Estimate strand temperature at 3 hr from Fig. 9.3.11: At 3 hr, carbonate aggregate,

u = 1.75 in.
θ_s = 925°F

(b) Determine $f_{pu\theta}$ from Fig. 9.3.9. For cold-drawn steel at 925°F

$f_{pu\theta}$ = 0.33(f_{pu}) = 89.1 ksi

(c) Determine $M_{n\theta}$ and w from Fig. 4.2.2.

$f_{ps\theta} = 89.1 \left[1 - \dfrac{0.28(1.224)(89.1)}{0.8(48)(6.25)(5)} \right] = 86.8$ ksi

$a_\theta = \dfrac{A_{ps} f_{ps\theta}}{0.85 f'_c b} = \dfrac{1.224(86.8)}{0.85(5)(48)} = 0.52$ in.

$M_{n\theta} = A_{ps} f_{ps\theta} (d - a_\theta/2)$
= 1.224(86.8)(6.25 − 0.52/2)/12
= 53.0 ft-kips

$w = \dfrac{8M}{b\ell^2} = \dfrac{8(53.0)(1000)}{(4)(25)^2} = 169$ psf

$w_\ell = w - w_d = 169 - 60 = 109$ psf

Example 9.3.4 Calculation of fire endurance of a rectangular beam

Given:

The 12RB24 shown.
Span = 30 ft
Dead load (including bm. wt.) = 1100 plf

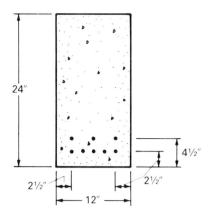

Live load = 1100 plf
Simple support, no restraint
Siliceous aggregate concrete, f'_c = 5000 psi
Stress-relieved ½ in. dia. strand, f_{ps} = 270 ksi

Problem:

Determine the necessary reinforcement for a 4-hr fire endurance rating.

Solution:

Try 8 strands
A_{ps} = 8(0.153) = 1.224 sq in.
y_s = [5(2.5) + 3(4.5)]/8 = 3.25 in.
d = 24 − 3.25 = 20.75 in.
$\bar{u}$ = [5(2.5) + 1(4.5) + 2(2.5)(0.5)]/8 = 2.44 in.

From Fig. 9.3.11, siliceous aggregate:
at 4 hr, strand temp. = 920°F

From Fig. 9.3.9:
$f_{pu\theta}$ = 0.34(270) = 91.8 ksi

From Fig. 4.2.2:

$f_{ps\theta} = 91.8 \left[1 - \dfrac{(0.4)}{(0.8)} \times \dfrac{(1.224)(91.8)}{(12)(20.75)(5)} \right] = 87.7$ ksi

$a_\theta = \dfrac{A_{ps} f_{ps\theta}}{0.85 f'_c b} = \dfrac{1.224(87.7)}{0.85(5)(12)}$
= 2.10 in.

$M_{n\theta} = A_{ps} f_{ps\theta} (d - a_\theta/2)$
= 1.224(87.7) (20.75 − 2.10/2)
= 2115 in.-kips = 176 ft-kips

Service load moment:
w = 2.20 kips/ft
M = $w\ell^2/8$ = 2.20(30)²/8
= 247.5 ft-kips > 176 ft-kips

Therefore, beam will not satisfy criteria for a 4-hr fire endurance.

Fig. 9.3.9 Strength-temperature relationships for various steels

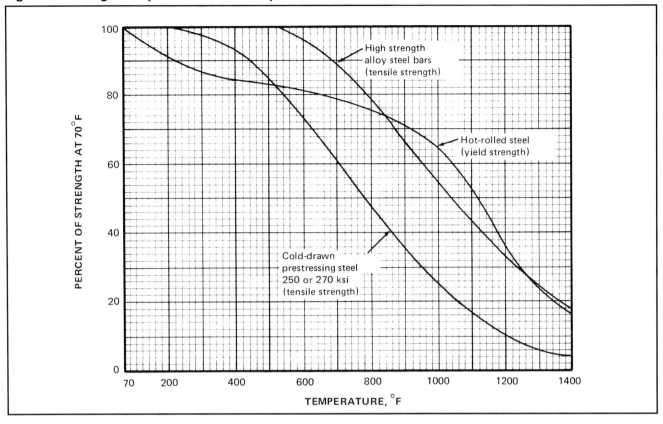

Fig. 9.3.10 Compressive strength of concrete at high temperatures

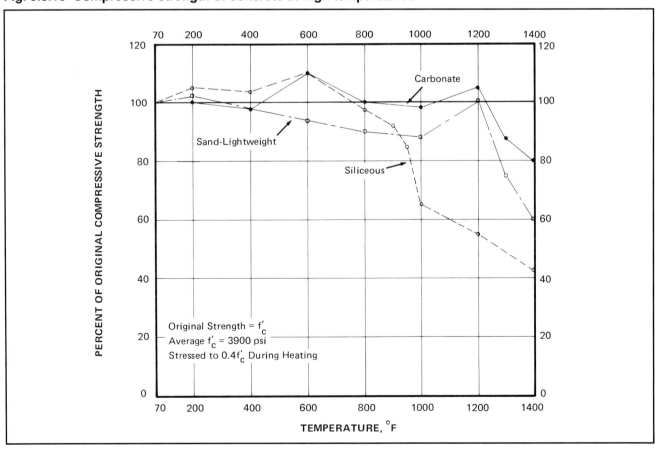

PCI Design Handbook/Fourth Edition

Fig. 9.3.11 Temperature within concrete slabs or panels during fire tests—normal weight concrete

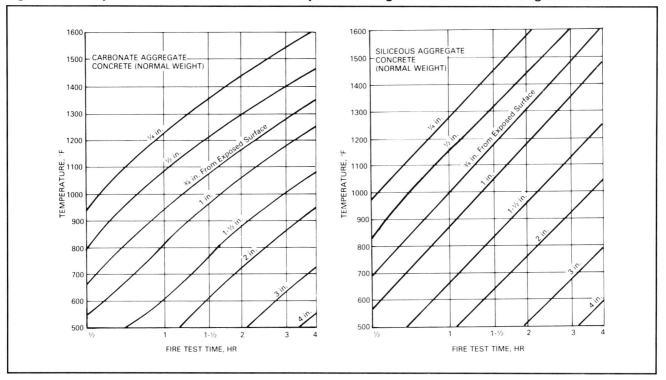

Fig. 9.3.12 Temperatures within concrete slabs or panels during fire tests—sand lightweight concrete

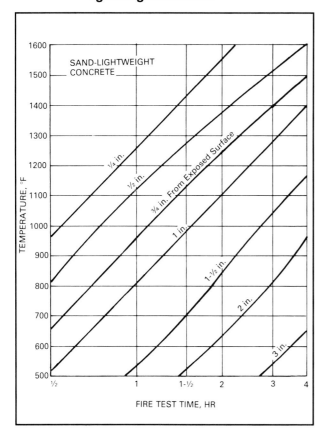

Solution No. 1:
Try adding 2 additional strands at 4½ in. from bottom:

A_{ps} = 10(0.153) = 1.53 sq in.

y_s = [5(2.5) + 5(4.5)]/10 = 3.5 in.

d = 24 − 3.5 = 20.5 in.

$\bar{u}$ = [5(2.5) + 1(4.5) + 2(4.25) + 2(2.5)(0.5)]/10
   = 2.80 in.

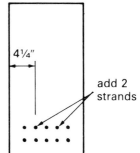

From Fig. 9.3.11:
  Strand temp. = 840°F

From Fig. 9.3.9:
  $f_{pu\theta}$ = 0.42(270) = 113.4 ksi

From Fig. 4.2.2:

$f_{ps\theta}$ = 113.4 $\left[1 - \dfrac{(0.40)}{(0.8)} \times \dfrac{(1.53)(113.4)}{(12)(20.5)(5)} \right]$ = 105.4 ksi

a_θ = $\dfrac{1.53(105.4)}{0.85(5)(12)}$ = 3.16 in.

$M_{n\theta}$ = 1.53(105.4)(20.5 – 3.16/2)
= 3051 in.-kips
= 254 ft-kips > 247.5 ft-kips OK

Solution No. 2:
Try adding 2 – #7 at 4.5 in. from bottom:
A_s = 1.20 sq in.
d(rebar) = 24 – 4.5 = 19.5 in.
u(rebar) = 4.25 in.

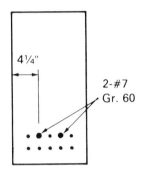

From Fig. 9.3.11:
Bar temp. at 4 hr = 500°F
From Fig. 9.3.9:
$f_{y\theta}$ = 0.83 f_y = 49.8 ksi
$a_\theta = \dfrac{1.224(87.7) + 1.20(49.8)}{0.85(5)(12)}$ = 3.28 in.
$M_{n\theta}$ (strand) = 1.224(87.7)(20.75 – 3.28/2)
= 2051 in.-kips = 171 ft-kips
$M_{n\theta}$ (bars) = 1.20(49.8)(19.5 – 3.28/2)
= 1067 in.-kips = 89 ft-kips
171 + 89 = 260 ft-kips > 247.5 OK

The calculation of fire endurance for stemmed units is similar to the above examples. Figures 9.3.3, 9.3.4, 9.3.5 and 9.3.6 give the temperatures on the vertical centerline of stemmed units for 1-hour, 2-hour, 3-hour and 4-hour exposures.

9.3.7.2 Continuous members

Continuous members undergo changes in stresses when subjected to fire. These stresses result from temperature gradients within the structural members, or changes in strength of the materials at high temperatures, or both.

Fig. 9.3.13 shows a two-span continuous beam whose underside is exposed to fire. The bottom of the beam becomes hotter than the top and tends to expand more than the top. This differential temperature effect causes the ends of the beam to tend to lift from their support thereby increasing the reaction at the interior support. This action results in a redistribution of moments, i.e., the negative moment at the interior support increases while the positive moments decrease.

Fig. 9.3.13 Moment diagram for a two-span continuous beam

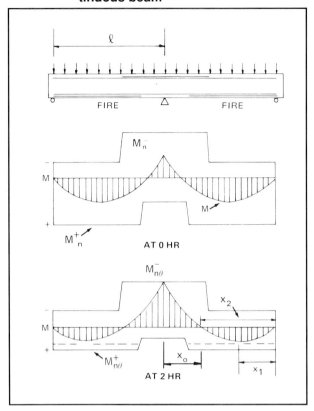

During a fire, the negative moment reinforcement (Fig. 9.3.13) remains cooler than the positive moment reinforcement because it is better protected from the fire. In addition, the redistribution that occurs is sufficient to cause yielding of the negative moment reinforcement. Thus, a relatively large increase in negative moment can be accommodated throughout the test. The resulting decrease in positive moment means that the positive moment reinforcement can be heated to a higher temperature before failure will occur. Therefore, the fire endurance of a continuous concrete beam is generally significantly longer than that of a simply supported beam having the same cover and the same applied loads.

It is possible to design the reinforcement in a continuous beam or slab for a particular fire endurance period. From Fig. 9.3.13 the beam can be expected to collapse when the positive moment capacity, $M_{n\theta}^+$ is reduced to the value of the maximum redistributed positive moment at a distance x_1 from the outer support.

Fig. 9.3.14 shows a uniformly loaded beam or slab continuous (or fixed) at one support and simply supported at the other. Also shown is the redistributed applied moment diagram at failure.

It can be shown that at the point of positive moment, x_1,

$$x_1 = \frac{\ell}{2} - \frac{M_{n\theta}^-}{w\ell} \qquad \text{(Eq. 9.3.5)}$$

at $x = x_2$, $M_x = 0$ and $x_2 = 2x_1$

$$x_o = \frac{2M_{n\theta}^-}{w\ell} \quad \text{(Eq. 9.3.6)}$$

$$M_{n\theta}^- = \frac{w\ell^2}{2} \pm w\ell^2 \sqrt{\frac{2M_{n\theta}^+}{w\ell^2}} \quad \text{(Eq. 9.3.7)}$$

Fig. 9.3.14 Uniformly loaded member continuous (or fixed) at one support

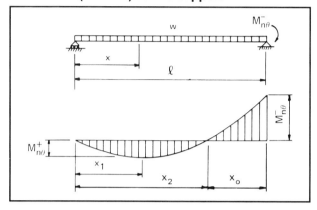

Fig. 9.3.15 Symmetrical uniformly loaded member continuous at both supports

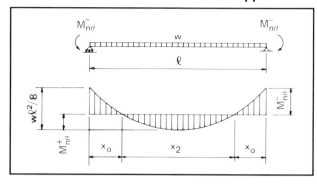

In most cases, redistribution of moments occur early during the course of a fire and the negative moment reinforcement can be expected to yield before the negative moment capacity has been reduced by the effects of fire. In such cases, the length of x_o is increased, i.e., the inflection point moves toward the simple support. If the inflection points moves beyond the point where the bar stress cannot be developed in the negative moment reinforcement, sudden failure may result.

Fig. 9.3.15 shows a symmetrical beam or slab in which the end moments are equal.

$$M_{n\theta}^- = \frac{w\ell^2}{8} - M_{n\theta}^+ \quad \text{(Eq. 9.3.8)}$$

$$\frac{wx_2^2}{8} = M_{n\theta}^+ \quad \text{(Eq. 9.3.9)}$$

$$x_2 = \sqrt{\frac{8M_{n\theta}^+}{w}} \quad \text{(Eq. 9.3.10)}$$

$$x_o = \tfrac{1}{2}(\ell - x_2)$$

$$= \frac{\ell}{2} - \frac{1}{2}\sqrt{\frac{8M_{n\theta}^+}{w}} \quad \text{(Eq. 9.3.11)}$$

To determine the maximum value of x_o, the value of w should be the minimum service load anticipated, and $(w\ell^2/8 - M_n^-)$ should be substituted for $M_{n\theta}^+$ in Eq. 9.3.11.

For any given fire endurance period, the value of $M_{n\theta}^+$ can be calculated by the procedures given in Sect. 9.3.7. Then the value of M_n^- can be calculated by the use of Eqs. 9.3.7 or 9.3.8 and the necessary lengths of the negative moment reinforcement can be determined from Eqs. 9.3.6 or 9.3.11. Use of these equations is illustrated in Example 9.3.5.

It should be noted that the amount of moment redistribution that can occur is dependent on the amount of negative moment reinforcement. Tests have clearly demonstrated that in most cases the negative moment reinforcement will yield, so the negative moment capacity is reached early during a fire test, regardless of the applied loading. The designer must exercise care to ensure that a secondary type of failure will not occur. To avoid a compression failure in the negative moment region, the amount of negative moment reinforcement should be small enough so that $\omega_\theta = A_s f_{y\theta}/b_\theta d_\theta f'_{c\theta}$ is less than 0.30, before and after reductions in f_y, b, d and f'_c are taken into account. Furthermore, the negative moment bars or mesh must be long enough to accommodate the complete redistributed moment and change in the inflection points. It should be noted that the worst condition occurs when the applied loading is smallest, such as the dead load plus partial or no live load. It is recommended that at least 20% of the maximum negative moment reinforcement extend throughout the span.

Example 9.3.5 Calculation of fire endurance for hollow-core slab with topping

Problem:

Design a floor using hollow-core slabs and topping for 22 ft span for 4 hr fire endurance. Service loads = 175 psf dead (including structure) and 150 psf live. Use 4 ft wide, 10 in. deep slabs with 2 in. topping, carbonate aggregate concrete. Continuity can be achieved at both ends. Use f'_c (precast) = 5000 psi, f_{pu} = 250 ksi, and f'_c (topping) = 3000 psi, sixteen ⅜ in., 250 ksi strands at u = 1.75 in. Provide negative moment reinforcement needed for fire resistance.

Solution:

A_{ps} = 16(0.080) = 1.28 sq in.
u = 1.75 in.
d = 12 − 1.75 = 10.25 in.

From Fig. 9.3.11, θ_s = 1010°F

From Fig. 9.3.9, $f_{pu\theta}$ = 0.24f_{pu} = 60 ksi

Using Table 4.10.2:*

$$\omega_{p\theta} = \frac{16(0.080)(60)}{48(10.25)(3)} = 0.052$$

K'_u = 136/0.9 = 151

$M_{n\theta}$ = K'_u bd²/12,000

= 151(48)(10.25)²/12,000

= 63.5 ft-kips/unit

= 15.9 ft-kips/ft

For simply supported members:

M = 0.325(22)²/8 = 19.7 ft-kips/ft

Req'd $M^-_{n\theta}$ = 19.7 − 15.9 = 3.8 ft-kips/ft

Assume d − a_θ/2 = 10.25 in., and f_y = 60 ksi

$$A^-_s = \frac{3.8(12)}{60(10.25)} = 0.074 \text{ in.}^2/\text{ft}$$

Use 20% of required A_s throughout span:

Try 6 x 6-W1.4 x W1.4 continuous plus

6 x 6-W2.9 x W2.9 over supports

A^-_s = 0.029 + 0.058 = 0.087 in.²/ft

Neglect concrete above 1400°F in negative moment region, i.e., from Fig. 9.3.11, neglect bottom ⅝ in. Also, concrete within compressive zone will be about 1350 to 1400°F, so use $f'_{c\theta}$ = 0.81f'_c (see Fig. 9.3.10) = 4.05 ksi.

Check $M^-_{n\theta}$, assuming that the temperature of the negative steel does not rise above 200°F. If greater then 200°F, steel strength should be reduced according to Fig. 9.3.9.

$$a_\theta = \frac{0.087(60)}{0.85(4.05)(12)} = 0.126 \text{ in.}$$

$M^-_{n\theta}$ = 0.087(60)(10.37 − 0.063)/12

= 4.48 ft-kips/ft

With dead load + ½ live load, w = 0.25 ksf, M = 15.12 ft-kips/ft, and M^-_n = 4.71 ft-kips/ft (calculated for room temperature)

M^+_{min} = 15.12 − 4.71 = 10.41 ft-kips/ft

From Eq. 9.3.11

$$\max x_o = \frac{22}{2} - \frac{1}{2}\sqrt{\frac{8(10.41)}{0.25}} = 1.87 \text{ ft}$$

Use 6 x 6-W1.4 x W1.4 continuous throughout plus 6 x 6-W2.9 x W2.9 for a distance of 3 ft from the support. Mesh must extend into walls which must be designed for the moment induced at the top.

*The values for K'_u in Table 4.10.2 include θ = 0.9. Since in the design for fire, θ = 1.0, the value of K'_u must be divided by 0.9.

9.3.7.3 Members restrained against thermal expansion

If a fire occurs beneath an interior portion of a large reinforced concrete slab, the heated portion will tend to expand and push against the surrounding part of the slab. In turn, the unheated part of the slab exerts compressive forces on the heated portion. The compressive force, or thrust, acts near the bottom of the slab when the fire test occurs but, as the fire progresses, the line of action of the thrust rises as the mechancial properties of the heated concrete changes. This thrust is generally great enough to increase the fire endurance significantly.

The effects of restraint to thermal expansion can be characterized as shown in Fig. 9.3.16. The thermal thrust acts in a manner similar to an external prestressing force, which, in effect, increases the positive moment capacity.

The increase in bending moment capacity is similar to the effect of added reinforcement located along the line of action of the thrust. It can be assumed that the added reinforcement has a yield strength (force) equal to the thrust. By this approach, it is possible to determine the magnitude and location of the required thrust to provide a given fire endurance.

The above explanation is greatly simplified because in reality restraint is quite complex, and can be likened to the behavior of a flexural member subjected to an axial force. Interaction diagrams similar to those for columns can be constructed for a given cross-section at a particular stage of a fire, e.g., 2 hr of a standard fire exposure.[5] The guidelines in ASTM E119 given for determining conditions of restraint are useful for preliminary design purposes. Most interior and many exterior bays of multi-bay floors or roofs can be considered to be restrained and the magnitude and location of the thrust are generally of academic interest only. In such cases, the fire endurance is governed by heat transmission rather than by structural considerations.

9.3.7.4 Shear resistance

Many fire tests have been conducted on simply supported reinforced or prestressed concrete elements as well as on elements in which restraint to thermal expansion occurred. Shear failures did not occur in any of those tests.

It should be noted that when beams which are continuous over one support (e.g., such as that shown in Fig. 9.3.13) are exposed to fire, both the moment and the shear at the interior support increase. Such a redistribution of moment and shear results in a severe stress condition. However, of the several fire tests of reinforced concrete beams in which that condition was simulated, failure occured only in one beam.[6] In that test, the shear reinforcement was inadequate, even for service load conditions without fire, as judged by the shear requirements of ACI 318-89. Thus, it appears from available test data that members which are designed for shear strength in accordance with ACI 318-89 will perform satisfactorily in fire situations, i.e., failure will not occur prematurely due to a shear failure.

9.3.8 Protection of Connections

Many types of connections in precast concrete construction are not vulnerable to the effects of fire, and consequently, require no special treatment. For example, gravity-type connections, such as the bearing between precast concrete panels and concrete footings or beams which support them, do not generally require special fire protection.

If the panels rest on elastomeric pads or other combustible materials, protection of the pads is not generally needed because deterioration of the pads will not cause collapse.

Connections that can be weakened by fire and thereby jeopardize the structure's load carrying capacity should be protected to the same degree as that required for the supported member. For example, an exposed steel bracket supporting a panel or spandrel beam will be weakened by fire and might fail causing the panel or beam to collapse. Such a bracket should be protected.

The amount of protection depends on (a) the stress-strength ratio in the steel at the time of the fire and (b) the intensity and duration of the fire. The thickness of protection materials required is greater as the stress level and fire severity increase.

Fig. 9.3.17 shows the thickness of various commonly used fire protection materials required for fire endurances up to 4 hr. The values shown are based on a critical steel temperature of 1000°F, i.e., a stress-strength ratio (f_s/f_y) of about 65%. Values in Fig. 9.3.17 (b) are applicable to concrete or dry-pack mortar encasement of structural steel shapes used as brackets or lintels.

9.3.9 Precast Concrete Column Covers

Steel columns are often clad with precast concrete panels or covers for architectural reasons. Such covers also provide fire protection for the columns.

Fig. 9.3.18 shows the relationship between the thickness of concrete column covers and fire endurance for various steel column sections. The fire endurances shown are based on an empirical relationship developed by Lie and Harmathy.[7]

The above authors also found that the air space between the steel core and the column covers has only a minor effect on the fire endurance. An air space will probably increase the fire endurance but only by an insignificant amount.

Most precast concrete column covers are 3 in. or more in thickness, but some are as thin as 2½ in. From Fig. 9.3.18, it can be seen that precast concrete column covers can qualify the column for fire endurances of at least 2½ hr, and usually more than 3 hr. For steel column sections other than those shown, including shapes other than wide flange beams, interpolation between the curves on the basis of weight per foot will generally give reasonable results.

For example, the fire endurance afforded by a 3-in. thick column cover of normal weight concrete for a 8 x 8 x ½ in. steel tube column will be about 3 hr 20 min (the weight of the section is 47.35 lb per ft).

Precast concrete column covers (Fig. 9.3.19) are made in various shapes such as (a) four flat panels with butt or mitered joints that fit together to enclose the steel column, (b) four L-shaped units, (c) two L-shaped units, (d) two U-shaped units, and (e) and (f) U-shaped units and flat closure panels. Type (a) would probably be most vulnerable to bowing during fire exposure while Type (f) would probably be the least vulnerable. There are, of course, many combinations to accommodate isolated columns, corner columns, and column walls.

To be fully effective the column covers must remain in place without severe distortion. Many types of connections are used to hold the column covers in place. Some connections consist of bolted or welded clip angles attached to the tops and bottoms of the covers. Others consist of steel plates embedded in the covers that are welded to angles, plates, or other shapes which are, in turn, welded or bolted

Fig. 9.3.16 Longitudinally restrained beam during fire exposure

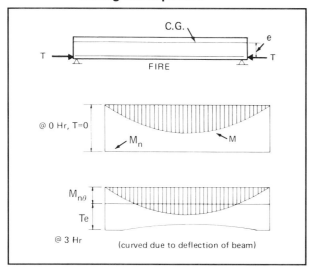

to the steel column. In any case, the connections are used primarily to position the column covers and as such are not highly stressed. As a result, temperature limits need not be applied to the steel in most column cover connections.

If restrained, either partially or fully, concrete panels tend to deflect or bow when exposed to fire. For example, for a steel column that is clad with four flat panels attached top and bottom, the column covers will tend to bulge at midheight thus tending to open gaps along the sides. The gap size decreases as the panel thickness increases.

With L, C, or U-shaped panels, the gap size is further reduced. The gap size can be further minimized by connections at midheight. In some cases, ship-lap joints can be used to minimize the effects of joint openings.

Joints should be sealed in such a way to prevent passage of flame to the steel column. A non-combustible material such as sand-cement mortar or ceramic fiber blanket can be used to seal the joint.

Precast concrete column covers should be installed in such a manner that, if they are exposed to fire, they will not be restrained vertically. As the covers are heated they tend to expand. Connections should accommodate such expansion without sub-

Fig. 9.3.17 Thickness of protection materials applied to connections consisting of structural steel shapes

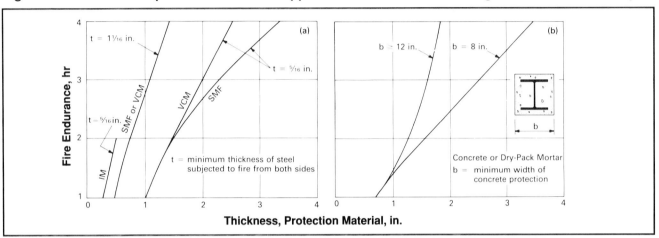

IM = intumescent mastic
SMF = sprayed mineral fiber
VCM = vermiculite cementitious material)

Fig. 9.3.18 Fire endurance of steel columns afforded protection by concrete column covers

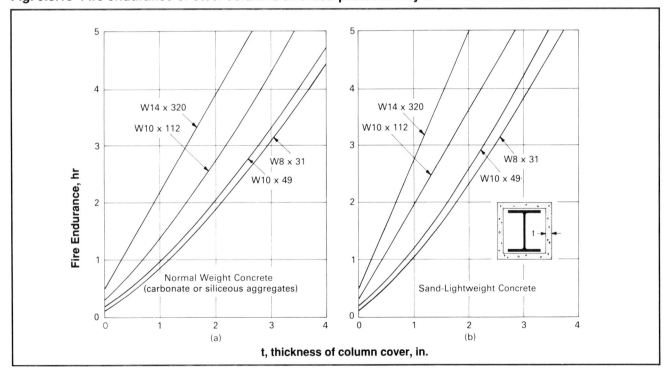

jecting the cover to additional loads.

Fire resistive compressible materials, such as mineral fiber safing, can be used to seal the tops or bases of the column covers, thus permitting the column covers to expand.

9.3.10 Code and Economic Considerations

An important aspect of dealing with fire resistance is to understand what the benefits are to the owner of a building in the proper selection of materials incorporated in his structure. These benefits fall into two areas: codes and economics.

Building codes are laws that must be satisfied regardless of any other considerations and the manner in which acceptance of code requirements is achieved is explained in the preceding pages. The designer, representing the owner, has no option in the code regulations, only in the materials and assemblies that meet these regulations.

Economic benefits associated with increased fire resistance should be considered by the designer/owner team at the time decisions are made on the structural system. Proper consideration of fire resistive construction through a life-cycle cost analysis will provide the owner economic benefits over other types of construction in many areas, e.g., lower insurance costs, larger allowable gross area under certain types of building construction, fewer stairwells and exits, increased value for loan purposes, longer mortgage terms, and better resale value. To ensure an owner of the best return on his investment, a life-cycle cost analysis using fire resistive construction should be prepared.

Beyond the theoretical considerations is the history of excellent performance of prestressed concrete in actual fires. Structural integrity has been maintained, fires are contained in the area of origin, and, in many instances, repairs consist of "cosmetic" treatment only, leading to early re-occupancy of the structure.

9.3.11 References

1. "Floor and Ceiling Assembly Consisting of Prestressed, Precast Concrete Double Tee Units with a Wallboard Ceiling," File R1319-131, February 21, 1973, Underwriters Laboratories, Inc., Northbrook, IL.

2. Abrams, M.S. and Gustaferro, A.H., "Fire Endurance of Two-Course Floors and Roofs," *Journal of the American Concrete Institute*, V. 66., No. 2, February 1969.

3. Gustaferro, A.H. and Abrams, M.S., "Fire Tests of Joints Between Precast Concrete Wall Panels: Effect of Various Joint Treatments," *PCI Journal*, V. 20, No. 5, September.-October 1975.

4. "Design for Fire Resistance of Precast Prestressed Concrete," Second Edition, MNL-124-89, Precast/ Prestressed Concrete Institute, Chicago, IL, 1989.

5. Abrams, M.S., Gustaferro, A. H. and Salse, E. A. B., "Fire Tests of Concrete Joist Floors and Roofs," RD 006B, Portland Cement Association, Skokie, IL, 1971.

6. "Fire Endurance of Continuous Reinforced Concrete Beams," RD 072B, Portland Cement Association, Skokie, IL.

7. Lie, T.T. and Harmathy, T.Z., "Fire Endurance of Concrete-Protected Steel Columns," *Journal of the American Concrete Institute*, V. 71, No. 1, January,1974.

8. Lie, T.T., "Contribution of Insulation in Cavity Walls to Propagation of Fire," *Fire Study No. 29*, Division of Building Research, National Research Council of Canada, Ottawa, Ontario, Canada.

9. "Fire Resistance Directory," Underwriters Laboratories, Inc., Northbrook, IL.

10. "Reinforced Concrete Fire Resistance," Concrete Reinforcing Steel Institute, Schaumburg, IL, 1980.

11. "Building Code Requirements for Reinforced Concrete," ACI 318-89, and "Commentary," ACI 318R-89, American Concrete Institute, Detroit, MI, 1989.

Fig. 9.3.19 Types of precast concrete column covers

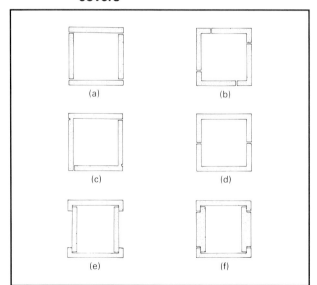

9.4 SANDWICH PANELS

9.4.1 General

Sandwich panels are composed of two concrete wythes separated by a layer of insulation. One of the concrete wythes may be a standard shape, as shown in Fig. 9.4.1, or any architectural concrete section produced for a single project. In place, sandwich panels provide the dual function of transferring load and insulating the structure. They may be used only for cladding, or they may act as beams, bearing walls or shear walls.

Fig. 9.4.1 Typical precast concrete load-bearing insulated wall panels

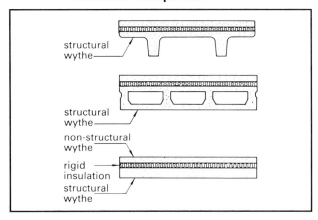

9.4.2 Structural Design

The structural design of sandwich panels is the same as design of other wall panels once the section properties of the panel have been determined. Three different assumptions may be used for the properties, depending on the construction:

1. Non-composite for full life cycle of panel: The two concrete layers act independently (Fig. 9.4.2a). The wythes are connected by ties and/or hangers which are flexible enough that they offer insignificant resistance to shrinkage and temperature movement. Positive steps may be taken to ensure that one wythe does not bond to the insulation. Typical techniques are by placing a sheet of polyethylene or reinforced paper over the insulation before the final concrete wythe is placed, by applying a retarder or form release agent to one side of the insulation, or by placing insulation in two layers (Fig. 9.4.9). One wythe is usually assumed to be "structural" and all loads are carried by that wythe, both during handling and in service, although, when designing for wind loads and slenderness effects, the panel may in some cases be designed as in (3) below.

2. Composite for full life cycle of panel: The two wythes act as a fully composite unit for the full life of the structure (Fig. 9.4.2b). In order to provide for composite behavior, positive measures must be taken to effect shear transfer between the wythes in the direction of panel span. This may be accomplished by rigid ties, longitudinal welded-wire trusses, or regions of solid concrete which join both wythes.

3. Composite during handling, but non-composite during service life: Composite action results largely because of early bond between insulation and adjacent wythe, in combination with flexible ties. The bond is considered unreliable for the long term, thus the panel is considered as non-composite for service life loads. Wind loads are distributed to each wythe in proportion to wythe stiffness. For effects of slenderness, the sum of the individual moments of inertia may be used.

For some machine-produced sandwich panels, tests and experience have demonstrated that a degree of composite action can be anticipated for the full life cycle. Manufacturer's recommendations for these panels should be followed.

The choice of panel type must consider the following.

A composite panel will have a stiffness significantly greater than a non-composite panel or partially composite panel of the same thickness. Thus, a composite panel can span greater heights. However, because there will be a significantly greater temperature gradient across the width of a composite panel, this panel type can be expected to bow more.

Panel size is limited only by stresses due to handling and service loads, and slenderness in load bearing panels, modified by the experience of the producer. Wythes should be no thinner than 2 in. or 3 times the maximum aggregate size. Panels with ¾ in. aggregate and 2 in. thick wythes have been successfully used.

The lifting points should be chosen to resist forces during stripping and handling without cracking, and the panel reinforced for service conditions. Prestressing in the long direction is particularly effective as a method of providing for crack-free panels; cracking due to restrained shrinkage in the short direction of the panel usually need not be considered, provided the dimension in that direction does not exceed about 12 ft. Prestressing should be concentric within each wythe which is stressed. The de-

Fig. 9.4.2 Non-composite and composite panels

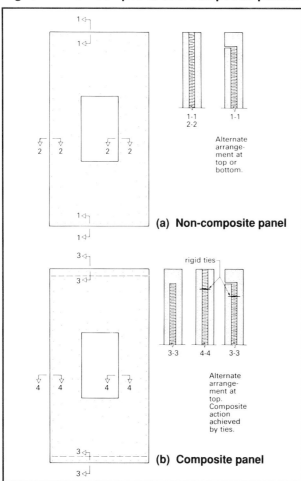

(a) Non-composite panel

(b) Composite panel

signer may consider prestressing the non-structural wythe to the same prestress level as the structural wythe, in order to minimize camber.

Example 9.4.1 Section properties of sandwich panels

Given:

The sandwich panel shown below.

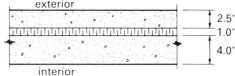

Problem:

Calculate the moment of inertia and section modulus if the section is (a) non-composite and (b) composite. Also find the load distribution for the non-composite case. Since the insulation is rigid, the lateral load applied to the non-composite panel will be resisted by each wythe in proportion to its stiffness.

Solution:

(a) The non-composite properties are as follows:
Interior (structural) wythe:

$I_i = bd^3/12 = 12(4)^3/12 = 64$ in.^{4}/ft width
$S_i = I/c = 64/2 = 32$ in.^{3}/ft width

Exterior (non-structural) wythe:

$I_e = 12(2.5)^3/12 = 15.6$ in.^{4}/ft
$S_e = 15.6/1.25 = 12.5$ in.^{3}/ft

Distribution:

$I_i + I_e = 64 + 15.6 = 79.6$ in.^{4}

Lateral load resisted by:

Interior: $64(100)/79.6 = 80\%$
Exterior: $15.6(100)/79.6 = 20\%$

(b) The composite properties are as follows:

	A	y	Ay	$\bar{y}$	$A\bar{y}^2$	I
Interior	48	2.00	96.0	1.63	127.5	64.0
Exterior	30	6.25	187.5	2.62	205.9	15.6
	78		283.5		333.4	79.6

$y_b = 283.5/78 = 3.63$ in.
$I_c = 333.4 + 79.6 = 413.0$ in.^{4}/ft width

9.4.3 Connections

9.4.3.1 Panel connections

Analysis and design of panel connections to resist normal and transverse wind and seismic shears as well as gravity and alignment loads are described in Chapters 3 and 6. Consideration must also be given to anticpated bowing, particularly of composite panels.

The effects of differential bowing between adjacent panels of similar span may be minimized by carefully aligning the panel during erection, and providing caulking on both the inside and outside of panel joints. Where adjacent panels are of significantly different stiffness, positive connections should be provided across the adjacent joint, to maintain alignment; such connections should be spaced not more than 15 ft on centers along the length of the common joint. The connection should provide for anticipated movement to prevent build-up of volumetric restraint forces.

When bowing does occur, it is generally outward. Chapter 3, Sect. 3.3.2, discusses the effects on connections. Recommendations stated there should be followed, in order to prevent the opening of joints and the resultant possibility of caulk failure. Special attention should also be paid to panels which are hung from adjacent panels. If a hung panel is supported by full height panels which are supported on a foundation, the assembly of panels can move as a unit, without distress to the panel or joint sealant. However, if the hung panel is also rigidly attached to the structure, this restraint may result in differential bowing of the hung panel with respect to the adjacent long panels. A detail to permit the hung panel to move with the adjacent panels should be provided. Attention should be paid to the position of the connection near the bottom of a panel supported on a foundation, in order to minimize the possibility of spalling due to thermal bowing, Fig. 9.4.3.

9.4.3.2 Wythe connectors

When one wythe is non-structural, its weight must be transferred to the structural wythe. This may be accomplished by using shear connectors or solid concrete ribs at the top or bottom of the panel. If the panel is non-composite, the shear connector should be a single element or a closely spaced pair of elements placed as near the center of rigidity of the panel as possible. This permits the non-structural wythe to contract and expand with the least restraint.

Shear anchors may be bent reinforcing bars, sleeve anchors, expanded metal, fiber connectors, or welded wire trusses, as illustrated in Fig. 9.4.4.

For ribbed panels, the shear connector is placed in the rib to assure proper embedment depth. In non-composite panels, it is preferable to have only one anchoring center. In a panel with two ribs, the shear connector can be positioned in either one of the ribs, and a flat anchor with the same vertical shear capacity is used in the other rib (Fig. 9.4.5). Since the flat anchor has little or no horizontal shear capacity, restraint of the exterior wythe is minimized. In a multi-ribbed panel, the shear connector is placed as near the center of rigidity as is possible, and flat anchors are used in the other ribs.

Fig. 9.4.3 Restraint at foundation

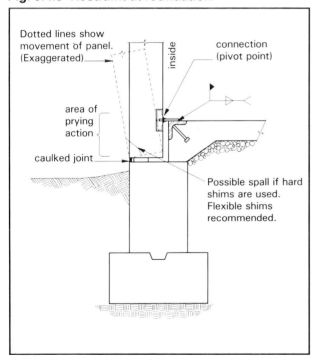

Fig. 9.4.4 Typical shear connectors

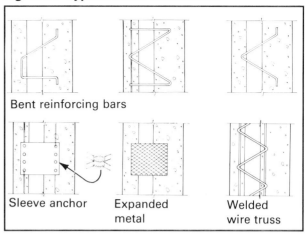

Fig 9.4.5 Anchorage for ribbed panels

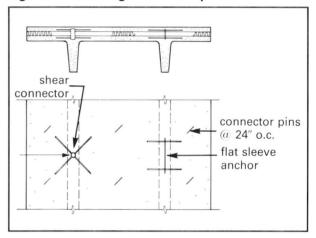

To complete the connection, metal tension/compression ties passing through the insulation are spaced at regular intervals to prevent the wythes from separating. Functions of wythe connectors are shown in Fig. 9.4.6. Typical tie details are shown in Fig. 9.4.7, and arrangements and spacing in Fig. 9.4.8. Wire tie connectors are usually 12 to 14 gauge, and preferably of stainless steel. Galvanized metal or plastic ties may also be acceptable. Ties of welded wire fabric and reinforcing bars are sometimes used. Shaped, crimped, or bent ties should be cold bent.

Tension/compression ties should be flexible enough so as not to resist temperature and shrinkage parallel to the panel surface, yet strong enough to resist a lifetime of stress reversals caused by temperature strains. Ties should be arranged, or coated, so that galvanic reaction between the tie and reinforcement will not occur.

The spacing of ties should be approximately 2 ft on centers, but not more than 4 ft, or at least 2 ties per 10 sq ft of panel area.

Fig. 9.4.6 Functional behavior of connectors

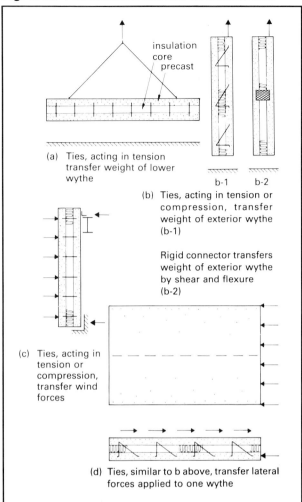

Fig. 9.4.7 Tension/compression ties

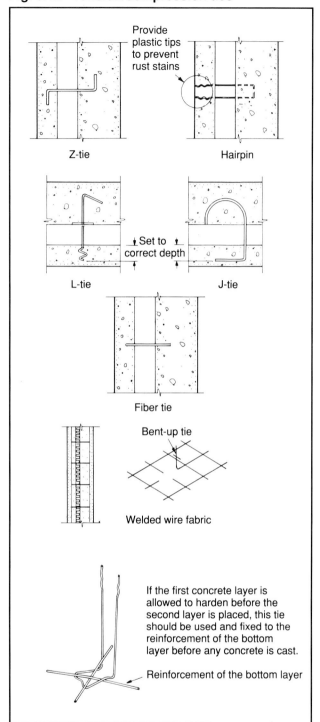

9.4.4 Insulation

Physical properties of insulation materials are listed in Table 9.4.1. Thermal properties are discussed in Sect. 9.1. The insulation should have low absorption or a water-repellent coating to minimize absorption of water from fresh concrete.

The thickness of insulation is determined as described in Sect. 9.1. A minimum of 1 in. is recommended. While there is no upper limit on the thickness, the deflection characteristics of the wythe connectors should be considered.

Openings in the insulation around connectors should be packed with insulation to avoid forming thermal bridges between wythes.

Using the maximum size of insulation sheets, consistent with the panel shape, is recommended. This will minimize joints and the resulting thermal links. Lapped abutting ends of single layer insulation, or staggered joints with double layer insulation, will effectively remove thermal links at joints (Fig. 9.4.9). Insulation may expand when subject to curing temperatures greater than 150°F; proper precautions should be taken, such as leaving room for the insulation to expand where ribs and insulation are in contact.

9.4.5 Thermal Bowing

Thermal bowing of exterior composite panels must be considered. Calculation of bow and the forces required to restrain it are discussed in Sect. 3.3.2. Differential movement between panels is sel-

Fig. 9.4.8 Arrangement of connectors between wythes

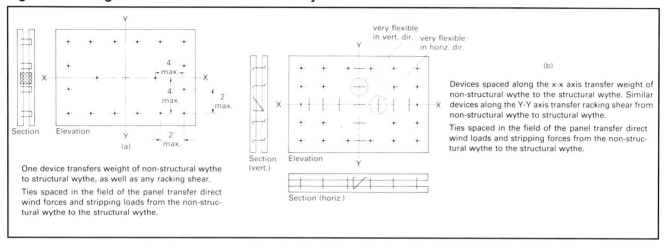

Table 9.4.1 Properties of insulation

	Expanded Polystyrene		Expanded Polystyrene (Extruded) (Molded)			Polyurethane Isocyanurate		Phenolic	Cellular Glass
Density (pcf)	1.0	2.0	1.4	2.0	4.0	2.0	6.0	2.0-3.0	8.5-12.0
Water absorption (% volume)	< 2.5		< 0.3			2-5		< 10.0	Non-absorptive
Comp. strength (psi)	10-14	25-33	15-25	25-50	115-125	25-35	100	10-35	100-210
Tensile strength (psi)	18	25	25	50	125	45	140	60	50
Linear coeff. of expansion	25-40		25-40			30-60		10-20	1.6-4.6
Shear strength (psi)	20	35	—	35	70	20	100	12	50
Flexural strength (psi)	25-30	50-75	40-70	50-100	100-140	50	210	25	80-100
Thermal conductivity (btu/in/hr)	0.30	0.23	0.20	0.20	0.20	0.18	0.13	0.12-0.23	0.35-0.57
Max. use temp.	165°F		165°F			250°F		300°F	900°F

dom a problem, except at corners or where abrupt changes in the building occur. Crazing or cracking is sometimes caused by bowing, but will be minimized if the recommendations of Sect. 9.4.2 are followed; such cracking is seldom of structural significance.

9.4.6 Typical Details

As with all precast concrete construction, satisfactory performance depends on proper detailing.

Some typical sandwich panel details are shown in Chapter 6.

Fig. 9.4.9 Preferred installation of insulation sheets

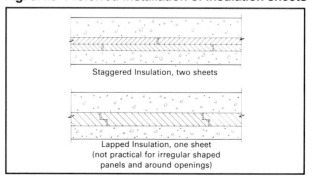

9.5 QUALITY CONTROL

9.5.1 Introduction

The use and application of precast and prestressed concrete demands a high level of quality in its design and production.

Each producer must have a total quality assurance program that ensures structural integrity and desired esthetic appearance in the final structure. However, the owner or the architect/engineer must be satisfied that materials, methods, products and quality control meet all the requirements of the project specifications. He can do this by conducting his own inspection of the plant, by engaging an independent agency to conduct inspections to prescribed criteria, or by specifying manufacturing facilities that are certified by the Precast/Prestressed Concrete Institute Plant Certification Program (see Chapter 10, Guide Specifications).

Precast and prestressed concrete products are sometimes site-cast by the general contractor or construction manager. In these instances, the same degree of quality should be expected by the owner or architect/engineer and the same rigid inspections and/or quality assurance, as listed above, must be provided. However, these expectations can only be achieved if clearly written and included in the project specifications.

9.5.2 Plant Certification

Plants certified under the PCI Plant Certification Program have a confirmed capability to produce quality precast or prestressed concrete products. This is based on three inspections per year by a structural engineering firm engaged by PCI; two of the inspections are unannounced, and one is a scheduled in-depth inspection requiring two or more days.

In addition, the program has included group and category certifications to identify whether the certified plant specializes in certain products or has general product capability. There are four groups: A1—architectural; A2—glass fiber reinforced concrete; B—bridges; and C—commercial (structural). Groups B and C each has four separate categories for product/inspection.

Inspection criteria and grading are based on the industry's standards for quality control. These standards are presented in PCI's three quality control manuals, one for structural precast and prestressed concrete products[1], one for architectural precast concrete products[2], and one for GFRC[3]. Inspections cover all phases of production including materials, production methods, product handling and storage, shipping, product appearance, testing, record keeping, personnel and safety practices. Failure to maintain a production facility at or above required minimum standards results in mandatory decertification.

9.5.3 PCI Quality Control Manuals

The three PCI manuals listed below contain industry-approved requirements for quality control for structural and architectural precast concrete products, and glass fiber reinforced concrete products. Any in-house quality assurance program should be based on these manuals. They should also be used by any outside inspection of production facilities. They form the basis for the PCI Plant Certification Program and the detailed grading system on which a plant can become certified.

9.5.4 References

1. "Manual for Quality Control for Plants and Production of Precast and Prestressed Concrete Products," Third Edition, MNL-116-85, Precast/Prestressed Concrete Institute, Chicago, IL, 1985.

2. "Manual for Quality Control for Plants and Production of Architectural Precast Concrete Products," MNL-117-77, Precast/Prestressed Concrete Institute, Chicago. IL, 1977.

3. "Manual for Quality Control for Plants and Production of Glass Fiber Reinforced Concrete Products," MNL-130-91, Precast/Prestressed Concrete Institute, Chicago, IL, 1991.

9.6 CONCRETE COATINGS AND JOINT SEALANTS

9.6.1 Coatings for Horizontal Deck Surfaces

Concrete surface sealers reduce moisture and salt (chloride) penetration, which is particularly important in parking structures. While these sealers can enhance the durability of any concrete topping, they do not substitute for basic durable concrete design, nor do they provide protection against penetration of moisture and chlorides through cracks. Research has shown that the performance of concrete sealers will vary, depending on the product and other variables. Products should be evaluated against the criteria established in the NCHRP 244 study.[3]

Sealers may be classified into two groups: penetrants and surface sealers.

Penetrants: These are generally silanes or siloxanes. They penetrate the surface, reacting with cementitious materials and making the concrete hydrophobic. They do not have crack-bridging capabilities. These materials do not appreciably affect the appearace or characteristics of the surface to which they are applied. While generally more expensive than other types of sealers they are typically longer lasting and less subject to wear under traffic or deterioration from exposure to sun.

Surface Sealers: These are generally polymer resins such as urethanes, epoxies, acrylics, or other proprietary blends. They provide protection by penetrating the surface slightly, and/or by providing a tough film over the surface; these materials also do not bridge cracks. Surface sealers are generally less expensive than penetrants, and performance characteristics of many of these products compare favorably with the penetrating sealers, as demonstrated by the NCHRP 244 criteria. These sealers are more likely to be subject to wear under traffic, and may be more slippery than the bare concrete surface.

9.6.2 Clear Surface Sealers for Architectural Precast Panels

Clear surface coatings or sealers are sometimes used on precast concrete wall panels to improve weathering qualities or to reduce attack of the concrete surface by airborne pollutants. Because of the quality of concrete normally achieved in plant-cast precast concrete, even with very thin sections, sealers are not required for waterproofing. Because the results are uncertain, use of sealers in locations having little or no air pollution is not recommended.

A careful evaluation should be made before deciding on the type of sealer. This includes consultation with the local precasters. In the absence of near-identical experience, it is desirable to test sealers on reasonably sized samples of varying age to verify performance over a suitable period of exposure or usage, based on prior experience under similar exposure conditions.

Sealers are usually applied to wall panels after erection to avoid problems with adhesion of joint sealants.

Any coating used should be guaranteed by the supplier or applicator not to stain, soil, or discolor the precast concrete finish. Also, some clear coatings may cause joint sealants to stain concrete. Consult manufacturers of both sealants and coatings or pretest before applying the coating.

Sealers should be applied in accordance with manufacturer's recommendations. Generally, good airless spray equipment is used for uniformity and to prevent surface rundown. Two coats are usually required to provide a uniform coating, because the first coat is absorbed into the concrete. The second coat does not penetrate as much and provides a more uniform surface color. Care must be taken to keep sealers off glass surfaces.

9.6.3 Joint Sealants

Joint sealants include viscous liquids, mastics or pastes, and tapes, gaskets and foams. Generally, a sealant is any material placed in a joint for the purpose of preventing the passage of moisture, air, heat or dirt into or through the joint.

Successful performance of a wall system is often dependent on good joint details and sealant selection. Too often, because the sealant is a relatively minor part of the total structure, the responsibility for selection is left to the contractor or erector who does not fully understand the task it is to perform.

The selection of the proper sealant is confused by the method of supply. Most of the raw materials are manufactured by a few major chemical companies. These basic materials are then compounded with other ingredients by literally hundreds of sealant manufacturers. The quality of finished product varies widely because the expertise of the formulators varies widely.

Viscous liquid sealants are often used in pourable form and normally are used in horizontal joints. Mastics are applied with a gun and are compounded to prevent sagging or flowing when used in vertical joints. Tapes are most often used around glass. Elastomeric gaskets are used in joints which experience considerable movement. Foams are used as air seals or backup for more durable surface seals.

Specification of sealants is difficult because many of the numerous formulations are not covered by standard specifications, either for materials or installation. Many of the "standard" specifications are also out of date, and often standard-writing agencies have specifications that do not agree.

Proven performance is still the procedure that is often relied upon. Refs. 5 through 9 will aid in the design and selection of joint sealants. Ref. 6 is also a very valuable aid for the design and detailing of architectural precast panel joints.

9.6.4 References

1. "A Guide to the Use of Waterproofing, Dampproofing, Protective, and Decorative Barrier Systems for Concrete," ACI 515.1R-79, *ACI Manual of Concrete Practice*, Part 5, American Concrete Institute, Detroit, MI.

2. Pfeifer, D.W. and Perenchio, W.F., "Coatings, Penetrants and Specialty Concrete Overlays for Concrete Surfaces," National Association of Corrosion Engineers Seminar, September 1982, Chicago, IL.

3. "Concrete Sealers for Protection of Bridge Structures," *NCHRP Report 244*, Transportation Research Board, Washington, DC.

4. Litvin, Albert, "Clear Coatings for Exposed Architectural Concrete," Development Department Bulletin D137, Portland Cement Association, Skokie, IL, 1968.

5. "Guide Specifications, Section 07900 (Sealants)", Sealant, Waterproofing and Restoration Institute, Kansas City, MO, 1982.

6. "Sealants: The Professionals' Guide", Sealant, Waterproofing and Restoration Institute, Kansas City, MO, 1990.

7. "Architectural Precast Concrete," Second Edition, MNL-122-89, Precast/Prestressed Concrete Institute, Chicago, IL, 1989.

8. "Guide to Joint Sealants for Concrete Structures," ACI 504R-77, *ACI Manual of Concrete Practice*, Part 5, American Concrete Institute, Detroit, MI.

9. Cook, J.P., "Construction Sealants and Adhesives," Wiley-Interscience, New York, NY.

9.7 VIBRATION IN CONCRETE STRUCTURES

9.7.1 Human Response to Building Vibrations

When the resonant frequency of a floor system is close to the frequency imparted by the occupants, and the deflection of the system is significant, motion will be perceptible and perhaps annoying. Perception is related to the activity of the occupant: a person at rest or engaged in quiet work will tolerate less vibration than a person performing an active function, such as dancing or aerobics. However, if a floor system dissipates the imparted energy in a very short period of time, the magnitude of motion is generally not perceived as annoying. Thus, the damping characteristics of the system contribute to acceptability.

Vibration is generally stated as a fraction of gravity acceleration. In an office environment, investigators report annoyance when vibration exceeds 0.005g; in an active environment, such as aerobics, investigators[10] report the participants will accept vibrations in the order of 0.05g. Thus, a design procedure would first establish the limiting degree of vibration for a particular floor usage, and from that determine the necessary minimum stiffness (natural frequency). Finally, the natural frequency of the floor is calculated, and compared to the minimum acceptable natural frequency. Ref. 10 is the basis for this section. Additional information is given there.

9.7.1.1 Minimum natural frequency

The minimum required stiffness of a floor system to offset a sense of disturbing vibration is a function of many variables, and not subject to exact calculation. A simplified method to estimate the minimum natural frequency of a floor system is given in Eq. 9.7.1:[10]

$$f_o \geq f\sqrt{1 + \left(\frac{1.3}{a_o/g}\right)\alpha\frac{w_p}{w}} \qquad \text{(Eq. 9.7.1)}$$

where:

- a_o/g = acceleration limit, % gravity
- f = forcing frequency, Hz
- f_o = natural frequency of the structural system
- w = total weight (DL + actual LL), psf
- w_p = actual LL, psf
- α = dynamic load factor
- g = acceleration constant, 386 in./sec.^{2}

Natural frequency

The natural frequency of a floor system will be a combination of the natural frequency of the framing members, including slab, beam and girders. For a precast beam supported on rigid supports, the natural frequency may be calculated as:

$$f_o = K\sqrt{\frac{gEI}{W\ell^3}} \qquad \text{(Eq. 9.7.2)}$$

where:

- g = 386 in./sec^2
- E = modulus of elasticity, psi
- I = moment of inertia of beam, in.^{4}
- ℓ = span, in.
- W = total supported weight on beam, lb
- K = 1.57 for simple beam, 0.56 for fixed end cantilever, and 3.50 for a fixed end beam.

Example 9.7.1 Gymnasium floor

An 8DT24+2 is to be used as a gymnasium floor and is supported on block walls. The design span is 50 ft. There are no adjacent quiet areas, so the acceleration fraction limit will be taken as 0.05 (see Table 9.7.1). The weight of the active participants is taken as 5 psf, and the floor system weighs 77 psf. The forcing frequency is estimated as 2.75. From Table 9.7.2, the dynamic load factor for the first harmonic is 1.5. From Eq. 9.7.1, the resonant frequency should be greater than:

$$f_o = 2.75\sqrt{1 + \frac{(1.3)(1.5)(5)}{(0.05)(82)}} = 5.05 \text{ Hz}$$

From Eq. 9.7.2

$$f_o = 1.57\sqrt{\frac{(386)(4.3 \times 10^6)(27,720)}{(82)(8)(50)(50 \times 12)^3}} = 4.00 \text{ Hz}$$

Since this is less than the required minimum natural frequency, the floor system is not acceptable. Possible solutions include increasing the depth of the floor, blocking the double tee flanges, or reducing the span.

Example 9.7.2 Stadium seats

The precast stadium seat shown is proposed to span 20 ft. Verify its acceptability with respect to vibration.

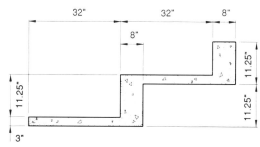

According to Ref. 10, the maximum forcing frequency for the first harmonic in a stadium can be assumed as 3 Hz with an associated dynamic load factor of 0.25, and the maximum forcing frequency for the second harmonic can be assumed as 5.5 Hz, with an associated dynamic load factor of approximately 0.05. The acceleration limits would fall within a range of 0.04 to 0.07 (Table 9.7.1). For the example, select an acceleration limit of 0.05, and an actual live load of 30 psf. The weight of the stadium seat is 388 lb./ft, or 65 psf. I_{min} = 3794 in.^{4}

In accordance with Eq. 9.7.1, the minimum required natural frequency, based on the first harmonic, is:

$$f_o = 3.0\sqrt{1 + \frac{(1.3)(0.25)(30)}{(0.05)(95)}} = 5.24 \text{ Hz}$$

and the minimum required natural frequency, based on the second harmonic, is:

$$f_o = 5.5\sqrt{1 + \frac{(1.3)(0.05)(30)}{(0.05)(95)}} = 6.53 \text{ Hz}$$

The natural frequency of the stadium seat is:

$$f_o = 1.57\sqrt{\frac{(386)(4.3 \times 10^6)(3794)}{(568)(20)(20 \times 12)^3}} = 9.94 \text{ Hz}$$

Thus the section is acceptable.

Table 9.7.1 Recommended acceleration limits for vibrations due to rhythmic activities

Occupancies affected by the vibrations	Acceleration limits % gravity
Office and residential	0.4 to 0.7
Dancing and weightlifting	1.5 to 2.5
Rhythmic activity only	4 to 7

Table 9.7.2 Recommended dynamic load factors for aerobics[11]

Harmonic	Forcing frequency, Hz	Dynamic load factor, α
1	2-2.75	1.5
2	4-5.5	0.6
3	6-8.25	0.1

9.7.2 Vibration Isolation for Mechanical Equipment

Vibration produced by equipment with unbalanced operating or starting forces can usually be isolated from the structure by mounting on a heavy concrete slab placed on resilient supports. This type of slab, called an inertia block, provides a low center of gravity to compensate for thrusts such as those generated by large fans.

For equipment with less unbalanced weight, a "housekeeping" slab is sometimes used below the resilient mounts to provide a rigid support for the mounts and to keep them above the floor so they are easier to clean and inspect. This slab may also be mounted on pads of precompressed glass fiber or neoprene.

The natural frequency of the total load on resilient mounts must be well below the frequency generated by the equipment. The required weight of an inertia block depends on the total weight of the machine and the unbalanced force. For a long-stroke compressor, five to seven times its weight might be needed. For high pressure fans, one to five times the fan weight is usually sufficient.

A floor supporting resiliently mounted equipment must be stiffer than the isolation system. If the static deflection of the floor approaches the static deflection of the mounts, the floor becomes a part of the vibrating system, and little vibration isolation is achieved. In general, the floor deflection should be limited to about 15% of the deflection of the mounts.

Simplified theory shows that for 90% vibration isolation, a single resilient supported mass (isolator) should have a natural frequency of about ⅓ the driving frequency of the equipment. The natural frequency of this mass can be calculated by:[9]

$$f_n = 188\sqrt{1/\Delta_i} \quad \text{(Eq. 9.7.3)}$$

where:

f_n = natural frequency of the isolator, CPM

Δ_i = static deflection of the isolator, in.

From the above, the required static deflection of an isolator can be determined as follows:

$$f_n = f_d/3 = 188\sqrt{1/\Delta_i} \text{ or}$$

$$\Delta_i = (564/f_d)^2 \quad \text{(Eq. 9.7.4)}$$

and:

$$\Delta_f \leq 0.15\,\Delta_i \quad \text{(Eq. 9.7.5)}$$

where:

f_d = driving frequency of the equipment, CPM

Δ_f = static deflection of the floor system caused by the weight of the equipment, including inertia block, at the location of the equipment, in.

Example 9.7.3 Vibration isolation

Given:

A piece of mechanical equipment has a driving frequency of 800 CPM.

Problem:

Determine the approximate minimum deflection of the isolator and the maximum deflection of the floor system that should be allowed.

Solution:

From Eq. 9.7.4:

Isolator, $\Delta_i = (564/800)^2 = 0.50$ in.

From Eq. 9.7.5:

Floor, $\Delta_f = 0.15(0.50) = 0.07$ in.

9.7.3 References

1. "Vibrations of Concrete Structures", *Special Publication SP-60*, American Concrete Institute, Detroit, MI, 1979.

2. Galambos, T.V., Gould, P.C., Ravindra, M.R., Surgoutomo, H. and Crist, R.A., "Structural Deflections—A Literature and State-of-the-Art Survey", *Building Science Series*, October 1973, National Institute of Standards and Technology, Washington, DC.

3. Murray, T.M., "Acceptability Criterion for Occupant-Induced Floor Vibration", *Sound and Vibration*, November 1979.

4. "Design and Evaluation of Operation Breakthrough Housing Systems", *NBS Report 10200, Amendment 4*, September 1970, U.S. Department of Housing & Urban Development, Washington, DC.

5. Wiss, J.F. and Parmelee, R. A., "Human Perception of Transient Vibrations", *Journal of the Structural Division*, ASCE, V. 100, No. ST4, April 1974.

6. "Guide to Floor Vibrations", *Steel Structures for Buildings (Limit States Design)*, CAN3-S16.1-M74, Appendix G, Canadian Standards Association, Rexdale, Ontario, Canada, 1974.

7. "Guide for the Evaluation of Human Exposure to Whole-Body Vibration", International Standard 2631, International Organization for Standardization, 1974.

8. Murray, T.M., "Design to Prevent Floor Vibration", *Engineering Journal*, Third Quarter, 1975, American Institute of Steel Construction, Chicago, IL.

9. Harris, C.M. and Crede, C.E., "Shock and Vibration Handbook," Second Edition, McGraw-Hill, New York, NY, 1976.

10. Allen, D.E. "Building Vibrations from Human Activities", *Concrete International*, V. 12, No. 6, June 1990.

11. Bachmann, H. and Ammann, W., "Vibrations in Structures Induced by Man and Machine," *Structural Engineering Documents*, No. 3, International Association for Bridge and Structural Engineering, Zurich, Switzerland, 1987.

9.8 CRACKING, REPAIR AND MAINTENANCE

9.8.1 Cracking

Minor cracking may occur in precast concrete without being detrimental, and it is impractical to impose specifications that prohibit all cracking. (Note: In other sections, the term "crack-free" design is used. This refers to a design in which concrete tension is kept within limits, and should not be construed as being a guarantee that there will be no cracking.) However, in addition to being unsightly, cracks are potential locations of concrete deterioration, and should be avoided if possible. Prestressing and proper handling procedures are two of the best methods of keeping cracks to a minimum. To evaluate the acceptability of a crack, the cause and service conditions of the precast unit should be determined.

Crazing, i.e., fine, random (commonly called "hairline") cracks, may occur in the cement film on the surface of concrete. The primary cause is the shrinkage of the surface with respect to the mass of the unit. Crazing has no structural significance and does not significantly affect durability, but may be visually accentuated if dirt settles in these minute cracks. Crazing should not be cause for rejection.

Tension cracks are sometimes caused by temporary loads during production, transportation, or erection of these products (see Chapter 5). These cracks may extend through to the reinforcement. If the crack width is narrow, the structural adequacy of the casting will remain unimpaired, as long as corrosion of the reinforcement is prevented. See Sect. 4.2.2.1 for recommended limits on crack width. The acceptability of cracks wider than recommended maximums should be governed by the function of the unit. Most can be effectively repaired and sealed.

Long-term volume changes can also cause cracking after the member is in place in the building, if the connections provide enough restraint to the member (see Sect. 3.3). Internal causes, such as corrosion of reinforcement or cement-aggregate reactivity, can also lead to long-term cracking and should be considered when materials are selected.

9.8.2 Repair and Maintenance

Products which are damaged prior to or during placement in the structure can usually be repaired, but major repairs should not be made until it is determined that the unit will be structurally sound.

Patching of architectural panels is an art requiring expert craftsmanship and careful selection and mixing of materials.

Concrete surfaces normally require little maintenance, except for those that are subject to harsh environments, such as parking structures in northern climates. Routine inspection and maintenance can do much to extend the life of such buildings.[8,9]

Small cracks (under 0.005 to 0.010 in.) may not need repair, unless failure to do so can cause corrosion of reinforcement. If repair is required for the restoration of structural integrity, cracks may be pressure injected with a low viscosity epoxy.

9.8.3 References

1. "Control of Cracking in Concrete Structures," ACI 224R-80, 1980, *ACI Manual of Concrete Practice*, Part 3, American Concrete Institute, Detroit, MI.

2. Abeles, P.W., Brown, E.T. and Morrow, J.W., "Development and Distribution of Cracks in Rectangular Prestressed Beams During Static and Fatigue Loading," *PCI Journal*, V. 13, No. 5, October 1968.

3. Mather, Bryant, "Cracking Induced by Environmental Effects," *Causes, Mechanism, and Control of Cracking in Concrete*, Special Publication SP-20, American Concrete Institute, Detroit, MI, 1968.

4. ACI Committee 201, "Guide to Durable Concrete," *Journal of the American Concrete Institute*, V. 74, No. 12, December 1977.

5. "Manual for Quality Control for Plants and Production of Architectural Precast Concrete Products," MNL-117-77, Precast/Prestressed Concrete Institute, Chicago, IL, 1977.

6. PCI Committee on Quality Control Performance Criteria, "Fabrication and Shipment Cracks in Prestressed Hollow-Core Slabs and Double Tees," *PCI Journal*, V. 28, No. 1, January-February 1983.

7. PCI Committee on Quality Control Performance Criteria, "Fabrication and Shipment Cracks in Precast or Prestressed Beams and Columns," PCI Committee on Quality Control Performance Criteria, *PCI Journal*, V. 30, No. 3, May-June 1985.

8. "Parking Structures: Recommended Practice for Design and Construction," MNL-129-88, Precast/Prestressed Concrete Institute, Chicago, IL, 1988.

9. "Parking Garage Maintenance Manual," National Parking Association, Wahington, DC.

9.9 PRECAST SEGMENTAL CONSTRUCTION

9.9.1 General

Segmental construction is a method of construction in which primary load carrying members are composed of individual segments post-tensioned together. This allows precast concrete to be used for long horizontal or vertical spans within size limitations imposed by manufacturing, transportation, and handling equipment. With proper planning and element selection, a large re-use of forms is possible, with the resulting economy.

The method is best known for a number of landmark bridges,[1,10] but has also been used in the construction of airport control towers, storage tanks for various solids and liquids, including natural gas, a long-span exterior Vierendeel truss on a high-rise building, an Olympic stadium and other non-bridge structures.[2-9]

Segmental construction requires that the designer give special consideration to:

1. Size and weight of the precast elements.
2. Configuration and behavior of the joints between elements.
3. Construction sequence, and the loads and deflections imposed at various stages.
4. The effect of normal tolerances and deviations upon the joints.

9.9.2 Joints

Joints may be either "open", to permit completion by a field placed grout, or "closed", where the joint is either dry or bonded by a thin layer of adhesive (Fig. 9.9.1).

9.9.2.1 Open joints

The individual segments are separated by an amount sufficient to place (usually by pressure) a grout mix, but not more than about 2 in. Prior to placing the segments, the joint surface is thoroughly cleaned and wire brushed or sand-blasted.

The perimeter of the joint is sealed with a gasket which is compressed by use of "come-alongs" or by a small amount of prestress. Gaskets are also provided around the post-tensioning elements to prevent leakage into the ducts, blocking passage of the tendons. Vents are provided at the top to permit escape of entrapped air during grouting. Prior to filling the joint, the surfaces should be thoroughly wetted, or coated with a bonding agent. Grout strength should be at least equal to that of the precast segments, but not less than 4000 psi.

After grouting, vents are closed and pressure increased to assure full grout intrusion. After a few days, the vent is reopened, and filled with grout as required.

9.9.2.2 Closed joints

If a closed joint is used, the segment is usually "match-cast", i.e., each segment is cast against its previously cast neighbor. A bond breaker is applied to the joint during casting. Thus, the connecting surfaces fit each other accurately, so that little or no filling material is needed at the joint. The sharpness of line of the assembled construction depends mainly on the accuracy of the manufacture of the segments.

Match cast segments are usually joined by coating the abutting surfaces with a thin layer of epoxy adhesive, and then using the post-tensioning to draw the elements together and hold them in position. In some structures, dry joints have been used successfully, in which case, the compression provided by the post-tensioning is relied upon for weatherproofing, as well as for transfer of forces.

Surface preparation of closed joints is extremely important. They should be sound and clean, free from all traces of form release agents, curing compounds, laitance, oil, dirt and loose concrete.

A small piece of foreign material in a joint, or imper-

Fig. 9.9.1 Types of joints

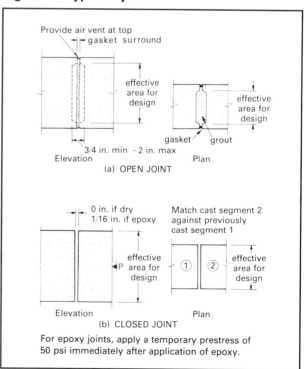

fect alignment will frequently cause the concrete to spall around the edges of the contact area of a closed joint. Thus, care in joint preparation and segment alignment cannot be over-emphasized.

9.9.3 Design

Analysis of precast segmental structures usually assumes monolithic behavior of the members under service loads, except that if dry joints are used, tension is not permitted between segments. Some designers also prefer to not allow tension in grouted joints, unless tests have indicated otherwise.

The behavior during construction is of particular importance in segmental structures. Stresses which may be caused by settlement and shortening of scaffolding, temperature changes, elastic shortening from post-tensioning and other construction related stresses and movements may need to be considered in the design.

Shear stresses across joints are resisted by epoxy adhesives or grout, sometimes in combination with shear keys. In dry joints, shear stresses are resisted by friction (and shear keys if present), with the post-tensioning providing the normal force. In the absence of test data, the coefficient of friction may be assumed to be 0.8 (Sect. 6.6).

Individual segments are designed for handling and erection stresses (Chapter 5). They may be pretensioned or reinforced with mild steel, depending on size and manufacturing procedure.

9.9.4 Post-Tensioning

Information on various post-tensioning systems and their applications is given in the *PTI Post-Tensioning Manual*.[11] Nearly any type of bonded system can be used.

9.9.4.1 Tendon and duct placement

In addition to the design parameters for in-service conditions, the tendon layout must consider the sequence of construction and the changes in load conditions during the various construction stages.

Ducts for the tendons are placed in the segments prior to concrete placement. In some cases, tendons are installed in the ducts before the segment is erected, or even before concrete is placed, and subsequently coupled at each joint. In others, the tendons are placed after the segments are erected, either full length of the member or in shorter lengths, again coupled at the joints. Special attention must be given to the alignment of the ducts, especially at the joints. They must be large enough in diameter to adequately place the tendon, allowing some tolerance in alignment, and receive the grout placed subsequent to stressing. Special attention must be given to the corrosion protection of the post-tensioning steel, if it is to be unbonded at any stage of construction. Also, drains should be installed in the ducts at low points of the tendon profile.

Post-tensioning tendons crossing joints should be approximately perpendicular to the joint surface in the smaller dimension of the segment, e.g., flange or web thickness, but may be inclined to the direction of the larger dimension (e.g., depth). This is to minimize any unbalanced shearing force which could lead to dislocation of edges at the joints.

9.9.4.2 Couplers

Couplers are designed to develop the full strength of the tendons they connect. Adjacent to the coupler, the tendons should be straight, or have very minor curvature for a minimum length of 12 times the diameter of the coupler. Adequate provisions must be made to ensure that the couplers can move during prestressing.

9.9.4.3 Bearing areas

Prestress force is transferred to the joined segments through bearing plates or other anchorage devices. This causes high load concentrations at the bearing areas, usually requiring vertical and horizontal reinforcement in several locations:

1. Under end surfaces, not more ¾ in. deep, to control possible surface cracking around anchorage.

2. Internally, to prevent splitting between individual anchors. Size and location of this area and the magnitude of splitting (bursting) force depends of the type of anchorage and the force in the post-tensioning tendon.

3. Internally, to prevent splitting between groups of anchors.

4. Adjacent to joint surfaces, to decrease the possibility of damage to segments during post-tensioning or handling.

Grouting of the ducts is provided for corrosion protection and to develop bond between the prestressing steel and the surrounding concrete. Grouting procedures should follow the *Recommended Practice for Grouting of Post-Tensioned Prestressed Concrete*, contained in Ref. 11.

9.9.5 References

1. Joint PCI-PTI Committee on Segmental Construction "Recommended Practice for Precast Post-Tensioned Segmental Construction," *PCI Journal*, V. 27, No. 1, January-February 1982.

2. Anderson, Arthur R., "World's Largest Prestressed LPG Floating Vessel," *PCI Journal*, V. 22, No. 1, January-February 1977.

3. Pery, William E., "Precast Prestressed Clinker Storage Silo Saves Time and Money," *PCI Journal*, V. 21, No. 1, January-February 1976.

4. Martynowicz, A. and McMillan, C.B., "Large Precast Prestressed Vierendeel Trusses Highlight Multistory Building," *PCI Journal*, V. 20, No. 6, November-December 1975.

5. Arafat, Z., "Giant Precast Prestressed LNG Storage Tanks at Staten Island," *PCI Journal*, V. 20, No. 3, May-June 1975.

6. Muller, Jean, "Ten Years Experience in Precast Segmental Construction," *PCI Journal*, V. 20, No. 1, January-February 1975.

7. Lamberson, E. A., "Post-Tensioned Structural Systems—Dallas-Ft. Worth Airport," *PCI Journal*, V. 18, No. 6, November-December 1973.

8. Zielinski, Z.A., "Prestressed Slabs and Shells Made of Prefabricated Components," *PCI Journal*, V. 9, No. 4, August 1964.

9. Zielinski, Z.A., "Prefabricated Building Made of Triangular Prestressed Components," *Journal of the American Concrete Institute*, V. 61, No. 4, April 1964.

10. "Precast Segmental Box Girder Bridge Manual," published jointly by the Precast/Prestressed Concrete Institute (Chicago, IL) and the Post-Tensioning Institute (Phoenix, AZ), 1978.

11. "Post-Tensioning Manual," Fifth Edition, Post-Tensioning Institute, Phoenix, AZ, 1990.

9.10 COORDINATION WITH MECHANICAL, ELECTRICAL AND OTHER SUB-SYSTEMS

9.10.1 Introduction

Prestressed concrete is used in a wide variety of buildings, and its integration with lighting, mechanical, plumbing, and other services is of importance to the designer. Because of increased environmental demands, the ratio of costs for mechanical and electrical installations to total building cost has increased substantially in recent years. This section is intended to provide the designer with the necessary perspective to economically satisfy mechanical and electrical requirements, and to describe some standard methods of providing for the installation of other sub-systems.

9.10.2 Lighting and Power Distribution

For many applications, the designer can take advantage of the fire resistance, reflective qualities and appearance of prestressed concrete by leaving the columns, beams, and ceiling structure exposed. To achieve uniform lighting free from distracting shadows, the lighting system should parallel the stems of tee members.

By using a reflective paint and properly spaced high-output flourescent lamps installed in a continuous strip, the designer can achieve a high level of illumination at a minimum cost. In special areas, lighting troffers can be enclosed with diffuser panels fastened to the bottom of the tee stems providing a flush ceiling (see Fig. 9.10.1). By using reflective paints, these precast concrete lighting channels can be made as efficient as conventional flourescent fixtures.

9.10.3 Electrified Floors

The increasing use of business machines, telephones, and other communication systems stresses the need for adequate and flexible means of supplying electricity and communication service. Since a cast-in-place topping is usually placed on prestressed floor members, conduit runs and floor outlets can be readily buried within this topping. Burying conduits in toppings of parking structures is not recommended because of the possibility of conduit corrosion. With shallow height electrical systems, a comprehensive system can be provided in a reasonably thin topping. The total height of conduits for these comprehensive electrical systems is as little as $1\frac{3}{8}$ in. Most systems can easily be included in a 2 to 4 in. thick slab. Voids in hollow-core slabs can also be used as electrical raceways (see Fig. 9.10.2).

When the system is placed in a structural composite slab, the effect of ducts and conduits must be carefully examined and their location coordinated with reinforcing steel. Tests on slabs with buried ductwork have shown that structural strength is not normally impaired by the voids.

Because of the high load-carrying capacity of prestressed concrete members, it is possible to locate high-voltage substations, with heavy transformers, near the areas of consumption with little or no additional expense. For extra safety, distribution feeds can also be run within those channels created by stemmed members. Such measures also aid the economy of the structure by reducing the overall story height and minimizing maintenance expenses.

9.10.4 Ductwork

The designer may also utilize the space within stemmed members or the holes inside hollow-core slabs for distribution ducts for heating, air-conditioning, or exhaust systems. In stemmed members three sides of the duct are provided by the bottom of the flange and the sides of the stems. The bottom of the duct is completed by attaching a metal panel to the tee stems in the same fashion as the lighting diffusers (see Fig. 9.10.1).

Connections can be made by several means, among them powder-activated fasteners, cast-in inserts or reglets. Field installed devices generally offer the best economy and ensure placement in the exact location where the connecting devices will be required. Inserts should only be cast-in when they can be located in the design stage of the job, well in advance of casting the precast members.

With hollow-core slabs, additional duct work can be eliminated. These members have oval, round, or rectangular voids of varying size which can provide ducts or raceways for the various systems. Openings core drilled in the field can provide access and distribution. The voids in the slabs are aligned and connected to provide continuity of the system. Openings can also be provided in intermediate supporting beams such as inverted tees, to allow duct continuity.

If high velocity air movement is utilized, the enclosed space becomes a long plenum chamber with uniform pressure throughout its length. Diffusers are installed in the ceiling to distribute the air. Branch runs, when required, can be standard ducts installed along the column lines.

When ceilings are required, proper selection of precast components can result in shallow ceiling spaces as shown in Fig. 9.10.3. This figure also illustrates the flexibility of space arrangements possi-

Fig. 9.10.1 Metal panels attached at the bottoms of the stems create ducts, and diffuser panels provide a flush ceiling

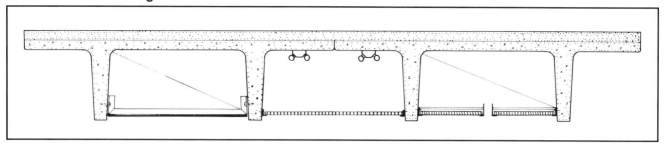

Fig. 9.10.2 Underfloor electrical ducts can be embedded within a concrete topping

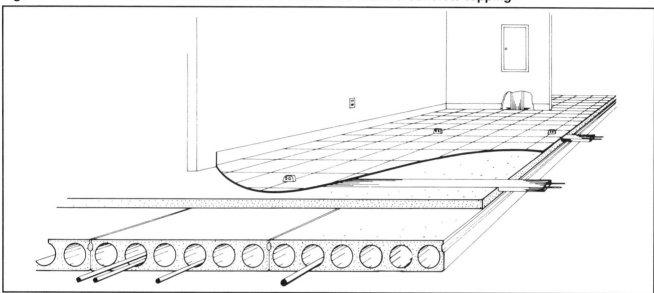

Fig. 9.10.3 Where ceilings are required, ducts, piping, and lighting fixtures can be accommodated within a shallow depth

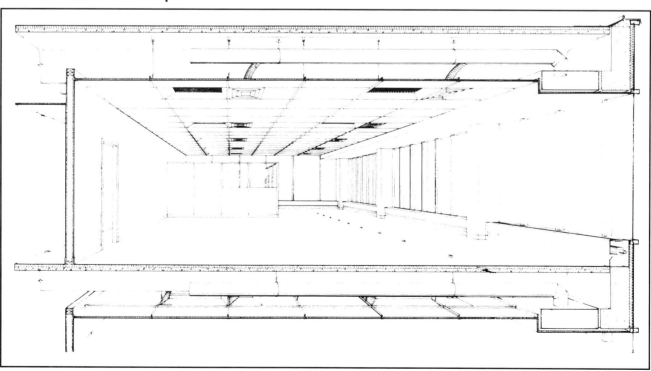

Fig 9.10.4 Large openings in floors and roofs are made during manufacture of the units; small openings are field drilled. Some common types of openings are show here.

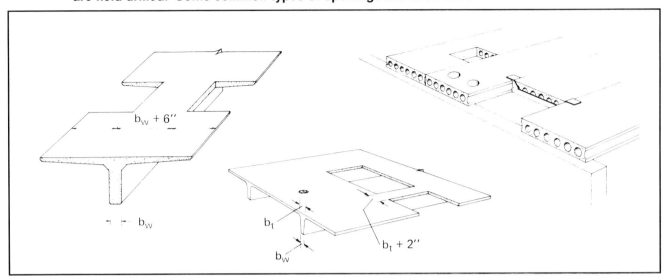

ble with long span prestressed concrete members.

Branch ducts of moderate size can also be accommodated by providing block-outs in the stems of tees or beams. To achieve best economy and performance in prestressed concrete members, particularly stemmed members, such block-outs should be repeated in size and location to handle all conditions demanded by mechanical, electrical or plumbing runs. While this may lead to slightly larger openings in some cases, the end result will probably be more economical. It should also be noted that sufficient tolerance should be allowed in sizing the openings to provide the necessary field assembly considerations (see Chapter 8).

Prestressed concrete box girders have been used to serve a triple function as air conditioning distribution ducts, conduit for utility lines and structural supporting members for the roof deck units. Conditioned air can be distributed within the void area of the girders and then introduced into the building work areas through holes cast into the sides and bottoms of the box girders. The system is balanced by plugging selected holes.

Vertical supply and return air trunks can be carried in the exterior walls, with only small ducts needed to branch out into the ceiling space. In some cases the exterior wall cavities are replaced with three or four sided precast boxes stacked to provide vertical runs for the mechanical and electrical systems. These stacked boxes can also be used as columns or lateral bracing elements for the structure.

In some cases it may be required to provide openings through floor and roof units. Large openings are usually made by blockouts in the forms during the manufacture; smaller ones (up to about 8 in.) are usually field drilled. When field drilling, care must be taken not to cut prestressing strand. Openings in flanges of stemmed members should be limited to the "flat" portion of the flange, that is, beyond 1 in. of the edge of the stem on double tees. Angle headers are often used for framing large openings in hollow-core floor or roof systems (see Fig. 9.10.4).

9.10.5 Other Sub-Systems

Suspended ceilings, crane rails, and other sub-systems can be easily accommodated with standard manufactured hardware items and embedded plates as shown in Fig. 9.10.5.

Architectural precast wall panels can be adapted to combine with pre-assembled window or door units. Door or window frames properly braced to prevent bowing during concrete placement, can be cast in the panels and then the glazing or doors can be installed prior to or after delivery to the job site. If the glazing or doors are properly protected, they can also be cast into the panel at the plant. When casting in aluminum window frames, particular attention should be made to properly coat the aluminum so that it will not react with the concete. It should be noted that repetition is one of the real keys to economy in a precast concrete wall assembly. Windows and doors should be located in identical places for all panels wherever possible.

Insulated wall panels can be produced by embedding an insulating material such as expanded polyurethane between layers of concrete. These "sandwich panels" are described in detail in Sect. 9.4. Sandwich panels are normally cast on flat beds or tilting tables. The inside surface of the concrete panel can be given a factory troweled finish followed by minor touch up work. The interior face is complet-

Fig. 9.10.5 Methods of attaching suspended ceilings, crane rails, and other sub-systems

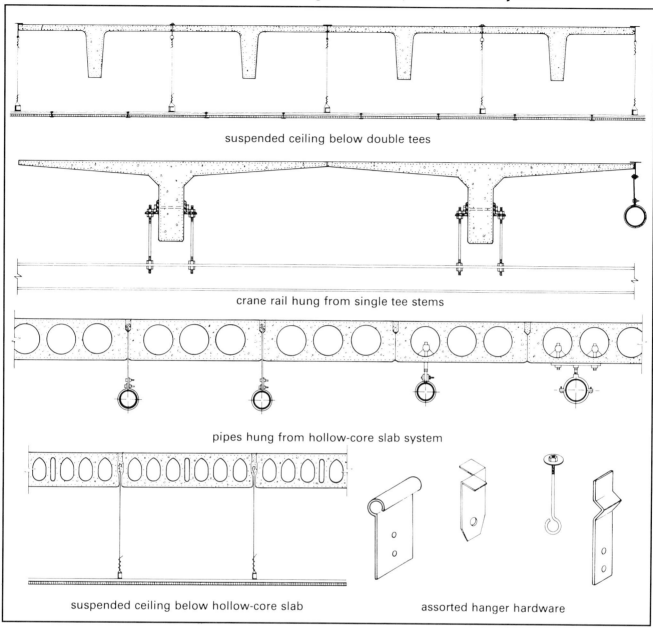

9.10.6 Systems Building

As more and more complete systems buildings are built with precast and prestressed concrete, and as interest in this method of construction increases, we can expect that more of the building sub-systems will be prefabricated and pre-coordinated with the structure.

This leads to the conculsion that those parts of the structure that require the most labor skills should logically be prefabricated prior to installation in the field. The prefabricated components can be pre-assemblies of basic plumbing systems or electrical/mechanical systems plus lighting.

For housing systems, electrical conduits and boxes can be cast in the precast wall panels. This process requires coordination with the electrical contractor; and savings on job-site labor and time are possible. The metal or plastic conduit is usually pre-bent to the desired shape and delivered to the casting bed already connected to the electrical boxes. It is essential that all joints and connections be thoroughly sealed and the boxes enclosed prior

Fig. 9.10.6 Kitchen/bathroom modules can be pre-assembled on precast prestressed slabs ready for installation in systems buildings

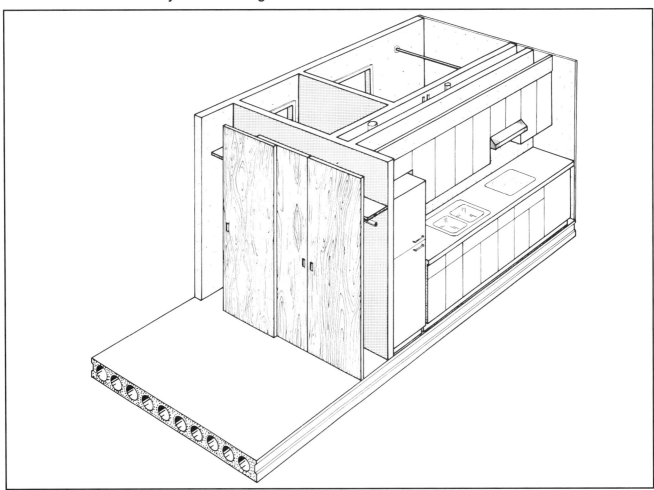

Fig. 9.10.7 Prefabricated wet-wall plumbing systems incorporate pre-assembled piping

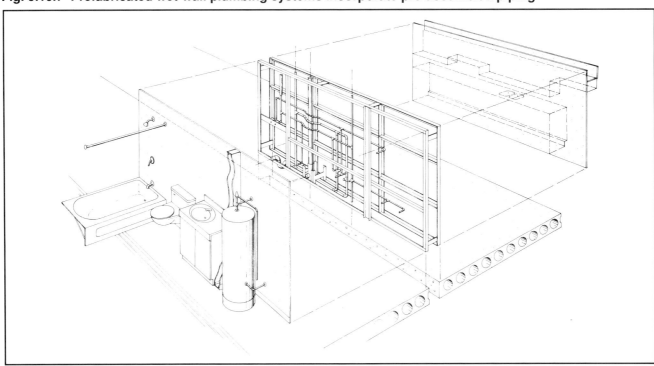

to casting in order to prevent the system from becoming clogged. The wires are usually pulled through at the job site. Television antennae and telephone conduits have also been cast-in using the same procedure.

To reduce on-site labor, prefabricated bathroom units or combination bathroom/kitchen modules have been developed (see Fig. 9.10.6). Such units include bathroom fixtures, kitchen cabinets and sinks, as well as wall, ceiling, and floor surfaces.

Plumbing units are often connected and assembled prior to delivery to the job sites. These bathroom/kitchen modules can be molded plastic units or fabricated from drywall components. To eliminate a double floor, the module can be plant built on the structural member or the walls of the unit can be designed strong enough for all fixtures to be wall hung. In the latter case, the units are placed directly on a precast floor and, in multi-story construction, are located in a stack fashion with one bathroom directly over the one below. A block-out for a chase is provided in the precast floor and connections are made from each unit to the next to provide a vertical plumbing stack. Prefabricated wet-wall plumbing systems, as shown in Fig. 9.10.7, incorporate pre-assembled piping systems using snap-on or no-hub connections made up of a variety of materials. These units only require a block-out in the prestressed flooring units and are also arranged in a stack fashion. Best economy results when bathrooms are backed up to each other, since a common vertical run can service two bathrooms.

Some core modules not only feature bath and kitchen components, but also HVAC components all packaged in one unit. These modules can also be easily accommodated in prestressed structural systems by placing them directly on the prestressed members with shimming and grouting as required.

CHAPTER 10
SPECIFICATIONS AND STANDARD PRACTICES

		Page No.
10.1	Guide Specification for Precast, Prestressed Concrete	10-2
10.2	Guide Specification for Architectural Precast Concrete	10-12
10.3	Code of Standard Practice for Precast Concrete	10-31
10.4	Recommendations on Responsibility for Design and Construction of Precast Concrete Structures	10-41

10.1 GUIDE SPECIFICATION FOR PRECAST, PRESTRESSED CONCRETE

This Guide Specification is intended to be used as a basis for the development of an office master specification or in the preparation of specifications for a particular project. In either case, this Guide Specification must be edited to fit the conditions of use.

Particular attention should be given to the deletion of inapplicable provisions. Necessary items related to a particular project should be included. Also, appropriate requirements should be added where blank spaces have been provided.

The Guide Specifications are on the left. *Notes to Specifiers are on the right.*

GUIDE SPECIFICATIONS	NOTES TO SPECIFIERS
1. GENERAL	
1.01 Description	
A. Work included:	*1.01.A This Section is to be in Division 3 of Construction Specifications Institute format. Verify that plans clearly differentiate between this work and architectural precast concrete if both are on the same job. One may need to list items such as beams, purlins, girders, lintels, columns, slab or deck members, etc. Some items such as prestressed wall panels could be included in either this or the Specification for Architectural Precast Concrete, depending on the desired finish and tolerance expectation.*
1. These specifications cover precast and precast, prestressed structural concrete construction, including product design not shown on contract drawings, manufacture, transportation, erection, and other related items such as anchorage, bearing pads, storage and protection of precast concrete.	
B. Related work specified elsewhere:	
1. Cast-in-place concrete: Section _____.	
2. Architectural precast concrete: Section _____.	
3. Post-tensioning: Section _____.	
4. Masonry bearing walls: Section _____.	
5. Structural steel: Section _____.	
6. Miscellaneous steel: Section _____.	
7. Waterproofing: Section _____.	
8. Flashing and sheet metal: Section _____.	
9. Sealants and caulking: Section _____.	
10. Painting: Section _____.	
11. Openings for other trades: Section _____.	
C. Work installed but furnished by others:	
1. Receivers or reglets for flashing: Section _____.	
2. Elevator guides: Section _____.	

| GUIDE SPECIFICATIONS | NOTES TO SPECIFIERS |

1.02 Quality Assurance

A. Manufacturer qualifications:

The precast concrete manufacturing plant shall be certified by the Precast/Prestressed Concrete Institute, Plant Certification Program, prior to the start of production.

** OR **

The manufacturer shall, at his expense, meet the following requirements:

1. Retain an independent testing or consulting firm approved by the architect/engineer and/or owner.

2. The basis of inspection shall be the Precast/Prestressed Concrete Institute *Manual for Quality Control for Plants and Production of Precast and Prestressed Concrete Products,* MNL-116, and *Manual for Quality Control for Plants and Production of Architectural Precast Concrete Products*, MNL-117.

1.02.A.2 The two quality control manuals form a general basis for the total inspection process. It is not intended, nor should it be expected unless so specified, that structural products must meet all the requirements of MNL-117 (see Sects. 2.03A and B).

3. This firm shall inspect the precast concrete plant at two-week intervals during production and issue a report, certified by a registered engineer, verifying that materials, methods, products and quality control meet all the requirements of the specifications, drawings, and MNL-116 and/or MNL-117. If the report indicates to the contrary, the architect/engineer, at the precaster's expense, will inspect and, at architect's option, may reject any or all products produced during the period of non-compliance with the above requirements.

* * * *

B. Erector qualifications:

Regularly engaged for at least _____ years in the erection of precast structural concrete similar to the requirements of this project.

1.02.B Usually 2 to 5 years.

C. Welder qualifications:

In accordance with AWS D1.1.

1.02.C Qualified within the past year.

D. Testing:

In general compliance with testing provisions in MNL-116, *Manual for Quality Control for Plants and Production of Precast and Prestressed Concrete Products.*

E. Requirements of regulatory agencies:

All local codes plus the following specifications, standards and codes are a part of these specifications:

1.02.E Always include the specific year edition of the specifications, codes and standards used in the design of the project and made part of the specifications.

GUIDE SPECIFICATIONS	NOTES TO SPECIFIERS

1. ACI 318—Building Code Requirements for Reinforced Concrete.
2. AWS D1.1—Structural Welding Code—Steel.
3. AWS D1.4—Structural Welding Code—Reinforcing Steel.
4. ASTM Specifications—As referred to in Sect. 2. Products, of this Specification.
5. AASHTO—Standard Specifications for Highway Bridges.

For projects in Canada, standards from the Canadian Standards Association should be listed in addition to, or in place of the U.S. Standards. Fire ratings are generally a code requirement. When required, fire-rated products should be clearly identified on the design drawings.

1.03 Submittals

A. Shop drawings:

1. Erection drawings:

    a. Member piece marks and completely dimensioned size and shape of each member.
    b. Plans and/or elevations locating and defining all products furnished by manufacturer.
    c. Sections and details showing connections, cast-in items and their relation to the structure.
    d. Relationship to adjacent material.

    1.03. A.1.d Details, dimensional tolerances and related information of other trades affecting precast concrete work should be furnished to precast concrete manufacturer.

    e. Joints and openings between members and between members and structure.
    f. Description of all loose, cast-in and field hardware.
    g. Field installed anchor location drawings.
    h. Erection sequences and handling requirements.

    1.03. A.1.h If the sequence of erection is critical to the structural stability of the structure, or for access to connections at certain locations, it should be noted on the contract plans and specified.

    i. All dead, live and other applicable loads used in the design.

2. Production drawings:

    a. Elevation view of each member.
    b. Sections and details to indicate quantities and position of reinforcing steel, anchors, inserts, etc.
    c. Lifting and erection devices.
    d. Dimensions and finishes.
    e. Prestress for strand and concrete strengths.
    f. Estimated cambers.
    g. Methods for storage and transportation.

1.03.A.2 Production drawings are normally submitted only upon request.

| GUIDE SPECIFICATIONS | NOTES TO SPECIFIERS |

B. Product design criteria:

1. Loadings for design:

   a. Initial handling and erection stress limits.

   b. All dead and live loads as specified on the contract drawings.

   c. All other loads specified for member, where applicable.

2. As directed on the contract drawings, design calculations of products shall be performed by a registered engineer experienced in precast, prestressed concrete design and submitted for approval upon request.

3. Design shall be in accordance with applicable codes, ACI 318, or AASHTO Standard Specifications for Highway Bridges.

C. Permissible design deviations:

1. Design deviations will be permitted only after the architect/engineer's written approval of the manufacturer's proposed design supported by complete design calculations and drawings.

2. Design deviations shall provide an installation equivalent to the basic intent without incurring additional cost to the owner.

1.03.B and C The contract drawings normally will be prepared using a local precast, prestressed concrete manufacturer's design data and load tables. Dimensional changes that would not materially affect architectural and structural properties or details are usually permissible.

Most precast, prestressed concrete is cast in continuous steel forms, therefore connection devices on the formed surfaces must be contained within the member since penetration of the form is impractical. Camber will generally occur in prestressed concrete members having eccentricity of the stressing force. If camber considerations are important, check with local prestressed concrete manufacturer to secure estimates of the amount of camber and of camber movement with time and temperature change. Architectural details must recognize the existence of camber and camber movement in connection with:

1. *Closures to interior non-load bearing partitions.*

2. *Closures parallel to prestressed concrete members (whether masonry, windows, curtain walls or others) must be properly detailed for appearance.*

3. *Floor slabs receiving cast-in-place topping. The elevation of top of floor and amount of concrete topping must allow for camber of prestressed concrete members.*

Designing for cambers less than obtained under normal design practices is possible, but this usually requires the addition of tendons or non-prestressed steel reinforcement and price should be checked with the local manufacturer.

As the exact cross section of precast, prestressed concrete members might vary somewhat from producer to producer, permissible deviations in member shape from that shown on the contract drawings might enable more manufacturers to quote on the project. Manufacturing procedures also vary between plants and permissible modifications to connection details, inserts, etc., will allow the manufacturer to use devices he can best adapt to his manufacturing procedure.

Be sure that loads shown on the contract drawings are easily interpreted. For instance, on members which are to receive concrete topping, be sure to state whether all superimposed dead and live loads on precast, prestressed members do or do not include the weight of the concrete topping.

It is best to list the live load, superimposed dead load, topping weight, and weight of the member, all

GUIDE SPECIFICATIONS

D. Test reports:

Reports of tests on concrete and other materials upon request.

1.04 Product Delivery, Storage, and Handling

A. Delivery and handling:

1. Precast concrete members shall be lifted and supported during manufacturing, stockpiling, transporting and erection operations only at the lifting or supporting points, as shown on the shop drawings, and with approved lifting devices. Lifting inserts shall have a minimum safety factor of 4. Reusable lifting hardware and rigging shall have a minimum safety factor of 5.

2. Transportation, site handling, and erection shall be performed with acceptable equipment and methods, and by qualified personnel.

B. Storage:

1. Store all units off ground.
2. Place stored units so that identification marks are discernible.
3. Separate stacked members by battens across full width of each bearing point.
4. Stack so that lifting devices are accessible and undamaged.
5. Do not use upper member of stacked tier as storage area for shorter member or heavy equipment.

2. PRODUCTS

2.01 Materials

A. Portland cement:

1. ASTM C150—Type I or III.

B. Admixtures:

1. Air-entraining admixtures: ASTM C260.
2. Water reducing, retarding, accelerating, high range water reducing admixtures: ASTM C494.
3. Calcium chloride or admixtures containing chlorides shall not be used.

C. Aggregates:

1. ASTM C33 or C330.

NOTES TO SPECIFIERS

as separate loads. Where there are two different live loads (e.g., roof level of a parking structure) indicate how they are to be combined.

2.01 Delete or add materials that may be required for the particular job.

2.01.B.3 Admixtures containing any chloride, other than impurities from admixture ingredients, are not to be used in prestressed concrete.

| GUIDE SPECIFICATIONS | NOTES TO SPECIFIERS |

D. Water:

Potable or free from foreign materials in amounts harmful to concrete and embedded steel.

E. Reinforcing steel:
  1. Bars:
     Deformed billet-steel: ASTM A615.
     Deformed rail-steel: ASTM A616.
     Deformed axle-steel: ASTM A617.
     Deformed low-alloy steel: ASTM A706.
  2. Wire:
     Cold-drawn steel: ASTM A82.
  3. Welded wire fabric:
     Welded plain steel: ASTM A185.
     Welded deformed steel: ASTM A497.

2.01.E.1 When welding of bars is required, weldability must be established to conform to AWS D1.4.

F. Strand:
  1. Uncoated, 7-wire strand:
     ASTM A416 — Grade 250 or 270.

G. Anchors and inserts:
  1. Materials:
     a. Structural steel: ASTM A36.
     b. Malleable iron castings: ASTM A47.
     c. Stainless steel: ASTM A666, type 304.
     d. Carbon steel plate: ASTM A283.
     e. Bolts: ASTM A307 or A325.
     f. Welded headed studs: AWS D1.1— Type B.
     g. Deformed bar anchors: ASTM A496.
  2. Finish:
     a. Shop primer: Manufacturer's standards.
     b. Hot dipped galvanized: ASTM A153.
     c. Zinc-rich coating: MIL-P-21035, self curing, one component, sacrificial.
     d. Cadmium coating: ASTM A165.

2.01.G.1.b Usually specified by type and manufacturer.

H. Grout:
  1. Cement grout: Portland cement, sand, and water sufficient for placement and hydration.
  2. Non-shrink grout: Premixed, packaged ferrous and non-ferrous aggregate shrink-resistant grout.

2.01.H Indicate required strengths on contract drawings.

2.01.H.2 Non-ferrous grouts with a gypsum base should not be exposed to moisture. Ferrous grouts should not be used where possible staining would

GUIDE SPECIFICATIONS	NOTES TO SPECIFIERS
	be undesirable. Non-shrink grouts are normally not used or required in the keyed joints between hollow-core floor and roof systems or hollow-core wall panels.
3. Epoxy-resin grout: Two-component mineral-filled epoxy-resin: ASTM C881 or FS MMM-A-001993.	2.01.H.3 Check with local suppliers to determine availability and types of epoxy-resin grouts.
I. Bearing pads:	
1. Chloroprene (Neoprene): Conform to Division II, Section 25 of AASHTO Standard Specifications for Highway Bridges.	2.01.I.1 AASHTO grade pads utilize 100 percent chloroprene as the elastomer. Less expensive commercial grade pads are available, but are not recommended.
2. Random oriented fiber reinforced: Shall support a compressive stress of 3000 psi with no cracking, splitting or delaminating in the internal portions of the pad.	2.01.I.2 Standard guide specifications are not available for random oriented, fiber reinforced pads. Proof testing of a sample from each group of 200 pads is suggested. Normal service load stresses are 1500 psi, so the 3000 psi test load provides a factor of 2 over service load stress. The shape factor for the test specimens should not be less than 2. If adequate test data are provided by the pad supplier, further proof testing may not be required.
3. Duck layer reinforced: Conform to Division II, Section 10.3.12 of AASHTO Standard Specifications for Highway Bridges or Military Specification MIL-C-882C.	
4. Plastic: Multi-monomer plastic strips shall be non-leaching and support construction loads with no visible overall expansion.	2.01.I.4 Plastic pads are widely used with concrete plank. Compression stress in use is not normally over a few hundred psi and proof testing is not considered necessary. No standard guide specifications are available.
5. Tetrafluoroethylene (TFE): Reinforced with glass fibers and applied to stainless or structural steel plates.	2.01.I.5 ASTM D2116 applies only to basic TFE resin molding and extrusion material in powder or pellet form. Physical and mechanical properties must be specified by naming manufacturer or other methods.
2.02 Concrete Mixes	
A. 28-day compressive strength: Minimum of _____ psi.	2.02.A and B Verify with local manufacturer. 5000 psi for prestressed products is normal practice, with release strength of 3500 psi.
B. Release strength: Minimum of _____ psi.	
2.03 Manufacture	
A. Manufacturing procedures shall be in general compliance with PCI MNL-116.	
B. Manufacturing tolerances shall comply with PCI MNL-116.	

| GUIDE SPECIFICATIONS | NOTES TO SPECIFIERS |

C. Finishes:

1. Standard underside: Resulting from casting against approved forms using good industry practice in cleaning of forms, design of concrete mix, placing and curing. Small surface holes caused by air bubbles, normal color variations, normal form joint marks, and minor chips and spalls shall be tolerated, but no major or unsightly imperfections, honeycomb, or other defects shall be permitted.

2.03.C.1 Other formed finishes that may be specified are:
Commercial Finish. Concrete may be produced in forms that impart a texture to the concrete (e.g., plywood or lumber). Fins and large protrusions shall be removed and large holes shall be filled. All faces shall have true, well-defined surfaces. Any exposed ragged edges shall be corrected by rubbing or grinding.
Grade B Finish. All air pockets and holes over ¼ in. in diameter shall be filled with a sand-cement paste. All form offsets or fins over ⅛ in. shall be ground smooth.
Grade A Finish. In addition to the requirements for Grade B Finish, all exposed surfaces shall be coated with a neat cement paste using an acceptable float. After thin pastecoat has dried, the surface shall be rubbed vigorously with burlap to remove loose particles. These requirements are not applicable to extruded products using zero-slump concrete in their process.

2. Standard top: Result of vibrating screed and additional hand finishing at projections. Normal color variations, minor indentations, minor chips and spalls shall be permitted. No major imperfections, honeycomb, or defects shall be permitted.

2.03.C.2 Other unformed finishes that may be specified are smooth, hard trowel finish, or scarified surface finish, using a rake or wire brush, for greater bond, if required, with composite topping.

3. Vertical ends: When exposed to view, strands shall be recessed a minimum of ½ in., the holes filled with grout and the ends of the member shall receive sacked finish.

2.03.C.3 When not exposed to view, normal plant practice is to cut strands flush and coat the ends of the member with bitumastic.

4. Special finish: If required, listed as follows:

2.03.C.4. Special finishes, (e.g., sandblasted or bushhammered), if required, should be described in this section of the specifications and noted on the contract drawings, pointing out which members require special finish. Such finishes will involve additional cost and consultation with the manufacturer is recommended. A sample of such finishes should be made available for review prior to bidding.

D. Openings:
Primarily on thin sections, the manufacturer shall provide for those openings 10 in. round or square or larger as shown on the structural drawings. Other openings shall be located and field drilled or cut by the trade requiring them after the precast, prestressed concrete products have been erected. Openings shall be approved by the architect/engineer before drilling or cutting.

2.03.D This paragraph requires other trades to field drill holes needed for their work, and such trades should be alerted to this requirement through proper notation in their sections of the specifications. Some manufacturers prefer to install openings smaller than 10 in. which is acceptable if their locations are properly identified on the contract drawings.

GUIDE SPECIFICATIONS

E. Patching:

Shall be acceptable providing the structural adequacy of the product and the appearance are not impaired.

F. Fasteners:

Manufacturer shall cast in structural inserts, bolts and plates as detailed or required by the contract drawings.

3. EXECUTION

3.01 Erection

A. Site access:

General contractor shall be responsible for providing suitable access to the building, proper drainage and firm, level bearing for the hauling and erection equipment to operate under their own power.

B. Preparation:

General contractor shall be responsible for:

1. Providing true, level bearing surfaces on all field placed bearing walls and other field placed supporting members.

2. Placement and accurate alignment of anchor bolts, plates or dowels in column footings, grade beams and other field placed supporting members.

3. All shoring required for composite beams and slabs.

C. Installation:

Installation of precast, prestressed concrete shall be performed by the manufacturer or a competent erector. Members shall be lifted by means of suitable lifting devices at points provided by the manufacturer. Temporary shoring and bracing, if necessary, shall comply with manufacturer's recommendations.

D. Alignment:

Members shall be properly aligned and leveled as required by the approved shop drawings. Variations between adjacent members shall be reasonably leveled out by jacking, loading, or any other feasible method as recommended by the manufacturer and acceptable to the architect/engineer.

NOTES TO SPECIFIERS

2.03.F Exclude this requirement from extruded products.

3.01.B Construction tolerances for cast-in-place concrete, masonry, etc., should be specified in those sections of the specifications.

3.01.D The following erection tolerance may be specified if other requirements do not control: Individual pieces are considered plumb, level and aligned if the error does not exceed 1:500 excluding structural deformations caused by loads.

GUIDE SPECIFICATIONS

3.02 Field Welding

A. Field welding is to be done by qualified welders using equipment and materials compatible to the base material.

3.03 Attachments

A. Subject to approval of the architect/engineer, precast, prestressed concrete products may be drilled or "shot" provided no contact is made with the prestressing steel. Should spalling occur, it shall be repaired by the trade doing the drilling or the shooting.

3.04 Inspection and Acceptance

A. Final inspection and acceptance of erected precast, prestressed concrete shall be made by the architect/engineer within a reasonable time after the work is completed.

NOTES TO SPECIFIERS

10.2 GUIDE SPECIFICATION FOR ARCHITECTURAL PRECAST CONCRETE

This Guide Specification is intended to be used as a basis for the development of an office master specification or in the preparation of specifications for a particular project. In either case, this Guide Specification must be edited to fit the conditions of use.

Particular attention should be given to the deletion of inapplicable provisions. Necessary items related to a particular project should be included. Also, appropriate requirements should be added where blank spaces have been provided.

The Guide Specifications are on the left. *Notes to Specifiers are on the right.*

GUIDE SPECIFICATIONS	NOTES TO SPECIFIERS
1. GENERAL	
1.01 Description	
A. Work included:	*1.01.A Local standard practice may indicate that responsibility for erection may not be included.*
1. The specifications establish general criteria for materials, production, erection and evaluation of precast concrete as required for subsequent related sections of these specifications. The work to be performed shall include all labor, material, equipment, related services, and supervision required for the manufacture and erection of the architectural precast concrete units shown on the contract drawings and schedules.	
B. Related work specified elsewhere:	
1. Concrete reinforcement: Section_____.	*1.01.B.1 Architectural precast concrete reinforcing steel requirements are different from cast-in-place reinforcement and should be specified in this section.*
2. Cast-in-place concrete: Section_____.	*1.01.B.2 For placement of anchorage devices in cast-in-place concrete for precast concrete panels.*
3. Precast, prestressed concrete: Section_____.	*1.01.B.3 For precast floor and roof slabs, beams, columns and other structural elements. Some items, such as prestressed wall panels on industrial buildings, could be included in either specification, depending on the desired finish and tolerance expectation.*
4. Structural steel framing: Section_____.	*1.01.B.4 For steel supporting structure, attachment of anchorage devices on steel for precast concrete panels, and sometimes loose anchors.*
5. Water repellent coatings: Section_____.	*1.01.B.5 For exposed face of panels. Delete when specified in this section.*

GUIDE SPECIFICATIONS	NOTES TO SPECIFIERS

    6. Insulation: Section_____.

1.01.B.6 *For insulation that is job-applied to precast concrete panels. Insulation cast in precast concrete panels during manufacture should be specified in this section.*

    7. Flashing and sheet metal: Section_____ .

1.01.B.7 *For counterflashing inserts and receivers, unless included in this section.*

    8. Sealants and caulking: Section_____ .

1.01.B.8 *For panel joint caulking and sealing.*

    9. Painting: Section_____ .

1.01.B.9 *For field touch-up painting. Delete when specified in this section.*

    10. Glass and glazing: Section_____.

1.01.B.10 *For glazing of precast concrete panels in plant. Delete when specified in this section.*

    11. Glazing accessories: Section _____.

1.01.B.11 *For reglets used with structural glazing gaskets. Delete when specified in this section.*

C. Work installed but furnished by others:

1.01.C *Delete when furnished by precast concrete manufacturer. Add additional items as may be required for the particular project.*

    1. Counterflashing receivers or reglets: Section _____ .

    2. Inserts or attachments for _____ : Section _____ .

1.01.C.2 *May include inserts/attachments for window or door frames, window washing equipment, etc.*

D. Testing agency provided by owner.

1.01.D *Delete when testing agency is provided by precast concrete manufacturer or general contractor. Coordinate with appropriate section of Division 1, General Conditions.*

1.02 Quality Assurance

A. Manufacturer qualifications:

The precast concrete manufacturing plant shall be certified by the Precast/Prestressed Concrete Institute, Plant Certification Program, prior to the start of production.

** OR **

The manufacturer shall, at his expense, meet the following requirements:

    1. Retain an independent testing or consulting firm approved by the architect/engineer and/or owner.

    2. The basis of inspection shall be the Precast/Prestressed Concrete Institute *Manual for Quality Control for Plants and Production of Architectural Precast Concrete Products,* MNL-117, and *Manual for Quality Control for Plants and Production of Precast and Prestressed Concrete Products,* MNL-116.

1.02.A *It is recommended that the architect establish procedures for approval of individual precast concrete manufacturers at least 10 days prior to bid date or identify approved bidders in the specification. The manufacture of architectural precast concrete requires a greater degree of craftsmanship than most other concrete products, and therefore requires some prequalification of the manufacturer. Manufacturers should have a minimum of 2 to 5 years of production experience in architectural precast concrete work of the quality and scope required on the project. Manufacturer should be able to show that the plant has experienced personnel, physical facilities, established quality control procedures, and a management capability sufficient to produce the required units without causing delay to the project. Manufacturer should demonstrate adequate financial responsibility, including ability to post bond, if required. When required by the architect, the manufacturer should submit written evidence of the above requirements.*

| GUIDE SPECIFICATIONS | NOTES TO SPECIFIERS |

3. This firm shall inspect the precast concrete plant at two-week intervals during production, and issue a report, certified by a registered engineer, verifying that materials, methods, products and quality control meet all the requirements of the specifications, drawings, and MNL-117 and/or MNL-116. If the report indicates to the contrary, the architect/engineer, at the precaster's expense, will inspect and, at architect's option, may reject any or all products produced during the period of non-compliance with the above requirements.

** OR **

Acceptable manufacturers:

1. _____ .
2. _____ .
3. _____ .

* * * *

B. Erector qualifications:

Regularly engaged for at least _____ years in erection of architectural precast concrete units similar to those required on this project.

1.02.B Usually 2 to 5 years.

C. Welder qualifications:

In accordance with AWS D1.1.

1.02.C Qualified within the past year. Delete when welding is not required.

D. Testing:

In general compliance with testing provisions in MNL-117, *Manual for Quality Control for Plants and Production of Architectural Precast Concrete Products.*

E. Testing agency:

1. Not less than _____ years experience in performing concrete tests of type specified in this section.

2. Capable of performing testing in accordance with ASTM E329.

3. Inspected by Cement and Concrete Reference Laboratory of the National Institute of Standards and Technology.

1.02.E Delete when provided by owner.

1.02.E.1 Usually 2 to 5 years.

F. Requirements of regulatory agencies:

Manufacture and installation of architectural precast concrete to meet requirements of _____ _____.

1.02.F Local building code or other governing code relating to precast concrete. For projects in Canada, standards from the Canadian Standards Association should be listed in addition to or in place of the U.S. standards.

GUIDE SPECIFICATIONS	NOTES TO SPECIFIERS
G. Allowable tolerances: 1. Manufacture and install wall panels so that each panel after erection complies with the dimensional tolerances listed in MNL-117.	*1.02.G Dimensional tolerances apply to both manufacturing and after manufacturing. The tolerances listed in PCI MNL-117 are also listed in Chapter 8 of this Handbook. Most manufacturers can meet closer tolerances, if required, but closer tolerances normally increase costs. The normal tolerances of the support system should also be recognized.*
H. Job mockup: 1. After standard samples are accepted for color and texture, submit full-scale unit meeting design requirements.	*1.02.H.1 Full-scale samples or inspection of the first production unit are normally required, especially when a new design concept or new manufacturing process or other unusual circumstance indicates that proper evaluation cannot otherwise be made. It is difficult to assess appearance from small samples.*
2. Mockup to be standard of quality for architectural precast concrete work, when accepted by architect/engineer.	*1.02.H.2 Use to determine range of acceptability with respect to color and texture variations, surface defects and overall appearance. Mockup should also serve as testing areas for remedial work. It should also be stated in the contract documents who the accepting authority will be.*
3. Incorporate mockup into work in a location reviewed by architect/engineer after keeping mockup in plant_____for checking purpose.	*1.02.H.3 Delete when mockup is not to be included in work. State how long unit should be kept. Mockup is normally incorporated in the building, at least for production units.*
I. Source quality control: 1. Quality control and inspection procedures to comply with applicable sections of MNL-117.	
2. Water absorption test on unit shall be conducted in accordance with MNL-117.	*1.02.I.2 Water absorption test is an early indication of weather staining (rather than durability). Verify the water absorption of the proposed face mix. For average exposures and based upon normal weight concrete (150 lbs per cubic foot), water absorption should not exceed 5% to 6% by weight. In order to establish comparable absorption figures for all materials, the current trend is to specify absorption percentages by volume. The stated limits for absorption would, in volumetric terms, correspond to 12% to 14% for average exposures and 8% to 10% for special conditions.*

1.03 Submittals

A. Samples:	*1.03.A Number of samples and submittal procedures should be specified in Division 1. All approved samples should be initialed by architect. Pre-bid samples should be submitted a minimum of 10 days prior to bid date.*

GUIDE SPECIFICATIONS	NOTES TO SPECIFIERS

1. Submit samples representative of finished exposed face showing typical range of color and texture prior to commencement of production.

2. Sample size: Approximately 12 in. x 12 in. and of appropriate thickness, representative of the proposed finished product.

1.03.A.1 If the back face of a precast concrete unit is to be exposed, samples of the workmanship, color, and texture of the backing should be shown as well as the facing.

B. Shop drawings:

1. Erection drawings:

    a. Member piece marks and completely dimensioned size and shape of each member.

    b. Plans and/or elevations locating and defining all products furnished by manufacturer.

    c. Sections and details showing connections, cast-in items and their relation to the structure.

    d. Relationship to adjacent material.

    e. Joints and openings between members and between members and structure.

    f. Description of all loose, cast-in and field hardware.

    g. Location, dimensional tolerances, and details of anchorage devices that are embedded in or attached to structure or other construction.

    h. Erection sequences and handling requirements.

    i. All dead, live and other applicable loads used in the design.

2. Production drawings:

    a. Member shapes (elevations and sections) and dimensions.

    b. Sections and details to indicate quantities and position of reinforcing steel, anchors, inserts, etc.

1.03.B State the number of copies required for approval or whether reproducibles are required. Current practice usually calls for two prints and one reproducible of shop drawing to be submitted for approval. General contractor should expedite submittal with architect/engineer to conform with allotted shop drawing approval time shown on the precast concrete supplier's order acknowledgment. When erection drawings contain all information sufficient for design approval, production drawings, except for shape drawings, need not be submitted for approval, except in special cases. However, record copies are frequently requested. Guidelines for the preparation of drawings are given in the PCI Drafting Handbook—Precast and Prestressed Concrete, Second Edition, MNL-119-90.

1.03.B.1.d Details, dimensional tolerances and related information of other trades affecting precast concrete work should be furnished to precast concrete manufacturer.

1.03.B.1.g Drawings normally prepared by precast concrete manufacturer and provided to general contractor for work by other trades.

1.03.B.1.h If the sequence of erection is critical to the structural stability of the structure, or for access to connections at certain locations, it should be noted on the contract plans and specified.

| GUIDE SPECIFICATIONS | NOTES TO SPECIFIERS |

      c. Lifting and erection devices.

      d. Finishes.

      e. Joint and connection details.

      f. Methods for storage and transportation.

   3. Shape drawings:

   For members with complex configurations, complete dimensions and details that also define mold shape.

C. Design calculations:

   Submit, on request, structural design calculations performed by a registered engineer experienced in the design of architectural precast concrete.

D. Design modifications:

   1. Submit design modifications necessary to meet performance requirements and field coordination.

   2. Variations in details or materials shall not adversely affect the appearance, durability or strength of units.

   3. Maintain general design concept without altering size of members, profiles and alignment.

E. Test reports:

   Submit, on request, reports on materials, compressive strength tests on concrete and water absorption tests on units.

1.03.E The number and/or frequency of each type of test should be clearly stated in the specifications by listing the required testing or by reference to applicable standards, such as PCI MNL 117. Schedule of required tests, number of copies of test reports, and how distributed are included in Testing Laboratory Services, Sect. _____.

1.04 Product Delivery, Storage, and Handling

A. Delivery and handling:

   1. Deliver all architectural precast concrete units to project site in such quantities and at such times to ensure continuity of erection.

   2. Handle and transport units in a position consistent with their shape and design in order to avoid stresses which would cause cracking or damage.

   3. Lift or support units only at the points shown on the shop drawings.

1.04.A Erector should coordinate arrival of precast concrete units and provide for possible storage and for erection in a safe manner within the agreed schedule and with due consideration for other trades. Handling procedures, including type and location of fastenings, should normally be left to the precaster, but the fastening devices should be located and identified on shop drawings.

GUIDE SPECIFICATIONS	NOTES TO SPECIFIERS

    4. Place non-staining resilient spacers of even thickness between each unit.

    5. Support units during shipment on non-staining shock-absorbing material.

    6. Do not place units directly on ground.

B. Storage at jobsite:

    1. Store and protect units to prevent contact with soil, staining, and physical damage.

    2. Store units, unless otherwise specified, with non-staining, resilient supports located in same positions as when transported.

    3. Store units on firm, level, and smooth surfaces.

    4. Place stored units so that identification marks are discernible, and so that product can be inspected.

1.04.B The ideal sequence of precast concrete erection is the unloading of units directly to their proper location on the structure without storing on the jobsite. If on-site storage is an absolute necessity to enable the erector to operate at the speed required to meet the established schedule, leaving the precast concrete units on the trailer eliminates extra handling or possible damage caused by improper on-site storage techniques.

2. PRODUCTS

2.01 Materials

A. Concrete:

    1. Portland cement:

        a. ASTM C150, type _____, _____ color.

2.01.A.1.a Type: [I(General use)], [III(High early strength)]. Color: (gray), (white), (buff). Gray is generally used for non-exposed backup concrete. Finish requirements will determine color selected for face mix.

        b. For exposed surfaces use same brand, type, and source of supply throughout.

2.01.A.1.b To minimize color variation. Specify source of supply when color shade is important.

    2. Air entraining agent: ASTM C260.

2.01.A.2 Delete if air entrainment is not required.

    3. Water reducing, retarding, accelerating, high range water reducing admixtures: ASTM C494.

2.01.A.3 Delete if water reducing, retarding, or accelerating admixtures are not required. Calcium chloride, or admixtures containing significant amounts of calcium chloride, should not be allowed. The selection of the particular admixture(s) should be left to the precast concrete manufacturer subject to approval by the architect/engineer.

    4. Coloring agent:

        a. Synthetic mineral oxide.

2.01.A.4 Investigate use of naturally colored fine aggregate in lieu of coloring agent. Delete if coloring agent is not required.

        b. Harmless to concrete set and strength.

2.01.A.4.b Consider effects upon concrete prior to final selection.

| GUIDE SPECIFICATIONS | NOTES TO SPECIFIERS |

      c. Stable at high temperature.

      d. Sunlight- and alkali-fast.

  5. Face mix aggregates:

      a. Provide fine and coarse aggregates for each type of exposed finish from a single source (pit or quarry) for entire job. They shall be clean, hard, strong, durable, and inert, free of staining or deleterious material.

2.01.A.5.a Approve or select the size, color and quality of aggregate to be used. Base choice on visual inspection of concrete sample and on assessment of certified test reports. Use same type and source of supply to minimize color variation. Fine aggregate is not always from same source as coarse aggregate.

      b. ASTM C33 or C330.

2.01.A.5.b Grading requirements are generally waived or modified.

      c. Material and color:_____.

2.01.A.5.c Specify type of stone desired such as crushed marble, quartz, limestone, granite, or locally available gravel as well as color. Some lightweight aggregates, limestones, and marbles may not be acceptable as facing aggregates. Omit where sample is to be matched.

      d. Maximum size and gradation: _____.

2.01.A.5.d State required sieve analysis. Omit where sample is to be matched.

  6. Backup concrete aggregates:

2.01.A.6 Delete when architectural requirements dictate that face mix be used throughout the unit.

      a. ASTM C33 or C330.

  7. Water: Free from deleterious matter that may interfere with the color, setting or strength of the concrete.

2.01.A.7 Potable water is ordinarily acceptable.

B. Reinforcing steel:

  1. Materials:

2.01.B.1 Grades of reinforcing steel are determined by the structural design of the precast concrete units. Panels are normally designed as crack free sections or with controlled cracking, thus the benefit of higher grade steel is not utilized.

      a. Bars:

2.01.B.1.a State uncoated, galvanized or epoxy coated. Use galvanizing or epoxy coating only where corrosive environment or severe exposure conditions justify extra cost. Availability of galvanized or epoxy coated bars should be verified.

        (1) Deformed steel: ASTM A615, grade 60.

        (2) Weldable deformed steel: ASTM A706.

2.01.B.1.a(2) Availability should be checked. When not available establish weldability in accordance with AWS D1.4.

GUIDE SPECIFICATIONS	NOTES TO SPECIFIERS

      (3) Galvanized reinforcing bars: ASTM A767.

2.01.B.1.a(3) Damage to the coating as a result of bending should be repaired with zinc-rich paint.

      (4) Epoxy coated reinforcing bars: ASTM A775.

2.01.B.1.a(4) Damage to the coating as a result of mishandling or field cutting should be repaired with epoxy paint.

   b. Welded wire fabric:

2.01.B.1.b Should be sheets, not rolls. State uncoated, galvanized or epoxy coated. Use galvanizing or epoxy coating only where corrosive environment or exposure conditions justify extra cost.

      (1) Welded plain steel: ASTM A185.

      (2) Welded deformed steel: ASTM A497.

      (3) Epoxy coated welded wire fabric: ASTM A884.

   c. Fabricated steel bar or rod mats: ASTM A184.

   d. Prestressing strand: ASTM A416, grade_____.

2.01.B.1.d Occasionally used in long and/or thin panels. Grades 250 or 270.

C. Cast-in anchors:

2.01.C Loose attachment hardware usually specified under Miscellaneous Metals.

  1. Materials:

   a. Structural steel: ASTM A36.

2.01.C.1.a For carbon steel connection assemblies.

   b. Stainless steel: ASTM A666, type 304, grade_____.

2.01.C.1.b Stainless steel anchors for use only when resistance to staining merits extra cost. (A), (B).

   c. Carbon steel plate: ASTM A283, grade_____.

2.01.C.1.c (A), (B), (C), (D).

   d. Malleable iron castings: ASTM A47, grade_____.

2.01.C.1.d Usually specified by type and manufacturer. Grades 32510 or 35018.

   e. Carbon steel castings: ASTM A27, grade 60-30.

2.01.C.1.e For cast steel clamps.

   f. Bolts: ASTM A307 or A325.

2.01.C.1.f For low-carbon steel bolts, nuts and washers.

   g. Welded headed studs: AWS D1.1—Type B.

   h. Deformed bar anchors: ASTM A496.

  2. Finish:

   a. Shop primer: FS TT-P-86, oil base paint, type I, or SSPC-Paint 14, or manufacturer's standard.

2.01.C.2.a For exposed carbon steel anchors.

| GUIDE SPECIFICATIONS | NOTES TO SPECIFIERS |

    b. Hot-dip galvanized: ASTM A153, electroplated or metallized.

2.01.C.2.b. For exposed carbon steel anchors where corrosive environment justifies the additional cost. Field welding should generally not be permitted on galvanized elements, unless the galvanizing is removed.

    c. Cadmium coating: ASTM A165.

2.01.C2.c Particularly appropriate for threaded fasteners.

    d. Zinc rich coating: MIL-P-21035, self curing, one component, sacrificial organic coating.

2.01.C.2.d 95% pure zinc in dried film. For field spot painting.

D. Receivers for flashing: 28 ga. formed _____, or polyvinyl chloride extrusions.

2.01.D (stainless steel), (copper), (zinc). Coordinate with flashing specification to avoid dissimilar metals. Delete when included in flashing and sheet metal section. Specify whether precaster or others furnish.

E. Sandwich panel insulation:_____.

2.01.E Specify type of insulation such as foamed plastic (polystyrene and polyurethane), glasses (foamed glass and fiberglass), foamed or cellular lightweight concretes, or lightweight mineral aggregate concretes. Thickness of sandwich panel insulation governed by wall U-value requirements.

F. Grout:

    1. Cement grout: Portland cement, sand, and water sufficient for placement and hydration.

2.01.F Indicate required strengths on contract drawings.

    2. Non-shrink grout: Premixed, packaged ferrous and non-ferrous aggregate shrink-resistant grout.

2.01.F.2 Grout permanently exposed to view should be non-oxidizing (non-ferrous).

    3. Epoxy-resin grout: Two-component mineral-filled epoxy-resin: ASTM C881 or FS MMM-A-001993.

2.01.F.3 Check with local suppliers to determine availability and types of epoxy-resin grouts.

G. Bearing Pads:

    1. Chloroprene (Neoprene): Conform to Division II, Section 25 of AASHTO Standard Specifications for Highway Bridges.

2.01.G.1 AASHTO grade pads having a minimum durometer hardness of 50 and utilizing 100 percent chloroprene as the elastomer. Less expensive commercial grade pads are available, but are not recommended.

    2. Random oriented fiber reinforced: Shall support a compressive stress of 3000 psi with no cracking, splitting or delaminating in the internal portions of the pad.

2.01.G.2 Standard guide specifications are not available for random oriented, fiber reinforced pads. Proof testing of a sample from each group of 200 pads is suggested. Normal service load stresses are 1500 psi, so the 3000 psi test load provides a factor of 2 over service stress. The shape factor for the test specimens should not be less than 2. If adequate test data are provided by the pad supplier, further proof testing may not be required.

| GUIDE SPECIFICATIONS | NOTES TO SPECIFIERS |

3. Duck layer reinforced: Conform to Division II, Section 10.3.12 of AASHTO Standard Specifications for Highway Bridges or Military Specification MIL-C-882C.

4. Plastic: Multi-monomer plastic strips shall be non-leaching and support construction loads with no visible overall expansion.

2.0.1.G.4 Compression stress in use is not normally over a few hundred psi and proof testing is not considered necessary. No standard guide specifications are available.

5. Tetrafluoroethylene (TFE): Reinforced with glass fibers and applied to stainless or structural steel plates.

2.0.1.G.5 ASTM D2116 applies only to basic TFE resin molding and extrusion material in powder or pellet form. Physical and mechanical properties must be specified by naming manufacturer or other methods.

2.02 Concrete Mixes

A. Concrete properties:

2.02.A The backup concrete and the surface finish concrete can be of one mix design, depending upon resultant finish, or the surface finish (face mix) concrete can be separate from the backup concrete. Clearly indicate specific requirements for each face of the product or allow manufacturer's option.

1. Water-cement ratio: Maximum 40 lbs. of water to 100 lbs. of cement.

2.02.A.1 Keep to a minimum consistent with strength and durability requirements and placement needs.

2. Air entrainment: Amount produced by adding dosage of air entraining agent that will provide 19% ± 3% of entrained air in standard 1:4 sand mortar as tested according to ASTM C185; or minimum 3%, maximum 6%.

2.02.A.2 Gradation characteristics of most facing mix concrete will not allow use of a given percentage of air. PCI recommends a range of air entraining be stated in preference to specified percentage.

3. Coloring agent: Not more than 10% of cement weight.

2.02.A.3 Amount used should not have any detrimental effects on concrete qualities. Delete if coloring agent is not required.

4. 28-day compressive strength: Minimum of 5000 psi when tested by 6 x 12 or 4 x 8 in. cylinders; or minimum 6250 psi when tested on 4 in. cubes.

2.02.A.4 Vary strength to match requirements. Strength requirements for facing mixes and back-up mixes may differ. Also the strength at time of removal from the molds should be stated if critical to the engineering design of the units. The strength level of the concrete should be considered satisfactory if the average of each set of any three consecutive cylinder strength tests equals or exceeds the specified strength and no individual test falls below the specified value by more than 500 psi.

B. Face mix:

2.02.B Delete if separate face mix is not used.

1. Minimum thickness of face mix after consolidation shall be at least one inch or a minimum of 1½ times the maximum size of aggregates used, whichever is larger.

2.02.B.1 Minimum thickness should be sufficient to prevent bleeding through of the backup mix and should be at least equal to specified minimum cover of reinforcement.

| GUIDE SPECIFICATIONS | NOTES TO SPECIFIERS |

    2. Water-cement and cement-aggregate ratios of face and backup mixes shall be similar.

2.02.B.2 Similar behavior with respect to shrinkage is necessary in order to avoid undue bowing and warping.

C. Design mixes to achieve required strengths shall be prepared by independent testing facility or qualified personnel at precast concrete manufacturer's plant.

2.02.C Proportion mixes by either laboratory trial batch or field experience methods using materials to be employed on the project for each type of concrete required. Tests will be necessary on all mixes including face, backup, and standard, which may be used in production of units. Water content should remain as constant as possible during manufacture.

2.03 Fabrication

A. Manufacturing procedures shall be in general compliance with PCI MNL-117.

B. Finishes:

2.03.B Finishing techniques used in individual plants may vary considerably from one part of the continent to another, and between individual plants. Many plants have developed specific techniques supported by skilled operators or special facilities.

    1. Exposed face to match approved sample or mockup panel.

2.03.B.1 Preferable to match sample rather than specify method of exposure.

        OR

    1. Smooth finish:

        a. As cast using flat, smooth, non-porous molds.

2.03.B.1.a Difficult to obtain uniform finish.

        OR

    1. Smooth finish:

        a. As cast using fluted, sculptured, board finish or textured form liners.

2.03.B.1.a Many standard shapes of plastic form liners are readily available.

        OR

    1. Textured finish:

        a. Achieve finish on face surface of precast concrete units by form liners applied to inside of forms.

        b. Distressed finish by breaking off portion of face of each flute.

2.03.B.1.b Delete if distressed finish is not desired.

        c. Achieve uniformity of cleavage by alternately striking opposite sides of flute.

        OR

    1. Exposed aggregate finish:

        a. Apply even coat of retardant to face of mold.

GUIDE SPECIFICATIONS

   b. Remove units from molds after concrete hardens.

   c. Expose coarse aggregate by washing and brushing or lightly sandblasting away surface mortar.

   d. Expose aggregate to produce a _____ exposure.

   OR

1. Exposed aggregate finish:

   a. Immerse unit in tank of acid solution.

   OR

   a. Pressure spray with acid and hot water solution.

   OR

   a. Treat surface of unit with brushes which have been immersed in acid solution.

   b. Protect hardware, connections and insulation from acid attack.

   c. Expose aggregate to produce a _____ exposure.

   OR

1. Exposed aggregate finish:

   a. Use power or hand tools to remove mortar and fracture aggregates at the surface of units (bushhammer).

   OR

1. Exposed aggregate finish:

   a. Hand place large facing aggregate, fieldstone or cobblestones in sand bed over mold bottom.

   b. Produce mortar joints by keeping cast concrete ½ in. to 1 in. from face of unit.

   OR

1. Sandblasted finish:

   a. Sandblast away cement-sand matrix to produce a _____ exposure.

NOTES TO SPECIFIERS

2.03.B.1.d (light) (medium) (deep). Finishes obtained vary from light etch to heavy exposure, but must relate to the size of aggregates. Matrix can be removed to a maximum depth of one-third the average diameter of coarse aggregate but not more than one-half the diameter of smallest sized coarse aggregate.

2.03.B.1.a Use reasonably acid resistant aggregates such as quartz or granite.

2.03.B.1.c (light) (medium) (deep).

2.03B.1.a Use with softer aggregates such as dolomite and marble.

2.03.B.1.a (light) (medium) (deep). Exposure of aggregate by sandblasting can vary from 1/16 in. or less to over 3/8 in. Remove matrix to a maximum depth of one-third the average diameter of coarse

| GUIDE SPECIFICATIONS | NOTES TO SPECIFIERS |

aggregate but not more than one-half the diameter of smallest sized coarse aggregate. Depth of sandblasting should be adjusted to suit the aggregate hardness and size.

OR

1. Honed or polished finish:

   a. Polish surface by continuous mechanical abrasion with fine grit, followed by special treatment which includes filling of all surface holes and rubbing.

2.03.B.1 Honing and polishing of concrete are techniques which require highly skilled personnel. Use with aggregates such as marble, onyx, and granite.

OR

1. Veneer faced finish:

   a. Cast concrete over tile, brick, terra cotta or natural stone placed in the bottom of the mold.

2.03.B.1.a Full scale mockup units with natural stone in actual production sizes, along with casting and curing of the units under realistic production conditions are essential for each new or major application or configuration of the natural stones.

   b. Connection of natural stone face material to concrete shall be by mechanical means.

2.03.B.1.b Provide a complete bondbreaker between the natural stone face material and the concrete. Ceramic tile, brick and terra cotta are bonded to the concrete.

* * * *

2. _____ back surfaces of precast concrete units after striking surfaces flush to form finish lines.

2.03.B.2 (Smooth float finish), (Smooth steel trowel), (Light broom), (Stippled finish). Use for exposed back surfaces of units.

C. Cover:

1. Provide at least ¾ in. cover for reinforcing steel.

2.03.C.1 Increase cover requirements when units are exposed to corrosive environment or severe exposure conditions. For exposed aggregate surfaces, the ¾ in. cover should be from bottom of aggregate reveal to surface of steel.

2. Do not use metal chairs, with or without coating, in the finished face.

2.03.C.2 If possible, reinforcing steel cages should be supported from the back of the panel, because spacers of any kind are likely to mar the finished surface of the panel. For smooth cast facing, stainless steel chairs may be permitted. The wires should be soft stainless steel and clippings should be completely removed from the mold.

3. Provide embedded anchors, inserts, plates, angles and other cast-in items with sufficient anchorage and embedment for design requirements.

D. Molds:

1. Use rigid molds to maintain units within specified tolerances conforming to the shape, lines and dimensions shown on the approved shop drawings.

GUIDE SPECIFICATIONS	NOTES TO SPECIFIERS

    2. Construct molds to withstand vibration method selected.

2.03.D.2 Molds for architectural precast concrete should be built to provide proper appearance, dimensional control and tightness. They should be sufficiently rigid to withstand pressures developed by plastic concrete, as well as the forces caused by consolidation. Unless otherwise agreed in the contract documents, the molds are the property of the precast concrete manufacturer.

E. Concreting:

    1. Convey concrete from the mixer to place of final deposit by methods which will prevent separation, segregation or loss of material.

    2. Consolidate all concrete in the mold by high frequency vibration, either internal or external or a combination of both, to eliminate unintentional colds joints, honeycomb and to minimize entrapped air on vertical surfaces.

2.03.E.2 The prime objective is to consolidate the concrete thoroughly, producing a dense, uniform product with fine surfaces, free of imperfections. Bonding between backup and face mix should be ensured if backup concrete is cast before the face mix has attained its initial set.

F. Curing:

2.03.F A wide variation exists in acceptable curing methods, ranging from no curing in some warm humid areas, to carefully controlled moisture-pressure-temperature curing. Consult with local panel manufacturers to avoid unrealistic curing requirements.

    1. Precast concrete units shall be cured until the compressive strength is high enough to ensure that stripping does not have an effect on the performance or appearance of the final product.

2.03.F.1 Stripping strength, which could be as low as 2000 psi, should be set by the plant based on the characteristics of the product and plant facilities. It is the responsibility of the precaster to verify and document the fact that final design strength has been reached.

G. Panel identification:

    1. Mark each precast panel to correspond to identification mark on shop drawings for panel location.

    2. Mark each precast panel with date cast.

H. Acceptance:

Architectural precast units which do not meet the color and texture range or the dimensional tolerances may be rejected at the option of the architect, if they cannot be satisfactorily corrected.

2.03. H It should be stated in the contract documents who the accepting authority will be—contractor, architect, engineer of record, owner or jobsite inspector.

GUIDE SPECIFICATIONS	NOTES TO SPECIFIERS

2.04 Concrete Testing

A. Make one compression test at 28 days for each day's production of each type of concrete.

    2.04.A This test should be only a part of an in-plant quality control program.

B. Specimens:

1. Provide two test specimens for each compression test.

    2.04.B.1 One test specimen may be used to check the stripping strength.

2. Obtain concrete for specimens from actual production batch.

3. 6 in. x 12 in. or 4 in. x 8 in. concrete test cylinder, ASTM C31.

OR

3. _____ sized concrete cube, _____ _____.

    2.04.B.3 Specify size. Cube specimens are usually 4 in. units, but 2 in. or 6 in. units are sometimes required. Larger specimens give more accurate test results than smaller ones. Source: (molded individually), (sawed from slab).

* * * *

4. Cure specimens using the same methods used for the precast concrete units until the units are stripped, then moist cure specimens until tested.

C. Keep quality control records available for the architect upon request for two years after final acceptance.

    2.04.C These records should include mix designs, test reports, inspection reports, member identification numbers along with date cast, shipping records and erection reports.

3. EXECUTION

3.01 Inspection

A. Before erecting architectural precast concrete, the general contractor shall verify that structure and anchorage inserts required to support panels are within tolerances.

B. Determine field conditions by actual measurements.

    3.0.1.B Any discrepancies between design dimensions and field dimensions which could adversely affect installation in accordance with the contract documents should be brought to the general contractor's attention. If such conditions exist, installation should not proceed until they are corrected or until design requirements are modified. Beginning of installation can mean acceptance of existing conditions.

| GUIDE SPECIFICATIONS | NOTES TO SPECIFIERS |

3.02 Erection

A. Clear, well-drained unloading areas and road access around and in the structure (where appropriate) shall be provided and maintained by the general contractor to a degree that the hauling and erection equipment for the architectural precast concrete products are able to operate under their own power.

B. General contractor shall erect adequate barricades, warning lights or signs to safeguard traffic in the immediate area of hoisting and handling operations.

C. Set precast units level, plumb, square and true within the allowable tolerances. General contractor shall be responsible for providing lines, center and grades in sufficient detail to allow installation. General contractor shall verify that bearing surfaces comply with specifications and, if not in compliance, shall make necessary corrections prior to start of erection.

D. Provide temporary supports and bracing as required to maintain position, stability and alignment as units are being permanently connected.

E. Tolerances for location of precast units shall be in accordance with Chapter 8 of this Handbook.

F. Set non-load bearing units dry without mortar, attaining specified joint dimension with steel or plastic spacing shims.

G. Fasten precast units in place by bolting or welding, or both, completing drypacked joints, grouting sleeves and pockets, and/or placing cast-in-place concrete joints as indicated on approved erection drawings.

3.02.A General contractor should coordinate delivery and erection of precast concrete products with other jobsite operations.

3.02.C Controlled reference lines should be used because the characteristics of precast concrete make a surface elevation difficult to define. Where thickness is not of exact concern, lines used in erection should be controlled from exposed exterior precast concrete surfaces.

3.02.F Shims should be near the back of the unit to prevent their causing a spall on face of unit when shim is loaded. The selection of the width and depth of field-molded sealants, for the computed movement in a joint, should be based on the maximum allowable strain in the sealant.

3.02.G The erector should protect units from damage caused by field welding or cutting operations and provide non-combustible shields as necessary during these operations. Structural welds should be made in accordance with the erection drawings which should clearly specify type, extent, sequence and location of welds. Adjustments or changes in connections, which could involve additional stresses in the products or connections, should not be permitted without approval by the architect/engineer. Precast concrete units should be erected in the sequence indicated on the approved erection drawings.

| GUIDE SPECIFICATIONS | NOTES TO SPECIFIERS |

H. Temporary lifting and handling devices cast into the precast concrete units shall be completely removed or, if protectively treated, left in place unless they interfere with the work of any other trade.

3.03 Repair

A. Repair exposed exterior surface to match color and texture of surrounding concrete and to minimize shrinkage.

3.03.A Repair is normally accomplished prior to final cleaning and caulking. It is recommended that the precaster execute all repairs or approve the methods proposed for such repairs by other qualified personnel. The precaster should be compensated for repairs of any damage for which he is not responsible. Repairs should be acceptable providing the structural adequacy of the product and the appearance are not impaired. All repairs and remedial work should be documented and kept in job record files.

B. Adhere large patch to hardened concrete with bonding agent.

3.03.B Bonding agent should not be used with small patches because of the greater likelihood of discoloring the patch.

3.04 Cleaning

A. After installation and joint treatment:_____
_____shall clean soiled precast concrete surfaces with detergent and water, using fiber brush and sponge, and rinse thoroughly with clean water in accordance with precast concrete manufacturer's recommendation.

3.04.A State whether erector or precaster should do cleaning under the responsibility of general contractor. Use cleaning materials or processes which will not change the character of exposed concrete finishes.

OR

A. After installation and joint treatment: Clean precast concrete panels with _____.

3.04.A (acid-free commercial cleaners), (steam cleaning), (water blasting), (sandblasting). Select cleaners with a non-chloride base. Use sandblasting only for units with original sandblasted finish. Ensure that materials of other trades are protected when cleaning panels.

* * * *

B. Use acid solution only to clean particularly stubborn stains after more conservative methods have been tried unsuccessfully.

C. Use extreme care to prevent damage to precast concrete surfaces and to adjacent materials.

D. Rinse thoroughly with clean water immediately after using cleaner.

3.05 Protection

A. All work and materials of other trades shall be adequately protected by the erector at all times.

| **GUIDE SPECIFICATIONS** | **NOTES TO SPECIFIERS** |

B. A fire extinguisher, of an approved type and in operating condition, shall be located within reach of all burning and welding operations at all times.

C. The erector shall be responsible for any chipping, spalling, cracking or other damage to the units after delivery to the jobsite unless damage is caused in site storage by others. After installation is completed, any further damage shall be the responsibility of the general contractor.

3.05.C After erection of any portion of precast concrete work to proper alignment and appearance, the general contractor should make provisions to protect all precast concrete from damage and staining.

10.3 CODE OF STANDARD PRACTICE FOR PRECAST CONCRETE

The precast concrete industry has grown rapidly and certain practices relating to the design, manufacture and erection of precast concrete have become standard in many areas of North America. This "Code of Standard Practice" is a compilation of these practices, and others deemed worthy of consideration, in the form of recommendations for the guidance of those involved with the use of structural and architectural precast concrete.

The goal of this Code is to build better understanding by suggesting standards which more clearly define procedures and responsibilities, thus resulting in fewer problems for everyone involved in the planning, preparation and completion of any project.

As the precast concrete industry continues to evolve, and it becomes apparent that additional practices have become standard in the industry or that current standards require modification, it is the intent of the Precast/Prestressed Concrete Institute to enlarge and revise this Code.

1. DEFINITIONS OF PRECAST CONCRETE

1.1 Structural Precast Concrete

Structural precast concrete usually includes beams, tees, joists, purlins, girders, lintels, columns, posts, piers, piles, slab or deck members, and wall panels. In order to avoid misunderstandings, it is important that the contract documents for each project list all the elements that are considered to be structural precast concrete.

1.2 Architectural Precast Concrete

Architectural precast concrete usually includes precast elements that require architectural finishes and/or exhibit decorative exposed surfaces. Typical architectural precast concrete elements include wall panels, window wall panels, mullions and column covers. In order to avoid misunderstandings, it is important that the contract documents for each project list all the elements that are considered to be architectural precast concrete.

1.3 Prestressed Concrete

Both structural and architectural precast concrete may be prestressed or non-prestressed. All structural precast concrete products referred to herein which are prestressed, are specifically referred to as prestressed concrete.

2. SAMPLES, MOCKUPS, AND QUALIFICATION OF MANUFACTURERS

2.1 Samples and Mockups

Samples, mockups, etc., are rarely required for structural prestressed concrete. If samples are required, they should be described in the contract documents and the samples should be manufactured in accordance with Sect. 3.2, *Architectural Precast Concrete*, Second Edition.*

2.2 Qualification of Manufacturer

Manufacture, transportation, erection and testing should be accomplished by a company, firm, corporation, or similar organization specializing in providing precast products and services normally associated with structural or architectural precast concrete construction.

The manufacturer may be requested to list similar and comparable work successfully completed, and demonstrate the adequacy of plant capability and facilities for performance of contract requirements.

Standards of performance are given in the PCI quality control manuals MNL-116-85 and MNL-117-77. Current certification under the PCI Plant Certification Program is normally accepted as fulfilling experience and plant capability requirements.

3. CONTRACT DOCUMENTS AND DESIGN RESPONSIBILITY

3.1 Contract Documents

Prior to initiation of the engineering-drafting function, the manufacturer should have the following contract documents at his disposal:

1. Architectural drawings.

2. Structural drawings.

3. Electrical, mechanical and plumbing drawings (if pertinent).

4. Specifications (complete with addenda).

*Available from Precast/Prestressed Concrete Institute.

Other pertinent drawings may also be desirable, such as approved shop drawings from other trades, roofing requirements, alternates, etc.

3.2 Design Responsibilities

It is the responsibility of the owner* to keep the manufacturer supplied with up-to-date documents and written information. The manufacturer should not be held responsible for problems arising from the use of outdated or obsolete contract documents. If updated documents are furnished, it may also be necessary to modify the contract.

The contract documents should clearly define the following:

1. Items furnished by the manufacturer.
2. Size, location and function of all openings, blockouts, and cast-in items.
3. Production and erection schedule requirements and restrictions.
4. Design intent including connections and reinforcement.†
5. Allowable tolerances. Normal field tolerances should be recommended by the manufacturer.
6. Dimension, material and quantity requirements.
7. General and supplemental general conditions.
8. Any other special requirements and conditions.
9. Site plan showing storage areas to be used, parking areas for trucks and equipment, etc.

Other design responsibility relationships are described in Sect. 10.4, page 10-41.

4. SHOP DRAWINGS

Shop drawings consist of erection and production drawings. Different areas of the country may use different terminology.

---

*The owner of the proposed structure or his designated representatives, who may be the architect, engineer, general contractor, public authority or others contracting with the precast manufacturer.

†When the manufacturer accepts design responsibility, the area or amount of responsibility must be clearly defined in the contract documents. The engineer or architect of record must be identified and it is understood that all designs are submitted through him for his approval and acceptance. The manufacturer's responsibility can be limited to member design only or it may include the entire structure. When the manufacturer is responsible for product design only, all loads which are to be applied to precast members, including forces developed by restraint, should be provided to the manufacturer by the owner unless otherwise agreed.

4.1 Erection Drawings

The information provided in the contract documents is used by the manufacturer to prepare erection drawings for approval and field use. They contain:

1. Plans and/or elevations locating and dimensioning all members furnished by the manufacturer.
2. Sections and details showing connections, finishes, openings, blockouts and cast-in items and their relationship to the structure.
3. Description of all loose and cast-in hardware including designation of who furnishes it.
4. Drawings showing location of anchors installed in the field.
5. Erection sequences and handling requirements.

4.2 Production Drawings

The contract documents are also used to prepare production drawings for manufacturing showing all dimensions together with locations and quantities for all cast-in materials (reinforcement, inserts, etc.) and completely defining all finish requirements.

Normal practices for the preparation of drawings for precast concrete are described in the PCI *Drafting Handbook—Precast and Prestressed Concrete*, Second Edition.

4.3 Discrepancies

When discrepancies or omissions are discovered on the contract documents, the manufacturer has the responsibility to check with the owner to resolve the problem. If this is not possible, the following procedures are normally followed:

1. Contract terms govern over specifications and drawings.
2. Specifications govern over drawings.
3. Structural drawings govern over architectural drawings.
4. Written dimensions govern over scale dimensions.
5. Sections govern over plans or elevations.
6. Details govern over sections.

Graphic verification should be requested for any unclear condition.

4.4 Approvals

Completed erection drawings, usually in reproducible form, should be submitted for approval. The exact sequence is dictated by construction schedules and erection sequences, and is determined when the contract is awarded.

Production drawings should preferably not be started prior to receipt of approved or approved as noted erection drawings. Production drawings should be submitted for approval only when so requested.*

Corrections should be noted on the reproducible erection drawings and copies made for distribution.

The following approval interpretations is normal practice:†

1. **Approved**—The approvers‡ have completely checked and verified the drawings for conformance with contract documents and all expected loading conditions. Such approval should not relieve the manufacturer from responsibility for his design when that responsibility is placed upon him by the contract. The manufacturer may then proceed with production drawings and production without resubmittal. Erection drawings may then be released for field use and plant use.

2. **Approved as Noted**—Same as above except that noted changes should be made and corrected erection drawings issued. Production drawings and production may be started after noted changes have been made.

3. **Not Approved**—Drawings must be corrected and resubmitted. Production drawings should not be started until "approved" or "approved as noted" erection drawings are returned.

5. MATERIALS

The relevant ASTM Standards that apply to materials for a project should be listed in the contract documents together with any special requirements that are not included in the ASTM Standards.

Note: Additional information regarding material specifications can be found in Sect. 10.1, "Guide Specification for Precast, Prestressed Concrete," page 10-2, and Sect. 10.2, "Guide Specification for Architectural Precast Concrete," page 10-12.

---

*When production drawings are the only drawings showing reinforcement, representative drawings should be submitted for approval.

†When production drawings are submitted, the same applies.

‡The contract should state who has approval authority.

6. TESTS AND INSPECTIONS

6.1 Tests of Materials

Manufacturers generally keep the test records required by the PCI manuals for quality control (MNL-116 and MNL-117). The contract documents may require the precast concrete manufacturer to make these records available for inspection by the owner's representative upon his request.

When the manufacturer is required to submit copies of test records to the owner and/or required to perform or have performed tests not required by MNL-116 and MNL-117, these special testing requirements should be clearly described in the contract documents along with the responsibility for payment.

6.2 Inspections

On certain projects the owner may require inspection of precast concrete products in the manufacturer's yard by persons other than the manufacturer's own quality control personnel. Such inspections are normally made at the owner's expense. The contract documents should describe how, when and by whom the inspections are to be made, the responsibility of the inspection agency, and who is to pay for them. Alternatively, the owner may accept plant certification in lieu of outside inspection, such as provided in the PCI Plant Certification Program.

6.3 Fire Rated Products

If the manufacturer is expected to provide a fire rated product and/or Underwriters Laboratories (U.L.) labels, these requirements should be clearly stated in the contract documents.

7. FINISHES

Finishes of precast concrete products, both structural and architectural, are probably the cause of more misunderstandings between the various members of the building team than any other question concerning product quality.

It is therefore extremely important that the contract documents describe clearly and completely the required finishes for all surfaces of all members, and that the erection drawings also include this information. When finish is not specified, the standard finish described in "Guide Specification for Precast, Prestressed Concrete" should normally be furnished.

For descriptions of the usual finishes for structural precast concrete, see Sect. 10.1 "Guide Specification for Precast, Prestressed Concrete," and for architectural precast concrete, see Sect. 3.5, *Architectural Precast Concrete*, Second Edition. Where special or critical requirements exist or where large expanses of exposed precast concrete will occur on a project, samples are essential and, if required, should be so stated and described in the contract documents.

8. DELIVERY OF MATERIALS

8.1 Manner of Delivery

The manufacturer should deliver the precast concrete to the erector* in a manner to facilitate the speed of erection of the building or as mutually agreed upon between the owner, manufacturer and erector. Special requirements of the owner for the delivery of materials or the mode of transport, should be stated in the contract documents.

8.2 Marking and Shipping of Materials

The precast concrete members should be separately marked in accordance with approved drawings in such a manner as to distinguish varying pieces and to facilitate erection of the structure. Any members which require a sequential erection should be properly marked.

The owner should give the manufacturer sufficient time to fabricate and ship any special plates, bolts, anchorage devices, etc., contractually agreed to be furnished by the manufacturer.

8.3 Precautions During Delivery

Special protection or precautions beyond that required in MNL-116 and MNL-117 should not be expected unless stated in the bid invitation or specifications. The manufacturer is not responsible for the product, including loose material, after delivery to the site unless required by the contract documents.

8.4 Access to Jobsite

Free and easy access to the delivery site should be provided to the manufacturer, including backfilling and compacting, drainage and snow removal, so that delivery trucks can operate under their own power.

8.5 Unloading Time Allowance

Delivery of product includes a reasonable unloading time allowance. Any delay beyond a reasonable time is normally paid for by the party which is responsible for the delay.

---

*The erector may be either the manufacturer or a subcontractor engaged by the manufacturer, or the general contractor.

9. ERECTION

9.1 Special Erection Requirements

When the owner requires a particular method or sequence of erection, this information should be stated in the contract documents. In the absence of such stated restrictions, the erector will proceed using the most efficient and economical method and sequence available to him, consistent with the contract documents.

9.2 Tolerances

Some variation is to be expected in the overall dimensions of any building or other structure. It is common practice for the manufacturer and erector to work within the tolerances recommended by the American Concrete Institute and the Precast/Prestressed Concrete Institute.

The owner, by whatever agencies he may elect, immediately upon completion of the erection, should determine if the work is plumb, level, aligned and properly fastened. Discrepancies should immediately be brought to the attention of the erector so that proper corrective action can be taken.

The work of the manufacturer and erector is complete once the precast product has been properly plumbed, leveled and aligned within the established tolerances. Acceptance for this work should be secured from the authorized representative of the general contractor (see Sect. 11.2).

9.3 Foundations, Piers, Abutments and Other Bearing Surfaces

The invitation to bid should state the anticipated time when all foundations, piers, abutments and other bearing surfaces will be ready and accessible to the erector.

9.4 Building Lines and Bench Marks

The precast erector should be furnished all building lines and bench marks at the site of the structure.

9.5 Anchor Bolts and Bearing Plates

9.5.1 The precast manufacturer normally furnishes but does not install anchor bolts, plates, etc. that are to be installed in cast-in-place concrete or masonry for connection with precast members. However, if this is to be the responsibility of the precast manufacturer, it should be so defined in the specifications. It is important that such items be installed true to line and grade, and that installation be completed in time to avoid delays or interference with the precast concrete erection.

9.5.2 Anchor bolts and foundation bolts are set by the owner in accordance with an approved drawing. They must not vary from the dimensions shown on the erection drawings by more than the following:

1. ⅛ in. center to center of any two bolts within an anchor bolt group, where an anchor bolt group is defined as the set of anchor bolts which receive a single fabricated shipping piece.

2. ¼ in. center to center of adjacent anchor bolt groups.

3. Maximum accumulation of ¼ in. per hundred feet along the established column line of multiple anchor bolt groups, but not to exceed a total of 1 in., where the established column line is the actual field line most representative of the centers of the as-built anchor bolt groups along a line of columns.

4. ¼ in. from the center of any anchor bolt group to the established column line through that group.

5. The tolerances of items 2, 3 and 4 apply to offset dimensions shown on the plans, measured parallel and perpendicular to the grid lines.

9.5.3 Erectors should check both line and grade in sufficient time before erection is scheduled to permit any necessary corrections. Proposed corrections should be submitted to the owner for approval. Corrections, if any, should be made by the general contractor before erection begins.

9.6 Utilities

Water and electricity should be furnished for erection and grouting operations by the owner.

9.7 Working Space

The owner should furnish adequate, properly drained, graded, and convenient working space for the erector and access for his equipment necessary to assemble the structure. The owner should provide adequate storage space for the precast concrete products to enable the erector to operate at the speed required to meet the established schedule. Unusual hazards such as high voltage lines, buried utilities, or areas of restricted access should be stated in the invitation to bid.

9.8 Materials of Other Trades

Other building materials or work of other trades should not be built up above the bearing of the precast concrete until after erection of the precast concrete members.

9.9 Correction of Errors

Corrections of minor misfits are considered a part of erection even if the precast concrete is not erected by the manufacturer. Any error in manufacturing which prevents proper connection or fitting should be immediately reported to the manufacturer and the owner so that corrective action can be taken.

9.10 Field Assembly

The size of precast concrete pieces may be limited by transportation requirements for weight and clearance dimensions. Unless agreed upon between the manufacturer and owner, the manufacturer should provide for such field connections that will meet required loads and forces without altering the function or appearance of the structure.

The manufacturer furnishes those items embedded in the precast members, and generally furnishes all loose materials for temporary and permanent connection of precast members. Temporary guys, braces, falsework, shims, and cribbing are the property of the erector and are removed only by the erector or with the erector's approval upon completion of the erection of the structure, unless otherwise agreed.

9.11 Blockouts, Cuts and Alterations

Neither the manufacturer nor the erector is responsible for the blockouts, cuts or alterations by or for other trades unless so specified in the contract documents. Whenever such additional work is required, all information regarding size, location and number of alterations is furnished by the owner prior to preparation of the precast production and erection drawings.

The general contractor is responsible for warning other trades against cutting of precast concrete members without prior approval of the engineer of record.

9.12 Temporary Floors and Access

The manufacturer or erector is not required to furnish temporary flooring for access unless so specified in the contract documents.

9.13 Painting, Caulking and Closure Panels

Painting, caulking and placing of closure panels between stems of flanged concrete members are services not ordinarily supplied by the manufacturer or erector. If any of these services are required of the manufacturer, it should be stated in the contract agreement.

9.14 Patching

A certain amount of patching of product is to be expected to repair minor spalls and chips. Patching should meet the finish requirements of the project and color should be reasonably matched. Responsibility for accomplishing this work should be resolved between the manufacturer and erector.

9.15 Safety

Safety procedures for the erection of the precast concrete members is the responsibility of the erector and must be in accordance with all local, state or Federal rules and regulations which have jurisdiction in the area where the work is to be performed, but not less than required in ANSI Standard A10.9, *Safety Requirements for Concrete Construction and Masonry Work.**

9.16 Security Measures

Security protection at the jobsite should be the responsibility of the general contractor.

10. INTERFACE WITH OTHER TRADES

Coordination of the requirements for other trades to be included in the precast concrete members should be the responsibility of the owner unless clearly defined otherwise in the contract documents.

The PCI manuals for quality control (MNL-116 and MNL-117) specify manufacturing tolerances for precast concrete members. Interfaces with other materials and trades must take these tolerances into account. Unusual requirements or allowances for interfacing should be stated in the contract documents.

11. WARRANTY AND ACCEPTANCE

11.1 Warranties

Warranties of product and workmanship have become a widely accepted practice in this industry, as in most others. Warranties given by the precast concrete manufacturer and erector should indicate that their product and work meet the specifications for the project.

*American National Standards Institute, New York, New York.

In no case should the warranty of the manufacturer and erector be in excess of the warranty required by the specifications. Warranties should in all instances include a time limit and it is recommended that this should not exceed one year.

In order to protect the interests of all parties concerned, warranties should also state that any deviations in the designed use of the product, modifications of the product by the owner and/or contractor or changes in other products used in conjunction with the precast concrete will cause said warranty to become null and void.

Warranty may be included as a part of the conditions of the contract agreement, or it may be presented in letter form as requested by the owner. A sample warranty follows:

> Manufacturer warrants that all materials furnished have been manufactured in accordance with the specifications for this project. Manufacturer further warrants that if erection of said material is to be performed by those subject to his control and direction, work will be completed in accordance with the same specifications.
>
> In no event shall manufacturer be held responsible for any damages, liability or costs of any kind or nature occasioned by or arising out of the actions or omissions of others, or for work, including design, done by others; or for material manufactured, supplied or installed by others; or for inadequate construction of foundations, bearing walls, or other units to which materials furnished by the precast manufacturer are attached or affixed.
>
> This warranty ceases to be in effect beyond the date of _____. Should any defect develop during the contract warranty period, which can be directly attributed to defect in quality of product or workmanship, precast manufacturer shall, upon written notice, correct defects or replace products without expense to owner and/or contractor.
>
> COMPANY NAME
>
> _____
> Signature   Title

11.2 Acceptance

Manufacturer should request approval and acceptance for all materials furnished and all work completed by him periodically and in a timely manner as deemed necessary in order to adequately protect the interests of everyone involved in the project. The size and nature of the project will dictate the proper intervals for securing approval and acceptance. Periodic approval in writing should be considered when it appears that such action will minimize possible problems which would seriously affect the progress of the project. A sample acceptance form follows:

FIELD INSPECTION REPORT

Project # _____

On this _____ day of _____ ,19_____

_____ of _____
Company Field Superintendent   Precast Manufacturer

and _____ of _____
General Contractor Superintendent   General Contractor

_____ have inspected _____

_____
portion of building being inspected

All of the work performed by the above indicated company in the above described portion of the project has been performed to the satisfaction of the above named General Contractor's Superintendent with the exception of the following:_____

_____

The General Contractor's Superintendent hereby releases the Precast Manufacturer of its responsibility to perform any other work in the above described portion of the project except as detailed herein.

The Precast Manufacturer in turn hereby releases the above described portion of the project to the General Contractor.

_____   _____
Precaster's Superintendent   Gen'l. Contr. Superintendent

Final inspection and acceptance of erected precast and prestressed concrete should be made by the architect/engineer within a reasonable time after the work is completed.

12. CONTRACT ADMINISTRATION

12.1 General Statement

Information relative to invoicing, payment, bonding and other data pertinent to a project or material sale should be specifically provided for in the major provisions of the contract documents or in the special terms and conditions applicable to all contractual agreements between manufacturer and owner.

Contract agreements may vary widely from area to area, but the objective should be the same in all instances. The contract agreement should be written to protect the interests of all parties concerned and at the same time, be specific enough in content to avoid misunderstandings once the project begins.

The intent of this section is to recommend those matters which ought to be considered, but not necessarily the form in which they should be expressed. The final statement of policies should be the result of careful consideration of all pertinent factors as well as of the normal practices in the area.

12.2 Retentions

Although retentions have been used for many years as a means of ensuring a satisfactory job performance, it is apparent that they directly contribute to the cost of construction, frequently lead to disputes, and often result in job delays. In view of the unfavorable consequences of retentions and possible abuse, it is recommended that the following procedure be followed:

1. Wherever possible, retentions should be eliminated and bonding should be used as the single, best source of protection. This should apply to prime contractors and subcontractors equally.

2. Where there are no bonding requirements, the retention percentage should be as low as possible. It is recommended that this be not more than 5 percent of the work invoiced.

3. The percentage level of any retention should be the same for subcontractors as for prime contractors on a job.

4. Release of retained funds and final payment, as well as computing the point of reduction of the retention, should be done on a line item basis, that is, each contractor or subcontractor's work considered as a separate item and the retention reduced by 50 percent upon substantial completion and the balance released within 30 days after final completion of that work.

5. Retained funds should be held in an escrow account with interest accruing to the benefit of the party to whom the funds are due.

6. When materials are furnished FOB plant or jobsite, it is recommended that there be no retentions.

12.3 Contract Agreement

1. Contract agreement should fully describe the project involved, including job location, project name, name of owner/developer, architect or other design professionals and all reference numbers identifying job relation information such as plans, specifications, addenda, bid number, etc.

2. Contract agreement should fully describe the materials to be furnished and/or all work to be completed by the seller.

3. All exclusions should be stated to avoid the possibility of any misunderstanding.

4. Price quoted should be stated to eliminate any possibility of misunderstanding.

5. Reference should be made to the terms and conditions governing the proposed contract agreement. The terms and conditions may best be stated on the reverse side of the contract form. Special terms or conditions should be stated in sufficient detail to avoid the possibility of misunderstanding.

6. The terms of payment should be specifically detailed so there is no doubt as to intent. Special care should be exercised where the terms of payment will differ from those normally in effect or where they deviate from the general terms and conditions appearing on the reverse side of the contract form.

7. A statement of policy should be made with reference to the inclusion or exclusion of taxes in the stated price.

8. The proposal form stating the full intent and conditions under which the project will be performed may contain an acceptance clause to be signed by the purchaser. At such time as said acceptance clause is signed, the proposal form then becomes the contract agreement.

9. Seller should clearly state the limits of time within which an accepted proposal will be recognized as a binding contract.

10. A statement indicating the classification of labor to perform the work in the field is advisable to eliminate later dispute over jurisdiction of work performed.

12.4 Terms and Conditions

The terms and conditions stated on the proposal contract agreement should include, but are not necessarily limited to, the following:

1. **Lien Laws**—Where the lien laws of a state specifically require advance notice of intent, it is advisable to include the required statement in the general terms and conditions.

2. **Specifications**—Seller should make a specific declaration of material and/or work specifications, but normally this should not be in excess of the specifications required by the contract agreement.

3. **Contract Control**—A statement should be made indicating that the agreement when duly signed by both parties supersedes and invalidates any verbal agreement and can only be modified in writing with the approval of those signing the original agreement.

4. **Terms of Payment**—Terms of payment should be specifically stated either on the face of the contract or in the general terms and conditions. Mode and frequency of invoicing should be so stated, indicating time within which payment is expected.

5. **Late Payment Charges**—The contract may provide for legal interest charges for late payments not made in accordance with contract terms, and if this is desired, it should be stated in the general terms and conditions. A statement indicating seller is entitled to reasonable attorney's fees and related costs should collection proceedings be necessary may also be included.

6. **Overtime Work**—Prices quoted in the proposal should be based on an 8-hour day and a 5-day week under prevailing labor regulations. Provisions should be included in the contract agreement to provide for recovery of overtime costs plus a reasonable markup when the seller is requested to provide such service.

7. **Financial Responsibility**—General terms and conditions may indicate the right of the seller to suspend or terminate material delivery and/or work on a project if there is a reasonable doubt of the ability of the purchaser to fulfill his financial responsibility.

8. **Payment for Inventory**
   a. It has become common practice to include in the contract terms and conditions provisions for the invoicing and payment of all materials stored at the plant or jobsite when deliveries or placement of said materials are delayed for more than a stipulated time beyond the originally scheduled date because of purchaser's inability either to accept delivery of materials or to provide proper job access.

   b. Under certain conditions, it may be necessary to purchase special materials or to produce components well in advance of job requirements to ensure timely deliveries. When job requirements are of such a nature, it is advisable to include provisions for payment of such raw and finished inventories stored in seller's plant or on jobsites on a current basis.

9. **Payment for Suspended or Discontinued Projects**—The terms and conditions should provide that in the event of a discontinued or suspended project, seller shall be entitled to payment for all material purchased and/or manufactured including costs, overhead and profit,

and not previously billed, as well as reasonable engineering and other costs incurred.

10. **Job Extras**—Requests for job extras should be confirmed in writing. Invoicing should be presented immediately following completion of the extra work with payment subject to the terms and conditions of the contract agreement, or as otherwise stated in the change order.

11. **Claims for Shortages, Damages or Delays**—Seller should, upon immediate notification in writing on the face of the delivery ticket of rejected material or shortage, acknowledge and furnish replacement material at no cost to purchaser. It is normal practice that the seller should not be responsible for any loss, damage, detention or delay caused by fire, accident, labor dispute, civil or military authority, insurrection, riot, flood or by occurrences beyond his control.

12. **Back Charges**—Back charges should not be binding on the seller, unless the condition is promptly reported in writing, and opportunity is given seller to inspect and correct the problem.

13. **Permits, Fees and Licenses**—Costs of permits, fees, licenses and other similar expenses are normally assumed by the purchaser.

14. **Bonds**—Cost of bonds is normally assumed by the purchaser.

15. **Taxes**—Federal, State, County or Municipal, occupation or similar taxes which may be imposed are normally paid by the purchaser and, in the case of sales taxes, through the seller. For tax exempt projects, purchaser issues to the seller a tax exempt certificate when the purchase agreement is finalized.

16. **Insurance**—Seller shall carry Workmen's Compensation, Public Liability, Property Damage and Auto Insurance and certificates of insurance will be furnished to purchaser upon request. Additional coverage required over and above that provided by the seller is normally paid by the purchaser.

17. **Services**—Heat, water, light, electricity, toilet, telephone, watchmen and general services of a similar nature are normally the responsibility of the purchaser unless specifically stated otherwise in the contract agreement.

18. **Safety Equipment**—The purchaser is normally responsible for necessary barricades, guard rails and warning lights for the protection of vehicular and pedestrian traffic and seller's equipment. Purchaser is also normally responsible for furnishing, installing and maintaining all safety appliances and devices required on the project under U.S. Department of Labor, *Safety and Health Standards for Construction Industry (OSHA 2207)*, as well as all other safety regulations imposed by other agencies having jurisdiction over the project.

19. **Warranty**—Seller should provide specific information relative to warranties given, including limitations, exclusions and methods of settlement. Warranties should not be in excess of warranty required by the specific project.

20. **Title**—Contract should provide for proper identification of title to material furnished. It is normal practice for title and risk of loss or damage to the product furnished to pass to the purchaser at the point of delivery, except in cases of FOB plant, in which event title to and risk of loss or damage to the product normally should pass to purchaser at plant pickup.

21. **Shop Drawing Approval**—Seller should prepare and submit to purchaser for approval all shop drawings* necessary to describe the work to be completed. Shop drawing approval should constitute final agreement to quantity and general description of material to be supplied. No work should be done upon material to be furnished by seller until approved shop drawings are in his possession.

22. **Delivery**—Delivery times or schedules set forth in contract agreements should be computed from the date of delivery to the seller of approved shop drawings. Where materials are specified to be delivered FOB to jobsite, the purchaser should provide labor, cranes or other equipment to remove the materials from the trucks and should pay seller for truck expense for time at the jobsite in excess of a specified time for each truck. On shipments to be delivered by trucks, delivery should be made as near to the construction site as the truck can travel under its own power. In the event delivery is required beyond the curb line, the purchaser should assume full liability for damages to sidewalks, driveways or other properties and should secure in advance all necessary permits or licenses to effect such deliveries.

23. **Builder's Risk Insurance**—Purchaser should provide Builder's Risk Insurance without cost to

---

*See Sect. 4 for definition of shop drawings.

seller, protecting seller's work, materials and equipment at the site from loss or damage caused by fire or the standard perils of extended coverage, including vandalism and malicious acts.

24. **Erection**—Purchaser should ensure that the proposed project will be accessible to all necessary equipment including cranes and trucks, and that the operation of this equipment will not be impeded by construction materials, water, presence of wires, pipes, poles, fences or framings. Purchaser should further indemnify and save harmless the seller and his respective representatives, including subcontractors, vendors, assigns and successors from any and all liability, fine, penalty or other charge, cost or expense and defend any action or claim brought against seller for any failures by purchaser to provide suitable access for work to be performed. Seller also reserves the right to discontinue the work for failure of purchaser to provide suitable access and the purchaser should be responsible for all expenses and costs incurred.

25. **Exclusions of Work to be Performed**—Unless otherwise stated in the contract, all shoing, forming, framing, cutting holes, openings for mechanical trades and other modifications of seller's products should not be performed by the seller nor are they included in the contract price. Seller should not be held responsible for modifications made by others to his product unless said modifications are previously approved by him.

26. **Sequence of Erection**—Sequence of erection should be as agreed upon between seller and purchaser and expressly stated in the contract agreement. Purchaser should have ready all foundations, bearing walls or other units to which seller's material is to be affixed, connected or placed, prior to start of erection. Purchaser should be responsible for the accuracy of all job dimensions, bench marks, and true and level bearing surfaces. Claims or expenses arising from the purchaser's neglect to fulfill this responsibility should be assumed by the purchaser.

27. **Arbitration**—In view of the difficulties and misunderstandings which may occur due to misinterpretation of contractual documents, it is recommended that the seller stipulate that all claims, disputes and other matters in question, arising out of or related to the contract, be decided by arbitration in accordance with the Construction Industry Rules of the American Arbitration Association then obtaining, or some other rules acceptable to both parties. The location for such arbitration should be stipulated.

28. **Contract Form**—Contract documents should stipulate policy governing acceptance of proposal on other then the seller's form. In the event purchaser does not accept the seller's proposal and/or contract agreement, but requires the execution of a contract on his own form, it is advisable that seller stipulate in writing on the contract agreement that the contract will be fulfilled according to his proposal originally submitted. All identifying information such as proposal number, dates, etc. should be included so there can be no question of the document referred to.

10.4 RECOMMENDATIONS ON RESPONSIBILITY FOR DESIGN AND CONSTRUCTION OF PRECAST CONCRETE STRUCTURES

1. INTRODUCTION

Design and construction of structures is a complex process. Defining the scope of work and the responsibilities of the parties involved in this process, by contract, is necessary to achieve a safe, high quality structure.

Besides the Owner, the parties involved in the design and construction of precast concrete structures or other structures containing precast concrete members may include the Engineer, Architect, Construction Manager, General Contractor, Manufacturer, Precast Engineer, Erector and Inspector. These and other terms related to precast concrete construction are defined in Sect. 2.0.

2. TERMINOLOGY

Terms used in engineering practice and the construction industry may have meanings that differ somewhat from ordinary dictionary definitions. The following definitions are commonly understood in precast concrete construction.

Approval (Shop Drawings or Submittals)

Action with respect to shop drawings, samples and other data which the General Contractor is required to submit, but only for conformance with the design requirements and compliance with the information given in the contract documents. Such action does not extend to means, methods, techniques, sequences or procedures of construction, or to safety precautions and programs incident thereto, unless specifically required in the contract documents.

Authority

The power, conferred or implied by contract, to exercise effective direction and control over an activity for which a party has responsibility.

Connection

A structural assembly or component that transfers forces from one precast member to another, or from one precast member to another type of structural member.

Contract Documents

The design drawings and specifications, as well as general and supplementary conditions and addenda, that define the construction and the terms and conditions for performing the work. These documents are incorporated by reference into the contract.

Contractor

A person or firm that enters into an agreement to construct all or part of a project.

Construction Manager

A person or firm engaged by the Owner to manage and administer the construction.

Design

(As a transitive verb) The process of utilizing the principles of structural mechanics and materials science to determine the geometry, composition, and arrangement of members and their connections in order to establish the composition and configuration of a structure. (As a noun) The product of the design process, as usually expressed by design drawings and specifications.

Design Drawings

Graphic diagrams with dimensions and accompanying notes that describe the structure.

Detail

(As a transitive verb) The process of utilizing the principles of geometry and the art of graphics to develop the dimensions of structural components.

(As a noun) The product of the detailing process, shown on either design or shop drawings, such as the graphic depiction of a connection.

Engineer/Architect

A person or firm engaged by the Owner or the Owner's representative to design the structure and/or to provide services during the construction process. In some cases the Engineer may be a Subcontractor to the Architect, or vice versa. The Engineer of Record is usually an individual employed by the Engineer or Architect.

Engineer of Record (EOR)

The registered professional engineer (or architect) who is responsible for developing the design drawings and specifications in such a manner as to meet all applicable requirements of governing state laws and of local building authorities. The EOR is commonly identified by the professional engineer's seal on the design drawings and specifications.

Erector

Usually the Subcontractor who erects the precast concrete components at the site. The General Contractor may also be the Erector.

General Contractor

A person or firm engaged by the Owner to construct all or part of the project. The General Contractor supervises the work of its Subcontractors and coordinates the work with other Contractors.

Inspector

The person or firm retained by the Owner or the Owner's representative to observe and report on compliance of the construction with the contract documents.

Manufacturer (Producer, Precaster, Fabricator)

The firm that manufactures the precast concrete components.

Owner

The public body or authority, corporation, association, firm or person for whom the structure is designed and constructed.

Precast Concrete

Concrete cast elsewhere than in its final position. Includes prestressed and non-prestressed components used in structural or nonstructural applications.

Precast Engineer

The person or firm who designs precast members for specified loads and who may also direct the preparation of the shop drawings. The Precast Engineer may be employed by the Manufacturer or be an independent person or firm to whom the Manufacturer subcontracts the work.

Responsibility

Accountability for providing the services and/or for performing the work required by contract.

Shop Drawings

Graphic diagrams of precast members and their connecting hardware, developed from information in the contract documents. They show information needed for both field assembly (erection) and manufacture (production) of the precast concrete. Shop drawings for precast concrete may be separated into erection and production drawings. Erection drawings typically describe the location and assembly details of each precast member at the construction site. Connection hardware is detailed on erection drawings and may be shown on production drawings. Production drawings contain all information necessary for the manufacturer to cast the member.

Specifications

Written requirements for materials and workmanship that complement the design drawings.

Subcontractor

A person or firm contracting to perform all or part of another's contract.

3. DESIGN PRACTICES

Practices vary throughout North America with respect to design of structures using precast concrete members. However, it is of fundamental importance that every aspect of the design be in accordance with the requirements of all state laws governing the practice of engineering and/or architecture, and also meet all requirements of local regulatory authorities. This generally requires that a registered professional engineer or architect accept responsibility as the Engineer of Record (EOR) for ensuring that these requirements are met. The EOR seals the contract documents. These documents constitute the structural design and are customarily submitted to regulatory authorities for a building permit. In addition, the EOR ordinarily approves shop drawings. The EOR may also have other ongoing responsibilities during construction to satisfy local authorities.

Structures are usually designed by an engineering firm that is retained by or on behalf of the Owner. A person within the firm is selected to prepare and/or to supervise the preparation of the contract documents. This person normally is registered to practice engineering in the state where the structure will be built and becomes the EOR when the design drawings and specifications are approved by the local regulatory authority. An individual in private practice may also be retained by an Owner to prepare the design and become the EOR. The design drawings and specifications become part of the contract documents used by contractors to construct the structure.

Procedures that allow consideration of alternative construction schemes are sometimes included in the contract documents. If a fully developed design is included in the contract documents, a contractor proposing an alternate for some part of the structure is expected to consider the effect of the alternate on all other parts of the structure, and to provide all necessary design changes. It is common to require that an alternate design be prepared under the direction of an engineer registered to practice in the state where the structure will be built. However, the EOR still has the responsibility to review and accept the alternate and to submit the alternate to the regulatory authorities.

Owners may also directly seek proposals from General Contractors who are willing to prepare the design. The General Contractor may use an employed registered professional engineer who is the EOR, or subcontract the design to a firm or individual who becomes the EOR. In some instances, a Manufacturer may be the General Contractor. Since the Owner will already have a contract with the selected General Contractor, any additional contracts will be between the General Contractor and Subcontractors. Under this arrangement, the Owner often retains an engineering consultant to review the proposals and the design related work of the selected General Contractor.

Normally the Manufacturer is a Subcontractor to the General Contractor. Most Manufacturers are willing to accept responsibility for component design of the members that they produce, provided that sufficient information is contained in the contract documents. This design work is commonly done by a Precast Engineer. The Manufacturer may also accept responsibility for design of the connections when the forces acting on the connections are defined by the EOR. At the time of bidding, if there is insufficient information in the contract documents to fully cover all the reinforcement requirements, the Manufacturer customarily assumes that industry minimum standards are acceptable.

Manufacturers usually have standard designs that fit common applications for their products and that are sufficient for design purposes. Manufacturers may design, produce and deliver a "structural frame" or a "structural shell" rather than a building ready for occupancy. In this case the Manufacturer may subcontract for parts of the work, such as the cast-in-place concrete or the precast erection.

Local regulatory authorities may approve design documents for starting construction without final design of the precast members. The design can be performed and submitted at a later time, often in conjunction with preparation of shop drawings. The EOR, if not employed by the Manufacturer, may require that a registered engineer seal the documents that depict the component design of the precast members. This does not relieve the EOR of responsibility to approve or accept the design as meeting the requirements of the contract documents, state laws and local authorities. However, some states accept designs made by these persons, relieving the EOR of some responsibility. For example, these persons have been called Specialty Engineers in the state of Florida.

4. RECOMMENDATIONS

The Precast/Prestressed Concrete Institute is keenly aware of the competitiveness in the marketplace for building systems. It believes precast concrete products provide a high quality structure. Along with quality, it is essential that a structure be both safe and serviceable, i.e., have structural integrity and perform as intended. Because the construction process involves many parties, it is essential that work assignments and responsibilities be clearly defined in the contractual arrangements. The Institute offers the following comments and recommendations towards achieving quality and integrity in precast concrete structures.

4.1 To the Owner

At the outset, the Owner must decide whether to enter into a contract with an Engineer or Architect to design and prepare contract documents for the structure which subsequently can be used to obtain bids by a General Contractor, or with a General Contractor (or Manufacturer willing to assume that responsibility) to both design and build the structure. These two arrangements may be summarized as follows:

1. Owner retains an Engineer or Architect who will be the EOR:
    a. To design and prepare contract documents sufficient for construction without further design by the General Contractor or Subcontractors.
    b. To prepare contract documents sufficient for construction with further design by the General Contractor or Subcontractor.

2. Owner contracts with a General Contractor for design and construction.

Under either arrangement, the Institute believes that the best control of structural integrity will be achieved when the EOR is given responsibility and authority for the entire structural design, including the connections, for review and approval of shop drawings, and for inspection during construction and acceptance of the finished construction. Under Arrangement 1b, the contract documents must establish the loadings and identify the criteria for design to be used by the contractor. Design work by any contractor should be submitted to and approved by the EOR. The EOR may require the design and shop drawings to be sealed by a registered professional engineer as a demonstration of qualifications. Under Arrangement 2, where the EOR will be engaged or employed by the Contractor, it may be desirable for the Owner to retain another Engineer or Architect for consultation on design criteria and verification that the design intent is achieved.

The Owner should consider the experience and qualifications of both the Engineer or Architect and the General Contractor with precast construction similar to that for the intended project. If the Owner plans to have an active role in the project, it is essential this role and lines of communication with the other parties be clearly expressed in written documents pertaining to the project. The Owner must allow sufficient time in the various phases of the construction process to achieve quality.

Changes initiated after the start of construction invariably add cost to the structure. A thorough review by the Owner after the design is completed, before construction is started, is essential.

The Precast/Prestressed Concrete Institute maintains a Plant Certification Program. The Program is recognized by the Council of American Building Officials (CABO) as a Quality Assurance Inspection Agency. CABO includes the following major building codes:

1. Building Officials & Code Administrators International (BOCA)
2. International Conference of Building Officials (ICBO)
3. Southern Building Code Congress International (SBCC)

PCI Plant Certification is also recognized in standard specifications by federal, state and local government agencies. The Owner or Engineer/Architect may choose to require that PCI Plant Certification be written into the project specifications.

No Manufacturer should be eliminated as a possible source for concrete products because they are not PCI certified. However, acknowledged confidence should be in order for the plant with PCI Certification credentials.

4.2 To the EOR (Engineer/Architect)

As discussed in Section 3, the role of the EOR (Engineer/Architect) varies significantly depending on whether this party is retained by the Owner or by other parties. If retained by the Owner, the EOR's lines of communication among the parties involved in the project should be established by written documents. Prebid and preconstruction conferences should be held, during which lines of communication and responsibilities are reviewed and the design requirements are discussed.

If retained or employed by the General Contractor or Manufacturer, the EOR must remain cognizant of the professional responsibility to satisfy state laws and local regulatory authorities. Procedures should be established to allow the EOR to discharge these responsibilities without restriction by other parties involved in the construction.

Review and approval of shop drawings by the EOR is essential to ensure the design intent is achieved. Timeliness of review may be crucial to the Manufacturer. The EOR should establish the schedule and define the responsibilities of the various parties involved in preparing shop drawings. These parties should accept that the approval of the EOR does not relieve them of their responsibilities.

The EOR should make clear whether the contract documents, specifications or drawings prevail in the event of conflicts. It is recommended that they prevail in the order given above.

Where the design involves a non-selfsupporting precast concrete frame, the contract documents should indicate which party is responsible for the design of the construction bracing and how long such bracing needs to remain in place.

Where erection procedures require special design and calculations, the contract documents should specify that a registered engineer perform these services. The EOR's review of this work should be limited to its effect on the integrity of the completed structure.

Interfaces between precast components and other construction materials require special attention. The EOR is responsible for considering these interface conditions during the design of the structure.

4.3 To the General Contractor

When the construction of a building that was independently designed for the Owner is undertaken, the General Contractor's responsibility is to build the structure in accordance with the contract documents. When design responsibility is accepted by the General Contractor, the professional nature of this function must be recognized and allowed to be achieved in a manner that does not compromise integrity or impair quality.

The responsibilities of the Manufacturer must be clearly defined. It must be understood that all design is submitted through the EOR for approval or acceptance. The Manufacturer's responsibility is usually limited to product design and preparation of shop drawings. However, it may include the design of the entire structure if agreed to by the Manufacturer.

When the Manufacturer is responsible only for product design, all loads which are applied to the precast members, including forces developed by restraint, must be provided by the EOR. The Manufacturer should be given responsibility and authority for properly implementing the design drawings, properly furnishing materials and workmanship, maintaining the specified fabrication and erection tolerances, and for fit and erectibility of the structure.

The Manufacturer must be given all of the drawings and specifications that convey the full requirements for the precast members. Drawings are commonly divided into architectural, structural, electrical and mechanical groupings depending on the size and scope of the project. Other pertinent drawings may also be desirable, such as approved shop drawings from other trades, roofing requirements, alternates, etc.

Timely review and approval of shop drawings and other pertinent information submitted by the Manufacturer is essential.

4.4 To the Manufacturer

After award of the contract, the Manufacturer and Precast Engineer should meet with the EOR to review design requirements. The Manufacturer should prepare shop drawings in accordance with the design information supplied in the contract documents and subsequent instructions from the EOR.

In those cases where the Manufacturer requests permission to revise certain connections to facilitate manufacture and/or erection, information supporting the revision should be submitted to the EOR for review and approval.

The Manufacturer and/or the Precast Engineer should have a direct channel of communication with the EOR as the project goes forward and should keep the General Contractor informed.

The Manufacturer should request clarification in writing from the EOR on special connections or unusual structural conditions not clearly defined by the design drawings or specifications.

CHAPTER 11
GENERAL DESIGN INFORMATION

	Page No.
11.1 Design Information	11-2
11.1.1 Dead Weights of Floors, Ceilings, Roofs, and Walls	11-2
11.1.2 Recommended Minimum Live Loads	11-3
11.1.3 Beam Design Equations and Diagrams	11-4
11.1.4 Camber (Deflection) and Rotation Coefficients for Prestress Force and Loads	11-12
11.1.5 Moments in Beams with Fixed Ends	11-13
11.1.6 Moving Load Placement for Maximum Moment and Shear	11-14
11.1.7 Moments, Shears, and Deflections in Beams with Overhangs	11-15
11.2 Material Properties	11-16
11.2.1 Table of Concrete Stresses	11-16
11.2.2 Concrete Modulus of Elasticity as Affected by Unit Weight and Strength	11-16
11.2.3 Properties and Design Strengths of Prestressing Strand and Wire	11-17
11.2.4 Properties and Design Strengths of Prestressing Bars	11-18
11.2.5 Idealized Stress-Strain Curve, 7-Wire Low-Relaxation Prestressing Strand	11-19
11.2.6 Transfer and Development Length for 7-Wire Uncoated Strand	11-20
11.2.7 Reinforcing Bar Data	11-21
11.2.8 Required Development Lengths for Reinforcing Bars	11-22
11.2.9 Common Stock Styles of Welded Wire Fabric	11-26
11.2.10 Special Welded Wire Fabric for Double Tee Flanges	11-27
11.2.11 Wires used in Welded Wire Fabric	11-27
11.3 Section Properties	11-28
11.3.1 Properties of Geometric Sections	11-28
11.4 Metric Conversion	11-31
11.4.1 Conversion to International System of Units (SI)	11-31

11.1 DESIGN INFORMATION

Design Aid 11.1.1   Dead weights of floors, ceilings, roofs, and walls

Floorings	Weight (psf)
Normal weight concrete topping, per inch of thickness	12
Sand-lightweight (120 pcf) concrete topping, per inch	10
Lightweight (90-100 pcf) concrete topping, per inch	8
⅞" hardwood floor on sleepers clipped to concrete without fill	5
1½" terrazzo floor finish directly on slab	19
1½" terrazzo floor finish with 1" mortar bed	30
1" terrazzo finish with 2" concrete bed	38
¾" ceramic or quarry tile with ½" mortar bed	16
¾" ceramic or quarry tile with 1" mortar bed	22
¼" linoleum or asphalt tile directly on concrete	1
¼" linoleum or asphalt tile with 1" mortar bed	12
¾" mastic floor	9
Hardwood flooring ⅞" thick	4
Subflooring (soft wood), ¾" thick	2½
Asphaltic concrete, 1½" thick	18

Ceilings	
½" gypsum board	2
⅝" gypsum board	2½
¾" plaster directly on concrete	5
¾" plaster on metal lath furring	8
Suspended ceilings	2
Acoustical tile	1
Acoustical tile on wood furring strips	3

Roofs	
Ballasted inverted membrane	16
Five-ply felt and gravel (or slag)	6½
Three-ply felt and gravel (or slag)	5½
Five-ply felt composition roof, no gravel	4
Three-ply felt composition roof, no gravel	3
Asphalt strip shingles (includes one layer of felt)	3
Rigid insulation, per inch	½
Gypsum, per inch of thickness	4
Insulating concrete (unit weight approx. 36 pcf), per inch	3

Walls	Un-Plastered	One side Plastered	Both sides Plastered
4" brick wall	40	45	50
8" brick wall	80	85	90
12" brick wall	120	125	130
4" hollow normal weight concrete block	28	33	38
6" hollow normal weight concrete block	36	41	46
8" hollow normal weight concrete block	51	56	61
12" hollow normal weight concrete block	59	64	69
4" hollow lightweight block or tile	19	24	29
6" hollow lightweight block or tile	22	27	32
8" hollow lightweight block or tile	33	38	43
12" hollow lightweight block or tile	44	49	54
4" brick 4" hollow normal weight block backing	68	73	78
4" brick 8" hollow normal weight block backing	91	96	101
4" brick 12" hollow normal weight block backing	119	124	129
4" brick 4" hollow lightweight block or tile backing	59	64	69
4" brick 8" hollow lightweight block or tile backing	73	78	83
4" brick 12" hollow lightweight block or tile backing	84	89	94
4" brick, steel or wood studs, ⅝" gypsum board	43		
Windows, glass, frame and sash	8		
4" stone	55		
Steel or wood studs, lath, ¾" plaster	18		
Steel or wood studs, ⅝" gypsum board each side	6		
Steel or wood studs, 2 layers ½" gypsum board each side	9		

DESIGN INFORMATION

Design Aid 11.1.2 Recommended minimum live loads*

UNIFORMLY DISTRIBUTED LOADS

Occupancy or Use	Live Load (psf)
Apartments (see Residential)	
Armories and drill rooms	150
Assembly halls and other places of assembly:	
Fixed Seats (fastened to floor)	60
Moveable seats	100
Platforms (assembly)	100
Stage floors	150
Balconies (exterior)	100
On one and two family residences only, and not exceeding 100 sq. ft.	60
Bowling alleys, poolrooms, and similar recreational areas	75
Corridors:	
First floor	100
Other floors, same as occupancy served except as indicated	
Dance halls and ballrooms	100
Decks (patio and roof), same as area served, or for the type of occupancy accommodated	
Dining rooms and restaurants	100
Dwellings (see Residential)	
Fire escapes	100
On single-family residential buildings only	40
Grandstands (see Stadiums and arena bleachers)	
Gymnasiums, main floors and balconies	100
Hospitals:	
Operating rooms, laboratories	60
Private rooms	40
Wards	40
Corridors, above first floor	80
Hotels (see Residential)	
Libraries:	
Reading rooms	60
Stack rooms (books & shelving at 65 pcf), not less than	150
Corridors, above first floor	80
Manufacturing:	
Light	125
Heavy	250
Marquees and canopies	75
Office buildings:	
Offices	50
Lobbies	100
File and computer rooms should be designed for loads based upon anticipated occupancy	
Parking structures:	
Passenger cars only	50
For trucks and buses use AASHTO lane loads(1)	
Penal institutions:	
Cell blocks	40
Corridors	100
Residential:	
Dwellings (one and two family):	
Uninhabitable attics without storage	10
Uninhabitable attics with storage	20

Occupancy or Use	Live Load (psf)
Residential (cont.)	
Habitable attics and sleeping areas	30
All other areas	40
Hotels and multifamily houses:	
Private rooms and corridors serving them	40
Public rooms and corridors serving them	100
Roofs**	
Schools:	40
Classrooms	40
Corridors above first floor	80
Sidewalks, vehicular driveways, and yards subject to trucking (2)	250
Stadiums and arena bleachers (3)	100
Stairs and exitways	100
Storage warehouse:	
Light	125
Heavy	250
Stores:	
Retail:	
First floor	100
Upper floors	75
Wholesale, all floors	125
Walkways and elevated platforms (other than exitways)	60

CONCENTRATED LOADS

Location	Load (lb)
Elevator machine room grating (on area of 4 sq. in.)	300
Finish light floor plate construction (on area of 1 sq. in.)	200
Parking structures(4)	
Office floors	2000
Scuttles, skylight ribs, and accessible ceilings	200
Sidewalks	8000
Stair treads (on area of 4 sq. in. at center of tread)	300

(1) American Association of State Highway and Transportation Officials.

(2) AASHTO lane loads should also be considered where appropriate.

(3) For detailed recommendations, see Assembly Seating, Tents and Air Supported Structures, ANSI/NFPA 102-1978 [Z20.3].

(4) Floors in parking structures or portions of buildings used for storage of motor vehicles should be designed for the uniformly distributed live loads shown or the following concentrated loads: (1) for passenger cars accommodating not more than nine passengers, 2,000 pounds acting on an area of 20 sq. in.; (2) mechanical parking structures without slab or deck, passenger cars only, 1,500 pounds per wheel; (3) for trucks or buses, maximum axle load on an area of 20 sq. in.

*Source: ASCE 7-88 (formerly ANSI A58.1). Local building codes take precedence; see local codes for live load reductions.
**See local codes or model building codes.

DESIGN INFORMATION

Design Aid 11.1.3 Beam design equations and diagrams

(1) Simple Beam—uniformly distributed load

$R = V = \dfrac{wl}{2}$

$V_x = w\left(\dfrac{l}{2} - x\right)$

M max. (at center) $= \dfrac{wl^2}{8}$

$M_x = \dfrac{wx}{2}(l - x)$

Δ max. (at center) $= \dfrac{5wl^4}{384\,EI}$

$\Delta_x = \dfrac{wx}{24\,EI}(l^3 - 2lx^2 + x^3)$

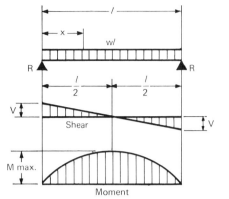

(2) Simple Beam—concentrated load at center

$R = V = \dfrac{P}{2}$

M max. (at point of load) $= \dfrac{Pl}{4}$

$M_x \left(\text{when } x < \dfrac{l}{2}\right) = \dfrac{Px}{2}$

Δ max. (at point of load) $= \dfrac{Pl^3}{48\,EI}$

$\Delta_x \left(\text{when } x < \dfrac{l}{2}\right) = \dfrac{Px}{48\,EI}(3l^2 - 4x^2)$

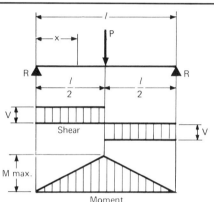

(3) Simple Beam—concentrated load at any point

$R_1 = V_1$ (max. when $a < b$) $= \dfrac{Pb}{l}$

$R_2 = V_2$ (max. when $a > b$) $= \dfrac{Pa}{l}$

M max. (at point of load) $= \dfrac{Pab}{l}$

M_x (when $x < a$) $= \dfrac{Pbx}{l}$

Δ max. $\left(\text{at } x = \sqrt{\dfrac{a(a+2b)}{3}} \text{ when } a > b\right) = \dfrac{Pab(a+2b)\sqrt{3a(a+2b)}}{27\,EI\,l}$

Δa (at point of load) $= \dfrac{Pa^2b^2}{3\,EI\,l}$

Δ_x (when $x < a$) $= \dfrac{Pbx}{6\,EI\,l}(l^2 - b^2 - x^2)$

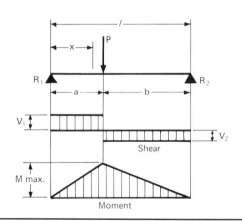

(4) Simple Beam—two equal concentrated loads symmetrically placed

$R = V = P$

M max. (between loads) $= Pa$

M_x (when $x < a$) $= Px$

Δ max. (at center) $= \dfrac{Pa}{24\,EI}(3l^2 - 4a^2)$

Δ_x (when $x < a$) $= \dfrac{Px}{6\,EI}(3la - 3a^2 - x^2)$

Δ_x (when $x > a$ and $< (l - a)$) $= \dfrac{Pa}{6\,EI}(3lx - 3x^2 - a^2)$

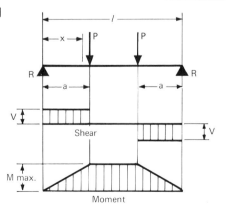

DESIGN INFORMATION

Design Aid 11.1.3 Beam design equations and diagrams (continued)

(5) Simple Beam—two unequal concentrated loads unsymmetrically placed

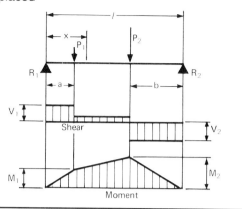

$R_1 = V_1 = \dfrac{P_1(l-a) + P_2 b}{l}$

$R_2 = V_2 = \dfrac{P_1 a + P_2(l-b)}{l}$

V_x (when $x > a$ and $< (l-b)$) $= R_1 - P_1$

M_1 (max. when $R_1 < P_1$) $= R_1 a$

M_2 (max. when $R_2 < P_2$) $= R_2 b$

M_x (when $x < a$) $= R_1 x$

M_x (when $x > a$ and $< (l-b)$) $= R_1 x - P_1(x-a)$

(6) Simple Beam—uniform load partially distributed

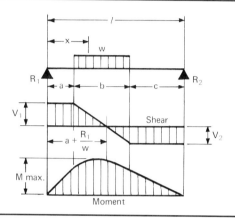

$R_1 = V_1$ (max. when $a < c$) $= \dfrac{wb}{2l}(2c + b)$

$R_2 = V_2$ (max. when $a > c$) $= \dfrac{wb}{2l}(2a + b)$

V_x (when $x > a$ and $< (a+b)$) $= R_1 - w(x-a)$

M max. $\left(\text{at } x = a + \dfrac{R_1}{w}\right) = R_1\left(a + \dfrac{R_1}{2w}\right)$

M_x (when $x < a$) $= R_1 x$

M_x (when $x > a$ and $< (a+b)$) $= R_1 x - \dfrac{w}{2}(x-a)^2$

M_x (when $x > (a+b)$) $= R_2(l-x)$

(7) Simple Beam—load increasing uniformly to one end (W is total load)

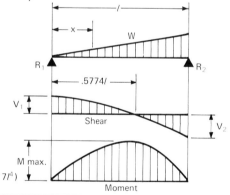

$R_1 = V_1 = \dfrac{W}{3}$

$R_2 = V_2$ max. $= \dfrac{2W}{3}$

$V_x = \dfrac{W}{3} - \dfrac{Wx^2}{l^2}$

M max. $\left(\text{at } x = \dfrac{l}{\sqrt{3}} = .5774 l\right) = \dfrac{2Wl}{9\sqrt{3}} = .1283\, Wl$

$M_x = \dfrac{Wx}{3l^2}(l^2 - x^2)$

Δ max. $\left(\text{at } x = l\sqrt{1 - \sqrt{\dfrac{8}{15}}} = .5193 l\right) = .01304 \dfrac{Wl^3}{EI}$

$\Delta_x = \dfrac{Wx}{180\, EI\, l^2}(3x^4 - 10 l^2 x^2 + 7 l^4)$

(8) Simple Beam—load increasing uniformly to center (W is total load)

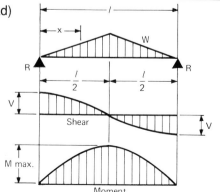

$R = V = \dfrac{W}{2}$

$V_x \left(\text{when } x < \dfrac{l}{2}\right) = \dfrac{W}{2l^2}(l^2 - 4x^2)$

M max. (at center) $= \dfrac{Wl}{6}$

$M_x \left(\text{when } x < \dfrac{l}{2}\right) = Wx\left(\dfrac{1}{2} - \dfrac{2x^2}{3l^2}\right)$

Δ max. (at center) $= \dfrac{Wl^3}{60\, EI}$

$\Delta_x = \dfrac{Wx}{480\, EI\, l^2}(5l^2 - 4x^2)^2$

DESIGN INFORMATION

Design Aid 11.1.3 Beam design equations and diagrams (continued)

(9) Beam overhanging one support—uniformly distributed load

$R_1 = V_1 = \dfrac{w}{2l}(l^2 - a^2)$

$R_2 = V_2 + V_3 = \dfrac{w}{2l}(l + a)^2$

$V_2 = wa$

$V_3 = \dfrac{w}{2l}(l^2 + a^2)$

V_x (between supports) $= R_1 - wx$

V_{x_1} (for overhang) $= w(a - x_1)$

M_1 $\left(\text{at } x = \dfrac{l}{2}\left[1 - \dfrac{a^2}{l^2}\right]\right) = \dfrac{w}{8l^2}(l + a)^2(l - a)^2$

M_2 (at R_2) $= \dfrac{wa^2}{2}$

M_x (between supports) $= \dfrac{wx}{2l}(l^2 - a^2 - xl)$

M_{x_1} (for overhang) $= \dfrac{w}{2}(a - x_1)^2$

Δ_x (between supports) $= \dfrac{wx}{24 EI l}(l^4 - 2l^2 x^2 + lx^3 - 2a^2 l^2 + 2a^2 x^2)$

Δ_{x_1} (for overhang) $= \dfrac{wx_1}{24 EI}(4a^2 l - l^3 + 6a^2 x_1 - 4a x_1^2 + x_1^3)$

(10) Beam overhanging one support—uniformly distributed load on overhang

$R_1 = V_1 = \dfrac{wa^2}{2l}$

$R_2 = V_1 + V_2 = \dfrac{wa}{2l}(2l + a)$

$V_2 = wa$

V_{x_1} (for overhang) $= w(a - x_1)$

M max. (at R_2) $= \dfrac{wa^2}{2}$

M_x (between supports) $= \dfrac{wa^2 x}{2l}$

M_{x_1} (for overhang) $= \dfrac{w}{2}(a - x_1)^2$

Δ max. $\left(\text{between supports at } x = \dfrac{l}{\sqrt{3}}\right) = \dfrac{wa^2 l^2}{18\sqrt{3}\, EI} = .03208 \dfrac{wa^2 l^2}{EI}$

Δ max. (for overhang at $x_1 = a$) $= \dfrac{wa^3}{24 EI}(4l + 3a)$

Δ_x (between supports) $= \dfrac{wa^2 x}{12 EI l}(l^2 - x^2)$

Δ_{x_1} (for overhang) $= \dfrac{wx_1}{24 EI}(4a^2 l + 6a^2 x_1 - 4a x_1^2 + x_1^3)$

(11) Beam overhanging one support—uniformly distributed load between supports

$R = V = \dfrac{wl}{2}$

$V_x = w\left(\dfrac{l}{2} - x\right)$

M max. (at center) $= \dfrac{wl^2}{8}$

$M_x = \dfrac{wx}{2}(l - x)$

Δ max. (at center) $= \dfrac{5wl^4}{384 EI}$

$\Delta_x = \dfrac{wx}{24 EI}(l^3 - 2lx^2 + x^3)$

$\Delta_{x_1} = \dfrac{wl^3 x_1}{24 EI}$

DESIGN INFORMATION

Design Aid 11.1.3 Beam design equations and diagrams (continued)

(12) Beam overhanging one support—concentrated load at any point between supports

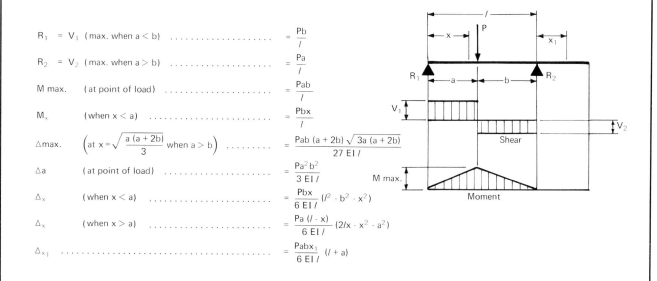

$R_1 = V_1$ (max. when $a < b$) $\ldots\ldots\ldots\ldots\ldots = \dfrac{Pb}{l}$

$R_2 = V_2$ (max. when $a > b$) $\ldots\ldots\ldots\ldots\ldots = \dfrac{Pa}{l}$

M max. (at point of load) $\ldots\ldots\ldots\ldots\ldots = \dfrac{Pab}{l}$

M_x (when $x < a$) $\ldots\ldots\ldots\ldots\ldots = \dfrac{Pbx}{l}$

Δ max. $\left(\text{at } x = \sqrt{\dfrac{a(a+2b)}{3}} \text{ when } a > b\right) \ldots\ldots = \dfrac{Pab(a+2b)\sqrt{3a(a+2b)}}{27\,EI\,l}$

Δa (at point of load) $\ldots\ldots\ldots\ldots\ldots = \dfrac{Pa^2 b^2}{3\,EI\,l}$

Δ_x (when $x < a$) $\ldots\ldots\ldots\ldots\ldots = \dfrac{Pbx}{6\,EI\,l}(l^2 - b^2 - x^2)$

Δ_x (when $x > a$) $\ldots\ldots\ldots\ldots\ldots = \dfrac{Pa(l-x)}{6\,EI\,l}(2lx - x^2 - a^2)$

$\Delta_{x_1} \ldots\ldots\ldots\ldots\ldots = \dfrac{Pabx_1}{6\,EI\,l}(l + a)$

---

(13) Beam overhanging one support—concentrated load at end of overhang

$R_1 = V_1 \ldots\ldots\ldots\ldots\ldots = \dfrac{Pa}{l}$

$R_2 = V_1 + V_2 \ldots\ldots\ldots\ldots\ldots = \dfrac{P}{l}(l + a)$

$V_2 \ldots\ldots\ldots\ldots\ldots = P$

M max. (at R_2) $\ldots\ldots\ldots\ldots\ldots = Pa$

M_x (between supports) $\ldots\ldots\ldots\ldots\ldots = \dfrac{Pax}{l}$

M_{x_1} (for overhang) $\ldots\ldots\ldots\ldots\ldots = P(a - x_1)$

Δ max. $\left(\text{between supports at } x = \dfrac{l}{\sqrt{3}}\right) \ldots = \dfrac{Pal^2}{9\sqrt{3}\,EI} = .06415\dfrac{Pal^2}{EI}$

Δ max. (for overhang at $x_1 = a$) $\ldots\ldots\ldots\ldots\ldots = \dfrac{Pa^2}{3\,EI}(l + a)$

Δ_x (between supports) $\ldots\ldots\ldots\ldots\ldots = \dfrac{Pax}{6\,EI\,l}(l^2 - x^2)$

Δ_{x_1} (for overhang) $\ldots\ldots\ldots\ldots\ldots = \dfrac{Px_1}{6\,EI}(2al + 3ax_1 - x_1^2)$

DESIGN INFORMATION

Design Aid 11.1.3 Beam design equations and diagrams (continued)

(14) Cantilever Beam—uniformly distributed load

$R = V = wl$

$V_x = wx$

M max. (at fixed end) $= \dfrac{wl^2}{2}$

$M_x = \dfrac{wx^2}{2}$

Δ max. (at free end) $= \dfrac{wl^4}{8EI}$

$\Delta_x = \dfrac{w}{24EI}(x^4 - 4l^3 x + 3l^4)$

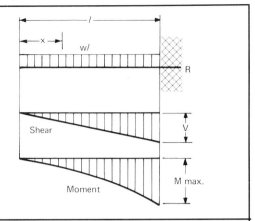

(15) Cantilever Beam—concentrated load at free end

$R = V = P$

M max. (at fixed end) $= Pl$

$M_x = Px$

Δ max. (at free end) $= \dfrac{Pl^3}{3EI}$

$\Delta_x = \dfrac{P}{6EI}(2l^3 - 3l^2 x + x^3)$

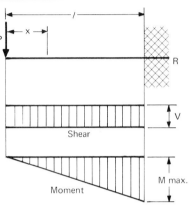

(16) Cantilever Beam—concentrated load at any point

$R = V = P$

M max. (at fixed end) $= Pb$

M_x (when $x > a$) $= P(x-a)$

Δ max. (at free end) $= \dfrac{Pb^2}{6EI}(3l - b)$

Δa (at point of load) $= \dfrac{Pb^3}{3EI}$

Δ_x (when $x < a$) $= \dfrac{Pb^2}{6EI}(3l - 3x - b)$

Δ_x (when $x > a$) $= \dfrac{P(l-x)^2}{6EI}(3b - l + x)$

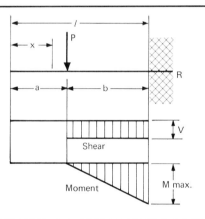

(17) Cantilever Beam—load increasing uniformly to fixed end

$R = V = W$

$V_x = W\dfrac{x^2}{l^2}$

M max. (at fixed end) $= \dfrac{Wl}{3}$

$M_x = \dfrac{Wx^3}{3l^2}$

Δ max. (at free end) $= \dfrac{Wl^3}{15EI}$

$\Delta_x = \dfrac{W}{60EI\,l^2}(x^5 - 5l^4 x + 4l^5)$

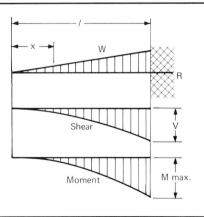

DESIGN INFORMATION

Design Aid 11.1.3 Beam design equations and diagrams (continued)

(18) Beam fixed at one end, supported at other—uniformly distributed load

$R_1 = V_1 = \dfrac{3wl}{8}$

$R_2 = V_2 \text{ max.} = \dfrac{5wl}{8}$

$V_x = R_1 - wx$

$M \text{ max.} = \dfrac{wl^2}{8}$

$M_1 \left(\text{at } x = \dfrac{3}{8}l\right) = \dfrac{9}{128} wl^2$

$M_x = R_1 x - \dfrac{wx^2}{2}$

$\Delta \text{max.} \left(\text{at } x = \dfrac{l}{16}(1+\sqrt{33}) = .4215l\right) = \dfrac{wl^4}{185\, EI}$

$\Delta_x = \dfrac{wx}{48\, EI}(l^3 - 3lx^2 + 2x^3)$

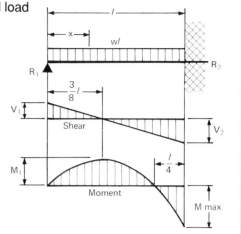

(19) Beam fixed at one end, supported at other—concentrated load at center

$R_1 = V_1 = \dfrac{5P}{16}$

$R_2 = V_2 \text{ max.} = \dfrac{11P}{16}$

$M \text{ max.} \quad (\text{at fixed end}) = \dfrac{3Pl}{16}$

$M_1 \quad (\text{at point of load}) = \dfrac{5Pl}{32}$

$M_x \left(\text{when } x < \dfrac{l}{2}\right) = \dfrac{5Px}{16}$

$M_x \left(\text{when } x > \dfrac{l}{2}\right) = P\left(\dfrac{l}{2} - \dfrac{11x}{16}\right)$

$\Delta \text{max.} \left(\text{at } x = l\sqrt{\dfrac{1}{5}} = .4472l\right) = \dfrac{Pl^3}{48\, EI\sqrt{5}} = .009317 \dfrac{Pl^3}{EI}$

$\Delta_x \quad (\text{at point of load}) = \dfrac{7Pl^3}{768\, EI}$

$\Delta_x \left(\text{when } x < \dfrac{l}{2}\right) = \dfrac{Px}{96\, EI}(3l^2 - 5x^2)$

$\Delta_x \left(\text{when } x > \dfrac{l}{2}\right) = \dfrac{P}{96\, EI}(x-l)^2 (11x - 2l)$

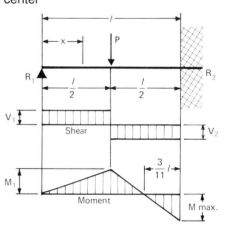

(20) Beam fixed at one end, supported at other—concentrated load at any point

$R_1 = V_1 = \dfrac{Pb^2}{2l^3}(a + 2l)$

$R_2 = V_2 = \dfrac{Pa}{2l^3}(3l^2 - a^2)$

$M_1 \quad (\text{at point of load}) = R_1 a$

$M_2 \quad (\text{at fixed end}) = \dfrac{Pab}{2l^2}(a + l)$

$M_x \quad (\text{when } x < a) = R_1 x$

$M_x \quad (\text{when } x > a) = R_1 x - P(x - a)$

$\Delta \text{max.} \left(\text{when } a < .414l \text{ at } x = l\dfrac{l^2 + a^2}{3l^2 - a^2}\right) = \dfrac{Pa}{3\, EI} \dfrac{(l^2 - a^2)^3}{(3l^2 - a^2)^2}$

$\Delta \text{max.} \left(\text{when } a > .414l \text{ at } x = l\sqrt{\dfrac{a}{2l + a}}\right) = \dfrac{Pab^2}{6\, EI} \sqrt{\dfrac{a}{2l + a}}$

$\Delta a \quad (\text{at point of load}) = \dfrac{Pa^2 b^3}{12\, EI l^3}(3l + a)$

$\Delta_x \quad (\text{when } x < a) = \dfrac{Pb^2 x}{12\, EI l^3}(3al^2 - 2lx^2 - ax^2)$

$\Delta_x \quad (\text{when } x > a) = \dfrac{Pa}{12\, EI l^3}(l-x)^2 (3l^2 x - a^2 x - 2a^2 l)$

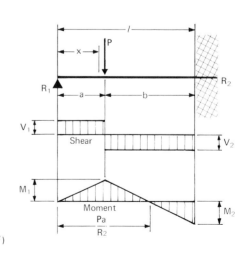

PCI Design Handbook/Fourth Edition

DESIGN INFORMATION

Design Aid 11.1.3 Beam design equations and diagrams (continued)

(21) Beam fixed at both ends—uniformly distributed load

$$R = V = \frac{wl}{2}$$

$$V_x = w\left(\frac{l}{2} - x\right)$$

$$M \text{ max.} \quad (\text{at ends}) = \frac{wl^2}{12}$$

$$M_1 \quad (\text{at center}) = \frac{wl^2}{24}$$

$$M_x = \frac{w}{12}(6lx - l^2 - 6x^2)$$

$$\Delta \text{max.} \quad (\text{at center}) = \frac{wl^4}{384\,EI}$$

$$\Delta_x = \frac{wx^2}{24\,EI}(l - x)^2$$

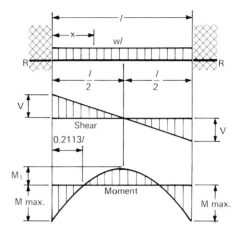

(22) Beam fixed at both ends—concentrated load at any point

$$R_1 = V_1 \quad (\text{max. when } a < b) = \frac{Pb^2}{l^3}(3a + b)$$

$$R_2 = V_2 \quad (\text{max. when } a > b) = \frac{Pa^2}{l^3}(a + 3b)$$

$$M_1 \quad (\text{max. when } a < b) = \frac{Pab^2}{l^2}$$

$$M_2 \quad (\text{max. when } a > b) = \frac{Pa^2 b}{l^2}$$

$$M_a \quad (\text{at point of load}) = \frac{2Pa^2 b^2}{l^3}$$

$$M_x \quad (\text{when } x < a) = R_1 x - \frac{Pab^2}{l^2}$$

$$\Delta \text{max.} \quad \left(\text{when } a > b \text{ at } x = \frac{2al}{3a + b}\right) = \frac{2Pa^3 b^2}{3\,EI\,(3a + b)^2}$$

$$\Delta a \quad (\text{at point of load}) = \frac{Pa^3 b^3}{3\,EI\,l^3}$$

$$\Delta_x \quad (\text{when } x < a) = \frac{Pb^2 x^2}{6\,EI\,l^3}(3al - 3ax - bx)$$

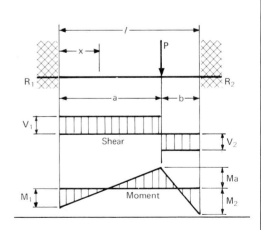

DESIGN INFORMATION

Design Aid 11.1.3 Beam design equations and diagrams (continued)

(23) Beam fixed one end—differential settlement of supports

$$V = {}^-R_1 = R_2 = \frac{3EI}{\ell^3}(\Delta_2 - \Delta_1)$$

$$M_{max} = \frac{3EI}{\ell^2}(\Delta_2 - \Delta_1)$$

$$M_x = M_{max}\left(1 - \frac{x}{\ell}\right)$$

$$\Delta_x = \Delta_1 + \frac{\Delta_2 - \Delta_1}{2}\left[3\left(\frac{x}{\ell}\right)^2 - \left(\frac{x}{\ell}\right)^3\right]$$

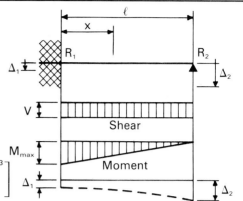

(24) Beam fixed one end—rotation of support

$$V = {}^-R_1 = R_2 = \frac{3EI}{\ell^2}\phi_1$$

$$M_{max} = \frac{3EI}{\ell}\phi_1$$

$$M_x = M_{max}\left(1 - \frac{x}{\ell}\right)$$

$$\Delta_{max} = \phi_1\left[\frac{\ell}{5.196}\right]$$

$$\Delta_x = \phi_1\left[-x + \frac{3x^2}{2\ell} - \frac{x^2}{2\ell^2}\right]$$

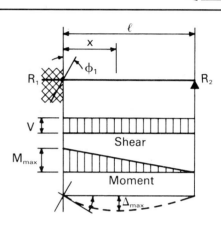

(25) Beam fixed both ends—differential settlement of supports

$$V = {}^-R_1 = R_2 = \frac{12EI}{\ell^3}(\Delta_2 - \Delta_1)$$

$$M_1 = {}^-M_2 = \frac{6EI}{\ell^2}(\Delta_2 - \Delta_1)$$

$$M_x = \frac{6EI}{\ell^2}(\Delta_2 - \Delta_1)\left(1 - \frac{2x}{\ell}\right)$$

$$\Delta_x = \Delta_1 + (\Delta_2 - \Delta_1)\left[3\left(\frac{x}{\ell}\right)^2 - 2\left(\frac{x}{\ell}\right)^3\right]$$

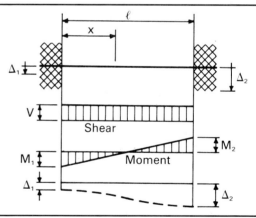

(26) Beam fixed both ends—rotation of support

$$V = {}^-R_1 = R_2 = \frac{6EI}{\ell^2}\phi_2$$

$$M_1 = \frac{2EI}{\ell}\phi_2$$

$$M_2 = \frac{4EI}{\ell}\phi_2$$

$$M_x = \frac{2EI}{\ell}\phi_2\left(1 - \frac{3x}{\ell}\right)$$

$$\Delta_{max}\ (\text{at}\ x = \frac{2}{3}\ell) = -\frac{4}{27}\ell\phi_2$$

$$\Delta_x = -\ell\phi_2\left[\left(\frac{x}{\ell}\right)^2 - \left(\frac{x}{\ell}\right)^3\right]$$

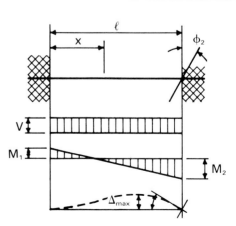

DESIGN INFORMATION

Design Aid 11.1.4 Camber (deflection) and rotation coefficients for prestress force and loads*

Prestress Pattern	Equivalent Moment or Load	Equivalent Loading	Camber (+ ↑)	End Rotation (∠+)	End Rotation (+∠)
(1) e, P, C.G., l	$M = Pe$	M at left support, simple beam span l	$+\dfrac{Ml^2}{16\,EI}$	$+\dfrac{Ml}{3\,EI}$	$-\dfrac{Ml}{6\,EI}$
(2) P, C.G., e, l	$M = Pe$	M at right support	$+\dfrac{Ml^2}{16\,EI}$	$+\dfrac{Ml}{6\,EI}$	$-\dfrac{Ml}{3\,EI}$
(3) e, P, C.G., e, P	$M = Pe$	M at both supports	$+\dfrac{Ml^2}{8\,EI}$	$+\dfrac{Ml}{2\,EI}$	$-\dfrac{Ml}{2\,EI}$
(4) P, C.G., e', P; l/2, l/2	$N = \dfrac{4Pe'}{l}$	Point load N at midspan	$+\dfrac{Nl^3}{48\,EI}$	$+\dfrac{Nl^2}{16\,EI}$	$-\dfrac{Nl^2}{16\,EI}$
(5) P, C.G., e', P; bl, bl	$N = \dfrac{Pe'}{bl}$	Two point loads N at bl from each support	$+\dfrac{b(3-4b^2)\,Nl^3}{24\,EI}$	$+\dfrac{b(1-b)\,Nl^2}{2\,EI}$	$-\dfrac{b(1-b)\,Nl^2}{2\,EI}$
(6) P, C.G., e', P; l/2, l/2	$w = \dfrac{8Pe'}{l^2}$	Uniform load w full span	$+\dfrac{5wl^4}{384\,EI}$	$+\dfrac{wl^3}{24\,EI}$	$-\dfrac{wl^3}{24\,EI}$
(7) P, C.G., e'; l/2, l/2	$w = \dfrac{8Pe'}{l^2}$	Uniform load w on left half	$+\dfrac{5wl^4}{768\,EI}$	$+\dfrac{9wl^3}{384\,EI}$	$-\dfrac{7wl^3}{384\,EI}$
(8) C.G., P, e'; l/2, l/2	$w = \dfrac{8Pe'}{l^2}$	Uniform load w on right half	$+\dfrac{5wl^4}{768\,EI}$	$+\dfrac{7wl^3}{384\,EI}$	$-\dfrac{9wl^3}{384\,EI}$
(9) P, C.G., e', P; bl, bl	$w = \dfrac{4Pe'}{(0.5-b)l^2}$; $w_1 = \dfrac{w}{b}(0.5-b)$	w_1 at ends, w in middle	$\left[\dfrac{5}{8} - \dfrac{b}{2}(3-2b^2)\right]\dfrac{wl^4}{48\,EI}$	$+\dfrac{(1-b)(1-2b)\,wl^3}{24\,EI}$	$-\dfrac{(1-b)(1-2b)\,wl^3}{24\,EI}$
(10) P, C.G., e'; bl	$w = \dfrac{4Pe'}{(0.5-b)l^2}$; $w_1 = \dfrac{w}{b}(0.5-b)$	w_1 at left end, w over l/2	$\left[\dfrac{5}{16} - \dfrac{b}{4}(3-2b^2)\right]\dfrac{wl^4}{48\,EI}$	$\left[\dfrac{9}{8} - b(2-b^2)\right]\dfrac{wl^3}{48\,EI}$	$-\left[\dfrac{7}{8} + b(2-b^2)\right]\dfrac{wl^3}{48\,EI}$
(11) C.G., P, e'; bl	$w = \dfrac{4Pe'}{(0.5-b)l^2}$; $w_1 = \dfrac{w}{b}(0.5-b)$	w_1 at right end, w over l/2	$\left[\dfrac{5}{16} - \dfrac{b}{4}(3-2b^2)\right]\dfrac{wl^4}{48\,EI}$	$\left[\dfrac{7}{8} - b(2-b^2)\right]\dfrac{wl^3}{48\,EI}$	$-\left[\dfrac{9}{8} + b(2-b)^2\right]\dfrac{wl^3}{48\,EI}$

*The tabulated values apply to the effects of prestressing. By adjusting the directional rotation, they may also be used for the effects of loads. For patterns 4-11, superimpose on 1, 2, or 3 for other C.G. locations.

DESIGN INFORMATION

Design Aid 11.1.5 Moments in beams with fixed ends

Loading	Moment at A	Moment at center	Moment at B
(1) Concentrated load P at center, span l	$-\dfrac{Pl}{8}$	$+\dfrac{Pl}{8}$	$-\dfrac{Pl}{8}$
(2) Concentrated load P at al from A	$-Pla(1-a)^2$		$-Pla^2(1-a)$
(3) Two loads P at third points	$-\dfrac{2Pl}{9}$	$+\dfrac{Pl}{9}$	$-\dfrac{2Pl}{9}$
(4) Three loads P at quarter points	$-\dfrac{5Pl}{16}$	$+\dfrac{3Pl}{16}$	$-\dfrac{5Pl}{16}$
(5) Uniform load W over full span	$-\dfrac{Wl}{12}$	$+\dfrac{Wl}{24}$	$-\dfrac{Wl}{12}$
(6) Uniform load W centered, with al unloaded at each end	$-\dfrac{Wl(1+2a-2a^2)}{12}$	$+\dfrac{Wl(1+2a+4a^2)}{24}$	$-\dfrac{Wl(1+2a-2a^2)}{12}$
(7) Two uniform loads W/2 of length al at each end	$-\dfrac{Wl(3a-2a^2)}{12}$	$+\dfrac{Wla^2}{6}$	$-\dfrac{Wl(3a-2a^2)}{12}$
(8) Uniform load W of length al at end A	$-\dfrac{Wla(6-8a+3a^2)}{12}$		$-\dfrac{Wla^2(4-3a)}{12}$
(9) Triangular load W, peak at center	$-\dfrac{5Wl}{48}$	$+\dfrac{3Wl}{48}$	$-\dfrac{5Wl}{48}$
(10) Triangular load W, peak at A	$-\dfrac{Wl}{10}$		$-\dfrac{Wl}{15}$

W = Total load on beam

PCI Design Handbook/Fourth Edition

DESIGN INFORMATION

Design Aid 11.1.6 Moving load placement for maximum moment and shear

(1) Simple Beam—one concentrated moving load

R_1 max. = V_1 max. (at $x = 0$) . . . . = P

M max. (at point of load, when $x = \dfrac{\ell}{2}$) . . . = $\dfrac{P\ell}{4}$

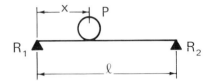

(2) Simple Beam—two equal concentrated moving loads

R_1 max. = V_1 max. (at $x = 0$) . . . . = $P\left(2 - \dfrac{a}{\ell}\right)$

M max. $\begin{cases} \begin{bmatrix} \text{when } a < (2 - \sqrt{2})\,\ell = .586\ell \\ \text{under load 1 at } x = \dfrac{1}{2}\left(\ell - \dfrac{a}{2}\right) \end{bmatrix} = \dfrac{P}{2\ell}\left(\ell - \dfrac{a}{2}\right)^2 \\[1em] \begin{bmatrix} \text{when } a > (2 - \sqrt{2})\,\ell = .586\ell \\ \text{with one load at center of span} \end{bmatrix} = \dfrac{P\ell}{4} \end{cases}$

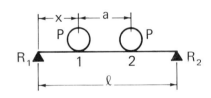

(3) Simple Beam—two unequal concentrated moving loads

R_1 max. = V_1 max. (at $x = 0$) . . . . = $P_1 + P_2 \dfrac{\ell - a}{\ell}$

M max. $\begin{cases} \left[\text{under } P_1, \text{ at } x = \dfrac{1}{2}\left(\ell - \dfrac{P_2 a}{P_1 + P_2}\right)\right] = (P_1 + P_2)\dfrac{x^2}{\ell} \\[1em] \begin{bmatrix} \text{M max. may occur with larger} \\ \text{load at center of span and other} \\ \text{load off span} \end{bmatrix} = \dfrac{P_1 \ell}{4} \end{cases}$

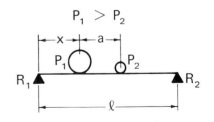

DESIGN INFORMATION

Design Aid 11.1.7 Moments, shears, and deflections in beams with overhangs

Loading and support	Reactions, and vertical shear	Bending moment M and maximum bending moment	Deflection y, maximum deflection, and end slope θ
Equal overhangs, uniform load $W = w\ell$	$R_1 = R_2 = \dfrac{W}{2}$ (A to B) $V = -\dfrac{W(c-x)}{\ell}$ (B to C) $V = W\left(\dfrac{1}{2} - \dfrac{x+c}{\ell}\right)$ (C to D) $V = \dfrac{W(c+d-x)}{\ell}$	(A to B) $M = -\dfrac{W}{2\ell}(c-x)^2$ (B to C) $M = -\dfrac{W}{2\ell}[c^2 - x(d-x)]$ $M = -\dfrac{Wc^2}{2\ell}$ at B and D $M = -\dfrac{W}{2\ell}\left(c^2 - \dfrac{d^2}{4}\right)$ at $x = \dfrac{d}{2}$ If $d > 2c$, $M = 0$ at $x = \dfrac{d}{2} \pm \left[\dfrac{d^2}{4} - c^2\right]^{1/2}$ If $c = 0.207\ell$, $M = -\dfrac{W\ell}{46.62}$ at $x = 0$ and $M = +\dfrac{W\ell}{46.62}$ at $x = \dfrac{d}{2}$ x is considered positive on both sides of the origin.	(A to B) $y = -\dfrac{Wx}{24EI\ell}[6c^2(d+x) - x^2(4c-x) - d^3]$ (B to C) $y = -\dfrac{Wx(d-x)}{24EI\ell}[x(d-x) + d^2 - 6c^2]$ $y = -\dfrac{Wc}{24EI}[3c^2(c+2d) - d^3]$ at A and D $y = -\dfrac{Wd^2}{384EI\ell}(5d^2 - 24c^2)$ at $x = \dfrac{d}{2}$ If $2c < d < 2.449c$, the maximum deflection between supports is $y = +\dfrac{W}{96EI\ell}(6c^2 - d^2)^2$ at $x = \dfrac{d}{2} \pm \left[3\left(\dfrac{d^2}{4} - c^2\right)\right]^{1/2}$ $\theta = +\dfrac{W}{24EI\ell}(6c^2d + 4c^3 - d^3)$ at A $\theta = -\dfrac{W}{24EI\ell}(6c^2d + 4c^3 - d^3)$ at D
Unequal overhangs, uniform load $W = w\ell$	$R_1 = \dfrac{W}{2d}(c+d-e)$ $R_2 = \dfrac{W}{2d}(d+e-c)$ (A to B) $V = -\dfrac{W}{\ell}(c-x)$ (B to C) $V = R_1 - \dfrac{W}{\ell}(c+x)$ (C to D) $V = -\dfrac{W}{\ell}(d+e-x)$	(A to B) $M = -\dfrac{W}{2\ell}(c-x)^2$ (B to C) $M = -\dfrac{W}{2\ell}(c+x)^2 + R_1 x$ (C to D) $M = -\dfrac{W}{2\ell}(e+d-x)^2$ $M = -\dfrac{Wc^2}{2\ell}$ at B $M = -\dfrac{We^2}{2\ell}$ at C M_{max} between supports $= \dfrac{W}{2\ell}(c^2 - x_1^2)$ at $x = x_1$ $= \dfrac{c^2 + d^2 - e^2}{2d}$ If $x_1 > c$, $M = 0$ at x $= x_1 \pm [x_1^2 - c^2]^{1/2}$ x is considered positive on both sides of the origin.	(A to B) $y = -\dfrac{Wx}{24EI\ell}\{[2d(e^2 + 2c^2) + 6c^2x - x^2(4c-x) - d^3]\}$ (B to C) $y = -\dfrac{Wx(d-x)}{24EI\ell}\{[2d(c^2 + e^2) - 2(c^2 + e^2) - \dfrac{2}{d}(e^2x + c^2(d-x))]\} \cdot x(d-x) + d^2 - 2(c^2 + e^2) - \dfrac{2}{d}(e^2x + c^2(d-x))$ (C to D) $y = -\dfrac{W(x-d)}{24EI\ell}[2d(c^2 + 2e^2) + 6e^2(x-d) - (x-d)^2(4e+d-x) - d^3]$ $y = -\dfrac{Wc}{24EI}[2d(e^2 + 2c^2) + 3c^3 - d^3]$ at A $y = -\dfrac{We}{24EI}[2d(c^2 + 2e^2) + 3e^3 - d^3]$ at D This case is too complicated to obtain a general expression for critical deflections between the supports. $\theta = +\dfrac{W}{24EI\ell}(4c^3 + 4c^2d - d^3 + 2de^2)$ at A $\theta = -\dfrac{W}{24EI\ell}(2c^2d + 4de^2 - d^3 + 4e^3)$ at D

PCI Design Handbook/Fourth Edition

11–15

11.2 MATERIAL PROPERTIES
CONCRETE

Design Aid 11.2.1 Table of concrete stresses

f'_c	$0.45 f'_c$	$0.6 f'_c$	$\sqrt{f'_c}$	$0.6\sqrt{f'_c}$	$2\sqrt{f'_c}$	$3.5\sqrt{f'_c}$	$4\sqrt{f'_c}$	$5\sqrt{f'_c}$	$6\sqrt{f'_c}$	$7.5\sqrt{f'_c}$	$12\sqrt{f'_c}$
2500	1125	1500	50	30	100	175	200	250	300	375	600
3000	1350	1800	55	33	110	192	219	274	329	411	657
3500	1575	2100	59	35	118	207	237	296	355	444	710
4000	1800	2400	63	38	126	221	253	316	379	474	759
4500	2025	2700	67	40	134	235	268	335	402	503	805
5000	2250	3000	71	42	141	247	283	354	424	530	849
5500	2475	3300	74	44	148	260	297	371	445	556	890
6000	2700	3600	77	46	155	271	310	387	465	581	930
6500	2925	3900	81	48	161	281	322	403	484	605	967
7000	3150	4200	84	50	167	293	335	418	502	627	1004
7500	3375	4500	87	52	173	303	346	433	519	650	1039
8000	3600	4800	89	54	179	313	358	447	537	671	1073

Design Aid 11.2.2 Concrete modulus of elasticity as affected by unit weight and strength

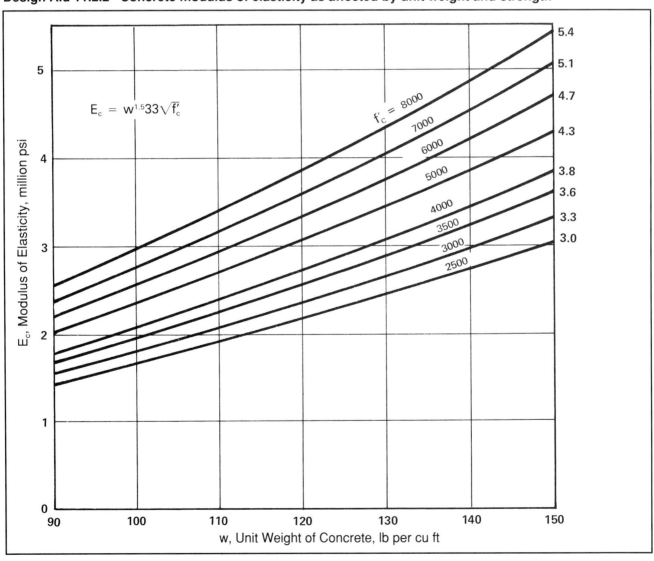

$E_c = w^{1.5} 33 \sqrt{f'_c}$

MATERIAL PROPERTIES
PRESTRESSING STEEL

Design Aid 11.2.3 Properties and design strengths of prestressing strand and wire

Seven-Wire Strand, f_{pu} = 270 ksi						
Nominal Diameter, in.	3/8	7/16	1/2	1/2 Special*	9/16	0.600
Area, sq. in.	0.085	0.115	0.153	0.167	0.192	0.217
Weight, plf	0.29	0.40	0.52	0.53	0.65	0.74
0.7 $f_{pu} A_{ps}$, kips	16.1	21.7	28.9	31.6	36.3	41.0
0.75 $f_{pu} A_{ps}$, kips	17.2	23.3	31.0	33.8	38.9	44.0
0.8 $f_{pu} A_{ps}$, kips	18.4	24.8	33.0	36.1	41.4	46.9
$f_{pu} A_{ps}$, kips	23.0	31.0	41.3	45.1	51.8	58.6

Seven-Wire Strand, f_{pu} = 250 ksi						
Nominal Diameter, in.	1/4	5/16	3/8	7/16	1/2	0.600
Area, sq. in.	0.036	0.058	0.080	0.108	0.144	0.216
Weight, plf	0.12	0.20	0.27	0.37	0.49	0.74
0.7 $f_{pu} A_{ps}$, kips	6.3	10.2	14.0	18.9	25.2	37.8
0.8 $f_{pu} A_{ps}$, kips	7.2	11.6	16.0	21.6	28.8	43.2
$f_{pu} A_{ps}$, kips	9.0	14.5	20.0	27.0	36.0	54.0

Three- and Four-Wire Strand, f_{pu} = 250 ksi				
Nominal Diameter, in.	1/4	5/16	3/8	7/16
No. of wires	3	3	3	4
Area, sq. in.	0.036	0.058	0.075	0.106
Weight, plf	0.12	0.20	0.26	0.36
0.7 $f_{pu} A_{ps}$, kips	6.3	10.2	13.2	18.6
0.8 $f_{pu} A_{ps}$, kips	7.2	11.6	15.0	21.2
$f_{pu} A_{ps}$, kips	9.0	14.5	18.8	26.5

Prestressing Wire										
Diameter	0.105	0.120	0.135	0.148	0.162	0.177	0.192	0.196	0.250	0.276
Area, sq. in.	0.0087	0.0114	0.0143	0.0173	0.0206	0.0246	0.0289	0.0302	0.0491	0.0598
Weight, plf	0.030	0.039	0.049	0.059	0.070	0.083	0.098	0.10	0.17	0.20
Ult. strength, f_{pu} ksi	279	273	268	263	259	255	250	250	240	235
0.7 $f_{pu} A_{ps}$, kips	1.70	2.18	2.68	3.18	3.73	4.39	5.05	5.28	8.25	9.84
0.8 $f_{pu} A_{ps}$, kips	1.94	2.49	3.06	3.64	4.26	5.02	5.78	6.04	9.42	11.24
$f_{pu} A_{ps}$, kips	2.43	3.11	3.83	4.55	5.33	6.27	7.22	7.55	11.78	14.05

*The 1/2 in. special strand has a larger actual diameter than the 1/2 in. regular strand. The table values take this difference into account.

MATERIAL PROPERTIES
PRESTRESSING STEEL

Design Aid 11.2.4 Properties and design strengths of prestressing bars

Smooth Prestressing Bars, f_{pu} = 145 ksi*						
Nominal Diameter, in.	¾	⅞	1	1⅛	1¼	1⅜
Area, sq. in.	0.442	0.601	0.785	0.994	1.227	1.485
Weight, plf	1.50	2.04	2.67	3.38	4.17	5.05
0.7 f_{pu} A_{ps}, kips	44.9	61.0	79.7	100.9	124.5	150.7
0.8 f_{pu} A_{ps}, kips	51.3	69.7	91.0	115.3	142.3	172.2
f_{pu} A_{ps}, kips	64.1	87.1	113.8	144.1	177.9	215.3

Smooth Prestressing Bars, f_{pu} = 160 ksi*						
Nominal Diameter, in.	¾	⅞	1	1⅛	1¼	1⅜
Area, sq. in.	0.442	0.601	0.785	0.994	1.227	1.485
Weight, plf	1.50	2.04	2.67	3.38	4.17	5.05
0.7 f_{pu} A_{ps}, kips	49.5	67.3	87.9	111.3	137.4	166.3
0.8 f_{pu} A_{ps}, kips	56.6	77.0	100.5	127.2	157.0	190.1
f_{pu} A_{ps}, kips	70.7	96.2	125.6	159.0	196.3	237.6

Deformed Prestressing Bars						
Nominal Diameter, in.	⅝	1	1	1¼	1¼	1⅜
Area, sq. in.	0.28	0.85	0.85	1.25	1.25	1.58
Weight, plf	0.98	3.01	3.01	4.39	4.39	5.56
Ult. strength, f_{pu}, ksi	157	150	160*	150	160*	150
0.7 f_{pu} A_{ps}, kips	30.5	89.3	95.2	131.3	140.0	165.9
0.8 f_{pu} A_{ps}, kips	34.8	102.0	108.8	150.0	160.0	189.6
f_{pu} A_{ps}, kips	43.5	127.5	136.0	187.5	200.0	237.0

Stress-strain characteristics (all prestressing bars):
For design purposes, following assumptions are satisfactory:

E_s = 29,000 ksi

f_y = 0.95 f_{pu}

*Verify availability before specifying.

MATERIAL PROPERTIES
PRESTRESSING STEEL

Design Aid 11.2.5 Idealized stress-strain curve, 7-wire low-relaxation prestressing strand

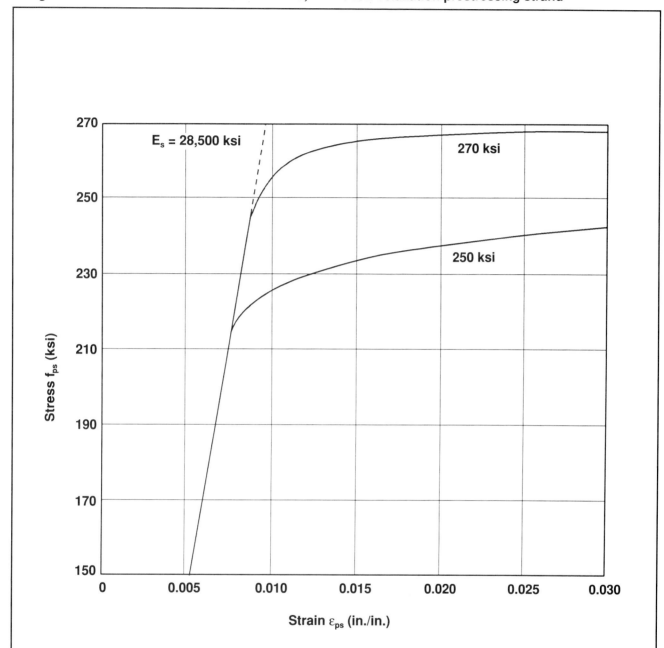

These curves can be approximated by the following equations:

250 ksi strand

$\varepsilon_{ps} \leq 0.0076$: $f_{ps} = 28{,}500\, \varepsilon_{ps}$ (ksi)

$\varepsilon_{ps} > 0.0076$: $f_{ps} = 250 - \dfrac{0.25}{\varepsilon_{ps}}$ (ksi)

270 ksi strand

$\varepsilon_{ps} \leq 0.0086$: $f_{ps} = 28{,}500\, \varepsilon_{ps}$ (ksi)

$\varepsilon_{ps} > 0.0086$: $f_{ps} = 270 - \dfrac{0.04}{\varepsilon_{ps} - 0.007}$ (ksi)

MATERIAL PROPERTIES
PRESTRESSING STEEL

Design Aid 11.2.6 Transfer and development lengths for 7-wire uncoated strand

The ACI 318-89 (Section 12.9.1) equation* for required development length may be rewritten as:

$$\ell_d = (f_{se}/3)d_b + (f_{ps} - f_{se})d_b$$

where:
ℓ_d = required development length, in.
f_{se} = effective prestress, ksi
f_{ps} = stress in prestressing steel at nominal strength, ksi
d_b = nominal diameter of strand, in.

The first term in the equation is the transfer length and the second term is the additional length required for the stress increase, $(f_{ps} - f_{se})$ corresponding to the nominal strength.

Transfer and development length⁺ in inches

Nominal Diameter, in.	f_{se} = 150 ksi					f_{se} = 160 ksi				
	Transfer Length	Development Length				Transfer Length	Development Length			
		f_{ps}, ksi					f_{ps}, ksi			
		240	250	260	270		240	250	260	270
⅜	18.8	52.5	56.3	60.0	63.8	20.0	50.0	53.8	57.5	61.3
⁷⁄₁₆	21.9	61.3	65.6	70.0	74.4	23.3	58.3	62.7	67.0	71.4
½	25.0	70.0	75.0	80.0	85.0	26.7	66.7	71.7	76.7	81.7
½ S‡	26.1	73.1	78.4	83.6	88.8	27.9	69.7	74.9	80.1	85.4
⁹⁄₁₆	28.1	78.8	84.4	90.0	95.6	30.0	75.0	80.6	86.3	91.9
0.600	30.0	84.0	90.0	96.0	102.0	32.0	80.0	86.0	92.0	98.0

* The ACI 318-89 equation was derived based on tests on ¼, ⅜ and ½ in. diameter strands. Its use to other sizes included in the table is based on inter/extrapolation as necessary. Considerable research is currently underway which may lead to revision to ACI 318-89 equation and thus the values given in the table. Research is also underway to produce recommendations on transfer and development lengths for epoxy coated strand.

+ The development length values given in the table must be doubled where bonding of the strand does not extend to the member end and the member is designed such that tension is produced under service loads (see ACI Code Sect. 12.9.3).

‡ The ½ in. special (½ S) strand has a larger nominal diameter than the ½ in. regular (½) strand. The table values for transfer and development length reflect this difference in diameters.

MATERIAL PROPERTIES
REINFORCING BARS

Design Aid 11.2.7 Reinforcing bar data

		ASTM STANDARD REINFORCING BARS		
Bar Size Designation	Weight (lb per foot)	NOMINAL DIMENSIONS		
		Diameter (in.)	Area (sq. in.)	Perimeter (in.)
#3	0.376	0.375	0.11	1.178
#4	0.668	0.500	0.20	1.571
#5	1.043	0.625	0.31	1.963
#6	1.502	0.750	0.44	2.356
#7	2.044	0.875	0.60	2.749
#8	2.670	1.000	0.79	3.142
#9	3.400	1.128	1.00	3.544
#10	4.303	1.270	1.27	3.990
#11	5.313	1.410	1.56	4.430
#14	7.650	1.693	2.25	5.320
#18	13.600	2.257	4.00	7.090

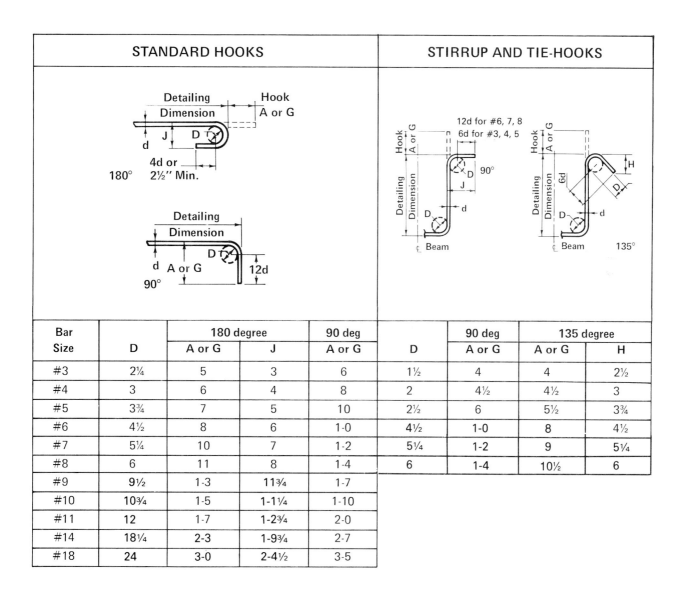

	STANDARD HOOKS			STIRRUP AND TIE-HOOKS			
Bar Size		180 degree		90 deg		90 deg	135 degree	
	D	A or G	J	A or G	D	A or G	A or G	H
#3	2¼	5	3	6	1½	4	4	2½
#4	3	6	4	8	2	4½	4½	3
#5	3¾	7	5	10	2½	6	5½	3¾
#6	4½	8	6	1-0	4½	1-0	8	4½
#7	5¼	10	7	1-2	5¼	1-2	9	5¼
#8	6	11	8	1-4	6	1-4	10½	6
#9	9½	1-3	11¾	1-7				
#10	10¾	1-5	1-1¼	1-10				
#11	12	1-7	1-2¾	2-0				
#14	18¼	2-3	1-9¾	2-7				
#18	24	3-0	2-4½	3-5				

MATERIAL PROPERTIES
REINFORCING BARS

Design Aid 11.2.8 Required development lengths for reinforcing bars

■ Bars in Tension:

ACI 318-89 gives the required development length as a basic development length multiplied by a number of modification factors to account for variables, such as spacing, cover epoxy coating. These provisions are formatted here first for $f_y = 60,000$ psi and $f'_c = 5,000$ psi and then adjusted with a factor, α_{mt} for different material properties.

$$\ell_d = \{[\{\overbrace{[(\ell_{db}) \times \alpha_A \times \alpha_B] \geq 25.5 d_b}^{(\ell_d)_1}\} \times (\alpha_c \leq 1.7) \times \alpha_D \times \alpha_E] \times \alpha_{mt}\} \geq 12 \text{ in.}$$

where:
- ℓ_d = required development length, in.
- d_b = diameter of bar, in.
- α = various modification factors
- ℓ_{db} = 34.0 A_b for #3 thru #11
- = 72.1 in. for #14
- = 106.1 in. for #18
- A_b = area of bar, in.²
- $(\ell_d)_1$ = intermediate value of development length, in. (See Table A)

The required development length is obtained as:

(1) Determine modification factors α_A and α_B from Figs. A and B, respectively.
(2) Read $(\ell_d)_1$ from Table A.
(3) Multiply table values successively with modification factors α_C, α_D, α_E, and α_{mt} given below.
(4) Compare the development length obtained in Step (3) with 12 in. to establish the required development length.

Modification factors α_C, α_D, α_E and α_{mt}:

α_C (ACI Sects. 12.2.4.1 and 12.2.4.3)
- Top bar 1.3
- Epoxy coated bar: cover < $3d_b$ or clear spacing < $6d_b$ 1.5
- otherwise 1.2
- Epoxy coated top bar: cover < $3d_b$ or clear spacing < $6d_b$ 1.7
- otherwise 1.56

α_D (ACI Sect. 12.2.4.2)
- Lightweight concrete 1.3

α_E (ACI Sect. 12.2.5)
- α_E = (A_s required)/(A_s provided)

$\alpha_{mt} = 1.18 f_y / \sqrt{f'_c}$, where f_y in ksi and f'_c in psi. (See Table B for values). Note for f_y = 60 ksi and f'_c = 5,000 psi, α_{mt} = 1.0

■ Bars in Compression:

The required development length may be written as:

$$\ell_d = \{[\overbrace{(1200.0 d_b / \sqrt{f'_c} \geq 18.0 d_b)}^{(\ell_d)_1} \times \alpha_E \times \alpha_F] \times \alpha_{mc}\} \geq 8 \text{ in.}$$

where:
- ℓ_d = required development length, in.
- d_b = diameter of bar, in.
- f'_c = 28-day concrete strength, psi
- $(\ell_d)_1$ = intermediate development length, in. (See Table A)
- α_E = modification factor per ACI Sect. 12.3.3.1
- = (A_s required)/(A_s provided)
- α_F = modification factor per ACI Sect. 12.3.3.2
- = 0.75, if reinforcement is enclosed within spiral ≥ ¼ in. diameter and ≤ 4 in. pitch or within #4 ties in conformance with ACI Sect. 7.10.5 and spacing ≤ 4 in., otherwise α_F = 1.0.

- α_{mc} = modification factor for f_y
- = f_y (ksi)/60

The required development length is obtained as:

(1) Read $(\ell_d)_1$ from Table A.
(2) Multiply the table value with modification factors α_E, α_F, α_{mc}.
(3) Compare Step (2) value with 8 in. to establish the required development length.

MATERIAL PROPERTIES
REINFORCING BARS

Design Aid 11.2.8 Required development lengths for reinforcing bars (continued)

■ **Bars with standard hooks:**

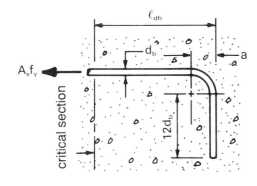

standard 90° hook

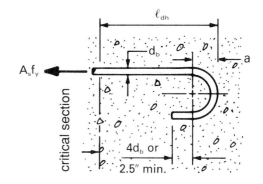

standard 180° hook

dimension a = $4d_b$ for #3 through #8,
              = $5d_b$ for #9, #10 and #11,
              = $6d_b$ for #14 and #18

$$\ell_{dh} = \{[(\ell_{hb}) \times \alpha_D \times \alpha_E \times \alpha_G \times \alpha_H] \times \alpha_{mt}\} \geq 8_{db} \text{ and 6 in.}$$

where:
- ℓ_{dh} = required development length, in.
- ℓ_{hb} = basic development length, in.
-    = 17.0 d_b for f_y = 60 ksi and f'_c = 5,000 psi (See Table A for values)
- α = various modification factors

The required development length is obtained as:

(1) Read $(\ell_d)_1$ from Table A. Note $(\ell_d)_1 = \ell_{hb}$
(2) Multiply table value successively with modification factors α_D, α_E, α_G, α_H, α_{mt}
(3) Compare the development length obtained in Step (2) with $8d_b$ and 6 in. to establish required development length.

Modification factors α_D, α_E, α_G, α_H, and α_{mt}:

α_D (ACI 12.5.3.5)
     Lightweight concrete.................................1.3

α_E (ACI 12.5.3.4)
     Excess steel $\dfrac{(A_s \text{ required})}{(A_s \text{ provided})}$

α_G (ACI 12.5.3.2)
     side cover ≥ 2.5 in. and end cover
     (90° hook only) ≥ 2 in. 0.7

α_H (ACI 12.5.3.3)
     Ties or stirrups (with #11 or smaller bars only)
     spaced at ≤ $3d_b$..0.8

α_{mt} = ...$1.18 f_y / \sqrt{f'_c}$
     where f_y is in ksi and f'_c is in psi. (See Table B for values).

MATERIAL PROPERTIES
REINFORCING BARS

Design Aid 11.2.8 Required development length for reinforcing bars (continued)

Figs. A and B— Modification Factors α_A and α_B

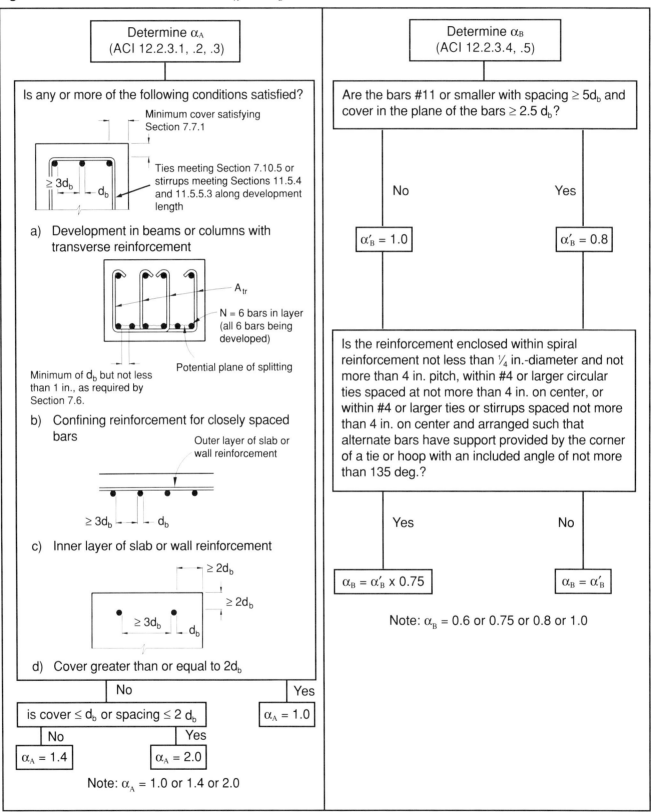

MATERIAL PROPERTIES
REINFORCING BARS

Design Aid 11.2.8 Required development lengths for reinforcing bars (continued)

Table A: Intermediate development length, $(\ell_d)_1$ in inches for $f_y = 60,000$ psi

	Bars in tension w/o hooks												Bars with standard hooks		Bars in compression			
f'_c psi	5000												f'_c psi	5000	f'_c psi	3000	4000	≥4445
α_A	1.0				1.4				2.0									
α_B	0.6	0.75	0.8	1.0	0.6	0.75	0.8	1.0	0.6	0.75	0.8	1.0	Bar#		Bar#			
Bar#																		
3	9.6	9.6	9.6	9.6	9.6	9.6	9.6	9.6	9.6	9.6	9.6	9.6	3	6.4	3	8.3	7.1	6.8
4	12.8	12.8	12.8	12.8	12.8	12.8	12.8	12.8	12.8	12.8	12.8	13.6	4	8.5	4	11.0	9.5	9.0
5	15.9	15.9	15.9	15.9	15.9	15.9	15.9	15.9	15.9	15.9	16.8	21.0	5	10.6	5	13.8	11.9	11.3
6	19.1	19.1	19.1	19.1	19.1	19.1	19.1	21.0	19.1	22.5	24.0	30.0	6	12.8	6	16.5	14.3	13.6
7	22.3	22.3	22.3	22.3	22.3	22.3	22.9	28.6	24.5	30.6	32.6	40.8	7	14.9	7	19.3	16.6	15.8
8	25.5	25.5	25.5	26.9	25.5	28.3	30.2	37.7	32.3	40.4	43.0	53.8	8	17.0	8	22.0	19.0	18.0
9	28.8	28.8	28.8	34.0	28.8	35.7	38.1	47.6	40.8	51.0	54.4	68.0	9	19.2	9	24.8	21.4	20.3
10	32.4	32.4	34.6	43.2	36.3	45.4	48.4	60.5	51.8	64.8	69.1	86.4	10	21.6	10	28.0	24.1	22.9
11	36.0	39.8	42.4	53.0	44.5	55.7	59.4	74.2	63.6	79.5	84.8	106.0	11	26.5	11	31.0	26.8	25.4
14	43.3	54.1	57.7	72.1	60.5	75.7	80.7	100.9	86.5	108.2	115.4	144.2	14	28.8	14	37.2	32.2	30.5
18	63.7	79.6	84.9	106.1	89.1	111.4	118.8	148.5	127.3	159.2	169.8	212.2	18	38.4	18	49.7	42.9	40.6

Table B: Material properties modification factor, α_{mt}

f_y, ksi	40						60					
f'_c, psi	3000	4000	5000	6000	7000	8000	3000	4000	5000	6000	7000	8000
α_{mt}	0.86	0.75	0.67	0.61	0.57	0.53	1.29	1.12	1.0	0.91	0.85	0.79

MATERIAL PROPERTIES
WELDED WIRE FABRIC

Design Aid 11.2.9 Common stock styles of welded wire fabric

Style Designation		Steel Area sq in. per ft		Approx. Weight
Former Designation (By Steel Wire Gage)	Current Designation (By W-Number)	Longit.	Trans.	lb per 100 sq ft
6x6-10x10	6x6-W1.4xW1.4	.029	.029	21
4x12-8x12**	4x12-W2.1xW0.9	.062	.009	25
6x6-8x8	6x6-W2.1xW2.1	.041	.041	30
4x4-10x10	4x4-W1.4xW1.4	.043	.043	31
4x12-7x11**	4x12-W2.5xW1.1	.074	.011	31
6x6-6x6*	6x6-W2.9xW2.9	.058	.058	42
4x4-8x8	4x4-W2.1xW2.1	.062	.062	44
6x6-4x4*	6x6-W4.0xW4.0	.080	.080	58
4x4-6x6	4x4-W2.9xW2.9	.087	.087	62
6x6-2x2*	6x6-W5.5xW5.5***	.110	.110	80
4x4-4x4*	4x4-W4.0xW4.0	.120	.120	85
4x4-3x3*	4x4-W4.7xW4.7	.141	.141	102
4x4-2x2*	4x4-W5.5xW5.5***	.165	.165	119

* Commonly available in 8 ft x 12 ft or 8 ft x 15 ft sheets.
** These items may be carried in sheets by various manufacturers in certain parts of the U.S. and Canada.
*** Exact W-Number size for 2 gage is 5.4.

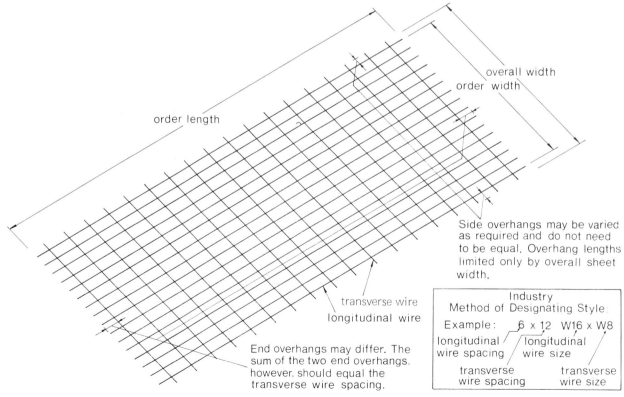

MATERIAL PROPERTIES
WELDED WIRE FABRIC

Design Aid 11.2.10 Special welded wire fabric for double tee flanges*

Application	Style Designation	Steel Area sq. in. per ft.		Approx. Weight lb per 100 sq. ft.
		Longit.	Trans.	
8-ft wide DT, 2-in. flange	12 X 6-W1.4 X W2.5	.014	.050	23
10-ft wide DT, 2-in. flange	12 X 6-W2.0 X W4.0	.020	.080	35
10-ft wide DT, 2½-in. flange	12 X 6-W1.4 X W2.9	.014	.058	27

*See "Standardization of Welded Wire Fabric," PCI Journal, July/Aug., 1976

Design Aid 11.2.11 Wires used in welded wire fabric

Wire Size Number		Nominal Diameter in.	Nominal Weight plf	Area — sq in. per ft of width Center to Center Spacing, in.						
Plain	Deformed			2	3	4	6	8	10	12
W31	D31	0.628	1.054	1.86	1.24	.93	.62	.465	.372	.31
W30	D30	0.618	1.020	1.80	1.20	.90	.60	.45	.36	.30
W28	D28	0.597	.952	1.68	1.12	.84	.56	.42	.336	.28
W26	D26	0.575	.934	1.56	1.04	.78	.52	.39	.312	.26
W24	D24	0.553	.816	1.44	.96	.72	.48	.36	.288	.24
W22	D22	0.529	.748	1.32	.88	.66	.44	.33	.264	.22
W20	D20	0.504	.680	1.20	.80	.60	.40	.30	.24	.20
W18	D18	0.478	.612	1.08	.72	.54	.36	.27	.216	.18
W16	D16	0.451	.544	.96	.64	.48	.32	.24	.192	.16
W14	D14	0.422	.476	.84	.56	.42	.28	.21	.168	.14
W12	D12	0.390	.408	.72	.48	.36	.24	.18	.144	.12
W11	D11	0.374	.374	.66	.44	.33	.22	.165	.132	.11
W10.5		0.366	.357	.63	.42	.315	.21	.157	.126	.105
W10	D10	0.356	.340	.60	.40	.30	.20	.15	.12	.10
W9.5		0.348	.323	.57	.38	.285	.19	.142	.114	.095
W9	D9	0.338	.306	.54	.36	.27	.18	.135	.108	.09
W8.5		0.329	.289	.51	.34	.255	.17	.127	.102	.085
W8	D8	0.319	.272	.48	.32	.24	.16	.12	.096	.08
W7.5		0.309	.255	.45	.30	.225	.15	.112	.09	.075
W7	D7	0.298	.238	.42	.28	.21	.14	.105	.084	.07
W6.5		0.288	.221	.39	.26	.195	.13	.097	.078	.065
W6	D6	0.276	.204	.36	.24	.18	.12	.09	.072	.06
W5.5		0.264	.187	.33	.22	.165	.11	.082	.066	.055
W5	D5	0.252	.170	.30	.20	.15	.10	.075	.06	.05
W4.5		0.240	.153	.27	.18	.135	.09	.067	.054	.045
W4	D4	0.225	.136	.24	.16	.12	.08	.06	.048	.04
W3.5		0.211	.119	.21	.14	.105	.07	.052	.042	.035
W3		0.195	.102	.18	.12	.09	.06	.045	.036	.03
W2.9		0.192	.098	.174	.116	.087	.058	.043	.035	.029
W2.5		0.178	.085	.15	.10	.075	.05	.037	.03	.025
W2.1		0.162	.070	.126	.084	.063	.042	.031	.025	.021
W2		0.159	.068	.12	.08	.06	.04	.03	.024	.02
W1.5		0.138	.051	.09	.06	.045	.03	.022	.018	.015
W1.4		0.135	.049	.084	.056	.042	.028	.021	.017	.014

11.3 SECTION PROPERTIES

Design Aid 11.3.1 Properties of geometric sections

SQUARE
Axis of moments through center

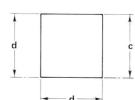

$A = d^2$
$c = \dfrac{d}{2}$
$I = \dfrac{d^4}{12}$
$S = \dfrac{d^3}{6}$
$r = \dfrac{d}{\sqrt{12}} = .288675\,d$

RECTANGLE
Axis of moments on diagonal

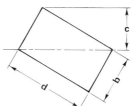

$A = bd$
$c = \dfrac{bd}{\sqrt{b^2 + d^2}}$
$I = \dfrac{b^3 d^3}{6(b^2 + d^2)}$
$S = \dfrac{b^2 d^2}{6\sqrt{b^2 + d^2}}$
$r = \dfrac{bd}{\sqrt{6(b^2 + d^2)}}$

SQUARE
Axis of moments on base

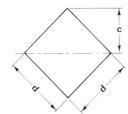

$A = d^2$
$c = d$
$I = \dfrac{d^4}{3}$
$S = \dfrac{d^3}{3}$
$r = \dfrac{d}{\sqrt{3}} = .577350\,d$

RECTANGLE
Axis of moments any line through center of gravity

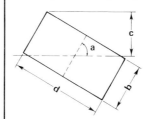

$A = bd$
$c = \dfrac{b\sin a + d\cos a}{2}$
$I = \dfrac{bd(b^2 \sin^2 a + d^2 \cos^2 a)}{12}$
$S = \dfrac{bd(b^2 \sin^2 a + d^2 \cos^2 a)}{6(b\sin a + d\cos a)}$
$r = \sqrt{\dfrac{b^2 \sin^2 a + d^2 \cos^2 a}{12}}$

SQUARE
Axis of moments on diagonal

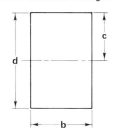

$A = d^2$
$c = \dfrac{d}{\sqrt{2}} = .707107\,d$
$I = \dfrac{d^4}{12}$
$S = \dfrac{d^3}{6\sqrt{2}} = .117851\,d^3$
$r = \dfrac{d}{\sqrt{12}} = .288675\,d$

HOLLOW RECTANGLE
Axis of moments through center

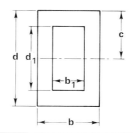

$A = bd - b_1 d_1$
$c = \dfrac{d}{2}$
$I = \dfrac{bd^3 - b_1 d_1^3}{12}$
$S = \dfrac{bd^3 - b_1 d_1^3}{6d}$
$r = \sqrt{\dfrac{bd^3 - b_1 d_1^3}{12 A}}$

RECTANGLE
Axis of moments through center

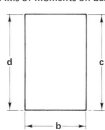

$A = bd$
$c = \dfrac{d}{2}$
$I = \dfrac{bd^3}{12}$
$S = \dfrac{bd^2}{6}$
$r = \dfrac{d}{\sqrt{12}} = .288675\,d$

EQUAL RECTANGLES
Axis of moments through center of gravity

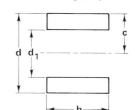

$A = b(d - d_1)$
$c = \dfrac{d}{2}$
$I = \dfrac{b(d^3 - d_1^3)}{12}$
$S = \dfrac{b(d^3 - d_1^3)}{6d}$
$r = \sqrt{\dfrac{d^3 - d_1^3}{12(d - d_1)}}$

RECTANGLE
Axis of moments on base

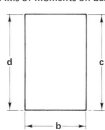

$A = bd$
$c = d$
$I = \dfrac{bd^3}{3}$
$S = \dfrac{bd^2}{3}$
$r = \dfrac{d}{\sqrt{3}} = .577350\,d$

UNEQUAL RECTANGLES
Axis of moments through center of gravity

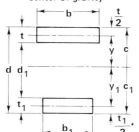

$A = bt + b_1 t_1$
$c = \dfrac{\tfrac{1}{2} bt^2 + b_1 t_1 (d - \tfrac{1}{2} t_1)}{A}$
$I = \dfrac{bt^3}{12} + bty^2 + \dfrac{b_1 t_1^3}{12} + b_1 t_1 y_1^2$
$S = \dfrac{I}{c} \qquad S_1 = \dfrac{I}{c_1}$
$r = \sqrt{\dfrac{I}{A}}$

SECTION PROPERTIES

Design Aid 11.3.1 Properties of geometric sections (continued)

TRIANGLE
Axis of moments through center of gravity

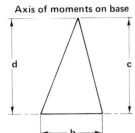

$A = \dfrac{bd}{2}$

$c = \dfrac{2d}{3}$

$I = \dfrac{bd^3}{36}$

$S = \dfrac{bd^2}{24}$

$r = \dfrac{d}{\sqrt{18}}$

HALF CIRCLE
Axis of moments through center of gravity

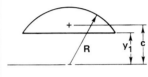

$A = \dfrac{\pi R^2}{2}$

$c = R\left(1 - \dfrac{4}{3\pi}\right)$

$I = R^4\left(\dfrac{\pi}{8} - \dfrac{8}{9\pi}\right)$

$S = \dfrac{R^3}{24}\dfrac{(9\pi^2 - 64)}{(3\pi - 4)}$

$r = R\dfrac{\sqrt{9\pi^2 - 64}}{6\pi}$

TRIANGLE
Axis of moments on base

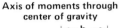

$A = \dfrac{bd}{2}$

$c = d$

$I = \dfrac{bd^3}{12}$

$S = \dfrac{bd^2}{12}$

$r = \dfrac{d}{\sqrt{6}}$

SEGMENT OF A CIRCLE
Axis of moments through circle center

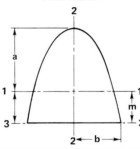

$I = \dfrac{\pi R^4}{8} + \dfrac{y_1}{2}\sqrt{(R^2 - y_1^2)^3}$

$\quad - \dfrac{R^2}{4}\left(y_1\sqrt{R^2 - y_1^2} + R^2 \sin^{-1}\dfrac{y_1}{R}\right)$

$A = \dfrac{\pi R^2}{2} - y_1\sqrt{R^2 - y_1^2}$

$\quad - R^2 \sin^{-1}\left(\dfrac{y_1}{R}\right)$

$c = \dfrac{2(R^2 - y_1^2)^{3/2}}{3} / A$

TRAPEZOID
Axis of moments through center of gravity

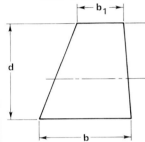

$A = \dfrac{d(b + b_1)}{2}$

$c = \dfrac{d(2b + b_1)}{3(b + b_1)}$

$I = \dfrac{d^3(b^2 + 4bb_1 + b_1^2)}{36(b + b_1)}$

$S = \dfrac{d^2(b^2 + 4bb_1 + b_1^2)}{12(2b + b_1)}$

$r = \dfrac{d}{6(b + b_1)} \times \sqrt{2(b^2 + 4bb_1 + b_1^2)}$

PARABOLA

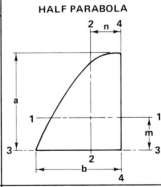

$A = \dfrac{4}{3}ab$

$m = \dfrac{2}{5}a$

$I_1 = \dfrac{16}{175}a^3 b$

$I_2 = \dfrac{4}{15}ab^3$

$I_3 = \dfrac{32}{105}a^3 b$

CIRCLE
Axis of moments through center

$A = \dfrac{\pi d^2}{4} = \pi R^2$

$c = \dfrac{d}{2} = R$

$I = \dfrac{\pi d^4}{64} = \dfrac{\pi R^4}{4}$

$S = \dfrac{\pi d^3}{32} = \dfrac{\pi R^3}{4}$

$r = \dfrac{d}{4} = \dfrac{R}{2}$

HALF PARABOLA

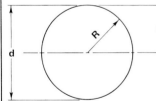

$A = \dfrac{2}{3}ab$

$m = \dfrac{2}{5}a$

$n = \dfrac{3}{8}b$

$I_1 = \dfrac{8}{175}a^3 b$

$I_2 = \dfrac{19}{480}ab^3$

$I_3 = \dfrac{16}{105}a^3 b$

$I_4 = \dfrac{2}{15}ab^3$

HOLLOW CIRCLE
Axis of moments through center

$A = \dfrac{\pi(d^2 - d_1^2)}{4}$

$c = \dfrac{d}{2}$

$I = \dfrac{\pi(d^4 - d_1^4)}{64}$

$S = \dfrac{\pi(d^4 - d_1^4)}{32d}$

$r = \dfrac{\sqrt{d^2 + d_1^2}}{4}$

COMPLEMENT OF HALF PARABOLA

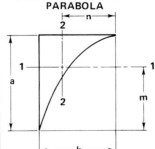

$A = \dfrac{1}{3}ab$

$m = \dfrac{7}{10}a$

$n = \dfrac{3}{4}b$

$I_1 = \dfrac{37}{2100}a^3 b$

$I_2 = \dfrac{1}{80}ab^3$

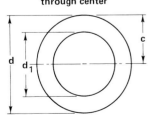

PCI Design Handbook/Fourth Edition

SECTION PROPERTIES

Design Aid 11.3.1 Properties of geometric sections (continued)

PARABOLIC FILLET IN RIGHT ANGLE

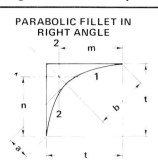

$a = \dfrac{t}{2\sqrt{2}}$

$b = \dfrac{t}{\sqrt{2}}$

$A = \dfrac{1}{6} t^2$

$m = n = \dfrac{4}{5} t$

$I_1 = I_2 = \dfrac{11}{2100} t^4$

*HALF ELLIPSE

$A = \dfrac{1}{2} \pi ab$

$m = \dfrac{4a}{3\pi}$

$I_1 = a^3 b \left(\dfrac{\pi}{8} - \dfrac{8}{9\pi} \right)$

$I_2 = \dfrac{1}{8} \pi ab^3$

$I_3 = \dfrac{1}{8} \pi a^3 b$

*QUARTER ELLIPSE

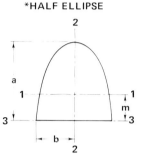

$A = \dfrac{1}{4} \pi ab$

$m = \dfrac{4a}{3\pi}$

$n = \dfrac{4b}{3\pi}$

$I_1 = a^3 b \left(\dfrac{\pi}{16} - \dfrac{4}{9\pi} \right)$

$I_2 = ab^3 \left(\dfrac{\pi}{16} - \dfrac{4}{9\pi} \right)$

$I_3 = \dfrac{1}{16} \pi a^3 b$

$I_4 = \dfrac{1}{16} \pi ab^3$

*To obtain properties of half circles, quarter circle and circular complement, substitute $a = b = R$.

*ELLIPTIC COMPLEMENT

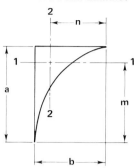

$A = ab \left(1 - \dfrac{\pi}{4} \right)$

$m = \dfrac{a}{6 \left(1 - \dfrac{\pi}{4} \right)}$

$n = \dfrac{b}{6 \left(1 - \dfrac{\pi}{4} \right)}$

$I_1 = a^3 b \left(\dfrac{1}{3} - \dfrac{\pi}{16} - \dfrac{1}{36 \left(1 - \dfrac{\pi}{4} \right)} \right)$

$I_2 = ab^3 \left(\dfrac{1}{3} - \dfrac{\pi}{16} - \dfrac{1}{36 \left(1 - \dfrac{\pi}{4} \right)} \right)$

REGULAR POLYGON
Axis of moments through center

$n =$ Number of sides

$\phi = \dfrac{180°}{n}$

$a = 2\sqrt{R^2 - R_1^2}$

$R = \dfrac{a}{2 \sin \phi}$

$R_1 = \dfrac{a}{2 \tan \phi}$

$A = \dfrac{1}{4} na^2 \cot \phi = \dfrac{1}{2} nR^2 \sin 2\phi = nR_1^2 \tan \phi$

$I_1 = I_2 = \dfrac{A(6R^2 - a^2)}{24} = \dfrac{A(12R_1^2 + a^2)}{48}$

$r_1 = r_2 = \sqrt{\dfrac{6R^2 - a^2}{24}} = \sqrt{\dfrac{12R_1 + a^2}{48}}$

BEAMS AND CHANNELS
Transverse force oblique through center of gravity

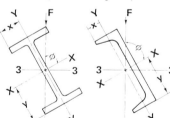

$I_3 = I_x \sin^2 \phi + I_y \cos^2 \phi$

$I_4 = I_x \cos^2 \phi + I_y \sin^2 \phi$

$f_b = M \left(\dfrac{y}{I_x} \sin \phi + \dfrac{x}{I_y} \cos \phi \right)$

where M is bending moment due to force F.

ANGLE
Axis of moments through center of gravity

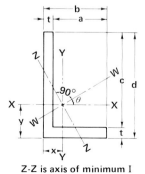

Z-Z is axis of minimum I

$\tan 2\theta = \dfrac{2K}{I_y - I_x}$

$A = t(b + c) \quad x = \dfrac{b^2 + ct}{2(b + c)} \quad y = \dfrac{d^2 + at}{2(b + c)}$

$K =$ Product of Inertia about X-X & Y-Y

$= \pm \dfrac{abcdt}{4(b + c)}$

$I_x = \dfrac{1}{3} \left(t(d - y)^3 + by^3 - a(y - t)^3 \right)$

$I_y = \dfrac{1}{3} \left(t(b - x)^3 + dx^3 - c(x - t)^3 \right)$

$I_z = I_x \sin^2 \theta + I_y \cos^2 \theta + K \sin 2\theta$

$I_w = I_x \cos^2 \theta + I_y \sin^2 \theta - K \sin 2\theta$

K is negative when heel of angle, with respect to c. g., is in 1st or 3rd quadrant, positive when in 2nd or 4th quadrant.

11.4 METRIC CONVERSION

Design Aid 11.4.1 Conversion to International System of Units (SI)

To convert from	to	multiply by
Length		
inch (in.)	millimeter (mm)	25.4
inch (in.)	meter (m)	0.0254
foot (ft)	meter (m)	0.3048
yard (yd)	meter (m)	0.9144
Area		
square foot (sq ft)	square meter (sq m)	0.09290
square inch (sq in.)	square millimeter (sq mm)	645.2
square inch (sq in.)	square meter (sq m)	0.0006452
square yard (sq yd)	square meter (sq m)	0.8361
acre (A)	hectare (ha) = 10,000 sq m	0.4047
Volume		
cubic inch (cu in.)	cubic meter (cu m)	0.00001639
cubic foot (cu ft)	cubic meter (cu m)	0.02832
cubic yard (cu yd)	cubic meter (cu m)	0.7646
gallon (gal) Can. liquid*	liter	4.546
gallon (gal) Can. liquid*	cubic meter (cu m)	0.004546
gallon (gal) U.S. liquid*	liter	3.785
gallon (gal) U.S. liquid*	cubic meter (cu m)	0.003785
Force		
kip	kilogram (kgf)	453.6
kip	newton (N)	4448.0
pound (lb)	kilogram (kgf)	0.4536
pound (lb)	newton (N)	4.448
Pressure or Stress		
kips/square inch (ksi)	megapascal (MPa)**	6.895
pound/square foot (psf)	kilopascal (kPa)**	0.04788
pound/square inch (psi)	kilopascal (kPa)**	6.895
pound/square inch (psi)	megapascal (MPa)**	0.006895
pound/square foot (psf)	kilogram/square meter (kgf/sq m)	4.882
Mass		
pound (avdp)	kilogram (kg)	0.4536
ton (short, 2000 lb)	kilogram (kg)	907.2
ton (short, 2000 lb)	tonne (t)	0.9072
grain	kilogram (kg)	0.00006480
tonne (t)	kilogram (kg)	1000
Mass (weight) per Length		
kip/linear foot (klf)	kilogram/meter (kg/m)	0.001488
pound/linear foot (plf)	kilogram/meter kg/m	1.488
pound/linear foot (plf)	newton/meter (N/m)	14.593

*One U.S. gallon equals 0.8321 Canadian gallon
**A pascal equals one newton/square meter

METRIC CONVERSION

Design Aid 11.4.1 Conversion to International System of Units (SI) (continued)

To convert from	to	multiply by
Mass per volume (density)		
pound/cubic foot (pcf)	kilogram/cubic meter (kg/cu m)	16.02
pound/cubic yard (pcy)	kilogram/cubic meter (kg/cu m)	0.5933
Bending Moment or Torque		
inch-pound (in.-lb)	newton-meter	0.1130
foot-pound (ft-lb)	newton-meter	1.356
foot-kip (ft-k)	newton-meter	1356
Temperature		
degree Fahrenheit (deg F)	degree Celsius (C)	$t_C = (t_F - 32)/1.8$
degree Fahrenheit (deg F)	degree Kelvin (K)	$t_K = (t_F + 459.7)/1.8$
Energy		
British thermal unit (Btu)	joule (j)	1056
kilowatt-hour (kwh)	joule (j)	3,600,000
Power		
horsepower (hp) (550 ft lb/sec)	watt (W)	745.7
Velocity		
mile/hour (mph)	kilometer/hour	1.609
mile/hour (mph)	meter/second (m/s)	0.4470
Other		
Section modulus (in.^{3})	mm^3	16,387
Moment of inertia (in.^{4})	mm^4	416,231
Coefficient of heat transfer (Btu/ft^2/h/°F)	W/m^2/°C	5.678
Modulus of elasticity (psi)	MPa	0.006895
Thermal conductivity (Btu-in./ft^2/h/°F)	Wm/m^2/°C	0.1442
Thermal expansion in./in./°F	mm/mm/°C	1.800
Area/length (in.^{2}/ft)	mm^2/m	2116.80

INDEX

Entry	Page No.
A-frames, transportation	5-17
American Association of State Highway Transportation Officials (AASHTO) grade pads	6-19
Absorption, water	10-15
Acceptability range	8-3
Acceptance	10-36
Acceptance, specification	10-11, 10-26
Access, jobsite	10-34
Accidental torsion	3-60
Acid cleaning	10-29
Acoustic design	9-20
Acoustical properties	9-20
Admixtures	1-27
Aggregates	10-6, 10-19
Air barriers	9-15
Air content	1-27
Air content, mortar	1-25
Air content, tolerance	1-27
Air entraining portland cement	1-27, 10-18
Airborne sound	9-20
Aircraft cable	5-10
Airport terminal	1-11
Alignment connections	6-48
Allowable compressive stress, bearing pads	6-21
Allowable loads, bolts	6-72
Allowable solid bearing	3-45
American Institute of Steel Construction (AISC)	6-16
American Society for Testing and Materials (ASTM)-E119	9-34
Anchor bolts	3-44, 3-47, 10-35
Anchor bolts, hooked	6-28
Anchorage, daps	6-34
Anchorage, grouted conduit	6-5
Anchorage, stone facing	7-17
Anchorage details, veneer panels	7-19
Anchorage seating loss	4-37
Anchorage tests, veneer panels	7-18
Angles, connection	6-20
Angles, shear strength	6-74
Angles, stiffeners	6-14
Approvals	10-33
Arbitration	10-40
Architectural concrete	1-15, 7-1
Architectural panel, earthquake analysis	3-81
Architectural panels	7-2, 8-5, 8-12
Arenas	1-17
Backup concrete	7-12, 10-19
Barricades	10-28
Bars, prestressing	11-18
Base connections, columns	6-27
Base plates	6-25
Base shear, seismic	3-60

Entry	Page No.
Bases, columns	3-44
Beam diagrams	11-3 to 11-11
Beam load tables	2-42
Beam-column construction	1-5
Beams	8-8
Beams with overhangs, diagrams	11-15
Bearing	8-13
Bearing, plain concrete	6-24
Bearing, reinforced concrete	6-24
Bearing connections	6-47
Bearing pads	6-18
Bearing pads, design recommendations	6-20
Bearing pads, minimum thickness	6-20, 6-21
Bearing pads, specification	10-8
Bearing piles	4-54
Bearing stress, anchor bolts	6-28
Bearing walls	1-4, 3-57, 3-61
Bearing walls, typical joints	6-51
Bench marks	10-34
Bilinear behavior	4-42
Boathouse	1-12
Bolts	6-15
Bolts, allowable loads	6-72
Bonded steel pads	6-18
Bowing	3-11, 8-5
Bowing, restrained	3-13, 3-17
Box modules	1-14
Box type building	3-61
Bracing	3-5
Bracing equipment	5-24
Bracket plate	6-14
Brackets, concrete	6-28
Brick	7-13
Bridges	1-19
Bridges, thermal	9-19
British thermal units (Btu)	9-3
Builders risk insurance	10-39
Building lines	10-34
Cable guys	5-24
Calcium chloride	1-28
Camber	4-40, 8-5
Camber, long time	4-44
Camber and deflection design aids	4-72
Camber coefficients	11-12
Canadian Standards Association	1-3
Cantilevers	4-56
Carbon equivalent	6-7
Caulking	7-18
Cazaly hanger	6-38
Ceiling weights	11-2
Cement content, minimum	1-27
Center of rigidity	3-30

Center of stiffness	3-49
Ceramic fiber felt	9-42
Ceramic tile	7-13
Chemical resistance	1-26
Chloride ions	1-28
Chloride penetration	1-26
Chloroprene	6-18, 10-8, 10-21
Chord forces	3-28
Cladding	7-2
Claims, contract	10-39
Clay products	7-13
Cleaning	10-29
Clearance, example	8-14
Clearances	8-13
Code of standard practice	10-31
Coefficient of friction	6-23
Coefficient of thermal expansion	1-24, 3-12
Coefficients, shear friction	6-24
Coil rods	6-72
Collins and Mitchell	4-32
Coloring agent	10-18, 10-22
Columns, eccentrically loaded	3-20
Columns, precast	2-50, 4-45
Columns, prestressed	2-50, 4-55
Column base plates, design	6-25
Column base plates, thickness	6-75
Column bases	3-44, 3-57
Column covers	7-11
Column covers, fire resistance	9-50
Column splices	3-57
Combined shear and tension, studs	6-13
Combined shear and torsion, weld groups	6-16
Component design	4-3
Composite members	4-19
Composite topping	3-28
Compression members	4-45
Compression reinforcement, effect on creep	5-16
Compressive strength, concrete	1-22
Compressive strength, effect of temperature	9-45
Computer models	3-56
Concentrated loads	4-13, 11-3
Concrete, curing	1-22
Concrete, material properties	1-22
Concrete, modulus of elasticity	1-23
Concrete, tensile strength	1-23
Concrete, unit weights	2-4
Concrete brackets	6-28
Concrete coatings	9-59
Concrete mixes, specification	10-8, 10-22
Condensation	9-15, 9-17
Confinement reinforcing, anchor bolts	6-28
Connection angles	6-20
Connection hardware	6-5
Connections	6-1
Connections, column base	6-27
Connections, design criteria	6-4
Connections, durability	6-5
Connections, fire protection	9-50
Connections, load factors	6-4
Connections, non-load bearing panels	7-4
Connections, sandwich panels	9-54
Connections, tolerances	8-13
Connections, typical-seismic loads	3-80
Construction loads	5-24
Construction loads, friction	6-23
Construction loads, safety factors	5-24
Continuity	4-56
Contract administration	10-37
Contract terms	10-38
Contracts	10-31
Corbels	6-28
Corbels, load tables	6-76
Corefloor	2-35
Corner separation, panel bowing	3-13
Corrosion resistance	6-16
Cotton duck fabric pads	6-18
Coupled shear walls	3-33
Couplers	6-7
Cover requirements	1-27, 10-25
Crack control	5-4
Crack widths	5-4
Cracking, repair and maintenance	9-64
Cracking, safety factor	5-3
Cracking, welded connections	6-16
Creep, correction factors	3-8
Creep strain, table	3-8
Critical section	4-7, 4-38
Crystallization	6-7
Cube specimens	1-22
Culvert	1-20
Curing, concrete	1-22, 10-26, 10-27
Cyclical movement, pads	6-20
Cylinders	1-22
Dapped ends	6-32
DBA (see deformed bar anchors)	
Dead weight, floors, ceilings, roofs, walls	11-2
Debonding	4-7
Decibel	9-20
Deck panels	1-19
Deflection	4-40
Deflection, long time	4-44
Deformed bar anchors (DBA)	6-15
Deformed prestressing bars, properties	11-18
Degree day	9-3
Degree of fixity, column base	3-48
Deicers	1-12, 1-29
Delivery	10-33
Design, responsibility	10-32, 10-42
Design analysis, architectural panels	7-3
Design calculations, specification	10-5, 10-17
Design criteria, connections	6-4
Development length, reinforcing bars	11-23
Development length, strand	4-22
Development length, strand, design aid	11-20
Dew point	9-3
Diagonal tension, daps	6-34
Diaphragm design	3-28, 3-77
Differential camber	8-5
Differential temperature	3-13
Direct shear, daps	6-33
Distressed finish	10-23
Docks	1-21
Double tee, partial prestress	4-11
Double tee flange strength	4-11

Double tee wall panels, interaction curves 2-52
Double tees load tables....................................2-7 to 2-26
Dowels, anchorage in conduit 6-5
Dowels, stainless steel 7-17
Drift..3-60
Dry connection 3-59
Drypack 1-24
Duck layer reinforcement 6-19, 10-8, 10-22
Ductility, connections..7-4
Ductwork 9-68
Durability 1-26
Durability, connections 6-5
Dynaspan 2-33

E119 9-34
Earthquake analysis 3-58
Eccentric bracket connections..................................6-17
Eccentric loading, weld groups................................6-17
Eccentrically loaded columns 3-20
Economic considerations, fire 9-52
Edge distance, punched holes 6-23
Effective length 3-26
Effective moment of inertia......................................4-43
Effective shear friction coefficient 6-23
Effective width of panels..4-50
Elastic deflections...4-42
Elastic deformation 3-16
Elastic stretch 5-26
Elastomeric pads 6-19
Elastomeric pads, fire 9-50
Electrified floors 9-68
Embedment length, dowels 6-5
End point criteria 9-35
Engineer of Record 10-43, 10-44
Engineer of Record, definition 10-42
Epoxy, veneer panel anchors....................................7-18
Epoxy coated reinforcement 1-27, 7-12, 10-20
Epoxy coated strand 1-29
Epoxy cement...1-24
Epoxy grout 1-25, 10-21
Equivalent volume change 3-49
Equivalent volume change strains, table..................3-51
Erection 5-17, 10-33
Erection, architectural panels 7-5
Erection, factors of safety 5-24
Erection, specification 10-10
Erection analysis, example.......................................5-27
Erection bracing 5-23
Erection drawings 10-32
Erection loads 5-24
Erection tolerances 8-7
Erector, definition 10-42
Errors, correction 10-35
Escorts, transportation 5-17
Estimating prestress losses.......................................4-37
Expansion inserts 6-15
Expansion joints 3-14
Extended end, daps 6-33

Fabrication, specification 10-23
Face mix...10-22

Factored cracking load 4-7
Factors of safety, construction loads 5-24
Factors of safety, handling and erection 10-6
Federal Construction Council 3-14
Fiber core 5-26
Field welding, specification......................................10-11
Fillet welds, allowable stresses 6-57
Finish, distressed 10-23
Finishes 10-23, 10-33
Fire endurance, sandwich panels..............................9-42
Fire extinguisher...10-30
Fire resistance 9-33
Fire tests 9-34, 9-35
Fixed end moments 11-13
Flame finish 7-16
Flanged elements 4-7
Flanking 9-30
Flatness tolerance, veneer panels 7-17
Flexibility coefficient, soil interaction 3-46
Flexible conduit, dowel anchorage 6-5
Flexicore 2-34
Flexural resistance coefficients 4-59
Floor members 8-9
Floor openings..4-55
Floor weights 11-2
Floors 8-9
Flow chart, flexural calculation 4-8
Form draft 5-3
Form liners 10-23
Form suction...5-3
Forms 7-5
Frame distortion, panels 3-16
Free edge, anchor bolts 6-28
Free edge, headed studs 6-8
Freeze-thaw 1-26
Friction 3-56, 6-23

Galvanized, welding 6-16
Galvanized reinforcement 1-28, 7-12
Gaskets, veneer panels...7-18
Geometric sections, properties.................................11-30
Glass Fiber Reinforced Concrete (GFRC) 1-16
Granite 7-15, 7-17
Ground motion 3-59
Grout 1-24
Grout keys 3-28
Grout keys, shear strength 3-28
Guide specifications, architectural precast
   concrete..10-12
Guide specifications, precast, prestressed
   concrete..10-2
Gusset plates 6-14
Gymnasium floor, vibrations 9-61
Gypsum wall board 9-39

Handling, connection hardware 5-10
Handling, specification 10-6, 10-17
Handling, with controlled cracking 5-4
Handling, without cracking 5-3
Handling considerations 5-6
Handling devices 5-8
Handling of panels, diagrams...................................5-7

Hanger connections	6-38
Hanger reinforcement, daps	6-34
Hanger steel, modification factor	6-39
Hardware, connections	6-5
Hardwood strips	6-20
Hauling (see transportation)	
Haunches, concrete (see corbels)	
Haunches, structural steel	6-30
Headed studs	6-7
Headed studs, combined loading	6-13
Headed studs, dimensions	6-61
Headed studs, free edge four adjacent sides	6-68
Headed studs, free edge one side	6-64
Headed studs, free edge three sides	6-67
Headed studs, free edge two adjacent sides	6-66
Headed studs, free edge two opposite sides	6-65
Headed studs, minimum member thickness	6-69
Headed studs, not near free edge	6-63
Headed studs, shear strength	6-71
Headed studs, strength limited by concrete	6-62
Headed studs, strength reduction of groups	6-70
Headed studs, tensile strength	6-61
Heat capacity	9-3
Heat flow	9-13
Heat transmittance	9-3
High strength threaded rods	6-15
Hollow-core	8-5
Hollow-core, load distribution	4-55
Hollow-core, openings	4-55
Hollow-core load tables	2-27 to 2-32
Hollow-core slabs, properties	2-33 to 2-35
Hollow-core wall panels, interaction curves	2-53
Hooked anchor bolts	6-28
Hooks	11-21
Horizontal alignment	8-3
Horizontal joints	6-51
Horizontal shear	4-29
Horizontal shear reinforcement	4-29
Hot dipped galvanized, welding	6-16
Human response to vibrations	9-61
Hybrid structures	3-57
Imaginary column	3-49
Impact factors, handling	5-3
Impact noise	9-21
Importance factor, seismic	3-60
Inclined lift lines	5-9
Initial camber	4-40
Inserts	6-15, 7-5
Inserts, wire capacity	6-72
Inspection	10-33
Inspection, specification	10-11, 10-27
Insulated panels	1-10
Insulation	9-56
Insurance	10-39
Intentional roughness	4-27
Interaction curve, column	2-5, 4-48
Interaction, shear wall-frame	3-56
Interface characteristics	8-18
Interface, other trades	10-36
Interfacing tolerances	8-17
Inverted tee beams, load tables	2-45 to 2-47

Jackson-Moreland alignment chart	3-26
Jails	1-14
Joints, segmental construction	9-65
Joints, thermal treatment	9-41
Joint clearance	8-13
Joint sealants	9-59
Justice facilities	1-14
Keyed joints, wall panels	6-50
L-shaped beams, load tables	2-43, 2-44
Large panel structures	3-57
Lateral load resisting systems	3-5
Lateral stability	5-14
Latex additives	7-14
Ledger beam, ledge design	6-36
Ledges, longitudinal bending	6-36
Ledges, transverse bending	6-36
Lift loops	5-10
Lifting loads, equalization	5-10
Light wall, parking structure	1-12
Lighting	9-68
Limestone panels	7-12, 7-17
Live loads, recommended minimum	11-3
Load and Resistance Factor Design (LRFD)	6-16
Load bearing panels, connections	6-46
Load distribution	4-55
Load distribution, assumed	4-57
Load factors, connections	6-4, 6-16
Load tables	2-7
Load tables, limiting criteria	2-3
Load transfer, architectural panels	7-3
Long term cambers-suggested multipliers	4-45
Long time camber/deflection	4-44
Loop inserts	6-15
Loov hanger	6-44
Loss of prestress	2-5, 4-36
Low boy trailers	5-17
Marble	7-15
Materials	10-33
Materials, specification	10-6, 10-18
Maximum camber variation	8-5
Maximum permitted computed deflections	4-41
Maximum strand diameter	5-5
Mechanical anchors, stone veneer	7-12
Mechanical systems, coordination	9-68
Metallic conduit	6-5
Metric conversion	11-32
Microcracking, granite	7-16
Minimum cement content	1-27
Minimum clearances	8-4
Minimum crack widths	4-16
Minimum eccentricity, columns	3-48
Minimum horizontal force	6-4
Minimum reinforcement, panels	4-16
Minimum shear reinforcement	4-27
Minimum thickness, bearing pads	6-20
Mixed building systems	8-7
Mockup, specification	10-15
Modeling of frames	3-48
Moderate exposure	1-27

Modulus of elasticity, clay products	7-13
Modulus of elasticity, concrete	1-23
Modulus of elasticity, design aid	11-16
Modulus of rupture	5-3
Moisture difference	3-12
Moment connections	6-45
Moment diagrams, design aids	11-4
Moment diagrams, fire	9-43, 9-47
Moment magnification	3-23, 3-24
Moment magnifier	3-24
Moment of inertia	4-43
Moment-resisting frames	3-5, 3-43, 3-49
Moments in panels, handling	5-7, 5-8
Mortar	1-24, 7-15
Moving loads	11-14
Mullions	7-11
Multimonomer plastic strips	6-19
Natural stone	7-12, 7-15
Neoprene (see chloroprene)	
Noise criteria	9-20, 9-25
Noise reduction	9-20, 9-21
Non-bearing panels	3-15
Non-load bearing panels, connections	6-46
Non-prestressed panels	4-17
Non-shrink grout	1-25
Office buildings	1-8
Other trades	10-35
Out of plane bending, beam ledges	6-37
Overjacking	4-37
Owner, definition	10-42
P-delta analysis	3-21
Panels, load bearing	3-17
Panels, non-load bearing	3-15, 6-46
Parking structures	1-12
Parking structure, earthquake analysis	3-73
Partial fixity	3-49
Partial penetration welds, allowable stresses	6-57
Partial prestress	4-11, 4-14
Patching	10-36
Payloads	5-17
PCI Committee on Prestressed Concrete Columns	4-45
PCI Hollow Core Producers Committee	4-55
Penetrating sealers	1-26, 9-59
Peripheral design force	3-57
Perm	9-3
Permeability	7-16, 9-3
Permeance	9-3
Pick-up points	5-9
Pigments	1-22
Pile caps	1-19
Piles	2-5, 4-52
Piles, allowable tension	4-53
Piles, section properties	2-56
Pipe braces	5-26
Plain concrete bearing	6-24
Plant certification	1-3, 9-58, 10-3, 10-13
Plastic bearing	6-18, 10-8, 10-22
Plastic coated tendons	5-6
Plastic deformation	3-16
Plastic section modulus	6-16, 6-73
Plow steel	5-26
Poisson's ratio	1-23
Polar moment of inertia, weld groups	6-19
Poles	1-21
Portland Cement Association	4-52
Post-tensioned wall panels	5-6
Post-tensioning, connections	6-18
Post-tensioning, segmental	9-66
Post-tensioning, wall connections	6-52, 6-53
Precast cladding	1-15
Precast columns, interaction curves	2-50, 2-51
Precast engineer, definition	10-42
Precast panels as forms	7-5
Precast wall panels, interaction curves	2-55, 4-45
Prefabricated modules	9-72
Premixed grout	1-25
Prestress losses	4-36
Prestressed columns, interaction curves	2-48, 4-45
Prestressed wall panels, handling	5-5
Prestressed wall panels, interaction curves	2-52, 4-45
Prestressing bars, properties	11-18
Prestressing steel, properties	11-17, 11-19
Prestressing tendons	1-25
Prestressing wire, properties	11-17
Primary control surface	8-3
Product delivery, specification	10-6, 10-17
Product handling	5-2
Product information	2-1
Product tolerances	8-4
Production considerations	7-5
Production drawings, specification	10-16
Properties of geometric sections	11-30
Properties of weld groups	6-19
Protection, specification	10-29
Protection of reinforcement	1-27
Prying action, panels	9-55
Quality assurance, specification	10-13
Quality control	9-58
R-values, insulating materials	9-40
Railroad ties	1-21
Raker beams	1-17
Random oriented fiber pads	6-18
Range of losses	4-37
Records	10-27
Rectangular beams, load tables	2-42
Reentrant corner, daps	6-34
Reinforced concrete bearing	6-24
Reinforcement, epoxy coated	1-27
Reinforcement, galvanized	1-27
Reinforcement, protection	1-27
Reinforcement limitations	4-7
Reinforcing steel, anchorage in grout	6-5
Reinforcing steel, typical welds	6-6
Reinforcing bar couplers	6-7
Reinforcing bars	1-26
Reinforcing bars, development length	11-23
Reinforcing bars, minimum weld lengths	6-58
Reinforcing bars, properties	11-21
Reinforcing bars, typical welds	6-6

Reinforcing bars, welding	6-7
Relative humidity	9-4
Relative humidity, correction factors	3-8
Relative humidity, map	3-7
Release reinforcement	4-20
Release strength	2-4, 10-8
Repair, specification	10-29
Repair of cracks	9-64
Residential buildings	1-6
Responsibility	8-2
Responsibility, erection	5-23
Responsibility, recommendations	10-41
Restrained members, fire	9-49
Restraint force build-up in beams, table	3-52
Reverberation	9-20
Ribbed members, handling stresses	5-9
Ribbed panels, fire resistance	9-39
Rigidity of shear walls	3-30
Roof members	8-9
Roof weights	11-2
Roofs, U-values	9-10
Rotation, pads	6-18, 6-21
Rotation, support beams	7-5
Rotation coefficients	11-12
Sabin	9-20
Safe superimposed loads	2-3
Safety equipment	10-39
Safety factor, cracking	5-3
Samples, pre-bid	10-15
Samples, veneer panels	7-18
Sand-cement mixtures	1-25
Sand-lightweight concrete	1-24
Sandblast	10-24
Sandstone	7-17
Sandwich panels	9-53
Sandwich panels, fire resistance	9-40
Sandwich panels, thermal	9-14, 9-19, 9-42
Sealers	1-26
Seasonal temperature change, map	3-7
Second order analysis	3-21
Secondary control surface	8-3
Section modulus, weld groups	6-19
Segmental construction	9-65
Seismic, design	3-58
Service load design	4-16
Service load stresses	4-50
Seven wire strand, properties	11-17
Severe exposure	1-27
Shaikh and Branson	4-44
Shape factor, bearing pads	6-20, 6-21
Shear and torsion design	4-32
Shear cone, headed studs	6-7
Shear depth ratio, daps	6-32
Shear design	4-22
Shear design aids	4-62
Shear-friction	6-23
Shear friction coefficients	6-24
Shear lag	3-44
Shear reinforcement	4-27
Shear resistance, bearing pads	6-21
Shear resistance, fire	9-49
Shear strength, angles	6-74
Shear strength, studs	6-9
Shear wall buildings	3-30
Shear wall deflections	3-31
Shear wall-frame interaction	3-56
Shear walls	3-5
Shear walls, coupled	3-33
Shear walls, rigidity	3-30
Shear walls, unsymmetrical	3-31
Shear walls with openings	3-33
Sheet piles	4-54
Sheet piles, section properties	2-57
Shop drawings	10-4, 10-16, 10-32
Shop drawings, approvals	10-39
Shop drawings, definition	10-42
Shop drawings, specification	10-16
Shop drawings, submittal	10-4, 10-16
Shore hardness, bearing pads	6-21
Shrinkage strain, table	3-8
Shrinkage-compensating cement	1-24
SI units	11-32
Site access, specification	10-10
Slate	7-16
Slenderness effects	4-50
Smoothness	8-7
Solid flat slab, load tables	2-36 to 2-41
Sound absorption	9-23
Sound absorption coefficient	9-20
Sound insulation	9-28
Sound pressure level	9-20
Sound Transmission Class (STC)	9-20, 9-21, 9-24
Sound transmission loss	9-20, 9-21
Sound wall	1-20
Span-to-depth ratio	3-5
Spancrete	2-35
Spandrels	3-18, 8-8
Special permits, transportation	5-17
Specifications	10-1, 10-2, 10-12
Splice sleeve	6-52
Stadiums	1-17
Stainless steel	6-16
Stainless steel dowels	7-17
Standard Building Code (SBC)	3-60
Standard cylinders	1-22
Standard hooks	11-21, 11-23
Standard practice	10-1
Standard practice, code	10-31
Standard time-temperature curve	9-34
Static load multipliers	5-5
Stemmed units, fire resistance	9-37
Stirrup hooks	11-21
Stone facing, anchorage	7-17
Stone veneer, sizes	7-16
Storage	5-16
Storage, specification	10-6, 10-17
Story drift	3-60
Strain compatibility	4-9
Strands	2-4, 8-5
Strand, epoxy coated	1-29
Strand, stress-strain curve	11-19
Strand diameter, panel thickness	5-5
Strand development length	4-22

Strand lifting loops	5-10
Strength design	4-6
Strength-temperature relationship	9-45
Stress block	4-7
Stress strain curve, design aid	11-19
Stripping forces	5-5
Stripping forces, example	5-10
Strongbacks, transportation	5-17
Structural integrity	3-53
Structural integrity, fire	9-42
Structural panels	8-5
Structural steel, connections	6-16
Structural steel, welding	6-16
Structural steel haunches	6-30
Structural steel haunches, design strength	6-81
Structural wall panels	8-11
Stud group	6-8, 6-10
Studs (see headed studs)	
Subgrade reaction	3-45
Submittal, specification	10-15
Sulphate attack	1-29
Sun effect, storage	5-16
Surface sealers	9-59
Suspended ceilings	9-70
Sweep	8-3
Swivel plate	5-10, 5-11
System building	9-71
Tanks	1-21
Teflon	6-20
Temperature change map	3-7
Temperature strains	3-9
Tempered hardwood strips	6-20
Temporary loading, stability considerations	5-25
Temporary loads, diagrams	5-25
Temporary supports	10-28
Tensile strength, concrete	1-23
Tensile strength, studs	6-8
Terms and conditions, contract	10-38
Terra cotta	7-13
Tests	10-33
Tests, anchorage of veneer panels	7-18
Tetrafluoroethylene (TFE)	6-20, 10-8
Thermal bowing	3-11
Thermal bowing, sandwich panels	9-56
Thermal bridges	9-19
Thermal conductance	9-4
Thermal expansion, concrete	1-23
Thermal fusible membrane (TFM)	9-16
Thermal loads, comparisons	9-12
Thermal mass	9-4
Thermal properties	9-3, 9-4, 9-7, 9-8
Thermal resistance	9-6
Thermal storage effects	9-13
Thin sections, studs	6-9
Threaded connectors	6-15
Threaded rods	6-15
Tie hooks	11-21
Tie-back connections	6-47
Ties, clay products	7-13
Tilt table	5-6, 5-9
Tilted beam, equilibrium	5-15

Tolerances	8-1
Torsion	4-30
Torsion, accidental	3-60
Torsion design aids	4-67 to 4-71
Trace elements	6-7
Traction steel	5-26
Transfer length	4-22
Transfer length, design aid	11-20
Transmittance	9-5
Transportation	5-17
Travertine	7-16
Tripping, diagrams	5-19
Truncated pyramid failure (shear cone)	6-8, 6-69
Tube structure	3-43
Tubes	6-30
Turning rigs	5-6
Type X gypsum wall board	9-39
Typical handling methods	5-4
U values	9-9
Ultraspan	2-35
Unbraced frame, analysis	3-45
Unbraced frame, stability analysis	3-45
Underwriters Laboratory (UL) Fire Resistance Directory	9-34
Undeveloped strand	4-61
Uniform Building Code (UBC)	3-60
Unsymmetrical shear walls	3-31
Utilities	10-35
Utility poles	1-21
Vapor retarder	9-15, 9-16
Veneer panel anchors, flexibility	7-18
Veneer jointing	7-18
Veneered panels	7-11
Vertical tension ties	3-57
Vibration	9-61
Vibration, inserts	6-15
Vibration isolation	9-62
Volume change, moment resisting frames	3-49
Volume change restraint forces, table	3-54
Volume change strains, typical building	3-10
Volume changes	1-24, 3-6
Volume-surface ratio, correction factors	3-9
Volume-surface ratios, typical members	1-24
Wall connections	3-73
Wall panels, effective width	4-51
Wall panels, grooved joints	6-49
Wall panels, mechanical connections	6-49
Wall panels, minimum reinforcement	5-4
Wall panels, post-tensioned	5-6
Wall panels, prestressed	5-5
Wall panel tripping, diagrams	5-19
Wall ties	7-14
Wall weights	11-2
Wall-beam connections	3-65
Wall-double tee connections	3-65, 3-67, 3-70
Wall-footing connection	3-67
Walls, U-values	9-9
Walls, vertical connections	3-57

Walnut Lane Memorial Bridge	1-2
Warpage, storage	5-16
Warping	8-5
Warranty	10-36
Waste water treatment plant	1-11
Water absorption	10-15
Water-cement ratio	1-27
Wedge action, inserts	6-15
Weight limitations, transportation	5-17, 5-18
Weld groups	6-16, 6-19
Weld groups, properties	6-19
Weld plates	8-5
Welded connections, cracking	6-16
Welded headed studs (see headed studs)	6-7
Welded wire fabric	1-26
Welded wire fabric, double tee flanges	11-27
Welded wire fabric, properties	11-26
Welding, hot dipped galvanized	6-16
Welding, reinforcing bars	6-7
Welding, structural steel	6-16
Welding rods	6-16
Welds, reinforcing bars	6-58, 6-59, 6-60
Welds, typical reinforcing bars	6-6
Wet connection	3-59
Wind load	3-17
Wind load, panels	3-17
Window wall	3-18
Wire, prestressing	11-17
Wire rope	5-10, 5-26
Wire rope, stretch	5-26
Wire rope core	5-26
Wire strand core	5-26
Wythe connectors	9-55
X-bracing	3-5
Z-factor	4-16, 7-21
Zia-McGee	4-30, 4-32